Methods in Enzymology

Volume 80
PROTEOLYTIC ENZYMES
Part C

METHODS IN ENZYMOLOGY

EDITORS-IN-CHIEF

Sidney P. Colowick   Nathan O. Kaplan

Methods in Enzymology

Volume 80

Proteolytic Enzymes

Part C

EDITED BY

Laszlo Lorand

DEPARTMENT OF BIOCHEMISTRY, MOLECULAR AND CELL BIOLOGY
NORTHWESTERN UNIVERSITY
EVANSTON, ILLINOIS

1981

ACADEMIC PRESS

A Subsidiary of Harcourt Brace Jovanovich, Publishers

New York   London
Paris   San Diego   San Francisco   São Paulo   Sydney   Tokyo   Toronto

ACADEMIC PRESS, INC.
111 Fifth Avenue, New York, New York 10003

United Kingdom Edition published by
ACADEMIC PRESS, INC. (LONDON) LTD.
24/28 Oval Road, London NW1 7DX

Library of Congress Cataloging in Publication Data
Main entry under title:

Proteolytic enzymes.

   (Methods in enzymology, v. 19, 45,
   Includes bibliographical references.
   Part B-   : Laszlo Lorand, editor.
   1. Proteolytic enzymes. I. Perlmann, Gertrude F.,
Date. II. Lorand, Laszlo, Date. III. Series:
Methods in enzymology; v. 19, etc.
QP601.M49 vol. 19, etc.   574.19'25s   75-26936
ISBN 0-12-181980-9 (v. 80) [574.19'256] AACR2

PRINTED IN THE UNITED STATES OF AMERICA

81 82 83 84   9 8 7 6 5 4 3 2 1

Table of Contents

Section I. Complement

Section II. Blood Clotting

Section III. The Plasmin System

Section IV. Enzymes Involved in Blood Pressure Regulation

Section V. Proteases of Diverse Origin and Function

Section VI. Inhibitors of Various Specificities

Contributors to Volume 80

Article numbers are in parentheses following the names of contributors.
Affiliations listed are current.

ALAN J. BARRETT (41, 42, 44, 54, 57), *Biochemistry Department, Strangeways Laboratory, Worts' Causeway, Cambridge CB1 4RN, England*

RALPH A. BRADSHAW (46, 53), *Department of Biological Chemistry, Washington University School of Medicine, St. Louis, Missouri 63110*

JOHN R. BROCKLEHURST (34), *Standard Telecommunication Laboratories Ltd., Harlow, Essex, England*

GEORGE J. BROZE, JR. (18, 19), *Division of Hematology/Oncology, Washington University School of Medicine at the Jewish Hospital of St. Louis, St. Louis, Missouri 63110*

RALPH J. BUTKOWSKI (23), *Hematology Research, Mayo Clinic, Rochester, Minnesota 55901*

WILLIAM M. CANFIELD (22), *Department of Biochemistry, University of Washington, Seattle, Washington 98195*

JOHN F. CANNON (34), *Department of Biochemistry, University of Wisconsin, Madison, Wisconsin 53706*

FRANCIS J. CASTELLINO (29), *Department of Chemistry, University of Notre Dame, Notre Dame, Indiana 46556*

TIM E. CAWSTON (52), *Rheumatology Research Unit, Addenbrooke's Hospital, Hills Road, Cambridge, United Kingdom*

DANA ČECHOVÁ (59), *Institute of Molecular Genetics, Czechoslovak Academy of Sciences, 160 20 Prague 6, Czechoslovakia*

ULLA CHRISTENSEN (28), *Chemistry Laboratory 4, University of Copenhagen, Universitetsparken 5, 2100 Copenhagen, Denmark*

CHIN HA CHUNG (50), *Department of Physiology and Biophysics, Harvard Medical School, Boston, Massachusetts 02115*

PATRICK L. COLEMAN (28, 33), *Department of Human Genetics, University of Michigan Medical School, Ann Arbor, Michigan 48109, and Clinical Systems Division, Photoproducts Department, Experimental Station, E. I. du Pont de Nemours & Co., Inc., Wilmington, Delaware 19898*

R. B. CREDO (27), *Department of Biology, Massachusetts Institute of Technology, Cambridge, Massachusetts 02139*

L. GAIL CROSSLEY (9), *Department of Biochemistry, University of Oxford, Oxford OX1 3QU, United Kingdom*

EARL W. DAVIE (13, 14, 16, 17, 26), *Department of Biochemistry, University of Washington, Seattle, Washington, 98195*

MICHAEL DOWNING (23), *Clinical Research, R/D, Alpha Therapeutic Corporation, Los Angeles, California 90032*

ARTHUR Z. EISEN (53), *Division of Dermatology, Department of Medicine, Washington University School of Medicine, St. Louis, Missouri 63110*

JACQUES ELION (23), *Institut de Pathologie Moléculaire, 24, rue du Fb. Saint-Jacques, 75014 Paris, France*

ERVIN G. ERDÖS (36, 37, 38), *Departments of Pharmacology and Internal Medicine, University of Texas Health Science Center, Dallas, Texas 75235*

NAOMI ESMON (31), *Laboratory of Protein Studies, Oklahoma Medical Research Foundation, Oklahoma City, Oklahoma 73104*

MANFRED EULITZ (60), *Institut für Hämatologie der Gesellschaft für Strahlen- und Umweltforschung, Abteilung Immunologie, 8000 Munich, Federal Republic of Germany*

MICHAEL T. EVERITT (45), *Department of Laboratory Medicine, University of Washington, Seattle, Washington 98195*

FRANZ FIEDLER (40), *Abteilung für Klinische Chemie und Klinische Biochemie in der Chirurgischen Klinik der Universität München, D-8000 Munich 2, Federal Republic of Germany*

EDWIN FINK (40), *Abteilung für Klinische Chemie und Klinische Biochemie in der Chirurgischen Klinik der Universität München, D-8000 Munich 2, Federal Republic of Germany*

HANS FRITZ (39, 40, 47, 60, 61), *Abteilung für Klinische Chemie und Klinische Biochemie in der Chirurgischen Klinik der Universität München, D-8000 Munich 2, Federal Republic of Germany*

KAZUO FUJIKAWA (16), *Department of Biochemistry, University of Washington, Seattle, Washington 98195*

J. GAGNON (11), *Medical Research Council Immunochemistry Unit, Department of Biochemistry, University of Oxford, Oxford OX1 3QU, United Kingdom*

REINHARD GEIGER (39), *Abteilung für Klinische Chemie und Klinische Biochemie in der Chirurgischen Klinik der Universität München, D-8000 Munich 2, Federal Republic of Germany*

ALFRED L. GOLDBERG (50, 51), *Department of Physiology and Biophysics, Harvard Medical School, Boston, Massachusetts 02115*

GREGORY A. GRANT (53), *Division of Dermatology, Department of Medicine and Department of Biological Chemistry, Washington University School of Medicine, St. Louis, Missouri 63110*

GEORGE D. J. GREEN (33, 62), *Biology Department, Brookhaven National Laboratory, Upton, New York 11973*

C. H. HAMMER (7), *Laboratory of Clinical Investigation, National Institute of Allergy and Infectious Diseases, National Institutes of Health, Bethesda, Maryland 20205*

R. A. HARRISON (7), *Medical Research Council Mechanisms in Tumour Immunity Unit, Cambridge CB2 2QH, England*

G. MICHAEL HASS (58), *Department of Bacteriology and Biochemistry, University of Idaho, Moscow, Idaho 83843*

RONALD L. HEIMARK (14), *Department of Biochemistry, University of Washington, Seattle, Washington 98195*

HIDEO IGARASHI (25), *Department of Microbiology, Tokyo Metropolitan Research Laboratory of Public Health, Shinjuku-ku, Tokyo 160, Japan*

SHIN-ICHI ISHII (64), *Department of Biochemistry, Faculty of Pharmaceutical Sciences, Hokkaido University, Sapporo 060, Japan*

SADAAKI IWANAGA (15, 24, 25), *Department of Biology, Faculty of Science, Kyushu University 33, Higashi-ku, Fukuoka-812, Japan*

CRAIG M. JACKSON (28), *Department of Biological Chemistry, Washington University School of Medicine, St. Louis, Missouri 63110*

KENNETH W. JACKSON (31), *Laboratory of Protein Studies, Oklahoma Medical Research Foundation, Oklahoma City, Oklahoma 73104*

J. JANATOVA (7), *Department of Pathology, School of Medicine, University of Utah, Salt Lake City, Utah 84132*

T. J. JANUS (27), *Department of Biochemistry, Molecular and Cell Biology, Northwestern University, Evanston, Illinois 60201*

DAVID JOHNSON (55), *Department of Biochemistry, East Tennessee State University, Quillen-Dishner College of Medicine, Johnson City, Tennessee 37601*

D. M. A. JOHNSON (11), *Medical Research Council Immunochemistry Unit, Department of Biochemistry, University of Oxford, Oxford OX1 3QU, United Kingdom*

VĚRA JONÁKOVÁ (59), *Institute of Molecular Genetics, Czechoslovak Academy of Sciences, 160 20 Prague 6, Czechoslovakia*

KEN-ICHI KASAI (64), *Department of Biochemistry, Faculty of Pharmaceutical Sciences, Teikyo University, Sagamiko, Kanagawa 199-01, Japan*

HISAO KATO (15), *Department of Biology, Faculty of Science, Kyushu University 33, Higashi-ku, Fukuoka-812, Japan*

JERRY A. KATZMANN (21), *Hematology Research, Mayo Clinic, Rochester, Minnesota 55901*

MICHAEL A. KERR (6, 8), *Department of Pathology, Ninewells Hospital and Medical School, Dundee, DD1 9SY, Scotland*

CHARLES KETTNER (63), *E. I. du Pont de Nemours & Co., Inc., Central Research and Development, Experimental Station, Building 328-112, Wilmington, Delaware 19898*

HEIDRUN KIRSCHKE (41), *Physiologisch-Chemisches Institut, Martin-Luther-Universität, 402 Halle (Saale), German Democratic Republic*

RICHARD J. KIRSCHNER (51), *Department of Physiology and Biophysics, Harvard Medical School, Boston, Massachusetts 02115*

WALTER KISIEL (22, 26), *Department of Biochemistry, University of Washington, Seattle, Washington 98195*

KOTOKU KURACHI (17), *Department of Biochemistry, University of Washington, Seattle, Washington 98195*

FREDERICK S. LARIMORE (50), *Eli Lilly Research Laboratories, Indianapolis, Indiana 46285*

STEVEN P. LEYTUS (34), *Department of Biochemistry, University of Illinois, Urbana, Illinois 61801*

HO-YUAN LIU (34), *Department of Biochemistry, University of Illinois, Urbana, Illinois 61801*

D. CAMPBELL LIVINGSTON (34), *Imperial Cancer Research Fund, Lincoln's Inn Fields, London WC2A 3PX, England*

L. LORAND (27), *Department of Biochemistry, Molecular and Cell Biology, Northwestern University, Evanston, Illinois 60201*

RICHARD LOTTENBERG (28), *Division of Hematology, University of Florida, College of Medicine, The J. Hillis Miller Health Center, Gainesville, Florida 32610*

WERNER MACHLEIDT (61), *Institut für Physiologische Chemie, Physikalische Biochemie und Zellbiologie der Universität München, D-8000 Munich 2, Federal Republic of Germany*

PHILIP W. MAJERUS (18, 19), *Division of Hematology/Oncology, Washington University School of Medicine, St. Louis, Missouri 63110*

WALTER F. MANGEL (34), *Department of Biochemistry, University of Illinois, Urbana, Illinois 61801*

KENNETH G. MANN (21, 23), *Hematology Research, Mayo Clinic, Rochester, Minnesota 55901*

RICHARD MELTON (10), *Department of Pathology, New York University Medical Center, New York, New York 10016*

JOSEPH P. MILETICH (18), *Department of Pathology, Division of Laboratory Medicine, Washington University School of Medicine, St. Louis, Missouri 63110*

MITSUYOSHI MORII (56), *Department of Biochemistry, University of Georgia, Athens, Georgia 30602*

TAKASHI MORITA (24, 25), *Department of Biology, Faculty of Science, Kyushu University 33, Higashi-ku, Fukuoka 812, Japan*

WERNER MÜLLER-ESTERL (47), *Abteilung für Klinische Chemie und Klinische Biochemie in der Chirurgischen Klinik der Universität München, D-8000 Munich 2, Federal Republic of Germany*

GILLIAN MURPHY (52), *Cell Physiology Department, Strangeways Research Laboratory, Worts' Causeway, Cambridge CB1 4RN, United Kingdom*

SHIGEHARU NAGASAWA (15), *Faculty of Pharmaceutical Sciences, Hokkaido University, Sapporo-060 Japan*

YALE NEMERSON (20), *Department of Med-*

icine, Mount Sinai School of Medicine, New York, New York 10029

MICHAEL E. NESHEIM (21, 23), *Hematology Research, Mayo Clinic, Rochester, Minnesota 55901*

HANS NEURATH (45, 48), *Department of Biochemistry SJ70, School of Medicine, University of Washington, Seattle, Washington 98195*

VICTOR NUSSENZWEIG (10), *Department of Pathology, New York University Medical Center, New York, New York 10016*

C. E. ODYA (38), *Medical Sciences Program, Pharmacology Section, Indiana University School of Medicine, Bloomington, Indiana 47401*

GARY A. PELTZ (34), *Stanford University School of Medicine, Stanford, California 94305*

STUART W. PELTZ (34), *Department of Biochemistry, University of Illinois, Urbana, Illinois 61801*

THOMAS H. PLUMMER, JR. (36), *Division of Laboratories and Research, New York State Department of Health, Albany, New York 12201*

R. R. PORTER (1), *Department of Biochemistry, Oxford University, Oxford OX1 3QU, United Kingdom*

JAMES R. POWELL (29), *Department of Chemistry, University of Notre Dame, Notre Dame, Indiana 46556*

R. PROHASKA (11), *Institut für Biochemie der Universität Wien, A-1090 Vienna, Austria*

A. REBOUL (5), *DRF/BMC, Centre d'Etudes Nucléaires de Grenoble, 85X, 38041 Grenoble Cedex, France*

K. B. M. REID (3, 11, 12), *Medical Research Council Immunochemistry Unit, Department of Biochemistry, University of Oxford, Oxford OX1 3QU, United Kingdom*

KENNETH C. ROBBINS (30), *Michael Reese Research Foundation, 530 East 31st Street, Chicago, Illinois 60616*

C. A. RYAN (58), *Institute of Biological Chemistry and Program in Biochemistry and Biophysics, Washington State University, Pullman, Washington 99164*

URSULA SEEMÜLLER (60), *Abteilung für Klinische Chemie und Klinische Biochemie in der Chirurgischen Klinik der Universität München, D-8000 Munich 2, Federal Republic of Germany*

ELLIOTT SHAW (62, 63), *Biology Department, Brookhaven National Laboratory, Upton, New York 11973*

R. B. SIM (2, 4, 5), *Medical Research Council Immunochemistry Unit, Department of Biochemistry, University of Oxford, Oxford OX1 3QU, United Kingdom*

EVE E. SLATER (35), *Department of Medicine, Harvard Medical School and Massachusetts General Hospital, Boston, Massachusetts 02114*

TESS A. STEWART (37), *Departments of Pharmacology and Internal Medicine, University of Texas Health Science Center, Dallas, Texas 75235*

LOUIS SUMMARIA (30), *Michael Reese Research Foundation, 530 East 31st Street, Chicago, Illinois 60616*

K. H. SREEDHARA SWAMY (50), *Ciba-Geigy Research Center, Goregan East, Bombay 400 063, India*

B. F. TACK (7), *Department of Pediatrics, Children's Hospital, and Harvard Medical School, Boston, Massachusetts 02115*

TAKAYUKI TAKAHASHI (43), *Laboratory of Protein Studies, Oklahoma Medical Research Foundation, Oklahoma City, Oklahoma 73104*

JORDAN TANG (31, 43), *Laboratory of Protein Studies, Oklahoma Medical Research Foundation, Oklahoma City, Oklahoma 73104*

KENNETH A. THOMAS (46), *Biochemistry Department, Merck Institute for Therapeutic Research, Merck Sharp and Dohme Research Laboratories, Rahway, New Jersey 07065*

M. L. THOMAS (7), *Department of Pathology, School of Medicine, University of Utah, Salt Lake City, Utah 84132*

PAULA B. TRACY (21), *Hematology Research, Mayo Clinic, Rochester, Minnesota 55901*

JAMES TRAVIS (55, 56), *Department of Biochemistry, University of Georgia, Athens, Georgia 30602*

HARALD TSCHESCHE (40), *Lehrstuhl für Biochemie, Universität Bielefeld, Postfach 8640, D-4800 Bielefeld 1, Federal Republic of Germany*

LLOYD WAXMAN (49), *Department of Physiology and Biophysics, Harvard Medical School, Boston, Massachusetts 02115*

JOHN A. WEARE (37), *Departments of Pharmacology and Internal Medicine, University of Texas Health Science Center, Dallas, Texas 75235*

BJÖRN WIMAN (32), *Department of Clinical Chemistry, Umeå University Hospital, S-901 85 Umeå, Sweden*

ROBERT C. WOHL (30), *Michael Reese Research Foundation, 530 East 31st Street, Chicago, Illinois 60616*

RICHARD G. WOODBURY (45), *German Cancer Research Center, D-6900 Heidelberg 1, Federal Republic of Germany*

GERT WUNDERER (61), *I. Frauenklinik und Hebammenschule der Universität München, D-8000 Munich 2, Federal Republic of Germany*

MARGALIT ZUR (20), *Department of Medicine, Mount Sinai School of Medicine, New York, New York 10029*

ROBERT ZWILLING (48), *University of Heidelberg, Zoological Institute, Division of Physiology, Federal Republic of Germany, and Department of Biochemistry, University of Washington, Seattle, Washington 98195*

Preface

This volume, as its companion Volumes XIX and XLV in the "Methods in Enzymology" series, developed from the renaissance of interest in proteolytic enzymes. Articles on naturally occurring inhibitors are a reminder of the control mechanisms that have evolved to regulate the activities of these enzymes. For the first time, considerable attention is devoted to the complement system which, in an interplay with the coagulation cascade and the fibrinolytic system of blood, constitutes such an important part of the general immune response. In addition, an effort has also been made to highlight developments in several representative areas such as blood pressure regulation, proteases in invertebrates, ATP-dependent enzymes, synthetic substrates, and to include the unexpected. Who would have predicted that one of the subunits of nerve growth factor might be a proteolytic enzyme?

Thanks are due to the authors who responded so willingly to aid in the organization of this volume.

LASZLO LORAND

METHODS IN ENZYMOLOGY

EDITED BY

Sidney P. Colowick and Nathan O. Kaplan

VANDERBILT UNIVERSITY
SCHOOL OF MEDICINE
NASHVILLE, TENNESSEE

DEPARTMENT OF CHEMISTRY
UNIVERSITY OF CALIFORNIA
AT SAN DIEGO
LA JOLLA, CALIFORNIA

METHODS IN ENZYMOLOGY

EDITORS-IN-CHIEF

Sidney P. Colowick   Nathan O. Kaplan

Section I

Complement

[1] The Proteolytic Enzymes of the Complement System

By R. R. PORTER

The complement system is a complex mixture of serum proteins that is activated sequentially, when antibodies in the blood of an immune animal interact with the corresponding antigens. Activation of complement causes lysis if the antigen is a cell, peptides are released that are chemotactic for phagocytes, and complement component products bound to the antibody–antigen complexes facilitate phagocytosis. The complement system is therefore an effective mechanism for destroying foreign substances and removing them from the blood.

The nature of the antigen is irrelevant and initiation of the activation is dependent on the aggregation of the antibody molecules caused by their binding to the antigen. The major feature of the activation is the conversion of proteolytic zymogens to active proteinases that catalyze the conversion of other zymogens to active enzymes down the pathway.

There are two pathways of activation, named the classical and the alternative pathway, both of which can be initiated by addition of antibody aggregates to serum. The alternative pathway can also be activated by the addition to serum of high-molecular-weight polysaccharides such as are found in bacterial or yeast cell walls. An activation scheme is shown in Fig. 1 and the proteolytic zymogens and proteinases are in heavy type. The proteinases are $C\bar{1}r$, $C\bar{1}s$, $C\bar{2}a$ in the classical pathway, and $\bar{D}$ and $\bar{Bb}$ in the alternative pathway. All are serine proteinases and $C\bar{1}r$ and $C\bar{1}s$ and $\bar{D}$ conform to the normal pattern with catalytic peptide chains of about 25,000 MW and, as far as is known, all have the expected N-terminal and active-site residues which are invariant or highly conserved in the catalytic peptide chains in other serine proteinases. They differ from each other in that $C\bar{1}r$ and $C\bar{1}s$ have large activation peptides of about 60,000 MW, similar to plasminogen, while $\bar{D}$ has only been found in serum as the active enzyme. An early report that $\bar{D}$ was also present in zymogen form has not been confirmed.

$\bar{Bb}$—and probably $C\bar{2}a$—are, however, the first examples of a new type of serine proteinase with catalytic peptide chains of about 60,000 MW. Bb was identified as a serine proteinase by its amino acid sequence, which showed that the invariant residues in and around the active site of serine proteases were present and in the same position relative to each other and to the C-terminal end of the chain. The typical N-terminal sequence is missing and is replaced by an additional 300 residues that so far have shown no homology to any other sequence. This suggests that the

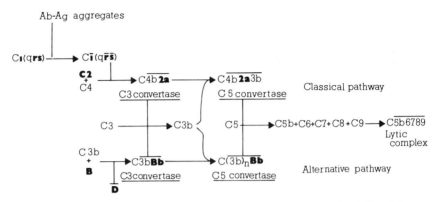

Fig. 1. A simplified scheme of the activation of complement by the classical and alternative pathways. The proteolytic zymogens and the active proteinases are shown in heavy type. The mechanism of activation of the alternative pathway is not clear but the present view is that activation is continuous and is controlled by the amount of C3b available.

active site of Bb is probably identical with that of other serine proteinases but that the activation mechanism is different.

Another unusual feature of the complement enzymes is that they usually function only in association with other proteins. Thus C1r is activated when the tetramer complex $C1r_2–Ca^{2+}–C1s_2$ binds to C1q on the aggregated antibody. C1s can be replaced by $\overline{C1s}$ inactivated by reaction with DFP but if C1s is omitted, C1r—though still able to bind—is not activated. Conversion of C1r to an active protease is the first step in the initiation of the complement cascade and evidence has been given that this occurs when the tetramer $C1r_2–Ca^{2+}–C1s_2$ binds to a C1q molecule bound to aggregated antibody. This noncovalent interaction leads to a conformational change in C1r which exposes a proteolytic site prior to peptide bond cleavage in a form of C1r designated C1r·. C1r· catalyzes conversion of C1r or C1r· to $\overline{C1r}$, the active enzyme.

C1r will activate C1s in solution as will $\overline{C1s}$ activate C4 and C2, but for a continuing sequence the $\overline{C1s}$ must be part of the C1 complex bound to aggregated antibody. The C3 and C5 convertases of both pathways are only active as complexes; $\overline{C2a}$ and $\overline{Bb}$ alone will not hydrolyze either C3 or C5. $\overline{D}$ will only activate B in the presence of C3b.

There are several explanations for this phenomenon. First, the noncatalytic proteins C1q, C4, and C3 serve as link proteins to bind the proteases to the activator, probably an essential feature when activation is taking place in plasma which is rich in proteolytic inhibitors. C3 and C4 bind on activation to antibodies and cell surfaces by a covalent reaction through a labile acyl group with a very short half-life and hence activation and binding must occur in close proximity. Further, it is probable that when, say, C3b + B interact a conformation change occurs, exposing a site vulnerable to $\overline{D}$, and this is also true for the C3 and C5 convertases though in these cases the interactions may cause exposure of the catalytic site rather than of the susceptible peptide chain in the substrate. Whatever the explanation, it is clear that association and assembly of the components is an essential feature of this complex proteolytic system. Study of the individual components will not in itself explain the activation mechanisms.

Finally, in this section will be found descriptions of some of the serum proteins that control the activation of complement either by inhibition of proteolysis or by proteolytic digestion of the different components. $\overline{C1}$ inhibitor ($\overline{C1}$ INH) is a glycoprotein that interacts so strongly with $\overline{C1}r$ and $\overline{C1}s$ that the $\overline{C1}$ INH is not dissociated in buffers containing sodium dodecyl sulfate and urea. The binding of $\overline{C1}$ INH to $\overline{C1}r$ and $\overline{C1}s$ in the $\overline{C1}$ complex causes them to dissociate from the complex. C3 inactivator (C3 INA) is a protease that splits one bond in the α'-chain of C3b when $\beta 1H$ is present and two bonds in the α'-chain of C4b if the C4b-binding protein is present. This is another example of proteolysis depending on the interaction of a third protein with either or both the enzyme and substrate. Neither the overall structure nor the pattern of inhibition have so far given any indication as to which class of proteinase the C3 INA belongs, but sequence studies have now shown that this enzyme also is a serine protease (L. M. Hsiung, J. Gagnon, and R. R. Porter, unpublished results).

The blood-clotting mechanism has many similarities to the complement system, but the latter has unusual features that, at present, appear to be unique. Much more structural information will be required before the mechanism of the proteolytic activation of the complement components are understood fully.

Detailed references are given in the articles that follow and descriptions of the complement system and its proteinases will be found in several recent reviews.[1-6]

[1] J. E. Fothergill and W. H. K. Anderson, Curr. Top. Cell. Regul. 13, 259 (1978).
[2] H. J. Müller-Eberhard, in "Molecular Basis of Biological Degradation Processes" (R. D. Berliner, H. Hermann, I. H. Lepow, and J. M. Tanzer, eds.). Academic Press, New York, 1978.
[3] P. J. Lachmann, in "The Antigens" (Sela, ed.), Vol. 5, p. 283. Academic Press, New York, 1979.
[4] R. R. Porter, Int. Rev. Biochem. 23, 177 (1979).
[5] R. R. Porter and K. B. M. Reid, Adv. Protein Chem. 33, 1 (1979).
[6] K. B. M. Reid and R. R. Porter, Annu. Rev. Biochem. 50, 433 (1981).

[2] The First Component of Human Complement—C1

By R. B. SIM

The first component of the complement system, C1, is a multiprotein complex responsible for the initiation of activation of the classical pathway of complement. C1 was first identified as a single functional component of the complement system by Ferrata.[1] It was shown to exist in plasma or serum in an unactivated form, which became activated specifically by binding to immune complexes. On activation, C1 acquired the ability to cleave the fourth (C4) and second (C2) components of complement, and to hydrolyze a number of amino acid esters.[2-5] The proteolytic and esterolytic activities could be inhibited[6] by diisopropyl phosphorofluoridate (DFP) and this suggested that activated C1 ($\overline{C1}$)[7] had serine protease activity. Lepow and colleagues[8] demonstrated in 1963 that C1 is composed of three types of subunits, designated subcomponent C1q, subcomponent C1r, and subcomponent C1s, which dissociate in the presence of EDTA, and reassociate on addition of Ca^{2+} ions.[8,9] The $\overline{C1s}$ subcomponent, which could be purified in unactivated (C1s) or activated ($\overline{C1s}$) form, was shown to mediate the DFP-sensitive esterolytic activity and the C4- and C2-cleaving activity of $\overline{C1}$.[8,10] Later studies demonstrated that the C1r subcomponent of C1 also exists in proenzyme (C1r) and activated ($\overline{C1r}$) forms.[11-13] The active form ($\overline{C1r}$) mediates proteolytic activation of proenzymic C1s, and has very weak esterolytic activity against a limited range of amino acid esters. These activities are also inhibited by DFP.[11,14,15]

[1] A. Ferrata, *Berl. Klin. Wochenschr.* **44**, 366 (1907).
[2] I. H. Lepow and L. Pillemer, *J. Immunol.* **75**, 63 (1955).
[3] I. H. Lepow, O. D. Ratnoff, and L. Pillemer, *Proc. Soc. Exp. Biol. Med.* **92**, 111 (1956).
[4] E. L. Becker, *J. Immunol.* **77**, 462 (1956).
[5] O. D. Ratnoff and I. H. Lepow, *J. Exp. Med.* **106**, 327 (1957).
[6] E. L. Becker, *J. Immunol.* **77**, 469 (1956).
[7] The convention in complement protein nomenclature is that activated components are designated by a superscript bar (e.g., $\overline{C1}$ = activated C1).
[8] I. H. Lepow, G. B. Naff, E. W. Todd, J. Pensky, and C. F. Hinz, Jr., *J. Exp. Med.* **117**, 983 (1963).
[9] G. B. Naff, J. Pensky, and I. H. Lepow, *J. Exp. Med.* **119**, 593 (1964).
[10] A. L. Haines and I. H. Lepow, *J. Immunol.* **92**, 468 (1964).
[11] G. B. Naff and O. D. Ratnoff, *J. Exp. Med.* **128**, 571 (1968).
[12] G. Valet and N. R. Cooper, *J. Immunol.* **112**, 1667 (1974).
[13] R. B. Sim and R. R. Porter, *Biochem. Soc. Trans.* **4**, 127 (1976).
[14] K. Sakai and R. M. Stroud, *Immunochemistry* **11**, 192 (1974).
[15] I. Gigli, R. R. Porter, and R. B. Sim, *Biochem. J.* **157**, 541 (1976).

The C1q subcomponent of C1 does not appear to possess any enzymatic activity, but mediates the attachment of the C1 complex to surfaces that activate complement, for example, immune complexes. The C1 macromolecule is thus an assembly of two serine protease proenzymes (C1r and C1s) and a nonenzymatic recognition protein (C1q).

The properties of individual isolated subcomponents of C1 are discussed elsewhere in this volume[16,17] and in recent reviews.[18,19] This article will describe the preparation and properties of the intact C1 macromolecule, and the interaction of its subcomponents.

Assay

C1, whether in proenzymic or activated form, is determined by hemolytic assay, that is by measuring the extent to which antibody-coated red blood cells are lysed by a mixture containing C1 plus an excess of all other complement components. This form of assay relies on the ability of C1 to adhere, via the C1q subcomponent, to antibody bound to red blood cells. Once bound, C1 becomes activated, and the $\overline{\text{C1s}}$ subcomponent cleaves C2 and C4, and this in turn leads to cleavage of C3 and C5 and finally to assembly of a lytic C5b–9 complex on the red cell membrane. This form of assay is suitable for determining C1 or $\overline{\text{C1}}$ in citrated plasma, serum, purified preparations, and also C1 or $\overline{\text{C1}}$ reconstituted from isolated subcomponents.

The reagents used for hemolytic assay of complement components are relatively complex and are time-consuming to prepare. Most are, however, available commercially, as indicated below.

Reagents

Buffers. Veronal (barbitone) buffers for complement assays are modified from those described by Mayer.[20] A stock concentrate (5 × physiological ionic strength) of veronal buffered saline (VBS) is made (41.5 g NaCl + 5.1 g sodium barbitone dissolved in 1 liter of water; pH adjusted to 7.5 with HCl). This can be stored for several months at 4°C. Buffers used in assays must be made up each day, as follows:

[16] K. B. M. Reid, this volume (section on C1q).
[17] R. B. Sim, this volume (section on C1r and C1s).
[18] R. M. Stroud, J. E. Volanakis, S. Nagasawa, and T. F. Lint, *in* "Immunochemistry of Proteins" (M. Z. Atassi, ed.), Vol. 3, p. 167. Plenum, New York, 1979.
[19] R. R. Porter and K. B. M. Reid, *Adv. Protein Chem.* 33, 1 (1979).
[20] M. M. Mayer, *in* "Experimental Immunochemistry" (E. A. Kabat and M. M. Mayer, eds.), 2nd ed., p. 133. Thomas, Springfield, Illinois, 1961.

DGVB^{++} (dextrose–gelatin–veronal buffer with divalent metal ions). To 100 ml of 5-fold VBS concentrate is added 1 g of gelatin + 25 g glucose dissolved together in 200 ml water. 0.5 ml of a stock 0.3 M CaCl$_2$–1.0 M MgCl$_2$ solution is added, and the buffer made up to 1 liter with water.

GVB–EDTA (gelatin–veronal buffer with EDTA). To 180 ml of 5-fold VBS concentrate is added 100 ml of 0.1 M trisodium EDTA, pH 7.5, and 1 g of gelatin dissolved in 200 ml water. The buffer is then made up to 1 liter with water.

EAC4^{hu} cells. Sheep red blood cells (E), sensitized with rabbit antibodies to sheep red blood cell stromata (A, hemolytic antibody or hemolysin[20]), and to which are attached the C4b fragment of human C4 (C4^{hu}) are prepared as described by Borsos and Rapp.[21] This reagent can be stored at 4°C for up to 14 days. Immediately before use, the cells are washed in DGVB^{++} and adjusted to a concentration of 1 × 10^8 cells/ml in DGVB^{++}. Estimation of cell concentration is described by Mayer.[20] EAC4^{hu} cells are available from Cordis Labs (Miami, Florida) and hemolysin from Cordis or Institut Pasteur (Paris).

C2. "Functionally pure" guinea pig C2 (Cordis Labs)—that is, a C2 preparation not significantly contaminated with other complement components—is prepared as described by Mayer.[20] This material is assayed for C2 activity[21,22] and stored at −70°C until required. Immediately before use, it is diluted in DGVB^{++} to approximately 2.5 × 10^9 effective C2 molecules/ml.

C–EDTA. Guinea pig serum (1 ml) is diluted to 20 ml in GVB–EDTA. Addition of EDTA to this serum serves to dissociate the guinea pig C1 present in the serum, and prevents formation of a C3 convertase in the guinea pig serum. This reagent therefore serves as a source of C3–C9.

Procedure

Serial 2-fold dilutions of the test material containing C1 or C$\overline{1}$ are made in DGVB^{++}. 1.0 ml of each dilution is mixed with 0.5 ml of EAC4^{hu} cells (10^8 cells/ml in DGVB^{++}). The samples are incubated for 1 hr at 30°C, with occasional shaking, then centrifuged (10 min, 1500 g). The supernatant is removed by aspiration, and the cells are resuspended in 0.5 ml of the C2 preparation (2.5 × 10^9 effective molecules/ml in DGVB^{++}) then incubated for 5 min at 37°C. GVB–EDTA (0.5 ml), followed by C–EDTA (1.0 ml) is

[21] T. Borsos and H. J. Rapp, *J. Immunol.* 99, 263 (1967).
[22] M. A. Kerr, this volume [6].

then added to each tube, and incubation is continued for 1 hr at 37°C. The samples are then diluted to a final volume of 4.4 ml with 145 mM NaCl and centrifuged (10 min, 1500 g). The release of hemoglobin and therefore the degree of lysis of the red blood cells is measured by reading the absorbance at 412 nm of the supernatant. Appropriate controls are included to determine spontaneous lysis of the cells (1.0 ml of DGVB^{++} buffer is added instead of a source of C1). The absorbance at 412 nm caused by 100% cell lysis is determined by diluting 0.5 ml cells with 3.9 ml water.

The C1 or C$\overline{1}$ activity present in a sample is generally expressed by determining from a graph the dilution of a C1 sample that is required to produce 50% lysis of the cells. The C1 activity is then expressed in 50% hemolysis (CH$_{50}$) units (e.g., if a C1 sample causes 50% cell lysis at a 1 : 20,000 dilution, the original undiluted sample is said to contain 20,000 CH$_{50}$ units/ml). The relationship between degree of lysis and the amount of C1 supplied is a sigmoidal curve that fits the von Krogh equation (see Ref. 20 for discussion). To convert this curve to a linear form, a logarithmic plot is used, in which $\log_{10}[y/100 - y]$ (where y is the percentage of cells lysed) is plotted against the reciprocal dilution (RD) of the C1 sample (RD is used to indicate 1/dilution factor—e.g., if the sample is diluted 200-fold, RD = 1/200). The reciprocal dilution corresponding to 50% lysis (where $y/[100 - y] = 1$) is read from the linear portion of this curve. Examples of this calculation method are given by Mayer.[20]

C1 titers are frequently reported in the form of "effective molecules/ml" rather than as CH$_{50}$ units. This form of calculation makes use of a plot of $-\ln[1 - Y]$ (where Y is the *fraction* of cells lysed) against reciprocal dilution of the sample. The basis of the calculation is described by Mayer[20] and Lachmann.[23]

Since the reagents required for C1 assay are complex and subject to variation, the CH$_{50}$ values found for C1 in a standard serum depend greatly on the condition of the reagents. For C1 in normal human serum, CH$_{50}$ values reported are generally in the range of 4×10^4 to 2×10^5 units/ml. The C1 hemolytic assay is essentially a comparative assay, and if comparison of a group of samples is required, it is preferable to assay all the samples on the same day, with the same batch of reagents. If groups of samples must be assayed on different days, it is essential to assay, with each group of samples, a standard material (e.g., pooled human serum). This allows assessment of day-to-day variation. Values obtained in different laboratories cannot be directly compared, as many variations in the assay procedure exist, and there is no system of standardization of reagents. In any one laboratory, provided that methods of preparation of

[23] P. J. Lachmann, *in* "Handbook of Experimental Immunology" (D. M. Weir, ed.), 2nd ed., Vol. 1, p. 5.1. Blackwell, Oxford, 1973.

reagents are well established and reproducible, the variation in the assay due to differences in batches of reagents and technical error is $\pm 15\%$.[24]

The assay procedure outlined above is often simplified by omitting the centrifugation step after incubation of EAC4[hu] cells + C1. This omission makes no significant difference in the assay of C1 in normal serum or plasma. However, in partially purified fractions, if this step is omitted, free subcomponent $C\bar{1}s$ alone can mimic the action of whole C1 or $C\bar{1}$ by causing cleavage and deposition of C2 on the EAC4[hu] cells. Thus to distinguish between the presence of $C\bar{1}s$ alone, and the presence of all three C1 subcomponents, the centrifugation step should be included.

Purification

The purification of C1 as an intact complex is made difficult by the relatively weak binding between C1q and the C1r and C1s subcomponents, and by the tendency of C1 to become activated during purification. A number of partial purification techniques have been developed in which a high degree of purification of C1 can be obtained in only one or two steps. These are based either on euglobulin precipitation or on the binding of C1 to immune complexes or to immunoglobulin G (IgG)–Sepharose.[1,8,15,25,26] In each case, however, final additional steps are required for purification to homogeneity, and these steps necessitate at least partial dissociation of C1.

With the exception of euglobulin precipitation at pH less than 6.0 (see discussion by Lepow et al.[27]), all methods for partial purification of C1 result in activation of the C1 unless protease inhibitors are present at all stages of the preparation. A partial purification method for C1,[13,15,28] which also forms the basis for isolation of C1r and C1s,[17] is described here. All steps are done at 4°C, unless stated otherwise.

Procedure

Citrated human plasma pooled from normal donors is taken and made approximately 20 mM CaCl$_2$ by addition of 20 ml of 1 M CaCl$_2$ per liter of plasma. The plasma is left to clot for 16 hr at 4°C in a glass vessel. The clot is removed by filtration through muslin, and the serum is stored at -20°C

[24] W. Opferkuch, V. Berger, S. Kapp, G. Melchers, W., Prellwitz, R. Ringelmann, and B. Wellek, in "Clinical Aspects of the Complement System" (W. Opferkuch, K. Rother, and D. R. Schultz, eds.), p. 4. Thieme, Stuttgart, 1978.

[25] D. H. Bing, this series, Vol. 34, p. 731.

[26] R. G. Medicus and R. M. Chapuis, J. Immunol. **125**, 390 (1980).

[27] I. H. Lepow, G. B. Naff, and J. Pensky, Ciba Found. Symp. [N.S.] **57**, 74 (1965).

[28] R. B. Sim, D. Phil. Thesis, University of Oxford (1976).

until required. Pooled outdated plasma is satisfactory as a starting material for this preparation. Plasma or serum containing high levels of immune complexes should not be used, and plasma containing heparin or EDTA as anticoagulant is not suitable for C1 preparation.

Step 1. Euglobulin Precipitation. One liter of serum is thawed and made 5 mM with DFP[29] (Sigma or Aldrich Chemical Co.) by addition of 2 ml of 2.5 M DFP. The serum is centrifuged for 30 min at 12,500 g. Precipitate and lipid film are removed. The serum is then adjusted to pH 7.4 by addition of 0.15 M HCl or 0.15 M NaOH, and again made 5 mM with DFP. A pH 7.4 euglobulin precipitate is then made by diluting the serum with 4 liters of 5 mM $CaCl_2$–0.12 mM iodoacetamide–0.2 mM 1,10-phenanthrolene.[30] The pH is again adjusted to 7.4, and the dilute serum stirred gently for 2 hr. At the end of the 2-hr period, DFP is again added to 1 mM, and the precipitate collected by centrifugation at 12,500 g for 45 min. The precipitate is resuspended by use of a hand homogenizer in 50 ml of 40 mM sodium acetate, 5 mM $CaCl_2$, pH 5.5, and the dispersion is made 5 mM DFP. The suspension is then centrifuged (45 min at 12,500 g) and the supernatant is discarded. This washing step, in low-ionic-strength acetate buffer, removes albumin and IgG from the precipitate. The precipitate is redissolved, by stirring with a glass rod, in 15–20 ml of 50 mM sodium acetate, 200 mM NaCl, 5 mM $CaCl_2$, pH 5.5. The solution is made 5 mM DFP, and centrifuged at 75,000 g for 30 min. Lipid film and insoluble residue are discarded.

Step 2. Gel Filtration. The redissolved euglobulins are then fractionated on a column (90 × 5 cm diameter) of Sepharose-6B (Pharmacia) equilibrated in 50 mM sodium acetate–200 mM NaCl–5 mM $CaCl_2$, pH 5.5. The column must be presaturated before its first use for fractionating euglobulins by passing 50 ml of human serum through it. This presaturation procedure prevents nonspecific binding of C1q to the Sepharose.[31] The elution profile is shown in Fig. 1. Fractions corresponding to the peak of C1 hemolytic activity are pooled, made 1 mM with DFP, and concentrated by ultrafiltration on a PM-10 membrane (Amicon) with very gentle stirring, to 20–30 ml. This material contains unactivated C1 and can be stored at pH 5.5 and 4°C for 2–3 months. No activation occurs under these conditions, although there is some loss of protein by aggregation.

To prepare the activated form of C1 (C$\overline{1}$), the procedure quoted above is modified in two ways: (a) DFP is added only while the serum is thaw-

---

[29] DFP is prepared as a stock 2.5 M solution by adding 1 g DFP to 1.2 ml anhydrous propan-2-ol. It is toxic, and precautions for use of DFP are described by the suppliers.

[30] Iodoacetamide and 1,10-phenanthrolene are used here as inhibitors of thiol- and metalloproteases, respectively.

[31] M. R. Mackenzie, N. Creevy, and M. Heh, *J. Immunol.* **106**, 65 (1971).

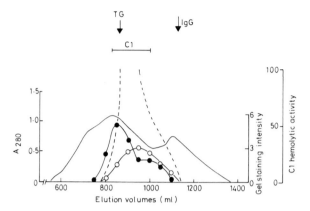

FIG. 1. Chromatography on Sepharose-6B. Conditions are as stated in the text. The elution positions of standard proteins, bovine thyroglobulin (TG) and rabbit immunoglobuliln G (IgG) are indicated. The elution positions of C1q, C1r, and C1s were found from scanning Coomassie blue-stained sodium dodecyl sulfate–polyacrylamide gels of column fractions. The elution profile of C1q (O——O) and of C1r and C1s, which always co-elute (●——●) is shown in arbitrary units of gel-staining intensity. C1 hemolytic activity (-----) was determined by assaying a 1 : 20,000 dilution of column fractions. A_{280} is shown as an unbroken line (——).

ing, and not subsequently; (b) The final concentrated pool from Sepharose-6B is adjusted to pH 7.4 by addition of 0.25 M Tris base, and incubated for 1 hr at 37°C in the presence of 5 mM CaCl$_2$. This material is then dialyzed against 50 mM sodium acetate–200 mM NaCl–5 mM CaCl$_2$, pH 5.5, and stored as described for the unactivated form.

Yield of activity at each step is shown in Table I. There is no difference in yield and purification factor between C1 and C$\bar{1}$ preparations. The final material contains only C1q, C1r, and C1s plus a major high-molecular-

TABLE I
PURIFICATION YIELDS

Material	Euglobulin precipitate	Pool from Sepharose-6B
Percent of total serum protein[a]	0.7–0.9	0.20–0.25
Percent of serum C1 hemolytic activity	75–87	45–55
Purification factor with respect to serum	83–124	180–275
Major proteins present[b]	C1q, C1r, C1s, C4c, C4bp, IgM, IgG, SAP	C1q, C1r, C1s, C4bp

[a] Percentages quoted are the acceptable range.
[b] SAP, serum amyloid P-component. C4bp, C4b-binding protein.

weight (HMW) contaminant now identified as C4b-binding protein. Traces of IgM may also be present. C4b-binding protein constitutes 15–25% of the total protein in the Sepharose-6B pool.

From the elution profile shown in Fig. 1, it is clear that C1 migrates on the Sepharose-6B column not as a macromolecular complex of C1q, C1r, and C1s, but in the form of C1q migrating independently but overlapping the elution position of C1r and C1s. The higher molecular weight peak containing most of the C1r and C1s is likely to represent a $C1r_2-C1s_2$ tetramer, while the lower molecular weight (LMW) peak of C1r and C1s may consist of dimers of C1r ($C1r_2$) and of C1s ($C1s_2$) migrating independently.

A recently published purification method,[26] involving a first step of affinity chromatography of C1 on an IgG–Sepharose column, followed by gel filtration on Ultrogel AcA 22, provides a yield of C1 activity equivalent to the method described above, while apparently eliminating most of the C4b-binding protein.

C1 in Other Species

C1 or $C\overline{1}$ have been partially purified, principally by euglobulin precipitation methods, from a large number of species, including mammals,[32] fish,[32] and amphibia.[32,33] $C\overline{1}$ from the frog is active in a standard hemolytic assay similar to that described above, and a hybrid $C\overline{1}$, containing frog C1q and human $C\overline{1}r$ and $C\overline{1}s$, is hemolytically active. Hybrid $C\overline{1}$ species can also be formed using combinations of human and other mammalian $C\overline{1}$ subcomponents (e.g., guinea pig[34] or cow[35]).

Properties of C1

The C1 macromolecule has been characterized both by studies on "intact" material (i.e., C1 in serum) and on C1 partially purified by techniques that aim to co-purify all the subcomponents. C1 has also been studied in highly purified systems in which C1 is reconstituted from subcomponents that have been isolated independently.[15,36,37] Earlier studies[8,9] indicated that C1 hemolytic activity in serum, when examined by sucrose density gradient centrifugation in physiological buffers, has a sedimentation coefficient of 18–19 S. C1 reconstituted from isolated C1q, C1r, and

[32] I. Gigli and K. F. Austen, *Annu. Rev. Microbiol.* **25**, 309 (1971).
[33] R. J. Alexander and L. A. Steiner, *J. Immunol.* **124**, 1418 (1980).
[34] F. G. Sassano, H. R. Colten, T. Borsos, and H. J. Rapp, *Immunochemistry* **9**, 405 (1972).
[35] R. D. Campbell, N. A. Booth, and J. E. Fothergill, *Biochem. J.* **183**, 579 (1979).
[36] R. J. Ziccardi and N. R. Cooper, *J. Immunol.* **118**, 2047 (1977).
[37] N. R. Cooper and R. J. Ziccardi, *J. Immunol.* **119**, 1664 (1977).

C1s, examined by the same method, migrates as an apparently smaller 15–16 S molecule.[36,38] This led to speculation that the C1 complex in serum might contain proteins additional to C1q, C1r, and C1s. Suggested involvement of a fourth protein, provisionally named "C1t" has been disproved,[39] however, and it is now generally agreed that C1 reconstituted only from C1q, C1r, and C1s possesses all of the properties of "native" C1 in serum. The discrepancy in sedimentation coefficient between C1 in serum and reconstituted C1 may be due to a weak interaction between the C1q subcomponent in serum C1 and monomeric IgG. This would lead to an increase in the average effective molecular weight (MW) of C1 in serum. Goers et al.[40] have demonstrated this effect for mixtures of monomeric IgG and C1q alone.

The stoichiometry of C1 in serum, and of the 16 S C1 reconstituted in solution, is C1q monomer : C1r monomer : C1s monomer = 1 : 2 : 2 on a molar basis.[15,36] The complex has MW 740,000–775,000.[18,26,36] The synthesis of the three subcomponents appears to be closely controlled such that, in normal sera, all the C1q, C1r, and C1s present circulate as macromolecular C1, with no subcomponent present in excess.[36,41] In various disease states, however, disequilibrium in subcomponent concentrations in serum is observed.[41]

C1 in serum dissociates readily into C1q plus C1r$_2$–C1s$_2$ complexes, and into other LMW complexes on dilution.[36,42] Serum C1 and purified 16 S C1 also dissociate on increase in ionic strength.[36]

Subcomponent Interactions

In solution, in the presence of Ca^{2+} ions, C1r and C1s bind strongly to each other to form a tetrameric C1r$_2$–C1s$_2$ complex.[12,19,28,43,44] The complex has MW 330,000–350,000,[19,43] a sedimentation coefficient of 8.7–10.0 S,[36,38,43] and a Stokes radius of 84–90 Å.[28] It is very asymmetric, and recent electron microscopy studies[43] indicate that it has an elongated S shape. The activated subcomponents, C̄1r and C̄1s form an equally stable C̄1r$_2$–C̄1s$_2$ complex.[45] Neither C1r nor C1s alone forms a strong complex

[38] G. Valet and N. R. Cooper, J. Immunol. 112, 339 (1974).
[39] R. H. Painter, J. Immunol. 119, 2203 (1977).
[40] J. W. Goers, R. J. Ziccardi, V. N. Schumaker, and M. M. Glovsky, J. Immunol. 118, 2182 (1977).
[41] A.-B. Laurell, U. Mårtensson, and A. G. Sjöholm, Acta Pathol. Microbiol. Scand., Sect. C 84, 45 (1976).
[42] M. Loos, H.-U. Hill, B. Wellek, and H.-P. Heinz, FEBS Lett. 64, 341 (1976).
[43] J. Tschopp, W. Villiger, H. Fuchs, E. Kilchherr, and J. Engel, Proc. Natl. Acad. Sci. U.S.A. 77, 7014 (1980).
[44] R. J. Ziccardi and N. R. Cooper, J. Immunol. 116, 496 (1976).
[45] G. J. Arlaud, S. Chesne, C. L. Villiers, and M. G. Colomb, Biochim. Biophys. Acta 616, 105 (1980).

with C1q in solution,[12,38] but when all three subcomponents are present, the Ca^{2+} ion-dependent 16 S $C1q-C1r_2-C1s_2$ complex (C1) is formed.[12,36]

Once C1 is bound to antibody–antigen complexes through the C1q subcomponent, the interactions between C1 subcomponents are altered.[28,44,46] Neither C1r nor C1s binds directly to immune complexes. Subcomponent C1r (or $C\overline{1}r$), however, binds directly and strongly to antibody–antigen–C1q complexes. C1s (or $C\overline{1}s$) does not bind to antibody–antigen–C1q complexes, but will bind only to antibody–antigen–C1q–C1r complexes. Thus it appears that, once C1q is bound to immune aggregates, a change in the mode of binding of the other subcomponents occurs, such that the C1q–C1r interaction becomes much stronger. This alteration in binding may be part of the sequence of events leading to C1 activation.

Experiments on the reassembly of C1 from isolated subcomponents in the presence of immune aggregates suggest that C1q, once fixed to antibody, may be able to bind two or more $C1r_2-C1s_2$ complexes. Maximal hemolytic activity of reconstituted C1 was observed when 4C1r and 4C1s monomers were supplied per molecule of C1q.[15] Electron-microscopic studies of reconstituted C1 have shown the presence of ringlike structures, which may contain two or three $C1r_2-C1s_2$ complexes.[43] These ring structures form only in the presence of C1q, but their mode of interaction with C1q is not yet clear.[43]

Activation of C1

C1 in serum is rapidly activated by incubation at 37°C, in the presence of Ca^{2+} ions, with immune aggregates, polyanion–polycation–C reactive protein complexes, lipid A component of bacterial walls, certain viruses, and subcellular particles.[18,19] Partially purified C1 preparations and C1 reassembled from isolated subcomponents are activated rapidly by these substances in a similar way to serum C1.[12,26,37,46]

The purified C1 preparations, however, are generally unstable, even in the absence of such activators. When restored to physiological pH and ionic strength, purified C1 undergoes slow activation on incubation at 37°C in the presence of Ca^{2+} ions.[26,27,37] The rates of "spontaneous" activation observed are very variable,[12,26,37,46] and it is not yet clear whether this activation is an inherent property of C1 or whether it is due to trace contamination with proteases (e.g., plasmin), which can cleave C1r or C1s. Since the "physiological" mechanism of C1 activation (i.e., rapid activation by immune aggregates) is suggested to involve conformational changes,[46] it is possible that simple reversible aggregation of C1, in the absence of other proteins, triggers activation at a slow rate.

[46] A. W. Dodds, R. B. Sim, R. R. Porter, and M. A. Kerr, *Biochem. J.* 175, 383 (1978).

The sequence of events during activation of C1 by antibody–antigen aggregates has been studied.[46] On binding of C1 to immune complexes, C1r is rapidly cleaved to form C$\bar{1}$r, and this is followed closely by cleavage of C1s to form C$\bar{1}$s. Cleavage of C1s is mediated by C$\bar{1}$r, but neither C$\bar{1}$s nor C$\bar{1}$r itself is capable of activating C1r.[46] It has been suggested that binding of C1 to antibody–antigen complexes causes a conformational change in C1q, and alteration of the mode of binding of C1r to C1q.[46] This in turn may induce autocatalytic activation of C1r, by formation of an "active zymogen" intermediate, C1r'. This type of mechanism has many similarities to the activation of plasminogen by streptokinase. On the basis of kinetic and differential inhibition studies, it appears likely that C1r cleaves a neighboring molecule of C1r to form C$\bar{1}$r, which in turn cleaves C1s.[46]

[3] Preparation of Human C1q, A Subcomponent of the First Component of the Classical Pathway of Complement

By K. B. M. Reid

Introduction

C1q is an unusual serum protein[1] since approximately half of its amino acid sequence is collagen-like, being composed of -Gly-X-Y- repeating triplets (where X is often proline and Y is often hydroxyproline or hydroxylysine). Further similarity to collagen is shown by the presence of glucosylgalactosyl disaccharide units on 80% of the hydroxylysine residues and by the susceptibility of the molecule to collagenase. The precise role of the collagen-like structures (thought to be present in the form of six triple helices) in C1q is not known but may involve the binding and activation of the proenzyme forms of C1r and C1s, the other two subcomponents of the first component. The C1r and C1s proenzymes both behave as single chains of molecular weight (MW) 83,000 in dissociating conditions, but under physiological conditions they form a calcium-dependent complex of the form $C1r_2–Ca^{2+}–C1s_2$. This complex or a dimer of it, in association with C1q, which has MW 410,000, yields C1, the first component of complement.[1-3] Since C1q does not appear to contain an enzy-

---

[1] R. R. Porter and K. B. M. Reid, Adv. Protein Chem. 33, 1 (1979).

[2] I. H. Lepow, G. B. Naff, E. W. Todd, J. Pensky, and C. F. Hind, J. Exp. Med. 117, 983 (1963).

[3] R. J. Ziccardi and N. R. Cooper, J. Exp. Med. 147, 385 (1978).

matic site, its function in the C1 complex is considered to involve the recognition and binding of activators of the classical pathway, thus allowing the activation of the C1r and C1s proenzymes to take place—which in turn leads to the sequential activation of the later components in the pathway. The C1 complex can be activated by the direct interaction of C1q with the activator,[4] that is, by an antibody-independent mechanism, or by the interaction of C1q with the Fc region of antibody in immune complexes containing IgM or certain subclasses of IgG.[1,5] There is a prerequisite for the presence of a large number of closely spaced binding sites on the activating particle, in both the antibody-dependent and antibody-independent activation mechanisms. For example, C1 binds strongly to aggregated immunoglobulins, which usually leads to activation, but weakly to monomeric immunoglobulin, which does not lead to activation.[5,6]

Serum Concentration

The best estimate of the concentration of C1q in normal adult serum would appear to be approximately 70 μg/ml,[7] despite the fact that values as high as 190 μg/ml have been reported using radial immunodiffusion. The lower value is probably more accurate since there are several factors that cause the overestimation of C1q levels by radial immunodiffusion,[7] and a value of 63 μg/ml was obtained by an entirely different procedure[8] involving the estimation of the hydroxyproline content of the euglobulin fraction of human serum (this fraction contains at least 95% of the C1q present in serum). The level of C1q in human serum is remarkably constant for individuals from 3 days old to 40 years old; above 40 years the level shows a steady increase.[9]

Assay

No enzymatic activity has been detected in C1q preparations, therefore estimation of its functional activity in partially purified samples can be made by use of assays that detect C1q binding alone or C1q binding

[4] R. M. Bartholomew, A. F. Esser, and H. J. Müller-Eberhard, *J. Exp. Med.* **147**, 844 (1978).
[5] W. Augener, H. M. Grey, N. R. Cooper, and Müller-Eberhard, *Immunochemistry* **8**, 1011 (1971).
[6] H. Metzger, *Contemp. Top. Mol. Immunol.* **7**, 119 (1978).
[7] R. J. Ziccardi and N. R. Cooper, *J. Immunol.* **118**, 2047 (1977).
[8] C. L. Rosano, N. Parhami, and C. Hurwitz, *J. Lab. Clin. Med.* **94**, 593 (1979).
[9] K. Yonemasu, H. Kitajima, S. Tanabe, T. Ochi, and H. Shinkai, *Immunology* **35**, 523 (1978).

followed by activation of the C1r–C1s complex. Binding of C1q can be estimated using an agglutination assay employing IgG-coated latex particles[10] or by a binding assay employing ^{125}I-labeled C1q and immune aggregates.[11] However, the most satisfactory assay procedure involves the hemolysis of erythrocytes, as this is a sensitive monitor of C1q functional activity since the molecule has to play both a binding role (to antibody coated on the erythrocytes) and an activation role (of the $C1r_2$–Ca^{2+}–$C1s_2$ complex). The hemolytic assay may be performed using functionally pure C1r, C1s, and C2 along with EAC4 cells and EDTA-treated guinea pig serum,[12,13] or, more simply, by using only one reagent, human serum lacking C1q, along with sheep erythrocytes and antibody to sheep erythrocytes.[14] Kolb *et al.*[14] have described the preparation of C1q-deficient serum and used this serum to detect C1q hemolytic activity at the 1-ng level.

Purification Procedures

Three different types of procedures have been used to prepare human C1q: (1) conventional purification procedures utilizing euglobulin precipitation, ion-exchange chromatography, and gel filtration; (2) precipitation procedures involving the use of DNA or low-ionic-strength buffers containing the chelating agents EGTA and EDTA; (3) affinity chromatography on IgG–Sepharose or immune aggregates.

Only one of the conventional purification procedures will be described in detail in this article. This procedure has consistently given good yields of human C1q (approximately 25 mg/liter of serum), which has allowed the isolation of large amounts of C1q for chain separation on a preparative basis[15]; preparation of large fragments by limited proteolysis[16]; functional studies[13,15]; physical studies[17]; and extensive amino acid sequencing.[18] The other procedures, while giving similar yields, involve the use of at least one or two of the column steps used in the conventional procedure.

1. Conventional Purification Procedure. Serum prepared from outdated human plasma has been used routinely in this laboratory to prepare C1q for functional and chemical studies. No significant increase in the final

[10] R. W. Ewald and A. F. Schubart, *J. Immunol.* 97, 100 (1966).
[11] C. Heusser, M. Boesman, J. H. Nordin, and H. Isliker, *J. Immunol.* 110, 820 (1973).
[12] M. A. Calcott and H. J. Müller-Eberhard, *Biochemistry* 11, 3443 (1972).
[13] K. B. M. Reid, R. B. Sim, and A. P. Faiers, *Biochem. J.* 161, 239 (1977).
[14] W. P. Kolb, L. M. Kolb, and E. R. Podack, *J. Immunol.* 122, 2103 (1979).
[15] K. B. M. Reid, D. M. Lowe, and R. R. Porter, *Biochem. J.* 130, 749 (1972).
[16] K. B. M. Reid, *Biochem. J.* 155, 5 (1976).
[17] B. Brodsky-Doyle, K. R. Leonard, and K. M. B. Reid, *Biochem. J.* 159, 279 (1976).
[18] K. B. M. Reid, *Biochem. J.* 179, 367 (1979).

yield of protein or activity has been detected when fresh or fresh-frozen plasma was used or when proteolytic inhibitors such as DFP, PMSF, phenanthroline monohydrate, or iodoacetamide were included in the early stages of the preparation. The inclusion of DFP or NPGB is necessary during the procedure if the proenzyme forms of C1r and C1s are to be prepared at the same time as C1q. Clotting of citrated plasma is achieved by the addition of $CaCl_2$ to a concentration of 20 mM and incubation at 37°C for 3 hr, followed by centrifugation and final separation of serum from the clot by the use of nylon gauze or muslin.

DEAE–Sephadex A-50 was obtained from Pharmacia Fine Chemicals, Uppsala, Sweden; CM-cellulose 32 was obtained from Whatman Ltd, Maidstone, Kent, United Kingdom; Bio-Gel A-SM was obtained from BioRad Laboratories, Richmond, California. A stirred ultrafiltration cell, fitted with a PM-10 membrane (Amicon Corporation, Lexington, Massachusetts) was used to concentrate the fractions containing C1q. All operations are performed at 4°C unless stated otherwise. The procedure given below is adapted from several publications[2,12,14,19] and unpublished observations.

The first step involves precipitation of the euglobulins at low ionic strength, which, for 2000 ml of serum, is conveniently performed in two 10-liter beakers. Human serum (2000 ml) is added, with constant stirring to 20 liters of pH 5.5 acetate buffer (78.3 ml of 4 N NaOH is added to approximately 8 liters of distilled water, the pH adjusted to 5.5 with glacial acetic acid—approximately 20.2 ml is required—and the solution made up to 10 liters with distilled water). The mixture is allowed to stand for 36 hr, and then most (approximately 18 liters) of the clear supernatant is siphoned off and the euglobulin centrifuged, in 1-liter buckets, at 1500 g for 30 min. The precipitate is resuspended in 200 ml of the pH 5.5 acetate buffer and centrifuged at 1500 g for 30 min. The washed precipitate is dissolved in 110 ml of 0.5 M NaCl–10 mM EDTA, pH 7.4, and centrifuged at 80,000 g for 90 min. The surface lipid layer is removed and the supernatant is dialyzed against 2 × 2000 ml of 20 mM NaCl–56 mM sodium phosphate–1 mM EDTA buffer, pH 7.4 (5.9 g NaCl + 33.4 g anhydrous Na_2HPO_4 + 7.80 g Na H_2PO_4 $2H_2O$ + 33.4 ml of 0.15 M EDTA pH 7.4 per 5 liters). The dialyzed sample (approximately 2100 $E_{280}^{1\ cm}$ units in 108 ml) is clarified by centrifugation at 1500 g for 30 min and applied to a column (7 × 27 cm) of DEAE–Sephadex A-50 that has been equilibrated with the 20 mM NaCl–56 mM sodium phosphate–1 mM EDTA buffer, pH 7.4. The column is eluted at a flow rate of 100 ml/hr with a hydrostatic head of 1 m and 25-ml fractions are collected. The C1q is not retained on the column and therefore all the protein eluted with the equilibrating buffer is pooled

[19] K. B. M. Reid, *Biochem. J.* **141**, 189 (1974).

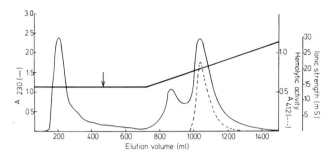

FIG. 1. Ion-exchange chromatography of C1q on CM-cellulose 32. The partially purified C1q (approximately 180 $E_{280}^{1\,cm}$ units in 60 ml) from DEAE–Sephadex A-50 was dialyzed against 0.23 M acetic acid–NaOH buffer, pH 5.2, and loaded onto a column (3.5 × 40 cm) of CM-cellulose 32 equilibrated in the same buffer. The column was developed as described in the text. C1q hemolytic activity eluted between 18 and 24 mS. ——, A_{230}; -----, hemolytic activity. The arrow marks the start of the gradient.

(approximately 180 $E_{280}^{1\,cm}$ units in 700 ml) and concentrated in an ultrafiltration cell to 60 ml using a PM-10 membrane. It is essential that the speed of the magnetic stirrer in the ultrafiltration cell be kept low in order to avoid irreversible denaturation of the C1q. Samples containing C1q can also be concentrated by the addition of ammonium sulfate at 50% of saturation at 4°C.[14] The concentrated sample (approximately 180 $E_{280}^{1\,cm}$ units in 60 ml) is then dialyzed against 2 × 2000 ml of 0.23 M acetic acid–NaOH buffer, pH 5.2 (55 ml of 4 M NaOH adjusted to pH 5.2 with acetic acid and made up to 1000 ml), and applied to a column (3.5 × 40 cm) of CM-cellulose 32 which is equilibrated with the same buffer. The column is washed with the starting buffer at 80 ml/hr until the A_{230} of the eluate is approximately 0.10, and then a linear gradient is developed using 750 ml of 0.23 M acetic acid–NaOH buffer, pH 5.2, and 750 ml of the same buffer containing 0.27 M NaCl. C1q is one of the last proteins to be eluted from the column, between 18 and 24 mS, at a position just over halfway through the salt gradient (Fig. 1). The peak eluted prior to the C1q peak, between 16 and 18 mS, contains activated properdin, therefore both major peaks eluted with the salt gradient are pooled (Fig. 2). If C1q hemolytic activity is not being measured it is useful to monitor the column fractions by SDS–polyacrylamide slab-gel electrophoresis in view of the very characteristic band patterns given by both proteins in reducing and nonreducing conditions.[14,20,21] Properdin has an apparent MW ≈ 60,000 in both reducing and onreducing conditions but may appear as a closely spaced doublet in the nonreducing conditions. C1q shows 2 bands, the A–B dimer and C–C

[20] K. B. M. Reid and R. R. Porter, *Biochem. J.* **155**, 19 (1976).
[21] K. Yonemasu and R. M. Stroud, *Immunochemistry* **9**, 545 (1972).

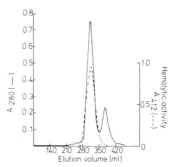

FIG. 2. Gel filtration of Clq on Bio-Gel A-5M. Partially purified Clq (approximately 25 $E_{280}^{1\ cm}$ units in 7.0 ml) from CM-cellulose 32 was dialyzed against 5 mM sodium phosphate–0.65 M NaCl–10 mM EDTA, pH 7.4, and after centrifugation, applied to a column (2.5 × 85 cm) of Bio-Gel A-5M equilibrated with the same buffer. The column was run at 10 ml/hr. ——— A_{280}; -----, hemolytic activity. This elution profile shows the final purification of half the sample derived from 2000 ml of human serum.

dimer, with apparent MWs 69,000 and 54,000 in nonreducing conditions and 3 bands, the separate A, B, and C chains, of apparent MWs 34,000, 32,000, and 27,000 in reducing conditions on SDS–polyacrylamide gel electrophoresis. The pool from the CM-cellulose column, containing the Clq (approximately 60 $E_{280}^{1\ cm}$ units in 300 ml) is concentrated by ultrafiltration. The only significant contaminating protein at this stage in the procedure is IgG (usually less than 10% of the material in the Clq pool). The sample from the CM-cellulose column after concentration to 14 ml is dialyzed against the buffer (2 × 1000 ml) used in the last step of the purification procedure, which involves gel filtration on a column (2.5 × 85 cm) of Bio-Gel A-5M[14] equilibrated with 5 mM sodium phosphate–0.65 M NaCl–10 mM EDTA, pH 7.4 (10.13 ml of 0.4 M Na$_2$HPO$_4$ + 5.95 ml of 0.4 M NaH$_2$PO$_4$ + 189.93 g NaCl + 333 ml of 0.15 M EDTA, pH 7.4). The elution profile in Fig. 2 shows the gel filtration of 7.0 ml (containing approximately 25 $E_{280}^{1\ cm}$ units) of the sample from CM-cellulose, that is, one-half of the sample. This step allows the separation of Clq from traces of IgM, C4-binding protein, and IgG (Fig. 2). The final yield of Clq is approximately 50 mg of Clq from 2 liters of serum (i.e., approximately 34 $E_{280}^{1\ cm}$ units—Clq having an $E_{1\ cm}^{1\%}$ of 0.68 (at 280 nm).

2. *Precipitation Procedures.* The procedure involving the precipitation of Clq by DNA,[11,22] followed by digestion of the DNA with DNAase, requires gel filtration and ion-exchange chromatography to free the Clq from contaminants, and thus this procedure is similar in some respects to the conventional procedure. In the precipitation method of Yonemasu and

[22] V. Agnello, R. J. Winchester, and H. G. Kunkel, *Immunology* **19**, 909 (1970).

Stroud[21] the C1q is precipitated three times by means of dialysis against low-ionic-strength buffers containing EDTA and EGTA. A high yield of C1q is obtained and a relatively small volumes (50 ml) of serum can be used without any lowering of the yield, therefore this method should be considered if only a small amount of serum is available or if only a small amount of C1q is required. However, human C1q prepared by this method has been reported to contain an inhibitor that greatly reduces the hemolytic activity of the preparation.[23] This inhibitor, which co-purifies with the C1g, may be separated from C1q by affinity chromatography on Con A–Sepharose; the inhibitor is not retained and the C1q can be eluted with α-methylglucopyranoside.[23]

3. *Affinity Chromatography.* In this method C1q is bound to IgG which is either covalently linked to Sepharose-4B[14,24,25] or present in preformed immune aggregates.[26] One recent successful application of the affinity procedure has been described by Kolb *et al.*,[14] who used a linear salt gradient to elute bound C1q from the IgG–Sepharose, rather than the stepwise conditions, involving 1 : 4 diaminobutane used in previous studies. As with the precipitation procedures, another step, such as gel filtration or ion-exchange chromatography, is necessary for the complete purification of C1q by the procedures involving binding to IgG and, in addition, the IgG–Sepharose affinity column,[14,24,25] or a large quantity of immune aggregates,[26] has to be prepared.

Yields

The conventional method yields 20–25 mg/liter of serum.[15,19] The DNA-precipitation method yields approximately 13 mg/liter of serum[11] while the precipitation procedure involving low-ionic-strength buffers and chelating agents yields, prior to removal of a C1q inhibitor, 50 mg/liter.[21] The affinity methods yield 17–50 mg/liter of serum.[14,25,26]

Preparation of C1q from Other Species

C1q has been prepared from the ox,[27] rabbit,[15,28,29] rat,[30] horse,[31] and frog.[32] Conventional procedures[15,27,28,31,32] or Yonemasu and Stroud's precipitation technique[29,30] have been used. The yields and properties of the

[23] J. D. Conradie, J. F. Volanakis, and R. M. Stroud, *Immunochemistry* 12, 967 (1975).
[24] C. R. Sledge and D. H. Bing, *J. Immunol.* 111, 661 (1973).
[25] S. N. Assimeh, D. H. Bing, and R. H. Painter, *J. Immunol.* 113, 225 (1974).
[26] G. J. Arlaud, R. B. Sim, A-M. Duplaa, and M. G. Colomb, *Mol. Immunol.* 16, 445 (1979).
[27] D. Campbell, N. A. Booth, and J. E. Fothergill, *Biochem. J.* 177, 531 (1978).
[28] S. M. Paul and P. A. Liberti, *J. Immunol. Methods* 21, 341 (1978).
[29] J. E. Volanakis and R. M. Stroud, *J. Immunol. Methods* 2, 25 (1972).

different C1q preparations indicate that a molecule very similar to human C1q is present at a level of approximately 70 mg/liter in these animals.

Stability

Repeated freezing and thawing of human C1q preparations, especially at neutral pH, leads to irreversible denaturation, as does vigorous stirring (e.g., in a stirred ultrafiltration cell). Samples stored at $-70°C$ in acetate buffer pH 5.3, containing 0.25 M NaCl, retain full hemolytic activity for at least 6 months. Kolb *et al.*[14] have recommended the precipitation of finally purified human C1q by ammonium sulfate at 50% saturation and 4°C, followed by resuspension in 50 mM NaCl containing 40% glycerol prior to storage at $-20°$ or $-70°C$. Rabbit C1q has been reported to show no evidence of aggregation on thawing after storage for 6 months at $-70°C$.[28]

Physical Properties and Composition

Human C1q has an extinction coefficient ($E_1^{1\%}{}_{cm}$) of 6.82 at 280 nm[15] and has an unusual amino acid composition for a serum protein since it contains 17% glycine, 5% 4-hydroxyproline, and 2.1% hydroxylysine.[12,15] C1q is a glycoprotein having a carbohydrate content of approximately 8.5%, and 6.0% of this carbohydrate is present as glucosylgalactosyl disaccharides, or galactose monosaccharides, linked to hydroxylysine residues in the collagen-like regions of the molecule.[33] The remaining 2.5% of the carbohydrate is composed of 6 asparagine-linked sugar chains, which are exclusively located in the C-terminal globular regions of the polypeptide chains of C1q.[34] In nondissociation conditions, C1q has MW 410,000 and an $S_{20,w}^0$, in acetate buffer pH 5.2, of 10.20 S, as estimated by analytical ultracentrifugation.[15] There are 18 polypeptide chains (6A, 6B, and 6C), each of MW $\approx$ 23,000, per molecule of C1q, and these chains are present as 9 disulfide-linked dimers (6 A–B dimers and 3 C–C dimers).[20] In each of the three types of chain, after a short N-terminal section of noncollagen-like amino acid sequence of 2–8 residues, there is a region of approximately 80 residues of typical -Gly-X-Y- repeating triplet, collagen-like sequence (the continuity of which is broken at one point in each chain), followed by approximately 110 residues of globular sequence which extends to the C-terminal (Fig. 3).[16,18] Evidence that these

---

[30] K. Höffken, P. J. McLaughlin, M. R. Price, V. E. Preston, and R. W. Baldwin, *Immunochemistry* **15**, 409 (1978).

[31] T. L. McDonald and D. Burger, *Immunology* **37**, 517 (1979).

[32] R. J. Alexander and L. A. Steiner, *J. Immunol.* **124**, 1418 (1980).

[33] H. Shinkai and K. Yonemasu, *Biochem. J.* **177**, 847 (1979).

[34] T. Mizuochi, K. Yonemasu, K. Yamashita, and A. Kobata, *J. Biol. Chem.* **253**, 7404 (1979).

```
          1            10             20              30              40          50
A chain  E D L C R A P D G K *K G E A G R P G R  R G R P G L K G E Q G E *P G A P G I R T G I Q — — G L K G D Q G E P *
B chain  E L S C T G P P *A I *P G I P G T  P G P D G Q P G T *P G I *K G E *K G L P — G L A G D H G E F G E *K G D P
C chain  N T G C Y G I P *G M P G L P G A P G K  D G Y D G L P G P *K G E P G I P A I K — G I R — — — G P P *G Q K G E P

         54       60              70              80              90             100         108
A chain  G P S G N *P G K V G Y P *P G P S  G P L G A R G I K G T P G S P G N I K D Q P P R P A F S A I R R N P P M G G
B chain  G I *P G D P G K V G P *K G P M  G P *K G P P G A P G A P *G P *K G E S G D Y K A T Q K I A F S A T R T I N V P L R
C chain  G L P G H *K G K D G P N G P P *G M P G V P *P G P M G I P G E P G E E G R Y K Q K F Q S V F
```

FIG. 3. Amino acid sequences of the collagen-like regions of the A, B, and C chains of human C1q. The numbering of the amino acid residues in all 3 chains is based on the B-chain sequence. The optimal alignment for showing the maximum homology between the 3 chains is obtained if a gap is left between positions 38 and 39 in the B- and C-chain sequences to allow for the "extra" threonine residue in the A-chain sequence. A triplet gap is left between residues 41 and 45 in the A- and C-chain sequences to allow for the "extra" triplet in the B-chain sequence. *, A hydroxylated residue. All except 3 (at positions B-50, B-65, and C-38), of the 14 hydroxylysine residues appear to be glycosylated. The data shown are taken from K. B. M. Reid, *Biochem. J.* 179, 367 (1979). Residues are given in the single-letter code: A, Ala; B, Asx; C, Cys; D, Asp; E, Glu; F, Phe; G, Gly; H, His; I, Ile; K, Lys; L, Leu; M, Met; N, Asn; P, Pro; Q, Gln; R, Arg; S, Ser; T, Thr; V, Val; W, Trp; X, unknown; Y, Tyr.

88	100	110	120	130

K D Q P R P A F S A I R R N P P M G G N V V I F D T V I T N Q E E P Y Q (N) H S G R F V

131	140	150	160	170

C T V P G Y Y Y F T F Q V L S Q W E I (N) L S I V S W S R G Q V R (R) / S L G F C D T T N

173	180	190	199	222

K G L F Q V V S G G E V L Q L E E G D Q V̄ X V E X D P (approx. 20 residues) F L A

FIG. 4. Partial amino acid sequence of the noncollagenous region of the A chain of human C1q. The identity of residues 124, 150, and 163 (which are given in parentheses) requires further confirmation and no sequence information is available for residues 200–219. /, No overlapping sequence has been established. Residues are given in the single-letter code as in Fig. 3. The data shown are taken from K. B. M. Reid, *Biochem. J.* 141, 189 (1974); 179, 367 (1979); and unpublished work.

collagen-like sequences combine to form triple-helical structures has been obtained by limited proteolysis of the molecule with collagenase,[15,35,36] pepsin,[16] circular dichroism studies,[17] and electron-microscopy studies.[17,37] Electron micrographs of C1q indicate that the molecule is divided into two distinct types of structure, that is, six globular "heads" each of which is connected by a strand to a central fibril-like region. A model has been proposed for C1q in which the six strands and central fibril-like region are considered to be composed of six collagen-like triple helices while each globular "head" is considered to be composed of the 110 residues of globular sequence from one A, one B, and one C chain.[20] A partial amino acid sequence of the noncollagenous region of one of the chains of human C1q is shown in Fig. 4.

The three types of chain found in C1q have proved unusually difficult to separate on a preparative basis, but this can be achieved by ion-exchange chromatography in solutions containing 9 M urea, after performic oxidation.[15] The globular "heads," which are considered to contain the binding sites for immunoglobulin, can be isolated after collagenase digestion of the whole molecule.[35,36] The collagen-like regions, which may be involved in the binding and activation of C1r and C1s and in the binding to receptors on lymphoid cells and platelets, can be isolated as a 176,000-MW fragment after limited proteolysis of C1q with pepsin.[16,37a]

[35] E. P. Paques, R. Huber, and H. Priess, *Hoppe-Seyler's Z. Physiol. Chem.* 360, 177 (1979).

[36] N. C. Hughes-Jones and B. Gardner, *Immunochemistry* 16, 697 (1979).

[37] H. R. Knobel, W. Villiger, and H. Isliker, *Eur. J. Immunol.* 5, 78 (1975).

[37a] Good recoveries of human C1q have been obtained in a simple two-step procedure involving Bio-Rex 70. A. J. Tenner, P. H. Lesarre, and N. R. Cooper, *J. Immunol.* 127, 648–653 (1981).

[4] The Human Complement System Serine Proteases C$\overline{1}$r and C$\overline{1}$s and Their Proenzymes

By R. B. SIM

The first component of human complement, C1, is a Ca^{2+}-dependent assembly of three types of subcomponents: C1q, C1r, and C1s. The C1r and C1s subcomponents are proenzymes of proteolytic enzymes, while C1q serves as a recognition and binding protein.[1,2] Activation of the classical pathway of complement is initiated by binding of C1, through the C1q subcomponent, to classical pathway activators, such as antibody–antigen complexes.[1,2] Binding of C1 to certain viral activators occurs through both C1s and C1q.[3] C1 binding is thought to cause conformational changes within C1, and result in autocatalytic activation of the proenzyme form of C1r. The activated C1r (C$\overline{1}$r)[4] subcomponent then activates the proenzyme form of C1s by limited proteolysis. Activated C1s (C$\overline{1}$s)[4] then cleaves and activates the second (C2) and fourth (C4) components of complement.[1,2] The properties of the C1 complex and of subcomponent C1q are described in other sections.[2,5]

Several procedures have been developed for isolation of C1r and C1s in both proenzymic and activated form. Substantial purification of C$\overline{1}$s was first achieved by Haines and Lepow,[6] and the proenzyme form was later isolated by Okamura *et al.*,[7] and in higher yield by Sakai and Stroud.[8] C1r has proved more difficult to isolate than C1s, probably because of lower solubility and tendency to aggregate. Isolation of C$\overline{1}$r was first reported by de Bracco and Stroud.[9] Conclusive proof that C1r occurred in proenzyme form was provided by Valet and Cooper,[10] who obtained the proenzyme in highly purified form. Isolation to homogeneity of proenzymic C1r was later reported by Ziccardi and Cooper[11] and by Porter and colleagues.[12,13]

---

[1] R. R. Porter and K. B. M. Reid, *Adv. Protein Chem.* **33**, 1 (1979).

[2] R. B. Sim, this volume (section on C1).

[3] R. M. Bartholomew and A. F. Esser, *Biochemistry* **19**, 2847 (1980).

[4] The convention in complement protein nomenclature is that activated components are designated by a superscript bar (e.g., C$\overline{1}$ = activated C1).

[5] K. B. M. Reid, this volume (section on C1q).

[6] A. L. Haines and I. H. Lepow, *J. Immunol.* **92**, 456 (1964).

[7] K. Okamura, M. Muramatu, and S. Fujii, *Biochim. Biophys. Acta* **295**, 252 (1973).

[8] K. Sakai and R. M. Stroud, *J. Immunol.* **110**, 1010 (1973).

[9] M. M. E. de Bracco and R. M. Stroud, *J. Clin. Invest.* **50**, 838 (1971).

[10] G. Valet and N. R. Cooper, *J. Immunol.* **112**, 1667 (1974).

[11] R. J. Ziccardi and N. R. Cooper, *J. Immunol.* **116**, 496 (1976).

[12] R. B. Sim and R. R. Porter, *Biochem. Soc. Trans.* **4**, 127 (1976).

[13] I. Gigli, R. R. Porter, and R. B. Sim, *Biochem. J.* **157**, 541 (1976).

Identification of C1r and C1s as serine protease zymogens rests on the inhibition of the activated forms, C̄1̄r and C̄1̄s, by diisopropyl phosphorofluoridate (DFP), on their overall structural similarities to larger serine proteases such as plasminogen/plasmin, and on limited sequence data, as discussed below.

I. Assay of C1r, C̄1̄r, C1s, and C̄1̄s

A. Hemolytic Assay

The most specific assays for C1r, C̄1̄r, C1s, and C̄1̄s are hemolytic assays, using a procedure very similar to that described for C1.[2] This assay method is based on that described by de Bracco and Stroud[9] and is also applicable to assay of C1q.[5]

The C1 subcomponent to be measured is incubated, in the presence of the other two subcomponent types, with EAC4 cells.[2] This results in reconstitution and activation of the C1 complex on the EAC4 cells. The quantity of C̄1̄ reconstituted and bound is, in the correct conditions, proportional to the concentration of the limiting subcomponent.

Procedure

Twofold serial dilutions (0.5 ml of each) of the subcomponent to be assayed are made in DGVB^{++},[2] and to each dilution is added 0.5 ml of DGVB^{++} containing an excess (2- to 5-fold on a molar basis over the initial concentration of the subcomponent being assayed) of the other two subcomponent types, plus 0.5 ml of EAC4^{hu} cells (10^8/ml in DGVB^{++}).[2] Cells and subcomponents are incubated for 1 hr at 30°C, then centrifuged (1500 g, 10 min). The supernatant is discarded, and the cells are washed twice with 1.0 ml of cold (4°C) DGVB^{++}. The cells are finally resuspended in buffer containing C2, and the assay completed by incubation with C2 and C–EDTA as in Ref. 2.

Subcomponent hemolytic assays of this type are subject to the same technical error and variation as the C1 hemolytic assay.[2] These assays are not suitable for determining the concentration of individual C1 subcomponents in normal serum or plasma, or in euglobulin precipitates from serum, since these materials contain all of the C1 subcomponents, and it is difficult to adjust an assay to a suitable excess of other subcomponent types. The assay is suitable for detection and comparative measure of each subcomponent type in partially purified material, and represents the most reliable method for verifying the activity of highly purified proenzymic C1r and C1s. Hemolytic assays can detect C1r, C̄1̄r, C1s, and C̄1̄s at the 1-10 ng level.

The assay method requires, in addition to the reagents required for C1 assay,[2] functionally pure C1q, C1r (C$\overline{1}$r), and C1s (C$\overline{1}$s) reagents. These can be prepared in a relatively crude form by DEAE–cellulose chromatography of an acid euglobulin precipitate of human serum.[14] It is preferable to use preparations of proenzymic C1r and C1s as reagents, since excess C$\overline{1}$s can mimic the action of whole C$\overline{1}$ if it remains nonspecifically bound to the EAC4 cells during centrifugation and washing.[2] The reagent preparations of C1q, C1r, and C1s must be extensively titrated against each other to establish the optimal quantity of each subcomponent to be used. Use of too great an excess of other subcomponents during an assay will result in loss of sensitivity due to partitioning of the subcomponent under test between cell-bound C$\overline{1}$ and C1 or other partial complexes in solution. Much of the preliminary cross-matching of reagents can be eliminated if pure C1q,[5] and C1r and C1s prepared as described below are available as reagents.

B. Esterolytic Assays

The activated subcomponents C$\overline{1}$r and C$\overline{1}$s can be assayed by hydrolysis of amino acid esters. These assays are not specific and should be used to measure only the very highly purified enzymes, or the quantity of C$\overline{1}$r or C$\overline{1}$s in certain well-defined serum fractions. The esterolytic specificities of C$\overline{1}$r and C$\overline{1}$s are discussed later in this section. The proenzymes, C1r and C1s, do not hydrolyze these substrates at a significant rate.

Assay of C$\overline{1}$r

C$\overline{1}$r cleaves the ester substrates acetyl-arginine methyl ester (AAME; Sigma Chemical Co.) and acetyl-glycine-lysine methyl ester (AGLME; Sigma). The hydrolysis is slow, and the assay is insensitive. It should be noted that C$\overline{1}$s cleaves both of these substrates faster than does C$\overline{1}$r, and other plasma proteases (e.g., plasmin, thrombin) also interfere. Great caution must be exercised in interpreting such assays. A procedure for C$\overline{1}$r assay by AAME hydrolysis is described by Ashgar et al.[15]

Assay of C$\overline{1}$s

C$\overline{1}$s has a wide esterolytic specificity and can be assayed readily using a number of amino acid esters, for example, N^α-carbobenzoxy-L-lysine p-nitrophenol ester (ZLNE; Sigma), N-carbobenzoxy-L-tyrosine p-nitro-

[14] I. H. Lepow, G. B. Naff, E. W. Todd, J. Pensky, and C. F. Hinz, Jr., J. Exp. Med. 117, 983 (1963).
[15] S. S. Ashgar, K. W. Pondman, and R. H. Cormane, Biochim. Biophys. Acta 317, 539 (1975).

phenol ester (ZTNE; Sigma) and acetyl-L-tyrosine ethyl ester (ATEE; Sigma).

1. ZLNE. Highly purified Cīs, obtained by "spontaneous" activation during purification,[12,13] cleaves ZLNE at a rate of 2700–3100 nmol ZLNE hydrolyzed/min/mg Cīs at 25°C.[16] The procedure for assay of Cīs by this method is reported elsewhere in this volume.[17] The assay is again not specific for Cīs, since ZLNE is a good plasmin substrate.[16] Cīr, however, does not cleave ZLNE, although occasional contamination of highly purified Cīr with a weak ZLNE esterase activity is observed.[16] In a neutral euglobulin precipitate of serum, prepared as in Ref. 2, and subsequently incubated (1 hr, 37°C in presence of Ca^{2+} ions) to activate all the C1 present, Cīs accounts for more than 95% of the total ZLNE esterase activity present.[18] Thus Cīs in neutral euglobulin fractions can be quantified by this method.

2. ZTNE. ZTNE provides a less sensitive assay for Cīs than does ZLNE. Plasmin also cleaves ZTNE, but Cīr, prepared as described here, does not.[16] In an activated neutral euglobulin precipitate of serum, as described above, Cīs accounts for only 50–60% of the total ZTNE esterase activity.[18] A procedure for Cīs assay by ZTNE has been described by Bing.[19]

3. ATEE. ATEE is regarded as being a relatively specific, though insensitive, substrate for Cīs.[20] Spectrophotometric[21] and pH-stat[15] assays for ATEE hydrolysis by Cīs have been described.

C. Other Assays

Cīs can be assayed by destruction of its natural substrate, C4.[13] Cīr may be measured by the rate at which it activates Cls.[13,22] Using an enzyme:substrate molar ratio of 1:10, at a Cls concentration of 450 μg/ml, Cīr activates 0.4–0.6 mg of Cls/min/mg Cīr at 37°C, pH 7.4, in isotonic saline in the presence of 5 mM EDTA.[18]

II. Purification of Clr, Cīr, Cls, and Cīs

Isolation of Clr and Cls and their activated forms has been achieved in several laboratories, using conventional precipitation, ion-exchange, and

---

[16] R. B. Sim, R. R. Porter, K. B. M. Reid, and I. Gigli, *Biochem. J.* **163**, 219 (1977).

[17] R. B. Sim and A. Reboul, this volume (section on Cī-Inh).

[18] R. B. Sim, D. Phil. Thesis, University of Oxford (1976).

[19] D. H. Bing, *Biochemistry* **8**, 4503 (1969).

[20] P. J. Lachmann, *in* "Defence and Recognition" (R. R. Porter, ed.), p. 361. Butterworth, London, 1973.

[21] G. Valet and N. R. Cooper, *J. Immunol.* **112**, 339 (1974).

[22] A. W. Dodds, R. B. Sim, R. R. Porter, and M. A. Kerr, *Biochem. J.* **175**, 383 (1978).

gel-filtration procedures.[7,8,11–13,21] A number of isolation procedures involving affinity chromatography on immunoglobulin G (IgG)–Sepharose,[23,24] or on antibody–antigen aggregates,[25,26] have been devised. The properties described for C1s (or C$\overline{1}$s) prepared by each of the methods cited are in very close agreement. For C1r and C$\overline{1}$r, however, several differences, principally in stability and activation properties, have been reported for material isolated in different ways. These differences are discussed in a later section.

The conventional isolation procedure described below gives a similar yield of C1r and C1s to that obtained with affinity-chromatography procedures.[24–26]

A. Isolation of C1r and C1s [12,13,18]

The isolation procedure for C1r and C1s continues directly from the partially purified C1 preparation described in Ref. 2. All steps are carried out at 4°C. The pool of purified C1, obtained from Sepharose-6B and concentrated to 20–30 ml, is taken and made 20 mM with EDTA by addition of one-quarter volume of 100 mM sodium EDTA, pH 5.5. Care should be taken that the pH does not fall below 5.2 on addition of EDTA. The preparation is then made 5 mM with DFP by addition of an appropriate volume of stock 2.5 M DFP solution.[2] The material is then dialyzed against 2 liters of 57 mM sodium phosphate–5 mM EDTA, pH 7.4. A light precipitate usually forms during dialysis, and this is removed by centrifugation (2600 g, 10 min), and discarded. The supernatant is made 2 mM with DFP.

DEAE–Cellulose Chromatography

The supernatant is loaded onto a column (16 × 2.5 cm diameter) of DEAE–cellulose (Whatman, DE-32 or DE-52) equilibrated in 57 mM sodium phosphate–5 mM EDTA, pH 7.4. The column is washed with 300–400 ml of starting buffer, at a flow rate of about 60 ml/hr, and then the bound proteins are eluted by a linear NaCl gradient made up of 250 ml of the starting buffer mixing with 250 ml of the starting buffer made 300 mM with NaCl. As shown in Fig. 1, C1q elutes in the starting buffer. Fractions eluted at 0.09–0.15 relative salt concentration (RSC) contain subcomponent C1r, and C1s is eluted at 0.16–0.25 RSC. The principal contaminant

[23] P. A. Taylor, S. Fink, D. H. Bing, and R. H. Painter, *J. Immunol.* **118**, 1722 (1977).
[24] R. M. Chapuis, H. Isliker, and S. N. Assimeh, *Immunochemistry* **14**, 313 (1977).
[25] G. J. Arlaud, R. B. Sim, A.-M. Duplaa, and M. G. Colomb, *Mol. Immunol.* **16**, 445 (1979).
[26] G. J. Arlaud, C. L. Villiers, S. Chesne, and M. G. Colomb, *Biochim. Biophys. Acta* **616**, 116 (1980).

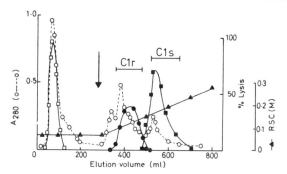

FIG. 1. DEAE–cellulose chromatography. Conditions are as described in the text. An arrow marks the start of gradient elution. The increase in ionic strength (▲) is shown as relative salt concentration (RSC). (A solution of RSC = 100 mM has the same conductivity as 100 mM NaCl.) The hemolytic activities of C1q (□), C1r (●), and C1s (■) were determined by assaying a 1 : 1000 dilution of column fractions. A_{280} is shown as open circles (○).

originally present in the C1 preparation, C4b-binding protein,[2] elutes with both C1q and C1r.

The peaks of C1r and C1s activity are pooled separately (Fig. 1). Each pool is made 2 mM with DFP, and diluted with twice its own volume of 93 mM sodium phosphate–200 mM NaCl, pH 5.3. Each pool is then concentrated to 10–12 ml by ultrafiltration with gentle stirring on a PM-10 membrane (Amicon).

Sephadex G-200 Chromatography

The concentrated pool of C1r is made 2 mM with DFP, and fractionated on a column (90 × 5 cm diameter) of Sephadex G-200 (Pharmacia) in 93 mM sodium phosphate–200 mM NaCl, pH 5.3. A typical elution profile is shown in Fig. 2A. The major contaminant in the C1r, C4b-binding protein, elutes in the void volume. Fractions containing C1r are pooled as indicated in Fig. 2A, and concentrated by ultrafiltration on a PM-10 membrane to a final concentration of about 250 μg C1r/ml. This material may be stored for 3–4 months at 4°C and pH 5.3 without loss of activity. Slight aggregation of the protein may occur on storage. Freezing of C1r preparations should be avoided, as this also tends to cause aggregation.

The concentrated C1s pool from DEAE–cellulose is also made 2 mM with DFP, and fractionated on Sephadex G-200 as described for the C1r pool (Fig. 2C). Small amounts of C4b-binding protein and C1r are removed by this procedure. Sephadex G-200 fractions containing C1s are pooled as indicated in Fig. 2C and concentrated as for C1r to a final concentration of 1–2 mg/ml C1s. This material is very stable, and may be stored for up to 6 months at 4°C and pH 5.3, or in physiological buffers at

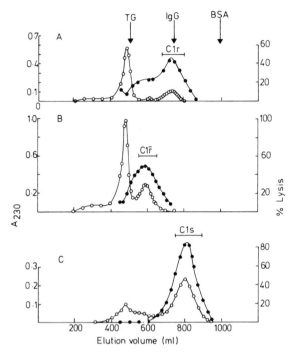

Fig. 2. Sephadex G-200 chromatography. Conditions are as described in the text. (A) Purification of C1r; (B) purification of C̄1r̄; (C) purification of C1s. Subcomponent hemolytic activities (●) were determined by assaying 1 : 1000 dilutions of column fractions. The elution positions of protein standards, bovine thyroglobulin (TG), rabbit immunoglobulin G (IgG), and bovine serum albumin (BSA) are indicated. A_{280} is shown as open circles (○).

pH 7.0–7.5. C1s, unlike C1r, is unaffected by freezing, and may be stored for more than 2 years at $-20°C$ in pH 7.0–7.5 isotonic buffers.

B. Isolation of C̄1r̄ and C̄1s̄[12,13,18]

Isolation of the activated forms, C̄1r̄ and C̄1s̄, is achieved by exactly the same procedure as for the proenzyme forms. The starting material is, however, the partially purified C̄1 pool from Sepharose-6B.[2] No DFP is added during processing of C̄1r̄ and C̄1s̄. The activated forms migrate on DEAE–cellulose exactly as do the proenzyme forms. C̄1s̄ also behaves identically to C1s on Sephadex G-200. C̄1r̄, however, behaves as a larger species than C1r on Sephadex G-200 (Fig. 2B). The basis of this difference is discussed later. An advantage of the different migration of C1r and C̄1r̄ at this pH is that traces of C̄1r̄ in C1r preparations can be removed. The activated forms, C̄1r̄ and C̄1s̄, are stored as described for C1r and C1s.

TABLE I
YIELDS AND PURIFICATION

Material	Total protein (mg)	Yield of activity (%)[a]		Purification factor	
		Clr	Cls	Clr	Cls
Serum	70,000	100	100	—	—
Euglobulin precipitate	520	79	79	106	106
C1 pool from Sepharose-6B	145	51	51	246	246
Cls pool from DEAE–cellulose	14	—	34	—	1700
Cls pool from Sephadex G-200	11	—	30	—	1910
Clr pool from DEAE–cellulose	55	30	—	382	—
Clr pool from Sephadex G-200	7.3	22	—	2110	—

[a] Yields of activity are based on C1 and C1-subcomponent hemolytic assays.

Purity and Yield

Final preparations of Clr, C$\bar{\mathrm{l}}$r, Cls, and C$\bar{\mathrm{l}}$s are homogeneous as judged by polyacrylamide-gel electrophoresis in the presence or absence of sodium dodecyl sulfate (SDS). The yields of activated subcomponents, C$\bar{\mathrm{l}}$r and C$\bar{\mathrm{l}}$s, are the same as yields of the proenzymes. A summary of typical purification stages is shown in Table I. The range of yields encountered in over 20 preparations is: Clr or C$\bar{\mathrm{l}}$r, 4.8–10.0 mg/liter of serum (average 7.6 mg); Cls or C$\bar{\mathrm{l}}$s, 7.9–16.0 mg/liter of serum (average 12.6 mg). The concentration of Clr in serum has been estimated at 34[27]–100[28] mg/liter, and of Cls as 22,[6] 31,[27] 33,[29] 80,[21] and 110[11] mg/liter. Estimates in this laboratory, based on recoveries of C1 and C1-subcomponent hemolytic activities during isolation indicate 30–35 mg/liter of Clr and Cls in serum made from standard anticoagulant-diluted plasma.

Other preparative methods using affinity methods[24–26] give similar yields of Clr and Cls to the procedure described above, and those described by Colomb and colleagues[25,26] are particularly rapid and convenient.

Clr and Cls in Other Species

C1 activity has been detected in a wide range of species, implying the presence of Clr and Cls subcomponents (for summary see Ref. 2). Rabbit

[27] R. J. Ziccardi and N. R. Cooper, J. Immunol. 118, 2047 (1977).
[28] M. M. E. de Bracco, C. L. Christian, and R. M. Stroud, Clin. Exp. Immunol. 16, 453 (1974).
[29] K. Nagaki and R. M. Stroud, J. Immunol. 105, 170 (1970).

$C\bar{1}r^{30}$ and $C\bar{1}s$,[31] bovine $C1s$ and $C\bar{1}s$,[32] and guinea pig $C1s^3$ have been isolated by methods similar to those used for human subcomponents. These other mammalian proteins are very much like the human proteins in structure and activity.

III. Properties of $C1r$, $C\bar{1}r$, $C1s$, and $C\bar{1}s$

A. Structural Properties

The proenzymic forms, $C1r$ and $C1s$, are both single polypeptide chains of MW 83,000–85,000, as determined by gel filtration in denaturing conditions[12,16] and ultracentrifugation in denaturing conditions.[33] Estimates of apparent molecular weights of $C1s$, and especially of $C1r$, by SDS–polyacrylamide gel electrophoresis are generally higher than 85,000, as summarized in Ref. 16. Both $C1r$ and $C1s$ are glycoproteins and are likely to show slightly anomalous migration on SDS–polyacrylamide gels. In nondenaturing conditions, $C1r$ exists as a noncovalently linked dimer of MW 166,000–188,000.[9-11,16,33] This dimer has a sedimentation coefficient of 6.7–7.9 S,[10,11,16,33,34] and is asymmetric, with an estimated axial ratio of about 10 : 1.[33] $C1s$, in nondenaturing conditions, exists as an MW 83,000–85,000 monomer, with sedimentation coefficient 4.1–4.7 S[16,21,33,35] and an estimated axial ratio of 6 : 1.[33] It is also generally found that $C1s$ forms a Ca^{2+} ion-dependent dimer of MW 170,000–176,000,[21,33] with sedimentation coefficient 5.7–6.0 S.[21,33] This species has been reported to dissociate at Ca^{2+} concentrations of less than 3 mM.[33] In addition to interactions between like monomers, $C1r$ and $C1s$ together form Ca^{2+}-dependent $C1r_2$–$C1s_2$ tetramers, which in turn bind to $C1q$. Interactions of this type are summarized in Ref. 2.

Partial specific volumes of 0.714 cm³/g for $C1r$ (or $C\bar{1}r$) and 0.717 cm³/g for $C1s$ (or $C\bar{1}s$) have been calculated from amino acid and carbohydrate compositions.[16] Extinction coefficients ($E_{1\,cm}^{1\%}$ at 280 nm) of 11.3[24]–11.7[16] for $C1r$, and 9.4[16]–11.7[24] for $C1s$ have been determined.

[30] Y. Mori, M. Koketsu, N. Abe, and J. Koyama, *J. Biochem. (Tokyo)* **85**, 1023 (1979).

[31] Y. Mori, E. Ueda, T. Takeuchi, S. Taniuchi, and J. Koyama, *J. Biochem. (Tokyo)* **87**, 1757 (1980).

[32] R. D. Campbell, N. A. Booth, and J. E. Fothergill, *Biochem. J.* **183**, 579 (1979).

[33] J. Tschopp, W. Villiger, H. Fuchs, E. Kilchherr, and J. Engel, *Proc. Natl. Acad. Sci. U.S.A.* **77**, 7014 (1980).

[34] G. J. Arlaud, S. Chesne, C. L. Villiers, and M. G. Colomb, *Biochim. Biophys. Acta* **616**, 105 (1980).

[35] G. J. Arlaud, A. Reboul, C. M. Meyer, and M. G. Colomb, *Biochim. Biophys. Acta* **485**, 215 (1977).

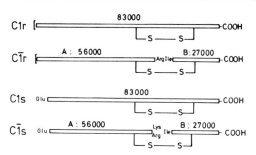

FIG. 3. Activation of C1s and C1r. The pattern of cleavage of C1r and C1s on activation is shown, and discussed in the text. Molecular weights of the polypeptide chains and known N- and C-terminal amino acids are shown. [indicates a blocked N-terminus.

On activation to C̄1r and C̄1s, the single polypeptide chains of C1r and C1s are both cleaved, probably at a single position.[16] As shown in Fig. 3, the point of cleavage is nearer the C-terminal end of the proenzyme polypeptide chain.[16,36,37] C̄1r and C̄1s monomers therefore each consist of 1 polypeptide chain of MW 56,000 (A chain) disulfide-linked to a 27,000-MW B chain.[16] The B chain contains the DFP-reactive site of the enzyme.[12,37,38] Molecular weights of 56,000 and 27,000 cited for the polypeptide chains of C̄1r and C̄1s are those estimated by gel filtration in denaturing conditions.[16] For the A chains of C̄1r and C̄1s, and for C̄1s B chain, these estimates agree with apparent molecular weights estimated by SDS–polyacrylamide gel electrophoresis.[16,23,35,39] However, C̄1r B chain migrates anomalously, with an apparent MW 35,000–36,000[16,25,39] on SDS–polyacrylamide gels. This behavior may again be caused by its relatively high carbohydrate content.[16]

The cleavage and activation pattern shown in Fig. 3 is typical of other serine proteases (e.g., coagulation factors and plasminogen) in that an initial cleavage of a single polypeptide chain proenzyme forms a 25,000–30,000-MW C-terminal fragment that contains the enzymic active site.

In nondenaturing conditions, C̄1r and C̄1s behave like the proenzyme forms, in that C̄1r normally exists as a noncovalent dimer,[10,16,34] and C̄1s as a monomer or Ca^{2+}-dependent dimer.[3,21,35] Arlaud and colleagues[34] have observed that the $C1r_2$ and $C̄1r_2$ dimers appear to be dissociated into 5.0 S monomers at low pH; dissociation of the proenzyme dimer is complete at pH 5.0, while more severe conditions (pH 4.0) are required to

[36] K. Takahashi, S. Nagasawa, and J. Koyama, *FEBS Lett.* **50**, 330 (1975).
[37] K. Takahashi, S. Nagasawa, and J. Koyama, *FEBS Lett.* **55**, 156 (1975).
[38] T. Barkas, G. K. Scott, and J. E. Fothergill, *Biochem. Soc. Trans.* **1**, 1219 (1973).
[39] R. J. Ziccardi and N. R. Cooper, *J. Immunol.* **116**, 504 (1976).

TABLE II
CARBOHYDRATE COMPOSITIONS OF C1r, C̄1r, C1s, C̄1s, AND THE SEPARATED CHAINS OF C̄1r AND C̄1s

Monosaccharide	Mol/mol protein							
		C̄1r				C̄1s		
	C1r	C̄1r	A chain	B chain	C1s	C̄1s	A chain	B chain
Fucose	0	0	0.8	0	0	0	0	0
Mannose	7.5	7.2	4.8	2.5	5.9	6.0	3.7	2.0
Galactose	9.2	9.5	6.9	2.8	5.6	5.3	3.6	1.8
N-Acetylglucosamine	15.1	15.8	5.5	8.3	8.4	8.2	5.4	2.6
N-Acetylgalactosamine	2.2	2.3	1.3	0.7	0.8	0.9	0.5	0.4
Sialic acid	5.2	5.1	2.9	1.7	7.5	7.2	4.6	2.5
Carbohydrate[a] as % total glycoprotein weight	9.4 }		7.4	12.4	7.1 }		6.5	7.5

[a] Hexosamines were calculated as N-acetylhexosamines and sialic acid as N-acetylneuraminic acid.

dissociate the $\overline{\text{C1r}}_2$ dimer. The differences in gel-filtration behavior of C1r and $\overline{\text{C1r}}$ at pH 5.3 (Fig. 2) are compatible with this type of dissociation.

Electron microscopy studies[33] suggest that each C1r and C1s monomer may have a two-domain structure. From radioiodination studies, Colomb and colleagues[26,34] have proposed that intermonomer interactions in the $\overline{\text{C1r}}_2$ and $\overline{\text{C1s}}_2$ dimers and in the $\overline{\text{C1r}}_2$–$\overline{\text{C1s}}_2$ complex are mediated principally by residues in the A-chain region of $\overline{\text{C1r}}$ and $\overline{\text{C1s}}$.

In standard pH 8.6 electrophoresis, in the presence of EDTA, C1s and $\overline{\text{C1s}}$ migrate as α_2-globulins,[6,21] and C1r and $\overline{\text{C1r}}$ as β-globulins.[9,39] $\overline{\text{C1r}}$ has also been reported to migrate as a γ-globulin in certain conditions.[39] A value of 4.9 has been reported for the isoelectric point of $\overline{\text{C1r}}$,[40] and this is likely to be similar for $\overline{\text{C1s}}$.

B. Chemical Properties

C1r and C1s are both glycoproteins, containing, respectively, about 40 and about 28 carbohydrate residues/molecule.[16] $\overline{\text{C1r}}$ and $\overline{\text{C1s}}$ have the same carbohydrate compositions as their proenzyme forms.[16] In $\overline{\text{C1r}}$, the A chain contains 22–23 sugar residues and the B chain, 16–17. In $\overline{\text{C1s}}$, the distribution of monosaccharides is 17–18 residues/A chain, and about 9 residues/B chain.[16,18] Full carbohydrate analyses[16,18] are shown in Table II. No information is yet available on the structure or attachment of the carbohydrate groups.

The amino acid compositions of C1r and of C1s are very similar,[12,16,21,39] and again the compositions of the activated forms are not significantly different from those of the proenzymes.[8,16,39] Full amino acid composition data for C1r, $\overline{\text{C1r}}$, C1s, and $\overline{\text{C1s}}$, and the A and B chains of $\overline{\text{C1r}}$ and $\overline{\text{C1s}}$ are shown in Table III. These data are from Refs. 12, 16, and 18, but are in close agreement with data provided by Cooper and colleagues.[21,39]

Limited amino acid sequence data for $\overline{\text{C1r}}$ and $\overline{\text{C1s}}$ are available,[16,18,36,37] and these are shown in Fig. 4. The N-terminal sequences of the B chains of $\overline{\text{C1r}}$ and $\overline{\text{C1s}}$ are homologous to each other and to the corresponding regions of other serine proteases. The $\overline{\text{C1s}}$ A chain N-terminal sequence shows no close relationship with other proteins. The A chains correspond to the N-terminal portion of the single polypeptide chains of the proenzyme.[16,36] $\overline{\text{C1r}}$ A chain and proenzyme C1r were found to have a blocked N-terminus,[16] although an earlier report suggests an Ile N-terminus for $\overline{\text{C1r}}$ A chain.[37] In a brief report, Barkas and colleagues[38] have indicated that the amino acid sequence around the active site of $\overline{\text{C1s}}$ is similar to the corresponding sequences in other serine proteases.

[40] K. Okamura and S. Fujii, *Biochim. Biophys. Acta* **534**, 258 (1978).

TABLE III

Amino Acid Composition of Subcomponents C1r, $\overline{C1r}$, C1s, $\overline{C1s}$, and the Separated Chains of $\overline{C1r}$ and $\overline{C1s}$

Amino acid			Residues/100 amino acid residues					
			$\overline{C1r}$				$\overline{C1s}$	
	C1r	$\overline{C1r}$	A chain	B chain	C1s	$\overline{C1s}$	A chain	B chain
Asp	10.3	10.2	10.3	9.6	10.9	10.9	11.0	10.7
Thr	4.8	4.9	4.9	5.3	4.7	5.1	4.9	5.7
Ser	6.5	6.3	6.0	7.1	6.5	6.3	6.6	5.4
Glu	12.3	12.1	12.3	11.7	11.2	10.9	11.6	10.1
Pro	5.9	5.9	6.5	4.5	7.1	7.1	7.0	6.9
Gly	10.5	10.4	9.8	12.0	10.8	10.2	10.0	10.7
Ala	5.5	5.3	5.0	7.2	6.0	5.9	5.5	7.2
½Cys	2.0	2.5	3.0	1.5	2.6	2.5	3.1	1.2
Val	5.6	5.3	4.9	6.6	6.9	7.2	6.7	8.5
Met	2.0	2.0	2.3	1.4	1.7	1.8	1.9	1.3
Ile	3.9	4.0	4.2	3.6	3.7	3.9	4.2	3.5
Leu	7.7	7.9	7.8	8.0	6.4	6.4	6.0	7.0
Tyr	4.1	4.2	3.9	4.2	4.0	4.1	4.0	4.1
Phe	4.5	4.7	4.9	4.0	4.6	4.7	5.0	4.1
Try	1.4	1.6	1.1	2.0	1.2	1.5	1.0	1.5
Lys	4.6	4.7	4.7	4.6	4.7	4.6	4.4	5.4
His	3.3	3.1	3.3	2.7	3.0	2.7	2.9	2.5
Arg	5.1	4.9	5.1	4.1	4.1	4.0	4.2	4.5

C. Enzymic Properties

The proenzymic forms of C1r and C1s have no demonstratable proteolytic or esterolytic activities, and they are not inactivated by DFP at the concentrations used during isolation procedures. The active forms, $\overline{C1r}$ and $\overline{C1s}$, are both very specific proteases. $\overline{C1r}$ cleaves a single Arg-Ile or Lys-Ile bond in proenzyme C1s[16,41] (Fig. 3). $\overline{C1r}$ does not cleave proenzymic C1r,[22,39] C4, or C2, and no other protein substrate has been found for $\overline{C1r}$.[1,22] $\overline{C1s}$ cleaves a single Arg-Ala bond in the sequence Leu-Gln-Arg-Ala-Leu-Glu in component C4,[42,43] and a single bond (likely to be Lys-Lys or Arg-Lys) in component C2.[44] $\overline{C1s}$ does not cleave C1r[39] or C1s,[7,13] and no other protein substrate has been found for C1s.[1,45]

[41] E. M. Press, personal communication (1978).
[42] J. P. Gorski, T. E. Hugli, and H. J. Müller-Eberhard, *Proc. Natl. Acad. Sci. U.S.A.* 76, 5299 (1979).
[43] E. M. Press and J. Gagnon, *Biochem. J.* 199, 352 (1981).
[44] M. A. Kerr, *Biochem. J.* 183, 615 (1979).
[45] O. D. Ratnoff and I. H. Lepow, *J. Exp. Med.* 106, 327 (1957).

$$\overline{\text{C}}\text{1s A chain}\quad \text{NH}_2\text{—E P T M Y G E I L S P N Y P Q A Y P X E V E K X W X I X V}$$

(5, 10, 15, 20, 25)

$$\overline{\text{C}}\text{1r B chain}\quad \text{NH}_2\text{—I I G G Q K A K M G N F P W Q V F T N Z X G X G X X X L L X X X X I L}$$

(5, 10, 15, 20, 25, 30, 35)

$$\overline{\text{C}}\text{1s B chain}\quad \text{NH}_2\text{—I I G G S D A D I K N F P W Q V F F D N}$$

(5, 10, 15, 20)

$$\text{Typical serine protease}\quad \text{NH}_2\text{—I V G G A G S P W Q V S L S G H F C G G L I}$$

FIG. 4. N-terminal amino acid sequences of $\overline{\text{C}}$1r and $\overline{\text{C}}$1s polypeptide chains. Sequences determined by automated Edman degradation of $\overline{\text{C}}$1s A, $\overline{\text{C}}$1r B chain, and $\overline{\text{C}}$1s B chain are shown. The B chain sequences are compared with the sequence of highly conserved residues in serine proteases [C. L. Young, W. C. Barker, C. M. Tomaselli, and M. O. Dayhoff, in "Atlas of Protein Sequence and Structure" (M. O. Dayhoff, ed.), Vol. 5, Suppl. 3, p. 73. National Biomedical Research Foundation, Washington, D.C., 1978].

$C\overline{1}r$ has a very limited esterolytic activity against acetyl-glycine-lysine-methyl ester (AGLME) and acetyl-arginine methyl ester (AAME).[46,47] The K_m for AGLME is $2 \times 10^{-2} M$. No other satisfactory ester or amide substrate has been found among a wide range tested.[16,18,46,47] N-Carbobenzoxy-L-tyrosine p-nitrophenol ester (ZTNE) was suggested to be a $C\overline{1}r$ substrate, but the statement has been withdrawn.[24] A recent publication has reported very high levels of ZTNE and N^α-carbobenzoxy-L-lysine p-nitrophenol ester (ZLNE) esterase activity in a $C\overline{1}r$ preparation, but this was not inhibited by known inhibitors of $C\overline{1}r$, and may have been due to contamination with a 26,000-MW ZTNE– and ZLNE–esterase found in an associated fraction.[48,49]

$C\overline{1}s$, in contrast to $C\overline{1}r$, hydrolyzes a wide range of methyl, ethyl and p-nitrophenol esters of basic and aromatic amino acids.[6,15,16,19,35,45,47] Representative K_m values include: acetyltyrosine ethyl ester (ATEE), $1.9 \times 10^{-2} M$[6]; ZTNE, $(5.6-7) \times 10^{-5} M$[16,19]; ZLNE, $8 \times 10^{-4} M$[16]; tosyl-arginine methyl ester (TAME), $(1.7-2.8) \times 10^{-3} M$[35]; benzoyl-arginine ethyl ester (BAEE), $(2.5-5.0) \times 10^{-3} M$.[35] In its esterolytic activity, $C\overline{1}s$ resembles plasmin, although as a protease its specificity is much more restricted than that of plasmin.[16]

The activities of $C\overline{1}r$ and $C\overline{1}s$ are inhibited irreversibly by $C\overline{1}$-inhibitor.[17] Other plasma protease inhibitors,[17] and soybean-trypsin inhibitor, lima bean-trypsin inhibitor, trasylol, ovomucoid, hirudin, heparin, and ϵ-aminocaproate do not inhibit $C\overline{1}r$ and $C\overline{1}s$.[45,46] Leupeptin is a competitive inhibitor of $C\overline{1}r$.[40] The enzymic activities of $C\overline{1}r$ and $C\overline{1}s$ are inhibited irreversibly by DFP[13,50] and by p-nitrophenyl-p'-guanidinobenzoate.[22] A large number of studies have documented reversible inhibition of $C\overline{1}r$ and $C\overline{1}s$ by amidines and guanidines. Such studies are cited in Refs. 15 and 19.

The proteolytic and esterolytic activities of $C\overline{1}r$ show broad pH optima, centered on pH 7.5–8.5.[18,46] Cleavage of C1s by $C\overline{1}r$ is very temperature dependent, having an activation energy of 23–32 kcal/mol.[18,46] C1s cleavage by $C\overline{1}r$ is much slower in the presence of Ca^{2+} ions than in the presence of EDTA,[13,46] and is inhibited strongly at ionic strength greater than 200 mM NaCl.[18,46,51]

The esterolytic activity of $C\overline{1}s$ against ATEE has a broad pH optimum centered on pH 7.0–8.0[6] and is relatively insensitive to ionic strength, being slightly and progressively inhibited if ionic strength is increased

[46] G. B. Naff and O. D. Ratnoff, *J. Exp. Med.* **128**, 571 (1968).

[47] J. E. Volanakis, R. E. Schrohenloher, and R. M. Stroud, *J. Immunol.* **119**, 337 (1977).

[48] J. M. Andrews and R. D. Baillie, *J. Immunol.* **123**, 1403 (1979).

[49] J. M. Andrews, *Fed. Proc., Fed. Am. Soc. Exp. Biol.* **37**, 1795 (abstr.) (1978).

[50] A. L. Haines and I. H. Lepow, *J. Immunol.* **92**, 468 (1964).

[51] G. J. Arlaud, A. Reboul, and M. G. Colomb, *Biochim. Biophys. Acta* **485**, 227 (1977).

above 200 mM NaCl. The presence of Ca^{2+} ions produces only very small changes in the esterolytic activities of C$\overline{1}$s.[6,35]

D. Stability and Activation

C1s exists as a stable proenzyme, and C$\overline{1}$s as a stable enzyme. Neither undergoes autolytic cleavage.[7,13,23,25] C1s and C$\overline{1}$s are relatively soluble and may be stored at concentrations of 1–2 mg/ml. Ca^{2+} ions have no effect on the solubility of either, although both bind Ca^{2+} ions (two relatively high-affinity binding sites per dimer).[52] Both tend to precipitate on exposure to pH 4.5–5.0, and C$\overline{1}$s is irreversibly inactivated by overnight exposure at 4°C to pH values below 5.0 and above 9.5.[50] The esterolytic and proteolytic activities of isolated C$\overline{1}$s are decreased by about 50% on heating at 56°C for 30 min.[50]

Proenzymic C1s can be activated by incubation with catalytic quantities of C$\overline{1}$r.[13,51] This activation can also be mimicked by trypsin,[21,23] although further degradation occurs. C$\overline{1}$s formed by cleavage of isolated proenzymic C1s by C$\overline{1}$r, and C$\overline{1}$s formed by "spontaneous" activation *during* isolation (as in the isolation procedure just described) appear to behave identically, although differences in specific enzymic activities have been recorded.[13]

C1r proenzyme has not been isolated in a completely stable form, and there is controversy as to whether isolated C1r undergoes autocatalytic activation. A type of autocatalytic activation is proposed as the mechanism of activation of C1r when it is bound within the C1 complex, in contact with C1 activators.[2] This latter activation occurs in response to a definite signal (i.e., binding of C1 to activators such as antibody–antigen aggregates) and requires the presence of C1q and C1s.[2,22] Valet and Cooper[10] initially reported substantial purification of C1r that was stable in solution. It was suggested, however, that this preparation contained traces of C$\overline{1}$-inhibitor, and on removal of this contaminant the isolated C1r was unstable, and activated rapidly on incubation at 37°C in the presence of EDTA.[11] The activation reaction was slowed by Ca^{2+} ions and inhibited by DFP.[11] The C$\overline{1}$r produced by this reaction did not accelerate activation of the proenzyme, and so an intermolecular autocatalytic process is unlikely. This species of C$\overline{1}$r, although able to cleave C1s, was unable to reconstitute whole C$\overline{1}$ hemolytic activity when mixed with C1q and C1s.[11] Assimeh *et al.*[53] reported similar activation of C1r prepared by an affinity method.[24] Arlaud and colleagues[26] have reported a slower, concentration-independent activation of proenzyme C1r. The activation in this case is about 50% complete after 20 min incubation at 37°C. This reaction was

[52] C. L. Villiers, G. J. Arlaud, R. H. Painter, and M. G. Colomb, *FEBS Lett.* **117,** 289 (1980).
[53] S. N. Assimeh, R. M. Chapuis, and H. Isliker, *Immunochemistry* **14,** 13 (1978).

completely inhibited by Ca^{2+} ions, but was relatively insensitive to DFP. An intramolecular autocatalysis (i.e., within the $C1r_2$ dimer) was proposed.[26] In contrast, Taylor et al.[23] have reported that although partially purified C1r activated as described by Ziccardi and Cooper,[11] the final isolated C1r preparation was stable. This suggests that activation of the less pure material was due to protease contamination. Proenzymic C1r isolated by the procedure described in detail above, although not perfectly stable, remains unactivated during incubation for up to 1 hr at 37°C.[22] The kinetics of the very slow activation that occurs in this case do not correspond to an autocatalysis mechanism, and the activation is thought to be a result of trace protease contamination. Valet and colleagues[54] have also produced evidence in support of protease contamination as the cause of apparent "autoactivation" of C1r.

C1r̄ produced by allowing activation to occur during isolation, as in the isolation procedure described in detail above, and also C1r̄ produced by binding and activation of purified C1 on antibody–antigen aggregates,[13] is able to reconstitute C1̄ hemolytic activity when mixed with C1s and C1q, in contrast to the material described by Ziccardi and Cooper.[11]

Several preparations of activated C1r̄ have been shown to undergo further cleavage on very prolonged (5–18 hr) incubation at 37°C in the presence of EDTA.[26,40,53] These proteolytic reactions result in cleavage of the A chain into fragments of about 35,000, 20,000, and 7000 MW.[26] It is again uncertain whether this is an autolytic mechanism[26] or the result of trace protease contamination.[54] It is not known whether C1r̄ prepared as described in the present article undergoes this type of degradation.

C1r and C1r̄ are both much less soluble than C1s, and tend to aggregate slowly if stored at concentrations greater than 250 $\mu g/ml$. Both C1r and C1r̄ have an increased tendency to aggregate in the presence of Ca^{2+}.[18,52] The $C1r_2$ dimer appears to have only one high-affinity Ca^{2+}-binding site, while the C1r̄$_2$ dimer has two to three.[52]

Proenzymic C1r can be activated rapidly by incorporation into an antibody–antigen $C1q$–$C1r$–$C1s$ complex.[2,22] Activation in this way results in cleavage of a single Arg-Ile bond in C1r.[41] Activation of C1r can be mimicked under certain conditions by digestion of C1r with trypsin[10,54] or plasmin.[55]

Acknowledgments

I thank Drs. G. J. Arlaud, E. Sim, and Mr. A. W. Dodds for helpful discussion. Dr. G. J. Arlaud, Professor M. G. Colomb, Professor J. Engel, and E. M. Press kindly communicated results prior to publication.

[54] J. Bauer, E. Weiner, and G. Valet, J. Immunol. 124, 1514 (abstr.) (1980).
[55] N. R. Cooper, L. A. Miles, and G. H. Griffin, J. Immunol. 124, 1516 (abstr.) (1980).

[5] Preparation and Properties of Human C1̄ Inhibitor

By R. B. SIM and A. REBOUL

Introduction

C1̄ inhibitor (C1̄-Inh) is a serum glycoprotein responsible for regulating the activities of the complement system serine proteases C1̄r and C1̄s. The presence of an inhibitor of C1̄s in human plasma was initially observed by Ratnoff and Lepow,[1] and substantial purification of the inhibitor was first reported by Pensky et al.[2] C1̄-Inh is in many respects a typical plasma protease inhibitor, and appears to act in a similar way to, for example antithrombin III, α_2-antiplasmin, and α_1-antitrypsin. C1̄-Inh is, however, the only plasma-protease inhibitor that interacts with C1̄r and C1̄s under physiological conditions.[3]

Methods for the purification and assay of C1̄-Inh have been described previously in this series.[4] Using information derived from earlier methods,[4] a simplified, higher-yield purification method has since been developed,[5] and further information is now available on the reaction of C1̄-Inh with C1̄r and C1̄s.

Assay

Determination of activity of C1̄-Inh in serum or plasma is of clinical importance for diagnosis of hereditary angioedema, a relatively frequent genetic defect characterized by total or partial absence of C1̄-Inh activity. (For discussion, see Ref. 4.)

C1̄-Inh is most conveniently assayed by determining the extent to which it inhibits the hydrolysis of amino acid esters by a standard preparation of C1̄s. A number of different amino acid esters are routinely used for this purpose, including N-acetyl-L-tyrosine ethyl ester (ATEE),[6] N^α-acetyl-L-lysine methyl ester (ALME),[4] N^α-tosyl-L-arginine methyl ester (TAME),[5] and N^α-carbobenzoxy-L-lysine p-nitrophenol ester (ZLNE).[7]

[1] O. D. Ratnoff and I. H. Lepow, J. Exp. Med. 106, 327 (1957).

[2] J. Pensky, L. R. Levy, and I. H. Lepow, J. Biol. Chem. 236, 1674 (1961).

[3] R. B. Sim, A. Reboul, G. J. Arlaud, C. L. Villiers, and M. G. Colomb, FEBS Lett. 97, 111 (1979).

[4] P. C. Harpel, this series Vol. 45, p. 751.

[5] A. Reboul, G. J. Arlaud, R. B. Sim, and M. G. Colomb, FEBS Lett. 79, 45 (1977).

[6] F. P. Schena, C. Manno, R. D'Agostino, G. Bruno, F. Cramarossa, and L. Bonomo, J. Clin. Chem. Clin. Biochem. 18, 17 (1980).

[7] R. B. Sim, G. J. Arlaud, and M. G. Colomb, Biochim. Biophys. Acta 612, 433 (1980).

METHODS IN ENZYMOLOGY, VOL. 80

All four methods cited provide similar sensitivity in that approximately 1 μg of C$\bar{1}$-Inh can be detected, but spectrophotometric assays using TAME and ZLNE are more rapid and require fewer reagents. An assay using ZLNE is described in detail here.

A standard preparation of C$\bar{1}$s is required for assaying C$\bar{1}$-Inh. For simple comparative assays, a neutral euglobulin precipitate from human serum, prepared as described by Linscott,[8] may be used as a crude C$\bar{1}$s preparation. This material should be incubated for 1 hr at 37°C and pH 7.5 after preparation to ensure complete activation of C$\bar{1}$s. To obtain higher sensitivity, however, it is necessary to use highly purified and fully activated C$\bar{1}$s. This is most easily obtained by using precipitation and DEAE–cellulose purification steps as described by Arlaud and colleagues.[9] For the purpose described here, a final affinity-chromatography step described by these authors, which removes traces of contaminant C$\bar{1}$r, may be omitted. The concentration of active C$\bar{1}$s in a crude or highly purified preparation can be estimated by direct assay with ZLNE.[10]

Reagents

C$\bar{1}$s. The stock C$\bar{1}$s preparation is dialyzed against 10 mM Tris-HCl–100 mM NaCl–1 mM EDTA, pH 8.0, and adjusted to a final concentration of 75–125 μg of active C$\bar{1}$s/ml.

Substrate. ZLNE (Sigma Chemical Co.) is dissolved to a final concentration of 3 mg/ml in a mixture of 9 parts by volume acetonitrile and 1 part by volume water. The stock substrate solution should be made up each day and kept on ice.

Assay buffer. Hydrolysis of ZLNE must be monitored at pH 6.0, as the substrate is unstable near neutral pH. The assay is done in 100 mM sodium phosphate–100 mM NaCl, 15 mM EDTA, pH 6.0.

C$\bar{1}$-Inh sample. The C$\bar{1}$-Inh samples to be tested should be at pH 7.0–8.0. The assay is suitable for titrating C$\bar{1}$-Inh in plasma or serum, as well as at various stages of purification. Plasma and serum samples should be diluted 5- to 10-fold with 10 mM Tris-HCl–100 mM NaCl–1 mM EDTA, pH 8.0.

Procedure

The C$\bar{1}$s preparation (50 μl) and C$\bar{1}$-Inh (up to 500 μl, 0–20 μg) are mixed and incubated for 15 min at 37°C. If the C$\bar{1}$-Inh samples are very dilute ($<$2 μg/ml), incubation should be performed for 30 min at 37°C. For

[8] W. D. Linscott, *Immunochemistry* **5**, 311 (1968).

[9] G. J. Arlaud, A. Reboul, C. M. Meyer, and M. G. Colomb, *Biochim. Biophys. Acta* **485**, 215 (1977).

[10] R. B. Sim, this volume [4].

each sample of C$\overline{1}$-Inh to be tested, a range of 3–5 different volumes should be incubated with C$\overline{1}$s. An appropriate control, containing 50 μl 10 mM Tris-HCl–100 mM NaCl–1 mM EDTA, pH 8.0, instead of C$\overline{1}$s, is set up for each volume of C$\overline{1}$-Inh used. The controls are treated throughout in the same way as test samples.

At the end of the incubation period, each C$\overline{1}$s + C$\overline{1}$-Inh mixture and each control is diluted to a final volume of 3.2 ml with the pH 6.0 assay buffer. Each 3.2-ml mixture is then transferred to a spectrophotometer cuvette and 50 μl of the stock substrate solution added. The contents of the cuvette are mixed by inversion, and the increase in absorbance at 340 nm due to p-nitrophenol release is then monitored for 2–5 min. Substrate hydrolysis may be determined at any temperature between 20°C and 37°C. Greater sensitivity and speed are obtained at higher temperatures. Increase in absorbance is linear up to an absorbance value of 0.35.

The rate of change of absorbance per minute (ΔA_{340}/min) is calculated for each sample. The ΔA_{340}/min value for each C$\overline{1}$s + C$\overline{1}$-Inh mixture is corrected by subtracting from it the ΔA_{340}/min value of the corresponding control. This correction takes account of spontaneous substrate hydrolysis and the possible presence of esterases in the C$\overline{1}$-Inh sample.[11]

For each set of assays, the activity of the uninhibited C$\overline{1}$s must be determined by measuring the ΔA_{340}/min for a mixture of 50 μl C$\overline{1}$s, 3.15 ml of pH 6.0 assay buffer, and 50 μl of substrate. This value is corrected for spontaneous hydrolysis of ZLNE by subtracting from it the ΔA_{340}/min developed in a mixture of 3.2 ml of pH 6.0 assay buffer and 50 μl substrate.

Corrected ΔA_{340}/min values for each C$\overline{1}$s + C$\overline{1}$-Inh mixture are expressed as a percentage of the corrected ΔA_{340}/min for C$\overline{1}$s alone, and finally the concentration of C$\overline{1}$-Inh in the sample is calculated from a graph of the type shown in Fig. 1. C$\overline{1}$-Inh concentration may be calculated in absolute terms provided that highly purified C$\overline{1}$s of known concentration is used. Alternatively, the concentration of both C$\overline{1}$s and C$\overline{1}$-Inh may be expressed as activity units or arbitrary units, as in Ref. 4.

C$\overline{1}$-Inh may also be titrated by any of several methods in which inhibition of C$\overline{1}$s or C$\overline{1}$ hemolytic activity is determined. Such methods may be up to 100-fold more sensitive than the esterase assays described previously, but they require relatively complex reagents. An assay of this type has been described by Gigli and colleagues.[12]

---

[11] Use of a split-beam spectrophotometer for these assays permits direct reading of test samples against controls, thus halving the number of individual assays to be performed. With experience in the assay method, a large number of similar C$\overline{1}$-Inh samples (e.g., plasmas) can be compared using only a single C$\overline{1}$-Inh dilution.

[12] I. Gigli, S. Ruddy, and K. F. Austen, *J. Immunol.* **100**, 1154 (1968).

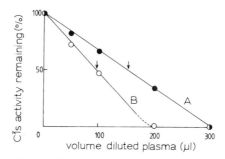

FIG. 1. Titration of C$\bar{1}$-Inh. C$\bar{1}$s (5 μg in 50 μl) was incubated as described in the assay procedure with varying volumes of two diluted human plasmas, A and B. Plasma dilution factor was 1 volume plasma: 9 volumes buffer. The titer of C$\bar{1}$-Inh is calculated from the volume of diluted plasma required to give 50% inhibition ($\downarrow$). For plasma A, this volume is 150 μl, and for B, 96 μl. The C$\bar{1}$-Inh concentration may be expressed in arbitrary units, where for example 1 unit = the quantity of C$\bar{1}$-Inh required to cause 50% inhibition of the C$\bar{1}$s present. Undiluted plasma A therefore contains [(1000/150) $\times$ plasma dilution] = 66.7 units/ml, and undiluted plasma B, 104.2 units/ml. Alternatively, from the known molecular weights of C$\bar{1}$s and C$\bar{1}$-Inh, and the 1 : 1 stoichiometry of inhibition, it can be calculated that inhibition of half the C$\bar{1}$s present requires 2.94–3.16 μg of C$\bar{1}$-Inh. Undiluted plasma A contains this quantity of C$\bar{1}$-Inh in 15 μl, and therefore contains 196–211 μg C$\bar{1}$-Inh/ml. Similarly, undiluted plasma B contains 306–329 μg/ml.

Purification

C$\bar{1}$-Inh can be purified from human plasma in three successive steps: (1) polyethylene glycol precipitation; (2) DEAE–cellulose chromatography; (3) concanavalin A–Sepharose chromatography.[5] Since C$\bar{1}$-Inh participates in inhibition of several proteases and can be degraded by others (e.g., plasmin[4]), it is essential to avoid activation of plasma proteases. To minimize protease activation and activity, coagulation contact-activation inhibitors and protease inhibitors are used during the preparation. The plasma used for purification of C$\bar{1}$-Inh should not have been in contact with glass, and contact with unsiliconized glass should be avoided during steps 1 and 2 of the preparation. Plasma should be kept frozen ($-20°$C) until it is required. Storage at 4°C for more than 3–4 hr should be avoided to limit cold-promoted activation of Factor VII.[13] Fresh plasma, containing acid citrate-dextrose (ACD) or phosphate-citrate-dextrose (PCD) anticoagulants is ideal for the preparation, but outdated ACD- or PCD-plasma is an adequate starting material, provided it has been stored at $-20°$C with no glass contact. All purification procedures are carried out at 2–4°C unless stated otherwise.

[13] E. A. van Royen, S. Lohman, M. Voss, and K. W. Pondman, *J. Lab. Clin. Med.* **92,** 152 (1978).

Protease Inhibitors

Diisopropyl phosphoroflurodate (DFP; Sigma or Aldrich Chemical Co.) is made up as a 2.5 M stock solution by adding 1 g DFP to 1.2 ml anhydrous isopropanol.[14]

Polybrene. Polybrene (Aldrich Chemical Co.) is made up as a 24 mg/ml solution in 0.4 M trisodium EDTA, pH 7.5.

Procedure

Human plasma (250 ml) is thawed at 37°C and centrifuged for 30 min at 10,000 g to remove debris. The plasma is then made 4 mM with DFP by addition of 400 μl of the DFP stock solution. The stock polybrene solution (13.2 ml) is then added to give final polybrene and EDTA concentrations of 1.2 mg/ml and 20 mM, respectively. The plasma is then stirred gently for 30 min at 25°C, and a further 400 μl of stock DFP solution is added.

Step 1. Polyethylene Glycol Precipitation. The plasma is now cooled to 4°C, and made 6% w/v polyethylene glycol (PEG) by addition of 36.4 ml of a 50% w/v solution of PEG 6000 (Sigma). The PEG solution is added slowly over a period of 20–25 min, with constant stirring. The plasma is stirred for a further 30 min at 4°C, then centrifuged (30 min at 10,000 g). The heavy yellow precipitate is discarded. A further 400 μl of stock DFP solution is added to the supernatant, and the supernatant is dialyzed to equilibrium against 5 liters of 20 mM sodium phosphate–50 mM NaCl–5 mM EDTA, pH 7.0.

The precipitation step results in removal of 35–40% of the total plasma protein with 75–85% recovery of C1̄-Inh activity, representing about 1.3-fold purification.

Step 2. DEAE-Cellulose Chromatography. The dialyzed supernatant is centrifuged (30 min, 10,000 g). A small gellike precipitate usually forms at this stage, and is discarded. The supernatant is loaded onto a column (20 × 5 cm diameter) of DEAE–cellulose (DE-32 or DE-52, Whatman), equilibrated in the dialysis buffer. Flow rate should be 80–100 ml/hr. Protein that does not bind to the column is discarded and the column is washed with the starting buffer until the absorbance at 280 nm of the eluate is <0.05. C1̄-Inh is then eluted in a linear gradient of NaCl formed by mixing 700 ml of the starting buffer with 700 ml of 20 mM sodium phosphate–300 mM NaCl–5 mM EDTA, pH 7.0. A typical elution profile

---

[14] Since DFP is volatile and highly toxic, some laboratories may prefer to use the less toxic inhibitor phenyl methyl sulphonyl fluoride (PMSF; Sigma Chemical Co.). This material is prepared as a stock 110 mM solution in ethanol and added to aqueous solutions to a final concentration of about 1 mM. It is less effective than DFP as a general serine-protease inhibitor.

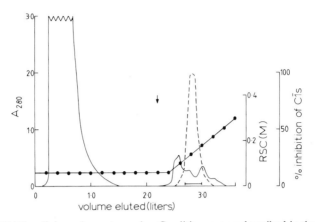

FIG. 2. DEAE–cellulose chromatography. Conditions are as described in the text. ——, A_{280}; - - - -, $C\bar{1}$-Inh activity, expressed as % inhibition of $C\bar{1}s$ in a standard assay by 100 μl of eluant fraction; ●——●, relative salt concentration (RSC) (a solution with RSC = 0.1 M has the same conductivity as 0.1 M NaCl). An arrow marks the start of gradient elution, and a bar indicates the fractions pooled for the next purification step.

is shown in Fig. 2. This step results in elimination of >99% of the total protein, with 50–55% recovery of $C\bar{1}$-Inh activity, representing 120- to 170-fold purification.

Step 3. Concanavalin A–Sepharose Chromatography. The pooled material from DEAE–cellulose chromatography is again made 4 mM with DFP and dialyzed against 5 liters of 20 mM Tris-HCl–100 mM NaCl, pH 8.0, and then applied to a column (8 × 5 cm diameter) of concanavalin A–Sepharose (Pharmacia), equilibrated in the same buffer. Flow rate should be about 20 ml/hr. The column is washed with the starting buffer until the absorbance at 280 nm of the eluate is <0.02. $C\bar{1}$-Inh is then eluted as a sharp peak (Fig. 3) by washing the column with 450 ml of the same buffer made 0.5% w/v with α-methylmannoside (Sigma). This purification step results in elimination of 30–50% of the remaining protein, with over 90% recovery of $C\bar{1}$-Inh activity, and therefore represents a 1.3- to 1.8-fold purification.

The pool of $C\bar{1}$-Inh is then dialyzed against 2 liters of 20 mM Tris-HCl–100 mM NaCl–1 mM EDTA, pH 8.0, to remove α-methyl-mannoside, and finally concentrated to about 1 mg/ml by ultrafiltration on a PM-10 membrane (Amicon). The purified protein is stable on storage at $-20°$C for more than 1 year.

The yield of $C\bar{1}$-Inh by this method is 19–36 mg of $C\bar{1}$-Inh (average over 14 preparations = 24 mg) from the 250 ml of starting material. The yield of activity is in the range 33–42%. The concentration of $C\bar{1}$-Inh in

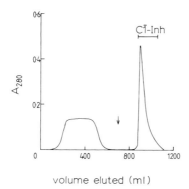

FIG. 3. Concanavalin A–Sepharose chromatography. Conditions are as described in the text. ——, A_{280}. An arrow marks the start of elution with α-methylmannoside, and a bar indicates the fractions pooled.

individual plasmas shows a relatively wide variation, and other investigators have suggested mean plasma concentrations of 235,[15] 185,[16] and about 130 mg/liter.[17] We have not systematically titrated C̄1-Inh in plasma from single donors, but single radial immunodiffusion and functional assays performed on pooled plasmas (8–15 donors) used for preparing C̄1-Inh consistently show an average C̄1-Inh concentration greater than 200 mg/liter.

Examination of C̄1-Inh purified by the above method on sodium dodecyl sulfate (SDS)–polyacrylamide gels reveals a single major component of 100,000 apparent molecular weight, with traces (generally <5%) of two other components of apparent MW 60,000 and 30,000. These are likely to be degradation products of C̄1-Inh.[5] Prolonged (over 1 month) storage of the isolated protein at 4°C in the absence of protease inhibitors leads to an increase in the quantity of the two minor components present. The final preparations are therefore likely to contain trace contamination with protease(s). Occasional contamination with transferrin and albumin is observed. Contaminating albumin can be removed by repeating step 3 of the preparation.

C̄1-Inh isolated by other methods[4,18] exhibited two major bands on Coomassie blue-stained SDS–polyacrylamide gels. One species had an apparent MW 9,000–10,000 lower than C̄1-Inh, and appeared to corre-

[15] N. Heimburger, *Cold Spring Harbor Conf. Cell Proliferation* 2, 367 (1975).
[16] F. S. Rosen, C. A. Alper, J. Pensky, M. R. Klemperer, and V. H. Donaldson, *J. Clin. Invest.* 50, 2143 (1971).
[17] R. J. Ziccardi and N. R. Cooper, *Clin. Immunol. Immunopathol.* 15, 465 (1980).
[18] D. P. Man and J. O. Minta, *Fed. Proc., Fed. Am. Soc. Exp. Biol.* 37, 591 (abstr. 1991) (1978).

spond to a major fragment produced by plasmin or trypsin digestion of $C\bar{1}$-Inh.[19] This degradation product is also observed in our preparations if protease inhibitors are not used during purification.

$C\bar{1}$-Inh in Other Species

$C\bar{1}$-Inh has been isolated from rabbit serum in low yield,[20] and has been shown to be very similar in molecular weight, temperature, and pH stability to the human protein. Rabbit $C\bar{1}$-Inh inhibits rabbit $C\bar{1}r$, $C\bar{1}s$, and plasmin, and also human $C\bar{1}s$.[20] Guinea pig $C\bar{1}$-Inh has also been purified, and although similar to human $C\bar{1}$-Inh in activity, was reported to have a much higher sedimentation coefficient (6.1 S).[21] Inhibitory activity against human $C\bar{1}s$ has been detected in the sera of numerous mammals, birds, and fish.[22,23] The sera of several primates, but not of lower mammals, contain protein that is antigenically related to human $C\bar{1}$-Inh.[23] Partial purifications of $C\bar{1}$-Inh from other mammalian species have been summarized by Gigli and Austen.[24]

Structural Properties

Isolated human $C\bar{1}$-Inh is a single polypeptide chain monomer of MW 100,000–105,000 as estimated by equilibrium centrifugation and by SDS–polyacrylamide gel electrophoresis.[4,5,25] It has a sedimentation coefficient of 3.7^{25}–4.0 S.[4] The isoelectric point is 2.7–2.8.[25] The amino acid and carbohydrate compositions of the protein have been determined by several groups and have been presented earlier in this series.[4] The amino acid composition is unremarkable, except for an unusually low content of glycine and sulfur-containing amino acids. The tyrosine and tryptophan content is low, consistent with the low extinction coefficient of the protein ($A_{1cm}^{1\%}$ 280 nm $= 4.5^{25}$). $C\bar{1}$-Inh is, however, very heavily glycosylated, containing 69 mol hexose, 47 mol hexosamine, and 51 mol sialic acid per mole of protein.[25] Consistent with the high carbohydrate content, the partial specific volume is low (0.667 ml/g[25]). No sequence data are available for $C\bar{1}$-Inh.

[19] P. C. Harpel and N. R. Cooper, *J. Clin. Invest.* **55**, 593 (1975).

[20] E. Ishizaki, Y. Mori, and J. Koyama, *J. Biochem. (Tokyo)* **82**, 1155 (1977).

[21] M. Loos, W. Opferkuch, and R. Ringelmann, *Z. Med. Microbiol. Immunol.* **156**, 194 (1971).

[22] L. R. Levy and I. H. Lepow, *Proc. Soc. Exp. Biol. Med.* **101**, 608 (1959).

[23] V. H. Donaldson and J. Pensky, *J. Immunol.* **104**, 1388 (1970).

[24] I. Gigli and K. F. Austen, *Annu. Rev. Microbiol.* **25**, 309 (1971).

[25] H. Haupt, N. Heimburger, T. Kranz, and H. G. Schwick, *Eur. J. Biochem.* **17**, 254 (1970).

Activity and Stability

Isolated C̄1-Inh loses activity on exposure to pH values outside the range 6.0–9.5.[2] The activity is totally destroyed by heating at 60°C for 30 min, but is stable to temperatures below 48°C.[2] Inhibitory activity is unaffected by neuraminidase treatment[18] but has been reported to be destroyed by exposure to 1 mM 2-mercaptoethanol.[18]

C̄1-Inh has been shown to inhibit, in vitro, plasma kallikrein,[26,27] plasmin,[27] activated Hageman Factor (FXII$_a$)[28] and its fragments,[29] and activated Factor XI (FXI$_a$)[28] as well as C̄1r[27] and C̄1s.[1] It has no significant inhibitory activity against trypsin or chymotrypsin.[2,25] It is degraded by trypsin and by excess plasmin.[19] Inhibition of plasmin by C̄1-Inh does not appear to be of physiological significance since, in whole plasma, α_2-antiplasmin and α_2-macroglobulin are more effective plasmin inhibitors.[30] C̄1-Inh may, however, be a major physiological inhibitor of plasma kallikrein,[31,32] of FXII$_a$ fragments,[13,29,32,33] and FXI$_a$.[32] It is the only plasma inhibitor of C̄1r and C̄1s.[3]

Inhibition of C̄1r, C̄1s, and plasmin by C̄1-Inh correlates with the formation of tightly bound complexes consisting of one molecule of inhibitor per molecule of protease.[7,19,34] These protease–(protease inhibitor) complexes are not dissociated by strong denaturants such as guanidine hydrochloride, urea, and SDS, or by low pH, and can be observed on SDS–polyacrylamide gels after staining with Coomassie blue.[3,5,7,19,34] The complexes are also resistant to reduction and alkylation. Patterns obtained on SDS–polyacrylamide gel electrophoresis of reduced and alkylated complexes of C̄1-Inh with C̄1r, C̄1s, or plasmin show that the inhibitor binds to the protease through the light "B" chains, which contain the active site.[3,19] Prior inactivation of C̄1r or C̄1s with DFP prevents subsequent binding of C̄1-Inh[34,35] and vice versa.[34] Thus it is likely that binding of C̄1-Inh to these proteins occurs via the catalytic active site.

Similar formation of denaturation-resistant equimolar complexes has been described for other proteases with plasma protease inhibitors, for

[26] I. Gigli, J. W. Mason, R. W. Colman, and K. F. Austen, *J. Immunol.* **104**, 574 (1970).

[27] O. D. Ratnoff, J. Pensky, D. Ogston, and G. B. Naff, *J. Exp. Med.* **129**, 315 (1969).

[28] C. D. Forbes, J. Pensky, and O. D. Ratnoff, *J. Lab. Clin. Med.* **76**, 809 (1970).

[29] A. D. Schreiber, A. P. Kaplan, and K. F. Austen, *J. Clin. Invest.* **52**, 1402 (1973).

[30] P. C. Harpel, *J. Exp. Med.* **146**, 1033 (1977).

[31] H. Fritz, E. Fink, and E. Truscheit, *Fed. Proc., Fed. Am. Soc. Exp. Biol.* **38**, 2753 (1979).

[32] H. Saito, G. H. Goldsmith, M. Moroi, and N. Aoki, *Proc. Natl. Acad. Sci. U.S.A.* **76**, 2013 (1979).

[33] P. S. Damus, M. Hicks, and R. D. Rosenberg, *Nature (London)* **246**, 355 (1973).

[34] G. J. Arlaud, A. Reboul, R. B. Sim, and M. G. Colomb, *Biochim. Biophys. Acta* **576**, 151 (1979).

[35] R. B. Sim, G. J. Arlaud, and M. G. Colomb, *Biochem. J.* **179**, 449 (1979).

example, Factor X_a–antithrombin III,[36] thrombin–antithrombin III,[36] trypsin–α_1-antitrypsin,[37] trypsin–antithrombin III,[37] and plasmin–α_2-antiplasmin.[30] Jesty[36] and Mahoney and colleagues[37] have recently shown that complexes of this type dissociate slowly on exposure to pH >9. Preliminary evidence (R. B. Sim, unpublished results; A. W. Dodds, unpublished results) indicates that this is also the case for plasmin–$\overline{C1}$-Inh and $\overline{C1}$s–$\overline{C1}$-Inh complexes. It is likely that the protease and inhibitor in such complexes are covalently associated, via formation of a tetrahedral intermediate.[37]

Interaction with $\overline{C1}$r and $\overline{C1}$s

The kinetics of interaction of $\overline{C1}$-Inh with isolated $\overline{C1}$r and $\overline{C1}$s have been studied.[7] Both proteases have a similar affinity for $\overline{C1}$-Inh (functional dissociation constant = $10^{-7}\,M$).[7] At 37°C, in the presence of EDTA, $\overline{C1}$r reacts 4- to 5-fold slower with $\overline{C1}$-Inh (second-order rate constant k_1 = $2.8 \times 10^3\,M^{-1}\,\mathrm{sec}^{-1}$) than does $\overline{C1}$s ($k_1 = 1.2 \times 10^4\,M^{-1}\,\mathrm{sec}^{-1}$).[7] In buffers containing Ca^{2+} ions, the rate of reaction of $\overline{C1}$r is reduced 2-fold, but the reactivity of $\overline{C1}$s toward $\overline{C1}$-Inh is unaffected.[7] The activation energy for interaction of $\overline{C1}$r and $\overline{C1}$-Inh (44.3 kcal/mol) is much higher than that for the $\overline{C1}$s–$\overline{C1}$-Inh interaction (11.7 kcal/mol), and thus at temperatures below 37°C the disparity in the reaction rates with $\overline{C1}$r and $\overline{C1}$s becomes much greater.[7] Isolated $\overline{C1}$r, which exists normally as a dimer,[10] forms large complexes with $\overline{C1}$-Inh, which are likely to have the composition $\overline{C1}r_2$–$\overline{C1}$-Inh_2.[3] Under certain conditions, however, the $\overline{C1}$r dimer appears to dissociate on interaction with $\overline{C1}$-Inh, and $\overline{C1}r_1$-$\overline{C1}$-Inh_1 complexes are formed.[38] Interaction of $\overline{C1}$r that is present in antibody–antigen–$\overline{C1}$ complexes, with $\overline{C1}$-Inh, also results in dissociation of the $\overline{C1}$r dimer.[39]

Inhibition of isolated $\overline{C1}$r and $\overline{C1}$s by $\overline{C1}$-Inh shows a pH optimum of 7.5–8.2 (R. B. Sim, unpublished results). The reaction of $\overline{C1}$s with $\overline{C1}$-Inh is insensitive to alteration in ionic strength, but the corresponding reaction with $\overline{C1}$r proceeds optimally at an ionic strength corresponding to 110–300 mM NaCl.[7]

Inhibition of $\overline{C1}$ by $\overline{C1}$-Inh is greatly enhanced in the presence of low concentrations of heparin.[7,40,41] This enhancement is due principally to an

[36] J. Jesty, *J. Biol. Chem.* **254**, 1044 (1979).

[37] W. C. Mahoney, K. Kurachi, and M. A. Hermodson, *Eur. J. Biochem.* **105**, 545 (1980).

[38] S. Chesne, G. J. Arlaud, C. L. Villiers, and M. G. Colomb, *Abstr., Int. Congr. Immunol., 4th, 1980* Abstract 15.1.05 (1980).

[39] R. J. Ziccardi and N. R. Cooper, *J. Immunol.* **123**, 788 (1979).

[40] R. Rent, R. Myrrmann, B. A. Fiedel, and H. Gewurz, *Clin. Exp. Immunol.* **23**, 264 (1976).

[41] K. Nagaki and S. Inai, *Int. Arch. Allergy Appl. Immunol.* **50**, 172 (1976).

acceleration of the rate of reaction of $\overline{\text{C1}}$-Inh with $\overline{\text{C1}}$s.[7,41] Heparin also increases the rate of $\overline{\text{C1}}$-Inh inactivation of $\overline{\text{C1}}$r[7] and plasmin (R. B. Sim and P. E. Gasson, unpublished results), but the enhancement is much smaller than that seen with $\overline{\text{C1}}$s. As discussed in Ref. 7, this enhancement is very similar to the effect of heparin on antithrombin III. The physiological significance of this effect is uncertain, as heparin has many inhibitory actions in the complement system.[42]

As discussed above, there is evidence that $\overline{\text{C1}}$-Inh interacts directly with the active sites of the proteases that it inhibits. It is therefore to be expected that $\overline{\text{C1}}$-Inh inhibits in parallel all of the esterolytic and proteolytic activities of each enzyme. Parallel inhibition by $\overline{\text{C1}}$-Inh of $\overline{\text{C1}}$s-dependent cleavage of C2, C4, and amino acid esters has been reported.[43] However other investigators have suggested differences in the sensitivities to $\overline{\text{C1}}$-Inh of the C2-cleaving, C4-cleaving, and various esterase activities of $\overline{\text{C1}}$s.[44,45] These discrepancies may be partly attributable to the diversity and complexity of the assay systems used, or to the use of insufficiently pure reagents. Nevertheless, this point has not been investigated systematically and requires further study.

Interaction with $\overline{\text{C1}}$

In the course of complement activation, $\overline{\text{C1}}$r and $\overline{\text{C1}}$s do not exist as free proteases in solution, but rather as Ca^{2+}-dependent macromolecular $\overline{\text{C1}}$r$_2$–$\overline{\text{C1}}$s$_2$ complexes bound to C1q, which is in turn bound to the activating surface (e.g., antibody–antigen complexes). $\overline{\text{C1}}$s bound within antibody–antigen–$\overline{\text{C1}}$ reacts with $\overline{\text{C1}}$-Inh at the same rate as does isolated $\overline{\text{C1}}$s,[7,35] but the reactivity of $\overline{\text{C1}}$r toward $\overline{\text{C1}}$-Inh is enhanced up to 4-fold when $\overline{\text{C1}}$r is bound within $\overline{\text{C1}}$.[7,35] Thus the reactivities of $\overline{\text{C1}}$r and $\overline{\text{C1}}$s toward $\overline{\text{C1}}$-Inh are much closer when $\overline{\text{C1}}$r and $\overline{\text{C1}}$s are within antibody–antigen–$\overline{\text{C1}}$ complexes than when the proteases are free in solution.[7,34,35] A number of studies[7,34,35,39,46] have established that interaction of $\overline{\text{C1}}$-Inh with antibody–antigen–$\overline{\text{C1}}$ leads, under physiological conditions, to dissociation of the $\overline{\text{C1}}$ complex. $\overline{\text{C1}}$-Inh reacts rapidly with the $\overline{\text{C1}}$s in antibody–antigen–$\overline{\text{C1}}$ complexes, and more slowly with the $\overline{\text{C1}}$r.[35] Binding of $\overline{\text{C1}}$-Inh to the $\overline{\text{C1}}$r leads to immediate dissociation of $\overline{\text{C1}}$r, $\overline{\text{C1}}$s, and $\overline{\text{C1}}$-Inh from the antibody–antigen–$\overline{\text{C1}}$ complexes.[35] The proteins released are in the form of a $\overline{\text{C1}}$r$_1$–$\overline{\text{C1}}$s$_1$–$\overline{\text{C1}}$-Inh$_2$ complex,[39] which has MW

[42] B. J. Johnson, *J. Pharm. Sci.* **66**, 1367 (1977).
[43] I. H. Lepow, G. B. Naff, and J. Pensky, *Ciba Found. Symp.* [N.S.] **57**, 74 (1965).
[44] K. Takahashi, S. Nagasawa, and J. Koyama, *Biochim. Biophys. Acta* **611**, 196 (1980).
[45] M. Kondo, I. Gigli, and K. F. Austen, *Immunology* **22**, 305 (1972).
[46] A.-B. Laurell, U. Johnson, U. Mårtensson, and A. G. Sjöholm, *Acta Pathol. Microbiol. Scand., Sect. C* **86**, 299 (1978).

330,000–382,000.[39] Specific assays for the $\overline{C1r}_1$–$\overline{C1s}_1$–$\overline{C1}$-Inh$_2$ complex in human serum have been developed by Laurell and colleagues[47] and these may be used as an indicator of C1 activation in patient plasma. A simple screening assay for $\overline{C1}$-Inh activity has been developed,[17] based on the alteration of antigenic properties of $\overline{C1r}$ when it is incorporated into the $\overline{C1r}_1$–$\overline{C1s}_1$–$\overline{C1}$-Inh complex.

Dissociation of the bound $\overline{C1}$ complex by reaction with $\overline{C1}$-Inh leaves most of the C1q still associated with the immune complex or other surface responsible for C1 activation. C1q "exposed" in this way may then interact with lymphoblastoid cells possessing C1q receptors.

[47] A.-B. Laurell, U. Mårtensson, and A. G. Sjöholm, *Acta Pathol. Microbiol. Scand., Sect. C* **87**, 79 (1979).

[6] The Second Component of Human Complement

By Michael A. Kerr

C2 is one of the least abundant of the complement components in human plasma, being present at a concentration of 15–20 mg/liter.[1,2] This and the extreme susceptibility of C2 to proteolytic digestion have in the past hampered its purification and molecular characterization. Functional studies using guinea pig and human preparations have, however, clearly identified the role of C2 in the classical pathway C3 and C5 convertases, and this has been reviewed extensively.[3–5]

The classical pathway C3 convertase is assembled from C2 and C4 on cleavage of these proteins by the $\overline{C1s}$ subcomponent of the $\overline{C1}$ complex. C4 is cleaved by $\overline{C1s}$ with the release of a small peptide, C4a, to form C4b (apparent MW 198,000)[6]: C2 is cleaved to C2a (apparent MW 74,000) and C2b (apparent MW 34,000).[7,8] For the generation of C3 convertase activity, C4 cleavage must precede C2 cleavage. The enzyme is not generated by the addition of C4 to a previously incubated mixture of C2 and $\overline{C1s}$, but

[1] M. A. Kerr and R. R. Porter, *Biochem. J.* **171**, 99 (1979).
[2] M. J. Polley and H. J. Müller-Eberhard, *J. Exp. Med.* **128**, 533 (1968).
[3] R. R. Porter and K. B. M. Reid, *Adv. Protein Chem.* **33**, 1 (1979).
[4] H. J. Müller-Eberhard, *in* "Molecular Basis of Biological Degradative Processes" (R. D. Berlin, H. Hermann, I. H. Lepow, and J. M. Tanzer, eds.), p. 65. Academic Press, New York, 1978.
[5] J. E. Fothergill and W. H. K. Anderson, *Curr. Top. Cell. Regul.* **13**, 259 (1979).
[6] R. A. Patrick, S. B. Taubman, and I. H. Lepow, *Immunochemistry* **7**, 217 (1979).
[7] M. A. Kerr, *Biochem. J.* **183**, 615 (1979).
[8] S. Nagasawa and R. M. Stroud, *Proc. Natl. Acad. Sci. U.S.A.* **74**, 2998 (1977).

is generated by the addition of C2 to a previously incubated mixture of C4 and C1̄s.[9,10] C4b and C2 have been shown to form a Mg^{2+}-dependent complex, and it is apparently the cleavage of C2 in this complex that results in C3 convertase activity. The interaction of C4b and C2 appears to be mediated by the C2b part of the C2 molecule.[8,9] C2a is the catalytic subunit of C3 convertase.

The C3 convertase once assembled is extremely unstable. In solution, the half-life of the enzyme at 37°C is less than 1 min. The decay of activity reflects the release of C2a from the C4b–C2b–C2a complex.[9,11] The C3 convertase can be stabilized by prior treatment of C2 with low concentrations of iodine.[12] The effect is to increase the affinity of C2a for C4b,[9] but the mechanism that is reported to involve the oxidation of adjacent SH groups is unclear. Whatever the mechanism, the treatment has been used routinely and has greatly facilitated study of C2. The C3 convertase cleaves C3 at a single bond to give C3a (apparent MW 9000) and C3b (apparent MW 185,000).[10,13] In the presence of excess C3b, C5 is also a substrate for the enzyme. C2a is the active subunit of this C5 convertase.[14]

A small as yet uncharacterized fragment of C2 may be released by the action of C1̄s and plasmin on C2 in the sera of patients showing genetic deficiency in C1̄ inhibitor (which is, in normal serum, inhibitory to both proteases). The fragment, termed C2 kinin, causes the increased vascular permeability in hereditary angioneurotic edema.[15]

Assay Procedure

Buffers

Calcium–magnesium stock: 30 mM calcium chloride, 100 mM magnesium chloride

Stock veronal buffer (5×): 25 mM sodium barbitone–0.7 M NaCl; 1 M HCl added to pH 7.4

GVB: 0.5 g gelatin (Difco Labs., Detroit, Michigan) dissolved in 100 ml water by boiling, added to 100 ml stock veronal buffer and brought up to 500 ml

[9] M. A. Kerr, Biochem. J. 189, 173 (1980).
[10] H. J. Müller-Eberhard, M. J. Polley, and M. A. Calcott, J. Exp. Med. 125, 359 (1967).
[11] R. M. Stroud, M. M. Mayer, J. A. Miller, and A. T. McKenzie, Immunochemistry 3, 163 (1966).
[12] M. J. Polley and H. J. Müller-Eberhard, J. Exp. Med. 126, 1013 (1967).
[13] V. A. Bokisch, H. J. Müller-Eberhard, and C. G. Cochrane, J. Exp. Med. 129, 1109 (1969).
[14] N. R. Cooper and H. J. Müller-Eberhard, J. Exp. Med. 132, 775 (1970).
[15] V. H. Donaldson, F. S. Rosen, and D. H. Bing, Trans. Assoc. Am. Physicians 90, 174 (1977).

DGVB^{++}: 250 ml GVB added to 250 ml 5% (w/v) glucose and then 2.5 ml Ca^{2+}–Mg^{2+} stock added

GVB–EDTA: 40 ml 0.1 M sodium EDTA pH 7.4 added to 60 ml GVB

Hemolytic Assay

Serial dilutions (0.1 ml) of component C2 in DGVB^{++} are incubated at 30°C with 0.1 ml of sheep erythrocytes (10^8 cells/ml) sensitized by antibody and components C1 and C4 (EAC14). The time of incubation is predetermined for each batch of cells such that maximum lysis is obtained: This time is usually 5 min. The hemolytic reaction is then completed by the addition of 0.3 ml of guinea pig serum diluted 1:30 with GVB–EDTA and incubation at 37°C for 1 hr. 1.0 ml of cold 0.15 M NaCl is then added, mixed well, and the suspension centrifuged at 1000 g for 10 min. The degree of lysis is measured from the A_{410} of the supernatant. 100% hemolysis and 0% hemolysis tubes are included. Results are expressed in 50% lysis units (CH$_{50}$ units) multiplied by the number of erythrocytes per ml (10^8). The preparation of EAC14 by the method of Borsos *et al.*[16] is described in several textbooks on complement.

Purification Procedures

Reagents

All buffers contain 0.04% w/v sodium azide as preservative.

0.4 M sodium phosphate buffer, pH 6.0. 69.9 g anhydrous disodium hydrogen orthophosphate, 547.3 g sodium dihydrogen orthophosphate dihydrate, and 4 g sodium azide are made up to 10 liters with water.

Veronal buffer (VBS). 5 mM veronal pH 8.5, 0.5 mM CaCl$_2$–2.0 mM MgCl$_2$–40 mM NaCl: 5.1 g sodium barbitone, 11.7 g sodium chloride, 2 g sodium azide added to 100 ml calcium, magnesium stock and made up to 5 liters.

Preparation of Chromatography Resins

a. *"Aged" CNBr-Activated Sepharose-4B.* Packed Sepharose-4B (100 ml) is washed well with water and resuspended in a minimal volume of 0.1 M sodium phosphate, pH 7.8. 13 g CNBr in 13 ml of dimethylformamide are added and the mixture stirred until the reaction is complete. Throughout this time the temperature is kept at 20°C by the addition of ice and at pH 10–11 by the addition of 4 M NaOH. When the reaction is complete the Sepharose is washed extensively with 0.1 M sodium phosphate (pH

---

[16] T. Borsos, H. J. Rapp, and H. R. Colten, *J. Immunol.* **105**, 1459 (1970).

7.8) and then stirred overnight at 4°C in 100 ml of the same buffer. This material is termed "aged" CNBr-activated Sepharose-4B.

b. Sepharose-4B-Bound (Anti-Factor B) Antibody. A partially purified rabbit (anti-Factor B) immunoglobulin fraction is prepared by precipitation of antiserum (60 ml) with 16% (w/v) Na_2SO_4. The precipitate is washed twice with 16% Na_2SO_4 and redissolved in 15 ml of 0.07 M sodium phosphate, pH 6.3. The sample is then dialyzed against the same buffer and loaded onto a column (12 × 2.5 cm) of DEAE–Sephadex A50 equilibrated in the buffer. Later fractions of the IgG peak, which is not retarded by the column, have anti-Factor B activity and these are pooled and stored at −20°C. 160 mg of this partially purified rabbit (anti-Factor B) IgG in 40 ml of 0.07 M sodium phosphate, pH 7.8, is coupled to 50 ml of Sepharose-4B that has been activated with CNBr (6.5 g in 6.5 ml dimethylformamide) as described above. The Sepharose-bound antibody is stirred overnight and then washed with 0.1 M sodium phosphate, pH 7.8.

Purification Procedure

Outdated human citrated plasma is made 20 mM in $CaCl_2$ and left to clot overnight at 4°C. The clot is then centrifuged down and the serum stored at −20°C.

Euglobulin Precipitation

Frozen serum (2000 ml) is thawed and, after the addition of 1 ml of 2.5 M iPr_2 P-F in propan-2-ol, is centrifuged at 23,000 g for 30 min at 4°C. The serum is brought to pH 5.5 with 1 M HCl and dialyzed against 8 liters of 5 mM EDTA (pH 5.5)–0.5% (w/v) benzamidine hydrochloride overnight. The dialysate is then centrifuged at 23000 g for 30 min and the supernatant decanted.

Chromatography on CM-Sephadex C-50

Fifty milliliters of 0.4 M sodium phosphate are added to the supernatant and after the pH is brought up to 6.0 with 1 N NaOH the mixture is loaded onto a column (10 × 8 cm) of CM-Sephadex C-50 equilibrated in 0.1 M sodium phosphate (pH 6.0)–0.5% (w/v) benzamidine hydrochloride. The column is then washed with 1 liter of 0.1 M sodium phosphate (pH 6.0)–0.5% (w/v) benzamidine hydrochloride and then developed with a linear gradient formed from 2 liters of 0.1 M sodium phosphate (pH 6.0) and 2 liters of 0.25 M sodium phosphate (pH 6.0), both 0.5% (w/v) benzamidine hydrochloride; all solutions being pumped through the column at a rate of about 1.5–2 liters/hr. C2 is eluted in the middle of the gradient.

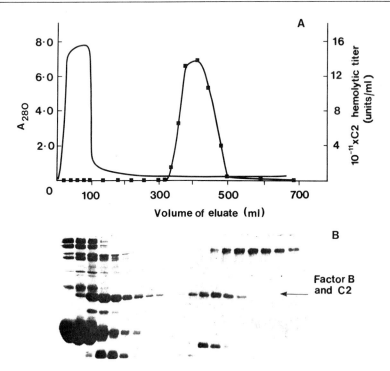

FIG. 1. Chromatography of C2 on "aged" CNBr-activated Sepharose-4B. The protein precipitated at 75% saturated $(NH_4)_2SO_4$, equilibrated with veronal buffer (VBS) and chromatographed on a column (15 × 3 cm) of "aged" CNBr-activated Sepharose-4B. C2 was eluted by sodium phosphate (pH 6.0) started at the point indicated by the arrow. —— A_{280}; —■—, C2 hemolytic activity. The fractions that were assayed were also subjected to SDS–polyacrylamide gel electrophoresis (shown in Fig. 1B). From M. A. Kerr and R. R. Porter, *Biochem. J.* **171**, 99 (1978).

Ammonium Sulfate Precipitation

Active fractions from the CM-Sephadex column are pooled (2 liters) and ammonium sulfate added to 50% saturation (291 g/liter). After stirring for 1 hr at 4°C the solution is centrifuged at 23,000 g for 30 min. The supernatant is decanted, filtered through Whatman No. 1 filter paper, and ammonium sulfate added (159 g/liter) to give 75% saturation. The suspension is then stirred and left to warm to room temperature (15–20°C) before centrifugation at 23,000 g for 90 min at 20°C. The pellet is redissolved in about 50 ml of 0.4 M sodium phosphate, pH 6.0, and stored overnight after the addition of 0.1 ml of 2.5 M iPr$_2$ P-F.

Chromatography on "Aged" CNBr-Activated Sepharose-4B

The cloudy suspension is centrifuged at 26,000 g for 30 min and the pellet discarded. The supernatant is passed through a column (30 × 4 cm) of Sephadex G-25 equilibrated in 5 mM veronal buffer (VBS) and then applied to a column (15 × 3 cm) of "aged" CNBr-activated Sepharose-4B equilibrated in the same buffer. The column is washed with the same buffer until A_{280} of the eluate is less than 0.05, and then the C2 is eluted with 250 ml of 0.4 M sodium phosphate, pH 6.0.

Chromatography on DEAE–Sepharose

The active fractions from the "aged" CNBr-activated Sepharose column (Fig. 1) are pooled and equilibrated with 5 mM veronal buffer (VBS) by passage through a column (50 × 5 cm) of Sephadex G-25 in that buffer. The protein is then applied to a column (30 × 2.5 cm) of DEAE–Sepharose equilibrated in the same buffer. The column is washed with 100 ml of the veronal buffer and developed with a linear gradient made from 500 ml of 5 mM veronal buffer (VBS) and 500 ml of 5 mM veronal buffer (pH 8.5)–0.5 mM CaCl$_2$–2.0 mM MgCl$_2$–80 mM NaCl.

Chromatography on Sepharose-4B-Bound Anti-Factor B Antibody

Although early fractions eluted by the gradient on the DEAE–Sepharose column (Fig. 2) are homogeneous C2, later fractions are contaminated with Factor B. This is conveniently removed by chromatography on Sepharose-4B-bound (anti-Factor B) antibody. All active frac-

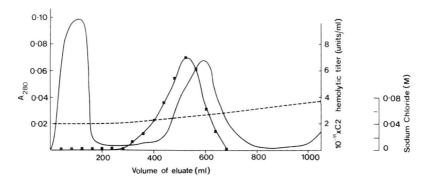

FIG. 2. Chromatography of C2 on DEAE–Sepharose. Active fractions from the "aged" CNBr-activated Sepharose-4B column, equilibrated with veronal buffer (VBS), were loaded onto a column (30 × 2.5 cm) of DEAE–Sepharose and the column developed with increasing sodium chloride concentration. ——, A_{280}; —■—, C2 hemolytic activity; ---, concentration of NaCl.

TABLE I
PURIFICATION OF C2 FROM HUMAN SERUM

Fraction	Volume (ml)	$10^{-13} \times$ Total activity (units)	Total protein[a] (mg)	$10^{-11} \times$ Specific activity (units/mg)	Activity yield (%)
Serum	2000	38	112000	0.034	100
CM-Sephadex pool	2200	26	n.d.[b]	n.d.	69.5
75% saturated (NH$_4$)$_2$SO4 pellet	55	26	980	2.65	68
Aged CNBr-activated Sepharose-4B pool	120	22	39	56.4	58
DEAE–Sepharose pool	250	16	18	88.9	41
Sepharose-4B-bound (anti-Factor B) antibody pool	70	14	9.0	155.6	37
Pool concentrated by ultrafiltration	11	13	8.8	147.7	34

[a] Protein determined by absorbance at 280 nm.
[b] Not determined.

tions from the DEAE–Sepharose column are pooled and then applied to a column (10×2 cm) of Sepharose-4B-bound (anti-Factor B) antibody equilibrated in 5 mM veronal buffer (VBS). The column is washed with 200 ml of the same buffer, and the C2, which is retarded because of the "aged" CNBr-activated Sepharose properties of the column, is eluted in a concentrated form by 200 ml of 5 mM veronal buffer (pH 8.5), 0.2 M NaCl. Factor B (inactive) is eluted and the column regenerated by washing with 1.0 M propionic acid and then reequilibrated with the veronal buffer.

The C2 fraction represents a purification of about 4000-fold from serum and the yield is about 8 mg from 2 liters of serum (Table I). The C2 is homogeneous by the criteria of SDS–polyacrylamide electrophoresis and N-terminal analysis. The final product is stable at neutral pH without the addition of proteolytic enzyme inhibitors and can be further concentrated by ultrafiltration and stored at $-70°C$. The purification procedure has been used routinely in our laboratory for 3 years to produce sufficient quantities of C2 for amino acid sequence analysis. The same purification scheme has been used recently for the purification of guinea pig C2,[17] the only modification being the requirement to increase the limit buffer for the DEAE–Sepharose column to 0.12 M NaCl. No traces of Factor B are

[17] M. A. Kerr and J. Gagnon, *Biochem. J.*, submitted for publication.

present in this preparation of the anti-Factor B column is therefore unnecessary.

A Note on the Use of "Aged" CNBr-Activated Sepharose-4B

The major difficulty in the purification of C2 is its separation from Factor B. Factor B, which is similar in molecular weight and charge to C2, but present in great excess over C2, is partially resolved from C2 by most chromatographic techniques. However, the repeated chromatography necessary to separate the two completely results in low yields of C2. It was during attempts to develop an affinity column that would allow rapid separation of C2 from Factor B that the curious property of C2 adsorbing on "aged" CNBr-activated Sepharose-4B was observed.[1] The nature of the interaction between C2 and the resin is unclear: It has, however, proved invaluable in the purification of C2, greater than 30-fold purification being achieved in a single step. Other proteins do however bind to the resin, the interaction being similar to an ion-exchange phenomenon in that it is decreased by higher salt concentrations and by a change in pH. The same separation however is not achieved on CM or DEAE–Sephadex or Sepharose under the same conditions. The capacity of "aged" CNBr-activated Sepharose is about 1 mg of protein/ml packed resin under the conditions used. Because of this relatively low capacity, it is essential that the earlier steps of the purification be carried out to remove most of the protein. Careful equilibration of column and sample buffer is also essential.

Precautions against Inactivation of C2

C2 is particularly susceptible to proteolysis during column chromatography, especially during earlier stages. To limit this, the $C\overline{1}s$ inhibitor benzamidine is added to buffers during early stages of the purification. The preparation is usually carried out in 1 week and all fractions stored where possible at pH 6.0. Chromatography columns are washed and repacked carefully before use to ensure removal of bound protein and to allow rapid flow rates: CM-Sephadex by washing with $0.4 M$ sodium phosphate (pH 6.0) and the "aged" CNBr-activated Sepharose and DEAE–Sepharose with 5 mM veronal buffer, $2 M$ NaCl (pH 8.5).

Molecular Properties

C2 comprises a single carbohydrate-containing polypeptide chain with an apparent molecular weight by SDS–polyacrylamide gel electrophoresis of 102,000; alanine is the N-terminal amino acid.[7,8] The molecule is rapidly cleaved by subcomponent $C\overline{1}s$ with the loss of hemolytic activity to yield

TABLE II

AMINO ACID COMPOSITION OF C2, C2a, AND C2b[a]

Amino acid	Amino acid composition (residues/100 residues)		
	C2	C2a	C2b
Cys	3.2	2.7	5.3
Asp	11.2	12.7	8.4
Thr	5.0	4.3	4.9
Ser	7.9	7.3	8.7
Glu	10.4	10.5	8.7
Pro	5.7	4.6	9.8
Gly	9.2	6.6	13.4
Ala	6.3	6.2	6.5
Val	6.8	6.7	7.0
Met	1.9	2.7	0.8
Ile	4.2	4.9	2.9
Leu	9.4	10.3	6.3
Tyr	2.7	1.6	3.4
Phe	4.4	4.4	4.0
His	2.7	2.8	2.1
Lys	5.1	5.9	2.5
Arg	5.3	5.5	5.3
Trp	n.d.[b]	n.d.	n.d.

[a] From M. A. Kerr, *Biochem. J.* 171, 99 (1978).
[b] Not determined.

two fragments, C2a with apparent MW 74,000 and C2b, 34,000. The two fragments are not linked by disulfide bonds and can easily be separated without the requirement for denaturing agents, suggesting they represent separate domains in the molecule. C2b is the N-terminal end of the molecule. The amino acid compositions of C2, C2a, and C2b are shown in Table II. The amino acid analysis of C2 shows no exceptional features other than its similarity to Factor B. C2a, as would be expected from its acidic behavior on electrophoresis, is rich in aspartic and glutamic acids, though no attempt has been made to assess the degree of amidation. C2b has a most unusual composition, being very rich in small uncharged amino acids—glycine (13.4%), proline (9.8%), and cysteine (5.3%)—but despite the high proline and glycine content, no hydroxylysine or hydroxyproline has been detected, suggesting that no collagen-like sequences such as those found in C1q are present.

N-terminal sequences of C2, C2a, and C2b are shown in Fig. 3. Cleavage of C2 by C$\overline{1}$s occurs at an Arg/Lys-Ile bond, the same bond cleaved in Factor B by Factor D. In spite of the similarity of properties of C2 and

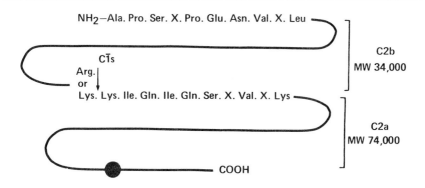

The activation of C2 by C̄1s

FIG. 3. The activation of C2. The cleavage of C2 by C̄1s showing amino-terminal sequences of the fragments formed. The active site is indicated to be in the C2a fragment (—●—).

Factor B and similar amino acid composition, this is the only homology in the small amount of sequence data that is available.[7]

Genetic Polymorphism

C2 from human serum is polymorphic, as judged by electrophoretic mobility.[18] The two common alleles C2[1] and C2[2] have gene frequencies of 0.96 and 0.04 and the structural gene is closely linked with the B locus of the major histocompatibility complex as is the gene for Factor B.[19,20] C2 deficiency is the most common of the human complement deficiencies; it is transmitted as an autosomal recessive trait. Of the more than 30 pedigrees that have been described, about half were healthy and half suffered some "immune complex disease."[21]

C2 as a Proteolytic Enzyme

C2a has been identified as the active subunit in the C3 convertase complex, C4b–(C2b) C2a, and in the presence of C3b as the active subunit of the C5 convertase complex. C3 is cleaved at a single Arg-Ser[22] bond

[18] T. Meo, J. P. Atkinson, M. Bernoco, O. Bernoco, and R. Ceppellini, *Proc. Natl. Acad. Sci. U.S.A.* **74**, 1672 (1977).
[19] S. M. Fu, H. G. Kunkel, H. P. Brusman, F. H. Allen, Jr., and M. Fotino, *J. Exp. Med.* **140**, 1108 (1974).
[20] C. A. Alper, *J. Exp. Med.* **144**, 1111 (1976).
[21] D. Glass, D. Raum, D. Gibson, J. S. Stillman, and P. H. Schur, *J. Clin. Invest.* **58**, 853 (1976).
[22] T. Hugli, *J. Biol. Chem.* **250**, 8293 (1975).

and C5 at a single Arg-X[23] bond. Isolated C2a is however unable to cleave either C3 or C5.

Both C2 and C2a have been shown to possess esterase activity against certain arginine and lysine esters, principally N-acetylglycine-lysine methyl ester.[24,25] C2 has also been reported to be inhibited by diisopropyl fluorophosphate,[26] though others were unable to detect this inhibition. In view of the structural and functional similarity of C2 to Factor B, which has recently been shown to be a serine protease by amino acid sequence analysis,[27] it is likely that C2 is a second member of this class of enzymes, mechanistically similar to typical trypsin-like enzymes but whose zymogens are activated by a mechanism unlike other zymogens of serine proteases.

[23] H. Fernandez and T. Hugli, *J. Immunol.* 117, 1688 (1976).
[24] N. R. Cooper, *Prog. Immunol.* 1, 568 (1971).
[25] N. R. Cooper, *Biochemistry* 24, 4245 (1975).
[26] R. G. Medicus, O. Götze, and H. J. Müller-Eberhard, *Scand. J. Immunol.* 5, 1049 (1976).
[27] D. L. Christie, J. Gagnon, and R. R. Porter, *Proc. Natl. Acad. Sci. U.S.A.* 77, 4923 (1980).

[7] The Third, Fourth, and Fifth Components of Human Complement: Isolation and Biochemical Properties

By B. F. Tack, J. Janatova, M. L. Thomas, R. A. Harrison, and C. H. Hammer

Introduction

The participation of the third, fourth, and fifth components of complement in inflammatory, immune surveillance, and immune-response pathways has been documented. Thus, the complement system is a principal effector of the humoral immune response and as such assumes an important role in host defense to infection. This system of proteins operates by two pathways, the classical and alternative, which are subject to intrinsic and extrinsic controls. Both pathways proceed by the sequential self-assembly of multimacromolecular enzyme complexes which have as their final substrates C3 and C5. The alternative pathway operates independently of C4, and C3 therefore represents the point of convergence. Activation proceeding by the classical pathway results in a selective and controlled fragmentation of each protein. Interactions of C3 and C4 fragments with surface receptors of the polymorphonuclear leukocyte, B lymphocyte, monocyte, and macrophage serve to modulate complement-dependent granulocyte and lymphocyte functions. The activation profile of C5, although similar to that of C3 and C4, elicits products that are different in two important respects: (1) the activation peptide, in addition to its vasoactive properties, also promotes chemotaxis; and (2) the macromolecular activation fragment initiates assemblage of the membranolytic (cytolytic) complex. While the low-molecular-weight activation peptides of each protein have been studied in considerable structural detail, the binding sites present in the macromolecular activation fragments of each protein that interact with membrane constituents have received less attention. At this time, the biological and structural properties of C3 are best understood and will be the focus of discussion in this article.

Methods of Assay and Isolation

A method for the isolation of multiple human complement components from fresh EDTA-plasma (2–11 liters) has been described.[1] By conventional ion-exchange and gel-permeation chromatography, milligram to gram amounts of several complement proteins were obtained in a state of high biochemical and functional purity with full hemolytic activity. To avoid unnecessary duplication, only those steps relevant to the isolation of C3, C4, and C5 will be presented here. This section will be followed by data relating to the physical and chemical properties, pathways of activation and inactivation, and active-site structures of these proteins.

Methods of Functional and Antigenic Assay

Guinea pig C1 and C9 were isolated as described by Nelson *et al.*[2] and Vroon *et al.*[3] Functionally pure guinea pig C5, C6, C7, and C8 were prepared by the method of Hammer *et al.*[4] Human C2 and C4 were purchased from Cordis Laboratories, Miami, Florida. Human C3, isolated by the method of Tack and Prahl,[5] was immunoadsorbed to remove trace amounts of C5, IgG, and IgA. Buffers for hemolytic assays were prepared according to Hammer *et al.*,[6] except that buffer D^{++} contained $0.20 M$ glucose. The $EAC\overline{14}$ cell intermediate used for C3, C5, C6, and C7 functional

[1] C. H. Hammer, G. H. Wirtz, L. Renfer, H. D. Greshman, and B. F. Tack, *J. Biol. Chem.* **256**, 3995 (1981).

[2] R. A. Nelson, J. Jensen, I. Gigli, and N. Tamura, *Immunochemistry* **3**, 111 (1966).

[3] D. H. Vroon, D. R. Schultz, and R. M. Zarco, *Immunochemistry* **7**, 43 (1970).

[4] C. H. Hammer, A. Nicholson, and M. M. Mayer, *Proc. Natl. Acad. Sci. U.S.A.* **72**, 5076 (1975).

[5] B. F. Tack and J. W. Prahl, *Biochemistry* **15**, 4513 (1976).

[6] C. H. Hammer, A. S. Abramovitz, and M. M. Mayer, *J. Immunol.* **117**, 830 (1976).

assays[7] was prepared with guinea pig C1 and human C4 as described by Borsos and Rapp.[8] Human C4 was assayed by the method of Gaither et al.[9] using EA cells and serum from guinea pigs deficient in C4 as the source of all other complement components. C8 and C9 were titrated with EAC-7 cells, as previously described.[7] C2, C1 esterase inhibitor (C1EI), C3b inactivator (C3bINA), and β1H were assayed according to Borsos et al.,[10] Gigli et al.,[11] and Gaither et al.[12] Titers for C2 through C9 are expressed as the averages of the product $Z \times$ reciprocal dilution obtained from a minimum of three independent measurements within the linear portion of the dose–response curve, where $Z = -\ln(1 - y)$, $y = \%$ lysis. One unit/ml corresponds to 1.5×10^7 effective molecules in a system containing 0.1 ml indicator cells (1.5×10^8/ml), 0.2 ml sample dilution, and 0.2 ml of a reagent containing the remaining complement components required for lysis.

The quantitation of antigenic levels of C3, C4, and C5 in plasma and purified fractions was done by radial immunodiffusion according to Mancini et al.[13] Ouchterlony analyses of column fractions were performed on diffusion plates prepared with phosphate-buffered saline containing 0.03% sodium azide, 10 mM EDTA, and 1% agarose. Antisera for these analyses were either obtained from commercial sources or by immunization of sheep, burros, or goats with the appropriate antigens. Animals, which were housed at the NIH animal farm, were immunized with 50–500 μg of protein emulsified in 50% Freund's complete adjuvant (2 ml), given intramuscularly at multiple sites in the hindlegs, or intradermally along the animal's back.

Antisera Fractionation and Immunoadsorbent Preparation

The large volumes of antisera required for these steps were obtained by plasmaphoresis. A procedure using octanoic acid and DEAE–cellulose[14] to obtain pure IgG antibody from mammalian sera was followed for fractionation of these antisera. Post-DEAE–cellulose antibody pools were concentrated by ultrafiltration using PM-30 membranes (Amicon Corp., Lexington, Massachusetts). Antibody at a final concentration

[7] C. H. Hammer, M. L. Shin, A. S. Abramovitz, and M. M. Mayer, J. Immunol. 119, 1 (1977).
[8] T. Borsos and H. J. Rapp, J. Immunol. 99, 263 (1967).
[9] T. A. Gaither, D. W. Alling, and M. M. Frank, J. Immunol. 113, 574 (1974).
[10] T. Borsos, H. J. Rapp, and M. M. Mayer, J. Immunol. 87, 310 (1961).
[11] I. Gigli, S. Ruddy, and K. F. Austen, J. Immunol. 100, 1154 (1968).
[12] T. A. Gaither, C. H. Hammer, and M. M. Frank, J. Immunol. 123, 1195 (1979).
[13] G. Mancini, A. O. Carbonara, and J. F. Heremans, Immunochemistry 2, 235 (1965).
[14] M. Steinbuch and R. Audran, Arch. Biochem. Biophys. 134, 279 (1969).

of 5 mg/ml was coupled to Sepharose-4B beads according to the method of March et al.[15] Unreacted groups were masked with a solution of 0.1 M ethanolamine (pH 9.0) following the coupling reaction. Sepharose-4B : IgG adsorbents were treated with 1 mM PMSF for 30 min at 37°C prior to use.

Preparation of Sepharose/L-Lysine Adsorbent

The Sepharose-4B/L-lysine adsorbent was prepared by coupling 200 g of L-lysine per liter of cyanogen bromide activated Sepharose-4B.[15,16] The capacity of 1.1 liters of this adsorbent was sufficient to deplete up to 3.5 liters of plasma of plasminogen.

Isolation Procedures (Part A)

All steps were performed at 4°C unless otherwise indicated. Centrifugations of polyethylene glycol (PEG) 4000 precipitates were carried out at 14,000 g for 25 min at a Sorvall RC2B centrifuge. The conductivity of buffers and solutions was measured at 0°C and reported in units of mS/cm. Stock buffers and solutions were Millipore-filtered prior to dilution and use. The data presented for purification of individual components through the DEAE–Sephacel (Pharmacia Fine Chemicals, Piscataway, New Jersey) step were obtained from the fractionation of a 2-liter pool of plasma. In some instances, 11-liter pools of plasma were processed and these preparations were used as the source of C3 and C5 which were then further purified.

1. Treatment of Plasma with Inhibitor Solutions. Platelet-free EDTA human plasma was obtained from four donors who were medication-free and had fasted prior to plasmaphoresis. Each unit of fresh plasma (about 500 ml) was diluted with a buffered inhibitor solution containing 1 M KH_2PO_4–0.2 M Na_2 EDTA–0.2 M benzamidine-HCl adjusted to pH 7.4 with NaOH (20 parts plasma to 1 part inhibitor). Following the addition of inhibitors the individual units were pooled, and 0.1 M PMSF in anhydrous isopropanol was added to a final concentration of 1 mM. The pH was adjusted to 7.4, if necessary.

2. Fractionation of Plasma with PEG 4000. The inhibitor-treated plasma pool was made 5% (w/v) in PEG 4000 by slow addition, with stirring, of the solid powder, and allowed to equilibrate for 1 hr. The precipitate that formed was removed by centrifugation. The 5% PEG supernatant containing the bulk of the plasma proteins and complement components was

[15] S. C. March, I. Parikh, and P. Cuatrecasas, *Anal. Biochem.* **60**, 149 (1974).
[16] D. G. Deutsch and E. T. Mertz, *Science* **170**, 1095 (1970).

adjusted to a specific conductivity of 12 mS/cm by the addition of solid NaCl prior to plasminogen depletion.

3. Plasminogen Depletion of the 5% PEG Supernatant. A glass column 13 cm in diameter was packed with 1.1 liters of Sepharose/L-lysine adsorbent and equilibrated in 50 mM K–Na phosphate buffer (pH 7.4; 11.8 mS/cm) containing 10 mM EDTA and 150 mM NaCl. The 5% PEG supernatant was applied and collected at a flow rate of 1300 ml/hr. Residual nonspecifically bound protein was removed from the column by washing with a 2-fold concentrated buffer and included in the adsorbed plasma pool.

Following several column washes with the 2-fold concentrated buffer as above, the bound plasminogen was eluted with concentrated buffer containing 200 mM ε-amino-caproic acid (EACA) and stored at −65°C. The adsorbent was regenerated by sequential treatment with 10 mM NaOH and 10 mM phosphoric acid. After washing with starting buffer, the gel was stored in the cold in the presence of 0.02% azide.

4. Concentration and Ionic-Strength Adjustment of the Plasminogen-Depleted PEG Supernatant. The post-Sepharose/L-lysine pool (13.2 mS/cm) was made 33 mM in EACA by the addition of solid inhibitor, and allowed to warm to about 15°C. A diluent solution (pH 7.4; 1.1 mS/cm), containing 6.7 mM Na$_2$ EDTA–6.7 mM benzamidine HCl–33.3 mM EACA, was made 1 mM in PMSF and warmed to 15°C. The ionic strength of the plasminogen-depleted PEG supernatant was rapidly reduced by four sequential additions of the 1.1 mS/cm diluent (to a total volume of 4 liters) and subsequent concentration of the adjusted pool to 2 liters with a Pellicon ultrafiltration system, at an average rate of 5 liters/hr. When the adjusted pool approached a specific conductivity of 1.35 mS/cm, the volume of the pool was reduced to 500 ml and transferred to a 1-liter graduate. The concentrator was rinsed with two 500-ml portions of ice-cold DEAE–Sephacel buffer described below, each concentrated to 200 ml, and these were added to the adjusted pool, now at 4°C. The specific conductivity and pH were adjusted with solid NaCl and 1 N NaOH to 1.35 mS/cm and 7.4, respectively.

5. DEAE–Sephacel Chromatography. The concentrated and plasminogen-depleted 5% PEG supernatant was applied at 220 ml/hr to a 6.5 × 146-cm (4850-ml bed volume) glass column of DEAE–Sephacel, equilibrated with a 3.2 mM K–sodium phosphate buffer (pH 7.4; 1.37 mS/cm) containing 6.4 mM Na$_2$ EDTA–6.4 mM benzamidine HCl–31.8 mM EACA. Following application of the sample, the column was washed with the same buffer at the above flow rate until 3 liters of effluent had been collected, the first 2400 ml of which were discarded, and the last 600

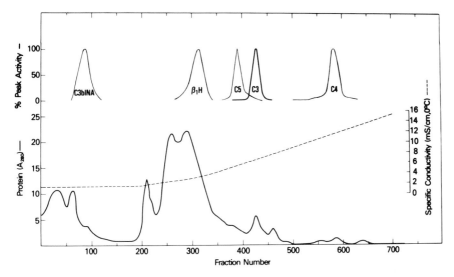

FIG. 1. DEAE–Sephacel chromatography of the 5% PEG supernatant of fresh human EDTA–plasma.

ml, enriched in IgG, pooled and stored frozen at −65°C. Fraction collection (22 ml/tube) was initiated and the column was developed with a 12-liter linear salt gradient. The limit buffer was identical in composition to the wash buffer, except that it also contained 300 mM NaCl (16 mS/cm). The flow rate was maintained at 200 ml/hr. Following application of the entire gradient, several liters of limit buffer were applied to complete the elution profile. Protein was monitored by absorbance at 280 nm after adjusting for interference by benzamidine.

The results of this column step are shown in Fig. 1. Complement-component activities were identified by functional assays on appropriate dilutions of column fractions, as described earlier. Where monospecific antisera were available, double-diffusion assays were run to identify regions of complement-component antigen prior to performing functional assays. Ceruloplasmin was detected by its blue color and monitored optically at 600 nm. In addition to C3bINA eluting at 1.35 mS/cm, C5 (5.5 mS/cm) and C4 (11.6 mS/cm) were individually resolved from other complement components and pooled as such. β_1H (3.4 mS/cm) overlapped with C8 (2.6 mS/cm) and C6 (2.8 mS/cm). C3 eluting at 6.6 mS/cm was effectively separated from C5, but overlapped with about 30% of the C1EI which eluted broadly at 7.4 mS/cm. Ceruloplasmin (7.8 mS/cm) was well resolved from C3.

Isolation Procedures (Part B)

1. C5 Purification: Gel Filtration on Sepharose CL-6B. The C5-containing fractions from a 10.2-liter plasma pool chromatographed on an 11 × 147-cm column of DEAE–Sephacel were pooled and the C5 was precipitated at pH 7.4 by the addition of solid PEG to 16% (w/v). Following equilibration for 1 hr at 4°C, the C5 precipitate was collected by centrifugation and resolubilized in DEAE–Sephacel wash buffer (pH 7.4) containing 500 mM NaCl. For gel filtration, 5.9% (15 ml) of the concentrated C5 pool (14.2 A_{280}/ml) was adjusted with PMSF to a final concentration of 1 mM and applied to a 5 × 85-cm column containing Sepharose CL-6B (Pharmacia Fine Chemicals). The column was equilibrated in 5 mM Na phosphate buffer (pH 7.4) containing 150 mM NaCl–5 mM EDTA. The flow rate was maintained at 125 ml/hr and 10-ml fractions collected, This step, shown in Fig. 2B, was effective in removing macromolecular protein and resulted in at 4.2-fold purification. The C5-containing fractions were pooled and concentrated by ultrafiltration on a PM-30 membrane to 2.6 A_{280}/ml. Immunochemical analyses revealed that the C5 pool contained C3, β_1H, HSA, IgG, and IgA.

2. C5: Hydroxylapatite Chromatography. The C5 pool concentrate was equilibrated with a 5 mM Na phosphate buffer (pH 7.4, 6.0 mS/cm) containing 100 mM NaCl and 5 mM Na$_2$EDTA. The sample was applied to a 1.5 × 26-cm hydroxylapatite (Gallard-Schlesinger Chem. Mfg. Corp., Carle Place, New York) column equilibrated with the same buffer and washed with 1 column volume at 20 ml/hr. The C5 was eluted by application of a linear salt gradient of 200 ml, collecting 2-ml fractions. The limit buffer composition was identical to the wash buffer except that it contained 1 M NaCl. Elution of residual protein from the column was completed by application of an additional linear salt gradient to 2 M NaCl, and a terminal kick buffer (pH 7.4) containing 125 mM Na phosphate–100 mM NaCl–5 mM EDTA. C5-containing fractions were pooled and concentrated to about 4 ml (4.9 A_{280}/ml) on a colloidion bag of 75,000-MW exclusion. Immunochemical tests indicated that this step was effective in removing β_1H, C3, and HSA.

3. C5: Immunoadsorption of IgG and IgA. Immunoglobulins present in the post-hydroxylapatite C5 pool were removed by adsorption on a Sepharose-4B : anti-β_1H-, IgG-, and IgA-containing immunoadsorbent. The C5 preparation was applied to a 1.5 × 8-cm column containing 14 ml of adsorbent equilibrated with 100 mM K–Na phosphate buffer (pH 7.4) containing 150 mM NaCl–5 mM Na$_2$ EDTA. The C5 drop-through was collected as a pool. The specific activity of the final product was 3.0×10^6

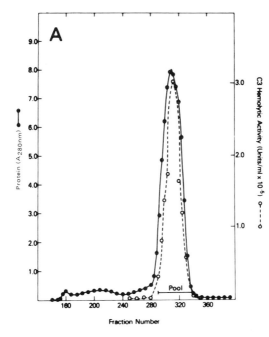

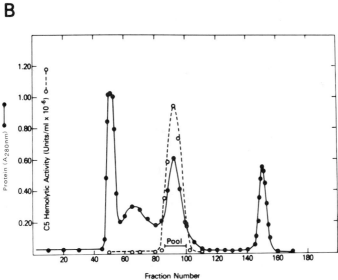

FIG. 2. (A) Gel filtration of the 16% PEG concentrated post-DEAE–Sephacel C3 pool on Sepharose CL-6B. (B) Gel filtration of the 16% PEG concentrated post-DEAE–Sephacel C5 pool (5.9%) on Sepharose CL-6B.

TABLE I
SUMMARY OF PURIFICATION[a]

	Protein		Activity			
					Specific activity	Purifi-
	Total A_{280}	Yield	Total units	Yield	units/A_{280}	cation
Step	($\times 10^{-3}$)	(%)	($\times 10^{-7}$)	(%)	($\times 10^{-3}$)	(-fold)
C5 component						
1. Pooled human plasma	620	100	138	100	2.23	1
2. DEAE–Sephacel 2nd. 16% PEG precipitate	3.62	0.85	63.8	46	176	79
3. Sepharose CL-6B	0.057	0.16[b]	4.29	53[b]	750	336
4. Hydroxylapatite	0.020	0.054	2.52	31	1290	577
5. Immunoadsorption	0.006	0.018	1.95	24	3000	1350
C3 component						
1. Pooled human plasma	718	100	20.9	100	0.29	1
2. DEAE–Sephacel 2nd. 16% PEG precipitate	6.10	0.85	18.2	87	29.8	102
3. Sepharose CL-6B	4.64	0.65	13.9	67	30.0	103
4. Immunoadsorption	0.422	0.59[b]	1.27	61[b]	30.0	103
C4 component						
1. Pooled human plasma	118	100	25.5	100	2.16	1
2. DEAE–Sephacel 2nd. 16% PEG precipitate	0.56	0.48	19.8	78	354	164

[a] These data are taken from C. H. Hammer, G. H. Wirtz, L. Renfer, H. D. Gresham, and B. F. Tack, *J. Biol. Chem.* **256**, 3995 (1981).
[b] Recovery adjusted for quantity used.

units/A_{280} and represented a 1350-fold purification. The recovery of C5 functional activity was 24% of that present in the starting material (Table I).

4. C3 Purification: Gel Filtration on Sepharose CL-6B. The C3-containing fractions from an 11.4-liter plasma pool chromatographed on a 12 × 90-cm column of DEAE–Sephacel were pooled, and the C3 precipitated by addition of solid PEG to 16% (w/v). After 1 hr of equilibration at 4°C, the C3 precipitate was collected by centrifugation and dissolved in a 100 mM K phosphate buffer (pH 7.4) containing 150 mM NaCl–5 mM Na$_2$ EDTA–50 mM EACA. This procedure was repeated one time to further concentrate and purify the C3 preparation. The second 16% PEG precipitation was effective in removing 5% of the protein, including trace amounts of ceruloplasmin. The C3 pool was applied to a 10 × 110-cm column containing Sepharose CL-6B, equilibrated with the above buffer, and eluted at a flow rate of 160 ml/hr, collecting 20-ml fractions. This step removed a small amount of high-molecular-weight material, as shown in Fig. 2A. Functional and antigenic analysis of the C3 pool (4.5 A_{280}/ml) indicated 4200 units/ml of C5, 0.24 mg/ml of IgG, and trace amounts of IgA.

5. *C3: Immunoadsorption of C5, IgG, and IgA.* Detectable amounts of C5 activity, IgG, and IgA antigen in the C3 preparation were removed by selective adsorption on a Sepharose-4B: anti-C5, -IgG, and -IgA immunoadsorbent. Ten percent of the post-Sepharose CL-6B C3 pool was applied at 4°C to a column containing about 30 ml of immunoadsorbent equilibrated with a 100 mM K–Na phosphate buffer (pH 7.4), containing 150 mM NaCl–5 mM Na$_2$ EDTA. The flow rate of the column was maintained at 60 ml/hr. Protein was monitored by absorbance at 280 nm and the C3 collected as a pool. The specific activity of the final product was 30,000 units/A_{280} and represented a 103-fold purification. The recovery of C3 functional activity was 61% of that present in the starting pool of plasma (Table I).

6. *C4 Purification: Precipitation with PEG.* The C4-containing fractions from a 2-liter plasma pool from the 6.5 × 146-cm column of DEAE–Sephacel were pooled, and the C4 precipitated by addition of solid PEG to 16% (w/v). After 1-hr equilibration at 4°C the precipitate was collected by centrifugation and dissolved in 50 mM K–Na phosphate buffer (pH 7.4), containing 10 mM Na$_2$ EDTA–150 mM NaCl. The biochemical purity of the post-DEAE C4 preparation was estimated to be about 90% based on the observed 164-fold purification (Table I) and the antigenic C4 concentration of the starting plasma (0.328 mg/ml).

Functional Purity of Isolated Components

The C5, C3, and C4 preparations (1.3 × 10⁶; 9.9 × 10⁴, and 2.1 × 10⁶ units/ml, respectively) were assayed for contamination by other complement components. The results of this evaluation are shown in Table II and demonstrate the high degree of functional purity of these proteins when assayed at concentrations up to 16 times that found in plasma.

Immunochemical Analysis

Double-diffusion analysis of the C5 and C3 preparations at 1.5 mg/ml and 3.3 mg/ml, respectively, against monospecific antisera to albumin,

TABLE II

FUNCTIONAL PURITY OF ISOLATED COMPONENTS[a]

Titrated complement component	Titer in pooled human plasma[b]	Titer in purified component[b]		
		C5	C3	C4
C2	2820	0	0	24
C3	18,400	0	99,000	32
C4	129,000	0	2	2,130,000
C5	135,000	1,300,000	1	39
C6	49,000	4	100	0
C7	68,200	1	29	26
C8	121,000	0	221	15
C9	95,000	0	36	180
C3bINA	5290	0	0	0
C1EI	18,700	1	9	2800
$\beta_1 H$	3400	0	0	5

[a] These data are taken from C. H. Hammer, G. H. Wirtz, L. Renfer, H. D. Gresham, and B. F. Tack, *J. Biol. Chem.* **256**, 3995 (1981).

[b] Titers are expressed as the reciprocal of the dilution producing $Z = 1$ (63% lysis); i.e., $Z = -\ln(1 - y)$, $y = \%$ lysis.

IgG, α_1-antitrypsin, α_2-macroglobulin, plasminogen, C1q, IgA, IgM, transferrin, α_2-HS-glycoprotein, ceruloplasmin, β_2-glycoprotein, C-reactive protein, C1EI, $\beta_1 H$, Factor B, properdin, C3bINA, C1s, C8, and C4-binding protein failed to reveal detectable levels of these plasma proteins.

Polyacrylamide-Gel Electrophoresis (PAGE)

The purified proteins revealed multiple chain structures when subjected to SDS–PAGE[17] under reducing conditions, as shown in Fig. 3. The C5 and C3 structures consisted of 2 chains each: α-chains with estimated MW 115,000 ± 12,000 and β-chains with estimated MW 75,000 ± 8000, in agreement with reported values.[5,18,19] The biochemical purity of the C4, as estimated earlier from measurement of the plasma-antigenic level and the observed fold purification, was consistent with the results of SDS–PAGE, which indicated bands of 93,000 ± 9000, 75,000 ± 8000, and 30,000 ± 3000. These are the expected molecular weights for the α-, β-, and γ-chains of C4.[20–22]

[17] J. V. Maizel, *in* "Methods in Virology" (K. Maramorosch and H. Koprowski, eds.), p. 179. Academic Press, New York, 1971.

[18] B. F. Tack, S. C. Morris, and J. W. Prahl, *Biochemistry* **18**, 1490 (1979).

[19] U. R. Nilsson, R. J. Mandle, and J. A. McConnell-Mapes, *J. Immunol.* **114**, 815 (1975).

[20] R. D. Schreiber and H. J. Müller-Eberhard, *J. Exp. Med.* **140**, 1324 (1974).

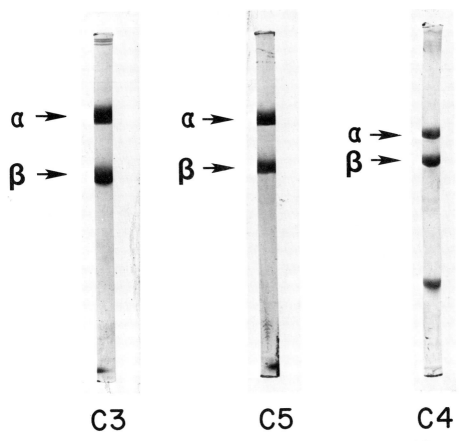

FIG. 3. The chain structure of purified complement components C3, C5, and C4 as assessed by SDS–PAGE. Each sample was reduced with β-mercaptoethanol prior to electrophoresis.

Comments on Purification Procedures

1. Pooled plasma that was adjusted with PEG to a final concentration of 5% (w/v) lost all C4 functional activity and more than 90% of the C2 activity. It was essential to mix individual units of plasma with EDTA and benzamidine to final concentrations of 10 mM each before they were pooled, adjusted to 1 mM in PMSF, and fractionated with PEG in order to obtain a stabilized supernatant solution for further fractionation.

2. In order to obtain a concentrated soluble pool of low ionic strength

[21] C. Bolotin, S. C. Morris, B. F. Tack, and J. W. Prahl, *Biochemistry* **16**, 2008 (1977).
[22] I. Gigli, I. von Zabern, and R. R. Porter, *Biochem. J.* **165**, 439 (1977).

for application to the DEAE–Sephacel column, it was essential to use the Pellicon ultrafiltration system as described. A direct rapid decrease in volume of the complement pool resulted in the preferential precipitation of C5 and β_1H due to elevated PEG levels.

3. The separation of C3 and C5 is dependent on the physical properties of the anion-exchange resin. The microgranular form of DEAE–cellulose (Whatman DE 52) does not effect a separation of these two proteins.[5] With a spherical form of cellulose (DEAE–Sephacel), however, total resolution was obtained. The separation between C3 and C4 was greatly enhanced by this resin as well.

4. The plasma levels of C5 and C3 were determined to be 67 and 1000 μg/ml, respectively, by radial immunodiffusion. The respective theoretical purification factors are therefore 910 and 63. The final specific activities of C5 and C3 were 1350 and 103. The final products, therefore, have significantly greater specific hemolytic activities (1.5- to 1.6-fold) than would have been predicted.

5. C5 is a relatively stable protein and can be maintained for several months at 4°C under sterile conditions. For longer periods it is recommended that the protein be stored at -70°C. Repeated freezing and thawing of C5 has not, in our experience, adversely affected the specific hemolytic activity. C3 and C4, however, are less stable. C3 solutions in K–Na phosphate buffer (pH 7.4) have been frozen as 100-μl drops in liquid nitrogen and subsequently maintained at -70°C for 2 years without a loss of specific activity greater than 50%. When C3 has been frozen in 10- to 100-ml volumes at -70°C, a sizable precipitate frequently forms on thawing. Recent studies have indicated that the preferred method of holding C3 is as a sterile solution at 4°C. At this time, there is no satisfactory way of maintaining C4 activity. This protein has been frozen in liquid nitrogen with full recovery of hemolytic activity on immediate thawing. Following storage at -70° for 6 months, however, this preparation had less than 10% of the original activity. Maintained as a sterile solution at 4°C, the anticipated half-life is 2–4 months.

Separation and Characterization of Hemolytically Active and Inactive Forms of C3 and C4

1. Chromatography on Activated Thiol Sepharose-4B. Fresh preparations of C3 and C4 frequently contain detectable amounts of hemolytically inactive forms, denoted C3(i) and C4(i). The inactive components in these samples have acquired a free sulfhydryl (SH) group which is reactive with Ellman's reagent [5,5'-dithiobis(2-nitrobenzoic acid); DTNB], 2,2'-dipyridyl disulfide (2-PDS), and [^{14}C]iodoacetamide.[23,24] SH-bearing mol-

[23] J. Janatova, B. F. Tack, and J. W. Prahl, *Biochemistry* **19**, 4479 (1980).
[24] J. Janatova and B. F. Tack, *Biochemistry* **20**, 2394 (1981).

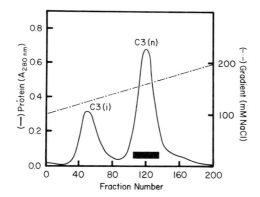

FIG. 4. Ion-exchange chromatography of a C3 sample following storage at $-70°C$ for 21 months. Twenty milliliters of protein (4.8 A_{280}/ml) in 20 mM Na phosphate buffer (pH 7.0) containing 2 mM Na_2EDTA and 100 mM NaCl was applied to a 1.4 × 15-cm column of QAE-Sephadex A-50 and DEAE–Bio-Gel A, mixed in a 3:1 v/v ratio. The column was developed with a linear NaCl gradient (600 ml) to a limit concentration of 200 mM. The flow rate was maintained at 25 ml/hr and 3-ml fractions collected. The C3(i) pool had an SH content of 0.89 mol/mol. The percentage of hemolytically active molecules in the C3(n) pool (indicated by the heavy bar) was evaluated by determination of the differential SH content before (0.18) and after (0.87) treatment with potassium bromide, and found to be 69%. All determinations of SH content were done with DTNB.

ecules can be removed from these preparations by batch adsorption or column chromatography on activated thiol Sepharose according to a protocol[24] similar to that described for papain.[25]

2. *Chromatography on QAE-Sephadex A-50/DEAE–Bio-Gel A.* The appearance of inactive components that either fail to react or only partially react with SH reagents have been observed on long-term storage of C3 and C4 samples at $-70°C$. Ion-exchange chromatography on QAE-Sephadex A-50 (Pharmacia Fine Chemicals) has been used to effectively resolve active and inactive components in these preparations.[23,24] The presence of DEAE–Bio-Gel A (BioRad Laboratories) in a QAE-Sephadex A-50 column, mixed in a 1:3 volume ratio, improves the physiochemical properties of the chromatographic column by increasing the flow rate and preventing excessive shrinkage on application of the NaCl gradient. Examples of such separations are shown in Figs. 4 and 5. Inactive forms that elute between C3(i) and C3(n) and between C4(i) and C4(n) are frequently detected and have a fractional SH-group content.

3. *Properties of Inactive Components.* All chromatographic forms exhibit a polypeptide-chain structure identical with that of the respective native protein (Fig. 3). The stage at which spontaneous inactivation af-

[25] K. Brocklehurst, J. Carlsson, M. P. J. Kierstan, and E. M. Crook, *Biochem. J.* **133**, 573 (1973).

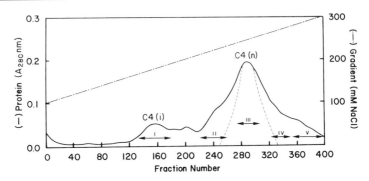

FIG. 5. Ion-exchange chromatography of a C4 sample following long-term storage at $-70°C$. Twelve milliliters of protein (6.7 A_{280}/ml) was applied to a 1.4 × 15.5-cm column of QAE-Sephadex A-50 and DEAE–Bio-Gel A mixed in a 3:1 v/v ratio. The protein solution and column were preequilibrated in 20 mM Na phosphate buffer (pH 7.0) containing 1 mM Na$_2$ EDTA–100 mM NaCl. A linear NaCl gradient (1200 ml) was developed to a limit concentration of 300 mM. Only pool III material had functional activity. The highest SH content was found in C4(i), pool I. (See also Fig. 6.)

fects the ability of C3 or C4 to sustain further complement activation has been determined. Only the native, hemolytically active protein is cleaved by its converting enzyme to give active fragments, C3b($\alpha'\beta$) or C4b ($\alpha'\beta\gamma$).[23,26] The failure of C4(i) to be cleaved by C1$\bar{s}$ is shown in Fig. 6A. A further observation that illustrates an inherent structural difference between active and inactive forms of these two proteins concerns an α-chain-specific fragmentation, which occurs when C3(n) and C4(n) are incubated in 0.2% SDS or 6 M guanidinium chloride at 37°C.[24,26–29] The fragmentation pattern for C4(n) is shown in Fig. 6B. The structural implications of this cleavage reaction will be discussed in a later section.

Compositional and Structural Analysis

A summary of presently known physical and chemical properties for C3, C4, and C5 is presented in Table III. The amino acid composition for each protein is shown in Table IV. The composition of C3[5] was calculated based on a weight average MW 187,650 ± 5650 determined by sedimentation equilibrium, which was corrected to a protein mass of 185,000 based on a 1.5% (w/w) carbohydrate content. The molecular weights of C4 and C5

[26] J. Janatova, P. E. Lorenz, A. N. Schechter, J. W. Prahl, and B. F. Tack, *Biochemistry* **19**, 4471 (1980).

[27] R. B. Sim and E. Sim, *Biochem J.* **193**, 129 (1981).

[28] J. B. Howard, *J. Biol. Chem.* **255**, 7082 (1980).

[29] J. P. Gorski and J. B. Howard, *J. Biol. Chem.* **255**, 10025 (1980).

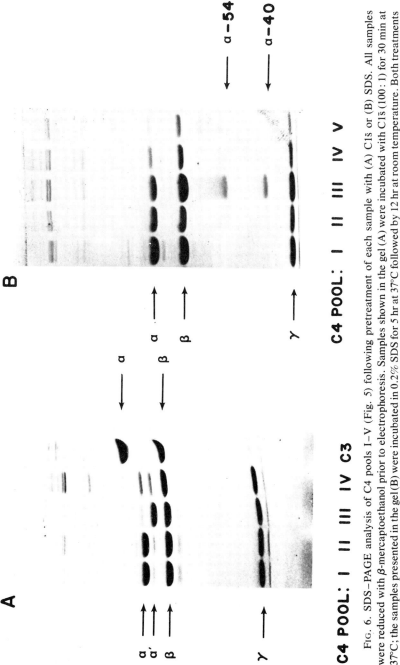

FIG. 6. SDS–PAGE analysis of C4 pools I–V (Fig. 5) following pretreatment of each sample with (A) C1s̄ or (B) SDS. All samples were reduced with β-mercaptoethanol prior to electrophoresis. Samples shown in the gel (A) were incubated with C1s̄ (100 : 1) for 30 min at 37°C; the samples presented in the gel (B) were incubated in 0.2% SDS for 5 hr at 37°C followed by 12 hr at room temperature. Both treatments were carried out in 20 mM Na phosphate buffer (pH 7.0) containing 2 mM Na$_2$ EDTA.

TABLE III
PROPERTIES OF C3, C4, AND C5

Property	C3	C4	C5
Plasma concentration[a]	1.0 mg/ml	0.33 mg/ml	† mg/ml
Sedimentation coefficient[b]	9.55	10.05	8.75
Partial specific volume[c]	0.736 ml/g	Not determined	Not determined
Molecular weight	187,650 ± 5650[c]	198,000 ± 20,000[d]	190,000 ± 19,000[e]
Chain structure	α β	α β γ	α β γ
Chain size	α-115,000 ± 12,000[f] β-75,000 ± 8000[f]	α-93,000 ± 9000[d] β-75,000 ± 8000[d] γ-30,000 ± 3000[d]	α-115,000 ± 12,000[e] β-75,000 ± 8000[e]
Extinction coefficient[e]	1.82×10^5 liter/mol/cm	Not determined	Not determined
Carbohydrate content (% by wt)	1.5[f]	6.9[g]	≥1.6[h]
Carbohydrate composition (% by wt)	0.3 fucose 0.8 hexose 0.3 hexosamine 0.5 N-acetylneuraminic acid	0.0 fucose 2.1 mannose 1.2 galactose 0.1 glucose 0.0 galactosamine 1.2 glucosamine 2.3 N-acetylneuraminic acid	Not determined

[a] C. H. Hammer, G. H. Wirtz, L. Renfer, H. D. Gresham, and B. F. Tack, *J. Biol. Chem.* **256**, 3995 (1981).
[b] H. J. Müller-Eberhard, *Annu. Rev. Biochem.* **44**, 697 (1975).
[c] B. F. Tack and J. W. Prahl, *Biochemistry* **15**, 4513 (1976).
[d] C. Bolotin, S. C. Morris, B. F. Tack, and J. W. Prahl, *Biochemistry* **16**, 2008 (1977).
[e] B. F. Tack, S. C. Morris, and J. W. Prahl, *Biochemistry* **18**, 1490 (1979).
[f] B. F. Tack, S. C. Morris, and J. W. Prahl, *Biochemistry* **18**, 1497 (1979).
[g] I. Gigli, I von Zabern, and R. R. Porter, *Biochem. J.* **165**, 439 (1977).
[h] H. N. Fernandez and T. E. Hugli, *J. Immunol.* **117**, 1688 (1976).

TABLE IV
AMINO ACID COMPOSITIONS OF C3, C4, AND C5[a]

Residue	C3	C4	C5
Lys	116.7	74.0	114.4
His	27.8	38.1	31.1
Arg	81.1	101.7	54.7
S-CMCys[b]	21.8	29.0	23.3
Asp	152.6	132.2	158.9
Thr	100.0	85.9	131.6
Ser	106.7	134.8	141.2
Glu	219.2	190.1	180.7
Pro	79.4	93.9	71.2
Gly	98.9	121.0	96.0
Ala	98.6	132.7	97.9
Val	146.2	127.6	121.9
Met	30.9	27.2	20.6
Ile	78.9	53.1	97.3
Leu	152.9	101.8	146.4
Tyr	55.6	50.5	78.3
Phe	60.8	61.1	67.8
Trp	12.0	18.2	9.4
Total residues	1640	1573	1643

[a] Reported as moles of amino acid per mole of protein.
[b] Determined as S-carboxymethylcysteine.

have not been determined with such precision. Rather, apparent molecular weights have been deduced from the molecular weights of the constituent polypeptide chains (Table III). Composition[21] of C4 was therefore based on MW 198,000 corrected to a protein mass of 184,000 for a 7% carbohydrate content. The composition reported for C5[18] is based on MW 190,000 corrected to a protein mass of 187,000 based on a minimum carbohydrate content of 1.6% (w/w). In general, the composition reported here for C3 is in agreement with that reported by others.[30,31] There are, however, major discrepancies in the values reported for cysteinyl, methionyl, and tryptophanyl residues, which have inherent problems with quantitation. When compared with the C4 compositions reported by Budzko and Müller-Eberhard[32], and Gigli *et al.*,[22] the data reported here are in general agreement, but show some divergence for lysyl, arginyl,

[30] D. B. Budzko, V. A. Bokisch, and H. J. Müller-Eberhard, *Biochemistry* 10, 1166 (1971).
[31] J. L. Molenaar, A. W. Helder, M. A. C. Muller, M. Goris-Mulder, L. S. Jonker, M. Brouwer, and K. Pondman, in "The Third Human Complement Component" (J. L. Molenaar, ed.), p. 72. Graduate Press, Amsterdam, 1974.
[32] D. B. Budzko and H. J. Müller-Eberhard, *Immunochemistry* 7, 227 (1970).

glutamyl, alanyl, cysteinyl, and phenylalanyl residues. The compositions reported for each protein must be considered to be only approximations, since genetic polymorphism has not been taken into account and rigorous carbohydrate analyses have not been done for each protein. Unusual amino acids such as hydroxyproline and hydroxylysine have been looked for but not found in these proteins. In view of the calcium-binding properties of C3, hydrolysates have been examined for 5,5'-[³H]dihydroxyleucine following reduction of the protein with [³H]diborane.[33] In agreement with the studies of Hauschka,[34] this residue was not found.

Polypeptide-Chain Isolation and Structure

1. Chain Separation. The polypeptide chains of C3 and C5 can be separated following total reduction and alkylation in 6 M guanidinium chloride on a 2.5 × 160-cm column of Sepharose CL-4B equilibrated in 0.1 M ammonium bicarbonate (pH 7.9), and containing 0.2% SDS.[18,35] The β-chain pool is frequently contaminated with α-chain, which can be removed by rechromatography. The α- and β-chain pools can be conveniently concentrated by lyophilization, thereby removing the bulk of ammonium bicarbonate, and the detergent then removed on an AG2-X10 resin column equilibrated in 0.1 M acetic acid containing 6 M deionized urea.[36] The reduction and alkylation of C4 in 250 mM Tris-HCl buffer (pH 8.0) containing 20 mM DTT, followed by gel filtration on a 2.5 × 160-column of G-200 equilibrated in 1.0 M acetic acid is effective in obtaining β- and γ-chain pools.[21] The α-chain, however, cannot be obtained free of β- and γ-chains by this procedure. The final purification of α-chain is achieved by filtration on Sepharose CL-6B in 0.1 M ammonium bicarbonate containing 0.2% SDS. An alternative method for separation of C4 chain is that described by Gigli *et al.*[22] This procedure involves the adsorption of a reduced and alkylated C4 sample on a hydroxylapatite column equilibrated in a 0.1 M Na phosphate buffer (pH 6.4) containing 0.1% SDS, and elution with a phosphate gradient between 0.2 and 0.5 M as described by Moss and Rosenblum.[37]

2. Compositional Analysis. The amino acid compositions of the polypeptide chains of these proteins are shown in Table V.[18,21,35] In general, there is fair agreement with the compositions of C3 and C4 chains determined by Taylor *et al.*[38] and Gigli *et al.*,[22] respectively. Major dis-

---

[33] G. Nelsestuen, personal communication.
[34] P. V. Hauschka, *Anal. Biochem.* **80**, 212 (1977).
[35] B. F. Tack, S. C. Morris, and J. W. Prahl, *Biochemistry* **18**, 1497 (1979).
[36] J. Lenard, *Biochem. Biophys. Res. Commun.* **45**, 662 (1971).
[37] B. Moss and E. N. Rosenblum, *J. Biol. Chem.* **247**, 5194 (1972).
[38] J. C. Taylor, I. P. Crawford, and T. E. Hugli, *Biochemistry* **16**, 3390 (1977).

TABLE V
AMINO ACID COMPOSITIONS OF THE POLYPEPTIDE CHAINS OF C3, C4, and C5[a]

Residue	C3α	C4α	C5α	C3β	C4β	C5β	C4γ
Lys	72.7	29.5	73.8	40.6	30.9	43.0	14.1
His	17.4	14.5	20.3	9.9	14.6	11.3	4.6
Arg	50.7	39.4	36.8	28.6	31.5	15.1	20.0
S-CMCys[b]	16.6	18.6	16.8	4.1	7.4	6.3	7.2
Asp	102.3	70.6	90.8	54.5	50.3	79.7	16.5
Thr	56.5	43.7	64.8	49.4	30.8	45.3	10.4
Ser	57.4	67.6	80.3	50.4	71.7	60.2	15.6
Glu	140.0	95.8	106.8	80.5	61.8	76.2	34.7
Pro	42.3	31.3	34.4	37.9	41.8	32.4	11.2
Gly	49.7	55.8	54.9	47.2	49.6	39.2	15.6
Ala	62.6	70.6	54.5	33.4	39.9	38.9	20.0
Val	58.6	50.9	73.2	71.6	63.1	52.8	20.7
Met	20.6	17.3	10.9	11.5	14.9	6.4	4.0
Ile	46.1	23.3	63.3	32.1	24.2	41.1	5.8
Leu	98.0	112.0	95.8	61.9	83.0	55.4	25.8
Tyr	33.2	14.5	44.0	23.6	19.3	37.2	11.4
Phe	36.0	19.6	39.4	25.0	26.1	29.5	10.5
Trp	10.0	9	7.0	2.4	—	1.7	8.5
Total residues	961	785	968	665	661	672	257

[a] Reported as moles of amino acid per mole of protein.
[b] Determined as S-carboxymethylcysteine.

crepancies, again, usually involve cysteinyl, methionyl, and tryptophanyl residues. These differences may be attributable in part to methods used for quantitation. Half-cystine values reported here were as S-carboxymethylcysteine or aminoethylcysteine and tryptophan as the free amino acid following hydrolysis in 3 M mercaptoethanesulfonic acid. Taylor et al.[38] determined the former as cysteic acid following performic acid oxidation and the latter by alkaline hydrolysis according to Hugli and Moore.[39]

3. Amino- and Carboxyl-Terminal Structures.[18,21,22,35] The α-, and β-chains of C3, C4, and C5 were subjected to automated Edman degradation using 50- to 100-nmol samples; the derived structures are shown in Fig. 7. PTH amino acids were identified and quantitated by either gas chromatography, back hydrolysis in HI, or high-pressure liquid chromatography. Protein samples were dialyzed against 1 mM N-ethylmorpholine containing 0.05% SDS prior to loading in a Joel 47K sequence analyzer. Introduction of these samples in SDS maintained solubility and resulted in a uniform film on drying in the spinning cup. Several attempts to sequence the isolated C5

[39] T. E. Hugli and S. Moore, J. Biol. Chem. 247, 2828 (1972).

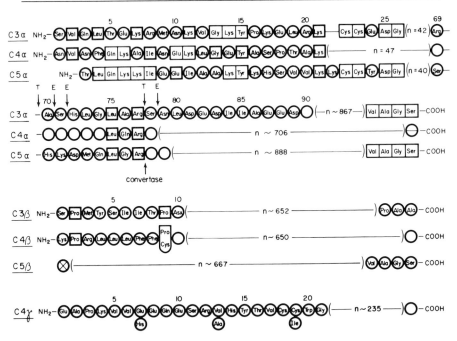

FIG. 7. The amino- and carboxyl-terminal structures of the polypeptide chains of C3, C4, and C5 aligned for maximal homology. Primary sites of cleavage in the amino-terminal region of the C3 α-chain with trypsin (T) and elastase (E) are indicated by arrows. Residues enclosed in squares indicate positions of homology and residues enclosed in circles represent differences. Open circles denote positions that have not been determined and the crossed circle an apparent pyrrolidone carboxylic acid residue.

β-chain have been unsuccessful. Since only a single sequence can be followed with intact C5 it has been concluded that the amino-terminus of the β-chain is blocked. However, the possibility remains that the β-chain has been rendered unsequenceable during isolation by cyclization of an N-terminal glutamine to pyrrolidone carboxylic acid.

Digestion of the C3 β-chain with carboxypeptidase Y (CpY) (as a suspension in $0.1\,M$ sodium acetate buffer, pH 6.0, at a substrate to enzyme molar ratio of 240 : 1) resulted in the release of 1.42 mol of alanine/mol and a trace of proline; when the amount of enzyme relative to substrate was increased to 50 : 1, 2.20 mol of alanine/mol and 1.14 mol of proline/mol were observed. A sequence of Pro-Ala-Ala is proposed for the C3 β chain from these data. The presence of a prolyl residue in the third position is assumed to be responsible for the observation that CpA was not effective in digesting beyond the carboxyl-terminal alanyl residue. Similar experiments with the

C3 α-chain have suggested a C-terminal sequence of (Ala, Val)-Gly-Ser. The C5 α- and C5 β-chains appear to share a carboxyl-terminal structure of (Ala, Val)-Ala-Gly-Ser. While digestion with CpA was unaffected by the presence of 0.1% SDS, CpY was only effective when SDS was omitted. The C-terminal structures of the C4 chains have not been determined.

These data, when compared with the N-terminal structures of C3a $(1-77)$,[40] C5a $(1-74)$[41] and C4a[42]; indicate that the activation peptides are derived from the amino-termini of their respective α-chains.

Fragmentation

The activation of each protein and subsequent modulation of its biological activity is effected by limited proteolysis involving highly specific enzyme systems.[43–45] These pathways are best understood for C3 and C4 and are briefly outlined below.

During complement activation, the α-chain of C3 is cleaved at a single site[46], either by a bimolecular complex of activated C4 and C2 (C$\overline{\text{4b2a}}$)[47] or by a complex of C3 and Factor B (C$\overline{\text{3iBb}}$)[48], to give C3a and C3b. The macromolecular fragment (C3b), comprised of an $\alpha'\beta$-chain structure, binds to Factor B, and this complex is subsequently converted to C$\overline{\text{3bBb}}$ by Factor D.[49] The bimolecular complex C$\overline{\text{3bBb}}$ is the major C3 convertase of the alternative pathway. Bb contributes the proteolytic-active site to this complex and binding of C3b is required for expression of its physiological activity. C$\overline{\text{3bBb}}$ activity is regulated by the plasma proteins C3bINA and β1H.[50] β_1H can effect the displacement of Bb from C3b and is a required cofactor for the C3bINA-dependent cleavage of C3b. This reaction, characterized by a double cleavage of the C3b α'-chain,[51] gives rise to C3bi, which is incapable of acting as a subcomponent in the C$\overline{\text{3bBb}}$

[40] T. E. Hugli, E. H. Vallota, and H. J. Müller-Eberhard, *J. Biol. Chem.* **250**, 1472 (1975).

[41] H. N. Fernandez and T. E. Hugli, *J. Immunol.* **117**, 1688 (1976).

[42] J. P. Gorski, T. E. Hugli, and H. J. Müller-Eberhard, *Fed. Proc., Fed. Am. Soc. Exp. Biol.* **38**, 1010 (1979).

[43] R. R. Porter and K. B. M. Reid, *Adv. Protein Chem.* **33**, 1 (1979).

[44] R. M. Stroud, J. E. Volanakis, S. Nagasawa, and T. F. Lint, *Immunochem. Proteins* **3**, 167 (1979).

[45] H. J. Müller-Eberhard and R. D. Schreiber, *Adv. Immunol* **29**, 1 (1980).

[46] U. R. Nilsson and J. Mapes, *J. Immunol.* **111**, 293 (1973).

[47] V. A. Bokisch, H. J. Müller-Eberhard, and C. G. Cochrane, *J. Exp. Med.* **129**, 1109 (1969).

[48] M. K. Pangburn and H. J. Müller-Eberhard, *J. Exp. Med.* **152**, 1102 (1980).

[49] H. J. Müller-Eberhard and O. Götze, *J. Exp. Med.* **135**, 1003 (1972).

[50] K. Whaley and S. Ruddy, *Science* **193**, 1011 (1976).

[51] R. A. Harrison and P. J. Lachmann, *Mol. Immunol.* **17**, 9 (1980).

complex. Further proteolytic action on C3bi leads to the generation of the immunochemically defined fragments, C3c and C3d.[51-53] An additional fragment, acidic in nature, has been termed C3e and is apparently derived from the C3c fragment on fluid-phase digestion of C3 with trypsin.[54] A further cleavage of surface-bound C3 described by Harrison and Lachmann[55] has been interpreted to result in the release of the C3e fragment, however, the protease responsible has not been identified.

An analogous series of reactions occurs with C4. The α-chain is cleaved by the C1$\bar{s}$ component[56] of activated C1, generating C4a and C4b.[32,57] C4b combines with C2, and C2 is subsequently cleaved by C1$\bar{s}$. C2a contributes the proteolytic-active site to the C$\overline{4b2a}$ complex.[58] The activity of this complex is regulated by the control proteins C3bINA and C4-binding protein.[59] The latter cofactor functions in a manner analogous to that described earlier for β_1H. An intermediate equivalent to C3bi has not been detected on cleavage of C4b by C3bINA. Rather, the products of this reaction appear to be C4c and C4d.

The requirements for C5 activation are more complex. Both C4b and C3b are required as cofactors for the C2a-dependent cleavage of the α-chain of C5 by the C$\overline{4b2a3b}$ complex.[60] For the corresponding convertase of the alternative pathway, at least two molecules of C3b are required as cofactors for Bb.[61-63] Activation results in two fragments, C5a and C5b. The macromolecular fragment C5b combines with C6 and C7,[64-66] thereby initiating the formation of the C5b–C9 membrane-attack complex.[67,68] Little is known about further events concerning C5b.

[52] M. K. Pangburn, R. D. Schreiber, and H. J. Müller-Eberhard, *J. Exp. Med.* **146**, 257 (1977).

[53] S. Nagasawa and R. M. Stroud, *Immunochemistry* **14**, 749 (1977).

[54] B. Ghebrehiwet and H. J. Müller-Eberhard, *J. Immunol.* **123**, 616 (1979).

[55] R. A. Harrison and P. J. Lachmann, *Mol. Immunol.* **17**, 219 (1980).

[56] I. H. Lepow, O. D. Ratnoff, F. S. Rosen, and L. Pillemer, *Proc. Soc. Exp. Biol. Med.* **92**, 32 (1956).

[57] R. A. Patrick, S. B. Taubman, and I. R. Lepow, *Immunochemistry* **7**, 227 (1970).

[58] N. R. Cooper, *Biochemistry* **14**, 4245 (1975).

[59] T. Fujita, I. Gigli, and V. Nussenzweig, *J. Exp. Med.* **148**, 1044 (1978).

[60] H. J. Müller-Eberhard, M. J. Polley, and M. A. Calcott, *J. Exp. Med.* **125**, 359 (1967).

[61] R. G. Medicus, O. Götze, and H. J. Müller-Eberhard, *J. Exp. Med.* **144**, 1076 (1976).

[62] M. R. Daha, D. T. Fearon, and K. F. Austen, *J. Immunol.* **117**, 630 (1976).

[63] W. Vogt, G. Schmidt, B. von Buttlar, and L. Dieminger, *Immunology* **34**, 29 (1978).

[64] N. R. Cooper and H. J. Müller-Eberhard, *J. Exp. Med.* **132**, 775 (1970).

[65] M. B. Goldlust, H. S. Shin, C. H. Hammer, and M. M. Mayer, *J. Immunol.* **113**, 998 (1974).

[66] P. J. Lachmann and R. A. Thompson, *J. Exp. Med.* **131**, 643 (1970).

[67] W. P. Kolb, J. A. Haxby, C. M. Arroyave, and H. J. Müller-Eberhard, *J. Exp. Med.* **135**, 549 (1972).

[68] M. M. Mayer, *Proc. Natl. Acad. Sci. U.S.A.* **69**, 2945 (1972).

Generation of C3b

A number of proteases are able to cleave C3 at, or proximal to, the site of action of C$\overline{4b2a}$ or C$\overline{3bBb}$. The inherent instabilities of the physiological enzymes have precluded their general use in solution-phase fragmentation studies. Two alternative procedures will be described below, one utilizing a cobra venom factor convertase (CVF)[69] and the other bovine pancreatic trypsin.[70]

1. CVF-Convertase Procedure. CVF is isolated from cobra venom (Naja naja kaonthis, Sigma) using DEAE–cellulose and Sephadex G-200 chromatography, as described by Ballow and Cochrane.[71] *Extreme caution* (use gloves and face mask and work in a fume hood) should be exercised in the handling of the crude snake venom and in discarding unwanted material (particularly the DEAE–cellulose drop-through fraction). The purified CVF is coupled to CNBr-activated Sepharose in 0.2 M sodium citrate buffer, pH 6.5, by the method of Cuatrecasas and Anfinsen.[72] After washing, the CVF–Sepharose is activated by incubation with freshly drawn human serum (equal volumes packed CVF–Sepharose and serum) containing 10 mM magnesium acetate for 30 min at 37°C. Following activation, the Sepharose is pelleted by centrifugation and washed at 22°C by repeated resuspension in 20 volumes of PBS. The washed CVF–Sepharose is immediately added to a 2 mg/ml solution of C3 (0.05 ml of a 50% suspension of CVF–Sepharose in PBS/mg C3) containing 5 mM magnesium acetate, and the mixture incubated, with gentle stirring, at 22°C for 12–16 hr. The CVF–Sepharose is then removed by centrifugation and can be reactivated after washing with 1.0 M NaCl. C3a can be resolved from C3b either by chromatography on hydroxylapatite[51], or as described below by gel filtration. The $\alpha'\beta$ chain structure of C3b prepared with CVF–Sepharose is shown in Fig. 9B (track 2).

2. Digestion with Bovine Pancreatic Trypsin. The studies of Bokisch *et al.*[47] and Budzko *et al.*[30] have shown that this enzyme can be used to specifically fragment human C3. The resultant C3a and C3b fragments were separated on G-100 in 0.15 M Na acetate buffer (pH 3.6), and reported to resemble the corresponding cleavage products obtained with the classical pathway convertase (C$\overline{4b2a}$), based on biological, chemical, and immunochemical analyses. The separation at pH 3.6, however, is problematic, since considerable precipitation of C3b occurs. The general approach, nevertheless, would be useful provided the separation of C3b and C3a could be done under milder conditions.

[69] J. D. Gitlin, F. S. Rosen, and P. J. Lachmann, *J. Exp. Med.* **141**, 1221 (1975).
[70] B. F. Tack and A. N. Schechter, unpublished observations.
[71] M. Ballow and C. G. Cochrane, *J. Immunol.* **103**, 944 (1969).
[72] P. Cuatrecasas and C. B. Anfinsen, this series, Vol. 22, p. 345.

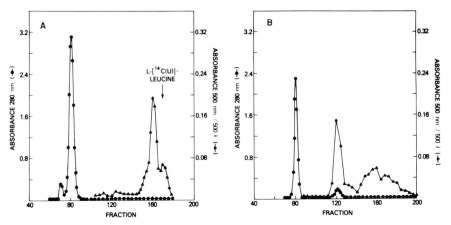

FIG. 8. (A) Gel filtration of a bovine trypsin digest of C3 on G-200. (B) Gel filtration of a porcine-elastase digest of C3b on G-200. Peptide material was detected by the Folin reaction as shown by solid triangles.

Trypsin, obtained from Worthington Biochemicals, was treated with L-1-tosylamide-2-phenylethyl chloromethyl ketone (TPCK) to inactivate possible chymotryptic activity. Digestion of 10 ml of C3 (10 mg/ml; pH 7.4) with TPCK : trypsin at a substrate to enzyme molar ratio of 130 : 1 for 1 min at 37°C effected complete cleavage to C3b. The digestion was terminated by the addition of 1 mM diisopropyl fluorophosphate (DFP) or a 10-fold molar excess of soybean-trypsin inhibitor (STI). Sufficient solid NaCl was added to the digest to give a final concentration of 1.0 M and the pH was reduced to 5.6 by the addition of glacial acetic acid. Separation of C3b and C3a was obtained by gel filtration on a G-200 column (2.5 × 160 cm) equilibrated in 0.05 M Na acetate buffer (pH 5.6) containing 1.0 M NaCl and 0.01 M EDTA (Fig. 8A; Fig. 9A, tracks 2 and 3). The α'-chain of C3b has been isolated, following reduction and alkylation in 6 M guanidinium chloride, by gel filtration on Sepharose CL-4B as earlier described for α- and β-chain separation. On automated Edman degradation, an amino-terminal sequence of Ser-Asn-Leu-Asp-Glu-Asp-Ile-Ile-Ala-Glu-Glu-Asp-Ile-Val was obtained.[35] The activation peptide (C3a) isolated by this procedure was shown on compositional and C-terminal analysis to be a mixture of C3a(1–69) and C3a(1–77). Analysis of trypsin digests for peptide material revealed only the presence of a peptide comprised of residues 70–77. Therefore, there are two primary sites of trypsin cleavage, the Arg-Ala bond at position 69–70, and the Arg-Ser bond tentatively positioned at 77–78. In view of our inability to detect inter-C3a/C3b peptide

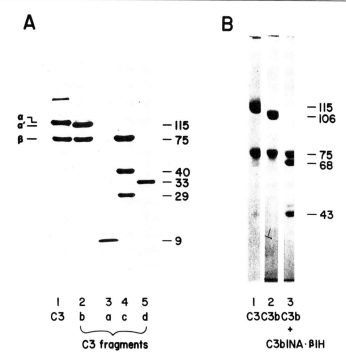

FIG. 9. (A) SDS–PAGE of C3 (1), C3b (2), C3a (3), C3c (4), and C3d (5) following reduction with β-mercaptoethanol. (B) SDS–PAGE of C3 (1), C3b (2), and C3bi (3) following reduction.

material, the carboxyl-termini of C3a and the amino-termini of the α'-chain of C3b are proposed to be contiguous.

Generation of C4b and C5b

Although we have no experience with the above fragments, others have used C1s̄ to generate C4b[59,73] and CVF–Sepharose to generate C5b.[74]

Generation of C3bi

C3bi can be generated using catalytic amounts of C3bINA and β1H. C3b (2 mg/ml) is incubated at 37°C for 12–16 hr in phosphate-buffered saline (pH 7.0) containing 1 mM magnesium acetate with C3bINA and β1H, 20 μg/ml each. The product has little, if any, of the 46,000-MW

[73] J. P. Gorski, T. E. Hugli, and H. J. Müller-Eberhard, *Proc. Natl. Acad. Sci. U.S.A.* **76**, 5299 (1979).
[74] R. G. Medicus, *Fed. Proc., Fed. Am. Soc. Exp. Biol.* **36**, 1244 (1977).

α-chain fragment,[51] both C3bINA-dependent scissions being essentially complete (Fig. 9B, track 3). As C3bi is highly susceptible to further proteolytic attack, pretreatment of all proteins with 1.0 mM DFP is desirable. This is particularly important with trypsin-generated C3b where the digestion has been stopped with STI. Some trypsin appears to be bound to C3b, probably through the covalent binding site, and fails to complex with STI. Therefore, non-DFP-treated C3b samples generated with trypsin are degraded beyond the C3bi level.

Generation of C3c and C3d with Elastase[70]

The activation fragments, C3c and C3d, can be prepared with porcine pancreatic elastase. This enzyme was chosen on the basis of studies by Taylor et al.,[38] which indicated that on digestion of C3 with human leukocyte elastase, the resultant cleavage products were indeed similar in their molecular weights and immunoelectrophoretic mobilities to the "physiological" fragments that have been isolated, albeit in poor yields, from activated sera.

Elastase (twice crystallized), obtained from Worthington Biochemicals, was solubilized and chromatographed on DEAE–Sephadex (A50) and Whatman CM-cellulose according to the method of Narayanan and Anwar.[75] Post G-200 C3b was concentrated to 10 mg/ml by ultrafiltration on a PM-30 membrane, and dialyzed against 0.02 M Tris-HCl buffer (pH 8.2) containing 0.5 mM CaCl$_2$–1.5 mM MgCl$_2$. Digestion of 10 ml of C3b (10 mg/ml) with elastase, at a molar ratio of substrate to enzyme of 13.3 : 1 for 240 min at 37°C, resulted in complete conversion of C3b to C3c and C3d. Following inactivation of elastase with 1 mM DFP or PMSF, the C3c and C3d fragments were resolved on the G-200 column previously used to separate C3b and C3a (Fig. 8B; Fig. 9A, tracks 4 and 5). The C3d fragment isolated by this method was a single-chain structure with MW 33,000 and was not disulfide bonded to other portions of the molecule. The C3c fragment, MW 145,000, was comprised of the intact β-chain and two α-chain subdomains with MW 40,000 and 29,000, which were disulfide-bridged to one another. The amino-terminal structure of C3d was determined by Edman degradation to be Ala-Gln-Met-Thr-Glu-Asp - Ala - Val - Asp - Ala - Glu - Arg - Leu - Lys - His - Leu - Ile - Val - Thr - Pro-Ser-Gly-Cys-Gly-Glu-Glu. The lack of identity between the amino-terminal structures of C3d and the α'-chain of C3b provides direct evidence that this α-chain subdomain (C3d) is not contiguous with C3a.

Treatment of C3 with elastase for shorter time periods can be used to generate C3b. The C3a pool from G-200 was a mixture of C3a (1–70) and C3a (1–71) due to a double cleavage at the Ala-Ser (position 70–71) and

[75] A. S. Narayanan and R. A. Anwar, Biochem. J. 114, 11 (1969).

Ser-His (position 71–72), respectively. The determination of the amino-terminal structure of the α'-chain of C3b prepared with elastase was expected to establish the overlap for the trypsin site at position 77–78. The structure of this chain, however, began at Asn (position 2 of the tryptic α'-chain), implying that the peptide bond Ser-Asn (position 78–79) was a further cleavage site for elastase.

Antisera to C3 Fragments[70]

Antibodies to C3, and to several of the recognized activation fragments isolated from trypsin and elastase digests, have been raised in rabbits and sheep. The animals were immunized with 1 mg of antigen emulsified in complete Freund's adjuvant. The rabbits were given multiple footpad injections and boosted twice with 0.5 mg/boost. The sheep were injected intramuscularly at multiple sites in the hindlegs, and boosted three times with 1.0 mg/boost. Double-diffusion analysis of each rabbit antiserum is shown in Fig. 10. The antiserum to C3 gave precipitin lines with each fragment (plate I); however, the line to C3d crossed over the line to C3c and fused with the C3 line. The antiserum to C3b failed to react with C3a, and the C3d line again crossed over the C3c line and fused with the line to C3 (plate II). Therefore, with both antisera tested, antigenic sites present on the C3d fragment are recognized as shared with C3 and C3b but absent on the C3c fragment. The antiserum to C3a was reactive with C3 and C3a only (plate III). The antiserum to C3c gave strong lines of identity with C3 and C3b, however it was unreactive with C3a and C3d (plate IV). The antiserum elicited to C3d gave precipitin lines with C3 and C3b that fused with the line to C3d (plate V). This antiserum was unreactive with C3a and C3c. Therefore, the antigenic structures recognized by the rabbit on challenge with C3d are equally reactive with the larger macromolecular fragments, C3 and C3b. The results of these diffusion analyses provide additional evidence for the purity of the subdomain structures isolated by the above procedures.

Tritium Labeling of C3 and Several Activation Fragments by Reductive Methylation

A general procedure for the tritiation of proteins to high specific activities has been described.[76] The method is based on the condensation reaction of α-amino groups of NH_2-terminal residues and ϵ-amino groups of lysyl residues with formaldehyde at pH 9.0.[77] The resultant Shiff's base is then reduced with tritiated sodium borohydride to give a stable

[76] B. F. Tack, J. Dean, D. Eilat, P. E. Lorenz, and A. N. Schechter, *J. Biol. Chem.* **255**, 8842 (1980).

[77] G. E. Means and R. E. Feeney, *Biochemistry* **7**, 2192 (1968).

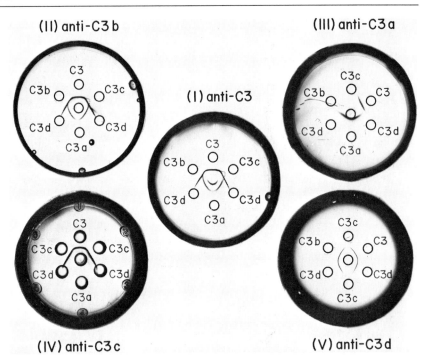

FIG. 10. Double-diffusion analysis of rabbit antisera to C3 and the activation fragments of C3. Undiluted antiserum was added to the center well of each plate (I–V). The concentration of each antigen was as follows: C3, C3b, and C3c, 1.25 mg/ml; C3d, 0.25 mg/ml; and C3a, 0.60 mg/ml. Each plate was developed for 48 hr. The dashed line drawn on plates I and II represents a fusion of the precipitin lines to C3 and C3d that was faint and therefore photographed poorly.

N-CH$_2$^{3}H adduct. The conditions used for labeling each protein and the observed specific activities are shown in Table VI. Double-diffusion analyses of [^{3}H]C3, [^{3}H]C4, and [^{3}H]C5 indicated that full antigenic reactivity was retained when compared with the corresponding unlabeled protein (Fig. 11). [^{3}H]C3 retained full hemolytic activity, and antigen-binding studies conducted with a sheep anti-human C3 antiserum and a burro anti-sheep IgG antiserum indicated that the labeled protein was 95% precipitable. A double-antibody radioimmunoassay for C3 was developed using the above antisera, and was sensitive in the 10^{-10} to 10^{-9} M range.

^{14}C-Methylated C3, C5, and C5a have been prepared using H^{14}CHO (53.3 mCi/mmol) and shown to retain full functional activity.[78] This procedure, however, results in significantly lower specific activities when com-

---

[78] D. L. Kreutzer, S. Kunkel, P. A. Ward, H. Showell, and R. H. McLean, *J. Immunol.* **123**, 2278 (1979).

TABLE VI

REACTANT CONCENTRATIONS USED FOR REDUCTIVE METHYLATION AND THE RESULTANT
SPECIFIC ACTIVITIES OF TRITIUM-LABELED PROTEINS

Sample	Protein (μM)	[^{3}H]NaBH$_4$ (mM)	HCHO (mM)	Specific activity[d] (Ci/mmol)	Modification[e] (%)
C3	48.5	4.6[a]	11.3	78.8	15.6
C3b	51.0	4.6[a]	11.3	62.8	13.5
C3c	64.9	4.6[a]	11.3	104.0	—
C3d	229.4	4.6[a]	9.0	21.1	—
C4	39.7	3.3[b]	10.3	358.0	14.1
C5	34.1	7.0[c]	43.5	141.0	21.2
β_1H	41.7	3.3[b]	10.3	290.0	—

[a] The specific activity of NaB^{3}H$_4$ was 8.6 Ci/mmol.
[b] The specific activity of NaB^{3}H$_4$ was 51.2 Ci/mmol.
[c] The specific activity of NaB^{3}H$_4$ was 11.4 Ci/mmol.
[d] Calculated from the observed count rates where tritium was counted with 40% efficiency.
[e] This calculation is based on a stoichiometric incorporation of 2 moles tritide per mole of lysine.

pared with the tritium-labeled proteins prepared with [^{3}H]NaBH$_4$ (8.6 Ci/mmol). [^{3}H]β_1H, labeled to a specific activity of 1.7×10^9 cpm/mg with [^{3}H]NaBH$_4$ (51.2 Ci/mmol), retains its cofactor activity for the C3bINA-dependent cleavage of C3b, and has also been used in cell-binding studies that established receptors for β_1H on human B lymphocytes.[79] The tritiated fragments of C3 (Table VI), labeled to high specific activities, have not been characterized with respect to retention of antigenic and/or functional activity. It is anticipated, however, that these labeled fragments will be useful for radioimmunoassay and, perhaps, for studies of cell-surface receptor: ligand interactions.

Labile Binding Site of C3 and C4

Interactions of nascent C3b with cell membranes and carbohydrate polymers proceed in part by the formation of covalent and hydrophobic bonds.[80] It has been established that the covalent component results from an acylation reaction. The acyl-group donor to the covalent bond is contributed by a residue present in the C3d fragment.[81] This bond is reactive with alkaline hydroxylamine and therefore may be an oxygen ester.[82]

[79] J. D. Lambris, N. J. Dobson, and G. D. Ross, *J. Exp. Med.* **152**, 1625 (1980).
[80] S. K. Law and R. P. Levine, *Proc. Natl. Acad. Sci. U.S.A.* **74**, 2701 (1977).
[81] S. K. Law, D. T. Fearon, and R. P. Levine, *J. Immunol.* **122**, 759 (1979).
[82] S. K. Law, N. A. Lichtenberg, and R. P. Levine, *J. Immunol.* **123**, 1388 (1979).

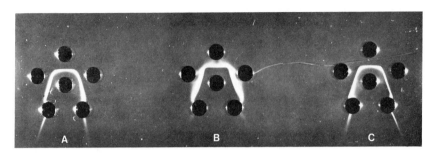

Fig. 11. Double-diffusion analysis of [³H]C3 (A), [³H]C4 (B), and [³H]C5 (C). The corresponding undiluted antiserum was added to the center well of each plate. The tritiated antigen at 1 mg/ml was placed in the top peripheral well and flanked by the corresponding unlabeled protein at 1 mg/ml. Following a 24-hr incubation each plate was photographed.

Nascent C4b is capable of covalent bonding to membrane components as well, and can be dissociated with hydroxylamine.[83] A further covalent interaction between C4b and IgG present in immune complexes has been described.[84] This bond, however, is more chemically resistant and perhaps amide in nature.[85] Our present knowledge of the active-site structures responsible for acyl-group activation is described below.

Evidence for an Amine-Sensitive Bond Comprised of a Thiol Component[23,24,26]

Treatment of C3 or C4 with nitrogen nucleophiles (hydroxylamine; hydrazine), chaotropes (potassium bromide; potassium thiocyanate), ionic detergents (SDS; deoxycholate), or denaturants (guanidinium chloride; urea) results in an irreversible loss of hemolytic function. While the inactivation by nucleophilic and chaotropic reagents is a nondegradative process, treatment with denaturants and ionic detergents results in an autolytic cleavage of the α-chain of each protein, giving 2 fragments (C3α, 46,000 and 70,000 MW; C4α, 40,000 and 54,000 MW). Any form of inactivation, whether it be spontaneous, chemical, or enzymatic in nature, is accompanied by the appearance of a stoichiometric SH group. The effects of 0.5 M hydroxylamine, 6 M guanidinium chloride, 0.2% SDS, and 1% (w/w) trypsin on the chain structure and SH content of C3 are shown in Fig. 12. All samples were radioalkylated with [¹⁴C]iodoacetamide prior to electrophoresis. On autoradiography, the SH group was localized in the α-chain of C3, the α'-chain of C3b, and the 46,000-MW α-chain fragment

[83] S. K. Law, N. A. Lichtenberg, F. H. Holcombe, and R. P. Levine, *J. Immunol.* 125, 634 (1980).
[84] J. W. F. Goers and R. R. Porter, *Biochem. J.* 175, 675 (1978).
[85] R. D. Campbell, A. W. Dodds, and R. R. Porter, *Biochem. J.* 189, 67 (1980).

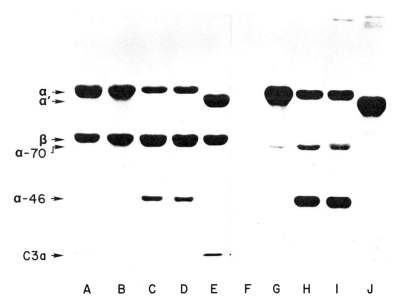

FIG. 12. SDS–PAGE of C3 samples pretreated with 0.5 M hydroxylamine (B). 6 M guanidinium chloride (C), 0.2% SDS (D), and 1% (w/w) trypsin (E). Prior to electrophoresis each sample was radioalkylated with [^{14}C]iodoacetamide and then reduced with β-mercaptoethanol. The left panel (A–E) represents the Coomassie blue-stained gel and the right panel (F–J) the corresponding autoradiogram. A C3 control was run in position A.

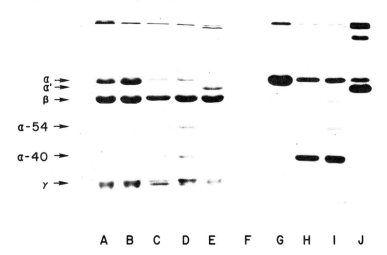

FIG. 13. SDS–PAGE of C4 samples pretreated with the same reagents and presented in the same order as C3 samples in Fig. 12, with the exception that C1s̄ was used in place of trypsin. The left panel (A–E) represents the stained gel and the right panel the corresponding autoradiogram. A C4 control was run in position A.

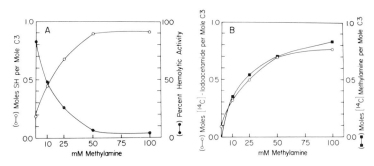

FIG. 14. (A) A plot of the loss of functional activity and acquisition of an SH group following treatment of C3 with the indicated concentrations of MA. Incubations were performed at a final protein concentration of 3.1 mg/ml in phosphate-buffered saline (pH 7.3) for 1 hr at 37°C. The thiol assay was carried out spectrophotometrically at 343 nm with 2-PDS. (B) A plot of radiolabel uptake for duplicate samples of C3 treated with [14C]MA or unlabeled MA followed by [14C]iodoacetamide.

of those samples preincubated in 6 M guanidinium chloride or 0.2% SDS. An identical set of data are shown in Fig. 13 for C4.

Identification of a Methylamine-Reactive Site

Treatment of native C3 or native C4 with [14C]methylamine (MA) results in a covalent and stoichiometric association of this reagent with either protein.[24,28,29,48,86] Incubation of C3 with increasing concentrations of MA results in the progressive loss of functional activity and the appearance of an SH group (Fig. 14A). The parallel nature of [14C]MA incorporation and SH-group appearance is illustrated in Fig. 14B. Hydrazine-inactivated C3 and C3b prepared with trypsin are unreactive with [14C]MA. Autoradiographic analyses of [14C]MA-inactivated C3 and MA-inactivated [1-14C]carboxamidomethylated C3 have shown a specific incorporation of each radiolabel into the α-chain.

An acyl group contributed by a glutamyl residue has been identified as the site of MA incorporation for C3.[28,86] The PTH-γ-glutamylmethylamide derivative obtained on Edman degradation was first isolated from MA-inactivated α_2-macroglobulin (α_2M) and definitively characterized by mass spectrometry.[87] The complete primary structure of a 35-residue tryptic peptide comprising the MA-reactive site of C3 has been determined.[88] When this sequence was compared with the amino-terminal structure of C3d, the tryptic peptide could be aligned commencing with the histidyl residue at position 15.[88] The cysteinyl residue at position 9 and

[86] B. F. Tack, R. A. Harrison, J. Janatova, M. L. Thomas, and J. W. Prahl, *Proc. Natl. Acad. Sci. U.S.A.* **77,** 5764 (1980).

[87] R. P. Swenson and J. B. Howard, *Proc. Natl. Acad. Sci. U.S.A.* **76,** 4313 (1979).

[88] M. L. Thomas, J. Janatova, W. R. Gray, and B. F. Tack, *Proc. Natl. Acad. Sci. U.S.A.* (in press).

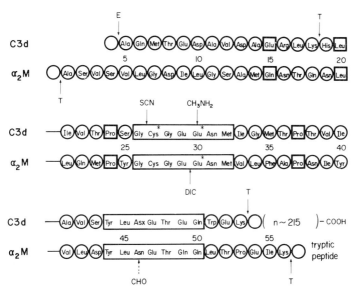

FIG. 15. The amino-terminal structure of C3d (positions 1–49) comprising the MA-reactive site, aligned for maximal homology with a 56-residue peptide from α_2M [R. P. Swenson and J. B. Howard, *J. Biol. Chem.* **255**, 8087 (1980)]. SCN denotes the cyanylation-induced cleavage site, Glu* the glutamyl residue reactive with MA (CH$_3$NH$_2$), Cys* the cysteinyl residue acquired on MA incorporation, and DIC the denaturant-induced cleavage site. CHO indicates a point of carbohydrate attachment in α_2M. T and E denote trypsin and elastase cleavage sites respectively. The position indicated by X in the C3d sequence has been tentatively identified as an aspartyl residue.

the MA-reactive glutamyl residue at position 12 in the tryptic peptide occupy positions 23 and 26, respectively, in the C3d fragment. The amino-terminal structure of C3d (positions 1–49) is shown in Fig. 15 and is aligned for maximal homology with the structure of the corresponding tryptic peptide from α_2M.[89] A 7-consecutive-residue identity in structure is evident at the MA-reactive site, followed by a further 7-residue region of identity toward the carboxyl-terminus. The asparaginyl residue at position 46 in the α_2M tryptic peptide has been identified as a point of carbohydrate attachment. The conservation in structure of the corresponding region in C3d suggests that this may be a site of carbohydrate attachment also. Further evidence for a shared amine-sensitive site between these two proteins has come from whole-plasma studies, where a MA-dependent incorporation of [¹⁴C]iodoacetamide into α_2M was observed.[86] The primary structure of the corresponding region in C4 has been recently reported.[90,91] The C4 sequence, in single letter code, is G S E G A L S P G G V A S L̲ L R

[89] R. P. Swenson and J. B. Howard, *J. Biol. Chem.* **255**, 8087 (1980).
[90] R. D. Campbell, J. Gagnon, and R. R. Porter, *Bioscience Reports* **1**, 423 (1981).
[91] R. A. Harrison, M. L. Thomas, and B. F. Tack, *Proc. Natl. Acad. Sci. U.S.A.* (in press).

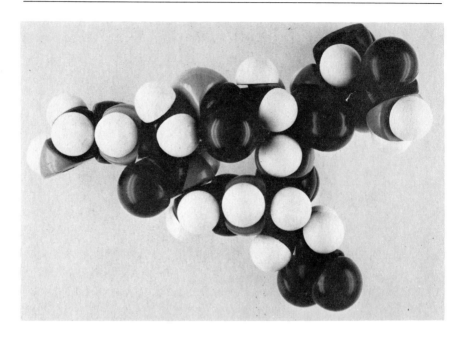

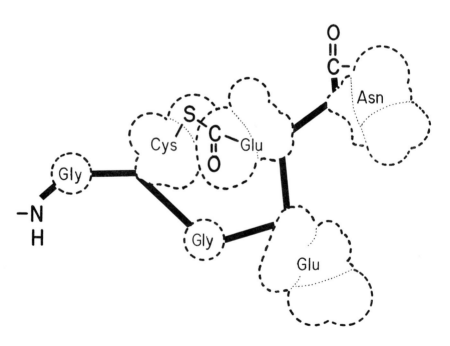

L P R G C G E E T M I Y L A P T L A A S R Y L D K T E Q. The residues underlined are those which show identity on comparison with the corresponding positions in C3d and α_2M (Fig. 15).

Treatment of C5 with MA does not result in a loss of hemolytic activity, expression of an SH group, or incorporation of this reagent.[24]

Proposed Nature and Function of the Labile Binding Site

These data indicate the presence of an internal thiolester bond in C3, C4, and α_2M. In each instance the presence of an amine-sensitive bond comprised of an acyl group and thiol component has been demonstrated. An alternative cyclic imide structure has been proposed for the MA-reactive site in α_2M.[87,89] This assignment was made, however, in the absence of data concerning the cysteinyl residue. In view of the nature of the covalent bond formed between C3b and C4b and membrane components, a transesterification mechanism is proposed. The acyl group of the glutamyl residue at position 12 of the C3 tryptic peptide is therefore thought to be transferred from the thiol of the cysteinyl residue at position 9 to an O— or N— nucleophilic group at the activating particle surface. The mechanism by which the thiolester is internally labilized following the conversion of C3 to C3b is unknown. It has been suggested, however, that a conformational change on cleavage of the Arg-Ser peptide bond at position 77–78 may position a histidyl residue for nucleophilic displacement at the carbonyl group of the thiolester, thereby effecting a transesterification reaction.[26] By analogy with the proposed mechanism for C3b and C4b binding, the covalent association of α_2M with trypsin and plasmin could proceed in an identical fashion. A space-filling model[90] of the active-site sequence Gly-Cys-Gly-Glu-Glu-Asn is shown in Fig. 16 and indicates that the cysteinyl residue at position 9 can bridge to the glutamyl residue at position 12 of the tryptic peptide.

[92] W. R. Gray, personal communication.

◄ FIG. 16. Possible conformation of thiolester loop. The model depicted is of the sequence Gly-Cys-Gly-Glu-Glu-Asn-, with a thiolester linkage between the side-chains of the Cys and the second Glu. Chain direction is from left to right, with the -Gly-Glu- residues displaced downward. A thiolester linkage cannot be formed when the residues are placed in α-helical conformation, since the side-chains are not long enough to bridge the gap. If a β-turn is constructed the bridge can be made, but only with excessive distortion. In contrast, the proposed conformation is essentially free from distortion of normal bond angles and distances, and the thiolester group is planar. It is presumed that the unmodified Glu is protruding into solvent, and that the main body of the protein lies above and behind the part of the molecule depicted here. The thiolester could then be partially shielded from solvent by nonpolar amino acid side-chains. Several minor variants of the conformation are also acceptable, and amino acids other than Gly can occupy the position following Cys. (This model was constructed and analyzed by W. R. Gray, Dept. of Biology, University of Utah, Salt Lake City, Utah.)

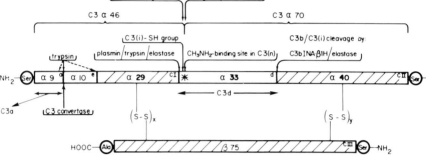

FIG. 17. A schematic representation of the C3 molecule. CI and CII represent α-chain subdomains that together with β-chain (CIII) comprise the C3c fragment. The positioning of an acidic fragment (C3e) with MW ≈ 10,000 at the amino-terminal end of the α'-chain is tentative. All other assignments, however, are based on direct structural determinations.

Denaturant-Induced Cleavage Reaction

The observation of a denaturant-induced cleavage (DIC) reaction for C3,[23,26–28] C4,[24,27,29] and $\alpha_2 M$[93,94] is almost certainly related to the presence of an internal thiolester. Any form of pretreatment of the native proteins that is disruptive of the thiolester totally prevents the cleavage reaction. The DIC site in $\alpha_2 M$ is located between the glutamyl residues at positions 29 and 30[94] as shown in Fig. 15. The larger carboxyl-terminal fragment of $\alpha_2 M$ has been isolated and characterized. The amino-terminus of this fragment is occupied by a pyrrolidone carboxylic acid residue derived from cyclization of the glutamyl residue at position 30.[94] The DIC site in C3 and C4 has not been fully characterized. Both proteins, however, acquire an SH group during the course of the cleavage reaction that is within the smaller 46,000-MW fragment of C3 and the 40,000-MW fragment of C4.[23,24] The 46,000-MW C3α fragment has been isolated and the amino-terminal structure shown to be identical with that of the α-chain.[95] Studies of the fragmentation patterns of S-cyanylated (SCN) forms of C3 and C4 have established that the site of autolytic cleavage is proximal to the position of the thiolester.[24,88] A further implication of the cyanylation-induced cleavage reaction is that the C3d fragment occupies a far more interior position in the α-chain than previously recognized. A schematic representation of the polypeptide chains, subdomain structure, and functional sites of C3 as presently known is presented in Fig. 17.

[93] P. C. Harpel, M. B. Hayes, and T. E. Hugli, *J. Biol. Chem.* **254**, 8669 (1979).
[94] J. B. Howard, M. Vermeulen, and R. P. Swenson, *J. Biol. Chem.* **255**, 3820 (1980).
[95] M. L. Thomas and B. F. Tack (unpublished observation).

Summary

C4,[96] C3,[97] and C5[98,99] are synthesized as single polypeptide-chain proforms that are processed into multiple-chain structures. C4 circulates in plasma as a 3-chain ($\alpha\beta\gamma$) structure; C3 and C5 are comprised of 2 chains each, α and β, of unequal size. Individual chains are held together by an undetermined number of disulfide bonds and by ionic and hydrophobic forces. The biological activities of C4, C3, and C5 are expressed and subsequently regulated through limited proteolysis.

Preparative procedures described in this chapter allow for the isolation of each protein in a state of high biochemical and functional purity. The native proteins, their constituent chains, and activation fragments have been subjected to primary sequence analysis. These studies have contributed to a fuller understanding of events germane to activation–inactivation pathways for each protein. The binding of nascent C3b and C4b to foreign cell surfaces, and subsequent interactions with specific receptors present on phagocytic cells provides for a positive control of infectious processes. In recent years, the nature and mechanism of the reaction responsible for the binding of C3b (and C4b) to cell surfaces has received considerable attention from investigators in several laboratories. It has been proposed that the bond between these two proteins and surface components on red cell membranes or bacterial cell wall products may be an oxygen ester. The mechanism of covalent attachment could therefore be one of transesterification. Support for this hypothesis has derived from the identification of an internal thiolester bond in native C3. In addition to this protein, data are available suggesting the presence of a thiolester in $\alpha_2 M$ and C4, and the absence of such a bond in C5. This site is proposed to (a) contribute to the stabilization of the native conformation prior to activation by the respective converting enzyme, (b) serve as the source of an activated acyl group for a transesterification reaction, and (c) sustain the DIC reaction of the C3α- and C4α-chains. Although there is considerable homology in size and structure between C5 and C3, including the existence of single polypeptide-chain proforms and similar activation mechanisms, there are substantial differences between these two proteins. Native C5 does not appear to contain an internal thiolester, and the association between C5b and cell-membrane structures is not covalent in nature.

[96] R. E. Hall and H. R. Colten, *Proc. Natl. Acad. Sci. U.S.A.* **74**, 1707 (1977).
[97] V. Brade, R. E. Hall, and H. R. Colten, *J. Exp. Med.* **149**, 759 (1977).
[98] F. Patel and J. O. Minta, *J. Immunol.* **123**, 2408 (1979).
[99] Y. M. Ooi and H. R. Colten, *J. Immunol.* **123**, 2494 (1979).

[8] Human Factor B

By Michael A. Kerr

Factor B was defined by early studies on the alternative pathway as a heat-labile protein necessary for the inactivation of C3 in serum by zymosan.[1] The identification of Factor B as a specific protein came from the study of a serum protein termed C3 proactivator which was able to complex with a protein from cobra venom to produce a C3-cleaving enzyme.[2] Factor B and C3 proactivator were shown to be the same protein[3] and both were subsequently shown to be identical to a protein of unknown function previously purified from human serum, called glycine-rich β-glycoprotein.[4-6] Factor B is present in normal human serum at concentrations of 120–300 μg/ml. It is an acute-phase reactant rising during "inflammatory" conditions. The role of Factor B in the alternative pathway has been reviewed extensively.[7-9]

The alternative-pathway C3 convertase is formed on cleavage of Factor B in a Mg^{2+}-dependent complex with C3b[10,11] (or the cobra venom factor subsequently identified as cobra C3b[2,12,13]) by Factor D. Factor B is cleaved into 2 fragments, Ba (apparent MW 30,000) and Bb (apparent MW 63,000). Bb is the catalytic subunit of the alternative-pathway C3 convertase, $\overline{C3b.B}$. In the presence of excess C3b, C5 is also a substrate for the enzyme. Bb is also the catalytic subunit of the alternative-pathway C5 convertase.[14,15] In vitro the alternative-pathway C3 convertase is extremely unstable, the activity decaying spontaneously with a half-life of several minutes. This decay reflects the release of Bb from the complex.[16,17] In serum this decay is retarded by the binding of properdin to the

[1] L. Pillemer, K. H. Lepow, and L. Blum, *J. Immunol.* **71**, 339 (1953).
[2] H. Götze and H. J. Müller-Eberhard, *J. Exp. Med.* **134**, 90s (1971).
[3] I. Goodkofsky and I. H. Lepow, *J. Immunol.* **107**, 1200 (1971).
[4] C. A. Alper, I. Goodkofsky, and I. H. Lepow, *J. Exp. Med.* **137**, 424 (1973).
[5] T. Boenisch and C. A. Alper, *Biochim. Biophys. Acta* **214**, 135 (1970).
[6] T. Boenisch and C. A. Alper, *Biochim. Biophys. Acta* **221**, 529 (1970).
[7] D. T. Fearon, *CRC Crit. Rev. Immunol.* **1**, 1 (1979).
[8] O. Götze, *Cold Spring Harbor Symp. Cell Proliferation* **2**, 255 (1975).
[9] R. R. Porter, *Annu. Rev. Biochem.* **23**, 177 (1979).
[10] H. J. Müller-Eberhard and O. Götze, *J. Exp. Med.* **135**, 1003 (1972).
[11] L. G. Hunsicker, S. Ruddy, and K. F. Austen, *J. Immunol.* **110**, 128 (1973).
[12] N. R. Cooper, *J. Exp. Med.* **137**, 451 (1973).
[13] C. A. Alper and D. Balavitch, *Science* **191**, 1275 (1976).
[14] M. R. Daha, D. T. Fearon, and K. F. Austen, *J. Immunol.* **117**, 630 (1976).
[15] W. Vogt, G. Schmidt, B. von Buttlar, and L. Diemenger, *Immunology* **34**, 29 (1978).
[16] D. T. Fearon, K. F. Austen, and S. Ruddy, *J. Immunol.* **111**, 1730 (1973).
[17] M. R. Daha, D. T. Fearon, and K. F. Austen, *J. Immunol.* **116**, 1 (1976).

METHODS IN ENZYMOLOGY, VOL. 80

complex.[18,19] Under certain conditions, C3b can in the absence of Factor D reversibly alter Factor B without cleavage to generate C3-cleaving activity.[20]

The C3 convertase cleaves C3 at a single bond to give the same fragments C3a and C3b formed by the classical pathway C3 convertase.[21] The C5 convertases of both pathways have also the same specificity.[22] There is therefore considerable structural and functional similarity between the enzymes of the alternative and classical pathways, Factor B being equivalent to C2, C3b to C4b. One major difference is that whereas C2 is cleaved by $C\overline{1}s$ even in the absence of C4b, Factor B is only cleaved by Factor D when it is complexed with C3b. It is however likely that only C2 that is complexed with C4b will form C3 convertase when cleaved by $C\overline{1}s$, the interaction of C2 and C4b being mediated by the C2b part of the molecule.[23,24] No such role for the analogous Ba fragment has been identified in the assembly of the alternative-pathway C3 convertase. The Ba fragment from guinea pig Factor B has been reported to stimulate polymorphonuclear leukocyte movement.[25] However, human Ba does not show this chemotactic activity.[26] Human Factor Bb induces the spreading of peripheral blood monocytes.[27]

Several purification schemes for human Factor B have been published.[26,28,29] All use a similar combination of ammonium sulfate precipitation, DEAE–ion-exchange chromatography, and cation-exchange chromatography. The methods all yield preparations homogeneous by the criterion of polyacrylamide-gel electrophoresis but with different yield. The purification scheme described was originally developed for the purification of C2.[30] It yields in addition to homogeneous Factor B, functionally pure C2 that can be used in the preparation of EAC43 cells used in the assay of Factor B and functionally pure Factor D also used in the assay.

[18] R. G. Medicus, R. D. Schreiber, O. Götze, and H. J. Müller-Eberhard, *Proc. Natl. Acad. Sci. U.S.A.* **73**, 612 (1976).
[19] D. T. Fearon and K. F. Austen, *Proc. Natl. Acad. Sci. U.S.A.* **73**, 3220 (1976).
[20] M. R. Daha, D. T. Fearon, and K. F. Austen, *Immunology* **31**, 789 (1976).
[21] T. E. Hugli, E. H. Vallota, and H. J. Müller-Eberhard, *J. Biol. Chem.* **250**, 1472 (1975).
[22] H. N. Fernandez and T. E. Hugli, *J. Biol. Chem.* **253**, 6955 (1978).
[23] M. A. Kerr, *Biochem. J.* **189**, 173 (1980).
[24] S. Nagasawa and R. M. Stroud, *Proc. Natl. Acad. Sci. U.S.A.* **74**, 2998 (1977).
[25] J. Hamuro, U. Hadding, and D. Bitter Suermann, *J. Immunol.* **120**, 438 (1978).
[26] P. H. Lesavre, T. E. Hugli, A. F. Esser, and H. J. Müller-Eberhard, *J. Immunol.* **123**, 529 (1979).
[27] O. Gotze, C. Bianco, and Z. A. Cohn, *J. Exp. Med.* **149**, 372 (1979).
[28] B. Curman, L. Sandberg-Tragardh, and P. Peterson, *Biochemistry* **16**, 5368 (1977).
[29] M. A. Niemann, J. E. Volanakis, and J. E. Mole, *Biochemistry* **19**, 1576 (1980).
[30] M. A. Kerr and R. R. Porter, *Biochem. J.* **171**, 99 (1978).

Assay Procedure

Buffers

  Calcium–magnesium stock: 30 mM calcium chloride–100 mM magnesium chloride

  Stock veronal buffer (5 ×): 25 mM sodium barbitone–0.7 M NaCl; 1 M HCl added to pH 7.4

  GVB: 0.5 g gelatin (Difco Labs, Detroit, Michigan) dissolved in 100 ml water by boiling, added to 100 ml stock veronal buffer, and brought up to 500 ml

  DGVB^{++}: 250 ml GVB added to 250 ml 5% (w/v) glucose and then 2.5 ml Ca^{2+}–Mg^{2+} stock added

  GVB–EDTA (6 : 4): 40 ml 0.1 M sodium EDTA of pH 7.4, added to 60 ml GVB

  GVB–EDTA (9 : 1): 10 ml 0.1 M sodium EDTA of pH 7.4, added to 90 ml GVB

Preparation of EAC43 Cells

  Sheep erythrocytes (1 × 10^8 cells/ml) sensitized with antibody and complement components C2 and C4,[31] (EAC14) are incubated with purified human C2 (400 ng/ml) and C3 (5 μg/ml) in DGVB^{++} for 5 min at 30°C and then chilled on ice. The cells are centrifuged at 1000 g for 5 min at 4°C and resuspended in GVB–EDTA (9 : 1). The cells are then washed once with the same buffer, incubated at 37°C for 15 min, and washed again with GVB–EDTA (9 : 1). After a second incubation at 37°C for 15 min the cells are washed three times in DGVB^{++} and resuspended in DGVB^{++} containing 40 units/ml streptomycin and penicillin before storage at 4°C.

Methods

  Factor B is assayed using EAC43 cells as follows: Factor B in 0.1 ml DGVB^{++} is incubated with 0.1 ml EAC43 cells (1 × 10^8 cells/ml in DGVB^{++}) and 0.1 ml Factor B (6 μg/ml in DGVB^{++}) for 30 min at 30°C. 0.3 ml guinea pig serum diluted 1 : 30 in GVB–EDTA (6 : 4) is added and the incubation continued for 1 hr at 37°C. After addition of 1.0 ml of 0.15 M NaCl the cells are pelleted by centrifugation at 1000 g, 4°C for 7 min and the degree of lysis determined from the absorption at 410 nm of the supernatant.

  Factor D is assayed by a similar procedure, the initial incubation containing Factor D in 0.1 ml DGVB^{++}, EAC43 cells (0.1 ml), and 0.1 ml Factor B (30 μg/ml in DGVB^{++}).

[31] T. Borsos, H. J. Rapp, and H. R. Colten, *J. Immunol.* **105**, 1439 (1970).

Purification Procedures

Reagents

All buffers contain 0.04 w/v sodium azide as preservative.

0.4 M Sodium phosphate buffer. 69.9 g anhydrous disodium hydrogen orthophosphate–547.3 g sodium dihydrogen orthophosphate dihydrate–4 g sodium azide, made up to 10 liters with water.

Veronal buffer (VBS). 5 mM veronal (pH 8.5)–0.5 mM CaCl$_2$–2.0 mM MgCl$_2$–40 mM NaCl is 5.1 g sodium barbitone–11.7 g sodium chloride–2 g sodium azide added to 100 ml calcium–magnesium stock and made up to 5 liters.

Preparation of "Aged" CNBr-Activated Sepharose-4B

Packed Sepharose-4B (100 ml) is washed well with water and resuspended in a minimal volume of 0.1 M sodium phosphate, pH 7.8. 13 g of CNBr in 13 ml of dimethylformamide is added and the mixture stirred until the reaction is complete. Throughout this time the temperature is kept at 20°C by the addition of ice, and pH 10–11 by the addition of 4 M NaOH. When the reaction is complete the Sepharose is washed extensively with 0.1 M sodium phosphate (pH 7.8) and then stirred overnight at 4°C in 100 ml of the same buffer. This material is termed "aged" CNBr-activated Sepharose-4B.

Purification Procedure

Outdated human citrated plasma is made 20 mM in CaCl$_2$ and left to clot overnight at 4°C. The clot is then centrifuged down and the serum stored at −20°C.

Euglobulin Precipitations

Frozen serum (2000 ml) is thawed and after the addition of 1 ml of 2.5 M iPr$_2$PF in propan-2-ol is centrifuged at 23,000 g for 30 min at 4°C. The serum is brought to pH 5.5 with 1 M HCl and dialyzed against 8 liters of 5 mM EDTA, (pH 5.5)–0.5% (w/v) benzamidine hydrochloride overnight. The dialysate is then centrifuged at 23,000 g for 30 min and the supernatant decanted.

Chromatography of CM-Sephadex C-50

Fifty milliliters of 0.4 M sodium phosphate are added to the supernatant and after the pH is brought up to 6.0 with 1 N NaOH the mixture is loaded onto a column (10 × 8 cm) of CM-Sephadex C-50 equilibrated in

0.1 M sodium phosphate (pH 6.0)–0.5% (w/v) benzamidine hydrochloride. The column is then washed with 1 liter of 0.1 M sodium phosphate (pH 6.0)–0.5% (w/v) benzamidine hydrochloride, and then developed with a linear gradient formed with 2 liters of 0.1 M sodium phosphate (pH 6.0) and 2 liters of 0.25 M sodium phosphate (pH 6.0), both 0.5% (w/v) benzamidine hydrochloride; all solutions are pumped through the column at a rate of about 1.5–2 liters/hr. Factor B is eluted in the middle of the gradient. A second gradient formed from 2 liters of 0.25 M sodium phosphate (pH 6.0) and 2 liters of 0.4 M sodium phosphate (pH 6.0), both 0.5% (w/v) benzamidine hydrochloride, can be used for the elution of functionally pure Factor D (see Article [11]).

Ammonium Sulfate Precipitation

Active fractions from the CM-Sephadex column are pooled (2 liters) and ammonium sulfate added to 50% saturation (291 g/liter). After stirring for 1 hr at 4°C the solution is centrifuged at 23,000 g for 30 min. The supernatant is decanted, filtered through Whatman No. 1 filter paper and ammonium sulfate added (159 g/liter) to give 75% saturation. The suspension is then stirred and left to warm to room temperature (15–20°C) before centrifugation at 23,000 g for 90 min at 20°C. The pellet is redissolved in about 50 ml of 0.4 M sodium phosphate (pH 6.0) and stored overnight after the addition of 0.1 ml of 2.5 M iPr$_2$ P-F.

Chromatography on "Aged" CNBr-Activated Sepharose-4B

The cloudy suspension is centrifuged at 26,000 g for 30 min and the pellet discarded. The supernatant is passed through a column (30 × 4 cm) of Sephadex G-25 equilibrated in 5 mM veronal buffer (VBS) and then applied to a column (15 × 3 cm) of "aged" CNBr-activated Sepharose-4B equilibrated in the same buffer. Factor B is eluted by washing the column with 500 ml of the same buffer. Functionally pure C2 can then be eluted with 250 ml of 0.4 M sodium phosphate, pH 6.0.

Chromatography on DEAE–Sepharose

The active fractions from the "aged" CNBr-activated Sepharose column (Fig. 1) are pooled and then applied to a column (30 × 2.5 cm) of DEAE–Sepharose equilibrated in 5 mM veronal buffer (VBS). The column is washed with 100 ml of the veronal buffer and developed with a linear gradient made from 500 ml of 5 mM veronal buffer (VBS) and 500 ml of 5 mM veronal buffer (pH 8.5)–0.5 mM CaCl$_2$–2.0 mM MgCl$_2$–100 mM NaCl.

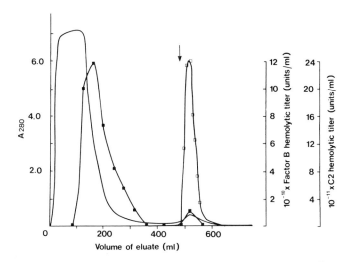

FIG. 1. Chromatography of Factor B on "aged" CNBr-activated Sepharose-4B. The protein precipitated at 75% saturated. $(NH_4)_2SO_4$ equilibrated with veronal buffer (VBS) and chromatographed on a column (15 × 3 cm) of "aged" CNBr-activated Sepharose-4B. ———, A_{280}; —■—, Factor B hemolytic activity; —□—, C2 hemolytic activity eluted by 0.4 M sodium phosphate started at the point indicated by the arrow. See also Article [6].

Rechromatography on "Aged" CNBr-Activated Sepharose-4B

Although most fractions eluted by the gradient on the DEAE–Sepharose column (Fig. 2) are homogeneous Factor B, early fractions are contaminated with another protein, probably hemopexin. This is conveniently removed by rechromatography on "aged" CNBr-activated Sepharose-4B. All active fractions from the DEAE–Sepharose column are pooled, dialyzed against veronal buffer (VBS), and then applied to a column (15 × 3 cm) of "aged" CNBr-activated Sepharose-4B, equilibrated in the same buffer. When the column is washed with 200 ml of the same buffer the contaminant passes through the column and the Factor B, which is retarded, is then eluted in a concentrated form by 200 ml of 0.1 M sodium phosphate, pH 6.0.

The Factor B fraction represents a purification of about 250-fold from serum and the yield is about 50–80 mg from 2 liters of serum. The Factor B is homogeneous by the criteria of SDS–polyacrylamide electrophoresis and N-terminal analysis. The final product is stable at neutral pH without the addition of proteolytic enzyme inhibitors and can be stored at 4°C without loss of activity. The purification procedure has been used routinely in our laboratory for 3 years to produce sufficient quantities of Factor B for amino acid sequence analysis.

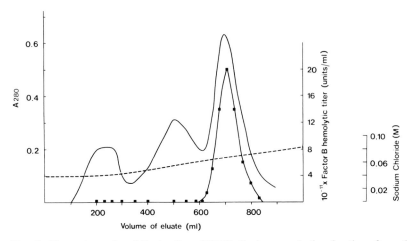

FIG. 2. Chromatography of Factor B on DEAE–Sepharose. Active fractions from the "aged" CNBr-activated Sepharose-4B column equilibrated with veronal buffer (VBS) were loaded onto a column (30 × 2.5 cm) of DEAE–Sepharose and the column developed with increasing sodium chloride concentration. ——, A_{280}; —■—, Factor B hemolytic activity; ----, concentration of NaCl.

Molecular Properties

Factor B is a single polypeptide chain with an apparent MW 86,000–93,000 by SDS–polyacrylamide gel electrophoresis.[26,28–30,32] This molecular weight has been confirmed by gel filtration, sedimentation equilibrium, and sedimentation velocity measurements.[28] The sedimentation coefficient is 5.9–6.2 S; partial specific volume, 0.72 ml/g; Stokes radius, 40 Å; diffusion constant ($D_{20,w}$), 5.4; and frictional ratio, 1.28.[28] The amino acid composition of Factor B has been determined in several studies and representative data are shown in Table I. Carbohydrate analyses have also been carried out.[28]

Factor B is rapidly cleaved by Factor D in the presence of C3b with the loss of hemolytic activity to yield 2 fragments, Ba with apparent MW 30,000, and Bb, 63,000. The 2 fragments are not linked by disulfide bonds and can be separated by gel filtration or by ion-exchange chromatography without the need for denaturing agents.[26,33] This suggests they represent separate domains within Factor B. Ba is the N-terminal part of the molecule.[29,32]

The amino acid composition of Ba and Bb are shown in Table II. Lesavre et al.[26] have reported Ba and Bb each to contain a single free SH

[32] M. A. Kerr, Biochem. J. 183, 615 (1979).

[33] D. L. Christie, J. Gagnon, and R. R. Porter, Proc. Natl. Acad. Sci. U.S.A. 77, 4923 (1980).

TABLE I
AMINO ACID ANALYSIS OF FACTOR B[a]

Amino acid	Residues/100 residues				
	Glycine-rich β-glycoprotein[a]	Curman et al.[b]	Kerr and Porter[c]	Lesavre et al.[d]	Niemann et al.[e]
Asp	10.8	10.2	10.2	10.0	10.4
Thr	5.3	5.4	5.5	5.6	5.6
Ser	6.2	7.5	6.9	7.6	6.9
Glu	11.5	11.7	11.0	11.5	12.1
Pro	5.9	5.2	5.2	6.0	6.7
Gly	8.4	8.7	8.3	8.5	9.1
Ala	5.1	4.9	4.6	4.7	4.7
Cys	2.5	3.2	2.9	2.8	—
Val	7.7	7.3	7.4	7.1	7.5
Met	1.2	1.4	1.1	1.7	1.6
Ile	4.7	4.5	4.9	4.6	4.7
Leu	7.1	6.7	7.2	7.0	7.5
Tyr	4.3	4.3	5.6	4.1	4.3
Phe	2.7	2.7	2.7	2.6	3.0
His	2.5	2.3	2.6	2.5	2.9
Lys	7.4	7.2	7.2	7.2	7.7
Arg	5.4	5.2	5.2	5.2	5.2
Trp	1.3	1.9	1.4	1.6	—

[a] T. Boenisch and C. A. Alper, *Biochim. Biophys. Acta* **221**, 529 (1970).
[b] B. Curman, L. Sandbergh-Tragardh, and P. A. Peterson, *Biochemistry* **16**, 5368 (1977).
[c] M. A. Kerr and R. R. Porter, *Biochem. J.* **171**, 99 (1978).
[d] P. H. Lesavre, T. E. Hugli, A. F. Esser, and H. J. Müller Eberhard, *J. Immunol.* **123**, 529 (1979).
[e] M. A. Niemann, J. E. Volanakis, and J. E. Mole, *Biochemistry* **19**, 1576 (1980).

group. The NH_2-terminal amino acid sequences of B, Ba, and Bb are shown in Fig. 3, together with extended sequence data on the COOH-terminal residues of Bb. Microheterogeneity has been reported in the sequence of Ba.[29]

The site of cleavage of Factor B by Factor D in the presence of C3b is an Arg-Lys bond at a site containing three basic residues Lys-Arg-Lys.[26] This is homologous to the site of cleavage of C2 by $C\bar{1}s$.[32] Although Factor B is only cleaved by Factor D in the presence of C3b, Factor B is cleaved in the absence of C3b by trypsin at the same site showing that the site is accessible. The Bb fragment is subsequently further degraded by trypsin whereas the Ba appears to be resistant.[29,32] Sequence analysis of the trypsin-produced Bb fragment showed a double sequence, the major sequence beginning with lysine representing 70% of the material and a minor

TABLE II
AMINO ACID COMPOSITION OF Ba AND Bb[a]

Amino acid	Glycine-rich γ-glycoprotein[b]	Bb			Ba	
		Niemann et al.[c]	Christie et al.[d]	Lesavre et al.[e]	Niemann et al.[c]	Lesavre et al.[e]
Asp	12.0	11.3	11.5	11.3	7.7	7.0
Thr	4.0	4.4	4.5	4.6	7.4	6.9
Ser	4.7	6.1	5.6	6.8	9.9	10.2
Glu	10.9	11.7	10.8	10.4	11.9	13.9
Pro	4.6	6.3	5.1	5.4	4.8	7.3
Gly	7.7	7.6	7.3	7.2	12.3	11.2
Ala	5.2	5.1	5.0	4.7	5.0	4.3
Cys	1.4	—	2.8	2.0	4.0	4.7
Val	9.0	8.0	9.5	8.6	4.9	4.0
Met	1.7	1.6	1.5	2.0	0.5	0.4
Ile	5.8	4.6	5.8	5.5	3.2	2.9
Leu	8.2	7.2	8.1	7.8	5.8	5.1
Tyr	4.2	4.1	3.9	3.9	4.2	4.9
Phe	3.0	3.1	2.7	2.6	2.5	2.5
His	2.6	2.6	2.6	2.7	2.6	2.2
Lys	9.1	8.8	9.3	9.8	3.1	2.7
Arg	3.7	3.8	4.1	4.2	8.1	8.1
Trp	2.4	—	—	0.8	—	2.1

[a] Values given as residues per 100 residues.
[b] T. Boenisch and C. A. Alper, Biochim. Biophys. Acta 214, 135 (1970).
[c] M. A. Niemann, J. E. Volanakis, and J. E. Mole, Biochemistry 19, 1576 (1980).
[d] D. L. Christie, J. Gagnon, and R. R. Porter, Proc. Natl. Acad. Sci. U.S.A. 77, 4923 (1980).
[e] P. H. Lesavre, T. E. Hugli, A. F. Esser, and H. J. Müller-Eberhard, J. Immunol. 123, 529 (1979).

sequence beginning with isoleucine having the same sequence displaced by one amino acid.[32]

Circular-dichroism spectra suggested 17% of the polypeptide chain of Factor B to be folded in α-helical arrangement. None of this α-helix is in Ba.[28]

Genetic Polymorphism

Factor B from human serum is polymorphic as judged by electrophoretic mobility.[6,34] The two common alleles F and S have gene frequencies of 0.28 and 0.72 in the Caucasian population. The structural gene

[34] C. A. Alper, T. Boenisch, and L. Watson, J. Exp. Med. 135, 68 (1972).

NH$_2$-Terminal B and Ba[a,b]

```
        1                   10                  20                  30
H₂N-T P W S L A R P Q G X C X L E G V E/V I K G G H F X L L X E X
        H           R
X A L E Y V – – – (≈250 residues)
```

NH$_2$-Terminal Bb[a–d]

```
        1                   10                  20                  30
H₂-N-K I V L D P S G S M N I Y L V L D G S D S I G A S X F T G A
                        40
K K C L V N L T E K – – – (≈250 residues)
```

Partial COOH-Terminal Half of Bb[d]

```
        1                   10                  20                  30
– – – V W E H R K G T D Y H K Q P W Q A K I S V I R P S K G X E S
                40                  50                  60
C M G A V V S E Y F V L T A A H C F T V D D K E H S K K V S
                70                  80                  90
V G G E K R D L E I E V V L F H P N Y N I N G K K E A G I P
                100                 110
E F Y D Y D V A L I K L K N K L K Y G Q
                                    220                 230
– – – (≈100 residues) – – – F L C T G G V S P Y A D P N T C R G D S
                240                 250                 260
G G P L I V H K R S R F I Q V G V I S W G V V D V C K N Q K
                270
R Q K Q V P A H A R – – – (≈20 residues) – – – F L — COOH
```

Fɪɢ. 3. Partial amino acid sequence of Factor B. Residues 4 and 9 of Ba have been identified differently by Kerr (1979) and Niemann et al. (1980). Residue 18 showed amino acid heterogeneity [Niemann et al. (1980)]. The sequence of Bb showing homology with serine protease active-site His, Asp, and Ser residues is underlined. [a]M. A. Kerr, Biochem. J. 183, 615 (1979). [b]M. A. Niemann, J. E. Volanakis, and J. E. Mole, Biochemistry 19, 1576 (1980). [c]P. H. Lesavre, T. E. Hugli, A. F. Esser, and H. J. Müller-Eberhard, J. Immunol. 123, 529 (1979). [d]D. L. Christie, J. Gagnon, and R. R. Porter, Proc. Natl. Acad. Sci. U.S.A. 77, 4923 (1980).

for Factor B is closely linked to the B locus of the major histocompatibility complex and to the structural gene for C2.[35,36] No well-documented cases of genetic deficiency of Factor B have been reported.

Factor B as a Proteolytic Enzyme

Bb has been identified as the active subunit in the C3 convertase complex, C3b Bb, and in the presence of excess C3b as the active subunit of the C5 convertase complex.[37] C3 is cleaved at a single Arg-Ser bond[21]

[35] C. A. Alper, J. Exp. Med. 144, 111 (1976).
[36] F. M. Allen, Jr., Vox Sang. 27, 382 (1974).
[37] W. Vogt, G. Schmidt, B. von Buttlar, and L. Dieminger, Immunology 39, 29 (1978).

and C5 at a single Arg-X bond.[22] Isolated Bb is however unable to cleave either C3 or C5. Bb has been shown to possess esterase activity against certain arginine and lysine esters, principally N-acetylglycine-lysine methyl ester.[2,38] Factor B has also been reported to be inhibited by diisopropyl fluorophosphate,[39] though others were unable to detect this inhibition.[40] Confirmation of the serine protease nature of Factor B has come recently from amino acid sequence studies (Fig. 3).[33] Although the catalytic chain Bb has twice the molecular weight of other serine proteases, the COOH-terminal part of the molecule shows strong homology with other serine proteases including the sequences around catalytically important charge-relay system residues. There is however no evidence of the characteristic NH_2-terminal sequence of other serine proteases that is believed to be important in the activity of these enzymes. The NH_2-terminal of Bb shows homology with C2a but with no other proteolytic enzyme.

[38] N. R. Cooper, *Prog. Immunol.* **1**, 567 (1971).
[39] R. G. Medicus, O. Götze, and H. J. Müller-Eberhard, *Scand. J. Immunol.* **5**, 1049 (1976).
[40] W. Vogt, W. Dames, G. Schmidt, and C. Dieminger, *Immunochemistry* **14**, 201 (1977).

[9] C3b Inactivator and β1H

By L. GAIL CROSSLEY

C3b inactivator (C3bINA), a β-globulin found in normal serum, was first identified in 1966 by its effects on the biological activities of cell-bound C3b. C3bINA abrogated the capacity of cells bearing C3b to participate in immune adherence, to undergo enhanced phagocytosis by polymorphonuclear leukocytes, and to react with subsequent components of the complement system leading to cytolysis.[1-3] A reactive site for bovine conglutinin was also revealed following the action of C3bINA on cell-bound C3b, giving rise to the early alternative name for this protein of "conglutininogen-activating factor (KAF)."[4] C3bINA was subsequently reported to inactivate not only C3b, but also C4b[5-7] and either β1H-

[1] R. A. Nelson, Jr., J. Jensen, I. Gigli, and N. Tamura, *Immunochemistry* **3**, 111 (1966).
[2] N. Tamura and R. A. Nelson, Jr., *J. Immunol.* **99**, 582 (1967).
[3] S. Ruddy and K. F. Austen, *J. Immunol.* **102**, 533 (1969).
[4] P. J. Lachmann and H. J. Müller-Eberhard, *J. Immunol.* **100**, 691 (1968).
[5] N. R. Cooper, *J. Exp. Med.* **141**, 890 (1975).
[6] S. Shiraishi and R. M. Stroud, *Immunochemistry* **12**, 935 (1975).
[7] M. K. Pangburn, R. D. Schreiber, and H. J. Müller-Eberhard, *J. Exp. Med.* **146**, 257 (1977).

globulin in the case of C3b,[7] or a 10 S serum protein, C4-binding protein, in the case of C4b,[8,9] were found to be essential cofactors for the action of C3bINA.

β1H was so named because of its electrophoretic mobility[10] and was later shown to potentiate the action of C3bINA on C3b. By binding stoichiometrically to C3b, β1H also prevented the interaction of other proteins with C3b and accelerated the rate of decay of the alternative-pathway convertase.[11–13] Since C3b plays a central role in the activation of both the pathways of complement, the regulation of its activities by C3bINA and β1H is one of the most important control mechanisms of the complement system.

Assay of C3bINA and β1H

Principle

The inactivation of cell-bound C3b provides the basis for hemolytic assays of C3bINA and β1H. However, if β1H alone is of interest, the alternative assays specific for β1H may be less time consuming. Cell-bound C3b on EAC43 cells incubated with Factors B and $\overline{D}$ in the presence of Mg^{2+} forms the alternative-pathway convertase. The subsequent addition of guinea pig serum diluted in buffer containing EDTA supplies the complement components C3–C9 and results in activation of the terminal components causing cell lysis. In the presence of limiting amounts of C3b on the EAC43 cells and an excess of Factors B and $\overline{D}$ and guinea pig serum, cell lysis is inhibited to a degree determined by the extent of cleavage of the cell-bound C3b by C3bINA and β1H. In a modification[14] of the original method,[15] assays of C3bINA are carried out in buffer containing a fixed concentration of β1H, or alternatively β1H is assayed in the presence of a fixed amount of C3bINA.[11,14]

Reagents

Buffers. The composition of solutions made daily is:
GVB. Veronal buffered saline, pH 7.5, containing 5 mM sodium barbital, 0.14 M NaCl, and 0.1% (w/v) gelatin

[8] S. Nagasawa, S. Shiraishi, and R. M. Stroud, J. Immunol. 116, 1743 (1976).
[9] T. Fujita and V. Nussenzweig, J. Exp. Med. 150, 267 (1979).
[10] U. R. Nilsson and H. J. Müller-Eberhard, J. Exp. Med. 122, 277 (1965).
[11] K. Whaley and S. Ruddy, J. Exp. Med. 144, 1147 (1976).
[12] J. M. Weiler, M. R. Daha, K. F. Austen, and D. T. Fearon, Proc. Natl. Acad. Sci. U.S.A. 73, 3268 (1976).
[13] D. H. Conrad, J. R. Carlo, and S. Ruddy, J. Exp. Med. 147, 1792 (1978).
[14] L. G. Crossley and R. R. Porter, Biochem. J. 191, 173 (1980).
[15] K. Whaley, P. H. Schur, and S. Ruddy, J. Clin. Invest. 57, 1554 (1976).

EDTA–DGVB, pH 6.5. 5% (w/v) glucose in water: GVB : (0.1 M EDTA, pH 7.5) in proportions 7.5 : 2.5 : 1 (v/v), adjusted to pH 6.5 with 1 M HCl

DGVB. Equal volumes of GVB and 5% (w/v) glucose

DGVB^{++}. DGVB containing 1.5×10^{-4} M Ca^{2+} and 0.5×10^{-3} M Mg^{2+}

GVB–EDTA. GVB : (0.1 M EDTA, pH 7.5) in proportions 9 : 1 (v/v) Guinea pig serum (stored at $-70°C$ in 1-ml aliquots) diluted 1 : 30 (v/v) with 0.04 M EDTA in GVB

Preparation of EAC43 Cells.[16] Sheep erythrocytes (E) sensitized by antibody (A) and components C1 and C4 (EAC14)[17] are incubated in DGVB^{++} (1×10^8 cells/ml) with human C2 (0.8 μg/ml) and C3 (5 μg/ml) for 5 min at 30°C, then chilled on ice for 10 min. The cells are centrifuged at 1000 g for 10 min at 4°C and resuspended in GVB–EDTA buffer. The latter step is repeated and the mixture incubated for 15 min at 37°C, centrifuged, resuspended in GVB–EDTA buffer and incubated for a further 15 min at 37°C. The resulting EAC43 cells are washed twice with DGVB^{++} and stored at 4°C in DGVB^{++} containing streptomycin and penicillin (40 units of each–ml).

Procedure

EAC43 cells in DGVB^{++} (1×10^8/ml) are pelleted by centrifugation at 1000 g for 10 min, washed twice by resuspension to the same concentration of cells in EDTA–DGVB (pH 6.5) buffer, and finally resuspended in EDTA–DGVB (pH 6.5) buffer containing either 10 μg β1H/ml for C3bINA assays, or 0.1 μg C3bINA/ml for β1H assays immediately before use. Dilutions of the sample to be assayed in EDTA–DGVB (pH 6.5) buffer (0.1 ml) are incubated at 37°C[15] for 30 min with 0.1 ml of the washed EAC43 cells. The inclusion of EDTA in the EDTA–DGVB (pH 6.5) buffer prevents the formation of C3b during assays of crude fractions, and the relatively low pH and ionic strength of the assay buffer have been found to be favorable for the action of C3bINA, which has a pH optimum of 6.[14,15] Ice-cold DGVB (0.5 ml) is added to stop the reaction and the cells washed twice with ice-cold DGVB and once with ice-cold DGVB^{++}. The pelleted cells are resuspended in DGVB^{++} (0.3 ml) containing Factors B (7 μg/ml) and $\overline{\text{D}}$ (0.4 μg/ml) and incubated at 30°C for 30 min. The amounts of Factors B and $\overline{\text{D}}$ are predetermined to give maximum lysis of EAC43 cells in the absence of C3bINA or β1H in the first incubation mixture. After addition of 0.3 ml of guinea pig serum diluted 1 : 30 with 0.04 M EDTA in GVB and further incubation at 37°C for 1 hr, 1 ml of ice-cold 0.15 M NaCl

[16] *See* Parkes, R. G. ViScipio, M. A. Kerr, and R. Prohaska. *Biochem. J.* **193**, 963 (1981).
[17] T. Borsos and H. J. Rapp, *J. Immunol.* **99**, 263 (1967).

is added and the mixture centrifuged at 1000 g for 10 min. The degree of lysis is determined from the absorption at 410 nm of the supernatant. As controls, the omission of C3bINA or β1H from the first incubation mixture measures the maximal lysis, while background lysis is determined by omission of Factors B and $\overline{D}$ from the second incubation mixture and 100% lysis is obtained by the addition of 1 ml of water instead of 0.15 M NaCl in the last step.

Unit of Activity

The degree of lysis is expressed as Z' hemolytic units, a unit being the reciprocal of the dilution of C3bINA or β1H that gives $Z' = 1$, where $Z' = -\ln (1 -$ proportion of cell lysis inhibited).[15]

Alternative Assay Methods

Alternative methods for assaying C3bINA include measurement by radial diffusion,[15] loss of immune adherence,[18] a hemolytic assay utilizing EAC423 cells and a C5–C9 reagent prepared from serum treated with hydrazine and immunochemically depleted of C3bINA,[7] or following the release of [^{125}I]C3c after trypsin treatment of EAC43 cells prepared with [^{125}I]C3 and previously incubated with C3bINA and β1H.[19]

β1H is conveniently assayed by measuring its ability to decay the properdin-stabilized amplification convertase by dissociation of Bb from cell-bound C3b.[12,20] A radioimmunoassay for β1H, which has the advantage of enhanced sensitivity, has also been reported,[20] in addition to quantitation by radial immunodiffusion[12] or the two-step trypsin-dependent radioassay described above.[19]

Purification of C3bINA

Principle

The relatively low concentration of C3bINA in serum (30–50 μg/ml),[7,21] together with difficulties encountered in the removal of the common contaminants transferrin and IgG, hampered early schemes for the purification of this protein, which was not obtained as a homogeneous preparation until 1977.[7,21] The most recent procedure affords a severalfold improved recovery of C3bINA compared with the previous methods.[14]

[18] W. D. Linscott, R. Ranken, and R. D. Triglia, *J. Immunol.* **121**, 658 (1978).
[19] T. A. Gaither, C. H. Hammer, and M. M. Frank, *J. Immunol.* **123**, 1195 (1979).
[20] K. Whaley, *J. Exp. Med.* **151**, 501 (1980).
[21] D. T. Fearon, *J. Immunol.* **119**, 1248 (1977).

C3bINA is present in significant amounts in both the euglobulin and pseudoglobulin fractions of serum, and initial fractionations using either ammonium sulfate or polyethylene glycol 4000 are also unsatisfactory, presumably because C3bINA associates with other proteins.[14]

Plasma depleted of plasminogen rather than serum is used as the starting material. Chromatography of plasma on the affinity adsorbent lysine–Sepharose gives quantitative recovery of C3bINA activity and prevents activation of the clotting factors. As an additional precaution against proteolysis, diisopropyl phosphofluoridate (DFP) is added repeatedly during the purification. During chromatography on QAE-Sephadex, C3bINA and transferrin, which serves as a visual marker, are separated from the bulk of bound protein including $\beta 1H$, by extended washing before application of the salt gradient. Removal of transferrin is achieved both by ammonium sulfate precipitation of C3bINA and by the adsorption of C3bINA to wheat germ agglutinin–Sepharose, which has a specificity for N-acetylglucosamine residues. Remaining contaminants are removed by chromatography on hydroxylapatite followed by gel filtration.[14]

Procedure

All procedures are carried out at 4°C and all buffers contain 0.02% (w/v) sodium azide. Tris buffers are prepared by dilution of 1 M Tris adjusted to the stated pH with HCl at 20°C. L-Lysine–Sepharose-4B (5 μmol lysine/ml of packed Sepharose) and wheat germ agglutinin–Sepharose-4B (5 mg/ml) are prepared by standard procedures[14,22] or may be obtained commercially. Plasminogen adsorbed to lysine–Sepharose from plasma is eluted with 0.2 M ϵ-aminocaproic acid in 0.3 M potassium phosphate buffer (pH 7.0), and the gel washed extensively with water. Each of the columns is used repeatedly after washing as described without repacking the column, with the exception of QAE-Sephadex, which is regenerated by washing successively with 0.1 M HCl, water, 0.1 M NaOH, and water.

Plasminogen Depletion of Plasma. Plasma (500 ml), to which 0.5 ml of 2.5 M DFP in propan-2-ol is added, is clarified by centrifugation at 1000 g for 30 min. The supernatant is applied to a column (5 × 15 cm) of lysine–Sepharose equilibrated with 0.1 M potassium phosphate (pH 7.0) containing 0.15 M NaCl and 15 mM EDTA at a flow rate of 200 ml/hr. The column is washed with buffer and the unadsorbed protein eluting after the void volume is collected until the absorption at 280 nm of the eluate is <1.0, and made 2.5 mM in DFP.

Chromatography on QAE-Sephadex. The plasminogen-depleted plasma is dialyzed against 20 mM Tris-HCl (pH 7.8)–60 mM NaCl, then passed

---

[22] D. G. Deutsch and E. T. Mertz, *Science* 170, 1095 (1970).

through a column (7 × 36 cm) of QAE-Sephadex equilibrated with the same buffer at a flow rate of 300 ml/hr. After washing the column with 1.5 column volumes of buffer, a linear gradient from 60 mM to 0.35 M NaCl in buffer (2 × 2 liters) is applied, followed by a step to buffer containing 1.0 M NaCl. The fractions containing C3bINA activity are combined, made 60% saturated (36 g/100 ml) in ammonium sulfate, and stirred for 2 hr. After centrifugation at 8000 g for 20 min, the precipitate is dissolved in 60 ml of 50 mM sodium phosphate (pH 7.0)–0.2 M NaCl, and DFP added to 2.5 mM.

Chromatography on Wheat Germ Agglutinin–Sepharose. The concentrated solution is dialyzed against 50 mM sodium phosphate (pH 7.0)– 0.2 M NaCl, and applied to a column (4 × 7.5 cm) of wheat germ agglutinin–Sepharose equilibrated with the same buffer at 15 ml/hr. After an initial wash with 1 column volume of buffer, the flow rate is increased to 30 ml/hr and when the adsorption at 280 nm of the eluate is <0.15, the adsorbed protein is eluted with buffer to which 100 mg/ml of N-acetyl-D-glucosamine is added. The active fractions are pooled and made 2.5 mM in DFP.

Chromatography on Hydroxylapatite. After dialysis against 25 mM potassium phosphate (pH 7.5), the solution is applied to a column (3.2 × 15 cm) of hydroxylapatite equilibrated with the dialysis buffer, at a flow rate of 35 ml/hr. The column is extensively washed with buffer, then a linear gradient (2 × 300 ml) of 25 mM to 0.2 M potassium phosphate (pH 7.5) is applied. The fractions containing C3bINA activity are concentrated by precipitation with ammonium sulfate (40 g/100 ml) as before and the precipitate redissolved in 3–5 ml of 25 mM Tris-HCl, pH 7.0.

Gel Filtration. The concentrated solution is passed directly through a column of Sephacryl-S200 (2.5 × 100 cm) equilibrated with 25 mM Tris-HCl (pH 7.0), containing 0.15 M NaCl at 12 ml/hr. The active fractions are combined, made 2.5 mM in DFP and concentrated by ultrafiltration with a Diaflo PM-10 membrane to 2–3 mg/ml and stored at 4°C.

Yield and Purity. The yield of C3bINA is 6 mg from 500 ml of plasma, with an overall recovery of hemolytic activity of 20% (Table I). The preparation appears 90–95% homogeneous on electrophoresis in polyacrylamide gels containing sodium dodecyl sulfate in the presence and absence of reducing agents, with a single impurity, probably IgG.

Properties of C3bINA

Chemical Composition

The molecular weight (MW) of C3bINA is 88,000[4,5,7] with a sedimentation rate of 4.5 S determined by sucrose-density gradient ultracentrifugal

TABLE I
PURIFICATION OF C3bINA FROM PLASMA[a]

Fraction	Volume (ml)	Total activity (units)	Total protein[b] (mg)	Specific activity (units/mg)	Activity yield (%)
Plasma	500	440,000	29,000	15	100
QAE-Sephadex pool	770	385,000	1600	240	88
Wheat germ agglutinin– Sepharose pool	223	152,000	150	1000	35
Hydroxylapatite pool	98	118,000	20	5900	27
Sephacryl-S200 pool	23	88,000	6	15,000	20

[a] From L. G. Crossley and R. R. Porter, *Biochem. J.* **191**, 173 (1980).
[b] Protein concentration determined by the Lowry method using bovine serum albumin as a standard.

analysis and a diffusion coefficient of 5.3×10^{-7} cm^2/sec calculated from gel filtration on Sephadex G-200.[5] Multiple bands are observed on alkaline-gel electrophoresis and isoelectric focusing of C3bINA, but a single band is observed on polyacrylamide-gel electrophoresis in the presence of dodecyl sulfate, suggesting that the heterogeneity is due to differences in charge rather than molecular weight.[21] Monospecific antisera have been raised against C3bINA,[7,21] and the isoelectric point falls in the pH range of 5.7–6.1.[21] The amino acid composition of C3bINA shown in Table II is similar to that of the other globulins except for an above average content of half-cystine.[7] The carbohydrate content of C3bINA is at least 10.7% (w/w), not including neuraminic acid, with 7.5% hexose and 3.2% glucosamine detected.[7]

C3bINA is composed of two polypeptide chains of MW 50K and 38K which are disulfide-bonded.[7,21] The smaller polypeptide chain contains relatively more carbohydrate than the larger chain as detected by periodic acid-Schiff staining of the two chains separated under reducing conditions on polyacrylamide gels.[7]

Stability

C3bINA is stable to heating at 56°C for 30 min and to exposure to $0.15 M$ hydrazine, 2 M guanidine-HCl and pH 2.2,[3,4,7] while treatment with 2-mercaptoethanol or potassium metaperiodate resulted in loss of C3bINA activity.[4,23] Purified C3bINA has been reported to be stable at 4°C at neutral pH for several weeks.[14]

[23] C. A. Alper, F. S. Rosen, and P. J. Lachmann, *Proc. Natl. Acad. Sci. U.S.A.* **69**, 2910 (1972).

TABLE II
AMINO ACID COMPOSITION OF C3bINA[a,b]

Amino acid	Residues[c]/molecule	Residues/1000 residues
Lysine	58	75
Histidine	17	21
Arginine	30	39
Aspartic acid	77	99
Threonine[d]	45	58
Serine[d]	61	78
Glutaminic acid	87	111
Proline	32	41
Glycine	73	94
Alanine	45	57
Half-cystine[e]	42	54
Valine[f]	51	65
Methionine[d]	11	14
Isoleucine[f]	35	45
Leucine	43	55
Tyrosine[d]	31	40
Phenylalanine	28	36
Tryptophan[g]	14	18
Total	780	1000

[a] From M. K. Pangburn, R. D. Schreiber, and H. J. Müller-Eberhard, *J. Exp. Med.* **146**, 257 (1977).
[b] Reported as moles of amino acid per mole of protein, assuming MW 88K.
[c] Average values from two replicates at each hydrolysis time (24, 48, and 72 hr).
[d] Extrapolated to zero time of hydrolysis.
[e] Determined after performic acid oxidation.
[f] Value for 72-hr hydrolysis.
[g] Determined after base hydrolysis according to T. E. Hugli and S. Moore, *J. Biol. Chem.* **247**, 2828 (1972).

Substrate Specificity

Soluble and Cell-Bound C3b. C3b is composed of an α'-chain (MW 105K) that is disulfide-bonded to a β-chain (MW 75K). The action of C3bINA on soluble C3b requires the presence of $\beta1H$[7] and results in the cleavage of the α'-chain only, to form the hemolytically inactive derivative, C3b$_i$. C3b$_i$ consists of two fragments with MW $\approx$ 67K and 43K, respectively, which are both disulfide-bonded to the intact β-chain.[7,24]

[24] S. Nagasawa and R. M. Stroud, *Immunochemistry* **14**, 749 (1977).

Recently, a second peptide bond in the α'-chain has been reported to be cleaved by C3bINA and β1H, with a fragment of MW $\approx$ 46K probably formed as a transient precursor of the stable 43K-MW fragment, in addition to the 67K-MW fragment. It is not known whether the putative 3K-MW peptide is cleaved from the N- or C-terminal end of the 46K-MW fragment.[9,25] Further cleavage of C3b$_i$, which is markedly more susceptible to proteolytic cleavage than its precursor C3b, requires exposure to trypsin or plasmin, with transient products being detected after very short exposure of C3b$_i$ to plasmin.[7,24,26] The stable products formed are two antigenically distinct fragments, C3c (MW 151K) and C3d (MW 27K), which were first isolated as breakdown products of C3 from aged serum or following trypsin treatment of C3b.[27] C3c is composed of two fragments of MW 29K and 43K, both disulfide-bonded to the intact β-chain, whereas C3d is a single polypeptide.[25]

Cell-bound C3b is attached by a covalent bond to the cell membrane via the α'-chain. Cleavage of the α'-chain by C3bINA and β1H forms two fragments of MW $\approx$ 60K and 40K, each linked to the intact β-chain by disulfide bonds. The resulting cell-bound C3b$_i$ remains covalently bound to the cell surface via the 60K-MW fragment and lacks the immune adherence and hemolytic activity of cell-bound C3b. Subsequent trypsin digestion releases a 130K-MW fragment (C3c), leaving a 40K-MW fragment covalently linked to the membrane, which is successively degraded to a 38K- and finally to a 32K-MW fragment (C3d).[28]

Soluble and Cell-Bound C4b. C4b is composed of three chains, α'(MW 90K), β(MW 80K), and γ(MW 30K), joined by interchain disulfide bonds. Either a 10 S serum protein, C4-binding protein,[8,9] or β1H, which is less efficient,[7,9] can act as the essential cofactor for the cleavage of soluble and cell-bound C4b by C3bINA. On cleavage, the biological activities of C4b are lost and two antigenically distinct fragments (C4c and C4d) are formed.[5,7] The α'-chain of soluble C4b is cleaved at two sites by C3bINA yielding three peptides of MW 47K, 25K, and 17K, while the β- and γ-chains remain intact. The 47K-MW fragment of the α'-chain, C4d, is dissociated from the remainder, C4c (MW 150K), consisting of the 25K- and 17K-MW fragments, which are both disulfide-bonded to the β- and γ-chains.[29]

Inactive C3 and C4. Although native C3 and C4 are not hydrolyzed by C3bINA,[3,5,7] the hemolytically inactive forms of both proteins formed on

[25] R. A. Harrison and P. J. Lachmann, *Mol. Immunol.* **17**, 9 (1980).

[26] J. D. Gitlin, F. S. Rosen, and P. J. Lachmann, *J. Exp. Med.* **141**, 1221 (1975).

[27] V. A. Bokisch, H. J. Müller-Eberhard, and C. G. Cochrane, *J. Exp. Med.* **129**, 1109 (1969).

[28] S. K. Law, D. T. Fearon, and R. P. Levine, *J. Immunol.* **122**, 759 (1979).

[29] T. Fujita, I. Gigli, and V. Nussenzweig, *J. Exp. Med.* **148**, 1044 (1978).

standing in the cold, on freezing and thawing, or on treatment with amines are cleaved.[14] In the case of inactive C3, the α-chain (MW 115K) is cleaved by C3bINA and β1H more slowly than the α'-chain (MW 105K) of C3b to yield two fragments of MW 75K and 43K.[14] The production of similar fragments from native C3 treated with C3bINA and β1H followed by prolonged exposure to immobilized F(ab')$_2$ has also been reported and may be due to a conformational change arising either from denaturation of the C3 or from the binding interaction with the antibody.[30]

Other Substrates. One other derivative of C3 produced by the action of an unidentified protease has been postulated to be a substrate for the action of C3bINA and β1H,[30] and the possibility that C5 derivatives may be cleaved has also been raised.[31] However, neither free C5b in the fluid phase[32] nor the complex C5b–6[33] is susceptible to proteolytic cleavage by C3bINA and β1H.

Effect of Protease Inhibitors

The proteolytic action of C3bINA is characterized by a high degree of substrate specificity, although several proteins, including β1H,[7] C4-binding protein,[8,9] and a glycoprotein isolated from human erythrocyte membrane,[34] are effective in the cofactor role. No synthetic substrates have yet been identified, and attempts to determine the class of protease to which C3bINA belongs have been unsuccessful to date. C3bINA is not inhibited by 10 mM DFP, 5 mM toluenesulfonyl lysyl chloromethyl ketone, 0.1 mM phenylmethylsulfonyl fluoride, 5 mM benzamidine, or by soybean- and lima bean-trypsin inhibitors or ovomucoid at 100 μg/ml, or by C$\bar{1}$ inactivator, α_2-macroglobulin, α_1-antitrypsin or inter-α-trypsin inhibitor at 5 μg/ml.[5,14] Chelating agents such as 1 mM 1,10-phenanthroline, 50 mM EDTA, and ethylene glycolbis(aminoethyl ether)tetraacetate (EGTA), or the addition of various divalent metal chlorides (1 mM) have no effect on C3bINA activity.[14] Similarly, the sulfhydryl reagents HgCl$_2$ (1 mM), iodoacetic acid (1 mM), iodoacetamide (0.02 M), and 4-chloromercuribenzoate (0.1 mM) do not inactivate C3bINA, although N-ethylmaleimide (1 mM) causes partial inhibition.[4,14] Strong inhibition of C3bINA is observed with 1 mM dithiothreitol or 10 mM 2-mercaptoethanol, presumably as a result of the reduction of interchain disulfide bonds in the substrate or enzyme.[4,5,14]

[30] R. A. Harrison and P. J. Lachmann, *Mol. Immunol.* **17**, 219 (1980).
[31] H. J. Müller-Eberhard, *Annu. Rev. Biochem.* **44**, 697 (1975).
[32] K. Yamamoto and H. Gewurz, *J. Immunol.* **120**, 2008 (1978).
[33] E. R. Podack and H. J. Müller-Eberhard, *J. Immunol.* **124**, 332 (1980).
[34] D. T. Fearon, *Proc. Natl. Acad. Sci. U.S.A.* **76**, 5867 (1979).

Purification of $\beta 1H$

Principle

Of the complement proteins, the concentration of $\beta 1H$ in serum (500 $\mu g/ml)^{12}$ is second only to that of C3. Two very similar procedures for the isolation of homogeneous $\beta 1H$ in four steps, including euglobulin precipitation, ion-exchange chromatography and gel filtration, have been reported.[11,12] Despite the relatively high serum concentration of $\beta 1H$, low overall recoveries of 1.5%[11] and 10%[12] respectively were obtained. Partially purified $\beta 1H$ is also readily available as a by-product of C3 purification. In the method of Tack and Prahl,[35] substitution of DEAE–Sephacel for DEAE–cellulose allows the separation of $\beta 1H$ from C3 and C5. Alternatively, gel filtration of the post-DEAE–cellulose pool on Sepharose-6B resolves high-molecular-weight material and $\beta 1H$ from C3 and C5, with pure $\beta 1H$ obtained following further chromatography on hydroxylapatite.[35,36] Harrison and Lachmann[37] have also reported high yields of $\beta 1H$ obtained as a by-product of their procedure for C3 purification.

Procedure

In the method of Weiler *et al.*,[12] serum is used as the starting material and $\beta 1H$ is detected by inhibition of the properdin-stabilized alternative-pathway convertase.

Chromatography on QAE-Sephadex. Serum (25 ml) is diluted with ice-cold water to a conductivity of 3.3 mS, adjusted to pH 7.5 with dilute HCl, and applied to a column (2.5 × 40 cm) of QAE-Sephadex equilibrated with 10 mM Tris-HCl–50 mM NaCl–2 mM EDTA, pH 7.5. After washing with 300 ml of buffer, a linear gradient of 2 liters from 50 mM to 0.33 M NaCl is applied and fractions containing $\beta 1H$ activity are concentrated to 18 ml by ultrafiltration.

Euglobulin Precipitation and Gel Filtration. The concentrated solution is adjusted to pH 5.5 with dilute HCl and dialyzed overnight against 6 liters of 0.1 M sodium acetate–2 mM EDTA, pH 5.5. The resulting precipitate is washed three times with dialysis buffer, dissolved in veronal-buffered 0.3 M NaCl containing 2 mM EDTA and applied to a column (2.5 × 80 cm) of Sephadex G-200 equilibrated with the same buffer. The fractions containing the peak of $\beta 1H$ activity, which coincides with the major protein peak, are combined (8 ml).

Chromatography on Hydroxylapatite. The solution containing $\beta 1H$ ac-

[35] B. F. Tack and J. W. Prahl, *Biochemistry* **15**, 4513 (1976).
[36] J. W. Prahl, unpublished results (1979).
[37] R. A. Harrison and P. J. Lachmann, *Mol. Immunol.* **16**, 767 (1979).

tivity is dialyzed overnight against 1 liter of phosphate buffer, 4 mS, at pH 7.9, and applied to a column (1.5 × 11 cm) of hydroxylapatite equilibrated with the same buffer. The column is washed with 50 ml of buffer and eluted with a linear phosphate gradient to 18 mS, of total volume 400 ml. β1H activity elutes at a conductivity of 8–9 mS and these fractions are pooled and concentrated in a dialysis bag against dry Sephadex G-200.

Yield and Purity. The yield of β1H is approximately 10% with a purification of 100-fold. A single band is observed on alkaline-gel electrophoresis of the preparation.[12]

Properties of β1H

Chemical Composition

The molecular weight of β1H is 150K determined by equilibrium ultracentrifugation and by polyacrylamide-gel electrophoresis in the presence of sodium dodecyl sulfate.[11] The sedimentation coefficients determined by analytical and sucrose-density gradient ultracentrifugation, 5.6 and 6.4 respectively, are consistent with MW 150K assuming a partial specific volume of 0.733.[11] In regard to gel filtration of β1H, it is noteworthy that an apparent MW 300K rather than 150K is observed on Sephadex G-200 in isotonic buffer and has been ascribed to asymmetry of the molecule.[11]

β1H is composed of a single polypeptide chain as evidenced by electrophoresis in the presence and absence of reducing agents and contains carbohydrate detected by the periodic acid-Schiff stain.[11] Monospecific antisera have been raised,[11,12] but as with C3bINA, little structural work has been done on β1H.

Stability

β1H is less heat stable than C3bINA, with 30% loss of activity after 30 min at 56°C. DFP, soybean-trypsin inhibitor, N-ethylmaleimide, and I_2 have no effect on β1H.[11]

Interaction between β1H and C3b Bound to Activating and Nonactivating Surfaces

β1H serves a multifunctional role in the regulation of the alternative pathway of complement, by dissociating the alternative-pathway convertase in addition to being required for the inactivation of C3b by C3bINA.[11,12,38,39] A high affinity is observed between β1H and C3b bound

[38] D. T. Fearon and K. F. Austen, *J. Exp. Med.* **146**, 22 (1977).
[39] M. K. Pangburn and H. J. Müller-Eberhard, *Proc. Natl. Acad. Sci. U.S.A.* **75**, 2416 (1978).

to nonactivating surfaces, resulting in ready inactivation of C3b by C3bINA and β1H. Conversely, the affinity between β1H and C3b bound to activating surfaces is decreased, and there is a corresponding decrease in the rate of C3b inactivation, allowing activation of the alternative pathway to proceed via generation of the alternative-pathway convertase.[40] Assessing the degree of interaction between β1H and membrane-bound C3b has served as a tool in studying specific natural and artificial constituents of activating and nonactivating surfaces, including sialic acid,[41] N-sulfated polysaccharides,[42] a human erythrocyte membrane protein,[34] and a bacterial lipopolysaccharide.[43]

[40] R. D. Schreiber, M. K. Pangburn, P. H. Lesavre, and H. J. Müller-Eberhard, *Proc. Natl. Acad. Sci. U.S.A.* **75**, 3948 (1978).
[41] M. D. Kazatchkine, D. T. Fearon, and K. F. Austen, *J. Immunol.* **122**, 75 (1979).
[42] M. D. Kazatchine, D. T. Fearon, J. E. Silbert, and K. F. Austen, *J. Exp. Med.* **150**, 1202 (1979).
[43] M. K. Pangburn, D. C. Morrison, R. D. Schreiber, and H. J. Müller-Eberhard, *J. Immunol.* **124**, 977 (1980).

[10] Human C4-Binding Protein (C4-bp)

By VICTOR NUSSENZWEIG *and* RICHARD MELTON

Introduction

C4-bp is a glycoprotein with a specific binding affinity for C4b, the major fragment resulting from the $C\overline{1}$-mediated cleavage of the fourth component of the complement system.[1,2] The main function of C4-bp is to inhibit the activity of the C3 convertase of the classical pathway, $C\overline{4b,2a}$.[3] By physicochemical, immunological, and functional criteria, C4-bp is identical to the C3b,C4b-inactivator cofactor described by Shiraishi and Stroud[4] and Nagasawa and Stroud.[5] C4-bp is present in human plasma in two forms, differing slightly in apparent molecular weight (590,000 and 540,000) and net charge, but whose functional properties are indistinguishable.[2] The nature of this polymorphism is unknown. In most individuals, C4-bp of higher molecular weight predominates (unpublished obser-

[1] A. Ferreira, M. Takahashi, and V. Nussenzweig, *J. Exp. Med.* **146**, 1001 (1977).
[2] J. Scharfstein, A. Ferreira, I. Gigli, and V. Nussenzweig, *J. Exp. Med.* **148**, 207 (1978).
[3] H. J. Müller-Eberhard, M. J. Polley, and M. A. Calcott, *J. Exp. Med.* **125**, 359 (1967).
[4] S. Shiraishi and R. M. Stroud, *Immunochemistry* **12**, 935 (1975).
[5] S. Nagasawa and R. M. Stroud, *Immunochemistry* **14**, 749 (1977).

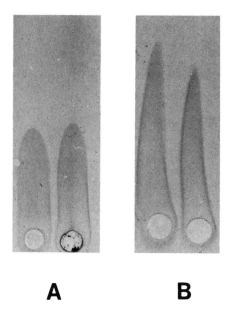

A   **B**

Fig. 1. Different shapes of C4-bp/anti-C4-bp rockets in the absence (A) and presence of saturating amounts of C4b (B).

vations). Both forms of C4-bp stain heavily with the periodic acid-Schiff reagent, suggesting high carbohydrate content.

It is very difficult to isolate C4-bp from a serum sample in which C4 has been activated and C4b/C4-bp complexes have been formed. For this reason, every precaution should be taken to prevent proteolysis of C4 during purification of C4-bp. Moreover, although the initial step of the fractionation (euglobulin precipitation) effectively separates most C4 from C4-bp, it is desirable during subsequent steps to identify and discard fractions that contain C4b.

Assay

Fractions should be monitored for the presence of C4-bp and C4. This can be done conveniently by rocket immunoelectrophoresis (RIE). This technique is rapid, quantitative, and, in addition, permits the identification of C4-bp complexed to degradation products of C4. The rockets formed between antibodies to C4-bp with free C4-bp have rounded tips, while the rockets formed between the same antibodies with complexed C4-bp have very sharp tips (Fig. 1). This morphological variation probably reflects changes in available antigenic determinants of C4-bp on binding of split products of C4.[2] RIE is performed by standard techniques in water-cooled electrophoretic cells.[6] We use 1.2% agarose in barbital buffer (0.023 M sodium barbital–0.0037 M barbituric acid–0.002 M EDTA, pH 8.6). The agarose gel, containing either antibody to C4 or to C4-bp, is cast on 5 × 20-cm gel bond films (Marine Colloids Division, FMC Corporation, Rockland, Maine). About 40 holes per film are punched into the agar and 5-μl samples of antigen are added. As standards, each film contains dilutions of the purified proteins (C4 and C4-bp). Dilutions of fresh serum can be used as standards provided that they contain 2 mM EDTA to prevent complement activation upon contact with the agarose. Electrophoresis is performed for 16 hr, at 3 V/cm. Antisera to C4 are available commercially. Samples of monospecific antisera to C4-bp may be provided by the authors upon request.

Purification can also be monitored by assaying for the ability of C4-bp to promote the decay of the C3 convertase of the classical pathway. The red cell intermediate EAC$\overline{1,4b,}$C2a is prepared as described elsewhere,[7] using 300 hemolytic units of C$\overline{1}$ and C4 and 1.5 hemolytic units of ^{oxy}C2. 100 μl of the fractions to be tested for the presence of C4-bp are incubated with 100 μl of EAC$\overline{1,4b,}$2a (1 × 10^8/ml) for 30 min at 30°C. Then 1.3 ml of guinea pig serum diluted 1:50 with a veronal buffer containing EDTA (EDTA–GVB, Ref. 7) is added and incubated for an additional 60 min at 37°C. The residual C3-convertase activity is calculated from the degree of hemolysis and compared to controls consisting of EAC$\overline{1,4b,}$2a cells incubated with buffer alone. We find that the inhibitory effect of C4-bp can be detected by this procedure at concentrations below 1 μg/ml (K. Iida and V. Nussenzweig, unpublished observations).

Methods of Isolation

Euglobulin Precipitation

Eight hundred fifty milliliters of fresh human serum are centrifuged at 4200 g for 30 min at 4°C. The supernatant is collected in a 6-liter beaker surrounded by ice and kept under the chemical hood. Add 150 ml of 100 mM EDTA, pH 7.1. Stir until the temperature drops to 4°C and slowly add 2 M diisopropyl fluorophosphate (DFP) to a final concentration of 4 mM. Continue stirring and slowly add 4 liters of cold wa   containing 1 mM EDTA, pH 7.1. Let stand at 4°C for 4 hr. Mix, and centɪifuge at 4200 g for 30 min at 4°C. Decant the supernatant, which contains most of the C4, and dry the interior of the bottles with paper towels. Dissolve the pellets with

---

[6] B. Weeke, in "A Manual of Quantitative Immunoelectrophoresis" (N. H. Axelsen, J. Krøll, and B. Weeke, eds.), p. 35. Universitetsforlaget, Oslo, 1973.

[7] L. G. Hoffman and M. M. Mayer, *Methods Immunol. Immunochem.* **4**, 166 (1977).

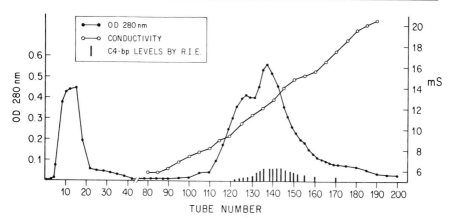

FIG. 2. Elution profile of the euglobulin fraction of human serum upon chromatography on DEAE–Sephadex. After washing the column with 0.01 M Tris-HCl buffer (pH 8.6) containing 0.075 M NaCl–0.005 M ε-aminocaproic acid–0.005 M EDTA (conductivity 5.8 mS at 0°C), a salt gradient is started. C4-bp is eluted at conductivities between 9 and 15 mS.

150 ml of 0.01 M Tris-HCl (pH 8.6), containing 0.075 M NaCl–0.005 M ε-aminocaproic acid–0.005 M EDTA, conductivity 5.8 mS at 0°C (buffer No. 1). Centrifuge at 4200 g for 15 min. Decant and add 0.15 ml of 2 M DFP to the supernatant, which is to be applied to the DEAE-A50 column.

Anion-Exchange Chromatography

Twenty grams of DEAE–Sephadex A-50 are swollen in water, and the fines are removed by sedimentation in a 2-liter cylinder. The gel is poured in a 2-liter glass-fritted funnel, and washed extensively with buffer No. 1. The gel is degassed and packed in a 2.5 × 40-cm column at 4°C, and washed with buffer No. 1 until equilibration. The euglobulin fraction is loaded, and the column washed with the same buffer until the optical density at 280 nm of the effluent is below 0.050. A linear salt gradient is then applied and 5-ml fractions collected at a rate of 40 ml/hr. The mixing chamber contains 500 ml of buffer No. 1, and the second flask contains an equal volume of the same buffer and solid NaCl (13.4 g/500 ml) to a final conductivity of 28 mS. Samples from selected tubes are assayed for the presence of C4-bp and C4 by rocket immunoelectrophoresis and also subjected to SDS–polyacrylamide gel electrophoresis performed under nonreducing conditions, as described by Laemmli.[8] The stacking and separating gels used are 3 and 5%, respectively. Results are shown in Figs. 2 and 3. C4-bp appears about 250–300 ml into the gradient at a conductivity of 9 mS. The low-molecular-weight form of C4-bp predomi-

[8] J. K. Lammeli, *Nature* (*London*) 227, 680 (1970).

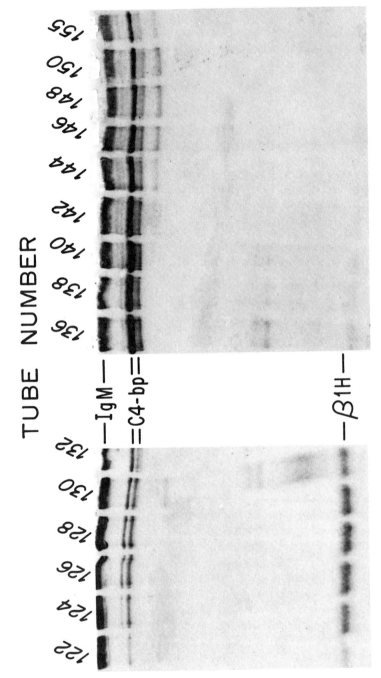

Fig. 3. SDS–PAGE under nonreducing conditions of selected fractions from the anion-exchange chromatography shown in Fig. 2. The stacking and running gels were 3 and 5%. C4-bp appears as 2 bands. Fractions eluted with higher salt concentrations contain increasing proportions of the C4-bp bands with higher apparent molecular weight. Two major contaminants are β 1H and IgM.

nates initially, but at higher conductivities the high-molecular-weight species is more abundant. Two of the major contaminants have been identified as IgM and $\beta 1H$. In this run, we could not detect the presence of C4b/C4-bp complexes. In other runs, these complexes were eluted at conductivities between 15 and 20 mS.

Cation-Exchange Chromatography

The contents of tubes 130–150 (Fig. 2) are pooled, the volume measured, and 2 M DFP added to a final concentration of 1 mM. To the pool, one-third volume of 33% polyethylene glycol dissolved in buffer No. 1 is added slowly, while stirring. After 5 hr at 4°C, the mixture is centrifuged at 4200 g for 15 min. The supernatant is discarded and the pellet resuspended and stirred for 30 min in 40 ml of buffer No. 2 (0.02 M sodium phosphate buffer, pH 7.2, conductivity 1.8 mS at 0°C). The suspension is centrifuged at 4200 g for 15 min. The supernatant contains most of the contaminant $\beta 1H$. The pellet is dissolved in 20 ml of buffer No. 2 to which solid NaCl (10.4 g/liter) is added, bringing the conductivity to 12 mS at 0°C. This mixture is stirred 10 min in the cold and centrifuged at 12,000 g for 15 min.

The supernatant is decanted, and its conductivity lowered to 8.0 with cold distilled water. It is immediately loaded onto a 2 × 15-cm column containing Bio-Rex 70 equilibrated with buffer No. 2 with added NaCl to a conductivity of 8.0 mS. The column is washed in the same buffer at a flow rate of 10 ml/hr. After a breakthrough peak that mainly or exclusively contains IgM, a gradient is started. The mixing chamber contains 250 ml of the starting buffer (conductivity 8 mS) and the other chamber, 250 ml of the same buffer with NaCl added, to a conductivity of 25 mS. C4-bp is eluted in fractions with conductivities between 8.8 and 14 mS (Fig. 4). The positive fractions are concentrated to a volume of about 5 ml in an Amicon chamber with an ultrafiltration XM-100 membrane. The major contaminant at this stage is IgM, but small amounts of $\beta 1H$ are also present (Fig. 5).

Molecular-Sieve Chromatography

This step is required to eliminate minor contaminants of low molecular weight. A column of 2.5 × 80 cm is filled with fined and degassed Sepharose-6B according to the instructions of the manufacturer. 250 mg of sucrose is added to 5 ml of the pooled fraction obtained from the Bio Rex 70 column and this is loaded onto the Sepharose-6B column. Elution proceeds with 0.02 M sodium phosphate buffer (pH 7.2) containing 0.15 M NaCl, with a flow rate of 15 ml/hr. The major OD peak contains C4-bp,

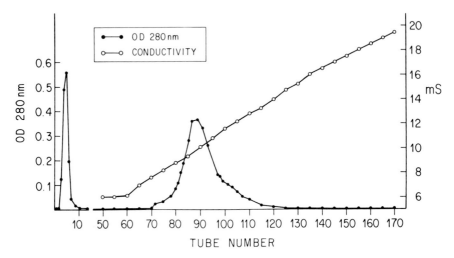

F<small>IG</small>. 4. Elution profile on Bio-Rex 70 of a serum fraction obtained after anion-exchange chromatography. The Bio-Rex 70 column had been equilibrated with 0.02 M sodium phosphate buffer (pH 7.2) with NaCl added, to a conductivity of 8 mS. After the breakthrough peak, and additional washing with the same buffer, a salt gradient is started. C4-bp is eluted at conductivities between 9 and 14 mS.

contaminating IgM, and small amounts of β1H. The pooled material is concentrated five times in an Amicon chamber as described above.

Removal of Contaminants by Affinity Chromatography in Solid Immunoabsorbents (Anti-IgM and Anti-β1H)

An antiserum to IgM can be prepared in rabbits using as antigen fractions 3–6 from the Bio-Rex column (Figs. 4 and 5). The IgG fraction of this antiserum is isolated by anion-exchange chromatography in DEAE–cellulose by standard procedures and immobilized on CnBr–Sepharose (Pharmacia Fine Chemicals, Piscataway, New Jersey), according to the instructions of the manufacturer. IgM is removed from the purified C4-bp by passage through a 1×5-cm column containing 1 g of Sepharose–anti-IgM in phosphate-buffered saline with a flow rate of 10 ml/hr. The eluted fraction is subsequently filtered through a Sepharose–anti-β1H column. The antiserum to β1H can be prepared by injection into rabbits of purified protein obtained from β1H-rich fractions isolated during the cation-exchange chromatography step of this procedure. The final recovery of C4-bp is about 20 mg, assuming that C4-bp at a concentration of 1 mg/ml has an absorption coefficient of 1 at 280 nm. Considering that the serum concentration of C4-bp is between 150 and 250 μg/ml, the final recoveries vary between 10 and 20%.

TUBE NUMBER

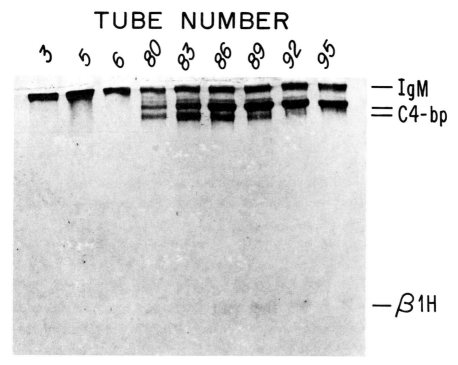

FIG. 5. SDS–PAGE under nonreducing conditions of selected fractions from the Bio-Rex 70 column (Fig. 4). The stacking and running gels were 3 and 5%. C4-bp appears on 2 bands, and the major contaminant is IgM. The breakthrough peak from the Bio-Rex 70 column (tubes 3–6) contains pure IgM.

Properties of Purified C4-bp

Purity. C4-bp is homogeneous in 5% SDS–PAGE, and free of IgM, IgG, β1H, or C4 antigenic determinants when analyzed by RIE with monospecific antisera.

Stability. C4-bp is stable for at least 2 years when kept frozen at $-70°$C. It loses activity when repeatedly frozen and thawed.

Structural Properties. C4-bp consists of several disulfide-bonded polypeptide chains with apparent molecular weight on SDS–PAGE of 70,000. Under nonreducing conditions the molecular weight of the two forms of the protein have been estimated to be 540,000 and 590,000. The amino acid composition is shown in the table. The N-terminal amino acid of both forms of C4-bp is glycine (E. Franklin, unpublished observation). The sedimentation coefficient estimated by sucrose-gradient centrifugation is 11 A.

AMINO ACID COMPOSITION OF C4-BINDING
PROTEIN[a]

Amino acid	Residues/1000 residues
Cys	36
Asp	83
Thr	56
Ser	94
Glu	118
Pro	72
Gly	97
Ala	38
Val	84
Met	30
Ile	39
Leu	60
Tyr	44
Phe	29
His	18
Lys	57
Arg	40

[a] Performed by Dr. E. Franklin.

Electrophoretic Mobility. C4-bp migrates as a slow β-globulin at pH 8.6 in buffers containing a chelating agent (EDTA). However, when the electrophoresis is performed in the presence of Ca^{2+} ions, C4-bp migrates as a slow γ-globulin. The mobility of C4-bp also changes markedly when complexed to C4b. The C4b/C4-bp complexes move much faster toward the anode than either protein. Furthermore, complex formation also increases the height and shape of the rockets formed after RIE.[2] The differences in properties between free and bound C4-bp were used to evaluate the stoichiometry of the reaction with C4b. It has been calculated that C4-bp is multivalent and that saturation of its combining sites is reached at ratios of 4–5 molecules of C4b per molecule of C4-bp.[2]

These observations help in part to clarify the conflicting results in the literature with regard to the electrophoretic patterns of C4 in human serum when precautions are not taken to prevent complement activation during runs in agarose. If C4 is activated and fragmented, C4b/C4-bp complexes are formed, and 3 bands containing C4 may be observed: native C4, free C4b, and C4b/C4-bp complexes.

Functional Properties. C4-bp binds specifically to C4b. The formation of C4b/C4-bp complexes does not require cofactors or the presence of divalent cations, and can be directly demonstrated by ultracentrifugation in sucrose gradients. Through its interaction with C4b, C4-bp controls the assembly and function of the classical pathway C3 convertase (C4b,2a). This enzyme, which mediates the cleavage of C3 into the C3a and C3b peptides, is labile at 37°C because of the decay of C2a. However, the enzymatic activity of C3 convertase can be fully restored following uptake by C4b of fresh C2 and its cleavage into C2a by C1s.

C4-bp can inhibit C3 convertase in two ways: alone or in conjunction with a serum enzyme, C3b/C4b inactivator. By itself, in a dose-dependent fashion, C4-bp prevents the assembly of C4b,2a, and promotes its decay. Both effects are most likely caused by a competition between C2a and C4-bp for binding sites on C4b.[9] In addition, C4-bp together with C3b/C4b inactivator, cleave the α'-chain of C4b into 3 fragments[10] in a two-step reaction,[11] and this leads to an irreversible loss of the ability of C4b to support the C3-convertase activity.

As a result of these activities, C4-bp considerably diminishes the consumption of C3 in serum following C1 activation and assembly of C4b,2a.[9]

An additional activity of C4-bp has been reported.[12] In a manner similar to β1H, it can serve as cofactor for the cleavage of *fluid-phase* C3b by C3b/C4b inactivator. However, important quantitative differences in this regard between the activities of β1H and C4-bp have been observed; that is, the cofactor activity of C4-bp is $\simeq$20 times smaller than β1H on a weight basis. When C3b is *cell bound,* its activity is not influenced by the presence of C4-bp even at very high concentrations. Considering that the serum concentration of β1H is higher than that of C4-bp, it is very likely that under physiological conditions the functions of C3b and C4b are specifically controlled by β1H and C4-bp, respectively.

Acknowledgment

This work was supported by NIH Grants #AI-08499 and CA-16247.

[9] I. Gigli, T. Fujita, and V. Nussenzweig, *Proc. Natl. Acad. Sci. U. S. A.* **76,** 6596 (1971).
[10] T. Fujita, I. Gigli, and V. Nussenzweig, *J. Exp. Med.* **148,** 1044 (1978).
[11] S. Nagasawa, C. Ishihara, and R. M. Stroud, *J. Immunol.* **125,** 578 (1980).
[12] T. Fujita and V. Nussenzweig, *J. Exp. Med.* **150,** 267 (1979).

[11] Preparation of Human Factor $\overline{D}$ of the Alternative Pathway of Complement

By K. B. M. Reid, D. M. A. Johnson, J. Gagnon, and R. Prohaska

Introduction

Factor $\overline{D}$, a component of the alternative pathway of human complement, is a serine esterase present in low concentration in the blood. It is not known whether factor $\overline{D}$ is synthesized first as a zymogen or in its active form. There is one report[1] that a small proportion of the factor $\overline{D}$ present in plasma is in a proenzyme form; however, recent studies[2] indicate that most, if not all, of the enzyme is present in plasma in its activated form. Factor B is another serine-esterase component of the alternative pathway that has clearly been shown to be present in plasma in its xymogen form. Proenzyme factor B forms a reversible complex with the C3b fragment of C3 in the presence of Mg^{2+}, and when it is in this complex it is readily split by factor $\overline{D}$ at a single Arg-Lys bond, thus yielding the Ba and Bb fragments.[3-5] This allows the formation of the enzymatically active $C\overline{3b,Bb}$ complex which can, *via* the active site in Bb, split C3, the most abundant complement component, to yield more C3b and C3a.

$$B + C3b \xrightleftharpoons{Mg^{2+}} B;C3b$$

$$B;C3b \xrightarrow{\overline{D}} Ba + C\overline{3b,Bb}$$

$$C3 \xrightarrow{C\overline{3b,Bb}} C3a + C3b$$

Factor $\overline{D}$ does not appear to be incorporated into the $C\overline{3b,Bb}$ complex,[2] unlike the situation found with most of the other complex proteases of the complement system where, usually, quite a strong interaction is observed between the activating component, or complex of components, and the next component in the pathway. Instead, factor $\overline{D}$ is considered to play a purely catalytic role, which would be consistent with its low serum concentration.

[1] D. T. Fearon, K. F. Austen, and S. Ruddy, *J. Exp. Med.* **139**, 355 (1974).
[2] P. H. Lesavre and H. J. Müller-Eberhard, *J. Exp. Med.* **148**, 1498 (1978).
[3] O. Götze, *Cold Spring Harbor Conf. Cell Proliferation* **2**, 255 (1975).
[4] P. H. Lesavre, T. E. Hugli, A. F. Esser, and H. J. Müller-Eberhard, *J. Immunol.* **123**, 529 (1979).
[5] M. A. Kerr, *Biochem. J.* **183**, 615 (1979).

Serum Concentration

The concentration of Factor $\overline{D}$ in human plasma, or serum, is in the range of 1.0–1.5 μg/ml as estimated by the recovery of hemolytic activity during the purification procedure.[4,6,7] A slightly higher value of 2 μg/ml has been reported by use of a radioimmunoassay procedure.[3]

Assay

The use of synthetic substrates to monitor the enzymatic activity of factor $\overline{D}$ is not possible since, to date, no synthetic substrate has been found that is readily hydrolyzed by highly purified factor $\overline{D}$ preparations.[2,6]

The hemolytic activity of factor $\overline{D}$ can be conveniently measured by the method of Lesavre et al.,[4] which involves the alternative pathway-dependent lysis of rabbit erythrocytes in solution, or by the method of Martin et al.,[8] which involves alternative pathway-dependent lysis of guinea pig erythrocytes on an agarose plate.

In the method of Lesavre et al.[4] a reagent depleted of factor $\overline{D}$ activity was obtained by passing 30 ml of fresh human serum, containing 0.01 M EDTA, through a column (2 × 20 cm) of Bio-Rex 70 equilibrated with isotonic veronal-buffered saline at 40, pH 7.4 (8.5 g sodium chloride + 0.375 g sodium barbital + 0.579 g barbitol per liter). The breakthrough fraction is collected and dialyzed against isotonic veronal-buffered saline, pH 7.4. In the assay, 25 μl of the factor $\overline{D}$-deficient reagent, made 5 mM with respect to $MgCl_2$ and EGTA, is mixed with 10^7 rabbit erythrocytes and 10 μl of the solution being tested for factor $\overline{D}$ hemolytic activity. Isotonic veronal-buffered saline (pH 7.4) containing 12 mM $MgCl_2$ and 0.1% gelatin, is used as a diluent and the final reaction volume is 150 μl. After 60 min at 37°C the reaction is stopped by the addition of 1.0 ml of isotonic veronal-buffered saline containing 10 mM EDTA and 0.1% gelatin. The factor $\overline{D}$ hemolytic activity present in the test sample is determined by the extent of hemolysis at 412 nm. In this assay approximately 50% of the cells are lysed on the addition of 3 ng of purified factor $\overline{D}$.

In the hemolytic diffusion-plate assay of Martin et al.,[8] guinea pig erythrocytes are incorporated into an agarose plate in the presence of EGTA, Mg^{2+}, and human serum specifically depleted of factor $\overline{D}$ by gel filtration on Sephadex G-75. On addition of samples containing factor $\overline{D}$ to

[6] A. E. Davis, C. Zalut, F. S. Rosen, and C. A. Alper, *Biochemistry* **18**, 5082 (1979).

[7] D. M. A. Johnson, J. Gagnon, and K. B. M. Reid, *Biochem. J.* **187**, 863 (1980).

[8] A. Martin, P. J. Lachmann, L. Halbwachs, and M. J. Hobart, *Immunochemistry* **13**, 317 (1976).

wells in the agarose plate, the alternative pathway is reconstituted by diffusion of the factor $\overline{D}$ through the gel, and activation of the alternative pathway takes place, with concomitant lysis of the unsensitized guinea pig cells by the "bystander lysis" mechanism. The hemolytic area is measured and compared with that given by a known amount of factor $\overline{D}$. In this assay readily measurable rings of lysis are obtained with as little as 2 ng of purified factor $\overline{D}$.

Purification Procedure

Outdated human plasma can be used to prepare satisfactory yields of Factor $\overline{D}$ but it is preferable to use fresh or fresh-frozen plasma since material having a similar molecular weight and similar ion-exchange properties may be generated on prolonged storage of plasma or even of the partially purified fractions during the purification procedure. Addition of protease inhibitors in the early stages of the procedure did not appear to increase the recovery of Factor $\overline{D}$. Plasma or serum may be used in the method given here, the yield and final specific hemolytic activity of factor $\overline{D}$ being the same in both cases. When serum was used it was prepared from the plasma by clotting overnight at 4°C on the addition of 1.0 M $CaCl_2$ to give a final concentration of 20 mM. The clot was removed by centrifugation and filtration through muslin.

CM-Sephadex C-50, Sephadex G-75, and concanavalin A–Sepharose were obtained from Pharmacia Fine Chemicals, Uppsala, Sweden. CM-cellulose 32 was obtained from Whatman, Maidstone, Kent, United Kingdom.

Human plasma, or serum (4000 ml), is dialyzed for 20 hr at 4°C against 40 liters of 5 mM EDTA, pH 5.4. (In this laboratory serum from outdated plasma has been used routinely in the preparation of factor $\overline{D}$, therefore the yields and elution profiles from one of these preparations have been taken for this article.) The euglobulin precipitate is removed by centrifugation at 2000 g for 1 hr. The conductivity of the supernatant is adjusted to 12–14 mS using 1 M sodium phosphate buffer, pH 6.0 (approximately 800 ml) and the solution is loaded onto a column (18 × 8 cm) of CM-Sephadex C-50, equilibrated in 0.2 M sodium phosphate buffer, pH 6.0 (27.38 g $NaH_2PO_4 \cdot 2H_2O$ + 3.5 g anhydrous Na_2HPO_4 per liter). A flow rate of 600 ml/hr is used during the 0.2 M sodium phosphate wash and a flow rate of 50 ml/hr is used during the elution of factor $\overline{D}$. The factor $\overline{D}$ hemolytic activity is eluted above 16 mS in two peaks (Fig. 1). The first peak of activity is probably due to enhancement of factor $\overline{D}$ activity by traces of factor B that had been left on the column after the 0.2 M sodium phosphate wash and is eluted immediately prior to the factor $\overline{D}$ with the 0.4 M

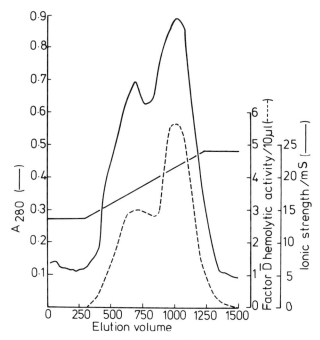

Fig. 1. Ion-exchange chromatography of factor $\overline{D}$ on CM-Sephadex C-50. 4000 ml of the supernatant ($\sim$208,000 A_{280} units) obtained after euglobulin precipitation was adjusted to 12–14 mS with 1 M sodium phosphate buffer (pH 6.0) and then applied to a column (18 × 8 cm) of CM-Sephadex C-50 equilibrated with 0.2 M sodium phosphate buffer, pH 6.0. The column was developed as described in the text. ———, A_{280}; ----, Factor $\overline{D}$ hemolytic activity. Factor $\overline{D}$, along with other proteins including traces of factor B, is eluted at about 16 mS.

buffer. No significant improvement in purification is obtained on eluting the factor $\overline{D}$ with a linear gradient instead of a stepwise wash.

The fractions containing factor $\overline{D}$ hemolytic activity are pooled and $(NH_4)_2SO_4$ is added to 50% saturation (291 g/liter). The suspension is stirred for 2 hr at 4°C, then centrifuged at 10,000 g for 2 hr. More $(NH_4)_2SO_4$ is added to the supernatant (125 g/liter) to give 70% saturation and the suspension is stirred for 2 hr at room temperature. The precipitate is dissolved in 40 ml of 0.1 M Tris-HCl–0.2 M NaCl–2 mM EDTA, pH 8.0, at 4°C and, after centrifugation, applied to a column (5 × 100 cm) of Sephadex G-75 equilibrated with the same buffer. The column is run at a flow rate of 30 ml/hr and Factor $\overline{D}$ hemolytic activity is eluted at a position on the column corresponding to an apparent molecular weight of 24,000 and thus is separated from a large number of proteins that have apparent molecular weights >30,000 (Fig. 2). The fractions containing factor $\overline{D}$ hemolytic activity are pooled and concentrated to 10 ml using an Amicon

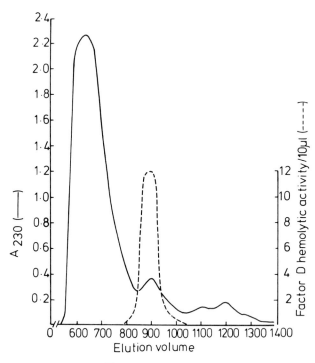

FIG. 2. Gel filtration of factor $\overline{D}$ on Sephadex G-75. The precipitate (~ 146 mg protein) obtained after $(NH_4)_2SO_4$ precipitation was dissolved in 40 ml of 0.1 M Tris-HCl–0.2 M NaCl–2 mM EDTA (pH 8.0) and, after centrifugation, applied to a column (5 × 100 cm) of Sephadex G-75 equilibrated with the same buffer. The column was run at 30 ml/hr. ——, A_{230}; ----, Factor $\overline{D}$ hemolytic activity.

Diaflo ultrafiltration cell fitted with a UM-2 membrane. Following the gel filtration on Sephadex G-75 the preparation is functionally pure with respect to factor $\overline{D}$ hemolytic activity but requires further ion-exchange chromatography and affinity chromatography for final purification.

Because of the small amount of protein (usually 15–20 mg) in the factor $\overline{D}$ pool from the Sephadex G-75 column it has been found convenient to combine the Sephadex G-75 factor $\overline{D}$ pools from two preparations (i.e., from 8000 ml of serum) before carrying out the final stages of purification. The partially purified factor $\overline{D}$ from Sephadex G-75 can be stored at $-70°$C until required. In the description of the purification of factor $\overline{D}$ given here, the material from the Sephadex G-75 steps from two different preparations is pooled and dialyzed against 2 × 500 ml of 0.23 M acetic acid–NaOH buffer, pH 5.2 (55 ml of 4 M NaOH adjusted to pH 5.2 with acetic acid and made up to 1000 ml) and applied to a column (1.5 × 20 cm)

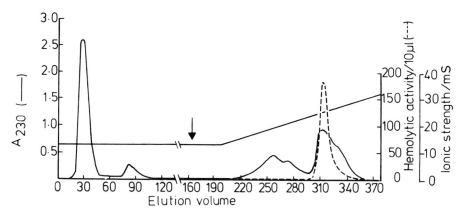

FIG. 3. Ion-Exchange chromatography of factor $\overline{D}$ on CM-cellulose 32. The partially purified factor $\overline{D}$ (~30 mg protein) from Sephadex G-75 was dialyzed against 0.23 M acetic acid–NaOH buffer (pH 5.2) and loaded onto a column (1.5 × 20 cm) of CM-cellulose 32 equilibrated in the same buffer. The column was developed as described in the text. Factor $\overline{D}$ hemolytic activity eluted between 22 and 27 mS. ——, A_{230}; ----, hemolytic activity. The arrow marks the start of the gradient.

of CM-cellulose 32 that is equilibrated with the same buffer. The column is washed with the starting buffer at 30 ml/hr until the A_{230} of the eluate is zero, and then a linear gradient is developed using 120 ml of 0.23 M acetic acid–NaOH buffer (pH 5.2) and 120 ml of the same buffer containing 0.3 M NaCl. The factor $\overline{D}$ hemolytic acitivity is eluted between 22 and 27 mS (Fig. 3). Another protein, which is completely removed by use of concanavalin A–Sepharose, is eluted between 25 and 28 mS in the gradient on CM-cellulose 32 (Fig. 3). This protein cannot be distinguished from factor $\overline{D}$ in nonreducing conditions on SDS–polyacrylamide gels, but in reducing conditions it has a higher apparent molecular weight compared to factor $\overline{D}$. No factor $\overline{D}$ hemolytic activity is associated with this contaminating protein, the concentration of which can vary considerably from preparation to preparation, sometimes comprising approximately 50% of the material in the factor $\overline{D}$ pool from CM-cellulose (Fig. 3). Thus the CM-cellulose 32 pool is routinely applied to a column of concanavalin A–Sepharose.

The fractions from the CM-cellulose 32 column, which contain factor $\overline{D}$ hemolytic activity, are pooled and concentrated by ultrafiltration using a UM-2 membrane to 10 ml, then dialyzed against 0.01 M Tris-HCl buffer (pH 8.0) containing 0.15 M NaCl, 1 mM MgCl$_2$ and 1 mM CaCl$_2$. After dialysis the sample (10 ml) is applied to a column (1 × 12 cm) of concanavalin A–Sepharose, equilibrated with the same buffer, and eluted at a flow rate of 10 ml/hr. The factor $\overline{D}$ hemolytic activity is not retarded on the

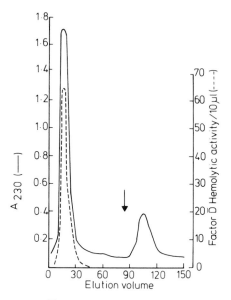

FIG. 4. Purification of factor $\overline{D}$ on concanavalin A–Sepharose. Partially purified factor $\overline{D}$ from the CM-cellulose 32 column ($\sim$3.4 mg protein) in 10 ml of 0.01 M (Tris-HCl/0.15 M NaCl, 1 mM MgCl$_2$ (pH 8.0) was applied to a column (1 × 12 cm) of concanavalin A–Sepharose equilibrated with the same buffer and eluted at a flow rate of 10 ml/hr. ——, A_{230}; - - - -, factor $\overline{D}$ hemolytic activity. The arrow marks the point at which elution with 2.5% (w/v) 1-O-methyl-α-D-glucopyranoside was begun.

TABLE I

PURIFICATION OF FACTOR $\overline{D}$ FROM 4 LITERS OF HUMAN SERUM

Fraction	Total protein (A_{280})	Total hemolytic activity × 10^{-4} (units)	Specific activity (units/A_{280})	Purification factor	Yield
Serum	221,000	38	1.73	1	100
Pseudoglobulin	209,000	35	1.69	0.98	92
CM-Sephadex C-50	422	32	758	440	84
50–70% (NH$_4$)$_2$SO$_4$ precipitate	144	26	1790	1040	68
Sephadex G-75	17	17.3	10,200	5890	45
CM-cellulose-32	3.2	9.2	29,600	17,100	24
Concanavalin A– Sepharose	1.8[a]	8.6	49,200	28,500	22

[a] The final yield of protein as estimated by amino acid analysis was 0.9 mg. This gives a final specific activity (units/mg of protein) for factor $\overline{D}$ of 103,000 and a purification factor of $\sim$60,000.

column but the contaminant, which has the same apparent molecular weight as factor $\overline{D}$ in nonreducing conditions on SDS–polyacrylamide gels, is bound to the column and can be eluted with 2.5% (w/v) 1-O-methyl-α-D-glucopyranoside dissolved in the same buffer used to equilibrate the column.

A summary of the purification of factor $\overline{D}$ is given in Table I. Similar purification procedures for factor $\overline{D}$ have also been reported by Lesavre *et al.*[4] and Davis *et al.*[6] Both these procedures use plasma as the starting material and Lesavre *et al.*[4] have reported that the Factor $\overline{D}$ present in 5 liters of plasma can be bound by a 1-liter column of Bio-Rex 70, which is useful if large quantities of plasma are to be processed. Davis *et al.*[6] reported that factor $\overline{D}$ binds to heparin–Sepharose and can be eluted with a linear salt gradient, which has proved useful in separating factor $\overline{D}$ from low-molecular-weight contaminants.

Yield

This purification procedure gives a yield of approximately 0.9 mg of factor $\overline{D}$ per 4000 ml of serum and a recovery of about 20% of the initial factor $\overline{D}$ hemolytic activity (Table I). Similar yields of factor $\overline{D}$ have been obtained by two other procedures.[4,6]

Stability

Human Factor $\overline{D}$ in 0.1 M Tris-HCl–0.2 M NaCl–2 mM EDTA (pH 8.0) containing 0.02% (w/v) NaN$_3$ is stable at 4°C for several months, with only a slight (approximately 10%) decrease in hemolytic activity. However, on storage at this temperature, protein aggregation has been observed. Storage at -70°C prevents this aggregation but freezing and thawing was found to cause a loss of approximately 30% of the original hemolytic activity.

Physical Properties and Composition

Human factor $\overline{D}$ is composed of a single polypeptide chain of apparent molecular weight of 24,000.[2,6,7,9] Its amino acid composition is unremarkable and the values published by two different laboratories[6,7] agree very closely but are significantly different from the amino acid composition reported in an earlier publication.[9] No estimate of the carbohydrate content of factor $\overline{D}$ has been made, but Lesavre *et al.*[4] And Davis *et al.*[6] found that their factor $\overline{D}$ preparations stained with Schiff stain after periodate

[9] J. E. Volanakis, R. E. Schrohenloher, and R. M. Stroud, *J. Immunol.* 119, 337 (1977).

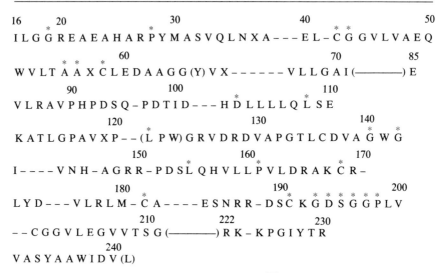

16 20 30 40 50
I L G G R E A E A H A R P Y M A S V Q L N X A - - - E L - C G G V L V A E Q
       60   70   85
W V L T A A X C L E D A A G G (Y) V X - - - - - - V L L G A I (————) E
    90   100   110
V L R A V P H P D S Q - P D T I D - - - H D L L L L Q L S E
        120   130   140
K A T L G P A V X P - - (L P W) G R V D R D V A P G T L C D V A G W G
       150   160   170
I - - - - V N H - A G R R - P D S L Q H V L L P V L D R A K C R -
       180   190   200
L Y D - - - V L R L M - C A - - - - E S N R R - D S C K G D S G G P L V
       210   222   230
- - C G G V L E G V V T S G (————) R K - K P G I Y T R
       240
V A S Y A A W I D V (L)

FIG. 5. Partial amino acid sequence of human Factor D̄. The residue numbering shown is that of the cow chymotrypsin A sequence to allow comparison of the factor D̄ sequence with other serine esterases. Residues are given in the single-letter code: A, Ala; B, Axs; C, Cys; D, Asp; E, Glu; F, Phe; G, Gly; H, His; I, Ile; K, Lys; L, Leu; M, Met; N, Asn; P, Pro; Q, Gln; R, Arg; S, Ser; T, Thr; V, Val; W, Trp; X, unknown; Y, Tyr. Brackets denote regions of unknown or uncertain sequence data. An asterisk above a residue denotes that it is highly conserved in other serine esterases. Gaps left to allow maximum homology of factor D̄ with chymotrypsin A are indicated by —. The data shown are taken from D. M. A. Johnson, J. Gagnon, and K. B. M. Reid, *Biochem. J.* **187**, 863 (1980), and D. M. A. Johnson, Ph.D. Thesis, University of Oxford (1980).

oxidation, indicating the presence of a glycoprotein. In the purification procedure used by Johnson *et al.*[7] it was observed that factor D̄ was not retained by a concanavalin A–Sepharose column and that 25 μg of factor D̄ prepared using this column did not show any significant stain with periodic acid-Schiff's reagent, under conditions where 0.3 μg of carbohydrate was readily detectable.[10]

Isoelectric focusing of purified factor D̄ gave two bands with isoelectric points at pH 7.0 and pH 6.6; both bands were hemolytically active and both were detected in normal human serum.[6] Factor D̄ appears to bind calcium ions, as judged by its electrophoretic behavior on agarose-gel electrophoresis in the presence of 1.8 mM calcium or in the presence of 5 mM EDTA. In the presence of calcium the factor D̄ had a γ-electrophoretic mobility, while in the presence of EDTA it had a β-mobility.[6] The factor D̄ in normal serum has the same electrophoretic mobility as isolated factor D̄ in the presence of calcium.[6] However, in the

[10] D. M. A. Johnson, Ph.D. Thesis, University of Oxford (1980).

presence of EDTA the factor $\overline{\text{D}}$ activity in normal serum showed an $\alpha_1-\alpha_2$ mobility, which suggested that it may be bound to another serum protein under these conditions.[6]

Factor $\overline{\text{D}}$ is a typical serine esterase, as is shown by its irreversible inhibition by DFP,[1,2] labeling with [1,3-^{14}C]DFP,[7] and the determination of the amino acid sequence around the serine residue that reacts with DFP.[7] Factor $\overline{\text{D}}$ splits factor B at a single Arg-Lys bond, thus indicating a trypsin-like specificity.[4,5] The presence of aspartic acid at position 189[7] (chymotrypsin numbering) is consistent with factor $\overline{\text{D}}$ belonging to a family of trypsin-like enzymes all of which have Asp[189] present in the substrate-binding pocket. Extensive N-terminal sequence analysis of the whole molecule[7,12] and sequence analysis of the CNBr-derived[9] and tryptic[10] peptides has provided data (Fig. 5) that emphasizes the homology between factor $\overline{\text{D}}$ and other serine esterases.

[11] T. Konno, Y. Katsuno, and H. Hirai, *J. Immunol. Methods* **21**, 325 (1978).

[12] J. E. Volanakis, A. S. Bhown, J. C. Bennett, and J. E. Mole, *Proc. Natl. Acad. Sci. U.S.A.* **77**, 1116(1980).

[12] Preparation of Human Properdin

By K. B. M. REID

Introduction

Properdin was first described by Pillemer *et al.*[1] in 1954, as a factor that, in the presence of magnesium and certain other nondialyzable serum factors, was involved in the inactivation of complement component C3 when the polysaccharide zymosan was incubated with human serum at 37°C. Pillemer *et al.*[1] found that properdin could be removed from serum by incubation with zymosan at 17°C, along with only a minimal destruction of C3 activity, and that the remaining C3 in the treated serum was resistant to inactivation on incubation with fresh zymosan at 37°C. This demonstration of the properdin, or alternative pathway of complement activation, helped to establish that the complement system could be activated by factors other than immunoglobulins. Properdin was isolated in a highly purified form and characterized by Pensky *et al.*[2] in 1968, who

[1] L. Pillemer, L. Blum, I. H. Lepow, O. A. Ross, E. W. Todd, and A. C. Wardlaw, *Science* **120**, 279 (1954).

[2] J. Pensky, C. F. Hinz, E. W. Todd, R. J. Wedgwood, J. T. Boyer, and I. H. Lepow, *J. Immunol.* **100**, 142 (1968).

showed that it had MW 220,000, a sedimentation coefficient of 5.2 S, and that it was distinct from the immunoglobulins and all the known complement components. The association of properdin with C3 or fragments of C3 in serum is well established,[3] but the exact stage, and to what degree, properdin enters the ongoing complement-reaction sequence is not completely clear. However, it is now considered that properdin does not play a central role in the activation of the alternative pathway since it has been shown that activation can take place in the absence of properdin and that the principal function of properdin appears to be its interaction with and stabilization of the complex proteases $\overline{C3bBb}$ or $\overline{C3b_nBb}$.[4-7] Thus, in contrast to the other control proteins associated with the complement system in normal serum, such as β1H and C3b inactivator properdin plays a stabilizing and amplifying role rather than a disruptive or degradative one.

Recent studies[6,8] have shown that properdin is present in serum in a native, or precursor form, which can be converted to an activated form that may be induced on its binding to the $\overline{C3b_nBb}$ complex. It is probable that most of the isolation procedures described, which involve euglobulin precipitation or the binding of the properdin to zymosan, yield the activated form. It has been reported that the two forms of properdin can only be distinguished by functional assays since chemically and antigenically they appear identical.[6] Activated properdin can be distinguished from native properdin by the following criteria: activated properdin causes consumption of the alternative-pathway components C3 and Factor B to take place when it is added to normal serum, while native properdin does not; activated properdin binds to surfaces coated with C3b in the absence of Factors B and D, while native properdin does not.[6]

The preparation of activated properdin, by a procedure that employs the same initial steps as used in the purification of C1q, will be described in this article. The activated properdin isolated by this method has been used in extensive amino acid sequence studies[9] and some functional studies.[9] A procedure for the isolation of native properdin has been described recently by Medicus et al.[6]

Serum Concentration. The concentration of properdin in fresh human serum is ~20 mg/liter. Of the other control proteins and components of

[3] J. Chapitis and I. H. Lepow, *J. Exp. Med.* 143, 241 (1976).
[4] D. T. Fearon and K. F. Austen, *J. Exp. Med.* 142, 856 (1975).
[5] R. D. Schreiber, M. K. Pangburn, P. H. Lesavre, and H. J. Müller-Eberhard, *Proc. Natl. Acad. Sci. U.S.A.* 75, 3948 (1978).
[6] R. G. Medicus, A. F. Esser, H. N. Fernandez, and H. J. Müller-Eberhard, *J. Immunol.* 124, 602 (1980).
[7] R. D. Schreiber and H. J. Müller-Eberhard, *J. Exp. Med.* 148, 1722 (1978).
[8] O. Götze, R. G. Medicus, and H. J. Müller-Eberhard, *J. Immunol.* 118, 525 (1977).
[9] K. B. M. Reid and J. Gagnon, *Mol. Immunol.* (submitted for publication).

the complement system, only Factor D (1.5 mg/liter) and C2 (15 mg/liter) are present in lower amounts than properdin, and they both play an enzymatic role in the system rather than the purely binding role that properdin is envisaged to play.

Assay

Activated properdin can be detected by a wide variety of techniques, such as the zymosan assay of Pillemer et al.[10]; radioimmunoassay[11]; stabilization of the C3bBb complex on sheep red blood cells[4,6]; measurement of the destruction of C3 and Factor B on the addition of activated properdin to serum depleted of properdin[12]; the enhancement of the alternative-pathway lysis of rabbit red blood cells by human serum that had been immunochemically depleted of properdin. The last assay procedure mentioned was used to detect properdin throughout the purification procedure described in this article.

Buffers

Magnesium stock. 100 mM magnesium chloride
Stock veronal buffer (5× concentrated). 42.5 g NaCl + 1.88 g sodium barbital + 2.88 g barbital per liter. The pH was adjusted to 7.4 using 1 M HCl.
VBS. 100 ml stock veronal buffer made up to 500 ml
GVB. 0.5 g gelatin (Difco Laboratories, Detroit, Michigan) dissolved in 100 ml water by boiling, added to 100 ml stock veronal buffer, and made up to 500 ml
Mg–DGVB. 250 ml GVB added to 250 ml 5% (w/v) glucose and then 7.5 ml of 100 mM magnesium chloride added
PBS. 8.48 g NaCl + 5.47 g Na$_2$HPO$_4$ + 1.73 g NaH$_2$PO$_4$ · 2H$_2$O, made up to 1000 ml to give a pH of 7.4
PBS–Mg–EGTA. PBS made 10 mM and 5 mM with respect to ethylene glycol bisaminotetraacetate and magnesium
Method. Rabbit red blood cells are collected from freshly drawn citrated blood, washed twice with PBS and twice with VBS, and then resuspended in Mg–DGVB to give 4 × 10^8 cells/ml.

To prepare serum depleted of properdin [i.e., R(P)], human serum (3.0 ml) was passed through a Sepharose antiproperdin column (0.6 × 6 cm) as described by Minta and Lepow[13] and then concentrated to its original volume by ultrafiltration in a stirred Amicon cell using a PM-10 mem-

[10] L. Pillemer, L. Blum, I. H. Lepow, L. Wurz, and E. W. Todd, *J. Exp. Med.* **103**, 1 (1956).
[11] J. O. Minta, I. Goodkofsky, and I. H. Lepow, *Immunochemistry* **10**, 431 (1975).
[12] O. Bötze and H. J. Müller-Eberhard, *J. Exp. Med.* **139**, 44 (1974).
[13] J. O. Minta and I. H. Lepow, *J. Immunol.* **111**, 286 (1973).

brane. The R(P) reagent (3.0 ml) was dialyzed for 16 hr against 1000 ml of PBS–Mg–EGTA.

A total assay volume of 160 μl is used. In a typical assay a mixture of 10^7 rabbit red blood cells in Mg–DGVB (25 μl), R(P) in PBS–Mg–EGTA (50 μl), sample (80–0 μl) in Mg–DGVB (or some other suitable buffer), and Mg–DGVB buffer (0–80 μl) is incubated at 37°C for 10 min, then ice-cold Mg–DGVB is added to stop the reaction and the tubes spun at 1000 g for 10 min at 2°C. The extent of hemolysis is estimated by reading the $E_{412}^{1\,cm}$ of the supernatant.

Fifty microliters of R(P) should give no, or little, hemolysis of the rabbit red blood cells and the addition of ~2 μg of activated properdin should cause over 90% lysis of the cells. The addition of approximately 10 μl of normal human serum alone should cause over 90% lysis of the cells.

Assay procedures for native properdin and mixtures of activated and native properdin have been described by Medicus et al.[6]

Purification Procedure

The procedure given here is similar to previously published procedures[2,14] in which euglobulin precipitation, ion-exchange chromatography, and affinity chromatography were used. The initial steps of this procedure are identical to those employed to prepare human C1q (this volume, Chapter 3) since activated properdin is eluted in good yield along with the C1q. Therefore for a full description of these initial steps, that is, preparation of 2 liters of human serum for euglobulin precipitation, ion-exchange chromatography on DEAE–Sephadex A-50 and CM-cellulose 32, reference should be made to the article cited.

The resolubilized euglobulin precipitate (approximately 2100 $E_{280}^{1\,cm}$ units) obtained from 2 liters of human serum was applied to a DEAE–Sephadex A-50 column, approximately 180 $E_{280}^{1\,cm}$ units passed straight through the column and were applied to a CM-cellulose 32 column. The peak eluted between 16 and 18 mS, immediately prior to the C1q peak, contains activated properdin (this volume, Chapter 3, Fig. 1). The fractions containing properdin activity (750–920 ml) are pooled and then concentrated by ultrafiltration in an Amicon stirred cell using a PM-10 membrane. The concentrated pool of protein from the CM-cellulose column (approximately 22 $E_{280}^{1\,cm}$ units in 6 ml) is dialyzed against 2 × 2000 ml of 0.15 M NaCl–12 mM sodium phosphate–0.2 mM EDTA buffer, pH 7.4 (8.85 g NaCl + 5 ml of .04 M NaH$_2$PO$_4$ + 23.4 ml of 0.4 M Na$_2$HPO$_4$ + 1.3 ml of 0.15 M EDTA, pH 7.4, per 1000 ml) and applied to a column (2.5 × 85 cm) of Sephacryl S-300 (Pharmacia Fine Chemicals), which is equilibrated with

[14] J. O. Minta and E. S. Kunar, J. Immunol. 116, 1099 (1976).

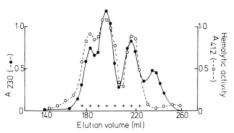

FIG. 1. Gel filtration of activated properdin on Sephacryl S-300. Partially purified proper-din (approximately 22 $E_{280}^{1\,cm}$ units in 6 ml) from CM-cellulose 32 was dialyzed against 0.15 M NaCl–12 mM sodium phosphate–0.2 mM EDTA buffer (pH 7.4), and after centrifugation, applied to a column (2.5 × 85 cm) of Sephacryl S-300 equilibrated with the same buffer. The column was run at 15 ml/hr. ●——●, A_{230}; ○----○, A_{412} (hemolytic activity). Reaction with antiproperdin is shown by +.

the same buffer. In this procedure, and under the conditions used for gel filtration, much of the activated properdin appears to form aggregates and properdin activity is eluted in three distinct peaks, between 160 ml and 225 ml (Fig. 1), which are separated from a fourth peak that contains IgG. If each of the three properdin peaks is pooled separately, concentrated and rerun on Sephacryl S-300, then each peak is again eluted in its original position. All three peaks contain properdin as judged by their hemolytic activities (the specific activity of each fraction being very similar; Fig. 1); their reaction with antiproperdin (Fig. 1); their identical behavior on SDS–polyacrylamide gel electrophoresis. All three properdin peaks from the Sephacryl S-300 column also yielded identical peptide patterns when treated with cyanogen bromide or subjected to limited proteolysis by trypsin. It is possible to monitor the elution of the properdin from the Sephacryl S-300 column by SDS–polyacrylamide gel electrophoresis rather than by activity measurements. Activated properdin appears as 2 very closely spaced bands of approximately 55,000 apparent molecular weight on 10% (w/v) SDS–polyacrylamide slab gels when run under non-reducing conditions and as a single band of 58,000 apparent molecular weight when reduced and then alkylated with iodoacetamide. Alkylation of the reduced properdin samples with iodoacetic acid rather than iodoacetamide prior to electrophoresis yields an apparent molecular weight of 69,000. The activated properdin from Sephacryl S-300 is pooled (175–215 ml, Fig. 1) and then concentrated by ultrafiltration to give approxi-mately 12 $E_{280}^{1\,cm}$ units in 10 ml, which is equivalent to 6.7 mg of properdin (as judged by amino acid analysis). This represents a yield of approximately 18% of that initially present in the serum.

Traces of IgG (and traces of C1q if it has not been completely removed

AMINO ACID COMPOSITION OF HUMAN PROPERDIN

	Residues/100 residues	
Amino acid	Minta and Lepow[a]	Reid and Gagnon[b]
Asx	4.70	5.03
Thr	5.70	6.41
Ser	7.26	6.97
Glx	15.30	13.53
Pro	11.60	11.60
Gly	12.74	11.89
Ala	5.58	5.53
Cys	4.96	9.31
Val	4.63	5.28
Met	1.43	0.90
Ile	1.32	1.56
Leu	5.07	5.26
Tyr	1.45	1.44
Phe	1.90	1.26
His	2.73	2.89
Lys	4.70	4.33
Arg	8.85	6.84
Trp[c]	—	—

[a] J. O. Minta and I. H. Lepow, *Immunochemistry* 11, 361 (1974).
[b] K. B. M. Reid and J. Gagnon, *Mol. Immunol.* 18, in press (1981).
[c] Not determined

by CM-cellulose chromatography) can be eliminated by affinity chromatography by applying the activated properdin to a Sepharose anti-R(P) column[8,13] or a Sepharose anti-human IgG column.[6]

Physical Properties and Composition

In nondissociating conditions, human properdin has a molecular weight of approximately 220,000[2] (although a value of 186,000 has also been reported)[15] and an $S^{\circ}_{20,w}$, in phosphate-buffered saline, pH 7.4, of 5.2 S as estimated by analytical centrifugation.[2,15,16] In dissociating conditions, properdin has an apparent molecular weight of 56,000–58,000 as estimated by SDS gel electrophoresis[6,9] and reduced and alkylated properdin has MW 58,500 as judged by gel filtration on Sephacryl S-300 in 6 M

[15] J. O. Minta and I. H. Lepow, *Immunochemistry* 11, 361 (1974).
[16] J. O. Minta, *J. Immunol.* 117, 405 (1976).

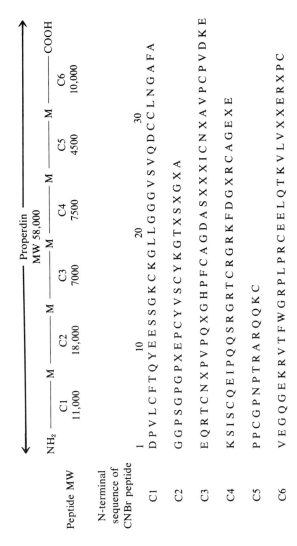

FIG. 2. N-terminal amino acid sequences and probable alignment of the major peptides produced by treatment of activated properdin with cyanogen bromide. The data shown are taken from K. B. M. Reid and J. Gagnon, *Mol. Immunol.* **18**, in press (1981) and are in agreement with the partial N-terminal sequence of 11 residues obtained, by R. G. Medicus, A. F. Esser, H. N. Fernandez, and H. J. Müller-Eberhard [*J. Immunol.* **124**, 602 (1980)], for both activated and native properdin. The apparent molecular weights of the cyanogen bromide peptides as estimated by gel filtration in $6 M$ guanidinium chloride are given. The single-letter code for amino acids is used: A, Ala; B, Asx; C, Cys; D, Asp; E, Glu; F, Phe; G, Gly; H, His; I, Ile; K, Lys; L, Leu; M, Met; N, Asn; P, Pro; Q, Gln; R, Arg; S, Ser; T, Thr; V, Val; W, Trp; X, unknown; Y, Tyr.

guanidinium chloride.[9] It thus appears probable that the intact molecule is composed of 4 polypeptide chains. Properdin is a glycoprotein having a carbohydrate content of approximately 9.8%, which is composed of 3.8% hexose, 3.8% sialic acid, 1.5% hexosamine, and 0.7% fucose.[15] An unusual amino acid composition has been reported for properdin (see table); only 4 amino acids—glutamic acid, proline, glycine, and half-cystine—account for 45% of the total amino acid composition.[9,15] However, amino acid sequence studies of the peptides produced by treatment with cyanogen bromide (Fig. 2) indicate that this unusual amino acid composition is evenly distributed throughout the molecule since no evidence of extended areas of repeating sequences has yet been found. The amino acid sequence studies[9] confirm and extend the N-terminal sequence data published by Medicus et al.[6] and show that it is very likely that the 4 polypeptide chains present in a molecule of properdin are identical. Medicus et al.[6] found that activated and native properdin have identical electrophoretic mobilities, apparent molecular weights on SDS gel electrophoresis, and amino- and carboxy-terminal amino acid sequences. They concluded that the transition from native to activated properdin was probably due to a small conformational change in the molecule since native properdin could be regenerated from activated properdin by the use of denaturing reagents.

Section II

Blood Clotting

[13] Introduction to Clotting in Blood Plasma

By EARL W. DAVIE

During the past 10 years, substantial progress has been made on the isolation and characterization of the various proteins involved in blood clotting and fibrinolysis. Indeed, most of the coagulation proteins have now been purified and well characterized from human as well as bovine plasma. Furthermore, excellent methods have recently been developed for two of the most difficult coagulation proteins to isolate, namely factor V[1-5] and factor VII.[6-9] These two proteins have been a problem to purify because of their considerable lability and low concentration in plasma. Some of the coagulation factors, however, have only been partially purified from human and bovine plasma at the present time. This is particularly true of factor VIII, which occurs in plasma as a trace protein. Indeed, the isolation of factor VIII continues to be one of the major challenges to the protein chemist working in the field of blood coagulation.

Some of the coagulation and fibrinolytic proteins and their inhibitors have become available in substantial amounts in recent years, making it possible to determine their primary structures. Thus far, the complete amino acid sequence has been established for human fibrinogen,[10,11] human and bovine prothrombin,[12-14] bovine factor IX,[15] bovine factor

[1] M. E. Nesheim, K. H. Myrmel, L. Hibbard, and K. G. Mann, *J. Biol. Chem.* **254**, 508 (1979).

[2] C. T. Esmon, *J. Biol. Chem.* **254**, 964 (1979).

[3] B. Dahlbäck, *J. Clin. Invest.* **66**, 583 (1980).

[4] W. H. Kane and P. W. Majerus, *J. Biol. Chem.* **256**, 1002 (1981).

[5] J. A. Katzmann, M. E. Nesheim, L. S. Hibbard, and K. G. Mann, *Proc. Natl. Acad. Sci. U.S.A.* **78**, 162 (1981).

[6] W. Kisiel and E. W. Davie, *Biochemistry* **14**, 4928 (1975).

[7] R. Radcliffe and Y. Nemerson, *J. Biol. Chem.* **250**, 388 (1975).

[8] G. J. Broze, Jr. and P. W. Majerus, *J. Biol. Chem.* **255**, 1242 (1980).

[9] S. P. Bajaj, S. I. Rapaport, and S. F. Brown, *J. Biol. Chem.* **256**, 253 (1981).

[10] A. Henschen, F. Lottspeich, E. Topfer-Petersen, M. Kehl, and R. Timpl, *Protides Biol. Fluids* **28**, 47 (1980).

[11] R. F. Doolittle, K. W. K. Watt, B. A. Cottrell, D. D. Strong, and M. Riley, *Nature (London)* **280**, 464 (1979).

[12] S. Magnusson, T. E. Petersen, L. Sottrup-Jensen, and H. Claeys, *Cold Spring Harbor Conf Cell Proliferation* **2**, p. 123 (1975).

[13] R. J. Butkowski, J. Elion, M. R. Downing, and K. G. Mann, *J. Biol. Chem.* **252**, 4942 (1977).

[14] D. A. Walz, D. Hewett-Emmett, and W. H. Seegers, *Proc. Natl. Acad. Sci. U.S.A.* **74**, 1969 (1977).

[15] K. Katayama, L. H. Ericsson, D. L. Enfield, K. A. Walsh, H. Neurath, E. W. Davie, and K. Titani, *Proc. Natl. Acad. Sci. U.S.A.* **76**, 4990 (1979).

METHODS IN ENZYMOLOGY, VOL. 80

X,[16,17] human antithrombin III,[18] and human plasminogen.[19-21] Also, the amino acid sequence of bovine high-molecular-weight kininogen is nearing completion.[22-24] Good progress has also been made on the carbohydrate structure of these proteins, including prothrombin,[25] factor X,[26] antithrombin III,[27] and high-molecular-weight (HMW) kininogen.[28]

During the isolation of prothrombin, factors VII, IX, and X, three new vitamin K-dependent proteins have been identified in plasma. These include protein C,[29] protein S,[30] and protein Z.[31] Protein C (MW 56,000) is composed of a heavy and a light chain, and these chains are held together by a disulfide bond(s). The primary sequence for bovine protein C has recently been completed.[32] It contains 11 residues of γ-carboxyglutamic acid, which are present in the light chain of the molecule. Protein C is converted to activated protein C by thrombin, trypsin, or a protease from Russell's viper venom.[33-35] Activated protein C is a serine protease that exhibits anticoagulant activity in the presence of phospholipid and cal-

[16] K. Titani, K. Fujikawa, D. L. Enfield, L. H. Ericsson, K. A. Walsh, and H. Neurath, *Proc. Natl. Acad. Sci. U.S.A.* **72**, 3082 (1975).

[17] D. L. Enfield, L. H. Ericsson, K. A. Walsh, H. Neurath, and K. Titani, *Proc. Natl. Acad. Sci. U.S.A.* **72**, 16 (1975).

[18] T. E. Petersen, G. Dudek-Wojciechowska, L. Sottrup-Jensen, and S. Magnusson, in "The Physiological Inhibitors of Blood Coagulation and Fibrinolysis" (D. Collen, B. Wiman, and M. Verstraete, eds.), p. 43. Elsevier/North-Holland Biomedical Press, Amsterdam, 1979.

[19] W. R. Groskopf, L. Summaria, and K. C. Robbins, *J. Biol. Chem.* **244**, 3590 (1969).

[20] B. Wiman and P. Wallen, *Eur. J. Biochem.* **58**, 539 (1975).

[21] L. Sottrup-Jensen, H. Claeys, M. Zajdel, T. E. Petersen, and S. Magnusson, *Prog. Chem. Fibrinolysis Thrombolysis* **3**, 191 (1978).

[22] Y. N. Han, M. Komiya, S. Iwanaga, and T. Suzuki, *J. Biochem. (Tokyo)* **77**, 55 (1975).

[23] Y. N. Han, H. Kato, S. Iwanaga, and T. Suzuki, *J. Biochem. (Tokyo)* **79**, 1201 (1976).

[24] H. Kato, Y. N. Han, S. Iwanaga, T. Suzuki, and M. Komiya, *J. Biochem. (Tokyo)* **80**, 1299 (1976).

[25] T. Mizuochi, K. Yamashita, K. Fujikawa, W. Kisiel, and A. Kobata, *J. Biol. Chem.* **254**, 6419 (1979).

[26] T. Mizuochi, K. Yamashita, K. Fujikawa, K. Titani, and A. Kobata, *J. Biol. Chem.* **255**, 3526 (1980).

[27] T. Mizuochi, J. Fujii, K. Kurachi, and A. Kobata, *Arch. Biochem. Biophys.* **203**, 458 (1980).

[28] Y. Endo, K. Yamashita, Y. N. Han, S. Iwanaga, and A. Kobata, *J. Biochem. (Tokyo)* **82**, 545 (1977).

[29] J. Stenflo, *J. Biol. Chem.* **251**, 355 (1976).

[30] R. G. DiScipio, M. A. Hermodson, S. G. Yates, and E. W. Davie, *Biochemistry* **16**, 698 (1977).

[31] C. V. Prowse and M. P. Esnouf, *Biochem. Soc. Trans.* **5**, 255 (1977).

[32] P. Fernlund and J. Stenflo, in "Vitamin K Metabolism and Vitamin K-Dependent Proteins" (J. W. Suttie, ed.), p. 84. University Park Press, Baltimore, Maryland, 1980.

[33] W. Kisiel, L. H. Ericsson, and E. W. Davie, *Biochemistry* **15**, 4893 (1976).

[34] C. T. Esmon, J. Stenflo, and J. W. Suttie, *J. Biol. Chem.* **251**, 3052 (1976).

[35] W. Kisiel, *J. Clin. Invest.* **64**, 761 (1979).

cium.[36,37] This is due to the proteolytic inactivation of factors V and VIII by activated protein C.

Protein S is a single-chain plasma protein with a molecular weight of approximately 64,000.[30,38] It contains 10 residues of γ-carboxyglutamic acid in its amino-terminal region. It occurs in plasma in a free form and in a complex with C4b-binding protein (C4bp).[39] The latter protein is involved in the regulation of the rate of complement activation.[40] No biological role has been established thus far for protein Z, another vitamin K-dependent protein present in plasma.

At the present time, the majority of the steps in the intrinsic and extrinsic pathways of blood coagulation are known. The initiation of these two pathways and their relationship to each other, however, are not well established. Some clarification of the initiation of the intrinsic pathway is beginning to emerge, however, at least as it occurs in the test tube.[41] Present evidence indicates that factor XII in the presence of a surface, such as kaolin or glass, will convert prekallikrein to kallikrein in a reaction stimulated by HMW kininogen. This reaction appears to be a substrate-induced catalysis by single-chain factor XII. The kallikrein thus generated then converts factor XII to factor XII_a in the presence of a surface and HMW kininogen. This reaction in turn is followed by the conversion of factor XI to factor XI_a by factor XII_a. This last reaction is also stimulated by HMW kininogen in the presence of a surface. These cascade reactions are illustrated as follows:

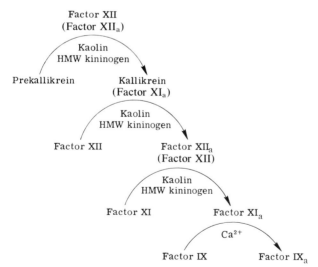

[36] W. Kisiel, W. M. Canfield, L. H. Ericsson, and E. W. Davie, *Biochemistry* **16**, 5824 (1977).
[37] G. A. Vehar and E. W. Davie, *Biochemistry* **19**, 401 (1980).
[38] R. G. DiScipio and E. W. Davie, *Biochemistry* **18**, 899 (1979).

The reactions in the middle phase of the intrinsic pathway of blood coagulation involve factors IX_a, VIII, X, V, and prothrombin. These reactions also require calcium and phospholipid or phospholipoprotein (PL). These reactions are as follows:

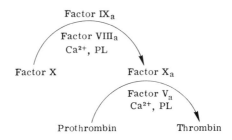

Factors IX_a and X_a are highly specific serine proteases, whereas factors VIII_a and V_a appear to function in these reactions as cofactors. Also, factors VIII and V must undergo limited proteolysis by a protease, such as thrombin, before they are active in these reactions. Kinetic studies indicate that factors VIII_a and V_a function primarily by increasing the V_{max} in their specific reactions.[42,43] In contrast, the phospholipid influences these reactions primarily by lowering the K_m. Under physiological conditions, the phospholipid or phospholipoprotein is provided by the platelets and becomes available during the formation of the platelet plug. This is formed at the site of the vascular injury and makes available binding sites for many of the coagulation proteins that interact during the coagulation process.

In the final phase of the coagulation process, fibrinogen is converted to fibrin in the presence of thrombin. The fibrin monomers then undergo side-to-side and end-to-end aggregation, leading to the fibrin gel. The fibrin gel is then crosslinked by covalent bonds in the presence of factor XIII_a.

[39] B. Dahlbäck and J. Stenflo, *Proc. Natl. Acad. Sci. U.S.A.* **78**, 2512 (1981).
[40] I. Gigli, T. Fujita, and V. Nussenzweig, *Proc. Natl. Acad. Sci. U.S.A.* **76**, 6596 (1979).
[41] R. L. Heimark, K. Kurachi, K. Fujikawa, and E. W. Davie, *Nature (London)* **286**, 456 (1980).
[42] J. Rosing, G. Tans, J. W. P. Govers-Riemslag, R. F. A. Zwaal, and H. C. Hemker, *J. Biol. Chem.* **255**, 274 (1980).
[43] G. Tans, J. Rosing, G. V. Dieyen, and H. C. Hemker, *in* "The Regulation of Coagulation" (K. G. Mann and F. B. Taylor, Jr., eds.), p. 173. Elsevier/North-Holland Biomedical Press, Amsterdam, 1980.

[14] Bovine and Human Plasma Prekallikrein

By RONALD L. HEIMARK and EARL W. DAVIE

Prekallikrein is the inactive precursor or zymogen form of kallikrein, a serine protease that releases kinin from kininogen. It is inactive or absent in the plasma of individuals with a coagulation disorder called Fletcher factor deficiency.[1] Fletcher factor deficiency is characterized by a prolonged partial thromboplastin time, which is due to a defect in the initiation of the intrinsic pathway of blood coagulation.[2-4] In addition, surface-mediated fibrinolysis and kinin formation have also been shown to be defective in Fletcher trait plasma.

The initiation of these closely related processes involves both plasma prekallikrein and factor XII. These two proteins are present in plasma as zymogens of serine proteases and have an unusual relationship to each other. Factor XII_a converts prekallikrein to kallikrein by the cleavage of an internal Arg-Ile bond[5] and, in a similar manner, kallikrein converts factor XII to factor XII_a by the cleavage of an internal Arg-Val bond.[6] A third plasma protein, high-molecular-weight (HMW) kininogen, stimulates both the activation of prekallikrein by factor XII_a and the activation of factor XII by kallikrein.[6-9]

Recently, it has been shown that the initial event in the activation of the intrinsic pathway of blood coagulation involves a substrate-induced catalysis of prekallikrein by the zymogen form of factor XII.[10] In this reaction, factor XII, bound to a surface such as kaolin, converts prekallikrein to kallikrein in the presence of HMW kininogen. The kallikrein thus generated converts factor XII to factor XII_a. Factor XII_a then converts factor XI to factor XI_a and this eventually leads to thrombin formation and a fibrin clot.

The purification of plasma prekallikrein has been difficult by conventional methods of protein fractionation, since its concentration in plasma

[1] W. E. Hathaway, L. P. Belhasen, and H. S. Hathaway, *Blood* **26**, 521 (1965).
[2] K. D. Wuepper, *J. Exp. Med.* **138**, 1345 (1973).
[3] H. Saito, O. D. Ratnoff, and V. H. Donaldson, *Circ. Res.* **34**, 641 (1974).
[4] A. S. Weiss, J. I. Gallin, and A. P. Kaplan, *J. Clin. Invest.* **53**, 622 (1974).
[5] R. L. Heimark, K. Fujikawa, and E. W. Davie, submitted for publication.
[6] K. Fujikawa, R. L. Heimark, K. Kurachi, and E. W. Davie, *Biochemistry* **19**, 1322 (1980).
[7] J. H. Griffin and C. G. Cochrane, *Proc. Natl. Acad. Sci. U.S.A.* **73**, 2554 (1976).
[8] H. L. Meier, J. V. Pierce, R. W. Colman, and A. P. Kaplan, *J. Clin. Invest.* **60**, 18 (1977).
[9] T. Sugo, N. Ikari, H. Kato, S. Iwanaga, and S. Fujii, *Biochemistry* **19**, 3215 (1980).
[10] R. L. Heimark, K. Kurachi, K. Fujikawa, and E. W. Davie, *Nature (London)* **286**, 456 (1980).

METHODS IN ENZYMOLOGY, VOL. 80

is low and it is very sensitive to protease digestion. A number of methods have been reported in recent years that have yielded preparations from human and animal plasmas.[11-17] In the isolation procedure to be described, benzamidine–agarose and agmatine– or arginine methyl ester–agarose column chromatography play an important role. Also, each purification step is performed in the presence of protease inhibitors to restrict activation of the precursor form.

Assay Methods

Coagulant Assay

Reagents

Phospholipid Suspension. One vial of rabbit brain cephalin (Sigma) is suspended in 100 ml of 0.15 M NaCl. The stock suspension is stored in aliquots of 1 ml at $-20°C$.

Kaolin Suspension. 50 mg of acid-washed kaolin (Fisher Scientific Co.) is suspended in 10 ml of 0.15 M NaCl.

Prekallikrein-Deficient Plasma. Human Fletcher factor-deficient plasma is obtained from George King Biomedical, Inc., and stored in small aliquots at $-70°C$.

For the coagulant assay of plasma prekallikrein, the test sample is diluted 50- to 200-fold with Michaelis buffer (0.036 M sodium acetate–0.036 M sodium barbital–0.145 M sodium chloride, pH 7.4) containing 0.1 mg/ml bovine serum albumin. The sample (0.05 ml) is then incubated at room temperature for 5 min with 0.05 ml of prekallikrein-deficient plasma and 0.05 ml of kaolin suspension. One-tenth milliliter of an equal mixture of the stock cephalin suspension and 0.025 M CaCl$_2$ is added and the clotting time determined by the tilting method. Activity is calculated from a calibration curve where the log of prekallikrein concentration is plotted against the log of the clotting time. One unit of activity is defined as the amount of activity present in 1.0 ml of pooled platelet-poor normal bovine plasma or human plasma.

[11] K. D. Wuepper and C. G. Cochrane, *J. Exp. Med.* **135,** 1 (1972).
[12] H. Takahashi, S. Nagasawa, and T. Suzuki, *J. Biochem. (Tokyo)* **71,** 471 (1972).
[13] R. Mandle and A. P. Kaplan, *J. Biol. Chem.* **252,** 6097 (1977).
[14] C. F. Scott, C. Y. Liu, and R. W. Colman, *Eur. J. Biochem.* **100,** 77 (1979).
[15] R. L. Heimark and E. W. Davie, *Biochemistry* **18,** 5743 (1979).
[16] B. N. Bouma, L. A. Miles, G. Beretta, and J. H. Griffin, *Biochemistry* **19,** 1151 (1980).
[17] K. Laake and A. M. Vennerod, *Thromb. Res.* **4,** 285 (1974).

Amidase Assay

Reagents

Trypsin Solution. A stock solution (0.2 mg/ml) is prepared employing TPCK-treated trypsin (Worthington), which is dissolved in 1 mM HCl containing 0.1 mM CaCl$_2$. It is stored in small aliquots at $-20°$C.

Ovomucoid Trypsin-Inhibitor Solution. Ovomucoid trypsin-inhibitor is obtained from Sigma (type II-0) and is dissolved at 10 mg/ml in 0.05 M Tris-HCl, pH 8.0.

Bz-Pro-Phe-Arg-pNA. The kallikrein chromogenic substrate, N-benzoyl-L-prolyl-L-phenylalanyl-L-arginine-p-nitroanilide (Pentapharm Ltd.), is prepared as a stock solution of 1 mM in distilled H$_2$O.

The amidase activity of kallikrein is measured by using the chromogenic kallikrein substrate Bz-Pro-Phe-Arg-p-nitroanilide after activation of prekallikrein by trypsin.[7] The sample (0.01–0.1 ml) is diluted to 0.9 ml with 0.05 M Tris-HCl buffer (pH 8.0) containing 0.15 M NaCl and incubated with 10 μl of the trypsin solution at 37°C. After 20 min, the trypsin is inactivated with 100 μg of the ovomucoid trypsin-inhibitor solution, and 0.1 ml of 1 mM Bz-Pro-Phe-Arg-p-nitroanilide is added. The mixture is incubated at 37°C for 3 min, and the reaction is stopped with 10 μl of glacial acetic acid. The absorbance at 405 nm is then read in a Gilford spectrophotometer.

Purification Procedures

Reagents

Heparin–Agarose. Heparin is coupled covalently to agarose beads by the method of Cuatrecasas.[18] Powdered cyanogen bromide (10 g) is rapidly added with vigorous stirring to 50 ml of a commercial preparation of agarose A-15m (100–200 mesh) or Sepharose-4B suspended in an equal volume of cold water. The pH of the suspension is maintained at 11 by continuous addition of 6 N NaOH, and the temperature is held at 18°C by the addition of crushed ice. After 20 min, the agarose suspension is poured onto a filter funnel and washed with a large excess of cold 0.1 M NaHCO$_3$, pH 8.3. The washed agarose is then mixed with 1 g of heparin (160 units/mg) in 50 ml of 0.2 M NaHCO$_3$, pH 8.3, and stirred slowly overnight at 4°C. It is washed extensively with 0.1 M Tris-HCl, pH 10.5, containing 1 M NaCl, followed by distilled water and 0.1 M sodium acetate, pH 4.0. It is then stored at 4°C in water containing 0.02% sodium azide.

---

[18] P. Cuatrecasas, *J. Biol. Chem.* **245**, 3059 (1970).

Arginine Methyl Ester–Agarose and Agmatine–Agarose. Agarose (50 ml) is activated as described above and is then mixed with 2.5 g of arginine methyl ester or agmatine in 50 ml of 0.2 M NaHCO$_3$ (pH 8.5) and stirred gently overnight. It is washed as described above and stored at 4°C in 0.02 M 2-(morpholino)ethanesulfonic acid (pH 6.0) containing 0.05 M NaCl and 0.02% sodium azide.

Benzamidine–Agarose. Benzamidine–agarose is prepared with an ε-aminocaproic acid spacer. In this procedure, the ε-aminocaproic acid is initially coupled to agarose beads (Bio-Gel A-15m, 100–200 mesh) by the procedure described above and then washed extensively with distilled water. Fifty milliliters of the ε-aminocaproyl-agarose is washed with 0.1 M 2-(morpholino)ethanesulfonic acid buffer (pH 4.75) and mixed with 50 ml of 0.1 M 2-(morpholino)ethanesulfonic acid buffer (pH 4.75) containing 1-cyclohexyl-3-(2-morpholinoethyl)carbodiimide metho-p-toluenesulfonate (100 mg/ml). The mixture is stirred gently at room temperature for 30 min. One gram of p-aminobenzamidine is then added, the pH is rechecked, and the mixture is allowed to react for 15 hr. The reaction is terminated by the addition of 1 g of glycine ethyl ester, and the benzamidine–agarose is washed with water and stored at 4°C.

Inhibitors. A stock solution of 1 M diisopropyl fluorophosphate (DFP) is carefully prepared by diluting 1 ml of pure DFP to 5.5 ml with anhydrous isopropanol. This solution is stored at -20°C in small aliquots. A polybrene stock solution is prepared by dissolving 10 g of polybrene (Aldrich) in 100 ml of H$_2$O.

Purification of Bovine Plasma Prekallikrein

Ammonium Sulfate Fractionation. Bovine blood is collected and mixed rapidly with 0.1 volume of anticoagulant solution (0.1 M sodium oxalate—heparin, 100 mg/liter–crude soybean-trypsin inhibitor, 100 mg/liter). The plasma is isolated at room temperature with a continuous-flow separator (DeLaval Model BLE519). Subsequent steps are performed at 4°C employing plastic containers, columns, and tubes. The plasma (14 liters) is stirred with barium sulfate (20 g/liter) for 30 min, and the slurry is centrifuged for 10 min at 7800 g in a Sorvall RC3 centrifuge. The supernatant is made 0.1 mM in EDTA, and solid ammonium sulfate is added slowly to 20% saturation. After the mixture is stirred for 30 min, the precipitate is removed by centrifugation for 15 min at 7800 g. The supernatant is then brought to 40% saturation with solid ammonium sulfate and stirred for 30 min, and the suspension is centrifuged for 40 min at 7800 g. The precipitate is dissolved in 4.5 liters of cold distilled water containing soybean-trypsin inhibitor (20 mg/liter), polybrene (50 mg/liter), and DFP (0.2 mM).

The solution is then dialyzed for 15 hr against 60 liters of distilled water, followed by dialysis against 80 liters of 0.03 M Tris-HCl buffer (pH 7.5) and 0.05 M NaCl for 14 hr. The recovery of prekallikrein is approximately 90% as shown in Table I.

Chromatography on DEAE–Sephadex. After dialysis, the conductivity of the sample is adjusted with cold 0.03 M Tris-HCl buffer (pH 7.5) to 5.0 mΩ at 4°C employing a Radiometer conductivity meter, type CDM2. Soybean-trypsin inhibitor, polybrene, and DFP are added at the same concentrations as noted above. The sample is divided in half, and each fraction is applied at a flow rate of approximately 400 ml/hr to a DEAE–Sephadex A-50 column (10 × 32.5 cm) previously equilibrated with the dialysis buffer. Each column is washed with 2 liters of 0.03 M Tris-HCl (pH 7.5) containing 0.05 M NaCl and 0.1 mM DFP. Prekallikrein passes directly through the DEAE–Sephadex column under these conditions, while factor XII remains adsorbed to the resin. This step is used in quantitation of prekallikrein in earlier steps to remove factor XII, which also corrects Fletcher factor deficiency.[19,20] The passthrough and wash fractions are then combined, and the pH is adjusted to 7.2 with 1 N HCl.

Chromatography on Heparin–Agarose. The pooled fraction is applied at a flow rate of 400 ml/hr to a heparin–agarose column (8 × 20 cm) previously equilibrated with 0.02 M Tris-HCl buffer (pH 7.2) containing 0.05 M NaCl. The column is washed with 3 liters of 0.02 M Tris-HCl buffer (pH 7.2) containing 0.055 M NaCl and 0.1 mM DFP, and the protein is eluted by a linear salt gradient formed by 3 liters of 0.055 M NaCl in 0.02 M Tris-HCl buffer (pH 7.2) and 3 liters of 0.55 M NaCl in 0.02 M Tris-HCl buffer, pH 7.2. Both solutions also contain polybrene (50 mg/liter) and 0.1 mM DFP. Fractions (200 ml) are collected in plastic bottles employing a Pharmacia preparative fraction collector. Prekallikrein is assayed with a kallikrein chromogenic substrate after activation with trypsin and appears in the leading edge of the main protein peak (Fig. 1). Polybrene (50 mg/liter), purified soybean-trypsin inhibitor (type I-S, Sigma) (20 mg), and DFP (0.1 mM) are added to the pooled fractions containing prekallikrein from the heparin–agarose column, and the sample is dialyzed overnight against 40 liters of 0.02 M sodium phosphate buffer, pH 8.0.

Rechromatography on DEAE–Sephadex. A precipitate that forms during dialysis is removed by centrifugation (15 min at 7800 g), and the sample is applied to a second DEAE–Sephadex A-50 column (6 × 35 cm) at a flow rate of 150 ml/hr. The DEAE–Sephadex column is previously equilibrated with 0.02 M sodium phosphate buffer, pH 8.0. After applica-

[19] H. Saito, O. D. Ratnoff, R. Waldmann, and J. P. Abraham, *J. Clin. Invest.* **55**, 1082 (1975).
[20] K. Fujikawa, K. Kurachi, and E. W. Davie, *Biochemistry* **16**, 4182 (1977).

TABLE I

PURIFICATION OF BOVINE PLASMA PREKALLIKREIN

Purification step	Volume (ml)	Total protein (mg)[a]	Total activity (units)[b]	Specific activity (units/mg)	Recovery (%)	Purification (-fold)
Plasma	14,000	9.8×10^5	2300	0.0023	100	1
(NH$_4$)$_2$SO$_4$ fractionation	6500	4.4×10^5	2070	0.005	90	2.2
First DEAE–Sephadex (pH 7.5)	10,500	1.3×10^5	1920	0.015	84	6.5
Heparin–agarose	1880	6.1×10^3	1250	0.21	55	90
Second DEAE–Sephadex (pH 8.0)	1050	820	850	1.0	37	450
CM-Sephadex	905	120	790	6.6	34	2900
Benzamidine–agarose	227	27.6	490	18	21	7700
Arginine methyl ester–agarose	181	5.5	320	58	14	25,000

[a] Protein concentration is determined by absorption employing $E_{280}^{1\%} = 10.0$ for plasma and subsequent steps up to the arginine methyl ester–agarose column, for which $E_{280}^{1\%} = 10.9$ is employed.

[b] Activity of prekallikrein is assayed with kaolin as described under Methods. Before assaying, the first three samples (1.0 ml) are dialyzed for 3 hr against 0.03 M Tris-HCl buffer (pH 7.5) and 0.05 M NaCl, and are passed through a DEAE–Sephadex column (0.7 × 4 cm), which is then washed with 4.0 ml of the same buffer.

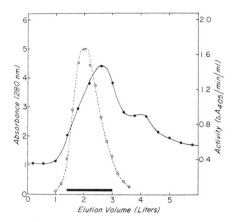

FIG. 1. Elution pattern for heparin–agarose column chromatography. Bovine prekalli-krein is eluted from the column (8 × 20 cm) with a linear gradient formed by 3 liters of 0.02 M Tris-HCl buffer (pH 7.2) containing 0.055 M NaCl and 3 liters of 0.02 M Tris-HCl buffer (pH 7.2) containing 0.55 M NaCl. Both solutions also contain polybrene (50 mg/liter) and 0.1 mM DFP. The flow rate is 400 ml/hr. Prekallikrein activity is measured after activation by trypsin as described under Methods. ●——●, Absorbance; ○----○, amidase activity. Fractions shown by the solid bar are pooled.

tion of the sample, the column is washed with 4 liters of equilibration buffer, and prekallikrein is eluted from the column by a linear salt gradient (Fig. 2A) formed by 1.5 liter of 0.02 M sodium phosphate buffer (pH 8.0) and 1.5 liter of 0.02 M sodium phosphate buffer (pH 8.0) containing 0.3 M NaCl. Each solution also contained 0.1 mM DFP.

Chromatography on CM-Sephadex. After addition of polybrene (100 mg/liter) and soybean-trypsin inhibitor (5 mg), the prekallikrein sample from the second DEAE–Sephadex column is dialyzed overnight against 40 liters of 0.05 M sodium acetate buffer (pH 5.6) containing 0.05 M NaCl. The sample is then applied to a CM-Sephadex C-50 column (5 × 25 cm) equilibrated with 0.05 M sodium acetate buffer (pH 5.6) containing 0.02 M NaCl. The flow rate is adjusted to 150 ml/hr. The column is then washed with 2 liters of 0.05 M sodium acetate buffer (pH 5.6) containing 0.1 M NaCl and polybrene (100 mg/liter), and prekallikrein is eluted from the column (Fig. 2B) with a linear gradient composed of 1.2 liters of 0.05 M sodium acetate buffer (pH 5.6) containing 0.1 M NaCl and polybrene (100 mg/liter) and 1.2 liters of 0.05 M sodium acetate buffer (pH 5.6) containing 0.35 M NaCl and polybrene (100 mg/liter).

Chromatography on Benzamidine–Agarose. Fractions containing pre-kallikrein from the CM-Sephadex column are combined and dialyzed overnight against 18 liters of 0.05 M imidazole hydrochloride (pH 6.0)

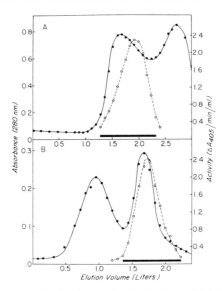

FIG. 2. Elution pattern for bovine prekallikrein from DEAE–Sephadex and CM-Sephadex column chromatography. (A) Elution pattern for the second DEAE–Sephadex column. Protein is eluted from the column (6 × 35 cm) with a linear gradient formed by 1.5 liter of 0.02 M sodium phosphate buffer (pH 8.0) and 1.5 liter of 0.02 M sodium phosphate buffer (pH 8.0) containing 0.3 M NaCl. Both solutions also contain 0.1 mM DFP. The flow rate is 150 ml/hr. (B) Elution pattern for the CM-Sephadex column. The fractions from the second DEAE–Sephadex column are dialyzed and applied to the CM-Sephadex column (5 × 25 cm), previously equilibrated with 0.05 M sodium acetate buffer (pH 5.6) containing 0.02 M NaCl. Protein is eluted from the column with a linear gradient formed with 1.2 liter of 0.05 M sodium acetate buffer (pH 5.6) containing 0.1 M NaCl and 100 mg/liter polybrene and 1.2 liter of 0.05 M sodium acetate buffer (pH 5.6) containing 0.35 M NaCl and 100 mg/liter of polybrene. The flow rate is 150 ml/hr. Prekallikrein is measured after conversion to kallikrein by trypsin as described under Methods. ●——●, Absorbance; ○----○, amidase activity. Fractions shown by the solid bar are pooled.

containing 0.025 M NaCl. The dialyzed solution is applied at a flow rate of 75 ml/hr to a benzamidine–agarose column (2.6 × 25 cm), which was previously equilibrated with 0.05 M imidazole hydrochloride buffer (pH 6.0) containing 0.025 M NaCl. The column is washed with 250 ml of 0.05 M imidazole hydrochloride buffer (pH 6.0) containing 0.025 M guanidine hydrochloride and 0.025 M NaCl. The prekallikrein is eluted by a linear gradient consisting of 250 ml of 0.05 M imidazole hydrochloride buffer (pH 6.0) containing 0.025 M guanidine hydrochloride and 0.025 M NaCl and 250 ml of 0.05 M imidazole hydrochloride buffer (pH 6.0), containing 0.75 M guanidine hydrochloride and 0.025 M NaCl. Prekallikrein is eluted with the ascending edge of the protein peak (Fig. 3A). When kallikrein contaminates the preparation, it appears as a second peak of activity in the descending portion of the protein peak.

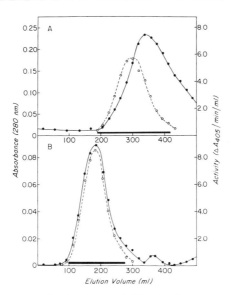

Fig. 3. Elution pattern for bovine prekallikrein from benzamidine–agarose and arginine methyl ester–agarose column chromatography. (A) Elution pattern from the benzamidine–agarose column for prekallikrein activity pooled from the CM-Sephadex column. Protein is eluted from the column (2.6 × 25 cm) with a linear salt gradient formed by 250 ml of 0.05 M imidazole hydrochloride buffer (pH 6.0) containing 0.025 M NaCl and 0.025 M guanidine hydrochloride, and 250 ml of 0.05 M imidazole hydrochloride buffer (pH 6.0) containing 0.025 M NaCl and 0.75 M guanidine hydrochloride. The flow rate is 75 ml/hr. (B) Elution pattern for the arginine methyl ester–agarose column. The fractions from the benzamidine–agarose column are dialyzed and applied to an arginine methyl ester–agarose column (1.6 × 30 cm) previously equilibrated with 0.05 M Tris-HCl buffer (pH 8.0) containing 0.05 M NaCl. Protein is eluted from the column with a linear gradient formed with 250 ml of 0.05 M Tris-HCl buffer (pH 8.0) containing 0.1 M NaCl and 250 ml of 0.05 M Tris-HCl buffer (pH 8.0) containing 1 M NaCl. Both solutions also contain 0.1 mM DFP. The flow rate is 40 ml/hr. Prekallikrein activity is measured after conversion to kallikrein by trypsin as described under Methods. ●——●, Absorbance; ○----○, amidase activity. Fractions shown by the solid bar are pooled.

Chromatography on Arginine Methyl Ester–Agarose. DFP (0.1 mM) is added to the pooled fractions of prekallikrein from benzamidine–agarose, and the sample is dialyzed against 4 liters of 0.05 M Tris-HCl buffer (pH 8.0) containing 0.05 M NaCl. The sample is applied to an arginine methyl ester–agarose column (1.6 × 30 cm) previously equilibrated with the dialysis buffer. The flow rate is adjusted to 40 ml/hr. The column is washed with 150 ml of 0.05 M Tris-HCl buffer (pH 8.0) containing 0.1 M NaCl and 0.1 mM DFP, and prekallikrein is eluted with a linear gradient consisting of 250 ml of 0.05 M Tris-HCl buffer (pH 8.0) containing 0.1 M NaCl and 0.1 mM DFP and 250 ml of 0.05 M Tris-HCl buffer (pH 8.0) containing 1.0 M NaCl and 0.1 mM DFP. Fractions containing prekallikrein (Fig. 3B) are pooled, and DFP is added to a final concentration of 0.1 mM. From 14 liters of bovine plasma, 5.5 mg of purified prekallikrein is obtained with a specific activity of 58 units/mg of protein as measured by the coagulant assay (Table I). The final preparation is dialyzed against 0.02 M 2-(morpholino)ethanesulfonic acid buffer (pH 6.0) containing 0.05 M NaCl, concentrated to ~15 ml, and stored at −70°C.

Purification of Human Plasma Prekallikrein

Ammonium Sulfate Fractionation and Chromatography on DEAE–Sephadex. Five liters of fresh-frozen human plasma are rapidly thawed at 37°C. After addition of 50 mg of heparin and 100 mg of soybean-trypsin inhibitor, the pooled plasma is stirred for 30 min at 4°C with 350 ml of 1 M barium chloride, and the precipitate formed is removed by centrifugation. The purification steps through chromatography on the second DEAE–Sephadex A-50 column are identical with those described in Chapter 16 in this volume for purification of human factor XII. The recovery and purification for these steps are shown in Table II.

Chromatography on the Second DEAE–Sephadex. The combined passthrough and wash fractions from the first DEAE–Sephadex column are dialyzed against the equilibration buffer and then applied to a second DEAE–Sephadex A-50 column (10 × 20 cm), which is previously equilibrated with 0.03 M Tris-HCl buffer (pH 7.5) containing 0.05 M NaCl. The column is washed with 2 liters of 0.03 M Tris-HCl buffer (pH 7.5) containing 0.05 M NaCl and 0.1 mM DFP. The passthrough and wash fractions are combined and the pH is adjusted to 7.2 with 1 N HCl. Soybean-trypsin inhibitor (100 mg) is then added. This sample is used for further purification of prekallikrein, and the fraction adsorbed to the column is employed for the purification of factor XII.

Chromatography on Heparin–Agarose. The pooled fraction is then applied to a heparin–agarose column (8 × 20 cm) previously equilibrated with 0.02 M Tris-HCl buffer (pH 7.2) containing 0.055 M NaCl. The column is washed and eluted as described for the bovine preparation. Prekallikrein appears early in the protein peak (Fig. 4A) and is well separated from factor XI, which elutes at the end of the protein peak.

Chromatography on CM-Sephadex. Polybrene (100 mg/liter) and soybean-trypsin inhibitor (50 mg) are then added to the prekallikrein sample from the heparin–agarose column and dialyzed overnight against 40 liters of 0.05 M sodium acetate buffer (pH 5.2) containing 0.05 M NaCl. This sample is applied to a CM-Sephadex C-50 column (4.5 × 19 cm) previously equilibrated with 0.05 M sodium acetate buffer (pH 5.2) containing 0.02 M NaCl. The flow rate is adjusted to 150 ml/hr. The column is

TABLE II

PURIFICATION OF HUMAN PLASMA PREKALLIKREIN

Purification step	Volume (ml)	Total protein (mg)[a]	Total activity (units)[b]	Specific activity (units/mg)	Recovery (%)	Purification (-fold)
Plasma	4600	2.42×10^5	4600	0.019	100	1
$(NH_4)_2SO_4$ fractionation	2250	1.02×10^5	4120	0.041	90	2.2
First DEAE–Sephadex (pH 5.8)	4500	7.2×10^4	3970	0.055	86	2.9
Second DEAE–Sephadex (pH 7.5)	8000	4.35×10^4	3750	0.086	82	4.5
Heparin–agarose	1330	3.28×10^3	3330	1.02	72	54
CM-Sephadex	1050	370	1940	5.2	42	280
Benzamidine–agarose	1350	260	1280	4.9	28	260
Concanavalin A–agarose	900	65	1080	16.6	24	880
Agmatine–agarose	124	8.5	583	68.6	13	3600

[a] Protein concentration is determined by adsorption employing $E_{280}^{1\%} = 10.0$.
[b] Activity of prekallikrein is assayed with the coagulant assay. Before assaying, the first three samples (1.0 ml) are dialyzed for 3 hr against 0.03 M Tris-HCl buffer (pH 7.5) and 0.05 M NaCl, and are passed through a DEAE–Sephadex column (0.7 × 4 cm), which is then washed with 4.0 ml of the same buffer.

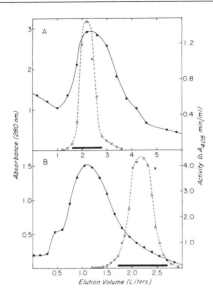

FIG. 4. Elution pattern for human prekallikrein from heparin–agarose and CM-Sephadex column chromatography. (A) Elution pattern from the heparin–agarose column for prekallikrein activity from the pooled passthrough and wash fractions from the DEAE–Sephadex column. Protein is eluted from the column (8 × 20 cm) with a linear gradient formed by 3 liters of 0.02 *M* Tris-HCl buffer (pH 7.2) containing 0.055 *M* NaCl and 3 liters of 0.02 *M* Tris-HCl buffer (pH 7.2) containing 0.55 *M* NaCl. Both solutions also contain polybrene (50 mg/liter) and 0.1 m*M* DFP. The flow rate is 400 ml/hr. (B) Elution pattern from the CM-Sephadex column. The fractions from the heparin–agarose column are dialyzed and applied to a CM-Sephadex column (4.5 × 19 cm) previously equilibrated with 0.05 *M* sodium acetate buffer (pH 5.2) containing 0.02 *M* NaCl. The column is eluted with a linear gradient formed with 1.5 liters of 0.05 *M* sodium acetate buffer (pH 5.2) containing 0.15 *M* NaCl and 100 mg/liter of polybrene and 1.5 liters of 0.05 *M* sodium acetate buffer (pH 5.2) containing 0.35 *M* NaCl and 100 mg/liter of polybrene. The flow rate is 150 ml/hr. Prekallikrein is measured after conversion to kallikrein by trypsin as described under Methods. ●——●, Absorbance; ○----○, amidase activity. Fractions shown by the solid bar are pooled.

washed with 4 liters of 0.05 *M* sodium acetate buffer (pH 5.2) containing 0.15 *M* NaCl and 100 mg/liter polybrene. The protein is eluted from the column with a linear salt gradient composed of 1.5 liters of 0.05 *M* sodium acetate buffer (pH 5.2) containing 0.15 *M* NaCl and polybrene (100 mg/liter) and 1.5 liters of sodium acetate buffer (pH 5.2) containing 0.35 *M* NaCl and polybrene (100 mg/liter). Prekallikrein appears in the descending portion of the main protein peak (Fig. 4B).

Chromatography on Benzamidine–Agarose. Fractions containing prekallikrein from the CM-Sephadex column are combined and dialyzed overnight against 18 liters of 0.02 *M* imidazole hydrochloride buffer, pH 6.0. The dialyzed solution is applied to a benzamidine–agarose column (2.6 × 28 cm) at a flow rate of 75 ml/hr. The column is previously equilibrated with 0.02 M imidazole hydrochloride buffer (pH 6.0) containing 0.02 M NaCl. The column is then washed with 250 ml of 0.02 M imidazole hydrochloride buffer (pH 6.0) containing 0.02 M NaCl and 0.02 M guanidine hydrochloride. Prekallikrein is present in the passthrough and wash fractions, and any contaminating kallikrein is adsorbed to the column. Kallikrein can be eluted from this column with 0.75 M guanidine hydrochloride in 0.02 M imidazole hydrochloride buffer (pH 6.0) containing 0.02 M NaCl.

Chromatography on Concanavalin A–Agarose. The combined fractions containing prekallikrein are dialyzed overnight against 18 liters of 0.02 M 2-(morpholino)ethanesulfonic acid buffer (pH 6.0) containing 0.05 M NaCl. The prekallikrein sample is applied at a flow rate of 100 ml/hr to a concanavalin A–agarose column (3.25 × 12 cm) equilibrated with the dialysis buffer. The column is washed with 2 liters of 0.02 M 2-(morpholino)ethanesulfonic acid buffer (pH 6.0) containing 0.25 M NaCl, and prekallikrein is eluted with 0.175 M α-methyl-D-glucoside in 0.02 M 2-(morpholino)ethanesulfonic acid buffer (pH 6.0) containing 0.25 M NaCl.

Chromatography on Agmatine–Agarose. The eluted fraction is dialyzed against 18 liters of 0.02 M imidazole hydrochloride buffer, pH 6.0. The prekallikrein sample is applied to an agmatine–agarose column (1.6 × 30 cm) previously equilibrated with the dialysis buffer at a flow rate of 50 ml/hr. The column is washed with 150 ml of 0.02 M imidazole hydrochloride buffer (pH 6.0) and is eluted with a linear gradient (Fig. 5)

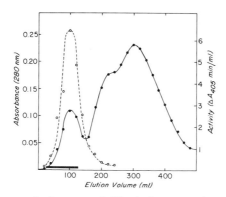

FIG. 5. Elution pattern for human prekallikrein from agmatine–agarose. Prekallikrein activity pooled from the concanavalin A–agarose column is dialyzed and applied to an agmatine–agarose column (1.6 × 30 cm) equilibrated with 0.02 M imidazole hydrochloride buffer, pH 6.0. The protein is eluted with a linear salt gradient formed with 250 ml of 0.02 M imidazole hydrochloride buffer (pH 6.0) and 250 ml of 0.02 M imidazole hydrochloride buffer (pH 6.0) containing 0.2 M NaCl. The flow rate is 50 ml/hr. Prekallikrein activity is measured after conversion to kallikrein by trypsin as described under Methods. ●——●, Absorbance; ○----○, amidase activity.

TABLE III
AMINO ACID AND CARBOHYDRATE COMPOSITIONS OF
PLASMA PREKALLIKREIN

Component	Residues/82,000 g of glycoprotein	
	Bovine[15]	Human[21]
Amino acid		
Lysine	38.3	39.1
Histidine	19.7	15.0
Arginine	30.1	24.4
Aspartic acid	54.0	51.8
Threonine	52.6	47.3
Serine	59.4	56.8
Glutamic acid	66.9	60.1
Proline	26.9	27.4
Glycine	55.7	54.7
Alanine	33.3	30.7
Half-cystine	38.8	32.0
Valine	29.8	37.4
Methionine	4.7	6.8
Isoleucine	40.3	30.6
Leucine	46.3	46.6
Tyrosine	16.5	20.3
Phenylalanine	26.1	27.4
Tryptophan	9.2	16.4
Carbohydrate		
Hexose	28.3 (6.2%)	37.3 (8.2%)
N-Acetylglucosamine	16.7 (4.5%)	18.5 (5.0%)
N-Acetylneuraminic acid	5.8 (2.2%)	5.8 (2.2%)
Protein (%)	87.1	84.6
Carbohydrate (%)	12.9	15.4

consisting of 250 ml of 0.02 M imidazole hydrochloride buffer (pH 6.0) and 250 ml of 0.02 M imidazole hydrochloride buffer (pH 6.0) containing 0.2 M NaCl. The purified prekallikrein (approximately 8 mg from 5 liters of plasma) is concentrated to ~15 ml and stored at −70°C.

General Properties of Plasma Prekallikrein

Bovine and human plasma prekallikrein are single-chain glycoproteins, each with a molecular weight of approximately 82,000. The amino acid and carbohydrate compositions are presented in Table III. Although

		1			5			
Human plasma prekallikrein	Gly	Cys	Leu	Thr	Gln	Leu	Tyr	Glu
Bovine plasma prekallikrein	Gly	Cys	Leu	Thr	Gln	Leu	Tyr	His

	10				15		
Asn	Ala	Phe	Phe	Arg	Gly	?	Asp
Asn	Ile	Phe	Phe	Arg	Gly	Gly	?

FIG. 6. Amino-terminal sequences of human[21] and bovine[15] plasma prekallikrein. Amino acids that are identical are shown in blocks, and those residues that are not known are shown as question marks.

similar in composition, minor differences are noted. The purified preparations migrate as a doublet with apparent molecular weights of 89,000 and 86,000 on sodium dodecyl sulfate–polyacrylamide gel electrophoresis.[15] Similar results have been observed by a number of groups for human plasma prekallikrein, which migrated with apparent molecular weights of 88,000 and 85,000.[13,16,21] A minimum molecular weight of 82,000 has been obtained by sedimentation equilibrium centrifugation of the bovine preparation.[15] The amino-terminal sequences of human and bovine plasma prekallikrein are shown in Fig. 6.[15] Since only a single glycine residue was observed in the first turn of the amino-terminal sequence analysis, it seems likely that the observed heterogeneity is due to heterogeneity in carbohydrate content or a modification at the carboxyl-terminal region of the polypeptide chain. The human and bovine preparations are nearly identical except for positions 8 and 10. Eight of the fifteen amino acid residues of human prekallikrein are homologous with the amino-terminal sequence of factor XI. The isoelectric points for the bovine and human preparations are 6.98[12] and 7.7,[11] respectively, as determined by isoelectric focusing.

Prekallikrein occurs in bovine plasma at a concentration of approximately 3 μg/ml of plasma. The extinction coefficient as determined by differential refractometry is 10.9.[15] The human protein appears to be more abundant in plasma with estimates ranging from 40 to 55 μg/ml of plasma.[16,22,23] However, the results presented here suggest a value of 15–20 μg/ml of plasma.

Several enzymes catalyze the activation of plasma prekallikrein to

[21] R. L. Heimark and E. W. Davie, in preparation.
[22] H. Saito, M. Poon, W. Vicic, G. H. Goldsmith, and J. E. Menitove, J. Lab. Clin. Med. 92, 84 (1978).
[23] B. N. Bouma, D. M. Kerbiriou, R. A. A. Vlooswijk, and J. H. Griffin, J. Lab. Clin. Med. 96, 693 (1980).

kallikrein including trypsin,[11] factor XII_a fragments,[7,13,14,16,24] factor XII_a,[15,25,26] and factor XII in the presence of certain negatively charged surfaces.[10,27] The conversion of prekallikrein to kallikrein involves the hydrolysis of an internal peptide bond resulting in a 2-chain structure held together by disulfide bond(s). Bovine kallikrein contains a heavy chain with an apparent molecular weight of 52,000 and a light chain of 38,000 or 33,000.[26] From human kallikrein, the heavy chain has an apparent molecular weight of 52,000 and the light chain of 36,000 or 33,000.[13,16] The cleavage site is apparently an arginine residue[25] resulting in generation of an identical amino-terminal sequence of Ile-Val-Gly-Gly-Thr-Asn-Ala-Ser-Trp-Gly for both of the light chains.[5] Antithrombin III in the presence of heparin rapidly inactivates plasma kallikrein,[5,28,29] and complete inhibition of bovine kallikrein activity was obtained when the molar ratio of enzyme to inhibitor was 1 : 1.[5] The light chain of plasma kallikrein contains the active site, as shown by labeling with radioactive DFP.[13]

[24] S. D. Revak, C. G. Cochrane, A. R. Johnston, and T. E. Hugli, *J. Clin. Invest.* **54**, 619 (1974).

[25] H. Takahashi, S. Nagasawa, and T. Suzuki, *FEBS Lett.* **24**, 98 (1972).

[26] R. L. Heimark, K. Fujikawa, and E. W. Davie, *Fed. Proc., Fed. Am. Soc. Exp. Biol.* **37**, 1587 (1978).

[27] C. G. Cochrane, S. D. Revak, and K. D. Wuepper, *J. Exp. Med.* **138**, 1564 (1973).

[28] A. M. Vennerod, K. Laake, A. K. Solberg, and S. Stromland, *Thromb. Res.* **9**, 457 (1976).

[29] B. Lahiri, A. Bagdasarian, B. Mitchell, R. C. Talamo, R. W. Colman, and R. D. Rosenberg, *Arch. Biochem. Biophys.* **175**, 737 (1976).

[15] HMW and LMW Kininogens

By Hisao Kato, Shigeharu Nagasawa, and Sadaaki Iwanaga

There exist at least two types of kininogens in mammalian blood plasmas, which differ in their molecular structures and biological functions. A kininogen with MW 50,000 was first isolated from bovine plasma by Habermann[1] in 1963 and subsequently by Suzuki *et al.*[2] in 1965. In 1966, Jacobsen[3] reported the presence of another kininogen with higher molecular weight in various mammalian plasmas. The higher-molecular-weight form of kininogen was isolated in pure form by Komiya *et al.*[4] from bovine plasma and by Habal *et al.*[5] from human plasma in 1974. Since then, the

[1] E. Habermann, *Biochem. Z.* **337**, 440 (1963).

[2] T. Suzuki, Y. Mizushima, T. Sato, and S. Iwanaga, *J. Biochem.* (*Tokyo*) **57**, 14 (1965).

[3] S. Jacobsen, *Br. J. Pharmacol. Chemother.* **26**, 403 (1966).

[4] M. Komiya, H. Kato, and T. Suzuki, *J. Biochem.* (*Tokyo*) **76**, 811 (1974).

[5] F. Habal, H. Z. Movat, and C. E. Burrows, *Biochem. Pharmacol.* **23**, 2291 (1974).

presence of two kinds of kininogens with different molecular weights was confirmed by many investigators and they were designated as low-molecular-weight (LMW) kininogen and high-molecular-weight (HMW) kininogen.[6,7]

In 1974, Wuepper et al.[8] found that a deficiency, Flaujeac trait, which showed a prolonged clotting time in vitro, lacks kininogen. Prior to this discovery, an unknown deficiency called Fitzgerald trait had been reported,[9] which subsequently was found to be due to a lack of HMW kininogen by Saito et al.[10] It has now been established that HMW kininogen is required as a cofactor in the contact phase of intrinsic blood coagulation.

Assay Methods

Assay of Kinin

Three kinds of kinins—bradykinin, kallidin, and methionyl-kallidin—are liberated from kininogen by various kinin-releasing enzymes, as shown in Table I. These kinins contract isolated smooth muscles, lower blood pressure, and increase capillary permeability. For the bioassay of kinins using these pharmacological activities, it should be kept in mind that the released kinins differ depending on the enzyme used, and that these kinins show quantitatively different pharmacological activities (Table I). For the quantitative estimation of kinin activity, guinea pig ileum and estrus rat uterus are ordinarily used, with the latter being more sensitive.

Reagent

De Jalon's solution: 9 g NaCl, 0.4 g KCl, 0.06 g $CaCl_2$, 0.15 g $NaHCO_3$, and 1 g glucose dissolved in 1 liter of deionized water

Procedures

A virgin Sprague-Dawley rat (120–200 g) is injected subcutaneously with estrogenic hormone (e.g., diethylstilbesterol, hexesterol, or estradiolbenzoate) 12–24 hr before assay. After the rat is stunned and ex-

[6] E. G. Erdos, ed., "Bradykinin, Kallidin and Kallikrein," Handb. Exp. Pharmacol., Vol. 25. Springer-Verlag, Berlin and New York, 1970.

[7] E. G. Erdös, ed., "Bradykinin, Kallidin and Kallikrein, Supplement," Handb. Exp. Pharmacol., Suppl. to Vol. 25. Springer-Verlag, Berlin and New York, 1979.

[8] K. D. Wuepper, D. R. Miller, and M. J. Lacombe, Fed. Proc., Fed. Am. Soc. Exp. Biol. 34, 859 (1975).

[9] R. Waldmann and J. P. Abraham, Br. J. Haematol. 44, 934 (1974).

[10] H. Saito, O. D. Ratnoff, R. Waldmann, and J. P. Abraham, J. Clin. Invest. 55, 1082 (1975).

TABLE I
KININS AND THEIR BIOLOGICAL ACTIVITY

Kinin	Kinin-releasing enzymes	Biological activity[a]			
		Contraction of rat uterus (0.03 ng/ml dose)	Contraction of guinea pig ileum (1 ng/ml dose)	Lowering of rabbit blood pressure (0.05 μg/kg dose)	Increase of guinea pig capillary permeability (1 ng dose)
Bradykinin Arg-Pro-Pro-Gly-Phe-Ser-Pro-Phe-Arg	Plasma kallikrein Plasmin Snake venom kininogenase	1	1	1	1
Kallidin Lys-Arg-Pro-Pro-Gly-Phe-Ser-Pro-Phe-Arg	Glandular kallikrein	0.6	0.3	1.9	1
Methionyl-kallidin Met-Lys-Arg-Pro-Pro-Gly-Phe-Ser-Pro-Phe-Arg	Pepsin[b] Unknown enzyme in plasma[b]	0.3	0.1–0.3	2–3	1

[a] Threshold doses of bradykinin are given in parentheses. The values are relative values of biological activity.

[b] No enzyme(s) in plasma that would liberate methionyl-kallidin has been identified. It has been reported that pepsin liberates the kinin from kininogen [V. Hial, H. R. Keiser, and J. J. Pisano, *Biochem. Pharmacol.* **25**, 2499 (1976)]. However, it remains to be established that the action of pepsin is directly on kininogen.

sanguinated, the estrus uterus is suspended in a 10-ml organ bath filled with De Jalon's solution. The bath is aerated and thermostatted at 28°C. The contraction of the uterus is recorded isotonically or isometrically using a transducer that is connected to a pen recorder. The contact time of a sample with the uterus is 90 sec and the interval between tests is 5 min. A standard curve is made by plotting the height of contractions against the amount of synthetic bradykinin, on semilogarithmic paper.

Assay method for kinin using guinea pig ileum has been described.[11]

Kinin can be also measured by radioimmunoassay[12–14] or by enzyme immunoassay.[15] For the radioimmunoassay, tyrosine[8]–bradykinin is labeled with ^{125}I by the chloramine-T method, and used as a radiolabeled tracer. Antiserum against bradykinin is prepared using bradykinin immunogen, which is obtained by coupling bradykinin to bovine serum albumin or to ovalbumin by toluene-2,4-diisocyanate or by water-soluble carbodiimide. After incubation of sample with $[^{125}I]Tyr^8$-bradykinin and antiserum, unbound antigen is separated using dextran–charcoal or polyethylene glycol and counted.

An enzyme immunoassay of bradykinin has been developed by using β-D-galactosidase as a labeling enzyme. Bradykinin is conjugated to β-D-galactosidase with N-(m-maleimide benzoyloxy)-succinimide. Antiserum is prepared from rabbits immunized with bradykinin linked to albumin with toluene-2,4-diisocyanate. After the incubation of sample with bradykinin–β-D-galactosidase and antiserum, precipitate was obtained by adding anti-rabbit IgG serum. Enzyme activity in the precipitate is estimated using 4-methylumbelliferyl-β-D-galactoside as substrate.

Assay of Kininogens

Kininogen can be estimated by measuring kinin biologically or immunologically following incubation with kinin-releasing enzymes. Differential assay for HMW kininogen and LMW kininogen is obtained by utilizing the specificity of kinin-releasing enzymes. Plasma kallikrein liberates bradykinin from HMW kininogen, whereas glandular kallikrein liberates kallidin from both of HMW and LMW kininogens. Trypsin and snake venom kininogenase liberate bradykinin from both kininogens. For the bioassay of kininogens, attention should be paid to the contamination of

[11] M. E. Webster and E. S. Prado, this series, Vol. 19, p. 680.
[12] R. C. Talamo, E. Haber, and K. F. Austen, *J. Lab. Clin. Med.* **74**, 816 (1979).
[13] O. A. Carretero, N. B. Oza, A. Piwonska, T. Ocholik, and A. G. Scicli, *Biochem. Pharmacol.* **25**, 2265 (1976).
[14] K. Shimamoto, T. Ando, T. Nakao, S. Tanaka, M. Sakuma, and M. Miyahara, *J. Lab. Clin. Med.* **91**, 721 (1978).
[15] A. Ueno, S. Oh-Ishi, T. Kitagawa, and M. Katori, *Adv. Exp. Med. Biol.* **120A**, 195 (1979).

kininase or inhibitors for enzymes in kininogen preparation, especially when partially purified kininogen or plasma is used. Kininase in the preparation can be inactivated by adding o-phenanthroline (3 mM) to the incubation mixture. For the measurement of kininogen content in blood plasma, a good method has been developed.[16] For total kininogen, plasma is pretreated at pH 2.0, and the bradykinin released by trypsin is separated and assayed on rat uterus. Under this condition, kininase is inactivated and the release of bradykinin-potentiating peptides is minimized. For HMW kininogen, plasma is incubated with glass powder in the presence of o-phenanthroline. For LMW kininogen, plasma is incubated with glass powder in the absence of o-phenanthroline and the HMW kininogen-depleted plasma is treated in the same way as for the total kininogen assay. By this method, HMW kininogen, LMW kininogen, and total kininogen levels in human plasma were determined to be 0.98 ± 0.05, 3.08 ± 0.09, 4.11 ± 0.10 μg bradykinin equivalents/ml plasma, respectively.

HMW kininogen can also be measured specifically by its ability to correct the kaolin-activated partial thromboplastin time of HMW kininogen-deficient plasma.[17] The method is essentially the same as for factor XII,[18] except for using HMW kininogen-deficient plasma in place of factor XII-deficient plasma. Immunological assay for HMW kininogen has been developed by Kleniewski and Donaldson.[19] Since HMW kininogen and LMW kininogen cross-react with regard to their antibodies, antiserum to human HMW kininogen is treated with an excess amount of LMW kininogen before a monospecific antibody to HMW kininogen is prepared. Using the antibody, human HMW kininogen is determined by inhibition of tanned red cell hemagglutination. A radioimmunoassay for human HMW kininogen has been recently reported, using ^{125}I-labeled HMW kininogen.[20] Specific antiserum to human HMW kininogen can be obtained with the use of light chain from human HMW kininogen as antigen.[21,22] The antiserum against human HMW kininogen does not cross-react with other mammalian plasma except monkey.[23]

Purification of Kininogens

Purification of human LMW kininogen has been reported by a variety of laboratories.[24–28] In general, purification of LMW kininogen requires

[16] Y. Uchida and M. Katori, *Thromb. Res.* **15**, 127 (1979).
[17] H. Saito, G. Goldsmith, and R. Waldmann, *Blood* **48**, 941 (1976).
[18] J. H. Griffin and C. G. Cochrane, this series, Vol. 45, p. 56.
[19] J. Kleniewski and V. H. Donaldson, *Proc. Soc. Exp. Biol. Med.* **156**, 113 (1977).
[20] D. Proud, J. V. Pierce, and J. J. Pisano, *J. Lab. Clin. Med.* **95**, 563 (1980).
[21] D. M. Kerbiriou, B. N. Bouma, and J. H. Griffin, *J. Biol. Chem.* **255**, 3952 (1980).
[22] K. Mori and S. Nagasawa, *J. Biochem.* (*Tokyo*) **89**, 1465 (1981).
[23] V. H. Donaldson and R. Breyler, *Proc. Soc. Exp. Biol. Med.* **160**, 134 (1979).

many steps and the recovery from plasma is very low. On the other hand, purification of HMW kininogen is very simple and the procedures reported by many investigators[7,8,29-31] include ion-exchange column chromatography on DEAE–Sephadex or QAE-Sephadex, CM-Sephadex or SP-Sephadex and gel filtration. However, a difficulty in the purification of HMW kininogen is the susceptibility of the kininogen to plasma kallikrein. Since HMW kininogen is present as a complex with prekallikrein in plasma, a trace of prekallikrein is always contaminating HMW kininogen during the purification procedures. Factor XII is easily activated on contact with a negatively charged surface, such as glass or CM-Sephadex, and activates prekallikrein into kallikrein. For these reasons, purified HMW kininogen preparations often contain a nicked kininogen or a kinin-free protein that underwent limited proteolysis by plasma kallikrein. Therefore, the purification of HMW kininogen requires a variety of inhibitors—such as hexadimethrine bromide, benzamidine, DFP, and soybean-trypsin inhibitor (SBTI)—to inhibit the activation of factor XII and to inactivate the activated factor XII and kallikrein. Although it is very difficult to separate a trace of prekallikrein from HMW kininogen, we have recently found that gel filtration in a buffer containing 1 M NaCl is very effective in removing prekallikrein.[32] Fujikawa et al.[33] also reported that DEAE–Sephadex column chromatography in the presence of 5 M urea is useful in removing prekallikrein from HMW kininogen. During purification, kininogen is measured quantitatively by the bioassay of kinin liberated by trypsin or other kinin-releasing enzymes, and qualitatively by immunodiffusion. HMW kininogen can be measured by its correcting ability regarding the prolonged clotting time of kininogen-deficient plasma. Attention should be paid to the fact that the immunodiffusion method and the clotting-time method do not discriminate between kininogen and kinin-free protein.

[24] J. V. Pierce and M. E. Webster, in "Hypotensive Peptides" (E. G. Erdös et al., eds.), p. 130. Springer-Verlag, Berlin and New York, 1966.

[25] J. Spragg and K. F. Austen, J. Immunol. 107, 1512 (1971).

[26] U. Hamberg, P. Elg, E. Nissinen, and P. Stelwagen, Int. J. Pept. Protein Res. 7, 261 (1975).

[27] S. Nagasawa and T. Nakayasu, Chem. Biol. Kallikrein-Kinin Syst. Health Dis. Fogarty Int. Cent. Proc., 27, 139 (1974).

[28] W. Sakamoto and O. Nishikaze, J. Biochem. (Tokyo) 86, 1549 (1979).

[29] R. E. Thompson, R. Mandle, Jr., and A. P. Kaplan, J. Exp. Med. 147, 488 (1978).

[30] T. Nakayasu and S. Nagasawa, J. Biochem. (Tokyo) 85, 249 (1979).

[31] D. M. Kerbiriou and J. H. Griffin, J. Biol. Chem. 254, 12020 (1979).

[32] T. Shimada, T. Sugo, H. Kato, and S. Iwanaga, Seikagaku 52, 734 (1980).

[33] K. Fujikawa, R. L. Heimark, K. Kurachi, and E. W. Davie, Biochemistry 19, 1322 (1980).

Purification Procedure for Bovine HMW Kininogen[4,32]

Nine volumes of bovine blood are mixed with one volume of anticoagulant (15 g of EDTA · Na$_2$, 50 g glucose, and 500 mg hexadimethrine bromide dissolved in 1 liter of deionized water). Plasma is obtained by centrifugation of blood at 4000 rpm for 30 min, using the swinging rotor of the Sorvall RC-3 centrifuge. Purification steps except step 1 are carried out in plastic ware or in siliconized glassware at 4°C. All buffers contain 50 μg of hexadimethrine bromide per milliliter, 1 mM benzamidine HCl, and 0.02% NaN$_3$.

Step 1. DEAE–Sephadex A-50 Column Chromatography. Bovine plasma (10 liters) is applied to a column (23 × 10.8 cm) of DEAE–Sephadex A-50, which was equilibrated with 0.02 M Tris-HCl buffer (pH 8.0) containing 0.05 M NaCl and 3 mM EDTA. After washing with 20 liters of the equilibration buffer, proteins are eluted by a linear gradient formed with 10 liters each of buffers containing 0.05 M NaCl and 0.6 M NaCl, respectively. Each 1-liter fraction is collected. HMW kininogen is eluted with 0.3 M NaCl. Prekallikrein is found both in the nonadsorbed and in the kininogen fractions. The HMW kininogen fractions are pooled and dialyzed against 20 liters of deionized water for 10 hr.

Step 2. CM-Sephadex C-50 Column Chromatography. The dialysate from step 1 is adjusted to pH 6.3 with 20% acetic acid. Conductivity of the dialysate is checked with a conductivity meter and adjusted to an ionic strength corresponding to 0.2 M NaCl by adding NaCl. To the solution, SBTI (100 mg) and DFP (final concentration of 10^{-3} M) are added. After standing for 1 hr with stirring, the solution is mixed with 1 liter bed volume of CM-Sephadex C-50, which was equilibrated with 0.05 M acetate buffer (pH 6.3), containing 0.2 M NaCl. Stirring is continued for 30 min and the slurry is used to prepare a column (9 × 16 cm) which is washed with 2 liters of the equilibration buffer. Proteins are eluted by a linear gradient, formed with 2 liters each of 0.05 M acetate buffer (pH 6.3) containing 0.2 M NaCl, and 0.8 M NaCl. As shown in Fig. 1, proteins are eluted in two broad peaks, the latter of which has HMW kininogen activity and cross-reacts with antiserum against HMW kininogen. Prekallikrein is eluted in the preceding fraction. Fractions of Nos. 100–140 are pooled and DFP is added to a final concentration of 10^{-3} M. After 30 min of stirring, the solution is dialyzed against 20 liters of deionized water for 5 hr and subjected to lyophilization.

Step 3. Sephadex G-150 Gel Filtration. The lyophilized material from step 2 is dissolved in deionized water and DFP is added to a final concentration of 10^{-3} M. After 30 min of stirring, the solution is dialyzed overnight against 0.02 M Tris-HCl buffer (pH 8.0) containing 1 M NaCl and

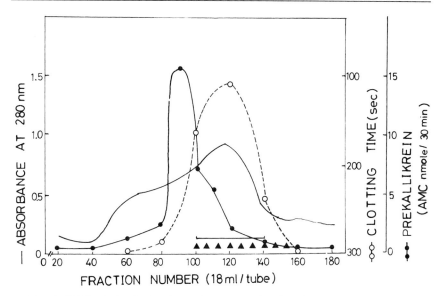

FIG. 1. CM-Sephadex C-50 column chromatography of bovine HMW kininogen. HMW kininogen in each fraction was assayed by the correction of kaolin-activated partial thromboplastin time of the Fitzgerald trait plasma. Prekallikrein was assayed using Z-Phe-Arg-MCA as substrate following activation by Factor XIIa, as described by T. Sugo, N. Ikari, H. Kato, S. Iwanaga, and S. Fujii [*Biochemistry* **19**, 3215 (1980)]. ▲, Cross-reacted with antiserum against HMW kininogen on immunodiffusion. Fractions as indicated by solid bar were collected.

3 mM EDTA. The dialyzed solution is applied to a column (4 × 141 cm) of Sephadex G-150, which was equilibrated with 0.02 M Tris-HCl buffer (pH 8.0) containing 1 M NaCl and 3 mM EDTA. Figure 2A shows that the bulk of prekallikrein is separated from HMW kininogen fraction by the gel filtration. Since prekallikrein is still contaminated with the HMW kininogen fractions, fraction Nos. 70–80 are subjected to gel filtration again using the same column. As shown in Fig. 2B, prekallikrein no longer contaminates the new fractions Nos. 70–80. The fractions are pooled and desalted using a column of Sephadex G-25, equilibrated with 0.2 M ammonium bicarbonate solution. Through these procedures, 220 mg of HMW kininogen is obtained, and the preparation shows a single band on SDS–polyacrylamide gel electrophoresis both in the presence and in the absence of 2-mercaptoethanol (see inset to Fig. 2B). Table II summarizes the purification for HMW kininogen from bovine plasma. The low recovery of HMW kininogen is due to the elimination of HMW kininogen fractions which contain prekallikrein.

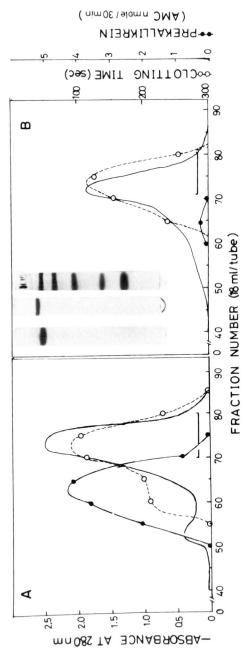

Fig. 2. First (A) and second (B) gel filtrations of bovine HMW kininogen. Inset of Fig. 2B shows patterns of SDS–polyacrylamide gel electrophoresis of purified HMW kininogen in the absence (1) and presence (2) of 2-mercapoethanol and marker proteins (3) (phosphorylase b, bovine serum albumin, ovalbumin, chymotrypsinogen, and cytochrome c). Fractions as indicated by solid bar were collected.

TABLE II
SUMMARY OF THE PURIFICATION OF BOVINE HMW KININOGEN[a]

Step	Total volume (ml)	Total A_{280}	Total units[b]	Recovery (%)
1. Plasma	10,200	719,100	10,200	100
2. DEAE–Sephadex A-50	5800	205,320	6960	68.2
3. CM-Sephadex C-50	860	1505	2150	21.1
4. Sephadex G-150	191	296	802	7.9
5. Sephadex G-150	160	166	464	4.5

[a] T. Shimada, T. Sugo, H. Kato, and S. Iwanaga, *Seikagaku* **52**, 734 (1980).

[b] HMW kininogen content was measured by its correcting activity of the abnormal clotting time of Fitzgerald trait plasma. One unit is defined as the amount of HMW kininogen in 1 ml of normal bovine plasma.

Purification Procedure for Human HMW Kininogen[30]

Step 1. QAE-Sephadex A-50 Chromatography. Human ACD-plasma (500 ml) is diluted by the addition of 250 ml of 0.02 M disodium phosphate, containing 100 mg of hexadimethrine bromide, 0.03 M EDTA-3Na, 0.3 M ε-aminocaproic acid (EACA), and 3 mM DFP, and adjusted to pH 8.0 with 1 N NaOH. The diluted plasma is kept at room temperature for 30 min and then cooled in an ice-water bath. To the cooled plasma is added about 300 ml of QAE-Sephadex A-50, which was equilibrated with 0.02 M sodium phosphate buffer (pH 8.0) containing 0.12 M NaCl, 0.1 M EACA, and 0.01 M EDTA-3Na. After stirring for 30 min in the cold room, the QAE-Sephadex slurry is poured into an 8-cm-diameter column to a height of 6 cm, and washed with 1 liter each of 0.12 M NaCl and 0.2 M NaCl in 0.02 M phosphate buffer (pH 8.0) containing 0.1 M EACA, 0.01 M EDTA-3Na, and 1 mM DFP, at a flow rate of 1 liter/hr. The column is then eluted with 1.5 liters of 0.02 M sodium phosphate buffer (pH 7.0) containing 0.4 M NaCl, 0.1 M EACA, 0.01 M EDTA-3Na, and 1 mM DFP. The flow rate is 1 liter/hr and 200-ml fractions are collected. Fractions in which kinin can be generated by plasma kallikrein are collected and a 600-ml pooled material is obtained.

Step 2. CM-Sephadex C-50 Chromatography. The pooled material is dialyzed overnight against 5 liters of 0.02 M sodium phosphate buffer (pH 6.0) containing 0.1 M NaCl and 0.01 M EACA. To the dialyzed sample is added 50 ml of CM-Sephadex C-50, which was equilibrated with 0.02 M sodium phosphate buffer (pH 6.0) containing 0.15 M NaCl, 0.1 M EACA, and 0.01 M EDTA-3Na. After stirring for 60 min, the CM-Sephadex slurry is poured into a 2.5-cm-diameter column to a height of 10 cm and washed

TABLE III

SUMMARY OF THE PURIFICATION OF HUMAN HMW KININOGEN[a,b]

Step	Protein (total A_{280})	Kininogen activity	
		Bradykinin equivalent (μg)	Specific activity (μg/A_{280})
1. QAE-Sephadex A-50	720	198	0.28
2. CM-Sephadex C-50	18.1	156	8.6
3. CM-Sephadex C-50	8.5	130	15.4

[a] T. Nakayasu and S. Nagasawa, *J. Biochem. (Tokyo)* **85**, 249 (1979).
[b] A typical procedure with 500 ml of human ACD-plasma.

with 100 ml of the equilibration buffer. The column is then eluted stepwise with 200 ml each of 0.3 M NaCl and 0.5 M NaCl with the equilibrating buffer, containing 1 mM DFP. The flow rate is 200 ml/hr and 15-ml fractions are collected. HMW kininogen is recovered in the 0.5 M NaCl eluate.

Step 3. CM-Sephadex C-50 Chromatography. The 0.5 M NaCl eluate is dialyzed overnight against 1 liter of 0.02 M sodium phosphate buffer (pH 6.0) containing 0.1 M NaCl and 0.1 M EACA, and applied to a column of CM-Sephadex C-50 (1 × 10 cm), equilibrated with the dialysis buffer. The column is eluted with a linear gradient of NaCl from 0.2 M to 0.6 M in the starting buffer, containing 1 mM DFP, at a flow rate of 20 ml/hr. 100 ml of each buffer is used. HMW kininogen is eluted as a single peak at a concentration of NaCl between 0.3 M and 0.4 M. The yield of a typical preparation is given in Table III.

Purification Procedure for Bovine LMW Kininogen

Although this method is essentially the same as the previous method,[34] minor modifications were made (unpublished results). In step 2, DEAE-Sephadex A-50 column chromatography was used and SP-Sephadex C-50 column was employed in step 3, instead of CM-cellulose. Since hydroxyapatite column was found to be very effective, it was added at the last step. Collection of blood is the same as the purification procedure for bovine HMW kininogen. Detection of LMW kininogen during the procedure can be made by the bioassay of kinin after incubation with snake venom kininogenase, or conveniently by immunodiffusion using antiserum either to HMW kininogen or LMW kininogen.

[34] M. Yano, H. Kato, S. Nagasawa, and T. Suzuki, *J. Biochem. (Tokyo)* **62**, 386 (1967).

Step 1. Zinc Acetate Precipitation. To 7.5 liters of plasma, ammonium sulfate is added to a final concentration of 25% with stirring. After 30 min, the precipitate is removed by centrifugation. To the supernatant, more ammonium sulfate is added to make a final concentration of 50%. After 1 hr, supernatant is obtained by centrifugation and dialyzed overnight against 40 liters of deionized water. To the dialysate (about 3 liters), zinc acetate is added to a final concentration of 20 mM, and the solution is adjusted to pH 8.0 with 0.1 M NaOH. The precipitate is removed by centrifugation and EDTA is added to the supernatant to make a final concentration of 50 mM. The solution is dialyzed overnight first against 40 liters of deionized water and then against 20 liters of deionized water and, finally, against 20 liters of 0.02 M Tris-HCl buffer (pH 8.0) containing 0.02 M NaCl.

Step 2. DEAE–Sephadex A-50 Column Chromatography. Half of the dialysate is applied to a column (4.3 × 40 cm) of DEAE–Sephadex A-50, which was equilibrated with 0.02 M Tris-HCl buffer (pH 8.0) containing 0.02 M NaCl. After washing with 2 liters of the equilibration buffer, protein is eluted by a linear gradient formed with 2 liters each of 0.02 M Tris-HCl buffer (pH 8.0) containing 0.02 M NaCl, and 0.4 M NaCl. Each 15-ml fraction is collected. Kininogen is eluted in the fractions between tube Nos. 130 and 170. The other half of the dialysate is chromatographed in the same way. The combined kininogen fraction is dialyzed overnight first against 20 liters of deionized water and then 20 liters of 0.05 M acetate buffer, pH 6.0.

Step 3. SP-Sephadex C-50 Column Chromatography. The dialysate from step 2 is applied to a column (4.3 × 40 cm) of SP-Sephadex C-50, which was equilibrated with 0.05 M acetate buffer, pH 6.0. Kininogen is found in the nonadsorbed fraction. After lyophilization, kininogen is dissolved in deionized water and dialyzed overnight against 10 liters of 0.02 M Tris-HCl (pH 8.0) containing 0.02 M NaCl.

Step 4. Sephadex G-100 Gel Filtration. The dialyzed solution from step 3 is applied to a column (5.5 × 135 cm) of Sephadex G-100, which was equilibrated with 0.02 M Tris-HCl buffer (pH 7.5) containing 0.02 M NaCl, and each 15-ml fraction is collected. Kininogen is found in the fractions between tube Nos. 58 and 78.

Step 5. DEAE–Sephadex A-50 Column Chromatography. The kininogen fraction from step 4 is applied to a column (4 × 40 cm) of DEAE–Sephadex A-50, which was equilibrated with 0.02 M Tris-HCl buffer (pH 8.0) containing 0.02 M NaCl. Chromatography is carried out the same way as in step 2. LMW kininogen is eluted in the second protein peak. The kininogen fraction is collected and dialyzed overnight against 20 liters of 0.01 M phosphate buffer, pH 8.0.

TABLE IV
PHYSICAL PROPERTIES OF HMW AND LMW KININOGENS

Property	HMW Kininogen			LMW Kininogen		
	Bovine[a]	Human[b]	Horse[c]	Bovine[d]	Human[e]	Rabbit[f]
Molecular weight	76,000[g] 80,000[h]	108,000[g] 110,000 − 120,000[h]	78,000[h]	48,000[g]	78,000[i] 78,000[h]	54,000[i]
Partial specific volume, V (ml/g)	0.718			0.688		
Isoelectric point	4.5	4.7		3.3	4.7	
$E_{1\,cm}^{1\,\%}$ (280 nm)	7.4	7.01	7.7	6.7		

[a] M. Komiya, H. Kato, and T. Suzuki, *J. Biochem.* (*Tokyo*) **76**, 811 (1974).
[b] R. E. Thompson, R. Mandle, Jr., and A. P. Kaplan, *J. Exp. Med.* **147**, 488 (1978); T. Nakayasu and S. Nagasawa, *J. Biochem.* (*Tokyo*) **85**, 249 (1979); D. M. Kerbiriou and J. H. Griffin, *J. Biol. Chem.* **254**, 12020 (1979); S. Schiffman, C. Mannhalter, and K. D. Tyner, *ibid.* **255**, 6433 (1980).
[c] T. Sugo, H. Kato, S. Iwanaga, and S. Fujii, *Eur. J. Biochem.* **115**, 439 (1981).
[d] H. Kato, T. Suzuki, K. Ikeda, and K. Hamaguchi, *J. Biochem.* (*Tokyo*) **60**, 643 (1966).
[e] S. Nagasawa and T. Nakayasu, *Chem. Biol. Kallikrein-kinin Syst. Health Dis. Fogarty Int. Cent. Proc.*, **27**, 139 (1974); W. Sakamoto and O. Nishikaze, *J. Biochem.* (*Tokyo*) **86**, 1540 (1979).
[f] T. P. Egorova, L. G. Makevnina, and T. S. Paskhina, *Biokhimiya* (*Moscow*) **41**, 1052 (1976).
[g] Sedimentation-equilibrium method.
[h] SDS–PAGE.
[i] Gel filtration.

Step 6. Hydroxyapatite Column Chromatography. The dialyzed solution from step 5 is applied to a column (1.5 × 21 cm) of hydroxyapatite Bio-Gel HTP, which was equilibrated with 0.01 M phosphate buffer, pH 8.0. Kininogen is found in the nonadsorbed fraction. The pooled material is subjected to lyophilization and desalted by a Sephadex G-25 column. Through these procedures, 360 mg of LMW kininogen, which shows a single band on SDS–polyacrylamide gel electrophoresis, is obtained.

Physical and Chemical Properties of HMW Kininogen

Physical properties of HMW kininogen in terms of molecular weight, partial specific volume, isoelectric point, and extinction coefficient are shown in Table IV. The molecular weight of bovine HMW kininogen has been extensively investigated by various methods.[4] A value of 76,000, calculated by the sedimentation equilibrium method, is in good agreement with the 80,000 obtained by gel filtration in 6 M guanidine hydrochloride. However, when bovine HMW kininogen was gel filtered on a column of

Bio-Gel P-300 in 0.02 M phosphate buffer (pH 7.0) containing 0.5 M NaCl, a value of 300,000 was calculated on the basis of calibration with standard proteins.[35] The abnormal behavior in gel filtration may be due to its glycoprotein nature, because the value reduces to 70,000 by using the sedimentation constant and Stokes radius based on the equation of Siegel and Monty.[4] On SDS–polyacrylamide gel electrophoresis, bovine HMW kininogen gave an apparent molecular weight of approximately 80,000 in the presence of 2-mercaptoethanol.[36] The molecular weight of human HMW kininogen, estimated by SDS–polyacrylamide gel electrophoresis is in the range of 110,000–120,000.[29–31,37] Kerbiriou and Griffin calculated 108,000 by sedimentation equilibrium.[31] Human HMW kininogen also shows an abnormal behavior on gel filtration similarly to bovine HMW kininogen. Habal et al.[38] calculated a value of 210,000 by gel filtration in nondenaturating buffer, and 108,000 in 4 M guanidine hydrochloride. From these results, they postulated that human HMW kininogen exists as a dimer in solution. However, this may be unlikely in view of the quoted results with the sedimentation-equilibrium method.[31]

Table V shows the amino acid composition and carbohydrate content of HMW kininogen. Since the carbohydrate content of human HMW kininogen has not been determined, the exact number of its amino acid residues can not be estimated. The values in Table V were calculated by assuming that the molecular weight of the polypeptide moiety in human HMW kininogen is 113,000. Amino acid compositions of HMW kininogens from bovine and horse plasma are very similar, but both differ from human HMW kininogen. However, the compositions of all three proteins become similar when values are expressed as mole percent of amino acid residues.

Fragmentation of HMW Kininogen by Plasma and Tissue Kallikreins

Bovine HMW Kininogen

The SDS-electrophoretic analysis of fragmentation of bovine HMW kininogen by plasma kallikrein indicates that HMW kininogen liberates initially a large peptide fragment, named fragment 1 · 2, together with bradykinin, and then this fragment is cleaved into fragment 1 and fragment 2. Fragment 2 is a histidine-rich peptide and contains a total of 41 amino acid residues.[39] On the other hand, fragment 1 is a glycopeptide containing

[35] M. Yano, S. Nagasawa, and T. Suzuki, J. Biochem. (Tokyo) 69, 471 (1971).
[36] M. Komiya, H. Kato, and T. Suzuki, J. Biochem. (Tokyo) 76, 823 (1974).
[37] S. Schiffman, C. Mannhalter, and K. D. Tyner, J. Biol. Chem. 255, 6433 (1980).
[38] F. N. Habal, B. J. Underdown, and H. Z. Movat, Biochem. Pharmacol. 24, 1241 (1975).
[39] Y. N. Han, M. Komiya, S. Iwanaga, and T. Suzuki, J. Biochem. (Tokyo) 77, 55 (1975).

TABLE V
CHEMICAL PROPERTIES OF HMW AND LMW KININOGENS

	HMW Kininogen			LMW Kininogen		
Carbohydrate	Bovine[a]	Horse[b]	Human[c]	Bovine[d]	Human[e,g]	Rabbit[f]
Hexose (%)	4.57	7.8	3.6	5.9	8.0	
Galactosamine (%)	1.53	1.9	—	ND[h]	3.9	
Glucosamine (%)	2.12	0.6	—	6.1		
Sialic acid (%)	4.35	3.6	—	4.4	4.9	
Total (%)	12.57	13.9	—	16.4	16.8	15.2

	HMW Kininogen						LMW Kininogen			
	Bovine[a]		Horse[b]		Human[c]		Bovine[d]	Human[e,g]		Rabbit[f]
Amino acid	Amino acid residue	Mole %	Amino acid residue	Mole %	Amino acid residue	Mole %	Mole %	Amino acid residue	Mole %	Mole %
Asp	63	10.8	64	11.0	116	11.6		37	9.8	10.8
Thr	41	7.1	39	6.7	79	7.8		24	8.4	7.7
Ser	49	8.5	28	4.8	74	7.4		28	8.7	6.3
Glu	64	11.1	74	12.7	137	13.7		46	14.4	10.5
Pro	39	6.6	53	9.1	64	6.4		26	5.6	6.3
Gly	39	6.7	48	8.3	72	7.2		20	6.4	7.7

Ala	31	5.4	31	5.3	46	4.6	26	7.0	7.4
½Cys	20	3.5	16	2.8	22	2.2	16	2.8	4.3
Val	38	6.6	31	5.3	45	4.5	30	6.1	5.4
Met	8	1.4	5	0.9	9	0.9	5	0.7	6.9
Ile	22	3.8	25	4.3	48	4.8	15	4.6	4.8
Leu	36	6.2	38	6.5	52	5.2	23	5.8	6.8
Tyr	19	3.3	13	2.2	31	3.1	13	3.6	2.8
Phe	20	3.5	19	3.3	39	3.9	14	4.3	4.3
His	28	4.9	32	5.5	48	4.8	9	1.4	3.0
Lys	43	7.3	39	6.7	73	7.3	26	6.8	6.0
Arg	13	2.3	22	3.8	32	3.2	11	3.7	2.8
Trp	8	1.3	4	0.7	12	1.2	4		
Total residues	581		581		999		373		

[a] M. Komiya, H. Kato, and T. Suzuki, J. Biochem. (Tokyo) 76, 811 (1974).
[b] T. Sugo, H. Kato, S. Iwanaga, and S. Fujii, Eur. J. Biochem. 115, 439 (1981).
[c] T. Nakayasu and S. Nagasawa, J. Biochem. (Tokyo) 85, 249 (1979).
[d] S. Nagasawa, Y. Mizushima, T. Sato, S. Iwanaga, and T. Suzuki, J. Biochem. (Tokyo) 60, 643 (1966).
[e] S. Nagasawa and T. Nakayasu, Adv. Exp. Med. Biol. 120B, 163 (1979).
[f] T. P. Egorova, L. G. Makevnina, and T. S. Paskhina, Biokhimiya (Moscow) 41, 1052 (1976).
[g] H. W. Sakamoto and O. Nishikaze, J. Biochem. 86, 1540 (1979).
[h] Not detectable.

TABLE VI

CHEMICAL PROPERTIES OF CLEAVAGE FRAGMENTS FROM HMW KININOGEN

	Heavy chain			Fragment 1 · 2		Light chain		Light chain, human[c]	Fragment, human[c]
	Bovine[a]	Horse[b]	Human[c]	Bovine[a]	Horse[d]	Bovine[a]	Horse[b]		
Molecular weight	48,500	49,000	62,000	14,500		16,000	15,000 20,000	43,000	8000
Carbohydrate content (%)									
Hexose	4.61			4.0		6.03			
Galactosamine	—			4.6		5.15			
Glucosamine	4.18			—		—			
Sialic acid	3.82			6.9		10.09			
Total	12.61			15.5		21.27			
Amino acid composition[e]									
Asp	45 (11.9)	37 (10.7)	65 (11.6)	8	6	16	15	46	7
Thr	26 (6.9)	27 (7.8)	44 (7.8)	4	6	15	9	30	5
Ser	32 (8.5)	18 (5.2)	42 (7.6)	7	5	13	9	33	5
Glu	44 (11.7)	40 (11.5)	85 (15.1)	11	21	14	12	51	13

Amino acid	(337)		(347)		(560)		(110)	(110)	(123)	(124)	(399)	
Pro	21	(5.6)	23	(6.6)	36	(6.4)	3	4	14	14	31	5
Gly	17	(4.5)	20	(5.8)	33	(5.8)	22	19	2	9	38	6
Ala	23	(6.1)	20	(5.8)	33	(5.8)	1	4	7	7	17	4
½Cys	17	(4.5)	14	(4.0)	15	(2.6)	—	—	1	2	3	1
Val	28	(7.4)	22	(6.3)	29	(5.2)	4.5	4	6	4	15	3
Met	4	(1.1)	2	(0.6)	5	(0.9)	2	2	2	2	5	1
Ile	17	(4.5)	18	(5.2)	29	(5.2)	1	4	5	4	17	3
Leu	24	(6.4)	23	(6.6)	31	(5.5)	5.5	5	11	10	23	4
Tyr	14	(3.7)	11	(3.2)	18	(3.2)	1	2	2	1	8	1
Phe	15	(4.0)	14	(4.0)	26	(4.7)	—	—	5	4	13	1
Lys	28	(7.4)	29	(8.4)	42	(7.6)	13	8	5	10	33	4
His	9	(2.4)	13	(3.7)	11	(2.0)	22	14	2	8	28	4
Arg	9	(2.4)	13	(3.7)	16	(2.9)	3	6	1	3	8	5
Trp	4	(1.1)	3	(0.9)	ND[f]		2	+[g]	2	1	ND[f]	
Total	337		347		560		110	110	123	124	399	

[a] H. Kato, Y. N. Han, S. Iwanaga, T. Suzuki, and M. Komiya, J. Biochem. (Tokyo) **80**, 1299 (1976).

[b] T. Sugo, H. Kato, S. Iwanaga, and S. Fujii, Eur. J. Biochem. **115**, 439 (1981).

[c] T. Nakayasu and S. Nagasawa, J. Biochem. (Tokyo) **85**, 249 (1979).

[d] T. Sugo, H. Kato, S. Iwanaga, and S. Fujii, Biochim. Biophys. Acta **579**, 474 (1979).

[e] Amino acid composition of each fragment is expressed as mol amino acid residues/mol of fragment. The values in parenthesis show mole percent of amino acid residues.

[f] Not determined.

[g] Detected after hydrolysis with 4N methanesulfonic acid.

also large amounts of histidine and it consists of a total of 69 residues.[40,41] The residual protein, named kinin- and fragment 1 · 2-free protein, which consists of 2 polypeptide chains named heavy and light chains, constitutes the NH$_2$- and the COOH-terminal portions of the parent HMW kininogen.[42] Table VI shows the amino acid and carbohydrate compositions of these fragments. A similar fragmentation of horse HMW kininogen with plasma kallikrein has been observed[43] and their compositions are also shown in Table VI. The chemical compositions of heavy and light chains and fragment 1 · 2 isolated from bovine and horse kininogens are quite similar to each other. The susceptibility of bovine HMW kininogen to other kininogenases has also been examined.[44] Although plasma kallikrein liberates fragment 1 · 2 very rapidly, the release of peptide fragments by hog pancreatic kallikrein, human urinary kallikrein, and snake venom kininogenase is very slow. Human urinary kallikrein hardly releases the peptide fragments and the final product is a kinin-free protein, which consists of heavy chain and fragment 1 · 2 light chain. Although plasmin has been known to be one of the kininogenases, its kinin-releasing activity is very weak. However, plasmin rapidly degrades bovine HMW kininogen into an altered kininogen with a molecular weight of about 50,000, cleaving off the fragment 1 · 2 portion and half of the light-chain portion.[44]

Human HMW Kininogen

On incubation of human HMW kininogen with human plasma kallikrein, a nicked kininogen and 2 kinin-free proteins (KFP), tentatively termed KFP-I and KFP-II, are produced.[22] At first, a single polypeptide chain of human HMW kininogen (120,000 MW) is cleaved by human plasma kallikrein into a nicked kininogen, composed of disulfide-linked polypeptide chains of MW 62,000 and 56,000. The nicked kininogen is then cleaved into a kinin and an intermediate KFP-I, which is composed of disulfide-linked 62,000- and 56,000-MW chains and is subsequently cleaved into stable KFP-II, composed of disulfide-linked 62,000- and 45,000-MW chains. The 56,000-MW chain, which was isolated by SP-Sephadex C-50 chromatography after reduction and alkylation of KFP-I, is cleaved into a 45,000-MW and an 8000-MW chain with plasma kallikrein. Therefore, the parent human HMW kininogen consists of 62,000-

[40] Y. N. Han, H. Kato, S. Iwanaga, and T. Suzuki, *J. Biochem.* (*Tokyo*) 79, 1201 (1976).
[41] Y. N. Han, H. Kato, S. Iwanaga, S. Oh-ishi, and M. Katori, *J. Biochem.* (*Tokyo*) 83, 213 (1978).
[42] H. Kato, Y. N. Han, S. Iwanaga, T. Suzuki, and M. Komiya, *J. Biochem.* (*Tokyo*) 80, 1299 (1976).
[43] T. Sugo, H. Kato, S. Iwanaga, and S. Fujii, *Biochim. Biophys. Acta* 579, 474 (1979).
[44] Y. N. Han, H. Kato, S. Iwanaga, and M. Komiya, *J. Biochem.* (*Tokyo*) 83, 223 (1978).

MW chain (heavy chain), kinin moiety, 45,000-MW chain (light chain), and 8000-MW fragment. The release of a fragment during the conversion of HMW kininogen into KFP-I has not been confirmed. A similar degradation pathway for human HMW kininogen with plasma kallikrein has been reported by Schiffman et al.[37] Table VI shows the amino acid composition of fragments isolated from human HMW kininogen. The compositions of human heavy and light chains are very similar to those of the heavy chain and the fragment 1 · 2-light chain from bovine and horse HMW kininogen. These results are also in good agreement with the data reported by Kerbiriou and Griffin.[31] The relatively high content of histidine in the light chain from human HMW kininogen indicates that it contains a histidine-rich portion corresponding to fragment 1 · 2 from bovine HMW kininogen, although the histidine-rich fragment is not liberated from human HMW kininogen by plasma kallikrein.

Amino Acid Sequence of HMW Kininogen

Figure 3 shows the partial amino acid sequence of bovine HMW kininogen.[39-42,45,46] The complete amino acid sequence of the carboxy-terminal portion of the bradykinin moiety in HMW kininogen has been determined. Fragment 1 · 2 has a characteristic amino acid sequence, containing 21 residues of histidine and glycine and 13 residues of lysine among 110 amino acid residues. In fragment 1 · 2, a repeating sequence of the type Gly-His-X or His-Gly-X appears many times, in addition to several sequences, most of which involved charged residues. Four oligosaccharide chains are attached to serine and threonine residues in the fragment 1 portion. The structures of the oligosaccharide chains were found to be trisaccharide units consisting of N-acetylneuraminyl-2-(2-3)-β-D-galactopyranosyl-(1-3)-N-acetylgalactosamine.[47] This structure also occurs as a tetrasaccharide variant, which has an additional branch in the form of an N-acetylneuraminyl residue attached by an α-ketosidic bond to carbon-6 of the N-acetylgalactosamine.[47] The light chain contains 123 amino acid residues and 7 oligosaccharide units. One characteristic of the light chain is the high content of 5 amino acids: Asp, Thr, Ser, Glu, and Pro. A sum of these amino acids amounts to 57% of the total residues. From sequence studies of bovine HMW kininogen, the protein was shown

---

[45] H. Kato, T. Sugo, N. Ikari, N. Hashimoto, I. Maruyama, Y. N. Han, S. Iwanaga, and S. Fujii, Adv. Exp. Med. Biol. 120B, 19 (1979).

[46] S. Iwanaga, H. Kato, T. Sugo, N. Ikari, N. Hashimoto, and S. Fujii, Colloq. Ges. Biol. Chem. 30, 243 (1979).

[47] Y. Endo, K. Yamashita, Y. N. Han, S. Iwanaga, and A. Kobata, J. Biochem. (Tokyo) 82, 545 (1977).

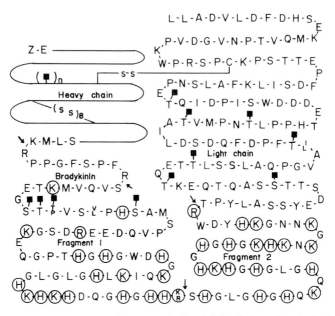

FIG. 3. Amino acid sequence of fragment 1 · 2 and light-chain regions in bovine HMW kininogen. Arrows show the sites cleaved by plasma kallikrein. ■, carbohydrate chains linked to threonine, serine, and asparagine residues.

to display microheterogeneity. Genetic variants exist in regard to 3 amino acid residues in fragment 1 · 2, that is, Pro to Thr, Val to Leu, and Lys to Arg. Although 4 oligosaccharide chains are attached to 2 threonine residues and 2 serine residues in fragment 1, some fragment 1 variants have oligosaccharide chains whereas others have none.[43] In addition to these, heterogeneity may be artificially introduced during the purification of HMW kininogen. On lyophilization or during steps under acidic conditions, dimers or polymers may form. Since HMW kininogen is highly susceptible to plasma kallikrein, a nicked kininogen (2-chain polypeptide) can be produced by traces of plasma kallikrein during the purification procedures.[36]

Figure 4 compares the structures of HMW kininogens from bovine, horse, and human plasmas. Those of bovine and horse HMW kininogens are quite similar; the similarities of amino acid compositions and molecular weights of heavy- and light-chain portions from both kininogens are also remarkable. Several peptide fragments analogous to the fragments 1 · 2, 1, and 2 of bovine HMW kininogen are liberated from horse HMW kininogen by plasma kallikrein. The amino acid sequence of horse HMW kininogen is only known for the peptide around the kinin moiety, and it

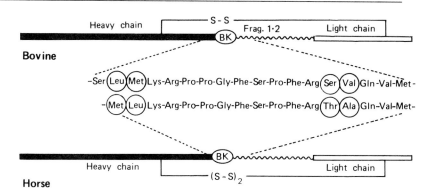

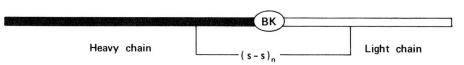

FIG. 4. Comparison of structures of HMW kininogens from bovine, horse, and human plasmas.

differs from that of bovine HMW kininogen in the residues circled in the figure. The structure of human HMW kininogen is different from those of bovine and horse HMW kininogens. The heavy-chain portion of human HMW kininogen is the largest.

Physical and Chemical Properties of LMW Kininogen

Bovine LMW kininogen is a glycoprotein with MW 50,000 (Table IV).[48,49] From this kininogen, tissue kallikrein liberates kallidin and snake venom kininogenase liberates bradykinin. No other peptide fragments are liberated.[42] In Fig. 5, the linear polypeptide sequences of HMW kininogen and LMW kininogen are shown for comparison. The large NH_2-terminal portions of both kininogens, that is the heavy-chain portions, are similar with respect to amino acid compositions, terminal residues, and molecular weights.[42,50] The similarity of these two heavy chains is also confirmed by

[48] S. Nagasawa, Y. Mizushima, T. Sato, S. Iwanaga, and T. Suzuki, *J. Biochem.* (*Tokyo*) 60, 643 (1966).
[49] H. Kato, T. Suzuki, K. Ikeda, and K. Hamaguchi, *J. Biochem.* (*Tokyo*) 62, 591 (1967).
[50] M. Komiya, H. Kato, and T. Suzuki, *J. Biochem.* (*Tokyo*) 76, 833 (1974).

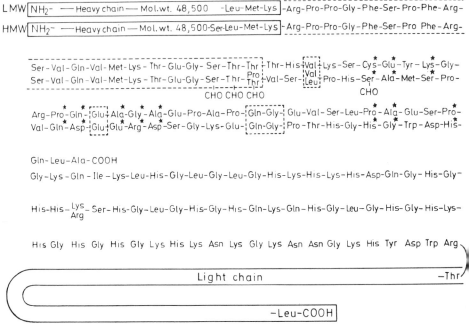

FIG. 5. Comparison of amino acid sequences of regions following the kinin moieties in HMW and LMW kininogens. Boxes indicate identical residues in the sequences, and CHO represents a carbohydrate unit. Residues marked by stars can be explained in terms of single-point mutations.

other techniques.[51] Tryptic peptide maps reflect the same results. The antibody against the heavy chain from HMW kininogen cross-reacts with the heavy chain from LMW kininogen. Comparing the structure of HMW kininogen and LMW kininogen, there exists a remarkable homology in the NH_2-terminal portion of the kinin moiety of both kininogens. On the other hand, there is a striking difference in the COOH-terminal portion of the kinin moiety of HMW and LMW kininogens, the latter of which consists of only 47 amino acid residues.[52] Moreover, there is no histidine-rich region corresponding to fragment 1 · 2 in HMW kininogen.

Function of Kininogen

Kininogen is a natural substrate of kallikrein, which is widely distributed in plasma and in various organs such as pancreas, kidney, salivary gland, intestine, and stomach.[6] Since kinin has a potent biological activ-

[51] M. Komiya, H. Kato, and T. Suzuki, *Biochem. Biophys. Res. Commun.* **49**, 1438 (1972).
[52] H. Kato, Y. N. Han, and S. Iwanaga, *J. Biochem. (Tokyo)* **82**, 377 (1977).

ity, kininogen is supposed to play an important role in these organs as substrate for kallikrein. The renal kallikrein–kinin system has been most intensively investigated, in close relation with the function of kidney, interacting with the renin–angiotensin–aldosterone–prostaglandin system.[53] Kidney kallikrein exists in the distal nephron as a membrane-bound enzyme and releases kinin from kininogen, which controls renal circulation. Although tissue kallikrein liberates kallidin from both HMW and LMW kininogens, the kinin is supposed to be derived from LMW kininogen, because no urinary kinins were detected in a subject with a congenital deficiency of LMW and HMW kininogens, whereas urinary kinins with normal amounts were found in a subject with a congenital deficiency of HMW kininogen (Fitzgerald trait).[53,54] Another function of kininogen is the acceleration of the intrinsic blood coagulation. It has been known that the kallikrein–kinin system is closely associated with intrinsic blood coagulation, since factor XII triggers the activation of both prekallikrein and factor XI. A relationship between the two systems has recently been disclosed by the discovery of asymptomatic congenital deficiencies of prekallikrein[55] and kininogen,[10,56–61] both of which show abnormal clotting time for kaolin-activated partial thromboplastin tests *in vitro*. Thus, two components of the kallikrein–kinin system, prekallikrein and HMW kininogen, can be classified as components of the blood-coagulation system. Table VII shows the kininogen contents of kininogen-deficient plasmas so far reported. In all cases the HMW kininogen content is nearly zero, but the LMW kininogen content varies. The abnormal clotting time *in vitro* can be corrected by adding human[58,59,61] or bovine HMW kininogen.[62,63] Variations of kininogen levels in physiological and pathological states have been reviewed.[64,65]

[53] O. A. Carretero and A. G. Scicli, *Am. J. Physiol.* **238**, F247 (1980).

[54] V. Hial, H. R. Keiser, and J. J. Pisano, *Biochem. Pharmacol.* **25**, 2499 (1976).

[55] K. D. Wuepper, *J. Exp. Med.* **138**, 1345 (1973).

[56] M. J. Lacombe, B. Veret, and J. P. Levy, *Blood* **46**, 761 (1975).

[57] K. D. Wuepper, D. R. Miller, and M. J. Locombe, *J. Clin. Invest.* **56**, 1663 (1975).

[58] R. W. Colman, A. Bagdasarian, R. C. Talamo, C. F. Scott, M. Seavey, J. A. Guimaraes, J. V. Pierce, and A. P. Kaplan, *J. Clin. Invest.* **56**, 1650 (1975).

[59] V. H. Donaldson, H. I. Glueck, M. A. Miller, H. Z. Movat, and F. Habal, *J. Lab. Clin. Med.* **87**, 327 (1976).

[60] C. L. Luteher, *Clin. Res.* **24**, 440A (1976).

[61] S. Oh-ishi, A. Ueno, Y. Uchida, M. Katori, H. Hayashi, H. Koya, K. Kitajima, and I. Kimura, *Adv. Exp. Med. Biol.* **120B**, 93 (1979).

[62] R. Waldmann, A. G. Scicli, R. K. McGregor, O. A. Carretero, J. P. Abraham, H. Kato, Y. N. Han, and S. Iwanaga, *Thromb. Res.* **8**, 785 (1976).

[63] R. T. Matheson, D. R. Miller, M. J. Lacombe, Y. N. Han, S. Iwanaga, H. Kato, and K. D. Wuepper, *J. Clin. Invest.* **58**, 1395 (1976).

[64] E. Habermann, *Handb. Exp. Pharmacol.* [N.S.] **25**, 251 (1970).

[65] R. W. Colman and P. Y. Wong, *Thromb. Haemostasis* **38**, 751 (1977).

TABLE VII
KININOGEN CONTENT OF HMW KININOGEN-DEFICIENT PLASMAS[a]

	HMW kininogen[b]	LMW kininogen[b]
Normal plasma	100	100
San Francisco	0.1	100
Williams	0.1	0
Dayton	0.1	0
Flaujeac	0.3	12
Fitzgerald	1.0	50
Reid	2.5	50–100
Detroit	3.5	100
Fujiwara[c]	0	1.3

[a] Each value represents the relative amounts of kininogen, taking the amounts in normal plasma as 100.
[b] D. Proud, J. V. Pierce, and J. J. Pisano *J. Lab. Clin. Med.* **95**, 563 (1980).
[c] S. Oh-Ishi, A. Ueno, Y. Uchida, M. Katori, H. Hayashi, H. Koya, K. Kitajima, and I. Kimura, *Adv. Exp. Med. Biol.* **120B**, 93 (1979).

Since discovery of the kininogen deficiency, the role of HMW kininogen in intrinsic blood coagulation has been studied by many investigators and it is now known that HMW kininogen accelerates the activation of factor XII, prekallikrein, and factor XI as cofactor.[66-68] Factor XII is activated by interacting with prekallikrein or factor XI in the presence of foreign surfaces such as kaolin and glass. Factor XII is also activated by plasma kallikrein and by the activated form of factor XI (XI_a). The activated factor XII (XII_a) activates prekallikrein and factor XI. Each reaction is accelerated by HMW kininogen, except the activation of factor XII by XI_a.[33,69-74] The effects of bovine HMW kininogen on these reactions have been analyzed in detail, using fluorogenic peptide substrates in the presence of kaolin.[75,76] The results indicate that at least five of the reac-

[66] A. P. Kaplan, H. L. Meier, and R. Mandle, Jr., *Semin. Thromb. Haemostasis* **3**, 1 (1976).
[67] J. H. Griffin and C. G. Cochrane, *Semin. Thromb. Haemostasis* **5**, 254 (1979).
[68] H. Z. Movat, *Handb. Exp. Pharmacol.* [N.S.] **25**, Suppl., 1 (1979).
[69] J. H. Griffin and C. G. Cochrane, *Proc. Natl. Acad. Sci. U.S.A.* **73**, 2554 (1976).
[70] H. L. Meier, J. V. Pierce, R. W. Colman, and A. P. Kaplan, *J. Clin. Invest.* **60**, 18 (1977).
[71] C. Y. Liu, C. F. Scott, A. Bagdasarian, J. V. Pierce, A. P. Kaplan, and R. W. Colman, *J. Clin. Invest.* **60**, 7 (1977).
[72] R. C. Wiggins, B. W. Bouma, C. G. Cochrane, and J. H. Griffin, *Proc. Natl. Acad. Sci. U.S.A.* **74**, 4636 (1977).
[73] H. Saito, *J. Clin. Invest.* **60**, 584 (1977).
[74] K. Kurachi, K. Fujikawa, and E. W. Davie, *Biochemistry* **19**, 1330 (1980).
[75] T. Sugo, N. Ikari, H. Kato, S. Iwanaga, and S. Fujii, *Biochemistry* **19**, 3215 (1980).
[76] T. Sugo, N. Ikari, H. Kato, S. Iwanaga, and S. Fujii, *Seikagaku* **51**, 589 (1979).

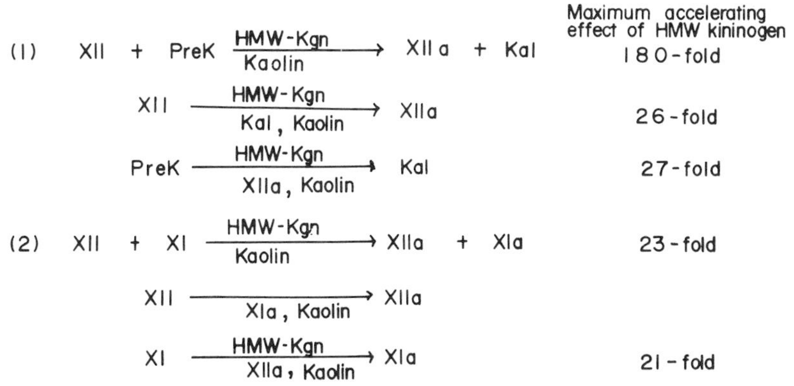

Maximum accelerating effect of HMW kininogen

(1)   XII + PreK $\xrightarrow{\text{HMW-Kgn}}$ (Kaolin) XIIa + Kal   180-fold

      XII $\xrightarrow{\text{HMW-Kgn}}$ (Kal, Kaolin) XIIa   26-fold

      PreK $\xrightarrow{\text{HMW-Kgn}}$ (XIIa, Kaolin) Kal   27-fold

(2)   XII + XI $\xrightarrow{\text{HMW-Kgn}}$ (Kaolin) XIIa + XIa   23-fold

      XII $\xrightarrow{}$ (XIa, Kaolin) XIIa

      XI $\xrightarrow{\text{HMW-Kgn}}$ (XIIa, Kaolin) XIa   21-fold

FIG. 6. Accelerating effect of bovine HMW kininogen on the kaolin-mediated activation of factor XII. (1) Activation of factor XII in the presence of prekallikrein. (2) Activation of factor XII in the presence of factor XI.

tions shown in Fig. 6 are accelerated by HMW kininogen. The functional site of HMW kininogen for these accelerating effects has also been studied by using derivatives from bovine HMW kininogen.[75] The results indicate that fragment 1 · 2 light chain has the same accelerating effect as HMW kininogen, whereas fragment 1 · 2 inhibits the activation of factor XII and the light chain has no effect. In the case of human HMW kininogen, the light chain that corresponds to bovine fragment 1 · 2 light chain, has been shown to have the same accelerating effect as HMW kininogen.[22,29,31,37] The accelerating effect of HMW kininogen as cofactor is mediated by interacting with kaolin and prekallikrein. Bovine HMW kininogen is efficiently adsorbed on kaolin and the adsorption is inhibited by prior treatment of kaolin with fragment 1 · 2 or with proteins containing the fragment 1 · 2 moiety.[77] Fragment 1 · 2 is extremely rich in basic amino acid residues, containing 21 histidines, 13 lysines, and 3 arginines in a total of 110 residues (Table VI). In general, it has been known that basic proteins such as protamine, lysozyme, and synthetic polylysine inhibit the activation of factor XII by negatively charged surfaces. Fragment 1 · 2 alone has also been shown to be a potent inhibitor of the contact activation of factor XII.[62,63,78] Therefore, the basic amino acid residues clustered in fragment 1 · 2 may be responsible for the accelerating effect of HMW kininogen or the inhibitory effect of fragment 1 · 2. On the other hand, human and bovine HMW kininogen has been known to form a complex with prekallikrein or factor XI in circulating plasma and in the purified sys-

---

[77] N. Ikari, T. Sugo, H. Kato, S. Iwanaga, and S. Fujii, *J. Biochem.* (*Tokyo*) **89**, 1699 (1981).
[78] S. Oh-ishi, M. Katori, Y. N. Han, S. Iwanaga, H. Kato, and T. Suzuki, *Biochem. Pharmacol.* **26**, 115 (1977).

tem.[72,74,77,79,80] The functional site of HMW kininogen that interacts with prekallikrein seems to be the light chain of the protein.[21,77] The association constant[81] of human HMW kininogen with prekallikrein is $3.4 \times 10^7 M^{-1}$. From the evidence presented thus far, the role of HMW kininogen in the kaolin-mediated activation of factor XII is suggested to be as follows. HMW kininogen is adsorbed on kaolin through its fragment 1 · 2 region, forming a complex with prekallikrein through the light-chain region. Since factor XII is also adsorbed to kaolin, this complex formation seems to provide the most favorable interaction between factor XII and prekallikrein. Studies on the susceptibility of HMW kininogen to proteases in plasma suggest that the cofactor activity of HMW kininogen is regulated by some proteases. Plasma kallikrein cleaves bovine HMW kininogen into a kinin-free protein with a full cofactor activity and then into kinin- and fragment 1 · 2-free protein, which has no cofactor activity. On the other hand, plasma kallikrein cleaves human HMW kininogen into a kinin-free protein. Therefore, the cofactor activity of HMW kininogen disappears in bovine plasma but not in human plasma, following the activation of prekallikrein. Furthermore, we have found that a very early cleavage product of the single-chain bovine HMW kininogen with bovine plasma kallikrein or human urinary kallikrein has an even more potent cofactor activity than intact HMW kininogen.[45] HMW kininogen is degraded very rapidly by plasmins into a product with no cofactor activity. Thus plasma kallikrein and plasmin seem to act as regulators of the cofactor activity of HMW kininogen.

[79] R. J. Mandle, R. W. Colman, and A. P. Kaplan, *Proc. Natl. Acad. Sci. U.S.A.* **73**, 4179 (1976).
[80] R. E. Thompson, R. Mandle, Jr., and A. P. Kaplan, *J. Clin. Invest.* **60**, 1376 (1977).
[81] R. E. Thompson, R. Mandle, Jr., and A. P. Kaplan, *Proc. Natl. Acad. Sci. U.S.A.* **76**, 4862 (1979).

[16] Human Factor XII (Hageman Factor)

By KAZUO FUJIKAWA and EARL W. DAVIE

Factor XII plays an important role in the contact activation of several biological systems, including blood coagulation, kinin generation, intrinsic fibrinolysis, and renin-angiotensin activation. Factor XII deficiency was first discovered in 1955 in a patient who had a prolonged clotting time.[1]

[1] O. D. Ratnoff and J. E. Colopy, *J. Clin. Invest.* **34**, 601 (1955).

METHODS IN ENZYMOLOGY, VOL. 80

Later, a prolonged lysis time and the absence of kinin formation and renin generation were also observed in factor XII-deficient plasma. However, individuals with factor XII deficiency are usually asymptomatic.[2-4]

Factor XII circulates in plasma as a zymogen to a serine protease. It is composed of a single polypeptide chain. In the test tube, factor XII is readily activated by plasma kallikrein in the presence of a negatively charged surface, such as kaolin, ellagic acid,[5-8] dextran sulfate, or sulfatide.[9] Factor XI_a and plasmin will also activate factor XII, but do so at a slower rate.[8] Another plasma protein, high-molecular-weight (HMW) kininogen, stimulates the activation of factor XII by kallikrein.[10-12] During the activation of factor XII, a specific internal peptide bond is cleaved by plasma kallikrein. Thus the activated form of factor XII (factor XII_a) is composed of 2 polypeptide chains held together by a disulfide bond(s). Factor XII_a, in turn, then activates plasma prekallikrein and factor XI in the presence of a negative surface. These two reactions are also stimulated by HMW kininogen (see Chapters 14 and 17 in this volume). Factor XII_a can also initiate the extrinsic system of coagulation through the activation of factor VII.[13-16]

Recently, the initial events in the surface activation of blood coagulation have been clarified in the bovine system. In these studies, it was shown that the zymogen form of factor XII can activate prekallikrein[17] or factor XI[18] on the surface of kaolin, dextran sulfate, or sulfatide. Furthermore, factor XII activates prekallikrein at the same reaction rate as factor XII_a.[17] Kaolin, celite, ellagic acid, and dextran sulfate function as negatively charged surfaces when these reactions are studied in the test

[2] A. P. Kaplan, *Prog. Hemostasis Thromb.* **4**, 127 (1978).

[3] J. H. Griffin and C. G. Cochrane, *Semin. Thromb. Hemostasis* **5**, 4 (1979).

[4] O. D. Ratnoff and H. Saito, *Curr. Top. Hematol.* **2**, 1 (1979).

[5] C. Y. Liu, C. F. Scott, A. Bagdasarian, J. V. Pierce, A. P. Kaplan, and R. W. Colman, *J. Clin. Invest.* **60**, 7 (1977).

[6] H. L. Meier, J. V. Pierce, R. W. Colman, and A. P. Kaplan, *J. Clin. Invest.* **60**, 18 (1977).

[7] R. C. Wiggins, B. N. Bouma, C. G. Cochrane, and J. H. Griffin, *Proc. Natl. Acad. Sci. U.S.A.* **74**, 4636 (1977).

[8] J. H. Griffin, *Proc. Natl. Acad. Sci. U.S.A.* **75**, 1998 (1978).

[9] K. Fujikawa, R. L. Heimark, K. Kurachi, and E. W. Davie, *Biochemistry* **19**, 1322 (1980).

[10] J. H. Griffin and C. G. Cochrane, *Proc. Natl. Acad. Sci. U.S.A.* **73**, 2554 (1976).

[11] T. Sugo, N. Ikari, H. Kato, S. Iwanaga, and S. Fujii, *Biochemistry* **19**, 3215 (1980).

[12] H. Saito, *J. Clin. Invest.* **60**, 584 (1977).

[13] K. Laake and R. Ellingsen, *Thromb. Res.* **5**, 539 (1974).

[14] K. Laake and B. Osterud, *Thromb. Res.* **5**, 759 (1974).

[15] H. Saito and O. D. Ratnoff, *J. Lab. Clin. Med.* **85**, 405 (1975).

[16] W. Kisiel, K. Fujikawa, and E. W. Davie, *Biochemistry* **16**, 4189 (1977).

[17] R. L. Heimark, K. Kurachi, K. Fujikawa, and E. W. Davie, *Nature (London)* **286**, 456 (1980).

[18] K. Kurachi, K. Fujikawa, and E. W. Davie, *Biochemistry* **19**, 1330 (1980).

tube. Sulfatide,[9] fatty acid,[19] and bacterial endotoxin[20] may be of importance as negatively charged surfaces under physiological conditions.

A number of purification methods have been described for human[21,22] and bovine[23,24] factor XII. In this chapter, a purification method of human factor XII is described that is reproducible and suitable for obtaining milligram amounts of protein.

Reagents

Stock Solution of Polybrene. Polybrene (Aldrich) is dissolved in water at 100 g/liter and stored at 4°C.

Stock Solution of Diisopropyl Fluorophosphate (DFP). DFP (1 M) is prepared by diluting 1 g of pure DFP to 5.5 ml with anhydrous isopropanol. This solution is stored at −20°C.

Phospholipid Suspension. One vial of rabbit brain cephalin (Sigma) is suspended in 100 ml of 0.15 M NaCl, and 1-ml aliquots are stored frozen.

Kaolin Suspension. Acid-washed kaolin (Fischer) is suspended in 0.15 M NaCl at 0.5 g/100 ml and 10^{-4} M NaN$_3$, and stored at room temperature.

Phospholipid–CaCl$_2$ Mixture. The phospholipid suspension (1 ml) is mixed with 1 ml of 0.025 M CaCl$_2$. This solution is prepared just before use.

Tris-Acetate Buffer. Tris-acetate buffer (0.05 M, pH 7.4) containing bovine serum albumin at 0.1 mg/ml and 10^{-4} M NaN$_3$ is stored in a refrigerator.

Benzamidine–Agarose with an ε-Aminocaproic Acid Spacer. Benzamidine–agarose is prepared according to Fujikawa et al.[23] with some modifications. Agarose-4B (Pharmacia), 250 ml of settled volume, is activated and coupled to 50 g of ε-aminocaproic acid in 500 ml of 0.1 M NaHCO$_3$ (pH 9.5). The agarose is gently stirred overnight at 4°C. It is then washed successively with water and 0.1 M 2(N-morpholino)-ethanesulfonic acid (MES, Sigma)–NaOH buffer (pH 4.75) and stored in sodium azide (10^{-4} M) at 4°C until use. ε-Aminocaproic acid–agarose (100 ml, settled volume) is suspended in 100 ml of 0.1 M MES–NaOH buffer (pH 4.75) and 10 g of 1-cyclohexyl-3-(2-morpholinoethyl)carbodiimide metho-p-toluenesulfonate (Aldrich) in 20 ml of 0.1 M MES–NaOH buffer

[19] H. L. Nossel, *in* "The Contact Phase of Blood Coagulation," p. 24. Davis, Philadelphia, Pennsylvania, 1964.

[20] D. C. Morrison and C. G. Cochrane, *J. Exp. Med.* **140**, 797 (1974).

[21] J. H. Griffin and C. G. Cochrane, this series, Vol. 45, p. 56.

[22] J. Y. C. Chan and H. Z. Movat, *Thromb. Res.* **8**, 337 (1976).

[23] K. Fujikawa, K. A. Walsh, and E. W. Davie, *Biochemistry* **16**, 2270 (1977).

[24] H. Claeys and D. Collen, *Eur. J. Biochem.* **87**, 69 (1978).

(pH 4.75) is added to the suspension. The mixture is then stirred for 30 min at room temperature. Five grams of p-aminobenzamidine-2 HCl(Vega or Sigma) is added to the suspension. The pH of the suspension is checked immediately after the p-aminobenzamidine is dissolved. If necessary, the pH is adjusted to 4.75 with 1 N NaOH and stirring is continued overnight. The residual activated groups are blocked by stirring for an additional 4 hr with 10 g of glycine ethyl ester. The final product is thoroughly washed with water and stored at 4°C in the presence of 10^{-4} M sodium azide.

D-*Prolyl-*L-*phenylalanyl-*L-*arginine-p-nitroaniline.*   D-Prolyl-L-phenyl-alanyl-L-arginine-p-nitroaniline-2 HCl (Kabi peptide substrate 2302) is dissolved in water at 6.1 mg/10 ml to make 1 mM stock solution.

Factor XII-Deficient Plasma. Blood is collected from a patient in one-tenth volume of 0.1 M sodium oxalate. Plasma is obtained by centrifugation of the blood at 12,000 g for 15 min. Small aliquots are stored at $-60°C$. Human factor XII-deficient plasma is also commercially available (George King Biomedicals).

Assay Procedures

Clotting Assay. The clotting activity of factor XII is determined by a kaolin partial thromboplastin time using human factor XII-deficient plasma. Fifty microliters of test sample (diluted 100-fold with the Tris-acetate buffer), 50 μl of the kaolin suspension, and 50 μl of factor XII-deficient plasma are mixed in siliconized glass test tubes (8 × 70 mm). The mixture is then incubated at room temperature for 2 min. The reaction mixture is transferred to a water bath at 37°C, and 100 μl of the phospholipid–CaCl$_2$ mixture is added. The clotting time is determined by continuous tilting of the tubes at 37°C. The units of factor XII activity are calculated from a standard calibration curve prepared by a serial dilution of pooled normal human plasma. The log of the factor XII concentration is plotted against the log of the clotting time in seconds. A straight line is obtained between 1/10 and 1/320 dilutions where the clotting times usually range from 160 to 340 sec. One unit of factor XII clotting activity is defined as that amount of factor XII present in 1 ml of normal human plasma.

Amidase Assay. Amidase activity of factor XII$_a$ is measured using the synthetic peptide substrate D-prolyl-L-phenylalanyl-L-arginine-p-nitro-aniline. The assay mixture contains 100 μl of 1 mM substrate (10 μmol), 0.1 mM Tris-HCl buffer (pH 7.6), and 2 to 4 μg of factor XII$_a$ in a final volume of 950 μl. The reaction mixture is incubated at 30°C for an appropriate time, and the reaction is terminated by the addition of 50 μl of 80% acetic acid. The amount of p-nitroaniline released is determined by the increase of absorbance at 405 nm. A standard p-nitroaniline solution of 0.05 mmol/liter gives an absorbance of 0.525 at 405 nm. Amidase activity is expressed as micromoles of p-nitroaniline released in 1 min per milligram of protein in the reaction mixture.

Purification Procedures

Ammonium Sulfate Fractionation. Fifteen liters of frozen outdated human plasma (collected in 0.01 M citrate) is thawed overnight at room temperature. After the addition of 200 mg of heparin (3.1 × 10⁴ units) and 500 mg of crude soybean-trypsin inhibitor (type II-S, Sigma), the plasma is stirred for 30 min at 4°C with 1050 ml of 1 M barium chloride. The suspension is centrifuged at 5000 rpm for 10 min. The supernatant is brought to 20% saturation with solid ammonium sulfate (107 g/liter) and stirred for 30 min at 4°C. The suspension is centrifuged and the precipitate is discarded. Additional solid ammonium sulfate is added to 40% saturation (129 g/liter) and the suspension is stirred for 30 min. The precipitate is removed by centrifugation and dissolved in 4.5 liters of chilled water containing 900 mg of polybrene, 7.8 g of benzamidine HCl, and 500 mg of crude soybean-trypsin inhibitor. This sample is divided into 3 equal fractions and stored frozen in plastic containers until use.

First DEAE–Sephadex Column Chromatography. DEAE–Sephadex A-50 (40 g) is swelled overnight in 4 liters of 0.05 M sodium-acetate buffer (pH 5.8) containing 0.06 M NaCl. After adjusting the pH of the DEAE–Sephadex suspension to pH 5.8 with 1 N NaOH, and DEAE–Sephadex is packed into a plastic column (7.5 × 30 cm) and the column washed with 2 liters of the same buffer in the cold. The frozen ammonium sulfate fraction, which corresponds to 5 liters of original plasma, is thawed and dialyzed for 2 days against two changes of 50 liters of 0.05 M sodium acetate buffer (pH 5.8) containing 0.06 M NaCl. After dialysis, the conductivity of the sample should be 6 mmho at 4°C using a conductivity meter (Radiometer, Type CDM 2e). If the conductivity is higher, the solution is adjusted to 6 mmho with chilled water. Any precipitate formed during dialysis is removed by centrifugation. Soybean-trypsin inhibitor (300 mg) is added and the sample is applied to a column of DEAE–Sephadex A-50 (7.5 × 22 cm), which is previously equilibrated with 0.05 M sodium acetate buffer (pH 5.8) containing 0.06 M NaCl. The column is washed with 2 liters of the same buffer, and the pass-through and wash fractions are combined.

Second DEAE–Sephadex Column Chromatography. DEAE–Sephadex A-50 (60 g) is swelled overnight in 4 liters of 0.03 M Tris-HCl buffer (pH 7.5) containing 0.05 M NaCl. After adjusting the pH of the DEAE–Sephadex suspension to 7.5 with 1 N NaOH, the DEAE–Sephadex is poured into a plastic column (10 × 35 cm) and washed with 2 liters of the same buffer in the cold. Soybean-trypsin inhibitor (300 mg) is added to the

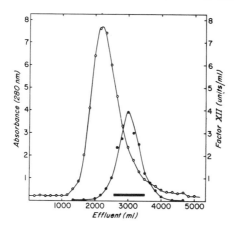

F<small>IG</small>. 1. Second DEAE–Sephadex column chromatography. The dialyzed sample of the first DEAE–Sephadex fraction starting with 5 liters of plasma is applied to a DEAE–Sephadex column (10 × 20 cm) in 0.03 M Tris-HCl buffer (pH 7.5) containing 0.05 M NaCl. The column is then extensively washed with 0.05 M Tris-HCl buffer (pH 7.5) containing 0.075 M NaCl and polybrene at 50 mg/liter, and the protein is eluted as described under Purification Procedures. The clotting activity is measured with samples diluted 100-fold with 0.05 M Tris-acetate buffer (pH 7.4) containing 0.1 mg/ml of bovine serum albumin. ○——○, Absorbance at 280 nm; ●——●, clotting activity of factor XII. Fractions shown by the solid bar were pooled.

first DEAE–Sephadex fraction and the sample is dialyzed overnight against 50 liters of 0.03 M Tris-HCl buffer (pH 7.5) containing 0.05 M NaCl. The conductivity of the dialyzed sample is adjusted when necessary to 5.0 mmho at 4°C with chilled water. The sample is then treated with 0.1 mM DFP for at least 1 hr and applied to a DEAE–Sephadex A-50 column (10 × 20 cm), which is previously equilibrated with the same buffer. The column is then washed with 2 liters of the same buffer. The pass-through and wash fractions are combined and used for the preparation of prekallikrein (see Chapter 14 in this volume). The DEAE–Sephadex column is extensively washed with about 18 liters of 0.03 M Tris-HCl buffer (pH 7.5) containing 0.075 M NaCl and polybrene at 50 mg/liter. The protein is eluted from the column with a linear gradient formed by 3 liters of 0.075 M NaCl in 0.03 M Tris-HCl buffer (pH 7.5) containing polybrene, and 3 liters of 0.5 M NaCl in 0.03 M Tris-HCl buffer (pH 7.5) containing polybrene. The fractions containing factor XII activity that are eluted in the trailing edge of the main protein peak are pooled, as shown in Fig. 1.

QAE-Sephadex Column Chromatography. QAE-Sephadex A-50 (12 g) is swelled overnight in 2 liters of 0.05 M Tris-HCl buffer (pH 8.4) containing 0.075 M NaCl. After adjusting the pH to 8.4, the QAE-Sephadex is poured

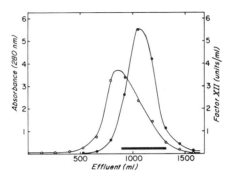

FIG. 2. QAE-Sephadex column chromatography. The fraction from the second DEAE–Sephadex column containing factor XII is dialyzed against 0.05 M Tris-HCl buffer (pH 8.4) containing 0.075 M NaCl, and applied to a QAE-Sephadex column (4.5 × 20 cm) in 0.05 M Tris-HCl buffer (pH 8.4) containing 0.075 M NaCl and polybrene at 50 mg/liter. After the sample is applied, the column is washed with 1 liter of 0.05 M Tris-HCl buffer (pH 8.4) containing 0.1 M NaCl and polybrene, and the protein is eluted as described under Purification Procedures. The sample is diluted 100-fold with Tris-acetate buffer (pH 7.4) containing 0.1 mg/ml of bovine serum albumin and measured for its clotting activity. O——O, Absorbance at 280 nm; ●——●, factor XII clotting activity. Fractions shown by the solid bar were pooled.

into a plastic column (4.5 × 23 cm) and the column is washed with 1 liter of the same buffer in the cold. Soybean-trypsin inhibitor (100 mg) is added to the second DEAE–Sephadex fraction, and the sample is dialyzed overnight against 40 liters of 0.05 M Tris-HCl buffer (pH 8.4) containing 0.075 M NaCl. DFP is added to the dialyzed sample at a final concentration of 0.1 mM. The sample is stirred for 1–2 hr and applied to a QAE-Sephadex A-50 column (4.5 × 20 cm), which is previously equilibrated with 0.05 M Tris-HCl buffer (pH 8.4) and containing 0.075 M NaCl and polybrene. After application of the sample, the column is washed with 1 liter of 0.05 M Tris-HCl buffer containing 0.1 M NaCl and polybrene. The protein is eluted with a linear gradient composed of 1 liter of 0.1 M NaCl and 1 liter of 0.5 M NaCl, both in 0.05 M Tris-HCl buffer. The fractions containing factor XII activity are pooled, as shown in Fig. 2.

CM-Cellulose Column Chromatography. Preswollen CM-cellulose (CM 52, Whatman) is suspended in 0.5 M sodium acetate buffer (pH 5.2) and stirred for 1 hr. The CM-cellulose is washed extensively on a funnel with 0.05 M sodium acetate buffer (pH 5.2) containing 0.06 M NaCl and then packed into a plastic column (3.2 × 10 cm). The column is washed with 500 ml of the same buffer in the cold. The QAE-Sephadex fraction is dialyzed overnight against 20 liters of 0.05 M sodium acetate buffer (pH 5.2) containing 0.06 M NaCl. This sample is then applied to a CM-cellulose column (3.2 × 10 cm), which is previously equilibrated with the

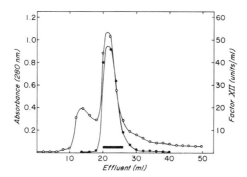

FIG. 3. CM-cellulose column chromatography. The active fractions pooled from the QAE-Sephadex column are dialyzed against 0.05 M sodium acetate buffer (pH 5.2) containing 0.06 M NaCl, and applied to a column (3.2 × 10 cm) of CM-cellulose equilibrated with 0.05 M sodium acetate buffer (pH 5.2) containing 0.06 M NaCl. After the column is washed with 0.5 liter of the same buffer, the protein is eluted as described under Purification Procedures. Aliquots are diluted 500-fold with Tris-acetate buffer (pH 7.4) containing 0.1 mg/ml bovine serum albumin and measured for clotting activity. O——O, Absorbance at 280 nm; ●——●, factor XII clotting activity. Fractions shown by the solid bar were pooled.

same acetate buffer. After washing the column with 500 ml of the buffer, the protein is eluted with a linear gradient containing 250 ml of 0.06 M NaCl and 250 ml of 0.5 M NaCl. Both solutions also contain 0.05 M sodium acetate buffer, pH 5.2 (Fig. 3). The fractions containing factor XII activity are tested for purity by sodium dodecyl sulfate (SDS)–polyacrylamide gel electrophoresis, and those fractions that give a single protein band are pooled. The CM-cellulose fraction usually gives a single protein band by SDS–polyacrylamide gel electrophoresis. The average yield of purified factor XII is about 35–40 mg from 5 liters of plasma. A summary of the purification procedure is shown in Table I. The final preparation is often contaminated by a small amount of factor XII_a, as judged by amidase activity. However, this factor XII_a activity is completely inhibited by treatment of the factor XII solution with 0.001 M DFP. If the preparation contains a substantial amount of factor XII_a, the zymogen and active enzyme can be separated by benzamidine–agarose column chromatography, as described below. The reduced sample of a mixture of factor XII and factor XII_a gives 3 protein bands with apparent molecular weights of 76,000, 52,000, and 28,000, as determined by SDS–polyacrylamide gel electrophoresis. A nonreduced sample appears as a single band on SDS–polyacrylamide gel electrophoresis. The prekallikrein and/or kallikrein contamination is not detectable in most preparations of factor XII when assayed by amidase activity against Bz-Pro-Phe-Arg-pNA. However, the clotting activity in prekallikrein-deficient plasma sug-

TABLE I
PURIFICATION OF HUMAN FACTOR XII

Purification step	Volume (ml)	Total protein (mg)	Total activity (units)	Specific activity (units/mg)	Recovery (%)	Purification (-fold)
Plasma	5000	258×10^3 [a]	5000[c]	0.019	100	1
Ammonium sulfate	2300	98×10^3 [a]	3335	0.034	66.7	1.8
First DEAE–Sephadex (pH 5.8)	4260	61×10^3 [a]	3067	0.050	61.3	2.6
Second DEAE–Sephadex (pH 7.5)	920	2.7×10^3 [a]	2484	0.920	49.7	48.4
QAE-Sephadex	618	1.25×10^3 [a]	2039	1.63	40.8	85.8
CM-cellulose	73	37.5[b]	1927	51.4	38.5	2705

[a] Protein concentrations are determined by absorbance at 280 nm assuming $E_{280}^{1\%} = 10.0$.
[b] Protein concentration is determined by absorbance at 280 nm assuming $E_{280}^{1\%} = 14.2$.
[c] The original factor XII activity is determined on an aliquot prior to the addition of various inhibitors.

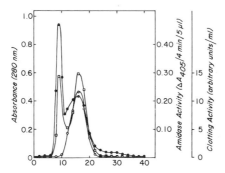

FIG. 4. Separation of factor XII and factor XII$_a$ by benzamidine–agarose column chromatography. A preparation from the CM-cellulose column, which is a mixture of factor XII and factor XII$_a$ (total protein 53 mg in 80 ml), is dialyzed overnight against 4 liters of 0.025 M MES–NaOH buffer (pH 6.0) containing 0.025 M NaCl. After removal of a small precipitate, the sample is applied to a benzamidine–agarose column (2 × 7 cm). The column is washed with 100 ml of the same buffer and the protein is eluted as described under Purification Procedures. Fractions are collected at 5 ml/tube with a flow rate of 30 ml/hr. Clotting activity is measured after the samples are diluted 500-fold. Amidase activity is assayed with 5-μl samples. ●——●, Absorbance at 280 nm; □——□, clotting activity; ○——○, amidase activity.

gests a prekallikrein contamination of 1%. This clotting activity is probably due to an intrinsic activity of factor XII in this assay, although this is not certain. Thus the contamination of prekallikrein in the final preparation should be carefully checked by amidase activity after treatment of the sample with trypsin, as described in Chapter 14 in this volume.

Separation of Factor XII and Factor XII$_a$ by Benzamidine–Agarose Column Chromatography. Some preparations of human factor XII (about 1 in 10) contain a substantial amount of factor XII$_a$.[1] Factor XII and factor XII$_a$, however, can be readily separated by benzamidine–agarose column chromatography. The preparation from the CM-cellulose column (containing about 50 mg of protein in a volume of approximately 10 ml) is dialyzed overnight against 4 liters of 0.025 M MES–NaOH buffer (pH 6.0) containing 0.025 M NaCl. A very small precipitate that forms during dialysis is removed by centrifugation and the sample is applied to a small benzamidine–agarose column (2 × 7 cm) at a flow rate of 30 ml/hr. The benzamidine–agarose column is previously equilibrated with 0.025 M MES–NaOH (pH 6.0) containing 0.025 M NaCl. The column is then washed with 100 ml of the same buffer and the protein is eluted with a linear gradient formed with 200 ml of 0.05 M MES–NaOH buffer (pH 6.0) containing 0.025 M NaCl and 200 ml of 1 M guanidine HCl (ultrapure grade, Schwarz/Mann) in 0.05 M MES–NaOH buffer (pH 6.0), plus 0.025 M NaCl. The flow rate is adjusted to 30 ml/hr. Both protein peaks have clotting activity, but there is no detectable amidase activity in peak 1 containing factor XII (Fig. 4). Factor XII$_a$, present in the second peak, is composed of 2 polypeptide chains with a trace amount of a third polypeptide chain. The factor XII and factor XII$_a$ fractions are dialyzed individually against 0.05 M Tris-HCl buffer (pH 7.5) containing 0.15 M NaCl to remove guanidine. These preparations can be stored for at least several months at $-20°C$ without any significant loss of activity.

Enzyme Activity of Human Factor XII and Factor XII$_a$

Clotting Activity. Purified human factor XII has a clotting activity of 40–50 units/mg. The clotting activity of factor XII$_a$ is about two times greater than that of factor XII.

Esterase and Amidase Activities. Factor XII$_a$ has substantial esterase and amidase activities against synthetic peptide substrates containing arginine. For instance, D-Pro-Phe-Arg-pNA is hydrolyzed at the rate of 10.0 μmol/min/mg at 30°C. The purified enzyme, however, does not hydrolyze benzoyl-Pro-Phe-Arg-pNA.

General Properties

Human factor XII is a glycoprotein composed of a single polypeptide chain. It contains 16.8% carbohydrate, including 4.2% hexose, 4.7% hexosamine, and 7.9% N-acetylneuraminic acid.[25] The amino acid and carbohydrate compositions of human factor XII are presented in Table II. Human factor XII migrates slightly slower than bovine factor XII on SDS–polyacrylamide gel electrophoresis. The molecular weight of the latter protein is 74,000, as determined by sedimentation equilibrium.[23] The molecular weight of human factor XII is about 76,000, as estimated by SDS–polyacrylamide gel electrophoresis. The amino-terminal sequence of human factor XII is Ile-Pro-Pro-Trp-Glu-Ala-Pro-Lys-Glu-His-Lys-Tyr,[25] and 7 of the first 12 residues are homologous to those of bovine factor XII. The latter protein has an amino-terminal sequence of Thr-Pro-Pro-Trp-Lys-Gly-Pro-Lys-Lys-His-Lys-Leu.[23]

Activation of factor XII by proteolysis takes place in the presence of negatively charged materials. It is catalyzed by plasma enzymes, such as kallikrein, factor XI$_a$, and plasmin.[5-11] Plasma kallikrein is probably the major activator under physiological conditions since it has the highest specific activity in this reaction.[8] Another plasma protein, HMW kinino-

---

[25] K. Fujikawa, B. McMullen, R. L. Heimark, K. Kurachi, and E. W. Davie, *Protides Biol. Fluids* **28**, 193 (1980).

[26] C. H. W. Hirs, this series, Vol. 11, p. 59.

[27] W. L. Bencze and K. Schmid, *Anal. Chem.* **29**, 1193 (1957).

TABLE II
AMINO ACID AND CARBOHYDRATE COMPOSITIONS
OF HUMAN FACTOR XII

	Factor XII (residues/76,000)
Amino acid	
Aspartic acid	35.8
Threonine	31.8
Serine	34.8
Glutamic acid	63.2
Proline	51.1
Glycine	44.6
Half-cystine[a]	29.6
Alanine	46.8
Valine	25.8
Methionine	3.6
Isoleucine	9.6
Leucine	51.1
Tyrosine	17.9
Phenylalanine	16.3
Tryptophan[b]	18.8
Histidine	25.9
Lysine	20.6
Arginine	36.7
Carbohydrate	
Hexose	17.7
Hexosamine	19.8
N-Acetylneuraminic acid	19.4
Protein (%)	83.2
Carbohydrate (%)	16.8

[a] Determined as cysteic acid according to Hirs.[26]
[b] Determined spectrophotometrically according to Bencze and Schmid.[27]

gen, stimulates the activation reaction severalfold.[10-12] The effect of HMW kininogen is variable, however, and depends on the concentration of enzyme, substrate, and the surface material. For instance, an inhibitory effect of HMW kininogen is observed in the activation of bovine factor XII by kallikrein when dextran sulfate is used as the surface material.[9]

During the activation of human factor XII by kallikrein, a specific internal peptide bond is initially cleaved, leading to the formation of an active enzyme. This enzyme is composed of a heavy chain (MW 52,000) and a light chain (MW 28,000), and these two chains are held together by a disulfide bond(s). The heavy chain is responsible for the binding of the

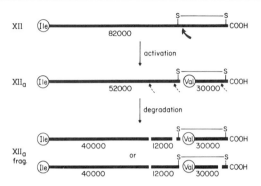

Fig. 5. Tentative mechanism for the activation and degradation of human factor XII. The solid arrow shows the initial cleavage site for the activation of factor XII by kallikrein. The dotted arrows show potential cleavage sites for the degradation reactions. The zymogen is shown with MW 82,000, representing the sum of the 3 smaller fragments. Accordingly, this value is somewhat larger than the value 76,000 obtained for the intact protein.

protein to negatively charged surfaces,[28] while the light chain contains the active site of the enzyme. The initial cleavage reaction is followed by a second peptide bond cleavage in the heavy chain. This occurs at a slower rate than the initial cleavage reaction.[10,25,29] Thus, human factor XII$_a$ is usually a mixture of two forms. One form is composed of 2 chains with MW 52,000 and 28,000, and a second form with MW 40,000, 10,000, and 28,000.[21] The specific activity of human factor XII$_a$ in a clotting assay remains constant after the initial cleavage, indicating that the second cleavage is a minor degradation step. It has been suggested that the 40,000-MW fragment is located at the amino-terminal region of the zymogen molecule and the 28,000-MW fragment is located at the carboxyl-terminal region of the factor XII molecule.[28] Recently, the alignment of the 3 fragments has been confirmed by sequence studies on the intact molecule and each fragment.[25] On prolonged incubation with kallikrein, a third cleavage occurs, and this peptide bond is probably split near the end of the carboxyl-terminal region of the 10,000-MW fragment or the 30,000-MW fragment. This leads to the formation of a 30,000-MW enzyme called factor XII$_a$ fragment. A tentative mechanism for the activation and degradation reactions is shown in Fig. 5.

The amino-terminal sequence of the 30,000-MW fragment is Val-Val-Gly-Gly-Leu-Val-Ala-Leu-Arg-Gly-. The sequence of Leu-Cys-Ala-Gly-Phe-Leu-Gly-Gly-Thr-Asp-Ala-Lys-Gln-Gly-Asp-SER-Gly-Gly-Pro- containing the active site of the enzyme is also present in the 30,000-MW

[28] S. D. Revak and C. G. Cochrane, *J. Clin. Invest.* **57**, 852 (1976).
[29] S. D. Revak, C. G. Cochrane, and J. H. Griffin, *J. Clin. Invest.* **59**, 1167 (1977).

fragment. DFP is also incorporated into the 30,000-MW fragment,[30] further indicating that the active-site serine is present in this polypeptide chain. DFP is incorporated into the zymogen form of factor XII, but this occurs at a much slower rate.[31] Furthermore, this incorporation is not accelerated by HMW kininogen and/or negatively charged surfaces.[31] These data indicate that human factor XII is a typical serine protease similar to the other coagulation factors.

Trypsin will activate human factor XII in the absence of a surface material. This, however, leads to a rapid degradation of the factor XII_a molecule and loss of clotting activity.[32,33] The specific esterase and amidase activities of activated factor XII remain constant, however, during the activation and degradation by trypsin. The factor XII_a fragment can be isolated by DEAE–Sephadex column chromatography,[34] and this fragment corresponds to the 30,000-MW fragment obtained by kallikrein activation. It is also identical to the factor XII_a fragment purified from human serum.[35-37]

Nonphysiological and physiological surface materials have been studied extensively for their effects on the activation of the human protein. These include kaolin, ellagic acid, celite, and dextran sulfate. Materials with a potential physiological role include sulfatide and galactoceramide 3-O-sulfate.[9] The latter is found in brain and red cells and most other organs. Fatty acids[19] and bacterial endotoxin[20] are also possible surface materials that may be active under physiological conditions.

[30] J. H. Griffin, Fed. Proc., Fed. Am. Soc. Exp. Biol. 36, 329 (1977).
[31] J. H. Griffin and G. Beretta, in "Kinin II" (T. Suzuki and H. Moriya, eds.), p. 39. Plenum, New York, 1979.
[32] C. G. Cochrane, S. D. Revak, and K. D. Wuepper, J. Exp. Med. 138, 1564 (1973).
[33] S. D. Revak, C. G. Cochrane, A. P. Johnston, and T. E. Hugli, J. Clin. Invest. 54, 619 (1974).
[34] K. Fujikawa and E. W. Davie, unpublished data.
[35] A. P. Kaplan and K. F. Austen, J. Immunol. 105, 802 (1970).
[36] A. M. Vennerod and K. Laake, Thromb. Res. 4, 103 (1974).
[37] A. P. Kaplan and K. F. Austen, J. Exp. Med. 133, 696 (1971).

[17] Human Factor XI (Plasma Thromboplastin Antecedent)

By KOTOKU KURACHI and EARL W. DAVIE

Factor XI is a glycoprotein (MW ~ 130,000) present in normal plasma.[1] It participates in the early stage of the intrinsic pathway of blood coagulation. Individuals with a congenital bleeding disorder called plasma

[1] E. W. Davie, K. Fujikawa, K. Kurachi, and W. Kisiel, Adv. Enzymol. 48, 277 (1979).

thromboplastin antecedent deficiency lack factor XI coagulant activity.[2,3] In normal plasma, factor XI is present in a precursor form and is converted to an enzyme (factor XI_a) by factor XII_a. Recently, the mechanism of activation of factor XI by factor XII_a has been reported.[4-6] This reaction involves the proteolytic cleavage of each of the two polypeptide chains in factor XI. Factor XI_a is a serine protease that converts factor IX to factor IX_a in the presence of calcium ions.[7,8] In this reaction, factor XI_a cleaves two specific peptide bonds in factor IX, converting it to factor IX_a. In these reactions, an activation peptide (MW 11,000) is released from the internal portion of the precursor molecule.

The purification of factor XI has been difficult, since its concentration in plasma is low and it is also very sensitive to degradation by proteolytic enzymes. Methods for the isolation of human factor XI,[4,5] rabbit factor XI,[9] and bovine factor XI[10] have been reported.

In the present procedure, heparin–agarose column chromatography, CM-Sephadex column chromatography, and benzamidine–agarose column chromatography were employed. Each purification step was performed in the presence of protease inhibitors in order to restrict activation of the precursor molecule.

Reagents

Factor XI-Deficient Plasma. The bovine factor XI-deficient plasma used for the assay was kindly provided by Dr. G. Kociba of Ohio State University, Columbus, Ohio. It was stored in small aliquots at $-70°C$. Human factor XI-deficient plasma is available from commercial sources, such as George King Biomedicals, Salem, New Hampshire.

Heparin–Agarose. Heparin–agarose is prepared as described elsewhere.[11] After every use, the heparin–agarose column is washed and reactivated as follows. One liter of heparin–agarose, packed in an 8 × 20-cm plastic column, is washed with 4 liters of 0.1 M Tris base containing 4 M NaCl, and then reequilibrated with 10 liters of 0.02 M Tris-HCl (pH

[2] R. L. Rosenthal, O. H. Dreskin, and N. Rosenthal, *Proc. Soc. Exp. Biol. Med.* **82**, 171 (1953).
[3] C. D. Forbes and O. D. Ratnoff, *J. Lab. Clin. Med.* **79**, 113 (1972).
[4] K. Kurachi and E. W. Davie, *Biochemistry* **16**, 5831 (1977).
[5] B. N. Bouma and J. H. Griffin, *J. Biol. Chem.* **252**, 6432 (1977).
[6] K. Kurachi, K. Fujikawa, and E. W. Davie, *Biochemistry* **19**, 1330 (1980).
[7] R. G. DiScipio, K. Kurachi, and E. W. Davie, *J. Clin. Invest.* **61**, 1528 (1978).
[8] K. Fujikawa, M. E. Legaz, H. Kato, and E. W. Davie, *Biochemistry* **13**, 4508 (1974).
[9] R. Wiggins, C. G. Cochrane, and J. H. Griffin, *Thromb. Res.* **15**, 475 (1979).
[10] T. Koide, H. Kato, and E. W. Davie, *Biochemistry* **16**, 2279 (1977).
[11] K. Fujikawa, A. R. Thompson, M. E. Legaz, R. G. Meyer, and E. W. Davie, *Biochemistry* **12**, 4938 (1973).

7.2) containing 0.05 M NaCl. After the column is used five times, the heparin–agarose is poured into a beaker and stirred with 3 liters of 0.1 M NaOH for 15 min at 4°C and rapidly filtered by suction. After extensive washing with cold water, the gel is packed into the column, washed, and equilibrated with the starting buffer as described above. For the second heparin–agarose column chromatography, the heparin–agarose is used after repeated washing with 0.02 M Tris-HCl (pH 7.2) containing 2 M NaCl, followed by equilibration with the starting buffer.

Benzamidine–Agarose. Benzamidine–agarose is prepared as described by Schmer[12] employing cyclohexyl(morpholinoethyl)carbodiimide as the coupling reagent. After the coupling reaction, the gel is washed extensively with 0.05 M imidazole HCl buffer (pH 6.3) containing 0.05 M NaCl, and stored at 4°C in the same buffer containing 0.02% sodium azide. The benzamidine–agarose is then packed in a plastic column (2.5 × 15 cm) and washed with 300 ml of 0.05 M imidazole HCl buffer (pH 6.3) containing 0.05 M NaCl before use. The benzamidine–agarose column is used after repeated washing with 300 ml of 0.05 M imidazole HCl (pH 6.3) containing 2 M guanidine HCl, followed by equilibration with the starting buffer.

Phospholipid Suspension. The stock suspension of phospholipid is prepared as follows. One vial (20 mg) of rabbit brain cephalin (Sigma Chemical Co.) is suspended in 100 ml of 0.02 M Tris-HCl (pH 7.4) containing 0.1 M NaCl, and briefly sonicated. The stock suspension is stored in aliquots of 2 ml at −20°C. Fresh aliquots are thawed each day for the assay.

Kaolin Suspension. Fifty milligrams of acid-washed kaolin (Fisher Scientific Co.) is suspended in 10 ml of saline and stored at room temperature.

Assay Procedures

Factor XI activity is routinely measured by the one-stage kaolin-activated partial thromboplastin time using bovine factor XI-deficient plasma as substrate. Samples to be assayed for factor XI activity are diluted with 0.02 M Tris-HCl (pH 7.4) containing 0.1 M NaCl and 1 mg bovine serum albumin/ml. Samples are diluted so that a clotting time between 40 and 150 sec is obtained. A 50-μl aliquot of the diluted sample is then incubated at 37°C for 10 min with 25 μl of bovine factor XI-deficient plasma and 50 μl of 0.5% kaolin suspension. After a 10-min preincubation period, 50 μl of 0.02% cephalin suspension and 50 μl of 0.05 M CaCl$_2$ solution are added to the mixture. A timer is activated on the addition of calcium, and the time required for clot formation as determined by the tilting method is recorded. The amount of factor XI is

---

[12] G. Schmer, *Hoppe-Seyler's Z. Physiol. Chem.* **353**, 810 (1972).

estimated from a standard calibration curve made by a serial dilution of normal human plasma. The log of the clotting time is plotted against the log of the concentration of platelet-poor normal human plasma. One unit of factor XI is defined as that amount of activity present in 1 ml of normal human plasma. Specific activity is expressed as units per absorbance at 280 nm. For the assay of factor XI_a, assay mixtures with or without kaolin are preincubated for only 30 sec. When no kaolin is used, 50 μl of saline is added instead.

Glass tubes (10 $\times$ 75-mm disposable culture tubes) are siliconized with PROSIL-28 (PCR Research Chemicals, Inc.) according to the instructions of the manufacturer, or with Dri film silicone (SC87) (General Electric Co.) as described elsewhere,[13] and used in the clotting assay.

Purification Procedures

Throughout the purification steps, plastic tubes and columns are used. All glassware is used after siliconizing as described above. All operations are performed at 4°C unless otherwise stated.

Human citrated whole plasma or cryosupernatant is used as the starting material.

Barium Chloride Adsorption and Ammonium Sulfate Precipitation. Barium chloride adsorption of the plasma is carried out to remove the vitamin K-dependent proteins (mainly prothrombin, factor X, factor IX, factor VII, protein C, and protein S) prior to ammonium sulfate precipitation. In this procedure, the human citrated plasma (15 liters) is mixed with 600 ml of 1 M barium chloride and the mixture is stirred for 30 min. The barium citrate pellet is removed by centrifugation at 5000 g for 15 min in a Sorvall RC3, and ethylenediamine tetraacetic acid is added to the supernatant to 0.1 mM. The barium chloride supernatant is then brought to 20% saturation by the slow addition of solid ammonium sulfate. After gentle stirring for 15 min, the precipitate is removed by centrifugation at 7800 g for 15 min. The supernatant is then brought to 45% saturation with solid ammonium sulfate, and after stirring for an additional 30 min, the suspension is centrifuged at 7800 g for 40 min. The precipitate is then dissolved in 4 liters of cold water containing polybrene (50 mg/liter), 0.2 mM diisopropyl fluorophosphate (DFP), and soybean-trypsin inhibitor (100 mg). The solution is dialyzed overnight against 100 liters of cold distilled water for 10–12 hr and then transferred to 80 liters of 0.02 M Tris-HCl (pH 7.2) to dialyze for an additional 8 hr.

First Heparin–Agarose Column Chromatography. After dialysis, a small precipitate is removed by centrifugation at 78,000 g for 15 min. At this

[13] M. E. Legaz and E. W. Davie, this series, Vol. 45, p. 83.

stage, the supernatant (about 6 liters) has a conductivity of 4 mmho at 4°C. The conductivity of the supernatant is then adjusted to 6 mmho by the addition of several grams of solid sodium chloride, and benzamidine (final concentration of 1 mM), DFP (final concentration of 0.2 mM), polybrene (final concentration of 50 mg/liter), and soybean-trypsin inhibitor (100 mg) are added.

The sample solution is then applied to a heparin–agarose column (8 × 20 cm), which is previously equilibrated with 0.02 M Tris-HCl (pH 7.2), containing 0.05 M NaCl. After application of the sample, the column is washed with 2.5 liters of 0.02 M Tris-HCl (pH 7.2) containing 0.05 M NaCl, 2 mM DFP, and polybrene at a concentration of 50 mg/liter. Factor XI is eluted from the column at a flow rate of 400 ml/hr employing a linear gradient formed with 3 liters of 0.02 M Tris-HCl (pH 7.2) containing 0.15 M NaCl and 3 liters of 0.02 M Tris-HCl (pH 7.2) containing 0.6 M NaCl (Fig. 1A). Each gradient buffer solution also contains 0.2 mM DFP and 50 mg of polybrene per liter. More than 95% of the contaminating protein passes through the column before the gradient elution is started. Factor XI is eluted after factor XII when the sodium chloride concentration approaches 0.42–0.45 M. It is not contaminated with appreciable amounts of factor XII. Factor XII is eluted at a sodium chloride concentration of about 0.3 M, and prekallikrein is eluted at 0.15–0.2 M sodium chloride. Factor XI is purified about 600-fold at this stage over the starting plasma (see table). Fractions containing factor XI activity, as shown by the bar in Fig. 1 are pooled, and benzamidine (final concentration, 1 mM), DFP (final concentration, 0.2 mM), polybrene (50 mg/liter), and soybean-trypsin inhibitor (100 mg) are added to the pooled fractions. The solution is then dialyzed overnight against 20 liters of 0.02 M phosphate (pH 7.2) containing 0.07 M NaCl. Polybrene (50 mg/liter) and DFP (final concentration of 0.2 mM) are added to the dialyzed sample, and a small precipitate in the pooled fractions is removed by centrifugation at 78,000 g for 15 min.

CM-Sephadex Column Chromatography. The dialyzed sample is mixed with CM-Sephadex C-50 (200 ml of settled volume), which is previously equilibrated with 0.02 M Tris-HCl (pH 7.2) containing 0.07 M NaCl. The suspension is gently stirred for 1.5 hr and allowed to settle for 20 min. The supernatant is removed by siphon. The CM-Sephadex is then poured into a plastic column (4.5 × 15 cm), and the column is washed with 1–1.5 liters of 0.02 M phosphate (pH 7.2) containing 0.07 M NaCl, polybrene (50 mg/liter), and DFP (final concentration of 0.2 mM) until the absorption at 280 nm is lower than 0.1. A linear gradient is then formed with 1 liter of 0.02 M phosphate buffer (pH 7.2) containing 0.07 M NaCl, and 1 liter of 0.02 M phosphate buffer (pH 7.2) containing 0.6 M NaCl. Both gradient solutions contain polybrene (50 mg/liter) and DFP (final concentration of

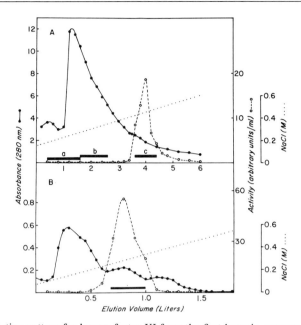

FIG. 1. Elution pattern for human factor XI from the first heparin–agarose column and the CM-Sephadex column. (A) The protein is eluted from the first heparin–agarose column (8 × 20 cm) with a linear gradient formed with 3 liters of 0.02 M Tris-HCl (pH 7.2) containing 0.15 M NaCl, and 3 liters of 0.02 M Tris-HCl (pH 7.2) containing 0.6 M NaCl, as described under Purification Procedures. Fractions (200 ml) are collected in bottles containing 12.5 mg polybrene. The flow rate is 400 ml/hr. Factor XI activity is determined by the one-stage method, as described under Assay Procedures. ●——●, Absorbance at 280 nm; ○----○, factor XI activity; · · · · · , concentration of NaCl (M). Fractions with factor XI activity, as shown by bar c, are pooled. Bars a and b indicate the elution positions of prekallikrein and factor XII, respectively. (B) The protein is eluted from the CM-Sephadex column (4.5 × 15 cm) with a linear gradient formed with 1 liter of 0.02 M phosphate buffer (pH 7.2) containing 0.07 M NaCl, and 1 liter of 0.02 M phosphate buffer (pH 7.2) containing 0.6 M NaCl, as described under Purification Procedures. Fractions (10 ml) are collected at a flow rate of 100 ml/hr. ●——●, Absorbance at 280 nm; ○----○, factor XI activity. Fractions shown by the bar are pooled.

0.2 mM). Factor XI is eluted from the column at 0.25–0.3 M sodium chloride (Fig. 1B). The flow rate is 100 ml/hr. Fractions containing factor XI, as shown by the bar, are pooled, and polybrene (50 mg/liter) and DFP (final concentration of 0.2 mM) are added.

Second Heparin–Agarose Column Chromatography. The pooled fractions from the CM-Sephadex column are concentrated to about 100 ml by ultrafiltration employing an Amicon Diaflo concentrator (PM-30 membrane), and dialyzed overnight against 4 liters of 0.05 M phosphate (pH 6.6) containing 0.15 M NaCl. The dialyzed sample is then applied to a

PURIFICATION OF HUMAN FACTOR XI

Purification step	Volume (ml)	Total protein (A_{280})	Total activity (units)	Specific activity (units/A_{280})	Recovery (%)	Purification (-fold)
Plasma	15,000	1.1×10^6	15,000	0.014	100	1
$(NH_4)_2SO_4$ fractionation	6000	5.7×10^5	13,200	0.023	88	1.7
First heparin–agarose	1050	1890	12,600	8.7	84	621
CM–Sephadex	350	67	10,850	162	72	11,570
Second heparin–agarose	185	24	5920	247	39	17,620
Benzamidine–agarose	135	9.4[a]	3920	417[b]	26	29,800

[a] This value is equivalent to 7.0 mg of factor XI.
[b] This value is equivalent to 559 units/mg of factor XI.

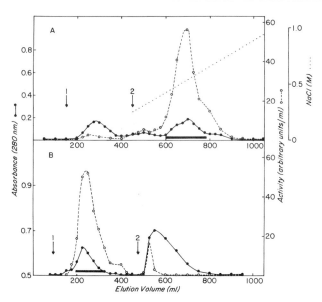

FIG. 2. Elution pattern for human factor XI from the second heparin–agarose column and the benzamidine–agarose column. (A) The heparin–agarose column (3.2 × 12 cm) is washed with 0.05 M phosphate buffer (pH 6.6) containing 0.2 M NaCl (arrow 1), and then eluted with a linear gradient formed with 300 ml of 0.05 M phosphate buffer (pH 6.6) containing 0.25 M NaCl, and 300 ml of 0.05 M phosphate buffer (pH 6.6) containing 1.0 M NaCl (arrow 2), as described under Purification Procedures. Fractions (5 ml) are collected at a flow rate of 100 ml/hr. ●——●, Absorbance at 280 nm; ○----○, factor XI activity; · · · · ·, concentration of NaCl (M). Fractions shown by the bar are pooled. (B) The benzamidine–agarose column (2.3 × 13 cm) is eluted with 0.05 M imidazole buffer (pH 6.3) containing 0.1 M NaCl (arrow 1), and then with 0.05 M imidazole buffer (pH 6.3) containing 0.1 M NaCl and 0.1 M guanidine hydrochloride (arrow 2), as described under Purification Procedures. Fractions (5 ml) are collected at a flow rate of 80 ml/hr. ●——●, Absorbance at 280 nm; ○----○, factor XI activity. Fractions shown by the bar are pooled.

heparin–agarose column (3.2 × 12 cm), which is previously equilibrated with 0.05 M phosphate (pH 6.6) containing 0.15 M NaCl. The column is washed with 300 ml of 0.05 M phosphate (pH 6.6) containing 0.2 M NaCl and 0.2 mM DFP. A linear gradient is formed with 300 ml of 0.05 M phosphate (pH 6.6) containing 0.25 M NaCl, and 300 ml of 0.05 M phosphate (pH 6.6) containing 1.0 M NaCl. Both solutions contain 0.2 mM DFP. Factor XI is eluted at a sodium chloride concentration of 0.45 M (Fig. 2A). Fractions containing factor XI activity are combined, and DFP is added to a final concentration of 0.2 mM. The pooled fraction is then concentrated to about 100 ml by ultrafiltration as described above.

Benzamidine–Agarose Column Chromatography. The pooled fraction from the second heparin–agarose column is then dialyzed overnight against 2 liters of 0.05 M imidazole HCl (pH 6.3) containing 0.15 M NaCl, and applied to a benzamidine–agarose column (2.3 × 13 cm). This column is previously equilibrated with 0.05 M imidazole HCl (pH 6.3) containing 0.05 M NaCl. The column is washed with 350 ml of 0.05 M imidazole HCl (pH 6.3) containing 0.1 M NaCl. In most preparations, more than 90% of the total factor XI activity is eluted in the wash and only a small portion of factor XI activity is retained on the column. The latter activity is eluted with 0.05 M imidazole buffer (pH 6.3) containing 0.1 M NaCl and 0.1 M guanidine hydrochloride (Fig. 2B). This elution profile differs somewhat from bovine factor XI, which binds rather tightly to the benzamidine–agarose column and can be eluted only with 0.05 M imidazole buffer (pH 6.3) containing 0.1 M NaCl and 0.1 M guanidine hydrochloride.

Some factor XI preparations contain a contaminant with MW 56,000 as estimated by sodium dodecyl sulfate (SDS)–polyacrylamide gel electrophoresis. These preparations are further purified by gel filtration on a Sephadex G-150 column (2.5 × 100 cm), which is previously equilibrated with 0.025 M Tris-HCl (pH 7.0) containing 0.15 M NaCl. Factor XI is eluted with the same buffer solution in the second protein peak. The contaminating protein, which is named protein K, forms an oligomer with a much higher molecular weight than factor XI and elutes from the Sephadex G-150 column prior to factor XI.

Fractions from the first peak in Fig. 2B are combined and concentrated to 0.5–1.0 mg/ml and stored at −70°C until further use. Human factor XI is purified about 30,000-fold by this method, with an overall yield of 20–25%. The isolation procedure takes about 9 days. A summary of the purification procedure is shown in Table I. The purified protein is homogeneous when examined by SDS–polyacrylamide gel electrophoresis and immunoelectrophoresis. Human factor XI is stable for many months when kept frozen at −70°C. Repeated freezing and thawing of factor XI leads to a gradual loss of activity.

General Properties and Activation of Factor XI

Human factor XI is a glycoprotein with MW ≈ 130,000.[4] It is composed of 2 identical polypeptide chains with MW ≈ 60,000, and these chains are linked by a disulfide bond(s). The protein contains about 5% carbohydrate, including hexose, hexosamine, and neuraminic acid. The amino acid and carbohydrate compositions of human and bovine factor XI have been published.[4]

The extinction coefficient of human factor XI is 13.4 as determined by differential refractometry,[4] and the isoelectric point is 8.6 as determined by electrofocusing.[14] The latter value appears to be considerably higher

---

[14] K. Kurachi, unpublished data.

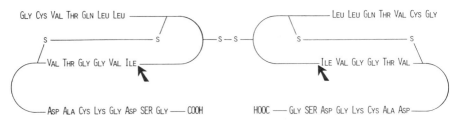

Fig. 3. Partial structure of human factor XI.[4] The 2 identical chains are held together by a disulfide bond(s). The active-site serine is shown in large caps. The two arrows indicate the site of cleavage in the 2 chains when the protein is converted to factor XI$_a$ by factor XII$_a$. The resulting 4 chains are held together by disulfide bonds. The number and exact location of the disulfide bonds are not known.

than that of bovine factor XI, since human factor XI passes through a DEAE–Sephadex column at pH 8.4, while bovine factor XI is well retained on the column. The amino-terminal sequence of the two subunits composing human factor XI[4] is Gly-Cys-Val-Thr-Gln-Leu-Leu-Lys-Asp-Thr-Cys-Phe-Glu-Gly-Gly-, and this sequence is homologous to that of bovine factor XI and bovine prekallikrein.[6] The subunits of human factor XI are indistinguishable by SDS–polyacrylamide gel electrophoresis. Bovine factor XI, however, is composed of two subunits that may be slightly different in their molecular weights.[6] The slight difference in the size of the subunits is probably due to a limited proteolysis of the intact subunit in the carboxyl-terminal region by an unknown protease other than factor XI$_a$ or factor XII$_a$.

Factor XI is activated by factor XII$_a$[4] or factor XII[6,14] in the presence of negatively charged surfaces such as kaolin. It is also activated by trypsin.[15] Upon activation by limited proteolysis, each subunit of human factor XI is cleaved at a specific peptide bond. This generates factor XI$_a$, a protein composed of 2 heavy chains (each with MW 35,000) and 2 light chains (each with MW 25,000) (Fig. 3). Bovine factor XI is also activated by a similar mechanism.[6] Factor XI$_a$ has esterase and amidase activity. The specific activity of human factor XI$_a$ is 18 μmol of tosyl-L-arginine methyl ester hydrolyzed per minute per milligram of enzyme. Factor XI$_a$ is inhibited by a number of protease inhibitors,[1] including antithrombin III,[4] DFP,[4] C1-inactivator,[16] pancreatic-trypsin inhibitor,[17] soybean-trypsin inhibitor,[17] α_1-antitrypsin inhibitor,[18] and Russell's viper venom inhibitor II.[6] Existence of two active sites in factor XI$_a$ is clearly shown by titration with antithrombin III.[4,6]

[15] C. Mannhalter, S. Schiffman, and A. Jacobs, *J. Biol. Chem.* **255**, 2667 (1980).

[16] C. D. Forbes, J. Pensky, and O. D. Ratnoff, *J. Lab. Clin. Med.* **76**, 809 (1970).

[17] K. D. Wuepper, *in* "Inflammation: Mechanisms and Control" (I. H. Lepow and P. A. Ward, eds.), p. 93. Academic Press, New York, 1972.

[18] L. W. Heck and A. P. Kaplan, *J. Exp. Med.* **140**, 1615 (1974).

[18] Purification of Human Coagulation Factors II, IX, and X Using Sulfated Dextran Beads

By Joseph P. Miletich, George J. Broze, Jr., and Philip W. Majerus

A major advance in the purification of the vitamin K-dependent coagulation factors occurred when it was reported that heparin coupled to agarose was useful in the isolation of bovine factor IX.[1] This matrix has since been used successfully in the purification of human factor IX,[2-4] and, under appropriate conditions, factors II and X as well.[5]

Heparin, a potent anticoagulant, is an acidic mucopolysaccharide that is as much as 40% ester sulfate by weight. Early attempts to produce an economical, synthetic anticoagulant centered around production of sulfuric acid esters of polysaccharides, and dextran sulfate was the most promising of these compounds produced.[6] Recently, dextran sulfate linked to agarose has also been employed in the purification of human factors II, IX, and X.[7]

Dextran already polymerized in bead form is readily available (Sephadex, Pharmacia), has efficient flow properties, and is relatively inexpensive. Here we describe the production of a highly sulfated derivative of these insoluble beads, and its use to isolate human plasma coagulation factors.

Sulfation of Bead-Polymerized Dextran[8]

Four hundred milliliters of chlorosulfonic acid were added very slowly to 2 liters of dry pyridine cooled in an ethanol–dry ice bath in a fume hood. Once this violently exothermic addition was complete, the mixture was transferred to a water bath maintained at 70°C to keep the pyridinium chlorosulfonate in solution. Three hundred grams of Sephadex G-50 (20–80 μm particle size) was then added with rapid mixing. The beads

[1] K. Fujikawa, A. R. Thompson, M. E. Leqaz, R. G. Meyer, and E. W. Davie, *Biochemistry* **12**, 4938 (1973).

[2] R. G. DiScipio, M. A. Hermodson, S. G. Yates, and E. W. David, *Biochemistry* **16**, 698 (1977).

[3] L. O. Anderson, H. Borg, and M. Miller-Andersson, *Thromb. Res.* **7**, 451 (1975).

[4] J. S. Rosenberg, P. W. McKenna, and R. D. Rosenberg, *J. Biol. Chem.* **250**, 8883 (1975).

[5] J. P. Miletich, C. M. Jackson, and P. W. Majerus, *J. Biol. Chem.* **253**, 6908 (1978).

[6] C. R. Ricketts, *Biochem. J.* **51**, 129 (1952).

[7] D. S. Pepper and C. Prowse, *Thromb. Res.* **11**, 687 (1977).

[8] From J. P. Miletich, G. J. Broze, and P. W. Majerus, *Anal. Biochem.* **105**, 304 (1980).

METHODS IN ENZYMOLOGY, VOL. 80

uniformly absorbed nearly all of the liquid after 20 min and further mixing was unnecessary. The reaction mixture was maintained at 70°C for 5 hr. Five hundred grams of sodium hydroxide dissolved in 5 liters of distilled water at 0°C was then rapidly added with mixing to neutralize the acid. The beads were collected on a sintered glass funnel and washed in 30 liters of 2 M sodium chloride adjusted to pH 10 with sodium hydroxide. The beads were collected and this step was repeated. Then the beads were washed four times with 40 liters of distilled water. This resulted in a suspension of approximately 35 liters, which was dehydrated by slow, successive addition of methanol to 50, 80, 90, and finally 95%. The dehydrated beads were again collected on a sintered glass funnel, washed with absolute methanol, and dried in a vacuum desiccator for 24 hr at room temperature.

Properties of Sulfated Dextran Beads

The procedure described above yields 800 g of dry sulfated dextran, free-flowing beads as the sodium salt, which theoretically corresponds to incorporation of 4.90 mol of sodium sulfate ester. Based on estimates of the degree of cross-linking of the dextran, there are 4.5 mol of unsubstituted hydroxyl residues available on the glucose residues and glyceryl bridges (plus a small contribution produced by side reactions during the crosslinking of the dextran with epichlorohydrin) in 300 g of Sephadex G-50. Thus the final product must contain a high degree of sulfate ester fixation.

One gram of dry beads swells to 35 ml in water. As anticipated with the substantial sulfate substitution in 0.15 M NaCl the packed volume decreases to 20 ml, and in 2.0 M NaCl to 5 ml.

The sulfated dextran beads can be recycled indefinitely. Materials that remain bound to the matrix following chromatography can be removed by washing with 1 M NaCl at pH ~10. The hydrated beads are stable for more than 2 years in buffered solutions at neutral pH. The hydrogen form is highly acidic and undergoes autohydrolysis, thus low pH (<2.0) must be avoided. The beads will deteriorate if stored in the dry form. When stored in aqueous solutions the usual precautions to prevent microbial growth must be observed.

Purification of Factors II, IX, and X

Starting Material. Ten liters of fresh frozen plasma was thawed at 37°C, pooled, and transferred to a cold room. All subsequent procedures were performed in plastic containers at 4°C unless otherwise indicated.

Barium Citrate Adsorption. Five hundred milliliters of 1.0 M BaCl$_2$ was added dropwise over 1 hr with stirring at 4°C. The precipitate was col-

lected by centrifugation at 5000 g for 20 min and the supernatant discarded. The precipitate was resuspended in 1.5 liters of 0.1 M NaCl, 0.01 M
BaCl$_2$, and then again collected by centrifugation. This precipitate was
finally resuspended in 0.02 M Tris-HCl (pH 7.0) to a final volume of 800 ml
and 1 g of diisopropyl fluorophosphate added with stirring in a fume hood.

Ammonium Sulfate Elution. Two hundred milliliters of 2.0 M ammonium sulfate (pH 7.0 with ammonium hydroxide) was added dropwise
over 1 hr with continuous stirring. The supernatant was collected after
centrifugation at 10,000 g for 30 min and diluted to 4 liters with water,
which reduced its conductivity to that of the buffer employed in the next
step. Benzamidine was added to a final concentration of 1 mM and the pH
of the solution adjusted to 7.0 if necessary.

DE-52 Cellulose Ion-Exchange Chromatography (Fig. 1). The solution

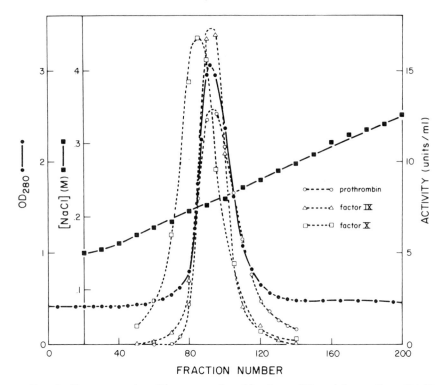

Fig. 1. Chromatography of human prothrombin, factor IX, and factor X on DE-52
cellulose. ●——●, Absorbance at 280 nm (no correction for the presence of benzamidine
has been made); ■——■, concentration of NaCl determined by conductivity; ○---○, prothrombin activity; △---△, factor IX activity; □---□, factor X activity. The units of activity
are relative, since the presence of benzamidine prevents exact measurement. Fraction size
was 17.5 ml. Coagulation factor activities were determined as previously described. From J.
P. Miletich, G. J. Broze, and P. W. Majerus, *Anal. Biochem.* **105,** 304 (1980).

was applied to a column (5 × 20 cm) of DE-52 cellulose (Whatman) equilibrated in 0.15 M NaCl–0.01 M sodium phosphate (pH 7.0)–0.001 M benzamidine, at a flow rate of 1.5 liters/hr, at room temperature. After the column was washed with 2 liters of starting buffer, the flow rate was decreased to 500 ml/hr and a linear gradient of 0.15–0.40 M NaCl in 0.01 M sodium phosphate (pH 7.0)–0.001 M benzamidine developed over 4 liters. Fractions 60–120 were pooled, the pH adjusted to 6.0 with HCl, and sufficient 0.01 M sodium phosphate (pH 6.0)–0.001 M benzamidine added to reduce the NaCl concentration to 0.15 M.

Dextran Sulfate Chromatography (Fig. 2). The solution was applied to a column (5 × 20 cm) of sulfated dextran beads equilibrated in 0.15 M NaCl–0.01 M sodium phosphate (pH 6.0)–0.001 M benzamidine at room

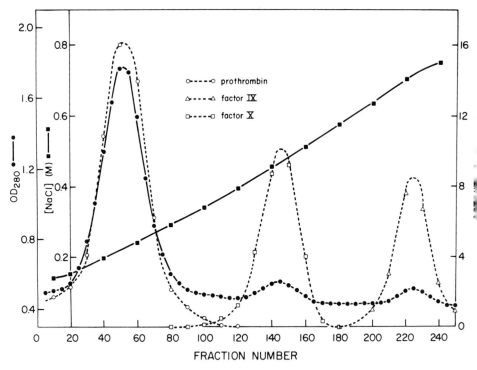

FIG. 2. Chromatography of DE-52 pool on sulfated dextran beads. ●——●, Absorbance at 280 nm (no correction for the absorbance of benzamidine has been made); ■——■, concentration of NaCl determined by conductivity; ○---○, prothrombin activity; △---△, factor IX activity; □---□, factor X activity. Units of activity are relative since the presence of benzamidine prevents exact measurement. Fraction size was 17.5 ml. Coagulation factor activities were determined as previously described. From J. P. Miletich, G. J. Broze, and P. W. Majerus, *Anal. Biochem.* **105,** 304 (1980).

temperature, at a flow rate of 400 ml/hr. The column was then washed with 1.5 liters of the same buffer and a 4-liter linear gradient developed from 0.15 to 0.75 M NaCl in 0.01 sodium phosphate (pH 6.0)–0.001 M benzamidine.

Concentration and Storage. Fractions containing the activities of factors II (24–78), X (125–168), or IX (201–243) were pooled and concentrated using PM-10 (Amicon) membranes. After 24-hr dialysis at 4°C against 100 volumes of 0.6 M NaCl–0.01 M sodium phosphate (pH 7.0)–0.001 M benzamidine, the factor pools were filtered (Gelman GN-6, 0.45-μm filters) and stored in 1-ml portions at −70°C.

Properties of the Final Product. The purification procedure is shown in the table; reduced and unreduced sodium dodecyl sulfate–polyacrylamide gel electrophoresis of the final products is shown in Fig. 3. The apparent molecular weights of the reduced proteins were 75,000 for prothrombin, 49,000 and 17,000 for the heavy and light chains of factor X, and 73,000 for factor IX. On unreduced gels the apparent molecular weights were 69,000, 68,000, and 58,000 for factors II, X, and IX, respectively.

Each purified factor was free of factor VII contamination. Factor VII is not absorbed by DE-52 cellulose under the conditions specified.

The prothrombin was free of factor IX or X activity, though in some samples 1–2% of the protein migrated as prothrombin intermediate 1 on polyacrylamide gels, presumably reflecting proteolysis by trace amounts of thrombin. Based on the recovery of activity and protein on three occasions, human plasma contains 85–95 μg of prothrombin/ml.

The factor X contained trace amounts (0.2–0.8%, w/w) of prothrombin as judged by thrombin assay following incubation with Taipan snake venom. Detection of trace factor IX activity was limited by the available assays, but was less than 0.1% if present. Although initially free of factor X_a activity, after several months' storage trace amounts (~0.2%) could be detected. Based on recovery of activity and protein, human plasma contains 7–8 μg of factor X/ml.

The factor IX was free of prothrombin, but assuming a specific activity of factor X of 165 U/ml, some preparations of factor IX contained as much as 1.0% (w/w) factor X. Human plasma contains 4–5 μg of factor IX/ml based on recoveries of activity and protein using this purification procedure.

Comments

The purification procedure described here permits the isolation of factors II, IX, and X in high yield and purity in 2 days. The products are very similar to those reported earlier by other investigators.

Purification of Human Prothrombin and Factors IX and X[a]

Material	Volume			Activity						Recovery (%)			Purification (-fold)		
				Units/ml			Units/mg[b]								
	II	IX	X	II	IX	X	II	IX	X	II	IX	X	II	IX	X
Plasma		10,000			1			0.017			100			1	
Barium sulfate eluate	—	960		—	9.7	10.1	—	3.7	3.9	—	93	97	—	220	230
DE-52 pool	—	1060		—	8.1	8.2	—	14.7	15.0	—	85	87	—	860	880
Sulfated dextran pools	990	808	778		5.12	7.10		242	167		41	55		14,250	9825
Final products	22.4	20.3	18.5	317	214	286	12.7	275	165	71	43	53	750	16,175	9700

[a] From J. P. Miletich, G. J. Broze, and P. W. Majerus, Anal. Biochem. **105**, 304 (1980).

[b] Protein concentrations were estimated from absorbance at 280 nm assuming: for plasma and the barium sulfate eluate, $E_{280}^{\%} = 10.0$; for the DE-52 pool and prothrombin fractions, $E_{280}^{\%} = 13.8$ (16); for factor IX, $E_{280}^{\%} = 13.3$ (1); and for factor X, $E_{280}^{\%} = 11.6$ (1).

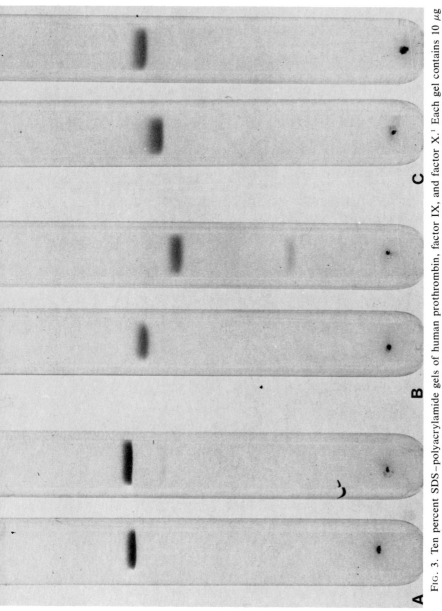

FIG. 3. Ten percent SDS–polyacrylamide gels of human prothrombin, factor IX, and factor X.[1] Each gel contains 10 μg of protein. (A) Prothrombin: left, unreduced; right, reduced. (B) Factor X: left, unreduced; right, reduced. (C) Factor IX: left, unreduced; right, reduced. The solid bar indicates the position of the tracking dye. From J. P. Miletich, G. J. Broze, and P. W. Majerus, *Anal. Biochem.* **105**, 304 (1980).

Sulfated dextran beads can also be used in the purification of protein C and antithrombin III, at steps where dextran sulfate–agarose or heparin–agarose are normally employed.[9] In the case of protein C, the addition of CaCl₂ (2.5 mM) to the buffer enhances the separation of protein C from prothrombin and factor X.[9] Sulfated dextran beads will also absorb factor VII,[9] and, most likely, other vitamin K-dependent proteins (e.g., protein S).

The mechanism of the interaction between the coagulation proteins and heparin–agarose, dextran sulfate–agarose, and dextran sulfate beads is obscure. Although heparin functions as an anticoagulant by promoting interaction of active factors and antithrombin III,[10] it is not clear that this is the same property that permits its use as an absorbant for the zymogen coagulation proteins. Apparently the presence of the sulfate group is essential, since cation-exchange chromatography using other materials including sulfopropyl Sephadex failed to separate the coagulation factors.

Acknowledgments

This research was supported by Grants HLBI 14147 (Specialized Center in Thrombosis), HL 07088, and HL 16634 from The National Institutes of Health, and in part by NIH Research Service Award GM 07200, Medical Scientist, from the National Institute of General Medical Sciences.

[9] G. J. Broze, unpublished observations (1980).
[10] R. D. Rosenberg and P. S. Damus, *J. Biol. Chem.* **248**, 6490 (1973).

[19] Human Factor VII

By GEORGE J. BROZE, JR., and PHILIP W. MAJERUS

Coagulation factor VII is a vitamin K-dependent plasma protein, which in the presence of tissue factor and calcium will activate factor X.[1,2] Bovine factor VII was isolated several years ago, and its properties have been studied extensively.[3-7] Purification of the human protein in sufficient

[1] Y. Nemerson, *Biochemistry* **5**, 601 (1966).
[2] W. S. Williams and D. G. Norris, *J. Biol. Chem.* **241**, 1847 (1966).
[3] R. Radcliffe and Y. Nemerson, *J. Biol. Chem.* **250**, 388 (1975).
[4] W. Kisiel and E. W. Davie, *Biochemistry* **14**, 4928 (1975).
[5] J. Jesty and Y. Nemerson, *J. Biol. Chem.* **249**, 509 (1974).
[6] J. Jesty, *Arch. Biochem. Biophys.* **185**, 165 (1978).
[7] W. Kisiel, K. Fujikawa, and E. W. Davie, *Biochemistry* **16**, 4189 (1977).

METHODS IN ENZYMOLOGY, VOL. 80

quantities for characterization was not accomplished, however, until much more recently.[8,9] The studies employing human factor VII have thus far shown its properties to be very similar to its bovine counterpart, with one notable exception (see below).

Assay Methods

Two methods are available for the determination of factor VII activity. A one-stage clotting assay is routinely used, employing tissue thromboplastin and factor VII-deficient plasma as substrate, and will be described below. Another assay employing a two-stage system in which the sample, tissue thromboplastin, and a source of factor X are incubated in the first stage followed by the addition of the chromagenic substrate S-2222 to assay the degree of factor X activation, has recently been described.[10,11] The specific activity of activated factor VII (factor VII_a) is considerably more than zymogen VII when assayed by the clotting method. However, in the chromogenic assay, factors VII and VII_a have the same apparent specific activity and it has been assumed that this is related to rapid activation of zymogen VII to VII_a by factor X_a during the first stage incubation.

Reagents

Human tissue thromboplastin. An extract from human brain can be prepared by a variety of methods.[8,12-14]

$CaCl_2 \cdot 2 H_2O$, 25 mM in H_2O

Assay buffer: 0.15 M NaCl–5 mM sodium citrate–0.05 M Tris-HCl (pH 7.4), and 5 mg/ml of bovine serum albumin.

Factor VII-deficient plasma. Congenitally deficient human plasma or bovine deficient plasma produced by the method of Nemerson and Clyne[15]

Procedure. The sample to be assayed is diluted appropriately in assay buffer. Sixty microliters of factor VII-deficient plasma, 60 μl of tissue thromboplastin, and 60 μl of the sample are incubated for 30 sec at 37°C.

---

[8] G. J. Broze and P. W. Majerus, *J. Biol. Chem.* **255**, 1242 (1980).

[9] S. P. Bajaj, S. I. Rapaport, and S. F. Brown, *J. Biol. Chem.* **256**, 253 (1981).

[10] U. Seligsohn, B. Osterud, and S. I. Rapaport, *Blood* **52**, 978 (1978).

[11] G. Avvisati, J. W. Ten Cate, E. W. Van Wijk, L. H. Kahte, and G. Mariani, *Br. J. Haematol.* **45**, 343 (1980).

[12] P. F. Hjort, *Scand. J. Clin. Lab. Invest.* **27**, Suppl., 17 (1957).

[13] P. A. Owren, *Scand. J. Clin. Lab. Invest.* **1**, 81 (1949).

[14] A. J. Quick, "Hemorrhagic Disease and Thrombosis," p. 438. Lea & Febiger, Philadelphia, Pennsylvania, 1966.

[15] Y. Nemerson and L. P. Clyne, *J. Lab. Clin. Med.* **83**, 301 (1974).

The reaction is then started by the addition of 60 μl of 25 mM CaCl$_2$ at 37°C.

Calibration. One unit of factor VII activity is defined as that amount in 1 ml of citrated pooled human plasma. Serial dilutions of normal citrated plasma are plotted against clotting times on log–log paper producing a linear relationship between approximately 0.025 and 0.00025 unit/ml of factor VII with clotting times of 25–100 sec.

Purification Procedure

Principle. Three major problems are encountered in the purification of human factor VII: (1) it is a trace protein in human plasma; (2) activation and subsequent degradation occur during the purification procedure; and (3) loss of activity occurs at protein concentrations less than approximately 30 μg/ml. To prevent proteolysis, soybean-trypsin inhibitor and high levels of benzamidine have been employed.

In the purification procedure, factor VII and the other vitamin K-dependent coagulation factors are separated from the bulk of unrelated plasma proteins by absorption to barium citrate. Factor VII in turn is separated from factors II, IX, X, and protein C by anion-exchange chromatography on DEAE–Sepharose. Much of the remaining contaminants are removed in a QAE-Sephadex anion-exchange step by taking advantage of the fact that factor VII elutes at a much lower ionic strength in the presence of Ca^{2+}. The remaining contaminants are removed by gel filtration on ACA-44.

Starting Material. Twenty liters of human citrated fresh-frozen plasma (with platelets removed) was thawed at 37°C, pooled, and transferred to the cold room. Benzamidine (156 g) and 500 mg of soybean-trypsin inhibitor (type I, Sigma, St. Louis, Missouri) were added and the plasma stirred for 30 min. All subsequent steps were performed at 4°C and in plastic containers.

Barium Citrate Adsorption and Elution and Ammonium Sulfate Fractionation. One liter of 1.0 M BaCl$_2$ was added dropwise over 45 min and the mixture stirred an additional 15 min. The barium citrate precipitate was collected by centrifugation at 3000 g for 15 min and the supernatant plasma decanted. The precipitate was washed twice with 6 liters of 0.2 M NaCl–0.05 M Tris-HCl (pH 8.3)–0.01 M benzamidine, and then resuspended in 3 liters of 0.15 M sodium citrate–0.1 M Tris-HCl (pH 8.3)–0.02 M benzamidine, and 25 mg/liter soybean-trypsin inhibitor. This mixture was stirred for 60 min to elute the adsorbed proteins, and the barium citrate removed by centrifugation at 3000 g for 30 min. Ammonium sulfate (442 g) was added to the supernatant slowly with stirring (25% saturation), and the precipitate removed by centrifugation at 6500 g for 20 min. An additional 772 g of ammonium sulfate was slowly added to the supernatant with stirring (65% saturation). The precipitate was collected by centrifugation at 8500 g for 30 min; and resuspended in 300 ml 0.2 M EDTA–0.1 M Tris-HCl (pH 8.3)–0.02 M benzamidine, and 25 mg/liter soybean-trypsin inhibitor. This solution was dialyzed against 0.05 M Tris-HCl (pH 7.5)–0.02 M benzamidine for 12 hr with three changes of dialysis buffer of 20 liters each. A fine precipitate that developed during dialysis was removed by centrifugation at 8500 g for 10 min.

DEAE–Sepharose Chromatography (Fig. 1). The solution was applied to a column (5 × 35 cm) of DEAE–Sepharose, equilibrated with 0.05 M Tris-HCl (pH 7.5)–0.02 M benzamidine at a flow rate of 35 ml/hr. The column was washed with 1 liter of the same buffer and then eluted with a 6-liter linear gradient from 0.0 to 0.5 M NaCl in 0.05 M Tris-HCl (pH 7.5)–0.02 M benzamidine. Fractions 232–250 were pooled (260 ml) and sufficient 0.05 M Tris-HCl (pH 7.5)–0.02 M benzamidine was added to adjust the conductivity to that of 0.15 M NaCl–0.05 M Tris-HCl (pH 7.5)–0.02 M benzamidine, and 6 mg of soybean-trypsin inhibitor was added.

QAE – Sephadex Chromatography. The solution was applied to a column (2.5 × 15 cm) of QAE–Sephadex equilibrated with 0.15 M NaCl–0.05 M Tris-HCl (pH 7.5)–0.02 M benzamidine at a flow rate of 100 ml/hr. The column was washed with 150 ml of the same buffer and then the factor VII activity was eluted with $CaCl_2$ containing buffer, which was prepared as follows: the starting buffer was made 0.005 M $CaCl_2$ by addition of 1.0 M $CaCl_2$, and then sufficient 0.05 M Tris-HCl (pH 7.5)–0.02 M benzamidine added to adjust the conductivity back to its original value. Under these conditions the factor VII is eluted from the column, while almost all the contaminating proteins remain bound. Four-milliliter fractions were collected in test tubes containing 200 μl of 0.25 M EDTA (pH 7.0), and those containing greater than 20 U/ml of factor VII activity were rapidly pooled and concentrated to 1.7 ml (PM-10 membrane, Amicon). The factor VII eluted as a sharp peak with tailing and the total volume of the eluant that was pooled was approximately 50 ml.

ACA-44 Gel Filtration (Fig. 2). The sample was then applied to a column (1.5 × 95 cm) of ACA-44 equilibrated with 0.15 M NaCl–0.02 M sodium phosphate (pH 7.0)–0.005 M benzamidine–0.001 M EDTA at a flow rate of 10 ml/hr, and 1.75-ml fractions were collected. Fractions 70–81 were pooled and concentrated (Amicon, PM-10) to approximately 0.5 mg/ml and stored at $-70°C$. For subsequent studies the factor VII is freed of benzamidine by gel filtration in a small Sephadex G-25 column or by extensive dialysis.

Comments on the Purification Procedure. The procedure just described (see Table I) is a modification of one previously described, which entailed

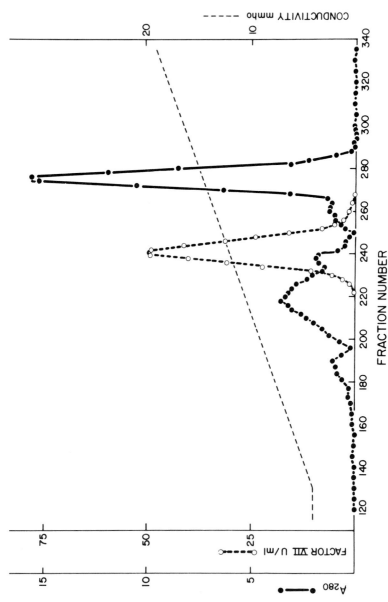

FIG. 1. Elution pattern of human factor VII from DEAE–Sepharose. Protein was eluted from the column (5 × 35 cm) with a linear gradient from 0.0 to 0.5 M NaCl in 0.05 M Tris-HCl (pH 7.5)–0.02 M benzamidine, over 6 liters. Fractions (15 ml) were collected at a flow rate of 350 ml/hr. The activities of prothrombin, factor X, and factor IX eluted under the larger protein peak following the factor VII. A_{280} was determined on samples diluted 20-fold from which the A_{280} of a 20-fold dilution of the elution buffer was subtracted. ●———●, Absorbance at 280 nm; O---O, factor VII activity; ----,

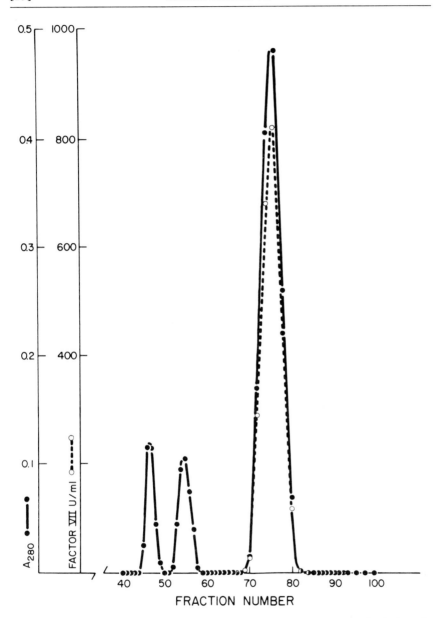

Fig. 2. Gel filtration on ACA-44. The factor VII after elution from QAE-Sephadex using $CaCl_2$ was concentrated to 1.7 ml and applied to a column (1.5 × 95 cm) of ACA-44. Fractions (1.75 ml) were collected at a flow rate of 10 ml/hr. A_{280} was determined on samples diluted 5-fold from which the A_{280} of a 5-fold dilution of elution buffer was subtracted. ●——●, Absorbance at 280 nm; ○---○, factor VII activity.

TABLE I
PURIFICATION OF HUMAN FACTOR VII

Purification step	Protein[a] (mg)	Activity (units)	Specific activity (units/mg)	Yield (%)	Purification (-fold)
Plasma	1.4×10^6	20,000	0.014	100	1
Barium citrate–ammonium sulfate	5.6×10^3	13,600	2.4	68	170
DEAE–Sepharose	205	11,300	55	57	3900
QAE-Sephadex	6.4	8100	1300	41	90,000
ACA-44	3.9	6600	1700	33	120,000

[a] Protein concentration was estimated by absorbance at 280 nm, assuming $E_{280}^{1\%} = 10$.

barium citrate absorption and elution, ammonium sulfate fractionation, QAE-Sephadex chromatography, QAE-Sephadex chromatography with $CaCl_2$ elution, and Sephadex G-100 gel filtration. The new method has produced somewhat greater yields of factor VII protein, of lower specific activity, presumably reflecting less contamination of zymogen VII with activated factor VII. Units of fresh-frozen plasma obtained from the Red Cross are frequently lipemic and, in the previous method, extensive washing of the barium citrate precipitate was required to remove sufficient lipoprotein contamination to allow absorption to a QAE-Sephadex column of reasonable size with good resolution. However, significant activation of factor VII appeared to occur during this extensive washing procedure. The use of DEAE–Sepharose in the method described here has allowed less washing of the barium citrate precipitate, and in addition more rapid chromatography. The use of a shorter and wider column for the second chromatography step has also decreased the time of chromatography and eliminated the need for prior concentration of the sample used in the previous purification method. Finally, we have found separation of factor VII from the last contaminants much improved when ACA-44 rather than Sephadex 100 gel filtration is used.

The final product is homogeneous when examined by SDS–polyacrylamide gel electrophoresis (see Fig. 3) and is without detectable factor IX, X, or prothrombin contamination. In addition, appropriate pools of the initial DEAE–Sepharose column can be used for the purification of factors IX, X, prothrombin, and protein C.

Properties

Chemical Composition. Human factor VII is a 48,000-MW glycoprotein containing a single polypeptide chain.[8] The amino acid composition is shown in Table II.[8] The amino-terminal sequence is Ala-Asn-Ala-Phe-

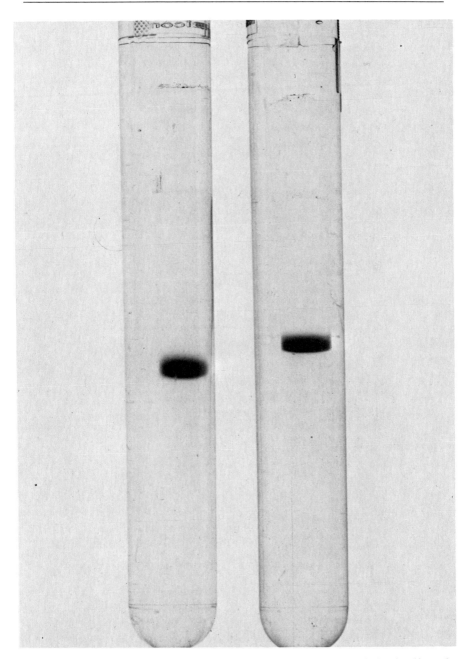

FIG. 3. Ten percent polyacrylamide-gel electrophoresis in sodium dodecyl sulfate of factor VII. Each gel contains 10 μg of protein: left, unreduced; right, following reduction with 5% 2-mercaptoethanol.

TABLE II
AMINO ACID COMPOSITION OF FACTOR VII

Amino acid	Residues/40,000 g of protein[a]
Aspartic acid	32
Threonine	20
Serine	31
Glutamic acid	44
Proline	21
Glycine	37
Alinine	21
Half-cystine	20
Valine	19
Methionine	2
Isoleucine	12
Leucine	32
Tyrosine	10
Phenylalanine	11
Histidine	10
Lysine	17
Arginine	20
Tryptophan	ND[b]

[a] Assumes that human factor VII has 40,000 g of protein, the remainder being carbohydrate.[4]
[b] Not determined.

Leu-(Gla)-(Gla)-Leu-Arg-Pro, where parentheses enclose γ-carboxyglutamic acid residues, which are suspected but not proven.[8,16] Factor VII appears to contain 9 Gla residues per molecule.[9]

Stability. Concentrated solutions of factor VII at neutral pH appear to be quite stable at temperatures ranging from −70°C to 25°C and can be frozen and thawed without detectable loss in activity. Dilute solutions of less than 30 μg/ml, however, require the presence of a carrier protein (e.g., bovine serum albumin) to prevent rapid loss of activity.

Proteolytic Activators. Single-chain factor VII can be activated by factor X_a or factor IX_a in the presence of phospholipids and calcium, and by thrombin or factor XII_a without additional cofactors.[8,17] This activation is associated with conversion of the single-chain protein to a 2-chain protein (28,000 and 20,000 MW)[9] when viewed on reduced SDS–polyacrylamide gel electrophoresis.[8]

Inhibitors. In contrast to bovine factor VII, relatively high concentrations of DFP (diisopropyl fluorophosphate) are required to inhibit human

[16] W. Kisiel, personal communication (1980).
[17] G. J. Broze and P. W. Majerus, *Clin. Res.* **27**, 459A (1979).

factor VII. Factor VII$_a$ is approximately 3-fold more sensitive to DFP than zymogen VII, and the presence of tissue factor and CaCl$_2$ accelerates the inactivation of both factor VII and VII$_a$ about 5-fold.[8] The time required for 50% inactivation by 15 mM DFP was as follows: factor VII, 160 min; factor VII$_a$, 60 min; factor VII plus tissue factor and CaCl$_2$, 26.5 min; and factor VII$_a$ plus tissue factor and CaCl$_2$, 11 min.[8]

Antithrombin III alone has no discernible effect on factor VII or VII$_a$; but in the presence of heparin, factor VII$_a$ is inactivated at a 25-fold faster rate than factor VII.[8] In a reaction mixture containing 11.5 μg/ml factor VII$_a$, heparin 10 μg/ml, and 200 μg/ml antithrombin III, 50% inactivation occurred at 11 min.[8]

Specificity. The two known substrates for factor VII/VII$_a$ are factor X and factor IX,[18] when tissue factor and CaCl$_2$ are present as cofactors. To date, detailed kinetic data describing these reactions have not been reported. It remains unclear whether single-chain zymogen factor VII possesses intrinsic coagulation activity. Its activity in native human plasma may be related to activation during the assay procedure itself, and studies using purified proteins are confounded by the probability of trace contamination by factor VII$_a$.[8] Preparations of factor VII isolated by the method described here have had a specific activity of 1.70–1.75 U/μg in the one-stage clotting assay. Conversion to the activated form (factor VII$_a$) is associated with an approximately 40-fold increase in activity.

Acknowledgments

Studies reported from the authors' laboratory were done during the tenure of a Clinician-Scientist Award from The American Heart Association and with funds contributed in part by The St. Louis Heart Association, and was also supported by Grants HLBI 14147 (Specialized Center in Thrombosis), HL 07088, and HL 16634 from The National Institutes of Health.

[18] B. Østerud and S. Rapaport, *Proc. Natl. Acad. Sci. U.S.A.* **74**, 5260 (1977).

[20] Radiometric Assays for Blood Coagulation Factors

By Margalit Zur and Yale Nemerson

Radiometric assays for blood coagulation factors were developed in order to quantitate precisely a number of coagulation reactions by directly monitoring a structural change associated with a particular reaction. Until recently, most assays were based on the acceleration of clotting time in

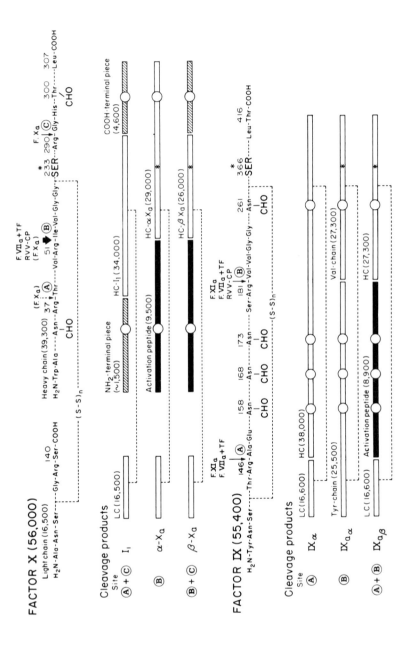

deficient plasma systems, employing complex biological reagents. These techniques, though in general sensitive and specific, suffer certain drawbacks: they permit limited adjustment of reactant concentrations, yield only rough approximations for reaction rates, and are inadequate for the study of many side and back reactions that may play critical roles in blood coagulation. During the last decade, the purification of most of the clotting factors has led to the clarification of the molecular details of their activation. Consequently, methods based on following a well-defined cleavage reaction that occurs concomitantly with the activation of a substrate zymogen could be designed and used for accurate kinetic studies of blood-coagulation components in purified systems as well as in plasma. This principle was applied to the detailed study of two key reactions that can initiate coagulation: the activation of factor X[1] and factor IX[2] by the "extrinsic," or tissue-factor, pathway of coagulation. Figure 1 depicts schematically the cleavage sites attacked by a number of activators of factors X and IX. When these two zymogens are converted to their enzymically active forms, they release approximately one-fifth of their mass as activation peptides that carry most of the carbohydrate content of the precursors.[3,4] The activation peptides are soluble in 5% trichloroacetic

[1] S. A. Silverberg, Y. Nemerson, and M. Zur, *J. Biol. Chem.* **252**, 8481 (1977).
[2] M. Zur and Y. Nemerson, *J. Biol. Chem.* **255**, 5703 (1980).
[3] K. Fujikawa, A. R. Thompson, M. E. Legaz, R. G. Meyer, and E. W. Davie, *Biochemistry* **12**, 4938 (1973).
[4] T. Mizuochi, K. Yamashita, K. Fujikawa, K. Titani, and A. Kobata, *J. Biol. Chem.* **255**, 3526 (1980).

◄ FIG. 1. Sites of cleavage and products formed during the activation of bovine factor X and factor IX by different mechanisms. The molecular weight approximations include the carbohydrate moiety, and vary slightly in reports from different laboratories. The diagram depicts relationships and is not drawn to scale. Activation peptides are shown as solid bars; shaded areas represent fragments that are not attached by disulfide bonds; circles mark carbohydrate; asterisks mark the location of the active serine. LC, light chain; HC, heavy chain; I, intermediate; RVV-CP, clotting protein from Russell's viper venom; CHO, carbohydrate. Following D. L. Enfield, L. H. Ericsson, K. Fujikawa, K. Titani, K. A. Walsh, and H. Neurath, *FEBS Lett.* **47**, 132 (1974); K. Fujikawa, M. H. Coan, M. E. Legaz, and E. W. Davie, *Biochemistry* **13**, 5290 (1974); K. Fujikawa, M. E. Legaz, H. Kato, and E. W. Davie, *ibid.* p. 4508; K. Fujikawa, K. Titani, and E. W. Davie, *Proc. Natl. Acad. Sci. U.S.A.* **72**, 3359 (1975); J. Jesty, A. K. Spencer, and Y. Nemerson, *J. Biol. Chem.* **249**, 5614 (1974); J. Jesty, A. K. Spencer, Y. Nakashima, Y. Nemerson, and W. Konigsberg, *ibid.* **250**, 4497 (1975); K. Katayama, L. H. Ericsson, D. L. Enfield, K. A. Walsh, H. Neurath, E. W. Davie, and K. Titani, *Proc. Natl. Acad. Sci. U.S.A.* **76**, 4990 (1979); P. A. Lindquist, K. Fujikawa, and E. W. Davie, *J. Biol. Chem.* **253**, 1902 (1978); K. Titani, K. Fujikawa, D. L. Enfield, L. H. Ericsson, K. A. Walsh, and H. Neurath, *Proc. Natl. Acad. Sci. U.S.A.* **72**, 3082 (1975). Reproduced from M. Zur and Y. Nemerson, *in* "Haemostasis and Thrombosis" (A. L. Bloom and D. P. Thomas, eds.). Churchill-Livingstone, Edinburgh and London (1981).

acid, which precipitates the larger zymogens and enzymes. Therefore when the zymogens are labeled by tritiation of their sialic acid residues, the rate of activation can be measured by counting the solubilized tritium-labeled activation peptides as a function of time. The increase in trichloroacetic acid-soluble tritiated peptide over time parallels the appearance of factor X_a (or of factor IX_a) as monitored by polyacrylamide-gel electrophoresis and by clotting assay.[1,2] The assay procedure consists of incubating the tritiated substrate with the appropriate enzyme and cofactor(s), removing subsamples into a solution of Ca^{2+}-chelator to stop the reaction, adding trichloroacetic acid, and separating the soluble tritiated activation peptide by centrifugation. Aliquots of the supernatant are counted for tritium in a liquid scintillation cocktail. The sensitivity of the technique is adequate for studying the coagulation reactions: using the procedures described below, 10–100 fmol of the activation peptides can be accurately determined.

Radiometric Assays in Purified Bovine Systems

The Activation of ^3H-Labeled Factor IX by Factor VII_a and Tissue Factor

The technique for measuring the activation of factor IX by the extrinsic pathway will be described in detail; the assay for factor X is identically performed. The basic technique is easily adapted, with minor changes, to many similar systems.

Protein Reagents. Coagulation factors are prepared by established procedures.[5] In our laboratory, bovine factors IX and VII are purified from barium citrate eluate of plasma. Factor IX, highly purified by the method of Fujikawa et al.,[3] is further freed from small amounts of prothrombin and factor VII by preparative acrylamide disc-gel electrophoresis[6]; its concentration is determined from $A_{280}^{1\%} = 14.9$,[7] taking a molecular weight value $M_r = 55,400$.[3] Factor VII is prepared and activated according to Radcliffe and Nemerson,[8] and its concentration determined from $A_{280}^{1\%} = 12.9$,[9] $M_r = 53,000$.[9] Thromboplastin is prepared from acetone-dried bovine brain powder by extraction with buffered saline according to Quick.[10]

[5] See this series, Vol. 45.
[6] F. Kalousek, W. Konigsberg, and Y. Nemerson, *FEBS Lett.* **50**, 382 (1975).
[7] K. Fujikawa, M. E. Legaz, H. Kato, and E. W. Davie, *Biochemistry* **13**, 4508 (1974).
[8] R. Radcliffe and Y. Nemerson, this series, Vol. 45, p. 49.
[9] R. Radcliffe and Y. Nemerson, *J. Biol. Chem.* **251**, 4797 (1976).
[10] A. J. Quick, "Hemorrhagic Diseases." Lea & Febiger, Philadelphia, Pennsylvania, 1959.

Tritiation of Substrates

Zymogen substrates are labeled by reductive tritiation of preoxidized sialic acid residues by the method of Van Lenten and Ashwell.[11] The procedure for tritiating factor IX is given as an example; identical conditions can be employed for labeling other coagulation glycoproteins.

Reagents

Bovine factor IX, 5 mg/ml in 3 ml of 0.1 M sodium acetate buffer (pH 5.8)–0.1 M NaCl

1 M Sodium periodate, freshly prepared

Ethylene glycol

0.2 M Sodium bicarbonate buffer (pH 8.5)–0.1 M NaCl

Sodium [^{3}H]borohydride, specific activity 5–10 Ci/mmol, a 100-mCi vial (New England Nuclear)

0.05 N NaOH

1 M Sodium borohydride, freshly dissolved in 0.05 N NaOH

Tris/saline buffer: 0.05 M Tris-HCl (pH 7.5)–0.1 M NaCl

50% Glycerol in Tris/saline buffer

0.5 M Diisopropyl phosphorofluoridate (DFP) in dry isopropanol

Aqueous liquid scintillation counting cocktail (we used satisfactorily the toluene-based New England Nuclear Formula 963)

Procedure. Bovine factor IX, 15 mg (0.27 μmol) in 3 ml of ice-cold acetate buffer, is oxidized in an ice bath using 40 μl of 1 M sodium periodate (40 μmol: 10-fold over the reported content of sialic acid, 15 mol/mol of protein),[7] and stirred gently (to avoid foaming) on ice for 30 min. Ethylene glycol (5 μl) is added to reduce excess reagent, and side products are removed by dialysis at 4°C against three changes of 1-liter bicarbonate buffer. After 4 hr, factor IX is tritiated in an ice bath by adding 100 mCi of sodium [^{3}H]borohydride dissolved in 100 μl of 0.05 N NaOH. The reaction mixture is stirred on ice for 25 min, and the unreacted groups are then reduced by adding 10 μl of 1 M unlabeled borohydride, and stirring for another 5 min at room temperature. The reaction products are separated on a Sephadex G-25 (medium) column (1 × 48 cm), equilibrated and eluted with Tris/saline buffer. Fractions (0.4 ml) are read for absorbance at 280 nm, and 1-μl samples are counted (in 3 ml of an aqueous liquid scintillation cocktail). Fractions of the same specific radioactivity are pooled. We have routinely obtained tritiated factor IX of specific radioactivity of 1.5 × 10^{10} cpm/μmol (at approximately 50% efficiency). The pooled tritiated factor IX is made 5 mM in

[11] L. Van Lenten and G. Ashwell, *J. Biol. Chem.* **246**, 1889 (1971).

DFP and stirred for 2.5 hr at room temperature, to reduce any trace enzymic activity. The protein is dialyzed at 4°C (for 3 hr, or overnight) against 50% glycerol in Tris/saline buffer. In practice, when stored at −18°C and carefully guarded against contamination, this material proved stable for at least 6–12 months. Less than 2.0% of the radioactivity was soluble in 5% trichloroacetic acid. Control assays, consisting of substrate and cofactors, but no exogenous enzyme, should show no increase over background during 30-min incubation at 37°C.

Tritiated factor X is prepared in a similar fashion, assuming 7 mol sialic acid/mol protein.[4] Specific radioactivity values are usually 20–30% lower.

The Radioassay

Equipment. An automatic diluter/dispenser (such as the Hamilton Digital Diluter, or the microprocessor-controlled Microlab-P) is a useful aid for the rapid processing of a large number of assays. Assays are conveniently run in 1.5-ml conical polypropylene centrifuge tubes, held at 37°C in water-filled glass tubes adapted to a small block Dri-bath. Capped polypropylene microcentrifuge tubes (400 μl) are used to stop the reaction and to precipitate TCA-insoluble material by high-speed centrifugation (e.g., Beckman model 152-B table-top microcentrifuge). High-efficiency counting is critical for maximal sensitivity and accuracy; we used 3.5-ml plastic counting vials, which fit glass counter adapters in a Beckman model LS 7000 or 7500 liquid scintillation counter.

Reagents

Reaction buffer: Tris/saline buffer containing 1 mg/ml bovine serum albumin (Sigma)

^{3}H-labeled factor IX, 1.5–10 μM in the reaction buffer; appropriate dilutions are periodically made from the stock solution in 50% buffered glycerol (usually of a concentration above 100 μM), and stored frozen in portions sufficient for 1 day's work

Factor VII$_a$ (0.4 mg/ml) is prepared by activation with factor X$_a$[12] (at 1 : 1500 ratio), in the presence of mixed brain phospholipids[13] (75 μg/ml) and CaCl$_2$ (5 mM). The activation is monitored by clotting assay,[14] and when the activity peak levels off (in 45–60 min), the reaction mixtures is made 2 mM in tetrasodium ethylenediaminetetraacetate (EDTA), to prevent degradation of factor VII$_a$ by factor X$_a$. The stock solution is stored frozen in small portions at

[12] R. Radcliffe and Y. Nemerson, *J. Biol. Chem.* **250**, 388 (1975).
[13] W. N. Bell and H. G. Alton, *Nature (London)* **211**, 880 (1954).
[14] Y. Nemerson and L. P. Clyne, *J. Lab. Clin. Med.* **83**, 301 (1974).

$-18°C$, and is stable for over 6 months. Once a month, dilutions (to 5–20 nM factor VII$_a$) are made in the reaction buffer, distributed in 100- μl aliquots for daily use, frozen, and are discarded a few hours after thawing. Thromboplastin is frozen in aliquots (0.4 ml) and discarded after thawing.

1.15 M CaCl$_2$

Stopping solution. 50 mM EDTA tetrasodium salt in 0.1% bovine serum albumin

15% Trichloroacetic acid (TCA)

Method. Pipette in a small tube, ^{3}H-labeled factor IX (2.5 μM, 50 μl), thromboplastin (50 μl), factor VII$_a$ (5 nM, 10–100 μl), and reaction buffer up to 250 μl. Withdraw a 20- μl sample and mix immediately with 80 μl cold stopping solution in a 400- μl capped microcentrifuge tube placed in an ice bath. This is the zero time blank. Allow the reaction mixture to equilibrate at 37°C for 1 min, then add 1.15 M CaCl$_2$ (1 μl), to start the reaction. At 1-min intervals, withdraw several 20-μl samples and treat each like the blank. Add to all samples ice-cold 15% TCA (50 μl), vortex well, and leave on ice for exactly 1 min. This time interval is adequate for precipitating the bulk proteins before any significant acid-catalyzed tritium exchange occurs. Spin for 4 min at 9000 g at room temperature. Take up duplicate 50-μl aliquots of the supernatant in 3 ml of aqueous liquid scintillation cocktail, and count for tritium. In practice, a period of rest in the dark (1–2 hr) is recommended, to allow the counts to stabilize and obtain accurate and reproducible results. Using the reagents described here, counts are reproducible with 2–17 hr of sampling. Total counts are determined by treating a subsample as described, but without centrifugation. From the total counts, the extent of the reaction can be estimated. In order to calculate the molar concentration of the product, two methods can be used. One method is to determine directly the solubility of the activation peptide in 5% TCA, and calculate the product concentration in the assay from the specific radioactivity.[1] The second involves allowing the reaction to run its full course to completion over several hours; calculation of the asymptotes obtained at high enzyme and a number of substrate concentrations provides the value of percent total counts released into solution on complete conversion of the substrate.[2] It is advisable to run such curves for each batch of tritiated substrate, since slight variations in the distribution of sialic acid residues on the molecule might occur. Initial velocities, as well as kinetic parameters, are most conveniently derived employing least-squares computer analyses. The choice of the proper time intervals for subsampling depends on the velocity (i.e., the enzyme, substrate, and cofactor concentrations) and is preferably made so as to give data points under 10% substrate hydrolysis. When increased sensitivity is desired—for example, to determine lower enzyme and cofactor concentrations, or under conditions where very low K_m values are indicated, or when using substrate preparations of low specific radioactivity—the composition of the sampling mixture can easily be adjusted by increasing the sample volume and diluting with more concentrated EDTA solutions. We have been able to determine, with excellent reproducibility, hydrolysis rates at substrate concentrations under 5 nM, and accurately detect femtomole quantities of factor VII$_a$ and tissue factor. In fact, the assay has been used to titrate tissue-factor preparations and assess the optimal conditions for reconstitution with lipids.[15]

Finally, some practical considerations should be mentioned. Care must be taken to avoid artifacts caused by spurious counts in solution. Two major sources for such errors are acid-catalyzed tritium exchange and cleavages catalyzed by the product, factor X$_a$ in particular (see Fig. 1). In order to counter the first contingency, it is not advisable to process more than 10 samples simultaneously. When slow reactions are monitored, and intervals between subsamples are longer than 10 min, better reproducibility is obtained when subsamples are processed with TCA and spun individually immediately on subsampling. Complications that result from a back reaction can be handled either by carrying out the assay in the presence of a suitable inhibitor (e.g., benzamidine[1] or antithrombin III), or by correcting the velocity calculation to account for the product activity, when the kinetic parameters governing the latter reaction under the same experimental conditions are known.

Applications of the Radioassay to Kinetic Studies

The radioassay has proved to be a powerful tool for analyzing the kinetic behavior of several components of the blood-coagulation system. The specificity and sensitivity of the technique have allowed its use in purified systems as well as in native or modified plasma preparations. It has been most extensively used for investigating the kinetics of the activation of factors X and IX by either the extrinsic or the intrinsic pathways, and elucidating in detail the roles of the enzymes and cofactors involved.

Kinetics of the Extrinsic Pathway-Initiated Activations

The radioassay was most systematically applied to the kinetic study of the initiation of the extrinsic pathway in the presence of factor VII$_a$ and tissue factor. The activation of purified bovine ^{3}H-labeled factor X in the presence of purified bovine factor VII$_a$ and bovine brain thromboplastin

[15] S. D. Carson and W. H. Konigsberg, *Science* **208**, 307 (1980).

was shown to be nearly absolutely dependent on tissue factor, which enhances the efficiency of the reaction by four orders of magnitude.[1] Used at saturating concentrations, thromboplastin increased the k_{cat} of the reaction nearly 3000-fold, and decreased the K_m 10-fold. The assay was performed in the presence of 10 mM benzamidine hydrochloride, to inhibit the feedback cleavages generated by factor X_a (see Fig. 1). This inhibitor was chosen because the K_i for the inhibition of factor VII$_a$ is approximately 400-fold higher than that for factor X_a.[16] Since the K_m is nearly twice the plasma concentration of factor X, it follows that the physiological level of factor X can influence the rate of formation of factor X_a via the extrinsic pathway. Further studies of the initial rates, done at low levels of hydrolysis in the absence of inhibitor, indicated a more complex dependence of the K_m on the concentration of tissue factor,[2] as shown in Table I. These data suggest that tissue factor acts as a buffer, but high concentrations of thromboplastin appear to suppress the activation of factor X more than the activation of factor IX. By using a purified tissue-factor preparation that permits the manipulation of lipid concentrations in the assay,[17] the K_m and k_{cat} values were seen to rise in proportion to the increase in phospholipids. These observations, correlated with the known differential avidity of factors IX and X for phospholipids,[18] formed the basis for a mathematical model[17] that accounts for the inhibitory effects of lipids as a form of substrate depletion. The two substrates for factor VII$_a$ (factors IX and X) each competitively inhibits the activation of the other, in purified systems[2] as well as in plasma,[19] and the K_i values are close to the K_m values obtained under similar conditions. Jesty and Silverberg[19] studied thromboplastin-induced activations in plasma artificially depleted of factors VIII, V, and fibrinogen, where either factor IX or factor X had also been removed by specific antibodies, so that their concentrations could be readjusted at will. They were able to calculate that at plasma levels of factors IX and X, factor X activation would be 16% inhibited by factor IX, whereas factor IX activation would be 54% inhibited by factor X.

The Activation of ^{3}H-Labeled Factor X by Components of the Intrinsic System

When factor X is activated in the presence of factor IX$_a$, factor VIII, and phospholipids, the hydrolysis products are identical with those

[16] M. Zur and Y. Nemerson, *J. Biol. Chem.* **253**, 2203 (1978).

[17] Y. Nemerson, M. Zur, R. Bach, and R. D. Gentry, *in* "The Regulation of Coagulation" (K. G. Mann and F. B. Taylor, eds.), p. 193. Elsevier/North-Holland Publ., Amsterdam and New York, 1980.

[18] G. L. Nelsestuen, W. Kisiel, and R. G. DiScipio, *Biochemistry* **17**, 2134 (1978).

[19] J. Jesty and S. A. Silverberg, *J. Biol. Chem.* **254**, 12337 (1979).

TABLE I

The Effects of Bovine Brain Thromboplastin on the Steady-State Parameters of Factor VII_a-Catalyzed Activations of ^3H-Labeled Factor IX and ^3H-Labeled Factor X[a]

Bovine brain thromboplastin (vol.%)	Factor IX				Factor X			
	K_m^b (nM)	k_{cat} (sec^{-1})	v_{calc}^c (nmol liter^{-1} sec^{-1})	k_{cat}/K_m (sec μM^{-1})	K_m (nM)	k_{cat} (sec^{-1})	v_{calc}^d (nmol liter^{-1} sec^{-1})	k_{cat}/K_m (sec μM^{-1})
2.5	87	0.15	0.007	1.7	88	2.2	0.165	25.1
10	177	0.40	0.011	2.3	230	5.2	0.242	22.7
20	191	0.42	0.011	2.2	300	3.8	0.150	12.7
50	243	0.26	0.006	1.1	433	2.6	0.081	5.2

[a] Reproduced from Zur and Nemerson.[2]
[b] Steady-state constants were computed using the nonparametric technique of Eisenthal and Cornish-Bowden.[16]
[c] Calculated velocity at the estimated plasma concentration of factor IX, 60 nM.[6]
[d] Calculated velocity at the estimated plasma concentration of factor X, 160 nM.[15]

produced via the tissue-factor pathway[20] (see Fig. 1). Therefore, the radioassay technique was applied by Hultin and Nemerson[21] to establish unambiguously that bovine factor IX_a alone (in the presence of phospholipids) can activate bovine factor X, but factor VIII (bovine or human) will not cleave ^3H-labeled factor X in the absence of factor IX_a. They showed further that factor VIII must be activated by thrombin in order to display measurable activity in the radioassay. By stabilizing the thrombin-activated factor VIII in the presence of 2 mM benzamidine hydrochloride, a specific and sensitive assay for factor VIII was designed. The presence of benzamidine in the assay also eliminates artifacts caused by the feedback reaction of factor X_a on ^3H-labeled factor X, factor VIII, or factor IX_a.

The Activation of Tritiated Factors IX and X by the Intrinsic Pathway in a Bovine Plasma System

Steinberg and Nemerson[22] used celite-activated factor X-depleted dilute bovine plasma to extract the Michaelis–Menten parameters for the activation of added ^3H-labeled factor IX by the product of contact activation (factor XI_a), and for the activation of added factor X by the plasma-generated factor IX_a, in the presence of factor VIII. They showed the K_m for the intrinsic activation of factor IX to be high (2 μM), nearly 30 times the plasma concentration of factor IX,[3] suggesting that the physiological rate of the activation would be a small fraction of the V_{max}, and very sensitive to factor IX concentration. The K_m value for the intrinsic activation of factor X (0.38 μM) is nearly twice the plasma concentration, and is close to the value derived for the extrinsic activation at tissue-factor saturation[1]; it also is independent of factor VIII at the concentrations studied. Increasing factor VIII concentration from 0.1 to 1 U/ml (plasma concentration) resulted in a 5-fold rate enhancement and nearly tripled the product level; that is, factor VIII regulates both the rate and the extent of the activation of factor X by factor IX_a. Since factor IX_a can also be formed by the extrinsic pathway, deficiencies in either factor VIII or factor IX (classic hemophilias) are thus seen to impair severely the concerted activation of factor X by both pathways. Indeed, in human plasma deficient in factors VIII and IX, the activation of added human ^3H-labeled factor X by thromboplastin also appears to be depressed.[23]

[20] R. D. Radcliffe and P. G. Barton, *J. Biol. Chem.* **248**, 6788 (1973).
[21] M. B. Hultin and Y. Nemerson, *Blood* **52**, 928 (1978).
[22] M. Steinberg and Y. Nemerson, submitted for publication.
[23] R. A. Marlar, A. H. Kleiss, and J. H. Griffin, *Thromb. Haemostasis* **42**, 166 (abstr.) (1979).

*The Activation of ³H-Labeled Factor X in Normal and
Deficient Human Plasmas*

Important insights on the modulation of factor VIII activity were obtained by Aronson and Bagley,[24] who studied the release of the activation peptide after the addition of tritiated human factor X to plasma activated in the presence of celite and phospholipids. In normal plasma, a short initial burst of peptide release (8%) could be repeatedly induced by the addition of factor VIII, and a similar response was obtained in factor VIII-deficient (hemophiliac) plasma, which initially requires external factor VIII supply to permit ³H-labeled factor X activation. A sustained production of the tritiated peptide was observed in factor V-deficient plasma, but no peptide was released when plasma was pretreated with thrombin (5 U/ml). The destruction of factor VIII by thrombin was thus inferred to be a major control of factor X production by the intrinsic pathway in plasma.

Assay of the Contact System in Human Plasma

The efficiency of a number of activators of the contact system can be assessed directly in plasma by assaying the activation product (factor XI_a) on ³H-labeled factor IX added to plasma in high excess.[25] The contact-activation capability of various materials can be thus compared, either by reference to a standard (e.g., celite), or by evaluating the concentration of activator ($K_{1/2}$) required to reach half-maximal velocity under defined conditions. This approach has potential usefulness in the search for biocompatible nonthrombogenic artificial surfaces and prostheses.

The Activation of ³H-Labeled Factor IX by Factor XI_a

Byrne *et al.*[26] used the radioassay to investigate the metal-ion specificity of the activation of bovine ³H-labeled factor IX by purified bovine factor XI_a. This reaction displays a high specificity of Ca^{2+}-binding to factor IX, and two classes of binding sites can be differentiated. Sr^{2+} is the only metal among several examined that could effectively substitute for Ca^{2+} in this reaction, and it interacts mainly with the weak class of Ca^{2+}-binding sites on factor IX. The use of the radioassay allowed the evaluation of the independent contributions of each class of Ca^{2+} sites to the overall activation. At low Ca^{2+} concentrations, when only the high-affinity sites on factor IX are saturated, peptide release does not occur, but it

[24] D. Aronson and J. Bagley, *Circulation* **58**, Suppl. II, 209 (abstr.) (1978).
[25] R. M. Berkowitz and Y. Nemerson, *Blood* **55**, 528 (1980).
[26] R. Byrne, G. W. Amphlett, and F. J. Castellino, *J. Biol. Chem.* **255**, 1430 (1980).

proceeds rapidly on the addition of suboptimal Sr^{2+} concentrations. These results suggest that the occupation of the weak Ca^{2+} sites, but not of the strong sites, is critical for this activation; however, the latter assist in producing maximal activation.

Kinetics of the Activation of [^3H]*Prothrombin*

Tritiated bovine prothrombin was first prepared by Butkowski *et al.*,[27] who have characterized the tritiated products and intermediates of the activation in the presence of a number of activators. Silverberg[28] has devised an assay to measure the proteolysis of tritiated prothrombin to fragment 1 and prethrombin 1 by thrombin, based on extracting the radiolabeled fragment into 4% *p*-toluenesulfonic acid. The steady-state parameters for this reaction suggest that, in the absence of other controls, plasma prothrombin will be rapidly hydrolyzed by thrombin, but 1 mM Ca^{2+} inhibits the reaction by over 90%.

Acknowledgment

This work was supported in part by grants from the National Institutes of Health (Grant # HL 22980).

[27] R. J. Butkowski, S. P. Bajaj, and K. G. Mann, *J. Biol. Chem.* **249**, 6562 (1974).
[28] S. A. Silverberg, *J. Biol. Chem.* **254**, 88 (1979).

[21] Factor V

By MICHAEL E. NESHEIM, JERRY A. KATZMANN,
PAULA B. TRACY, and KENNETH G. MANN

Introduction

Factor V is a high-molecular-weight single-chain plasma protein that, when activated, functions as a cofactor in the factor X_a-catalyzed conversion of prothrombin to the blood-clotting enzyme thrombin.[1-3] Factor V was first identified as a functional entity in 1943 by Owren,[4,5] who deduced its presence from the *in vitro* clotting behavior of plasma of a patient with

[1] K. G. Mann, this series, Vol. 45, p. 123.
[2] E. W. Davie and K. Fujikawa, *Annu. Rev. Biochem.* **44**, 799 (1975).
[3] J. W. Suttie and C. M. Jackson, *Physiol. Rev.* **57**, 1 (1977).
[4] P. A. Owren, *Lancet* **1**, 446 (1947).
[5] P. A. Owren, *Acta Med. Scand., Suppl.* **194**, 1 (1947).

a bleeding diathesis. Subsequently, numerous workers[6-11] provided partial purification of factor V and described its ability to augment considerably the rate of prothrombin activation in reconstructed systems. In addition, the potential of factor V to be converted to the more active species, factor V_a, was recognized in numerous studies.[5-14]

Further elucidation of the properties of both factor V and factor V_a followed the isolation of bovine factor V to apparent homogeneity.[15,16] The molecular weight of the single-chain circulating molecule was shown to be 330,000.[15] The molecule displays unusual hydrodynamic behavior, with a frictional coefficient of 2.01[15] and a Stokes radius determined both by ultracentrifugation and gel filtration of about 91–93 Å.[15-17] Its activation was shown to occur by discrete proteolysis by thrombin to give a multiple-subunit protein.[16,18,19] The activated protein augments the rate of prothrombin conversion by several orders of magnitude.[6,20-22] This augmentation can be rationalized based on the ability of factor V to bind both phospholipid[23] and factor X_a[21] with high affinity, thereby promoting the formation of a complex of phospholipid-bound factor V_a and factor X_a. Because of its Ca^{2+} and phospholipid-binding properties, prothrombin can efficiently interact with the enzymatic complex.[1] In addition to providing binding determinants for assembly of the prothrombinase complex, factor V_a enhances by a factor of 3000 or more the turnover rate (V_{max}) of prothrombin conversion to thrombin.[22] Current evidence suggests that in vivo platelets comprise the locus of assembly of the prothrombinase complex.[24-26] Studies of (1) the binding of factor X_a to human or bovine

[6] M. P. Esnouf and F. Jobin, *Biochem. J.* **102**, 660 (1967).
[7] P. G. Barton and D. J. Hanahan, *Biochim. Biophys. Acta* **133**, 506 (1967).
[8] R. W. Colman, *Biochemistry* **8**, 1438 (1969).
[9] F. A. Dombrose and W. H. Seegers, *Thromb. Diath. Haemorrh.* **57**, Suppl., 241 (1974).
[10] T. Chulkova and G. Hernandez, *Clin. Chim. Acta* **62**, 21 (1975).
[11] C. M. Smith and D. J. Hanahan, *Biochemistry* **15**, 1830 (1976).
[12] D. Papahadjopoulous, C. Hougie, and D. J. Hanahan, *Biochemistry* **3**, 264 (1964).
[13] G. Philip, J. Moran, and R. W. Colman, *Biochemistry* **9**, 2212 (1970).
[14] S. P. Bajaj, R. J. Butkowski, and K. G. Mann, *J. Biol. Chem.* **250**, 2150 (1975).
[15] M. E. Nesheim, K. H. Myrmel, L. Hibbard, and K. G. Mann, *J. Biol. Chem.* **254**, 508 (1979).
[16] C. T. Esmon, *J. Biol. Chem.* **254**, 964 (1979).
[17] K. G. Mann, M. E. Nesheim, and P. B. Tracy, *Biochemistry* **20**, 28 (1981).
[18] M. E. Nesheim and K. G. Mann, *J. Biol. Chem.* **254**, 1326 (1979).
[19] S. Saraswathi, R. Rawala, and R. W. Colman, *J. Biol. Chem.* **253**, 1024 (1978).
[20] C. T. Esmon, W. G. Owen, and C. M. Jackson, *J. Biol. Chem.* **249**, 8045 (1974).
[21] M. E. Nesheim, J. B. Taswell, and K. G. Mann, *J. Biol. Chem.* **254**, 10952 (1979).
[22] J. Rosing, G. Tans, J. W. P. Govers-Riemslag, R. F. A. Zwaal, and H. C. Hemker, *J. Biol. Chem.* **255**, 274 (1980).
[23] J. W. Bloom, M. E. Nesheim, and K. G. Mann, *Biochemistry* **18**, 4419 (1979).
[24] P. B. Tracy, J. M. Peterson, M. E. Nesheim, F. C. McDuffie, and K. G. Mann, *J. Biol. Chem.* **254**, 10354 (1979).

platelets[25,27]; (2) the binding of bovine factor V and factor V_a to bovine platelets[24]; and (3) the coordinate binding of both factor X_a and factor V_a to platelets with concurrent assessment of the formation of prothrombin converting activity,[26] are all consistent with the notion that factor V_a constitutes a binding site for factor X_a on platelets.

This article constitutes a report on the procedures currently used in this laboratory to isolate unactivated factor V from bovine plasma. In addition, a method using conventional procedures for the partial purification of factor V from human plasma is presented. This is followed by a description of an affinity technique, based on the use of an immobilized monoclonal hybridoma antibody, that yields electrophoretically homogeneous human factor V. The immobilized antibody has properties such that it tightly and specifically interacts with human factor V under conditions of physiological ionic strength, but dissociates at elevated ionic strength, thus allowing its use as an affinity adsorbant.

Further information concerning factor V is available in another article in *Methods in Enzymology* by Colman and Weinberg[28]; in a comprehensive review by Colman[29] and references cited therein; plus more recently published articles by Saraswathi *et al.*,[19] Esmon,[16] and various publications from this laboratory.[15,17,18,21,23,24,26]

Assay of Factor V Activity

Assays for factor V are based on its ability to enhance the rate of conversion of prothrombin to thrombin. The catalyst responsible for this conversion is a complex consisting of the serine protease factor X_a, factor V_a (which functions as a cofactor), phospholipid, and calcium ion. Deletion of activated factor V (factor V_a) from this complex at physiological levels of prothrombin will lead to a decrease in the rate of the reaction by at least four orders of magnitude.[21] Incremental replacement of factor V up to levels that are stoichiometric with factor X_a will restore the reaction rate correspondingly. Although levels of factor V_a in excess of factor X_a will not contribute further to the reaction rate, the rate of prothrombin conversion can be calibrated in terms of factor V (factor V_a) levels over a range of concentrations less than or equal to the factor X_a concentration. The methods employed to determine rates of prothrombin conversion utilize either (*a*) factor V-deficient plasma as a source of prothrombin, factor X (the precursor of factor X_a), and fibrinogen (as substrate for the

[25] J. P. Miletich, C. M. Jackson, and P. W. Majerus, *J. Biol. Chem.* **253**, 6908 (1978).

[26] P. B. Tracy, M. E. Nesheim, and K. G. Mann, *J. Biol. Chem.* **256**, 743 (1981).

[27] B. Dahlbäck and J. Stenflo, *Biochemistry* **17**, 4938 (1978).

[28] R. W. Colman and R. M. Weinberg, this series, Vol. 45, p. 107.

[29] R. W. Colman, *Prog. Hemostasis Thromb.* **3**, 109 (1976).

thrombin derived from prothrombin); or (*b*) systems of purified components consisting of prothrombin, factor X_a, and fibrinogen (or synthetic thrombin substrates). In the assay utilizing plasma deficient in factor V, factor X_a is produced *in situ* by the addition of thromboplastin and Ca^{2+} to both activate endogenous factor X and supply the requisite phospholipid. The rate of thrombin formation is indicated by the time required to form a fibrin clot, which is decreased at increased levels of factor V. With assays based on purified components, prothrombin activation is initiated by adding purified factor X_a, Ca^{2+}, and phospholipid to purified prothrombin. The rate of thrombin formation is determined either by including fibrinogen in the assay so that clot formation occurs, or by removing aliquots at intervals and determining thrombin levels in a second assay based on a suitable thrombin substrate.

The assay based on factor V-deficient plasma has the advantage that it is convenient, rapid, and does not require isolation and purification of the required components. The complex kinetics of the coupled processes leading to clot formation, however, do not provide a linear relationship between factor V and clotting time. In addition, if clot formation is detected visually, a somewhat subjective assessment of the end point in the assay is required. Furthermore, other plasma components or processes that conceivably might influence the complex reactions leading to clot formation cannot be ruled out as interferences in the assay. Nonetheless, the strength of the factor V-deficient plasma assay is its speed and convenience. It generally is the assay of choice for monitoring activity levels throughout purification procedures. A more carefully controlled assessment of activity can be obtained using assays based on purified components in which the relationship between factor V levels and rates of prothrombin conversion approaches linearity, and responses of the assay can be attributed unambiguously to factor V.

One feature of either type of factor V assay is that the product of the reaction, thrombin, activates factor V by proteolysis, and thus the activity of the added sample may be altered during the course of the assay itself. Because of this, assessment of the relative content of factor V and factor V_a of the sample cannot be unambiguously determined. This may constitute a serious disadvantage in efforts to isolate unactivated factor V or to estimate factor V levels in samples that are activated to various degrees. This positive feedback may be overcome by employing an assay developed in this laboratory that uses purified components plus the reversible fluorescent inhibitor of thrombin, dansylarginine *N*-(3-ethyl-1,5-pentanediyl)amide (DAPA). DAPA inhibits newly formed thrombin, thus attenuating feedback activation of factor V. The inhibitor also provides a continuous monitor of thrombin production by virtue of the enhanced fluorescence of the DAPA–thrombin complex.

For the purposes of this article, two assays will be described in detail. The first is based on factor V-deficient plasma; the second is based on a system of purified components utilizing DAPA to measure rates of thrombin formation. Other one- and two-stage assay systems may be found in other publications and references cited therein.[25,28]

Assay with Factor V-Deficient Plasma

Thromboplastin. Rabbit brain thromboplastin is prepared by the method of Bowie *et al.*[30] Rabbit brains are obtained from Pel-Freez Biologicals, Inc., Rogers, Arkansas. The meninges and blood vessels are removed under cold running water and the brains are quick-frozen in polyethylene bags for storage. To prepare thromboplastin reagent, two brains are placed on the plate-glass shelf of a bacteriological incubator and triturated with a heavy spatula to form a thin, even film. The film is dried at 35–37°C for 18–20 hr. It is then shaved onto a sheet of clean paper with a glass microscope slide. If drying is adequate, curls similar to wood shavings should be obtained. Inadequate drying produces sticky masses; overdrying yields a powder. The dried brain may be stored in sealed jars at 4°C. Thromboplastic activity is extracted by weighing approximately 0.5 g dried brain into a 50-ml glass centrifuge tube. To this tube is added 10 ml of 0.154 *M* NaCl warmed to 45–48°C. The suspension is placed in a bath at 45–48°C and gently stirred with a glass rod for 20 sec. Vigorous stirring is avoided to prevent disintegration of the shavings. After 6 min the suspension is again stirred for 10 sec. After a total of 12 min of incubation, the suspension is cooled to room temperature in a water bath and stirred a final time for 10 sec. The suspension is then centrifuged for 2 min at 175 *g*. The supernatant is then recovered and used as thromboplastin reagent. It may be stored at 4°C and used for 5–7 days without loss of activity. Alternatively, commercially available thromboplastin reagents (e.g., Simplastin, from General Diagnostics Division, Warner-Lambert Co., Morris Plains, New Jersey) may be used.

Factor V-Deficient Plasma. The activity of factor V depends on a tightly associated calcium ion incorporated in the molecule.[31] The removal of Ca^{2+} renders the molecule inactive. This property can be utilized to prepare a substrate plasma deficient in factor V activity. The procedure has been described in detail by Bloom *et al.*[32] Human plasma prepared from blood collected in acid–citrate–dextrose anticoagulant is treated with EDTA to a final concentration of 10 m*M*. The pH is adjusted to 7.5, and

[30] E. J. W. Bowie, J. H. Thompson, Jr., P. Didisheim, and C. A. Owen, Jr., "Mayo Clinic Laboratory Manual of Hemostasis." Saunders, Philadelphia, Pennsylvania, 1971.
[31] L. S. Hibbard and K. G. Mann, *J. Biol. Chem.* **255**, 638 (1980).
[32] J. W. Bloom, M. E. Nesheim, and K. G. Mann, *Thromb. Res.* **15**, 595 (1979).

the plasma is incubated at 37°C. At intervals, a 50- μl aliquot is removed and added to 50 μl of 0.02 M imidazole HCl–0.15 M NaCl, pH 7.4. Thromboplastin (50 μl) is added, and after approximately 15 sec at 37°C, 50 μl of 0.025 M CaCl$_2$ is added and the time required for clot formation is determined visually at 37°C. Prior to the addition of EDTA, normal fresh plasma will clot in about 20 sec. Incubation is continued until the clot time (prothrombin time) exceeds 120 sec. Usually 5 hr of incubation is sufficient; if not, incubation is continued overnight. The incubation is terminated by the addition of CaCl$_2$ to 10 mM final concentration, and dialysis is performed at 4°C with one change against 50 volumes of 0.01 M sodium oxalate–0.02 M Tris-HCl–0.15 M NaCl, pH 7.4. The plasma is then dispensed in glass tubes in aliquots of 2.5–3.0 ml, stoppered, and stored at −20°C.

Assay Procedure. The assay is performed in a 12 × 75-mm disposable glass tube. The sample is prepared by dilution of the sample to 0.005–0.1 unit/ml in buffer (0.02 M imidazole HCl–0.15 M sodium chloride, pH 7.4). Fifty microliters each of diluted sample, factor V-deficient plasma, and thromboplastin are pipetted into the tube. The solution is briefly warmed (5–10 sec) in a 37°C water bath. A 50-μl aliquot of 0.025 M calcium chloride is added with mixing to start the assay, and a timer with 0.1-sec graduations is started simultaneously. The tube is gently rocked back and forth in the bath at approximately 1-sec intervals while the contents of the tube are continuously monitored visually for the formation of a clot. Clot formation constitutes the end point of the measurement. The useful range of clot times typically extends from 25 to 75 sec. Corresponding factor V levels in the sample range from 0.1 to 0.005 unit/ml. The formation of the clot is signaled by an abrupt change in the consistency of the reaction mixture from a fluid to a gel. The time required to form a clot shortens with increasing levels of factor V. A standard curve is prepared using serial dilutions of freshly prepared citrated human or bovine plasma. One unit of activity is defined as the activity present in 1 ml of the standard plasma before activation by thrombin. Bovine plasma typically contains three to five times the activity of human plasma, although either is a suitable standard. Log–log plots of clot time versus factor V levels are linear over two orders of magnitude of factor V content. A typical standard curve is shown in Fig. 1.

Since factor V is susceptible to activation, two assays are essential to assess fully the factor V content of a sample. The first assay is carried out as described above and measures primarily factor V that has been unintentionally activated. For the second assay, an aliquot of the sample is treated with α-thrombin (1.0 NIH unit/ml, final concentration) for 1 min at 37°C to activate the factor V to factor V$_a$ fully. The activated sample is

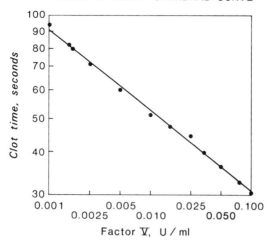

FACTOR V ASSAY STANDARD CURVE

FIG. 1. Standard curve for factor V assay using factor V-deficient plasma. The units on the horizontal axis are units of factor V/ml sample, with one unit defined as the level assayed in fresh bovine plasma (citrate) prior to activation by thrombin. (From Ref. 15.)

then diluted further in buffer (typically 10- to 50-fold), and the diluted sample is immediately assayed as above. The second assay provides a measure of the total factor V content of the sample. The ratio of activity measured after and before deliberate activation by thrombin provides an estimate of the degree to which the sample had been unintentionally activated. Ratios of measured activities (activation quotients) greater than or equal to 20 usually indicate relatively little unintentional activation of the factor V in the sample.

Assay with DAPA

Materials. Dansylarginine N-(3-ethyl-1,5-pentanediyl)amide (DAPA) is synthesized as described in Chapter 23 by Mann *et al.* in this volume. This method is also described by Nesheim *et al.*[33] Stock solutions are prepared in H_2O, stored at 4°C in the dark, and used within 2 weeks. Bovine prothrombin and factor X are prepared as described by Bajaj and Mann,[34] and factor X_a is prepared from factor X as described by Downing *et al.*[35] Vesicles of phosphatidylserine–phosphatidylcholine (1/3, w/w) (PCPS) are

[33] M. E. Nesheim, F. G. Prendergast, and K. G. Mann, *Biochemistry* **18**, 996 (1979).
[34] S. P. Bajaj and K. G. Mann, *J. Biol. Chem.* **248**, 7729 (1973).
[35] M. R. Downing, R. J. Butkowski, M. M. Clark, and K. G. Mann, *J. Biol. Chem.* **250**, 8897 (1975).

prepared as described by Barenholz *et al.*[36] They are stored at 4°C and used within 1 week of preparation.

Procedure. A solution of substrate is prepared having the following composition: prothrombin (0.1 mg/ml, 1.39 μM, $E^{1\%}_{280}$ = 1.44); DAPA (3.0 μM); PCPS (20 μM); CaCl$_2$ (2.0 mM). This solution is prepared (at least 10 min prior to use) in 0.02 M Tris-HCl–0.15 M NaCl (pH 7.4) and may be maintained at room temperature for several hours. Substrate (typically 1.5 ml) is pipetted into a cuvette and fluorescence intensity (λ_{ex} = 335 nm, λ_{em} = 565 nm) is determined. The sensitivity of the instrument is adjusted to approximately 60% of full scale. This signal is then offset to about 5% of full scale and the sensitivity of the instrument is increased by a factor of two. With these settings, complete conversion of prothrombin to thrombin will result in an increase in fluorescence intensity of approximately 80% of full scale. Fluorescence measurements may be made with instruments equipped with monochromators (e.g., Perkin Elmer MPF-44A) or employing filter arrangements (e.g., Aminco-fluorocolorimeter). Excitation may be at 280 nm, which measures thrombin by virtue of energy transfer.[37] Under these conditions the absolute intensity obtained is greater than that obtained with excitation at 335 nm, and the differential before and after production of thrombin is considerably greater. With excitation at 280 nm, however, the protein concentration resulting from sample addition must not be sufficiently high to absorb the majority of the excitation radiation.

After addition of the substrate solution and adjustment of instrumental sensitivity, an aliquot of the sample is added. The reaction is initiated by the addition of factor X$_a$ to a final concentration of 1 × 10^{-8} (0.55 μg/ml) and the reaction course is recorded continuously by virtue of enhanced fluorescence intensity. Up to a concentration of factor V (factor V$_a$) of about 3 × 10^{-9} M, initial rates of the reaction are linear with respect to the factor V$_a$ concentration. The sample is assayed both before and after deliberate activation by thrombin as described for the assay with factor V-deficient plasma. Because DAPA greatly attenuates feedback by thrombin during the assay, a more positive assessment of endogenous factor V$_a$ is possible than with the factor V-deficient plasma assay.

Other than purified factor V, no ideal standard exists. Fresh bovine plasma, however, diluted 1/100 and then treated with thrombin for 1 min at 37°C, will yield a solution in which factor V is fully active at a level of approximately 1 × 10^{-9} M. This solution will possess activity at a level of 0.25–0.50 unit/ml, when a unit is defined as the activity per milliliter of bovine plasma, as measured by factor V-deficient plasma *prior to activa-*

[36] Y. Barenholz, D. Gibbs, B. J. Litman, J. Gull, T. E. Thompson, and F. D. Carlson, *Biochemistry* **16**, 2806 (1977).

[37] D. L. Straight, M. E. Nesheim, K. G. Mann, and R. L. Lundblad, *Circulation, Part II* **62**(4), Abstract No. 1064 (1980).

tion by thrombin. Using this definition of a unit of activity, fully active factor V at $2 \times 10^{-9} M$ (0.66 μg/ml) corresponds to a solution with activity of approximately 1 unit/ml.

Isolation of Bovine Factor V

Numerous procedures for the isolation of bovine factor V have been published. Only three recently reported procedures have provided analysis of purity by gel electrophoresis. Saraswathi *et al.* report an isolate exhibiting primarily 2 bands on electrophoresis in SDS,[19] while Esmon[16] and Nesheim *et al.*[15] have reported isolates that electrophorese as a single component in SDS. The procedure to be described below is routinely used in this laboratory to isolate bovine factor V.

Materials. QAE-cellulose is obtained from Schleicher and Schuell and washed initially according to the procedure described for DEAE–cellulose by Sophianopoulous and Vestling.[38] Sepharose CL-4B and octyl-Sepharose are obtained from Pharmacia. Octyl-Sepharose is prepared for initial use and between subsequent uses by washing in order with: 5% Triton X-100, 90% methanol (degassed), and H_2O (degassed). Between uses it is stored in 90% methanol. Ammonium sulfate (ultrapure) is obtained from Schwarz/Mann, Orangeburg, New York. Tris base and soybean-trypsin inhibitor are obtained from Sigma Chemical Corp., St. Louis, Missouri. Heparin (beef lung) is obtained from Upjohn, Kalamazoo, Michigan. Benzamidine HCl is obtained from Aldrich, Milwaukee, Wisconsin. Cibacron Blue dye is obtained from Polysciences Inc., Warrington, Pennsylvania. Cibacron Blue Sepharose is prepared as described by Travis *et al.*[39] Alternatively, the resin may be obtained from Pharmacia. Bovine blood for the isolations described here was obtained from the Rock Dell Meat Shop, Rock Dell, Minnesota. All other reagents and chemicals are of analytical grade.

Collection of Blood and Preparation of Plasma. Factor V is extraordinarily sensitive to proteolysis and can be both activated[15,16] and inactivated[40,41] by enzymes that may accompany the coagulation process and subsequent events. Thus "proper" collection of blood is essential for the isolation of the native undegraded form of factor V. The method of collection that optimizes the likelihood of successful isolation of factor V is venipuncture into a combination of coagulation inhibitors. With suitable precautions, however, blood may also be collected by exsanguination into

[38] A. J. Sophianopoulous and C. S. Vestling, *Biochem. Prep.* **9**, 102 (1962).

[39] J. Travis, J. Bowen, D. Tewksbury, D. Johnson, and R. Purnell, *Biochem. J.* **157**, 301 (1976).

[40] W. Canfield, M. Nesheim, W. Kisiel, and K. G. Mann, *Circulation, Part II* **58**(4), Abstract No. 816 (1978).

[41] F. J. Walker, P. W. Sexton, and C. T. Esmon, *Biochim. Biophys. Acta* **571**, 333 (1979).

the same inhibitors. For the purpose of this article, the latter procedure will be described in detail. The conditions employed by this laboratory for collection by venipuncture have been described previously.[15]

At the slaughterhouse the animal is suspended head down and an incision is made leading to exsanguination in approximately 1–2 min. The first 10% of the outflow is discarded, and the subsequent stream is collected into a plastic pail containing 1 liter of anticoagulant. The anticoagulant consists of 2.85% trisodium citrate, 10 mM benzamidine, 0.01% soybean-trypsin inhibitor, and heparin (2.5 units/ml). Blood is collected until the total volume (blood and anticoagulant) is 10 liters. If the flow of the blood becomes slow before 10 liters are collected, the collection is terminated in order to eliminate blood that has spent relatively long periods of time in contact with the tissues surrounding the wound. Samples that amount to less than 10 liters but at least 8 liters are suitable for further workup. During collection, efforts are made to assure complete mixing of blood with anticoagulant. Immediately following collection, droplets of blood on the wall of the pail are washed with anticoagulated blood into the bulk of the sample and thus are not allowed to stand for prolonged periods in the absence of anticoagulant. For the isolation to be described below, the blood of two animals is collected in the above fashion, yielding ultimately about 10 liters of plasma. Plasma is prepared from the anticoagulated blood within 1.5 hr, or more rapidly, if possible. Plasma is obtained by centrifugation at 4000 g at 4°C for 20 min, or as routinely accomplished in this laboratory, through the use of a DeLaval Cream Separator operated at a blood flow rate of about 1 liter/min.

Treatment with BaCl$_2$ and PEG 6000. Over a period of 20 min, 1 M BaCl$_2$ is added to the plasma (80 ml/liter) at 4°C with stirring. This is followed by the addition of 50% (w/v) polyethylene glycol 6000 to a final concentration of 4%. The addition is carried out with stirring at 4°C over a period of about 10–20 min. After an additional 30 min of stirring, the suspension is centrifuged at 6000 g at 4°C for 30 min, and the supernatant is retained for the next step. If the recovery of the vitamin K-dependent plasma proteins is desired, the suspension may be centrifuged prior to the addition of PEG 6000. The resulting pellets are then retained for the recovery of the vitamin K-dependent proteins, PEG 6000 is added to the supernatant, and the isolation of factor V is continued.

Chromatography on QAE-Cellulose. Packed QAE-cellulose (equilibrated in 0.025 M Tris-HCl–5.0 mM CaCl$_2$, 1.0 mM benzamidine, pH 7.5) is added to the briskly stirring supernatant of the previous step (approximately 100 ml packed resin per liter of sample). A suitable vessel for this step is a 50-liter plastic garbage can fitted with an overhead mechanical stirrer operated at a rate sufficient to ensure uniform dispersion of the

QAE-cellulose throughout the sample. Volumes of H_2O and 0.025 M Tris-HCl–5.0 mM CaCl$_2$, 1.0 mM benzamidine (pH 7.5), each equal to the initial volume of the sample, are added at 4°C and stirring is continued at 4°C. After 30 min of stirring, the resin is allowed to settle over a period of 2 hr. The bulk of the fluid (80%) over the settled resin is then removed by siphon and the remainder is transferred with the resin into a 2-liter glass funnel fitted with a coarse sintered-glass filter and placed on a 4-liter side-arm Erlenmeyer flask. Flow is aided by partial vacuum. Once the resin is packed into the funnel, it is overlaid with a piece of Whatman No. 1 filter paper and washed with 2 liters of 0.02 M imidazole HCl–5.0 mM CaCl$_2$ (pH 6.5), and then 2 liters of the same buffer with 0.1 M NaCl. Factor V is then eluted with the same buffer containing 0.3 M NaCl. The eluate is collected in fractions of about 500 ml each. Factor V elutes sharply and can be collected in 1 or 2 such fractions. The washing and elution are carried out at 4°C. Solid ammonium sulfate (0.361 g/ml) is added to the eluate and, after 30 min stirring at 4°C, the suspension is centrifuged in two 250-ml bottles in a swinging bucket rotor (Sorvall HS-4) at 10,000 g for 20 min at 4°C. The supernatants are discarded and additional suspension is added to the pellets and centrifuged. This is repeated as necessary until all the suspended material has been collected in two equal pellets. The tightly packed pellets may be stored overnight at 4°C awaiting the next step of purification. After chromatography, the QAE-cellulose is washed with 2 liters of 1.0 M NaCl in 0.02 M imidazole HCl–5.0 mM CaCl$_2$, pH 7.5; partially dried by applying vacuum to force excess fluid from the resin; and then stored at 4°C in 1.0 M NaCl. Prior to subsequent use it is again washed with 1.0 M NaCl, and equilibrated in 0.025 M Tris-HCl–5.0 mM CaCl$_2$–1.0 mM benzamidine, pH 7.5.

Chromatography on Octyl-Sepharose. No benzamidine is included in this step, since it interferes with the hydrophobic chromatography. The pellet obtained with QAE-cellulose by precipitation with (NH$_4$)$_2$SO$_4$ is dissolved at 22°C in about 100 ml of 0.01 M Tris-borate, 1.0 M NaCl, 1.0 mM CaCl$_2$. The solution is then applied to a column (4 × 32 cm) of octyl-Sepharose equilibrated in the same buffer at 22°C. When the sample has been applied to the column, flow is halted for 20 min and then commenced with the same buffer at a flow rate of about 7 ml/min. Flow is continued until the absorbance of the eluate is 0.1 or less. Factor V is then eluted with the same buffer without NaCl (Fig. 2). Peak fractions may be pooled and applied directly onto the next column. Conductivity should be checked, however, and a value equivalent to less than 0.1 M NaCl is required if direct application to the next column is desired. Alternatively, solid (NH$_4$)$_2$SO$_4$ (0.435 g/ml) may be added to the pooled fractions and, after stirring at 4°C for 30 min, factor V can be pelleted by centrifugation

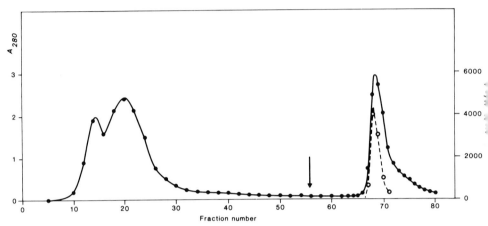

FIG. 2. Chromatography at 22°C of bovine factor V on octyl-Sepharose (4 × 32 cm). The sample is obtained from QAE-cellulose and is applied in 0.01 M Tris-borate–1.0 mM CaCl$_2$–1.0 M NaCl, pH 8.3. Elution is accomplished with the buffer minus NaCl. Fractions of 20 ml are collected at a flow rate of about 7 ml/min. ●——●, Absorbance at 280 nm; ○----○, factor V activity after activation by thrombin.

at 10,000 g for 20 min at 4°C. The sides of the container should be carefully wiped to remove excess (NH$_4$)$_2$SO$_4$. The pellets obtained in this fashion may be stored overnight at 4°C.

Chromatography on Cibacron Blue Sepharose. The sample for this step consists of either the pooled peak fractions obtained from the previous step, or the pellets obtained from that peak after precipitation with (NH$_4$)$_2$SO$_4$. If the material had been pelleted from (NH$_4$)$_2$SO$_4$, it is dissolved in 100 ml of 0.01 M Tris-borate–1.0 mM CaCl$_2$. The sample is applied at 22°C to a column (2.4 × 20 cm) of Cibacron Blue Sepharose equilibrated in the same buffer. The column is washed with the same buffer until the initial peak has passed and the absorbance of the eluate drops to baseline. Elution of factor V is then accomplished with an 800-ml linear gradient starting in the same buffer and ending in buffer plus 0.6 M NaCl. Factor V elutes about two-thirds of the way through the gradient (Fig. 3). Peak fractions are pooled, chilled to 0°C and solid (NH$_4$)$_2$SO$_4$ (0.435 g/ml) is added with stirring. Stirring is continued at 4°C for 30 min and then factor V is pelleted by centrifugation in a swinging bucket rotor at 10,000 g for 20 min at 4°C. The pellet is carefully drained to remove excess (NH$_4$)$_2$SO$_4$ and then dissolved in 5–10 ml of 0.01 M Tris-borate–1.0 mM CaCl$_2$–50% glycerol. The sample of factor V is then stored at −20°C. Typical results of the isolation of bovine factor V are shown in Table I.

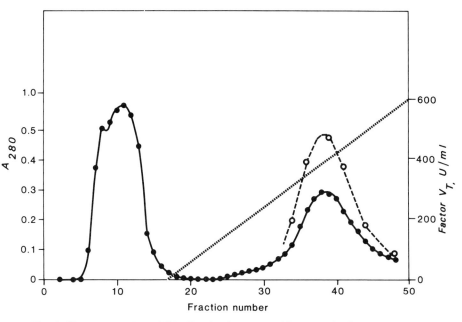

Fig. 3. Chromatography at 22°C of bovine factor V on Cibacron Blue Sepharose (2.4 × 20 cm). The sample is obtained from octyl-Sepharose and is applied to the column equilibrated in 0.01 M Tris-borate–1.0 mM CaCl$_2$, pH 8.3. Elution is accomplished with an 800-ml linear gradient starting in buffer and terminating in buffer plus 0.6 M NaCl. Fractions of 20 ml are collected at a flow rate of 3.5–4.0 ml/min. ●——●, Absorbance at 280 nm; ○----○, factor V activity after activation by thrombin; ·····, linear gradient of NaCl.

Isolation of Human Factor V

Partial Purification of Human Factor V. Three units (600–700 ml) of human plasma prepared from blood collected in acid–citrate–dextrose anticoagulant is processed in a fashion identical to the procedures described for the isolation of bovine factor V through the steps including QAE-cellulose chromatography and subsequent precipitation with ammonium sulfate. The pellet is then dissolved in 0.01 M Tris-borate–1.0 mM CaCl$_2$–5.0 μM DAPA, to a final volume of 5–10 ml. This is applied to a column (2.8 × 100 cm) of Ultrogel 22 (LKB Instruments, Inc., Rockville, Maryland) equilibrated in the same buffer at 4°C. Chromatography is carried out at a rate of 30 ml/hr, and fractions of 15 ml are collected. Factor V activity typically elutes in fractions 15–20. These are pooled and precipitated with ammonium sulfate at a final concentration of 70% saturation (0.435 g/ml) followed by centrifugation at 4°C, 10,000 g, for

TABLE I
ISOLATION OF BOVINE FACTOR V

Step	Volume (liters)	Total protein (mg)	Total[a] activity (units)	Activation[b] quotient	Specific activity (units/A_{280})	Purification factor	Yiel (%
Plasma	10.0	500,000	605,000	44.6	1.21	1.00	10(
Barium citrate– 4% PEG 6000 supernatant	10.5	290,000	515,000	40.5	1.78	1.47	85.
QAE-cellulose eluate	0.80	3620	251,000	25.3	69.3	57.2	41.
Octyl-Sepharose eluate	0.12	216	181,000	32.2	838	693	29.
Cibacron Blue Sepharose eluate	0.22	90.7	144,000	33.9	1590	1310	23.
Buffered 50% glycerol solution	0.0075	75.0	119,000	31.0	1600	1320	19.

[a] Activity measured with factor V-deficient plasma after activation by thrombin.
[b] Ratio of activity after and before activation.

20 min in a swinging bucket rotor. The supernatant is poured off, excess ammonium sulfate is carefully wiped from the sides of the tube, the pellet is taken up in approximately 20 ml of 0.01 M Tris-borate–1.0 mM CaCl$_2$ (pH 8.3) and applied to a 1.5 × 5-cm column of Cibacron Blue Sepharose operated at 22°C. The column is washed with the same buffer, then the same buffer plus 0.1 M NaCl. Factor V is finally eluted with the same buffer, 0.5 M NaCl. Peak fractions are pooled, ammonium sulfate is added to 70% saturation (0.435 g/ml), and, after 30 min at 0°C, the pellet is collected by centrifugation at 10,000 g at 4°C for 20 min. The pellet is dissolved in 2–5 ml of 50% glycerol/water and stored at −20°C. Factor V is typically 20–40% pure at this point and appears to be a mixture of factor V and the intermediates and end products that arise during conversion of factor V to factor V_a.

Monoclonal Hybridoma Antibody to Human Factor V. The partially purified human factor V was used to immunize mice, and hybridoma techniques were used to obtain a cell line producing a monoclonal antibody to factor V. The generation of this cell line has been described.[42] This hybridoma antibody is an IgGk immunoglobulin and is monoclonal as

[42] J. A. Katzmann, M. E. Nesheim, L. S. Hibbard, and K. G. Mann, *Proc. Natl. Acad. Sci. U.S.A.* **78**, 162 (1981).

detected by isoelectric focusing. The monoclonal antibody partially inhibits human factor V with a $K_d = 3 \times 10^{-9} M$. Since $1.2 M$ NaCl will disrupt the antibody–antigen complex, the binding appears to be ionic. The cell line is grown as an ascitic tumor in mice, and ascitic fluids contain 5–10 mg of monoclonal antibody/ml (40–50 mg/mouse).

Monoclonal antibody is purified by chromatography on Ultrogel AcA-34. Ascitic fluid (20 ml) is made 50% saturated in ammonium sulfate, stirred at 4°C for 30 min, and centrifuged to obtain the precipitate. The pellet is dissolved in 7 ml of buffer (0.01 M NaPi–0.15 M NaCl, pH 7.4) and chromatographed on a column of Ultrogel 34 (2.5×100 cm) equilibrated in the same buffer. Tubes containing the antibody are pooled and dialyzed against 0.2 M sodium citrate, pH 6.5. The antibody is then coupled to agarose.[43] To a 1-volume slurry of agarose (4%, Biorad) and water (1 : 1) is added 1 volume of 2 M Na_2CO_3. This mixture is stirred vigorously as 0.05 volume of cyanogen bromide in acetonitrile (2 g/ml) is added. Stirring is continued for 2 min and the mixture is then filtered on a sintered-glass filter funnel and washed successively with 5–10 volumes of 0.1 M $NaHCO_3$ (pH 9.5), 5–10 volumes H_2O, and 5–10 volumes 0.2 M sodium citrate (pH 6.5). The activated agarose is filtered to a cake; 1 volume of 0.2 M sodium citrate (pH 6.5) is added; and 1 mg of antibody for each milliliter of resin is added. This suspension is incubated overnight at 4°C with gentle stirring. More than 90% of the antibody is coupled to the resin, as judged by the absorbance of the supernatant of the slurry. Unreacted groups on the resin are blocked by a 4-hr, room-temperature incubation after the addition of 1 M glycine (1/10, v/v). The resin is then filtered and washed successively with 20 volumes of 0.1 M sodium acetate–0.5 M NaCl (pH 4), 20 volumes of 2.0 M urea–0.5 M NaCl, 20 volumes of 0.1 M $NaHCO_3$–0.5 M NaCl (pH 10), and 20 volumes of 0.01 M sodium phosphate–0.15 M NaCl (pH 7.4).

Isolation of Human Factor V Using a Hybridoma Antibody. Fresh human plasma (440 ml) or fresh-frozen human plasma (no more than 1 day old) obtained from blood collected in acid–citrate–dextrose is treated by the addition of 1.0 M barium chloride (80 ml/liter of starting plasma) with stirring at 4°C. Polyethylene glycol 6000 (50%, w/v) is added to this suspension (at 4°C, with stirring) to a final concentration of 4%. After stirring for 30 min, the sample is centrifuged at 10,000 g at 4°C for 20 min. The supernatant is transferred to a fresh centrifuge bottle, and polyethylene glycol 6000 (50%, w/v) is added (at 4°C, with stirring) to a final concentration of 10%. After 30 min at 4°C, the sample is centrifuged at 4°C, 10,000 g, for 20 min. The supernatant is discarded and the pellet is dissolved in

[43] S. C. March, I. Parikh, and P. Cuatrecasas, *Anal. Biochem.* **60**, 149 (1974).

50–100 ml of 0.02 M Tris–0.15 M NaCl (pH 7.4) at 4°C and centrifuged at 4°C, 10,000 g, for 10 min to clarify. The solution is applied to a column (2.5 × 4 cm, plastic syringe) of immobilized monoclonal antibody to factor V. The sample is added to the column at 4°C over a period of about 2.5 hr. The column is washed with the sample buffer and then transferred to the laboratory at room temperature and washed with a buffer consisting of 0.02 M imidazole HCl–0.1 M NaCl, pH 6.5. Washing is continued until absorbance at 280 nm of the eluate is 0.01. Factor V is then eluted with the same buffer made 1.2 M in NaCl, and typically elutes in 3–5 fractions, 20 ml each. The fractions are pooled, chilled in an ice bath, and applied to a 1.6 × 5-cm column of phenyl-Sepharose equilibrated at 4°C in the same buffer. The phenyl-Sepharose column is washed with the same buffer until absorbance values are at baseline. Elution of factor V is then accomplished with the same buffer without NaCl, at 22°C. Approximately 4–5 ml is collected per fraction. Peak fractions are pooled (typically 2 fractions) in a 20-ml glass vial, and solid ammonium sulfate (Schwarz/Mann) is added to a final concentration of 75% saturation (0.476 g/ml). After the ammonium sulfate is dissolved, the suspension of factor V is stored at 4°C. The concentration of the suspended factor V is typically 80–120 μg/ml. Recovery of factor V from the suspension is accomplished by centrifugation at 10,000 g, 4°C, for 15 min, removal of the supernatant, and solubilization of the pellet in the desired buffer. The isolation of human factor V is summarized in Table II. Other procedures yielding apparently homogeneous human factor V have been reported.[44,44a]

Immunochemical Techniques for Factor V Analyses

Preparation of Antibodies. Purified bovine and human factor V preparations are injected into burros in 10 weekly subcutaneous injections of 0.2 ml of either 0.05 mg of bovine factor V or 0.1 mg of human factor V in a 1 : 1 emulsion with complete Freund's adjuvant. Prior to the sixth, and all subsequent injections, serum is obtained by venipuncture and assessed for antibody titer in the radioimmunoassay described below. One week following the tenth injection, the animals are anesthetized and exsanguinated by jugular venipuncture. The purity and specificity of the antisera is determined by double immunodiffusion in agar[45] against either human or bovine plasma and purified factor V. In each case single precipitin lines of identity are obtained with both plasma and purified factor V. Antisera to normal burro IgG is obtained in an analogous procedure by weekly injections of a goat with 0.2 mg of burro IgG.

[44] B. Dahlbäck, *J. Clin. Invest.* **66**, 583 (1980).
[44a] W. H. Kane and P. W. Majerus, *J. Biol. Chem.* **256**, 1002 (1981).
[45] A. B. Auernheimer and F. O. Atchley, *Am. J. Clin. Pathol.* **38**, 548 (1962).

TABLE II
ISOLATION OF HUMAN FACTOR V

Step	Volume (ml)	Total A_{280}	Total[a] units	Activation[b] quotient	Units/ A_{280}	Purification (-fold)	Yield (%)
Plasma	440	24,000	13,000	70	0.54	1.0	100
Barium citrate–4% PEG supernatant	510	18,400	11,300	64	0.62	1.1	87.2
10% PEG pellet[c]	75	10,500	9200	103	0.88	1.6	70.8
αHFV[d] eluate	80	4.97	4690	82	940	1740	36.1
φ-Sepharose[e] eluate	10	2.00	3690	53	1840	3410	28.4

[a] Activity measured with factor V-deficient plasma after activation by thrombin.
[b] Ratio of activity after and before activation by thrombin.
[c] Pellet obtained with 10% PEG 6000 dissolved in 0.02 M Tris-HCl–0.15 M NaCl, pH 7.4.
[d] Immobilized monoclonal antibody to human factor V.
[e] Phenyl-Sepharose.

Radiolabeling of Factor V. Human and bovine factor V are radiolabeled with ^{125}I using Bolton–Hunter reagent.[46] The proteins (1.0 ml, 0.5 mg/ml) are dialyzed against 0.1 M sodium borate (pH 8.5) for 3 hr and then treated with 1 mCi of Bolton–Hunter reagent for 20 min at 0°C. The iodination reaction is quenched with 50 μl of 2 M glycine in 0.1 M sodium borate, pH 8.5. The ^{125}I-labeled factor V is diluted with 3 volumes of Tris-borate buffer (0.01 M Tris-borate–1 mM CaCl$_2$, pH 8.3) and applied to a 1-ml Cibacron Blue Sepharose column equilibrated with the same buffer in order to separate the iodinated factor V from other products of the conjugation reaction. The column is washed with 30 ml of Tris-borate buffer, followed by 30 ml of the same buffer containing 0.1 M NaCl. The ^{125}I-labeled factor V is eluted with the Tris-borate buffer containing 0.5 M NaCl. Typically, 50–60% of the protein is recovered in three 2-ml fractions. The fractions are pooled, dialyzed against 50% glycerol · H$_2$O (v/v), and stored at −20°C. This procedure results in the incorporation of 10–50% of the radioactivity into factor V with specific radioactivities of 250–1000 cpm/ng.

Double Antibody Precipitation Radioimmunoassay. All reagents for this assay are diluted in assay buffer (0.075 M Tris–0.075 M NaCl, pH 7.0) containing 1% bovine serum albumin and 0.2% Triton X-100. To plastic

[46] A. E. Bolton and W. M. Hunter, *Biochem. J.* **133,** 529 (1973).

tubes (1 × 6 cm) are added 0.2 ml of ^{125}I-labeled factor V (20 ng), 0.2 ml of appropriate plasma dilutions, and 0.2 ml of a 1 : 8000 dilution of burro anti-bovine factor V antisera or a 1 : 4000 dilution of burro anti-human factor V antisera. The dilutions of antisera are made in dilute normal burro serum (1 : 40) to provide sufficient IgG to form a manageable precipitate after the addition of goat anti-burro antisera in the following step. Following a 1-hr incubation at 37°C, 0.2 ml of goat anti-burro antisera is added and the tubes are incubated for an additional 16 hr at 4°C. The tubes are then centrifuged; the precipitate is washed twice with assay buffer and then assayed for radioactivity. The anti-factor V antibody dilutions used in this assay are all dependent on the antibody titer; therefore the determination of the antibody titer of the antisera is now described, using a slight modification of the radioimmunoassay.

A 1 : 40 dilution of the antisera precipitates 100% of the radiolabeled factor V (20 ng) when assessed in the above assay in the absence of added plasma. This dilution of the burro anti-factor V antisera is serially diluted with 1 : 40 dilutions of normal burro serum. This dilution series is used to determine the antibody dilution that precipitates 40–50% of the radiolabeled factor V, since that is the dilution of choice for use in all the radioimmunoassays.

A standard curve is prepared for each assay with known amounts of purified factor V. The amount of purified, unlabeled factor V added to each assay tube ranges from 8 to 1024 ng. Since this is a competition assay, the amount of radioactivity found in the precipitate is inversely proportional to the amount of unlabeled factor V added. Typically, the assay is linear from 16 to 1024 ng of factor V per tube when the amount of unlabeled factor V is plotted on a log scale. The linear region of the assay can be made more sensitive by the addition of less radiolabeled factor V.

Physical Properties of Bovine Factor V

The gross physical properties of bovine factor V are summarized in Table III. The protein circulates at approximately 35 μg/ml of plasma. The factor V concentration in human plasma is about 7 μg/ml. The molecular weight of bovine factor V has been determined by sedimentation equilibrium analysis both under native and denaturing conditions and by sedimentation velocity studies of the reduced random coil.[15] A striking feature of the protein is its apparent deviation from globular structure. It sediments with $S_{20,w}^{0}$ = 9.19 S, which together with the molecular weight yields a frictional ratio of 2.01. Factor V has a Stokes radius of 91–93 Å as determined both by sedimentation equilibrium plus sedimentation velocity measurements and by gel filtration on 6% agarose.[15-17] When the molecule

TABLE III
GROSS PHYSICAL PROPERTIES OF
BOVINE FACTOR V[a]

Property	Value
Molecular weight	330,000
$S_{20,w}^0$	9.19 S
$\bar{v}$	0.712 ml/g
Stokes radius[b]	91–93 Å (51 Å)[d]
Stokes radius[c]	95 Å (59 Å)[d]
f/f_{min}	2.01
$E_{280,1\ cm}^{1\%}$	9.6
Neutral sugars (%)	11.9
NH$_2$-terminal sequence	Ala-Lys-Leu-Arg

[a] Ref. 15.
[b] Calculated from molecular weight and $S_{20,w}^0$, also determined by gel filtration.[17]
[c] From Ref. 16.
[d] Stokes radius of activated factor V (factor V$_a$).

is modeled as a prolate ellipsoid[47] with 0.2 g water/g protein, an axial ratio of 25 : 1 is obtained. The apparent asymmetry of factor V does not permit accurate estimation of molecular weight by gel filtration based on standards of spherical configuration.

The amino acid composition of bovine factor V is presented in Table IV. The protein contains a relative abundance of aspartic acid plus asparagine and glutamic acid plus glutamine (592 mol/mol of factor V) relative to the basic amino acids lysine plus arginine (258 mol/mol of factor V). Factor V is a glycoprotein containing glucosamine and 12% neutral sugars by weight. The extinction coefficient of a 1% solution at 280 nm is 9.6.

Activation and Inactivation of Factor V and Factor V$_a$

Factor V circulates as a relatively inactive procofactor in plasma. In the presence of catalytic amounts of thrombin, however, the activity of the molecule increases dramatically. Our best estimates indicate that the relative activity increases at least 400-fold. This increase was observed in prethrombin-1 activations carried out in the presence of the inhibitor DAPA to attenuate the feedback activation of factor V by thrombin.[21] The conversion of bovine factor V to factor V$_a$ proceeds by virtue of

[47] J. L. Oncley, *Ann. N.Y. Acad. Sci.* **41**, 121 (1951).

TABLE IV
AMINO ACID COMPOSITION OF BOVINE FACTOR V[a]

Amino acid	Mol/mol[b]
Aspartic acid	229
Threonine	121
Serine	204
Glutamic acid	293
Proline	199
Cysteine	24
Glycine	146
Alanine	131
Valine	108
Methionine	43
Isoleucine	124
Leucine	240
Tyrosine	94
Phenylalanine	88
Lysine	148
Histidine	71
Arginine	110
Tryptophan	21
Glucosamine	65
Neutral sugars	11.9%

[a] From Ref. 15.
[b] Moles of amino acid per mole of factor V MW 330,000.

proteolysis of the native molecule initially to two intermediates, one of about 150,000 and the other of 205,000 apparent molecular weight, and proceeds to end products of approximately 94,000, 74,000, 71,000, and 31,000 apparent molecular weights.[18] Analysis of the activation of bovine factor V by electrophoresis in sodium dodecyl sulfate (SDS) is shown in Fig. 4; a model of the process is indicated in Fig. 5.

The integrity of activated factor V is dependent on calcium ion, as shown by Esmon.[16] In the presence of EDTA, the cofactor activity is lost, and the constituent polypeptides of factor V_a can be separated by ion-exchange chromatography. Activity can be reconstituted by a recombination of isolated peptides in the presence of calcium ion,[16] indicating that activated factor V (factor V_a) is a multiple subunit protein; the exact nature of factor V_a has not been deduced to date. Factor V can be activated also by a component of Russell's viper venom. As with thrombin, activation proceeds by proteolysis, but in a manner apparently different from that seen with thrombin. A single cleavage of the molecule to give

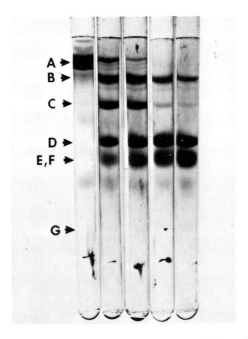

FIG. 4. Time course of the activation of bovine factor V (7.0 × 10^{-7} M) by bovine thrombin (2.0 × 10^{-8} M) at 37°C, pH 7.4. Analysis is by electrophoresis in 0.1% SDS–5% polyacrylamide. Gels from left to right correspond to 0, 10, 20, 60, and 180 sec after the addition of thrombin. At 180 sec the sample was fully active. Component A is factor V with MW 330,000 as measured by sedimentation equilibrium. Indicated intermediates (B, C) and end products (D–G) have apparent molecular weights estimated by gel electrophoresis of 205,000 (B); 150,000 (C); 94,000 (D); 74,000 (E); 71,000 (F); and 31,000 (G). (From Ref. 18.)

peptides of approximate molecular weights of 220,000 and 80,000 are sufficient to generate factor V$_a$ activity from the venom protein.[48] Other activators also have been described.[28,29] In addition, a putative factor V activator from platelets has been reported by Østerud et al.[49]

Recent studies have indicated that activated factor V can be inactivated through proteolysis catalyzed by activated protein C (APC),[40,41] an enzyme derived from a vitamin K-dependent precursor, protein C. The activation of protein C has been shown to occur by proteolysis by thrombin.[50] The inactivation of factor V$_a$ by APC is marked by further proteolysis of the 94,000- and 74,000-MW peptides of activated factor

[48] C. T. Esmon, in "The Regulation of Coagulation" (K. G. Mann and F. B. Taylor, Jr., eds.), p. 137. Elsevier/North-Holland Publ., Amsterdam and New York, 1980.
[49] B. Østerud, S. I. Rapaport, and K. K. Lavine, Blood 49, 819 (1977).
[50] W. Kisiel, L. H. Ericsson, and E. W. Davie, Biochemistry 15, 4893 (1976).

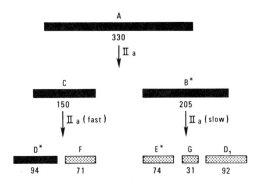

FIG. 5. Model of the activation of bovine factor V by thrombin. Data such as that of Fig. 4 and, for example, Ref. 18, indicate that activation of factor V proceeds through two intermediates as indicated. The majority of expressed activity appears to correlate with the cleavage of component C. Components C and D are shown properly located with respect to NH_2-terminal sequence and component B has an NH_2-terminal sequence different from components A, B, D. Components A, C, and D have the NH_2-terminal sequence: Ala-Lys-Leu-Arg, whereas the NH_2-terminal sequence of component B is Ser-Phe-Tyr-Pro (unpublished observations). The asterisks indicate minimum peptides which constitute actor V_a. The solid bars represent peptides placed in sequence by NH_2-terminal amino acid analysis.

V.[40,41] APC will also cleave unactivated factor V, but at a rate considerably slower than factor V_a. In addition, factor X_a (for which factor V_a serves a cofactor function) will protect factor V_a from inactivation by APC.[51,52]

Binding Properties of Factor V and Factor V_a

Interactions of factor V (factor V_a) with Ca^{2+}, phospholipid, factor X_a, prothrombin, and platelets have been studied and reported. The results of some of these studies are tabulated in Table V. Equilibrium dialysis indicated that bovine factor V will bind two equivalents of Ca^{2+} with $K_d = 5.9 \times 10^{-5} M$. In addition, a bound Ca^{2+} with $K_d < 10^{-8} M$ was also detected. Factor V and factor V_a were shown to bind phospholipid (vesicles of 75% phosphatidylcholine–25% phosphatidylserine) (PCPS) with about 10 times the avidity of prothrombin or factor X for similar vesicles. From alterations in the kinetics of either prothrombin activation or hydrolysis of the tetrapeptide substrate Ile-Glu(OR)-Gly-Arg-pNO$_2$ anilide (S2222), an interaction between factor X_a and phospholipid vesicle-bound factor V_a has been inferred. This interaction of factor X_a and phospholipid-bound

[51] M. E. Nesheim, L. S. Hibbard, P. B. Tracy, J. W. Bloom, K. H. Myrmel, and K. G. Mann, in "The Regulation of Coagulation" (K. G. Mann and F. B. Taylor, Jr., eds.), p. 145. Elsevier/North-Holland Publ., Amsterdam and New York, 1980.

[52] J. Stenflo and B. Dahlbäck, in "The Regulation of Coagulation" (K. G. Mann and F. B. Taylor, Jr., eds.), p. 225. Elsevier/North-Holland Publ., Amsterdam and New York, 1980.

TABLE V

BINDING INTERACTIONS OF PROTHROMBINASE COMPONENTS

rant	Fixed components[a]	Interaction	K_d (M)	Stoichiometry[b]	Method[c]	Ref.
2+	V	Ca^{2+} + V	5.9×10^{-5}	2.0 (Ca^{2+}/V)	E	31
2+	V	Ca^{2+} + V	$<1.0 \times 10^{-8}$	1.0 (Ca^{2+}/V)	E	31
	PCPS, Ca^{2+}	V + PCPS	4.6×10^{-8}	327 (PCPS/V)	E	23
	PCPS, Ca^{2+}	V_a + PCPS	4.4×10^{-7}	76 (PCPS/V_a)	E	23
PS	V_a, Ca^{2+}, X_a, S2222	V_a + PCPS	3.0×10^{-7}	83 (PCPS/V_a)	K	59
	PCPS, Ca^{2+}, X_a, II	X_a + V(PCPS)	2.9×10^{-9}	0.94 (V/X_a)	K	21
	PCPS, Ca^{2+}, X_a, II	X_a + V_a (PCPS)	7.3×10^{-10}	0.87 (V_a/X_a)	K	21
	PCPS, Ca^{2+}, X_a, S2222	X_a + V_a (PCPS)	6.1×10^{-10}	0.90 (V_a/X_a)	K	59
	Bovine platelets	V + platelets	3.1×10^{-9}	870 (V/platelet)	E	24
	Bovine platelets	V_a + platelets	3.7×10^{-10}	830 (V_a/platelet)	E	24
	Bovine platelets	V_a + platelets	3.3×10^{-9}	3400 (V_a/platelet)	E	24
	Bovine platelets, X_a, Ca^{2+}, II	V_a + platelets	1.9×10^{-10}	920 (V_a/platelet)	K	26
	V_a, Ca^{2+}, PCPS, S2222	X_a + V_a (PCPS)	9.3×10^{-10}	0.99 (V_a/X_a)	K	59
	V_a, Ca^{2+}, PCPS, II	X_a + V_a (PCPS)	7.1×10^{-10}	0.61 (X_a/V_a)	K	21
	V_a, Ca^{2+}, bovine platelets	X_a + V_a (platelets)	6.0×10^{-10}	1.04 (V_a/X_a)	E	26
	Bovine platelets	X_a + platelets	1.9×10^{-10}	360 (X_a/platelet)	E	27
	Human platelets	X_a + platelets	3.0×10^{-11}	160 (X_a/platelet)	E	25
	PCPS, Ca^{2+}	II + PCPS	2.3×10^{-6}	104 (PCPS/II)	E	23
	V_a, X_a, PCPS, Ca^{2+}	II + "prothrombinase"	1.0×10^{-6}	—	K	21
	PCPS, Ca^{2+}	X + PCPS	2.5×10^{-6}	48 (PCPS/X)	E	23
	PCPS, Ca^{2+}	X_a + PCPS	2.7×10^{-6}	66 (PCPS/X_a)	E	60[e]

[a] All experiments were performed by varying the concentration of one of the components (titrant) in the presence of fixed levels of the remaining components.

[b] Stoichiometries are expressed as levels at saturation. Formal stoichiometries with PCPS (vesicles of 25% phosphatidylserine–75% phosphatidylcholine) are expressed in terms of total phospholipid (monomer) per mole of protein.

[c] E, K represent binding constants determined by equilibrium-binding measurements or inferred from kinetics, respectively.

[d] The K_d represents the formal apparent K_m of the interaction of prothrombin with the prothrombinase complex.

[e] Determined by fluorescence polarization of an active-site dansylated derivative of factor X_a. See Ref. 60.

factor V_a is characterized by 1 : 1 stoichiometry and $K_d = 7.0 \times 10^{-10} M$. An interaction between the proteins in the presence of bovine platelets (also with 1 : 1 stoichiometry and $K_d = 6.0 \times 10^{-10} M$) has been deduced from equilibrium-binding measurements.

The binding of ^{125}I-labeled bovine factor V and factor V_a to bovine platelets has been reported.[24] The data indicate two classes of binding sites for factor V_a, one with $K_d = 3.4 \times 10^{-10} M$ and the other with $K_d = 3.4 \times 10^{-9} M$. Factor V_a can displace all bound factor V, but factor V will displace only the less tightly bound factor V_a, suggesting a binding domain selective for factor V_a. Studies with both bovine[26] and human platelets[53] indicate that the binding of factor X_a to platelets is dependent on the binding of factor V_a, such that the latter constitutes the equivalent of a factor X_a receptor.

Studies with immobilized prothrombin indicated that factor V_a, but not factor V, interacts measurably with prothrombin.[54]

Functional Properties of Activated Factor V

Activated factor V functions as a cofactor in the factor X_a-catalyzed activation of the blood-clotting zymogen, prothrombin. Although Ca^{2+}, phospholipid, and factor V_a have no known intrinsic enzymatic capacities, they form a complex with factor X_a (prothrombinase complex), which catalyzes prothrombin activation orders of magnitude more rapidly than factor X_a alone. The influence of various combinations of the nonenzymatic components on the rate of activation of physiological levels of prothrombin is illustrated in Table VI. Rates are expressed relative to the rate observed with factor X_a alone. The absolute rate obtained with factor X_a alone is 0.0044 mol thrombin produced per minute per mole of factor X_a. The addition of all components, other than factor V_a, increases the rate by a factor of 22, whereas the inclusion of factor V_a leads to an overall rate enhancement of 278,000. The fully competent prothrombinase complex, consisting of optimal levels of factor X_a, Ca^{2+}, PCPS, and factor V_a, converts prothrombin to thrombin with a V_{max} of 2100 mol/min/mol of factor X_a.[21,22] This value is observed whether the complex is on the surface of either phospholipid vesicles or platelets.[21]

Model of Prothrombinase[51]

A summary of the binding and functional properties of prothrombinase components is depicted in Fig. 6. The curved surface represents a section of a phospholipid vesicle. A molecule of factor V_a is shown tightly associated with the vesicle bound in 1:1 stoichiometry with factor X_a. The fissure in the factor V_a molecule suggests proteolytic activation of the cofactor. The combination of factor V_a, factor X_a, and phospholipid (plus Ca^{2+}) comprises the enzymatic complex (prothrombinase). Indicated also are molecules of the substrate, prothrombin, with their prothrombin

[53] W. H. Kane, M. J. Lindhout, C. M. Jackson, and P. W. Majerus, *J. Biol. Chem.* **255**, 1170 (1980).
[54] C. T. Esmon, W. G. Owen, D. L. Duiguid, and C. M. Jackson, *Biochim. Biophys. Acta* **310**, 289 (1973).

TABLE VI
INFLUENCE OF PROTHROMBINASE COMPONENTS
ON RATE OF PROTHROMBIN ACTIVATION[a]

Components[b]	Relative rate
X_a	1.0^d
X_a, PCPS[c]	1.0
X_a, Ca^{2+}	2.3
X_a, Ca^{2+}, PCPS	22
X_a, Ca^{2+}, PCPS, V_a	278,000
X_a, Ca^{2+}, V_a	356

[a] See Ref. 21.
[b] Present at levels sufficient to maximize the rate of prothrombin conversion in a system of all components (X_a, Ca^{2+}, PCPS, V_a). Prothrombin at $1.39 \times 10^{-6} M$ (0.1 mg/ml).
[c] Vesicles of phosphatidylserine–phosphatidylcholine (1/3, w/w).
[d] Absolute rate (22°C) is 0.0044 mol thrombin/ min/mol factor X_a.

fragment 1, prothrombin fragment 2, and prethrombin-2 domains.[55] The interaction of the substrate with phospholipid is suggested through Ca^{2+} bridging mediated by the γ-carboxyglutamate residues of the fragment 1 domain. A dimer of prothrombin[56] mediated by Ca^{2+} bridging as proposed by Bloom and Mann[57] is shown in solution. An interaction between prothrombin and factor V_a through the fragment 2 domain is shown in accordance with studies of the kinetics of prothrombin activation[14,58] plus chromatography of factor V_a on immobilized prothrombin.[54] In this model, factor V_a constitutes the equivalent of a receptor for factor X_a in a manner analogous to its proposed role in the assembly of prothrombinase on platelets.[24-26] The marked enhancement of rate at physiological concentrations of prothrombin (278,000-fold greater than factor X_a alone)[21] can be accounted for quantitatively by assembly of the enzymatic components.[59] The binding of substrate to the vesicles constitutes a great increase in its local concentration. This effect alone, given co-concentration of the enzyme (factor X_a), contributes an increase in rate by an approximate factor of 90. Factor V_a by virtue of its high affinity for both factor X_a and phospholipid allows co-concentration of factor X_a and subsequent

[55] J. W. Bloom and K. G. Mann, *Biochemistry* **18**, 1957 (1979).
[56] F. G. Prendergast and K. G. Mann, *J. Biol. Chem.* **252**, 840 (1977).
[57] J. W. Bloom and K. G. Mann, *Biochemistry* **17**, 4430 (1978).
[58] C. T. Esmon and C. M. Jackson, *J. Biol. Chem.* **249**, 7791 (1974).
[59] M. E. Nesheim, S. Eid, and K. G. Mann, *J. Biol. Chem.* **256**, 9874 (1981).

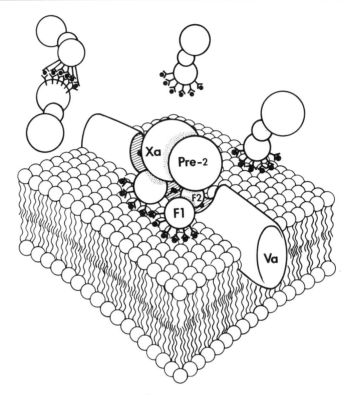

Fig. 6. Model of prothrombinase.[51] The enzyme prothrombinase is depicted as a stoichiometric (1 : 1), Ca^{2+}-dependent complex of phospholipid-bound factor V_a and factor X_a. Formation of the complex enhances the activation of prothrombin at 1.39×10^{-6} M (0.1 mg/ml) by a factor of 278,000[21] relative to the rate with factor X_a alone. The increase can be accounted for by concentration of substrate and enhanced turnover.[2,59] The assembled complex at 22°C catalyzes prothrombin conversion with apparent $K_m = 1.0$ μM and $V_{max} = 2100$ mol thrombin/min/mol factor X_a.[21] The V_{max} of prothrombinase assembled on platelets is indistinguishable from that assembled on phospholipid vesicles.

expression of the 90-fold increase in rate. Factor V_a, by an as yet unknown mechanism, also enhances the V_{max} of the reaction by a factor of 3100. This increase, along with the 90-fold increase resulting from concentration of the substrate, yields a calculated overall increase of 280,000, in good agreement with the 278,000-fold increase observed experimentally.[21]

Acknowledgments

This work was supported by grants from the National Institutes of Health HL 16150 and HL 17430 and by the Mayo Foundation.

[60] M. E. Nesheim, C. Kettner, E. Shaw, and K. G. Mann, *J. Biol. Chem.* **256**, 6537 (1981).

[22] Snake Venom Proteases That Activate Blood-Coagulation Factor V

By WALTER KISIEL[1] and WILLIAM M. CANFIELD

The venoms of *Vipera russelli* and *Bothrops atrox* each contain a serine protease that increases the coagulant activity of factor V severalfold.[2,3] The presence of a component in Russell's viper venom (RVV) that could increase the activity of factor V was initially described by Hjort.[2] Subsequently, Schiffman *et al.*[4] reported the separation of the factor V activator from the factor X activator by gel filtration of RVV in Sephadex G-200. The factor V-activating enzyme, designated RVV-V, has now been purified to homogeneity,[5-7] and some of its molecular properties have been reported.[7,8]

The factor V activator from *B. atrox* venom (thrombocytin) has only recently been isolated and characterized.[9,10] Structural changes in factor V associated with its activation by thrombocytin have been reported as well.[3]

The procedures for the isolation of RVV-V and thrombocytin used in this laboratory are virtually identical and include an initial gel filtration of the crude venom followed by final purification of the enzyme by SP-Sephadex ion-exchange chromatography. Since both of these enzymes are highly cationic, plasticware is recommended in the collection and storage of these proteins.

Assay Methods

Principle. Factor V is a high-molecular-weight plasma protein that augments the proteolytic activity of factor X_a toward prothrombin in the presence of phospholipid and calcium ions. Factor V circulates in plasma

---

[1] An Established Investigator of the American Heart Association.
[2] P. G. Hjort, *Scand. J. Clin. Lab. Invest.* **9**, 119 (1957).
[3] R. Rawala, S. Saraswathi, and R. W. Colman, *Circulation* **28**, II-209 (1978).
[4] S. Schiffman, I. Theodor, and S. I. Rapaport, *Biochemistry* **8**, 1397 (1969).
[5] C. T. Esmon, Ph.D. Thesis, Washington University, St. Louis, Missouri (1973).
[6] C. M. Smith and D. J. Hanahan, *Biochemistry* **15**, 1830 (1976).
[7] W. Kisiel, *J. Biol. Chem.* **254**, 12230 (1979).
[8] C. T. Esmon and C. M. Jackson, *Thromb. Res.* **2**, 509 (1973).
[9] E. P. Kirby, S. Niewiarowski, K. Stocker, C. Kettner, E. Shaw, and T. M. Brudzynski, *Biochemistry* **18**, 3564 (1979).
[10] S. Niewiarowski, E. P. Kirby, T. M. Brudzynski, and K. Stocker, *Biochemistry* **18**, 3570 (1979).

METHODS IN ENZYMOLOGY, VOL. 80

as an inactive, or marginally active, procofactor that is converted to a reactive cofactor (factor V_a) by minor proteolysis.[11,12] In the assay to be described, the venom protease is incubated with factor V substrate for a defined period. Following the incubation period, factor V is assayed by its ability to correct the clotting time of human factor V-deficient plasma following addition of tissue factor and calcium ions. The clotting times in this assay are proportional to the concentration of factor V_a present in the incubation mixture, which in turn is directly related to the concentration of the venom factor V activator. In the assay of RVV-V, barium sulfate-adsorbed plasma is a satisfactory source of factor V. In the assay of thrombocytin, however, partially purified factor V (free of fibrinogen) is required in that *B. atrox* venom contains a thrombin-like protease (batroxabin) that coelutes with thrombocytin in gel-filtration columns.

Reagents

Diluent Buffer. 50 mM Tris-HCl (pH 7.5) containing 100 mM NaCl and 1 mg/ml bovine serum albumin

Barium Sulfate-Adsorbed Plasma. Bovine plasma, stirred for 30 min with barium sulfate (40 mg/ml), followed by centrifugation at 10,000 g for 10 min

Factor V. Substrate factor V purified from bovine plasma through the octyl-Sepharose step as described by Nesheim *et al.*[13]

Factor V-Deficient Plasma. Factor V-deficient plasma is prepared from normal oxalated human plasma by aging at 37°C, essentially as described by Colman.[14] The one-stage prothrombin time of plasma prepared in this manner is routinely greater than 150 sec. The factor V-deficient plasma is stored in plastic tubes at -20°C.

Thromboplastin. Human brain thromboplastin is prepared by a modification of the procedure described by Quick.[15] Human brain acetone powder is filtered through a 100-mesh Tyler screen (W.S. Tyler, Cleveland, Ohio). The material that passed through the screen (100-mesh fraction) is used for the preparation of thromboplastin. Fifteen grams of the 100-mesh brain powder is extracted with 250 ml of 0.15 M NaCl at 48–50°C for 20 min. The extract is centrifuged for 10 min at 2000 rpm and the supernatant

[11] C. T. Esmon, *J. Biol. Chem.* **254**, 964 (1979).
[12] M. E. Nesheim and K. G. Mann, *J. Biol. Chem.* **254**, 1326 (1979).
[13] M. E. Nesheim, K. H. Myrmel, L. Hibbard, and K. G. Mann, *J. Biol. Chem.* **254**, 508 (1979).
[14] R. W. Colman, this series, Vol. 45B, p. 107.
[15] A. J. Quick, *in* "Hemorrhagic Disease and Thrombosis," p. 438. Lea & Febiger, Philadelphia, Pennsylvania, 1966.

is stored in small aliquots in plastic tubes at $-20°C$. Thromboplastin prepared in this manner produced clotting times of 12–15 sec in the one-stage prothrombin time assay.

Procedure. The reaction mixture, consisting of 0.1 ml of diluted enzyme and 0.1 ml of factor V substrate (0.1–1 U/ml) is incubated at 37°C for 1 min in a plastic culture tube. An aliquot of the reaction mixture is then diluted 50- to 100-fold and 0.1 ml of the diluted sample transferred to a glass culture tube (10 × 75 mm). The following reagents are added sequentially to this factor V sample: 0.1 ml of factor V-deficient plasma, 0.1 ml of thromboplastin, and 0.1 ml of 25 mM $CaCl_2$. An electric timer is started with the addition of the calcium chloride, and the clotting time is determined. A dilution of the venom protease is usually chosen to yield clotting times between 20 and 50 sec. The amount of venom factor V activator is estimated from a standard calibration curve made by serial dilution of crude venom using the same factor V substrate. One unit of factor V-activator activity is arbitrarily defined as the amount of activity present in 1 μg of crude venom. A straight line is obtained when the log of the clotting time is plotted against the log of venom factor V-activator concentration.

Protein is determined spectrophotometrically at 280 nm using an extinction coefficient of 10.0 for impure preparations of each enzyme. For homogeneous preparations of RVV-V and thrombocytin, extinction coefficients of 15.6[7] and 6.35,[9] respectively, are employed.

Purification Procedure

RVV-V

Step 1. Chromatography on Sephadex G-150. One gram of lyophilized *V. russelli* venom (Sigma Chemical Co., St. Louis, Missouri) is dissolved in 5 ml of 0.05 M sodium acetate (pH 5.0) containing 0.5 M NaCl. The solution is clarified by centrifugation and applied to a Sephadex G-150 column (2.6 × 95 cm) previously equilibrated at room temperature with the same buffer. Protein is eluted from the column by ascending flow at a rate of 35 ml/hr and fractions of 3.5 ml are collected in plastic tubes. Crude venom protein is resolved into three major protein peaks by Sephadex G-150 column chromatography, and RVV-V activity elutes in the ascending portion of the second major protein peak.

Step 2. SP-Sephadex Column Chromatography. SP-Sephadex C-50 (10 g) is allowed to hydrate at room temperature for 1–2 days in excess 0.05 M sodium acetate (pH 5.0) containing 0.5 M NaCl. Resin fines are removed from the swollen ion exchanger by repeated elutriation, and the

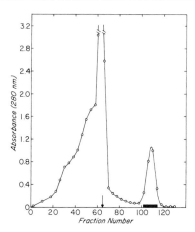

FIG. 1. Elution pattern of RVV-V from SP-Sephadex C-50 column. Protein was eluted from the column (1.6 × 35 cm) with a linear gradient of NaCl generated from 150 ml each of 0.4 *M* and 1.0 *M* NaCl in 0.05 *M* sodium acetate (pH 5.0) as described under Purification Procedure. Fractions (3 ml) were collected at a flow rate of 30 ml/hr. The arrow indicates where the gradient was started. Fractions indicated by the solid bar were pooled.

resin is then equilibrated with 0.05 *M* sodium acetate (pH 5.0) containing 0.25 *M* NaCl. The RVV-V fractions from step 1 are pooled, diluted with an equal volume of 0.05 *M* sodium acetate (pH 5.0) and applied to an SP-Sephadex C-50 column (1.6 × 35 cm) previously equilibrated at room temperature with 0.05 *M* sodium acetate (pH 5.0)–0.25 *M* NaCl. The column is washed with about 150 ml of 0.05 *M* sodium acetate (pH 5.0) containing 0.4 *M* NaCl. RVV-V is eluted from the column with a linear gradient of NaCl arising from 150 ml of 0.05 *M* sodium acetate (pH 5.0)– 0.4 *M* NaCl and 150 ml of 0.05 *M* sodium acetate (pH 5.0)–1.0 *M* NaCl. Fractions of 3 ml are collected in plastic tubes at a flow rate of 30 ml/hr. Figure 1 shows a typical elution pattern. A symmetrical protein peak containing RVV-V activity eluted in the gradient at approximately 0.6 *M* NaCl. The overall yield of RVV-V from 1 g of crude venom is about 25 mg.

Thrombocytin

Thrombocytin is isolated from *B. atrox* venom (Sigma Chemical Co.; Lot No. 47C-0114) by Sephadex G-150 column chromatography and SP-Sephadex ion-exchange column chromatography. The experimental conditions and buffers are identical to those used in the isolation of RVV-V with the exception of the sodium chloride concentration in the wash and gradient buffers of the SP-Sephadex chromatography. *B. atrox* venom is

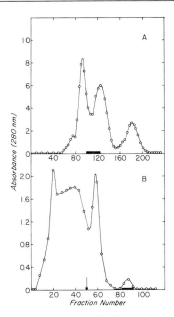

Fɪɢ. 2. Elution patterns of thrombocytin. (A) Pattern from a Sephadex G-150 column. Commercial *B. atrox* venom (1 g) was dissolved in 5 ml of 0.05 *M* sodium acetate (pH 5.0)–0.5 *M* NaCl and centrifuged. The supernatant was applied to a Sephadex G-150 column (2.6 × 95 cm) and eluted with the above buffer. Fractions (3.5 ml) were collected at a flow rate of 35 ml/hr. Fractions bound by the solid bar were pooled and applied to the SP-Sephadex column. (B) Pattern from an SP-Sephadex C-50 column. Protein was eluted from the column (1.6 × 35 cm) with a linear gradient of NaCl arising from 150 ml each of 0.5 *M* and 1.5 *M* NaCl in 0.05 *M* sodium acetate (pH 5.0) as described under Purification Procedure. Fractions (3 ml) were collected at a flow rate of 30 ml/hr. The arrow indicates where the gradient was started. Fractions bound by the solid bar were pooled.

resolved into three major protein peaks by gel filtration and thrombocytin activity eluted between the first and second major protein peaks (Fig. 2A). Thrombocytin (fractions 100–124) from the Sephadex G-150 column are pooled, diluted with an equal volume of 0.05 *M* sodium acetate (pH 5.0), and applied to an SP-Sephadex C-50 column equilibrated with 0.05 *M* sodium acetate (pH 5.0)–0.25 *M* NaCl. The column is washed with about 50 ml of 0.05 *M* sodium acetate (pH 5.0) containing 0.5 *M* NaCl. Thrombocytin is eluted from the column with a linear gradient of NaCl generated from 150 ml each of 0.5 *M* NaCl and 1.5 *M* NaCl in 0.05 *M* sodium acetate (pH 5.0). Thrombocytin elutes in the gradient at approximately 1 *M* NaCl (Fig. 2B). The overall yield of thrombocytin from 1 g of lyophilized *B. atrox* venom is 8–10 mg based on an extinction coefficient of 6.35 for the purified enzyme.[9]

Properties of RVV-V

Physicochemical. The factor V activator from Russell's viper venom is a single-chain protein with MW 27,200 as determined by sedimentation equilibrium under denaturing conditions.[7] The purified preparation is homogeneous by SDS-gel electrophoresis[16] and disc-gel electrophoresis at pH 4.3.[17] The apparent molecular weight of RVV-V by SDS–gel electrophoresis is 28,000 for the unreduced protein and 29,000 for the reduced protein. The amino-terminal sequence of RVV-V is Val-Val-Gly-Gly-Asp-Glu-Cys-Asn-Ile, which is homologous to that found in the NH_2-terminal end of several classical serine proteases.[7] RVV-V is a glycoprotein consisting of 94% protein and 6% carbohydrate. The amino acid and carbohydrate compositions of the enzyme are shown in Table I.

Stability. The procoagulant activity of purified preparations of RVV-V is remarkably stable at temperatures ranging from -70 to $50°C$. The procoagulant activity of RVV-V remains constant for several months at $4°C$ when the enzyme is stored in 50 mM Tris-HCl (pH 7.5) containing 0.02% sodium azide at protein concentrations of 0.5 to 1.5 mg/ml. Furthermore, repeated freezing and thawing has little if any effect on the clotting activity of the enzyme at the above concentrations.[7] When incubated at $85°C$ for 60 min at a concentration of 0.2 mg/ml, RVV-V retains approximately 50% of its coagulant and esterase activity.[8]

Specificity. Earlier studies revealed that RVV-V could increase the coagulant activity of factor V preparations 2- to 10-fold.[6,18,19] Furthermore, this activation process led to an apparent decrease in the molecular weight of factor V as measured by gel-filtration column chromatography.[6,19] More recent studies using homogeneous preparations of bovine factor V and RVV-V clearly demonstrate that the venom protease converts single-chain factor V to factor V_a by the cleavage of a single internal peptide bond in the factor V molecule. The resulting factor V_a is composed of a heavy chain ($M_r = 230,000$) and a light chain ($M_r = 80,000$) as determined by SDS–gel electrophoresis.[20–23] The heavy and light chains of factor V_a are noncovalently associated in solution. Calcium ions apparently mediate their interaction, so that the two chains are readily separated by gel-filtration[24] or ion-exchange chromatography[21,22] in the presence of EDTA.

[16] K. Weber and M. Osborn, *J. Biol. Chem.* **244**, 4406 (1969).

[17] R. A. Reisfeld, U. J. Lewis, and D. E. Williams, *Nature (London)* **195**, 281 (1962).

[18] R. W. Colman, *Biochemistry* **8**, 1445 (1969).

[19] D. J. Hanahan, M. R. Rolfs, and W. C. Day, *Biochim. Biophys. Acta* **286**, 205 (1972).

[20] M. J. Lindhout and C. M. Jackson, *Thromb. Hemostasis* **42**, 491 (1979).

[21] C. T. Esmon, *in* "The Regulation of Coagulation" (K. G. Mann and F. B. Taylor, eds.), p. 137. Elsevier/North-Holland Publ., Amsterdam and New York, 1980.

[22] W. M. Canfield and W. Kisiel, *Circulation* **60**, II-246 (1979).

TABLE I
AMINO ACID AND CARBOHYDRATE
COMPOSITIONS OF RVV-V[a]

Components	RVV-V (residues/27,200 residues)
Amino acid	
Lysine	12
Histidine	10
Arginine	16
Aspartic acid	23
Threonine	15
Serine	15
Glutamic acid	11
Proline	15
Glycine	21
Alanine	12
Half-cystine	12
Valine	16
Methionine	3
Isoleucine	18
Leucine	15
Tyrosine	6
Phenylalanine	6
Tryptophan	6
Carbohydrate	
Fucose	1
Galactose	2
Mannose	1
N-Acetylglucosamine	3
N-Acetylneuraminic acid	1
Protein (%)	94
Carbohydrate (%)	6

[a] From W. Kisiel, *J. Biol. Chem.* **254**, 12230 (1979).

Aside from human and bovine factor V, no other protein substrate for RVV-V has been identified to date. Prolonged incubation of RVV-V with human[4] or bovine[25] factor VIII, bovine fibrinogen,[8] bovine prothrombin,[8] and bovine factor X[8] are without effect on either the covalent structures or activities of these proteins. Rabbit factor V, however, is not activated by RVV-V.[4]

[23] The molecular weight values reported for the heavy and light chains of factor V_a obtained through either RVV-V or thrombin proteolysis vary somewhat depending on the SDS gel electrophoresis system and reference proteins employed.

[24] W. M. Canfield, W. Kisiel, and E. W. Davie, in preparation.

[25] G. Vehar, unpublished data.

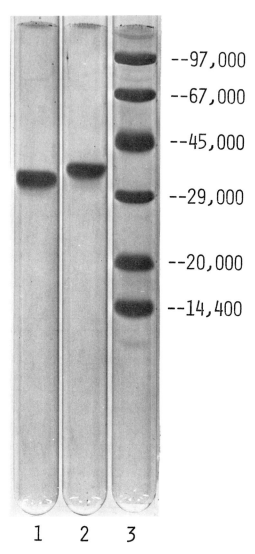

--97,000

--67,000

--45,000

--29,000

--20,000

--14,400

1   2   3

FIG. 3. Sodium dodecyl sulfate–polyacrylamide gel electrophoresis of thrombocytin (TCN) from *B. atrox* venom. Sample 1, 20 μg of TCN; sample 2, 20 μg of reduced TCN; sample 3, mixture of reduced standard proteins including 10 μg each of phosphorylase b (97,000), bovine serum albumin (67,000), ovalbumin (45,000), carbonic anhydrase (29,000), soybean-trypsin inhibitor (20,000), and α-lactalbumin (14,400).

		5						10					15			
Thrombocytin	Val	Ile	Gly Gly	Arg	Val	Cys	Lys	Ile	Asn	Lys	His	Arg	Ser	Leu Val	Leu	Leu	Phe -
RVV-V	Val	Val	Gly Gly	Asp	Glu	Cys	Asn	Ile	Asn	Glu	His	Pro	Ile	Leu Val	Ala	Leu	Tyr -

FIG. 4. Comparison of the amino-terminal sequences of thrombocytin and RVV-V.

RVV-V hydrolyzes arginine esters, such as benzoylarginine ethyl ester[8] and tosylarginine methyl ester.[7] It also possesses amidase activity toward the chromogenic substrate, phenylalanylpipecolylarginyl-p-nitroanilide (S-2238).[7]

Inhibitors. The coagulant and esterase activities of RVV-V are inhibited in parallel by the serine-protease inhibitor, diisopropyl fluorophosphate (DFP).[7,8] Covalent binding of the inhibitor to the enzyme has been confirmed with [³H]DFP.[7] Bovine antithrombin III, in the presence or absence of heparin, has no apparent effect on the esterase or coagulant activity of the enzyme.[7]

Properties of Thrombocytin

Physicochemical. Thrombocytin is a single-chain glycoprotein with MW 36,000 as determined by SDS–gel electrophoresis for the reduced protein.[9] The unreduced protein exhibits MW 33,000 by this technique (Fig. 3). The amino-terminal sequence of thrombocytin is Val-Ile-Gly-Gly-Arg-Val-Cys-Lys,[26] which is homologous to that found in the amino-terminal end of RVV-V (Fig. 4). The amino acid and carbohydrate compositions of thrombocytin are shown in Table II. Thrombocytin isolated in this laboratory from commercial lots of *B. atrox* venom of unspecified variety appears to be more highly glycosylated than the preparation described by Kirby *et al.*[9] obtained from *B. atrox* (*marajoensis*) venom. Interestingly, our preparation of thrombocytin contains no detectable galactose by gas chromatography and mass spectrometry,[27] but contains significant levels of glucose and *N*-acetylgalactosamine. It is uncertain whether the glucose is an intrinsic component of thrombocytin or is derived from the Sephadex column used during its isolation. The presence of *N*-acetylgalactosamine would suggest simple mucin-type sugar units *O*-glycosidically linked to the protein in addition to the more prevalent glycosyl moieties *N*-glycosidically linked to the protein via *N*-acetylglucosamine and asparagine residues.

Stability. The amidolytic and proteolytic activities of purified preparations of thrombocytin are stable for several weeks at 4°C when the en-

[26] W. Kisiel, unpublished data.
[27] H. J. Yang and S. I. Hakomori, *J. Biol. Chem.* **246**, 1192 (1971).

TABLE II
AMINO ACID AND CARBOHYDRATE COMPOSITIONS
OF THROMBOCYTIN[a]

Components	Thrombocytin	
	1[b]	2[c]
Amino acid		
Lysine	25	22
Histidine	7	8
Arginine	16	14
Aspartic acid	31	25
Threonine	16	14
Serine	22	21
Glutamic acid	16	12
Proline	23	21
Glycine	27	22
Alanine	18	16
Half-cystine	16	13
Valine	20	17
Methionine	4	3
Isoleucine	20	20
Leucine	27	24
Tyrosine	5	3
Phenylalanine	5	5
Tryptophan	10	12
Carbohydrate		
Fucose		3
Mannose	4[d]	8
Glucose		5[e]
N-Acetylglucosamine	5[f]	11
N-Acetylgalactosamine		4
N-Acetylneuraminic acid	1	4
Protein (%)	94.4	82.9
Carbohydrate (%)	5.6	17.1

[a] Compositions expressed as residues/36,000 g of glycoprotein.
[b] From E. P. Kirby, S. Niewiarowski, K. Stocker, C. Kettner, E. Shaw, and T. M. Brudzynski, *Biochemistry* **18**, 3564 (1979); nearest integer values.
[c] From W. Kisiel, unpublished data.
[d] Reported as neutral hexose.
[e] Sample dialyzed for 36 hr against 4 liters of 50 mM ammonium bicarbonate and lyophilized prior to carbohydrate analyses.
[f] Reported as N-acetylhexosamine.

zyme is dissolved in 0.05 M sodium acetate (pH 5.0)–1 M NaCl–0.02% sodium azide at protein concentrations of 100–200 μg/ml. The enzyme, however, begins to lose activity after storage for several months under these conditions. Alternatively, the enzyme can be stored at $-80°C$ for several months with no significant loss of activity.

Specificity. The specificity of thrombocytin for protein substrates closely parallels that observed for thrombin. In the activation of single-chain factor V by thrombin, at least two peptide bonds in factor V are cleaved sequentially to produce functional factor V_a composed of a heavy chain ($M_r = 94,000$) and a light chain ($M_r = 80,000$) that are associated by noncovalent bonds.[11,12,23]

Several other protein substrates for thrombocytin have been identified by Niewiarowski *et al.*[10] and these include fibrinogen, prothrombin, factor XIII, and factor VIII. In general, these protein substrates are also good substrates for thrombin. As in the activation of factor V, thrombocytin appears to act on these substrates at a significantly lower rate than that observed for thrombin.

Thrombocytin will also induce platelet aggregation and release of platelet constituents.[10] In addition, thrombocytin has a weak ability to induce hemagglutination activity in washed platelets.[28]

Thrombocytin possesses amidolytic activity against several synthetic paranitoanilides, which include Bz-Ile-Glu-Gly-Arg-pNA,[9] Bz-Phe-Val-Arg-pNA,[10] Tos-Gly-Pro-Arg-pNA,[9] and Phe-Pip-Arg-pNA.[24]

Inhibitors. Thrombocytin is inhibited by diisopropyl fluorophosphate[9,24] by the active-site titrant, p-nitrophenyl-p-guanidobenzoate,[9] and by D-phenylalanyl-L-prolyl-L-arginylchloromethyl ketone.[9] The latter reacts with a reactive histidine in the enzyme.[9] The platelet-aggregating and amidolytic activities of thrombocytin are inhibited simultaneously when the enzyme is incubated with either soybean-trypsin inhibitor or anti-thrombin III in the presence of heparin.[10] Hirudin and α_1-antitrypsin are, however, without any significant effect on thrombocytin.[10]

[28] T. K. Gartner, D. R. Philips, and D. C. Williams, *FEBS Lett.* **113**, 196 (1980).

[23] Prothrombin

By KENNETH G. MANN, JACQUES ELION, RALPH J. BUTKOWSKI,
MICHAEL DOWNING, and MICHAEL E. NESHEIM

The isolation of human and bovine prothrombin and the fragments produced during their activation have been described in an earlier volume.[1,2] This chapter is intended to be an update on prothrombin structure.

Immunochemical assessment of prothrombin levels in human plasma indicates that the proenzyme circulates at a concentration of about 150 mg/liter (2 μM) with a range of 110–212 mg/liter.[3] Detailed analysis has not been provided for the bovine proenzyme; however, plasma concentrations appear to be similar to those of the human based on activity measurements.

The primary structures of the human[4-8] and bovine molecule[9,10] have been elucidated, and these are compared in Fig. 1. The sites of cleavage of the prothrombin molecule by factor X_a and by thrombin are also identified in Fig. 1, as are the disulfide-bond cross-links. The latter are indicated as superscripts above each cysteine residue. When compared to human prothrombin as a standard, bovine prothrombin contains a 1-residue insertion among the first 4 residues. We have chosen to number the sequence represented in Fig. 1 according to the human prothrombin structure. All cleavages designated for the bovine molecule therefore will have the residue identified by the location appropriate for the human molecule. The reader should be aware, however, that if one were to use a

[1] K. G. Mann, this series, Vol. 45, p. 123.

[2] R. L. Lundblad, H. S. Kingdon, and K. G. Mann, this series, Vol. 45, p. 156.

[3] F. C. McDuffie, C. Griffin, R. Niedringhaus, K. G. Mann, C. A. Owen, Jr., E. J. W. Bowie, J. Peterson, G. Clark, and G. G. Hunder, *Thromb. Res.* **16**, 759 (1979).

[4] R. J. Butkowski, Ph.D. Thesis, University of Minnesota, Minneapolis (1974).

[5] J. Elion, R. J. Butkowski, M. R. Downing, and K. G. Mann, *Circulation* **53-54**, Suppl. 2, 118 (1976).

[6] D. A. Walz, D. Hewett-Emmett, and W. H. Seegers, *Proc. Natl. Acad. Sci. U.S.A.* **74**, 1969 (1977).

[7] R. J. Butkowski, J. Elion, M. R. Downing, and K. G. Mann, *J. Biol. Chem.* **252**, 4942 (1977).

[8] M. R. Downing, R. J. Butkowski, M. M. Clark, and K. G. Mann, *J. Biol. Chem.* **250**, 8897 (1975).

[9] C. M. Heldebrant, C. Noyes, H. S. Kingdon, and K. G. Mann, *Biochem. Biophys. Res. Commun.* **54**, 155 (1973).

[10] S. Magnusson, T. E. Petersen, L. Sottrup-Jensen, and H. Claeys, *Cold Spring Harbor Conf. Cell Proliferation* **2**, 123 (1975).

bovine prothrombin numbering system, all bovine cleavage sites would occur at a numerical position one beyond that identified for the human molecule. The carbohydrate content of the bovine molecule has been reported by a number of laboratories.[1,11-13] There appear to be 3 carbohydrate side-chains, 2 of which are located in prothrombin fragment 1, attached to Asn^{76} and Asn^{100}. The third chain is located in the thrombin B-chain region to Asn^{375}.

The prothrombin molecule can be considered to be composed of three domains that are somewhat independent with respect to noncovalent structure.[10,14] The prothrombin fragment 1 domain corresponds to the sequence from the NH_2-terminal up to residue 155. The structure from residues 156 to 273 corresponds to the prothrombin fragment 2 domain. The prothrombin 2 domain corresponds to the sequence containing residue 274 to residue 581. Prethrombin 2 is the precursor of α-thrombin. Cleavage of the Arg^{322}-Ile^{323} bond in prethrombin 2 gives rise to α-thrombin, which is composed of the A chain (residues 274–322) and the B chain (residues 323–581). One additional cleavage occurs in human, but not bovine, α-thrombin. The Arg^{256}-Thr^{257} bond is rapidly cleaved by thrombin, and thus the stable form of human α-thrombin is 13 residues shorter than the stable form of bovine α-thrombin.

During prothrombin activation, factor X_a appears to cleave the two bonds required for activation of the molecule in a sequential fashion.[8,13,15-18] The Arg^{273}-Thr^{274} bond is cleaved first, and this cleavage liberates the entire "pro" end of the prothrombin molecule, prothrombin fragment 1 · 2 (residues 1–273) and prethrombin 2. Factor X_a then cleaves the Arg^{322}-Ile^{323} bond to give rise to α-thrombin. The cleavage that releases prothrombin fragment 1 is caused by thrombin cleavage of the Arg^{155}-Ser^{156} bond. This occurs in both the bovine and the human molecules. When prothrombin is cleaved at Arg^{155}, the remaining segment (residues 156–581) is termed prethrombin 1. The other thrombin cleavage between Arg^{286}-Thr^{287} occurs only in the human molecule.[8]

The amino acid compositions of each of the domains as obtained from the sequence data are contained in Table I. The carbohydrate distribution

[11] S. Magnusson, *Ark. Kemi* **23**, 285 (1965).
[12] B. G. Hudson, C. M. Heldebrant, and K. G. Mann, *Thromb. Res.* **6**, 215 (1975).
[13] W. G. Owen, C. T. Esmon, and C. M. Jackson, *J. Biol. Chem.* **249**, 594 (1974).
[14] J. W. Bloom and K. G. Mann, *Biochemistry* **10**, 1957 (1979).
[15] C. M. Heldebrant, R. J. Butkowski, S. P. Bajaj, and K. G. Mann, *J. Biol. Chem.* **248**, 7149 (1973).
[16] K. S. Stenn and E. R. Blout, *Biochemistry* **11**, 4502 (1972).
[17] H. Pirkle and I. Theodor, *Thromb. Res.* **5**, 511 (1974).
[18] W. Kisiel and D. J. Hanahan, *Biochem. Biophys. Res. Commun.* **59**, 570 (1974).

1   10

Human Ala Asn Thr   Phe Leu GLA GLA Val Arg Lys Gly Asn Leu GLA
Bovine   Lys Gly

Human Leu GLA Ser Ser Thr Ala Thr   Asp Val Phe Trp Ala Lys Tyr Thr
Bovine   Leu Ser

                                70   CHO
            143 |

Human Glu Gly Asn Cys Ala Glu Gly Leu Gly Thr Asn Tyr Arg Gly Asn
Bovine   Val   Met
                                   CHO
                                   |

Human Tyr Pro His Lys Pro Glu Ile   Asn Ser Thr Thr His Pro Gly Ala
Bovine

                              130
           86   *114*

Human Pro Trp Cys Tyr Thr Thr Asp Pro Thr Ala Arg Arg Gln Glu Cys
Bovine   Ser   Leu   Glu

               a
               ↓

Human Arg Ser Glu Gly Ser Ser Val Asn Leu Ser Pro Pro Leu Glu Gln
Bovine   Gly   Thr Thr Ser Gln   Leu   Thr

                     190
                         231

Human Thr His Gly Leu Pro Cys Leu Ala Trp Ala Ser Ala Gln Ala Lys
Bovine   Ser   Ser Arg   Ser   Glu

                                          230
           243   *191*

Human Asn Phe Cys Arg AsnPro Asp Gly Asp Glu Glu Gly Val Trp Cys
Bovine   Ala

           250
          170

Human Cys Glu Glu Ala Val Glu Glu Glu Thr Gly Asp Gly Leu Asp Glu
Bovine   Pro   Asp Gly Asp Leu   Arg   Gly

                                       c   290

Human Tyr Gln Thr Phe Phe Asn Pro Arg↓Thr Phe Gly Ser Gly Glu Ala
Bovine Phe   Pro   Glu Lys   Ala

      310   *d*
                                               ↓

Human Thr Glu Arg Glu LeuLeu Glu Ser Tyr Ile Asp Gly Arg Ile Val
Bovine   Gln Lys   Phe   Glu

FIG. 1. The primary structures of human and bovine prothrombin are presented, with the numbering system based on the human molecule. The numbers appearing in italics above the Cys residues correspond to the identification of Cys pairs in disulfide bridges. The arrows labeled *a, b, c,* and *d* correspond to the sites of cleavage for both molecules by factor X_a (at positions *b* and *d*; by thrombin for both molecules at position *a*; and for human prothrombin

                                                                    30
        22   17
Arg GLA Cys Val GLA GLA Thr Cys Ser Tyr GLA GLA Ala Phe GLA Ala
            Leu   Pro   Arg
                    50
    60   47
Ala Cys Glu Thr Ala Arg Thr Pro Arg Asp Lys Leu Ala Ala Cys Leu
            Ser   Asn   Glu   Asn Glu
                                                            90
                                        126
Val Ser Ile Thr Arg Ser Gly Ile Glu Cys Gln Leu Trp Arg Ser Arg
        Val
        110
                        138
Asp Leu Gln Glu Asn Phe Cys Arg Asn Pro Asp Ser Ser Ile Thr Gly
        Arg   Gly
                    65   150
Ser Thr Pro Val Cys Gly Gln Asp Gln Val Thr Val Met Val Thr Pro
    Val   Arg   Glu   Ile
170
248
Cys Val Pro Asp Arg Gly Gln Gln Tyr Gln Gly Arg Leu Ala Val Thr
                        Arg Glu   Arg
                                        210

Ala Leu Ser Lys His Gln Asp Phe Asn Ser Ala Val Gln Leu Val Glu
            Asp   Pro   Pro   Ala
                                        219
Tyr Val Ala Gly Lys Pro Gly Asp Phe Gly Tyr Cys Asp Leu Asn Tyr
            Asp Gln   Glu
                            270   b
Asp Ser Asp Glu Glu Arg Ala Ile Glu Gly Arg ↓Thr Ala Thr Ser Glu
    Pro   Pro Asp Ala   Ser Glu Asp His
441
Asn Cys Gly Leu Arg Pro Leu Phe Glu Lys Lys Ser Leu Glu Asn Lys
Asp   Gln Val Gln Asp Glu
                    330
Glu Gly Ser Asn Ala Glu Ile Gly Met Ser Pro Trp Gln Val Met Leu
        Gln Asp   Val   Leu

(continued)

by thrombin at position *c*). The active-site residues characteristic of members of the mammalian serine-protease family are encircled, and carbohydrate side-chains are identified by the
notations CHO. An octapeptide sequence that is repeated twice and is totally conserved in
both the bovine and human molecules is underlined.

							350 366							
Human Phe	Arg	Lys	Ser	Pro	Gln	Glu	Leu	Leu	Cys	Gly	Ala	Ser	Leu	Ile
Bovine														

CHO
|

Human Trp	Asn	Lys	Asn	Phe	Thr	Glu	Asn	Asp	Leu	Leu	Val	Arg	Ile	Gly
Bovine	Asx					Val	Asp							
							410								

Human Leu	Glu	Lys	Ile	Tyr	Ile	His	Pro	Arg	Tyr	Asn	Trp	Arg	Glu	Asn
Bovine	Asp											Lys		

					295									
Human Ser	Asp	Tyr	Ile	His	Pro	Val	Cys	Leu	Pro	Asn	Arg	Glu	Thr	Ala
Bovine									Asp	Lys	Gln			
				470										

Human Asn	Leu	Lys	Ser	Thr	Val	Thr	Ala	Asp	Val	Gly	Lys	Gly	Gln	Pro
Bovine	Arg	Arg	Glu		Trp		Thr	Ser		Ala	Glu	Val		
													510 495	

Human Lys	Asp	Ser	Thr	Arg	Ile	Arg	Ile	Thr	Asp	Asn	Met	Phe	Cys	Ala
Bovine	Ala							Asn	Asp					
			530											

Human (Ser)	Gly	Gly	Pro	Phe	Val	Met	Lys	Ser	Pro	Phe	Asn	Asn	Arg	Trp
Bovine								Tyr						
									570					

Human Lys	Tyr	Gly	Phe	Tyr	Thr	His	Val	Phe	Arg	Leu	Lys	Lys	Trp	Ile
Bovine														

FIG. 1 (continued)

as presented in Table I is based on the bovine composition data, and presumes a symmetrical 3-chain distribution: that is, two-thirds of the carbohydrate mass in prothrombin fragment 1 and one-third of the carbohydrate in prethrombin 2.

Prothrombin Fragment 1

Prothrombin fragment 1 is the vitamin K-dependent portion of the prothrombin molecule.[19–21] The 10 GLAs at positions 6, 7, 14, 16, 19, 20, 25, 26, 29, and 32 correspond to γ-carboxyglutamic acid, which arises from

[19] J. Stenflo, P. Fernlund, W. Egan, and P. Roepstorff, *Proc. Natl. Acad. Sci. U.S.A.* **71**, 2730 (1974).
[20] G. L. Nelsestuen, T. H. Zytkovicz, and J. B. Howard, *J. Biol. Chem.* **249**, 6347 (1974).
[21] S. Magnusson, L. Sottrup-Jensen, T. E. Petersen, H. R. Morris, and A. Dell, *FEBS Lett.* **44**, 189 (1974).

370

350
Ser Asn Arg Trp Val Leu Thr Ala Ala (His) Cys Leu Leu Tyr Pro Pro
　　Asp
　　　　390

Lys His Ser Arg Thr Arg Tyr Glu Arg Asn Ile　Glu Lys Ile Ser Met
　　　　　　　　　　　　　　Lys Val
　　　　　　　　　　　　　　　　　430

Leu (Asp) Arg Asp Ile Ala Leu Met Lys Leu Lys Lys Pro Val Ala Phe
　　　　　　　　　　Leu　　　　Arg　　Ile Glu Leu
450

Ala Ser Leu Leu Gly Ala Gly Tyr Lys Gly Arg Val Thr Gly Tyr Gly
　　Lys　　　His　　　Phe　　　　　　　　　Trp
　　　　　　　　　　　　490

509
Ser Val Leu Gln Val Val Asn Leu Ala Leu Val Gln Arg Pro Val Cys
　　　　　　　　　　　　Pro　　　　Glu

553
Gly Tyr Lys Pro Asp Glu Gly Lys Arg Gly Asp Ala Cys Glu Gly Asp
　　　　　　Gly
　　　　　　　　550

523
Tyr Gln Met Gly Ile Val Ser Trp Gly Glu Gly Cys Asp Arg Asp Gly
　　　　　　　　　　　　　　　　　　　　Asn

Gln Lys Val Ile Asp Gln Phe Gly Glu
　　　　　　　Arg Leu　Ser

Fig. 1 (continued)

glutamic acid by virtue of a postribosomal synthetic event prior to prothrombin secretion.[22-26] The GLA residues confer on the prothrombin
fragment 1 segment of the molecule the ability to bind calcium ions,[25,27-29]
and through this bound Ca^{2+} to bind acidic phospholipids.[30] The signifi-

[22] P. O. Ganrot and J. E. Nilehn, Scand. J. Clin. Lab. Invest. 22, 23 (1968).
[23] F. Josso, J. M. LaVergne, M. Gouault, O. Prou-Wartelle, and J. P. Soulier, Thromb. Diath. Haemorrh. 20, 88 (1968).
[24] A. J. Quick, M. Stanley-Brown, and F. W. Bancroft, Am. J. Med. Sci. 190, 501 (1935).
[25] G. L. Nelsestuen and J. W. Suttie, J. Biol. Chem. 247, 8176 (1972).
[26] J. W. Suttie, G. A. Grant, C. T. Esmon, and D. V. Shah, Mayo Clin. Proc. 49, 933 (1974).
[27] S. P. Bajaj, R. J. Butkowski, and K. G. Mann, J. Biol. Chem. 250, 2150 (1975).
[28] R. Benarous, J. Elion, and D. Labie, Biochimie 58, 391 (1976).
[29] R. A. Henriksen and C. M. Jackson, Arch. Biochem. Biophys. 170, 149 (1975).
[30] S. N. Gitel, W. G. Owen, C. T. Esmon, and C. M. Jackson, Proc. Natl. Acad. Sci. U.S.A. 70, 1344 (1973).

TABLE I
AMINO ACID COMPOSITIONS FOR HUMAN AND BOVINE PROTHROMBIN[a]

	Prothrombin	Prothrombin fragment 1	Prothrombin fragment 2	Prethrombin 2
Aspartic acid	29 (35)	6 (4)	9 (13)	14 (18)
Asparagine	31 (25)	8 (10)	6 (5)	17 (10)
Threonine	36 (29)	18 (10)	3 (5)	15 (14)
Serine	37 (36)	10 (11)	8 (9)	19 (16)
Glutamic acid	40 (43)	7 (11)	13 (11)	20 (21)
Glutamine	22 (18)	5 (2)	8 (4)	9 (12)
γ-Carboxy glutamate	10 (10)	10 (10)	— (—)	— (—)
Proline	30 (35)	9 (10)	6 (9)	15 (16)
Cysteine	24 (24)	10 (10)	6 (6)	8 (8)
Glycine	48 (48)	9 (11)	12 (12)	27 (25)
Alanine	37 (34)	12 (10)	10 (10)	15 (14)
Valine	34 (35)	8 (9)	8 (5)	18 (21)
Methionine	8 (6)	1 (1)	— (—)	7 (5)
Isoleucine	21 (20)	4 (4)	1 (1)	16 (15)
Leucine	43 (46)	8 (10)	9 (9)	26 (27)
Tyrosine	22 (19)	5 (4)	4 (4)	13 (11)
Phenylalanine	20 (20)	4 (4)	3 (3)	13 (13)
Lysine	29 (31)	4 (5)	3 (2)	22 (24)
Histidine	9 (9)	2 (2)	2 (—)	5 (7)
Arginine	39 (45)	12 (15)	5 (8)	22 (22)
Tryptophan	12 (14)	3 (3)	2 (2)	7 (9)
Total	581 (582)	155 (156)	118 (118)	308 (308)
Carbohydrate (%)[b]	8.1 (8.2)	18.1 (18.1)	— (—)	(5.2)(5.2)

[a] Bovine composition is presented in parentheses.
[b] Human carbohydrate content is estimated based on similarity to the bovine molecule.

cance of the γ-carboxyglutamic acid containing fragment 1 region of the molecule to metal-ion binding and lipid binding is a matter of current investigation by a number of laboratories. This domain of the molecule can be shown to bind a number of different metal ions with moderate affinity, and with a variety of stoichiometries.[31–33] On binding of metal ion, conformational changes[32,34–36] and dimerization[32,37] of the prothrom-

[31] B. C. Furie, K. G. Mann, and B. Furie, J. Biol. Chem. 251, 3235 (1976).
[32] F. G. Prendergast and K. G. Mann, J. Biol. Chem. 252, 840 (1977).
[33] S. P. Bajaj, T. Nowak, and F. J. Castellino, J. Biol. Chem. 251, 6294 (1976).
[34] G. L. Nelsestuen, J. Biol. Chem. 251, 5648 (1976).
[35] J. W. Bloom and K. G. Mann, Biochemistry 17, 4430 (1978).
[36] K. A. Koehler, M. M. Sarasua, P. Robertson, Jr., and R. C. Hiskey, in "The Regulation of Coagulation" (K. G. Mann and F. B. Taylor, Jr., eds.), p. 75. Elsevier-North-Holland Publ., Amsterdam, 1980.
[37] C. M. Jackson, C.-W. Peng, G. M. Brenckle, A. Jonas, and J. Stenflo, J. Biol. Chem. 254, 5020 (1979).

TABLE II
METAL-ION BINDING TO PROTHROMBIN
FRAGMENT 1

Metal ion	K_d (M)	No. of sites
Calcium[27]	6.3×10^{-4}	5–6
Calcium[29]	6.3×10^{-4}	10–12
Calcium[28]	1.3×10^{-4}	5
Magnesium[32]	2.0×10^{-4}	—
Manganese[33]	2.2×10^{-5}	2
Gadolinium[31]	1.6×10^{-7}	2

bin fragment 1 region have been observed, and though there is not unanimity with respect to the significance of these events to phospholipid binding, there is reasonably good agreement that metal-ion binding is a prerequisite to association of the prothrombin fragment 1 region of the molecule with phospholipid vesicles. Table II[38–40] presents current data with respect to metal-ion binding by prothrombin fragment 1, and Table III illustrates current quantitative data for binding to phospholipid vesicles.

Prothrombin Fragment 2

The prothrombin fragment 2 domain has weak calcium-binding ability,[27] and appears to be related somehow to factor V (factor V_a) interactions of the prothrombin molecule during activation.[27,41] There is a good deal of internal homology between prothrombin fragment 1 and prothrombin fragment 2, with two major exceptions: (1) the γ-carboxyglutamic acid-containing NH_2-terminal portion of the prothrombin fragment 1 has no equivalent in prothrombin fragment 2; (2) prothrombin fragment 2 possesses 13 additional residues at the COOH-terminal, which do not appear to have an equivalent homologous structure in prothrombin fragment 1. The remaining structure, however, is remarkably similar among the 2 prothrombin fragments. Twenty-four percent of the residues are identical, and a sequence of 8 amino acids (residues 111–118 in prothrombin fragment 1, and 216–233 in prothrombin fragment 2) are identical in the 2 fragments from both the human and the bovine species. The cysteine residues also occupy identical positions in both bovine and human pro-

[38] G. L. Nelsestuen and T. K. Lim, *Biochemistry* **16**, 4164 (1977).
[39] F. A. Dombrose, S. N. Gitel, K. Zawalich, and C. M. Jackson, *J. Biol. Chem.* **254**, 5027 (1979).
[40] J. W. Bloom, M. E. Nesheim, and K. G. Mann, *Biochemistry* **18**, 4419 (1979).
[41] C. T. Esmon and C. M. Jackson, *J. Biol. Chem.* **249**, 7791 (1974).

TABLE III
PROTHROMBIN BINDING TO ACIDIC PHOSPHOLIPID IN THE
PRESENCE OF CALCIUM

Phospholipid vesicle components[a]	[Ca^{2+}] (mM)	K_d	n^b
25% PS–75% PC[38]	1	0.9×10^{-6}	9.2
25% PS–75% PC[39]	2	2.3×10^{-6}	13.3
50% PG–50% PC[40]	1	1.6×10^{-6}	19.0

[a] PS, phosphatidylserine; PC, phosphatidylcholine; PG, phosphatidylglycerol.
[b] PS per prothrombin.

thrombin, and the 3 disulfide bridges found in prothrombin fragment 2 are homologous with the last 3 disulfide bridges of prothrombin fragment 1.[10]

Prethrombin 2 and α-Thrombin

The third domain in prothrombin corresponds to prethrombin 2, the immediate precursor of thrombin. During prothrombin activation, prethrombin 2 is cleaved at Arg[322]. This cleavage gives rise to the A chain-B chain structure of thrombin, and permits expression of the active site of the enzyme. Thrombin is homologous in many structural features with the other members of the mammalian serine-protease family.[42,43] Twenty-eight percent of the residues are identical when one compares thrombin and chymotrypsin. Similar comparisons indicate a 24% identity with bovine trypsin, and a 19% identity with porcine elastase. However, trypsin and elastase have no equivalent to the A chain of thrombin. The familiar catalytic residues identified in all serine proteases are found in thrombin at His[365], Asp[419], and Ser[527] (corresponding respectively to residues 57, 102, and 195 in chymotrypsinogen). The primary binding site of trypsin is highly homologous to that of thrombin. A comparison of the similarities between chymotrypsin and thrombin reveals 6 major insertions (Table IV) in the thrombin molecule, which may represent structures responsible for a more elaborate binding site in the enzyme, or its function in other biological systems. The explicit functions for these insertions have not been identified.

A direct comparison of human and bovine thrombin shows the expected homologies between the species. 15.9% of the residues represent

[42] B. S. Hartley, *Philos. Trans. R. Soc. London, Ser.* B **257**, 77 (1970).
[43] J. Elion, M. R. Downing, R. J. Butkowski, and K. G. Mann, in "Chemistry and Biology of Thrombin" (R. L. Lundblad, J. W. Fenton II, and K. G. Mann, eds.), p. 97. Ann Arbor Sci. Publ., Ann Arbor, Michigan, 1977.

TABLE IV
MAJOR INSERTIONS IN HUMAN THROMBIN AS
COMPARED TO CHYMOTRYPSIN

NH$_2$-Terminal extension	287(274)[a]–294
A Loop	306–320
B Loop	370–382
C Loop	469–476
D Loop	487–494
E Loop	512–520

[a] Bovine thrombin.

substitutions. The highest rate of substitution occurs in the A-chain of the molecule, and most of these are near the NH$_2$-terminal of the A-chain. Of particular significance is the substitution of an Arg for a Lys at residue 286. This substitution makes this particular bond thrombin-sensitive; hence, human α-thrombin, as purified, has this 13-residue segment deleted from the NH$_2$-terminal of the A-chain. Each of the thrombin cleavage sites in the human and bovine prothrombin molecule are preceded by the sequence Pro-Arg, and are followed by the sequence Ser (or Thr), ——, Gly, Ser. In contrast, the factor X$_a$ cleavage sites are preceded by the sequence Ile, Glu (or Asp), Gly, Arg.

β and γ-Thrombin

Degraded forms of human and bovine α-thrombin have been observed by a variety of investigators.[44–46] Bovine β-thrombin has been carefully studied and the cleavage sites of this derivative form of the molecule have been identified.[47,48] In the formation of bovine β-thrombin from α-thrombin, the A-chain has been cleaved at two points corresponding to Lys[285] and Lys[303]. The B-chain of bovine thrombin is cleaved at Lys[387] and Lys[396]. This results in a 3-chain bovine β-thrombin structure, which contains a disulfide-bridged A-chain remnant linked to a B$_2$-chain (residues 397–581); and a noncovalently associated B$_1$-chain fragment

[44] K. G. Mann and C. W. Batt, *J. Biol. Chem.* **244**, 6555 (1969).

[45] R. D. Rosenberg and D. F. Waugh, *J. Biol. Chem.* **245**, 5049 (1970).

[46] J. W. Fenton, II, B. H. Landis, D. A. Walz, and J. S. Finlayson, *in* "The Chemistry and Biology of Thrombin" (R. L. Lundblad, J. W. Fenton II, and K. G. Mann, eds.), p. 43. Ann Arbor Sci. Press, Ann Arbor, Michigan, 1977.

[47] H. S. Kingdon, C. M. Noyes, and R. L. Lundblad, *in* "The Chemistry and Biology of Thrombin" (R. L. Lundblad, J. W. Fenton II, and K. G. Mann, eds.), p. 91. Ann Arbor Sci. Press, Ann Arbor, Michigan, 1977.

[48] R. L. Lundblad, C. Noyes, K. G. Mann, and H. S. Kingdon, *J. Biol. Chem.* **254**, 8524 (1979).

(residues 323–387), which contains the active-site histidine (His[365]). The cleavage site in human β-thrombin has been tentatively identified as Arg[395] in the human B-chain, whereas the cleavage site giving rise to human γ-thrombin has been tentatively identified as Lys[476] in the human B-chain.[46] The deletion of peptide regions in human thrombin that correspond to the deletions resulting from cleavages at Lys[303] and Lys[387] in bovine β-thrombin have not been reported. However, it is likely that these cleavages have also occurred because these two Lys bonds and their surrounding sequences are conserved when one compares the human and bovine enzymes.

It is unlikely that these proteolyzed forms of thrombin are relevant structures in blood-coagulation processes or that they arise from autolysis per se. They do represent useful derivatives for analysis of thrombin structure and function, since most of the clotting activity is lost in these two derivatives, whereas the catalytic site and ability to hydrolyze small substrates is still present in these derivative forms of the enzyme.[46,49,50]

Two additional modified thrombin species have been reported that correspond to cleavage of prothrombin at Arg[322] by a snake venom protease obtained from *Echis carinatus* venom.[51,52] This cleavage results in a thrombin derivative in which the whole "pro" end of the molecule remains covalently attached by virtue of disulfide bonds to the catalytic B-chain unit. In the absence of inhibitors, this product is not stable, since the *E. carinatus* thrombin produced will autolytically cleave the fragment 1 region by cleaving after Arg[155]. These two snake venom-derived species then correspond to the 2 chain molecules: Meizo thrombin,[53] residues 1–322 disulfide bridged to the B chain; and Meizo thrombin *des* fragment 1 with a light chain corresponding to residues 156–322.

Procedures for the isolation of each of the prothrombin domains and α- and β-thrombin have been described in detail in Volume 45 of this series.[1,2] Complete knowledge of the sequences, however, allow an update of the physical properties of the prothrombin molecule. These are presented in Table V.[32,54–58] For Table V, it should be noted that the total

[49] K. G. Mann, C. M. Heldebrant, and D. N. Fass, *J. Biol. Chem.* **246**, 5994 (1971).

[50] R. L. Lundblad, L. C. Uhteg, C. N. Vogel, H. S. Kingdon, and K. G. Mann, *Biochem. Biophys. Res. Commun.* **66**, 482 (1975).

[51] B. R. Franza, D. L. Aronson, and J. S. Finlayson, *J. Biol. Chem.* **250**, 7057 (1975).

[52] T. Morita, S. Iwanaga, and T. Suzuki, *J. Biochem.* (*Tokyo*) **79**, 1809 (1976).

[53] Nomenclature Committee, *Int. Soc. Thromb. Hemostasis, 1975* (1975).

[54] J. S. Ingwall and H. A. Scheraga, *Biochemistry* **8**, 1860 (1969).

[55] F. Lamy and D. F. Waugh, *J. Biol. Chem.* **203**, 489 (1953).

[56] W. H. Seegers, E. Marciniak, R. K. Kipfer, and K. Yasunaga, *Arch. Biochem. Biophys.* **121**, 372 (1967).

[57] A. C. Cox and D. J. Hanahan, *Biochim. Biophys. Acta* **207**, 49 (1970).

[58] D. J. Winzor and H. A. Scheraga, *J. Phys. Chem.* **68**, 338 (1964).

TABLE V
PHYSICAL PROPERTIES OF PROTHROMBIN AND ITS ACTIVATION FRAGMENTS

	Molecular weight (g/mol)	Partial specific volume (ml/g)	$S_{20,w}^{0}$ [a]	Extinction[a] coefficient ($E_{280}^{1\%}$)	Peptide molecular weight (g/mol)	Mean residue weight (g/residue)
·othrombin						
Human	71,600	0.713	—	14.7[57]	65,685	113.1
Bovine	72,100	0.714	4.6 – 4.8[54,55]	14.4	66,183	113.7
·othrombin fragment 1						
Human	21,700	0.691	—	11.9	17,759	114.6
Bovine	22,000	0.697	2.3[32]	10.5	17,995	115.4
·othrombin fragment 2						
Human	12,866	0.705	—	—	12,866	109.0
Bovine	12,791	0.702	—	12.5	12,791	108.4
·rethrombin 1						
Human	49,900	0.722	—	17.8	47,944	112.5
Bovine	50,200	0.722	3.9[56]	16.4	48,206	113.2
·rethrombin 2						
Human	37,000	0.727	—	17.3	35,096	113.9
Bovine	37,400	0.729	—	19.5[58]	35,433	115.0

[a] All values, except those otherwise cited, are from Ref. 1.

molecular weights reported are based on the known sequences and reported (partial) carbohydrate information. In the absence of explicit information with respect to the carbohydrate structures present on prothrombin fragment 1 and on the thrombin B-chain, some small error is possible in the total molecular weights reported.

Assays for Prothrombin and Thrombin

The classical assay for prothrombin is based on the conversion of the proenzyme to thrombin, and subsequent assay for thrombin using fibrinogen as a substrate.[59] It is to this assay that all subsequent methods must be compared. The procedures for these assays have been detailed in Volume 45 of this series.[1,2] More recently, a number of synthetic substrates have been synthesized based on knowledge of the specificity of thrombin-cleavage sites in natural products, which have proven to be of use for quantitation of thrombin.

Some of the more popular substrates include S2160 (Bz-Phe-Val-Arg-pNA-HCl[60,61]), S2238 (H-D-Phe-Pip-Arg-pNA-2HCl[60,61]), Chromozym TH

[59] A. G. Ware and W. H. Seegers, *Am. J. Clin. Pathol.* **19**, 471 (1949).
[60] KABI Peptide Research, Mölndal, Sweden.
[61] Ortho Diagnostics, Inc., Raritan, New Jersey.

(CBz-Gly-Pro-Arg-pNA-HCl[62]   or   Tos-Gly-Pro-Arg-pNA-HCl[63]),   and sarcosyl-Pro-Arg-pNA-HCl.[64] These substrates show reasonably good specificity for α-thrombin when compared to factor X_a, but should be used with caution for two reasons. The first of these is related to our experience with S2238. This substrate is hydrolyzed only very slowly by factor X_a. However, it is an uncompetitive inhibitor of factor X_a with K_i approximately 10^{-5} M.[65] Therefore, S2238 should not be present in the reaction mixture during the conversion step of prothrombin to thrombin, since factor X_a will be inhibited. We have not analyzed the other synthetic substrates for similar properties, and caution is advised in this matter. The second problem area has to do with the specificity of the substrates and mixtures of α, β-, and γ-thrombin. S2160 and S2238 are hydrolyzed at appreciable rates[66] by β- and γ-thrombin, while these two degraded forms of thrombin clot fibrinogen only very slowly.[46,47] Thus evaluation of the clotting activity in a thrombin preparation cannot be easily assessed by use of the synthetic substrates.

Most of our experience has been with the use of S2160 and S2238 in assays of thrombin itself, and in the production of thrombin from prothrombin. Of these two substrates, the reagent of choice is S2238, because of its better solubility properties. In a typical experiment in which one wishes to observe the generation of thrombin from prothrombin, aliquots are withdrawn from the reaction mixture, and added to a cuvette containing 0.05 mM S2238–5 mM EDTA–0.02 M Tris-HCl–0.15 M sodium chloride, pH 7.4. The progress of the reaction is observed by monitoring the change in absorbance at 405 nm. The EDTA present in the reaction mixture quenches the influence of calcium on prothrombin conversion by the prothrombinase complex, thus providing for a reasonably static assessment of the amount of thrombin present at any point in the reaction progress. At physiological concentrations of prothrombin and factor X_a, the rate of thrombin production is altered by approximately 280,000-fold when saturating levels of the cofactors factor V_a, phospholipid, and calcium, are added to the reaction mixture. Since EDTA destroys the influence of all cofactors, the reaction rate is enormously attenuated.

We have recently developed a new procedure for monitoring the progress of prothrombin-activation reactions by use of the reagent DAPA

---

[62] Boehringer Mannheim Biochemical, Indianapolis, Indiana.
[63] Pentapharm., Ltd., Basel, Switzerland.
[64] Abbott Laboratories, North Chicago, Illinois.
[65] M. E. Nesheim and K. G. Mann, unpublished material.
[66] E. F. Workman, Jr., L. C. Uhteg, H. S. Kingdon, and R. L. Lundblad, in "The Chemistry and Biology of Thrombin" (R. L. Lundblad, J. W. Fenton II, and K. G. Mann, eds.), p. 23. Ann Arbor Sci. Press, Ann Arbor, Michigan, 1977.

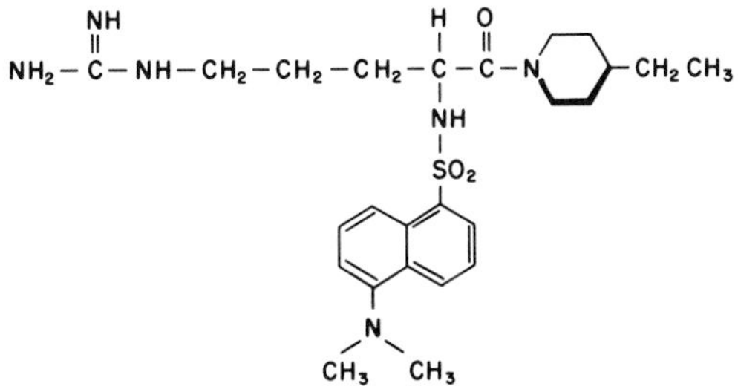

FIG. 2. Structural representation of DAPA: dansylarginine N-(3-ethyl-1,5-pentanediyl)-amide.

[dansylarginine N-(3-ethyl-1,5-pentanediyl)amide[67,68]]. The structure of DAPA is presented in Fig. 2. DAPA is a fluorescent competitive inhibitor of thrombin that binds to the active site of the enzyme in 1:1 stoichiometry. The dissociation constant for the inhibitor binding to the bovine enzyme has been determined to be $4.3 \times 10^{-8} M$.[68] In addition to inhibiting the enzyme, the bound probe exhibits enhanced fluorescence intensity, fluorescence lifetime, and altered excitation and emission spectra. As a consequence of various processes, including energy transfer from a tryptophan in the active site of thrombin,[69] the bound DAPA exhibits enhanced fluorescence when compared to the free inhibitor. And, as one would anticipate, the fluorescence polarization value obtained for the bound probe is also increased. In addition to these properties, DAPA does not inhibit factor X_a, nor has it been observed to inhibit other blood-coagulation serine proteases to any significant extent. Although the studies are preliminary, the enzymes tested thus far include Hageman factor, kallikrein, and factor XI_a[70]; factor IX_a, and factor VII_a.[71] As a consequence of these properties, DAPA provides an elegant method for examining the kinetics of prothrombin conversion to thrombin; since the reagent, when present in a prothrombin-converting reaction mixture, will bind to the thrombin as it is produced and display enhanced fluorescence intensity that can be monitored in a direct analysis of the number of thrombin

[67] S. Okamoto, A. Hijikata, K. Kinjo, R. Kikumoto, K. Ohkubo, S. Tonomura, and T. Kobe, *J. Med. Sci.* **21**, 43 (1975).

[68] M. E. Nesheim, F. G. Prendergast, and K. G. Mann, *Biochemistry* **18**, 996 (1979).

[69] D. L. Straight, M. E. Nesheim, K. G. Mann, and R. L. Lundblad, *Circulation* **62**, Suppl. 2, 1064 (1980).

[70] A. Kaplan, SUNY Health Science Center, Stonybrook, New York (personal communication).

[71] Y. Nemerson, Mt. Sinai School of Medicine, New York, New York (personal communication).

active sites produced.[72] Thus in the presence of DAPA, the progress of zymogen activation to enzyme can be monitored on a continuous basis.

Unfortunately, no commercial source of DAPA is presently available, and the compound must be synthesized. Fortunately, however, the synthesis is rather straightforward.

Synthesis of DAPA

Materials. Dansylarginine is obtained from Pierce Chemical, Rockford, Illinois. N,N'-carbonyldiimidazole is obtained from Sigma Chemical Company, St. Louis, Missouri, and is purified by sublimation. The material is heated to 90°C at a pressure of 20 μm until about 10% of the mass is sublimed. This material is discarded. Heating is then continued at 115°C at the same pressure until approximately all but 10% of the material remaining from the previous step has sublimed. The sublimate, collected by heating at 115°C is stored desiccated at 22°C in the dark. 4-Ethylpiperidine was obtained as a gift from Dr. William Sowers of Reilly Tar Chemical Co., Indianapolis, Indiana, as this material is not available commercially. 4-Methylpiperidine is obtained from Aldrich Chemical Co., Milwaukee, Wisconsin. This compound may be used in place of 4-ethylpiperidine to yield an inhibitor with properties similar to DAPA. Dimethyl sulfoxide, acetonitrile, and ethyl acetate are obtained from Burdick and Jackson, Muskegon, Michigan. Dimethyl sulfoxide is dried over molecular sieves (No. 4, Aldrich) prior to use. Heptane is obtained from Fisher Chemicals. The hydrophobic resin used for isolation of the compound is Mallinckrodt XAD-2, obtained from Curtin Scientific, Minneapolis, Minnesota. This resin must be cleaned prior to its initial use. It is initially suspended in water and allowed to settle; fines are removed by decantation. This is repeated until the supernatant is clear. The resin is then washed as follows: initially a mixture of hot chloroform–acetone (1 : 2) is used. This is followed by a second wash with hot acetone. The resin is then washed three times with water followed by 0.5 N sodium hydroxide, again followed by water, then 0.5 N HCl, and then water once more. It is then washed with warm acetic acid and finally, with copious quantities of water. The material is then stored suspended in water. For subsequent uses it can be regenerated with H_2O, followed by glacial acetic acid, and again H_2O in that order.

Seventy-five milligrams of dansylarginine and 180 mg of N,N'-carbonyldiamidizole are dissolved at 22°C in 0.6 ml of dried dimethyl sulfoxide at 22°C. The solution is incubated for 10 min and 180 μl of 4-ethylpiperidine (or 4-methylpiperidine) are added, and incubation is continued at 22°C for 4 hr. Then 2 ml of 0.15 M sodium chloride is added,

[72] M. E. Nesheim, J. B. Taswell, and K. G. Mann, *J. Biol. Chem.* **254**, 10952 (1979).

and the mixture is extracted with 3 ml, and then 1 ml of ethyl acetate. The phases are separated by gentle centrifugation if necessary. The combined ethyl acetate extract (upper phase) is backwashed three times with water, and phase separation is facilitated by centrifugation if necessary. The ethyl acetate extracts are taken to dryness with a stream of nitrogen. The resulting oil is dried under vacuum (20 μm) for 2–3 hr to remove traces of water. The residue is dissolved in 3 ml of ethyl acetate, and 6 ml of heptane is added slowly with stirring, resulting in the appearance of a solid to semisolid material. Precipitation of this material may be facilitated by chilling to 0°C, overnight if necessary. When the liquid phase has clarified, it is decanted and the remaining solid is washed with 5 ml of heptane. Residual solvent in the residue is removed by a stream of nitrogen. The relatively dry residue is dissolved in 10 ml of 0.15 M HCl, and added over a period of approximately 30 min at 22°C to a column (1 × 10 cm) of XAD-2 hydrophobic resin equilibrated in water. Nonhydrophobic materials, including excess hydrochloric acid, are removed by washing the column with 100–200 ml of water. The dansylarginine 4-ethylpiperidine amide hydrochloride is then eluted at 22°C from the column with approximately 40 ml acetonitrile : water (4 : 1, v/v). The column eluate is then subjected to rotary flash evaporation in a water bath at about 40°C, and the concentrated solution is then lyophilized to dryness. This procedure typically results in 75 mg of dry water-soluble yellowish powder. Care must be exercised in the process of flash evaporation to prevent bumping as the last traces of acetonitrile are removed. Occasionally, the resulting aqueous solution obtained after flash evaporation is somewhat turbid. If this is the case, the material may be extracted three times with 2 volumes of diethyl ether with separation of phases aided by gentle centrifugation. This procedure clarifies the aqueous phase without extracting appreciable amounts of DAPA. The clarified aqueous phase is then lyophilized to dryness. The dry material can be stored desiccated at −20°C indefinitely.

This procedure yields the monohydrochloride salt of the inhibitor, and solutions may be prepared easily at concentrations of 10 mM in distilled water. The molecular weight of the monohydrochloride salt is 539.14, the extinction coefficient (mM) at 330 nm; 1 cm path length is 4.01. The purity of the synthesized material may be assessed by thin-layer chromatography on polyamide sheets (F1700 micropolyamide, obtained from Schleicher and Schuell, Keene, New Hampshire) in the solvent system acetonitrile–0.02 M sodium acetate, pH 4.5 (9 : 1, v/v). DAPA migrates with an R_f value of approximately 0.86. The starting material, dansylarginine, chromatographs with R_f 0.38.

DAPA interacts reversibly with thrombin with a dissociation constant of 4.3 × $10^{-8}M$, and 1 : 1 stoichiometry. When bound, it yields an approximate 2.5-fold increase in fluorescence intensity with excitation at 335 nm and emission at 565 nm. Since the process of DAPA binding to the enzyme thrombin produced is a stoichiometric process, the sensitivity of thrombin analysis using DAPA is inherently less sensitive than the use of a substrate that is cleaved by thrombin. Thus DAPA is not appropriate for measuring very small amounts of thrombin, or very slow rates of thrombin production. DAPA is best suited to measure the rapid formation of thrombin under conditions of near-optimal concentrations of cofactor and near-physiological concentrations of substrate (approximately 10^{-6} M). We have used both a dual monochrometer fluorometer (Perkin-Elmer MPF44) and a filter fluorometer (Amino Filter Colorimeter). The primary requirement for the fluorescence instrumentation is that there is adequate facility to subtract baseline fluorescence, since a solution of DAPA at $3 \times 10^{-6} M$ (corresponding to the reaction mixture) is intensely fluorescent. When the cuvette containing the reaction mixture is placed in the fluorometer, the full-scale deflection observed should be approximately 60%. This signal is then offset to about 5% of full scale, and the overall sensitivity of the instrument is increased by a factor of two. Under this set of conditions, complete activation of prothrombin (at 0.1 mg/ml) will result in an increase of intensity equivalent to about 80% of full scale. The reaction is initiated, generally, by the addition of factor X_a to a final concentration of about 2×10^{-9} M. At this concentration of enzyme, cofactor, and substrate, the overall reaction will be complete in about 2 min at 22°C. In a typical experiment, a fluorescence cuvette would be prepared containing $1.39 \times 10^{-6} M$ prothrombin (0.1 mg/ml), and $1.5 \times 10^{-8} M$ factor V_a, $1.2 \times 10^{-5} M$ phospholipid (20% phosphatidylserine–80% phosphatidylcholine), 2 mM calcium chloride, 0.02 M Tris-HCl, 0.15 M sodium chloride (pH 7.4), and DAPA, 3×10^{-6} M. The reaction could be initiated by the addition of any constituent, with factor X_a (2×10^{-9} M) being the preferred reagent. The excitation monochromator of the fluorometer should be set either at 280 or 335 nm, with the emission monochromator set at 565 nm.

A variety of procedures can be used to estimate the rate of the process, either from the initial slopes or by taking into account the whole reaction progress. The end point of the reaction corresponds to complete conversion of the added prothrombin (up to ~0.2 mg/ml) to thrombin. This maximum level of intensity can be used to estimate the concentration of active prothrombin in the system. In addition, since the overall rate of the reaction is so strongly influenced by cofactor, calcium, and phospholipid, as well as factor X_a, the effective concentrations of any of these constituents can be estimated on the initial rate of the reaction, provided a calibration curve has been established prior to the determination.

Acknowledgments

This work was supported by grants from the National Institutes of Health HL 16150 and HL 17430 and by the Mayo Foundation.

[24] Prothrombin Activator from *Echis carinatus* Venom

By TAKASHI MORITA and SADAAKI IWANAGA

The venom of *Echis carinatus* (saw-scaled viper) contains procoagulants. One of the most well-known procoagulants is a prothrombin-activating enzyme[1,2] (ECV-prothrombin activator),[3] which catalyzes the conversion of prothrombin to α-thrombin.[4] The activator is also capable of generating α-thrombin from abnormal prothrombin produced by treating mammals with vitamin K antagonists.[5] The process of activation of prothrombin by the venom activator differs significantly from that by factor X_a.[6-8]

Assay Method

Principle. The prothrombin activator of *E. carinatus* venom (ECV) cleaves only a single arginyl-isoleucyl bond linking the thrombin A and B chains in the zymogen molecule, forming a meizothrombin that consists of 2 polypeptide chains. This active meizothrombin is converted autocatalytically to meizothrombin 1, and meizothrombin 1 generates α-thrombin. Therefore, three kinds of thrombin derivatives appear in the process of prothrombin activation by ECV-prothrombin activator (Fig. 1). Although the clotting activity of meizothrombin 1 is less than 50 NIH thrombin units/mg of protein, it shows 1.5 times higher TAME-esterase activity than that of α-thrombin.

One of the assay methods used for ECV-prothrombin activator is to determine its ability to form thrombin from bovine prothrombin; the other

---

[1] The existence of another procoagulant factor X-activating component has been reported [D. A. Walz, H. Dene, and W. H. Seegers, *Fed. Proc., Fed. Am. Soc. Exp. Biol.* **37**, 372 (1978)]. This activator, with MW 25,000, also activates factor IX and protein C.

[2] A. Schieck, F. Kornalik, and E. Habermann, *Naunyn-Schmiedeberg's Arch. Pharmacol.* **272**, 402 (1972).

[3] Abbreviations used are: ECV, *E. carinatus* venom; TAME, N^α-toluenesulfonyl-L-arginine methyl ester; SDS, sodium dodecyl sulfate; NPGB, *p*-nitrophenyl *p*-guanidinobenzoate; PCMB, sodium *p*-chloromercuribenzoate; BAEE, N^α-benzoyl-L-arginine ethyl ester; AGLME, N^α-acetylglycyllysyl methyl ester; Bz, benzoyl; Tos, toluenesulfonyl; PNA, *p*-nitroanilide; DFP, diisopropyl phosphorofluoridate.

[4] F. Kornalik, *Folia Haematol.* **80**, 73 (1963).

[5] G. L. Nelsestuen and J. W. Suttie, *J. Biol. Chem.* **247**, 8176 (1974).

[6] T. Morita, S. Iwanaga, and T. Suzuki, *Seikagaku* **46**, 569 (1974).

[7] B. R. Franza, D. L. Aronson, and J. S. Finlayson, *J. Biol. Chem.* **250**, 7057 (1975).

[8] T. Morita, S. Iwanaga, and T. Suzuki, *J. Biochem. (Tokyo)* **79**, 1089 (1976).

METHODS IN ENZYMOLOGY, VOL. 80

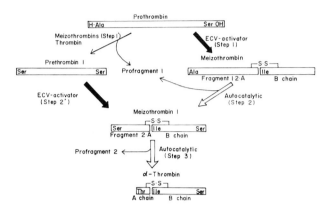

Fig. 1. Pathways of bovine prothrombin activation by ECV-prothrombin activator. Reproduced from T. Morita, S. Iwanaga, and T. Suzuki, *J. Biochem.* (*Tokyo*) **79**, 1089 (1976).

is to estimate the TAME-esterolytic activity of meizothrombins (thrombin) formed during the activation of prothrombin.

Method A

Reagents

Bovine fibrinogen dissolved in 0.09 M Tris-HCl buffer (pH 7.2) containing 0.06 M NaCl, to give a final concentration of 0.6%

Bovine prothrombin, 1 mg/ml in 0.15 M Tris-HCl buffer, pH 7.2

Procedure. The assay mixture contains 0.1 ml of 0.6% fibrinogen, 0.1 ml of the activator solution (diluted with 0.15 M Tris-HCl buffer, pH 7.2), and 0.1 ml of prothrombin solution containing 1 mg/ml of 0.15 M Tris-HCl buffer, pH 7.2. The first two components are mixed at 37°C in a polyethylene fibrino-cup and prothrombin solution is added 20 sec later. The clotting time is measured on a Fibrometer from Biochemical Lab., Maryland. One unit of the activator activity is arbitrarily defined as the amount of activity present in 1 mg of the crude venom. A calibration curve is drawn by graphing the clotting time versus the venom concentration on log–log paper. Samples are diluted to give clotting times between 20 and 60 sec.

Method B

Reagents

Tris-HCl buffer, 0.4 M, pH 8.5

Substrate solution. 0.1 M TAME (378.9 mg of TAME-HCl dissolved in 10 ml of distilled water), stored at 4°C

Alkaline hydroxylamine solution. 2 M (13.9%) aqueous hydroxylamine hydrochloride mixed with an equal volume of 3.5 M NaOH just before use and stored at 4°C

Trichloroacetic acid (TCA) solution. 6 g of TCA dissolved in 100 ml of 3 M HCl

Ferric chloride solution. 14.87 g of $FeCl_3 \cdot 6H_2O$ dissolved in 500 ml of 0.04 N HCl and stored in a dark bottle

Procedure. A sample is incubated with 10–20 μg purified prothrombin and 0.4 M Tris-HCl buffer (pH 8.5) in a total volume of 0.8 ml. After incubation for 10 min at 37°C, a mixture containing 0.1 ml of 0.1 M TAME and 0.1 ml of 0.05 M EDTA, adjusted to pH 8.5, is added and incubated further for 10 min. Then the reaction is terminated by adding 0.5 ml of 6% TCA, followed with 1.0 ml of alkaline hydroxylamine solution.[9] After the mixture has stood at room temperature for 30 min, 2.0 ml of ferric chloride solution is added and the color developed is measured at 500 nm within 30 min. Formation of gas bubbles in the colorimeter cell is largely prevented if the solution is read against a reagent blank prepared by substituting buffer for enzyme and TAME. One unit of the esterolytic activity is defined as the amount of enzyme that hydrolyzes 1.0 μmol of the substrate per minute.

Purification[10]

Step 1. Gel Filtration of Sephadex G-150. The lyophilized venom of *E. carinatus* purchased from Sigma Chemical Co., St. Louis, Missouri (988 mg, total absorbance at 280 nm = 1192) is dissolved in 10 ml of 0.05 M Tris-HCl buffer (pH 8.0) and allowed to stand for 30 min. The clear yellow supernatant after centrifugation at 6000 rpm for 15 min is applied to a column of Sephadex G-150 (3.5 × 122 cm) equilibrated with 0.05 M Tris-HCl buffer, pH 8.0. Elution is carried out with the equilibration buffer, as shown in Fig. 2. Prothrombin-activator activity is detected in the fractions with tube numbers 64–76, and the activator is separated from TAME esterase, which is originally contained in the venom. However, a proteinase with caseinolytic activity is not separable from the activator on this column. The activator fractions indicated by a solid bar in Fig. 2 are pooled.

Step 2. Chromatography on DEAE–Sephadex A-25. The pooled fraction from step 1 is applied to a column of DEAE-Sephadex A-25 (2.5 × 25 cm) equilibrated with 0.05 M Tris-HCl buffer, pH 8.0. After washing the column with the equilibration buffer, proteins are eluted with a linear 1-liter concentration gradient of Tris-HCl buffer (pH 8.0) from 0.05 M to

[9] P. S. Robert, *J. Biol. Chem.* **232**, 285 (1978).
[10] T. Morita and S. Iwanaga, *J. Biochem. (Tokyo)* **83**, 559 (1978).

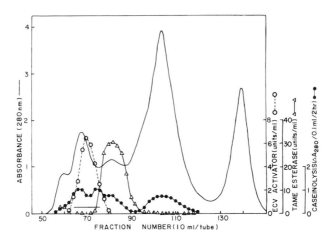

FIG. 2. Separation of prothrombin activator from the venom of *E. carinatus* by gel filtration on a Sephadex G-150 column (step 1). Elution is carried out with the equilibration buffer at a flow rate of 50 ml/hr. The fractions indicated by the solid bar are pooled.

0.5 *M*. The results are shown in Fig. 3. Prothrombin activator appears in the third peak and the caseinolytic activity is found in the first peak. The fractions with tube numbers 70 to 85 in Fig. 3 are pooled, concentrated with an Amicon ultrafiltration apparatus with a PM-10 membrane, and dialyzed overnight against 2 liters of 0.05 *M* acetate buffer, pH 6.0.

Step 3. Chromatography on DEAE–Cellulose. The dialyzed activator fraction from step 2 is chromatographed on a column of DEAE-cellulose (2.5 × 30 cm) equilibrated with 0.05 *M* acetate buffer, pH 6.0. After washing the column with 60 ml of the equilibration buffer, proteins are eluted with a linear 200-ml concentration gradient of sodium acetate (pH

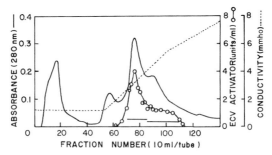

FIG. 3. DEAE-Sephadex A-25 column chromatography of ECV-prothrombin activator (step 2). The gradient system consists of two 500-ml mixing chambers connected at the bottom. The flow rate is 50 ml/hr.

6.0) from 0.05 M to 0.25 M. After elution of an inactive protein, the activator appears at a salt concentration between 0.15 M and 0.18 M. The fractions with activator activity are combined and concentrated with the Amicon ultrafiltration apparatus.

Step 4. Gel Filtration of Sepharose-6B. The concentrated activator fraction from step 3 is gel-filtered through a column of Sepharose-6B (1.3 × 140 cm) equilibrated with 0.05 M Tris-HCl buffer, pH 8.0. Elution is carried out with the equilibration buffer. Two protein peaks appear, and the first major peak contains the activator activity. These fractions are pooled and concentrated to 5 ml by ultrafiltration.

Step 5. Rechromatography on DEAE–Cellulose. The concentrated activator fraction from step 4 is applied to a column of DEAE-cellulose (1 × 13 cm) equilibrated with 0.05 M Tris-HCl buffer, pH 8.0. The activator is eluted with a linear 300-ml concentration gradient of Tris-HCl buffer (pH 8.0) from 0.05 M to 0.2 M. The results are shown in Fig. 4. The activator appears ahead of the major protein peak, and the protein in these fractions shows a single band on SDS-gel electrophoresis. The activator fractions indicated by a solid bar in Fig. 4 are pooled, concentrated by ultrafiltration, and preserved at −70°C.

A summary of the procedures for purification of prothrombin activator is shown in Table I. The yield of purified protein is about 8%, and about 57-fold purification over the activator activity of the crude venom is achieved.

Properties[10]

Purity. The purified activator give a single band on disc-gel electrophoresis at pH 8.3. Furthermore, a single peak with the activator activity is observed at the same position as the protein band stained with Coomassie brilliant blue R-250. On SDS-gel electrophoresis with and without 2-mercaptoethanol, the preparation gives a single molecular species, suggesting that the protein consists of a single polypeptide chain. Also, the material shows a strong color reaction with a periodic acid-Schiff reagent. The preparation is not contaminated by endopeptidase or plasmin-like enzymes, which are found in the crude venom. Moreover, no TAME-esterolytic activity can be detected.

Molecular Weight. The molecular weight of the purified activator is estimated to be 55,000 by molecular sieving on a Sepharose-6B column in the presence of 6 M guanidine, and a single symmetrical peak is obtained on this column. The molecular weight is also calculated, using a Ferguson plot, to be 56,000 ± 1000.

Isoelectric Point. The isoelectric point of the activator was determined

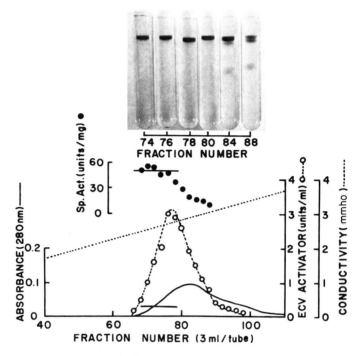

FIG. 4. Rechromatography of ECV-prothrombin activator on a DEAE-cellulose column (step 5). The pooled fraction from step 4 (described in text) is applied to the column and eluted with a linear gradient consisting of 150 ml each of 0.05 M and 0.2 M Tris-HCl buffer, pH 8.0. The flow rate is 21 ml/hr. SDS-gel patterns of unreduced samples from the fractions indicated are inserted in the figure.

by isoelectric focusing with carrier ampholytes, pH range 3–10. The activity is focused as a single symmetrical peak with an isoelectric point of pH 4.5 ± 0.1.

Amino Acid Composition. Table II shows the amino acid composition of the purified preparation. The material contains relatively large amounts of aspartic and glutamic acids; together they constitute nearly 25% of the total amino acid residues. Hexosamines appear to be present, because two peaks corresponding to authentic galactosamine and glucosamine are detectable on the chart of the amino acid analyzer.

Stability. On heating for 10 min at 50°C, the activator is stable in the pH range of 6.5–8.5 and is rather unstable below pH 5 and above pH 10. The thermal stability of the activator was examined by heating a solution in 0.05 M Tris-HCl buffer (pH 8.0) at various temperatures. The activator is very sensitive to heat treatment and its activity is completely lost on

TABLE I
SUMMARY OF THE PURIFICATION PROCEDURE OF ECV-PROTHROMBIN ACTIVATOR[a]

Step	Total A_{280}	Total activity (units)	Specific activity	Recovery (%)	Purification (-fold)
Crude venom (988 mg)	1193	988	0.83	100	1
1. Sephadex G-150	173	768	4.44	78	5.4
2. DEAE–Sephadex A-25 (pH 8.0)	70	743	10.61	75	12.8
3. DEAE–cellulose (pH 6.0)	26.9	360	13.38	36	16.1
4. Sepharose-6B	12.5	178	14.24	18	17.2
5. DEAE–cellulose (pH 8.0)	1.7	80	47.06	8	56.7

[a] T. Morita and S. Iwanaga, *J. Biochem. (Tokyo)* **83**, 559 (1978).

TABLE II
AMINO ACID COMPOSITION OF
ECV-PROTHROMBIN ACTIVATOR[a,b]

Amino acid	Mol%
Lys	5.69
His	2.48
Arg	6.75
Asp	15.24
Thr	5.63
Ser	5.61
Glu	9.26
Pro	5.06
Gly	7.72
Ala	6.20
Val	4.92
Met	2.90
Ile	4.59
Leu	6.43
Tyr	4.42
Phe	3.28
Cys	3.80
Trp	—

[a] From T. Morita and S. Iwanaga, *J. Biochem.* (*Tokyo*) **83**, 559 (1978).
[b] Average values estimated on samples of 24- and 48-hr hydrolysates, using a JEOL JLC-5AH amino acid analyzer.

heating at 60°C for 10 min; the activity is not regained on cooling. Although the purified preparation is stable at -70°C for at least 3 months, overnight storage at 4°C causes a considerable loss of activity.

Optimal pH. The maximum activity is observed at pH 8.0–8.5 using Tris-HCl, phosphate, and glycine-NaOH buffers. The relative activity at pH 8.5 is highest in Tris-HCl buffer; the activity in Tris-HCl buffer is about twice those in phosphate and glycine-NaOH buffers.

Inhibitors. The effects of chelating agents and SH-group blocking agents. At 5 to 10 mM concentration, EDTA, o-phenanthroline, glutathione, cysteine, 2-mercaptoethanol, and dithiothreitol each inhibit the enzyme activity completely. PCMB, iodoacetic acid, and N-ethylmaleimide have no effect. The effect of divalent cations. At a final concentration of 10 mM, Ca^{2+}, Cu^{2+}, and Mg^{2+} are found to have no effect, although Co^{2+}, Zn^{2+}, Fe^{2+}, Cd^{2+}, Ni^{2+}, Mn^{2+}, and Hg^{2+} show a strong inhibitory effect. The activator is insensitive to DFP, benzamidine, and NPGB. The effects of proteinase inhibitors. The following proteinase inhibitors from plants, animals, and microbes have no effect: soybean, lima bean, Bowman-Birk, ovomucoid, snake venom and pancreatic Kunitz-type trypsin inhibitors, hirudin, plasminostreptin, leupeptin, antipain, chymostatin, and pepstatin.

Substrate Specificities.[10] The activator activates bovine[3,5] and human[4] prothrombin by cleaving only a single arginyl-isoleucyl bond linking the thrombin A and B chains in the zymogen molecule. For the activation of prothrombin, it does not require any Ca^{2+}, phospholipids, and factor V. The apparent Michaelis constants (K_m) for prothrombin and prethrombin 1 are $1.03 \times 10^{-6} M$ and $1.05 \times 10^{-6} M$, respectively. The purified prothrombin activator does not activate various zymogens, such as factor X, factor IX, plasminogen, prekallikrein, and trypsinogen from a bovine source. These results indicate an extremely high specificity of the activator toward prothrombin. No hydrolysates of TAME, BAEE, AGLME, L-lysine ethyl ester, and L-leucine ethyl ester are observed. Furthermore, the activator does not liberate p-nitroaniline, when the specific substrates Bz-Phe-Val-Arg-pNA and Tos-Val-Ile-Pro-Arg-pNA for thrombin, Tos-Ile-Glu-Gly-Arg-pNA for factor X_a, and Bz-Pro-Phe-Arg-pNA for kallikrein are incubated with the enzyme at 37°C for 10 min.

[25] Staphylocoagulase

By Takashi Morita, Hideo Igarashi, and Sadaaki Iwanaga

Staphylocoagulase is an extracellular protein produced by *Staphylococcus aureus*.[1,2] It reacts with a component of plasma, named "coagulase-reacting factor," to produce a clot. Coagulase-reacting factor is now understood to be prothrombin. Staphylocoagulase is known to

[1] J. P. Soulier, S. Lewi, A. M. Pauty, and O. Prou-Wartelle, *Rev. Fr. Etud. Clin. Biol.* **12**, 544 (1967).

[2] M. Zajdel, Z. Wegrzynowicz, J. Sawecka, J. Jeljszewicz, and G. Pulverer, *in* "Staphylococci and Staphylococcal Diseases" (J. Jeljszewicz, ed.) p. 549. Fischer, Stuttgart, 1976.

METHODS IN ENZYMOLOGY, VOL. 80

coagulate human plasma, but not bovine plasma,[3] by forming an active molecular complex with prothrombin.[4]

Assay Methods

Principle. Staphylocoagulase forms an active molecular complex with human prothrombin. This complex has an ability to convert fibrinogen to a fibrin clot (Method A), to hydrolyze both chromogenic (Method B) and fluorogenic peptide substrates (Method C).

Method A [5]

Reagents

1 : 10 diluted rabbit plasma. Normal rabbit plasma is diluted with the buffer, which consists of 2% polypepton, 1% sodium citrate, 0.85% NaCl, 0.05% NaN_3, and 0.01% thimerosal (w/v).

Procedure. Twofold or tenfold serial dilutions of a culture filtrate or a fraction of each purification step, in 0.2-ml amounts, are mixed with an equal volume of 1 : 10 diluted rabbit plasma, and the highest dilution of staphylocoagulase is recorded, in which the plasma clots within 1 hr at 37°C in a water bath.

Method B [6]

Reagents

Tris-HCl buffer. 0.15 M Tris-HCl (pH 8.4) containing 0.01% bovine serum albumin

Human prothrombin solution. 100 μg/ml in Tris-HCl buffer, pH 8.4

Staphylocoagulase solution. A test sample or purified staphylocoagulase dissolved in 0.1 M ammonium bicarbonate, pH 8.0

Substrate solution. 0.065 mM tosyl-Gly-Pro-Arg-p-nitroanilide (Chromozym-TH, Pentapharm AG, Basel) dissolved in water

Procedure. Human prothrombin (100 μl) and staphylocoagulase (100 μl) are mixed with 600 μl of 0.15 M Tris-HCl buffer (pH 8.4) and incubated at 37°C for 2 min. Then, 400 μl of the substrate solution is added.

[3] Staphylocoagulase from some strains activate bovine prothrombin [M. Zajdel, A. Wegrzynowicz, J. Jeljszewicz, and G. Pulverer, *in* "Staphylococci and Staphylococcal Infections" (J. Jeljszewicz ed.), p. 364. Polish Med. Publ., 1973].

[4] H. C. Hemker, B. M. Bass, and A. D. Muller, *Biochim. Biophys. Acta* **379**, 180 (1975).

[5] H. Igarashi, T. Morita, and S. Iwanaga, *J. Biochem. (Tokyo)* **86**, 1615 (1979).

[6] H. Igarashi, M. Takahashi, T. Morita, and S. Iwanaga, *Jpn. J. Bacteriol.* **34**, 227 (1979); **35**, 300 (1980).

Absorbance change at 405 nm is measured at 37°C. The end-point method instead of the above mentioned initial velocity measurement is available by adding acetic acid to give 10% concentration, to terminate the reaction.

Method C[7]

Reagents

Prothrombin. Purified human prothrombin prepared by the method of Mann[8] or Miletich *et al.*,[9] and protein concentration determined based on the absorbance at 280 nm using $E^{1\%}_{280}$ value of 15.5[8]

Buffer. 0.05 M Tris-HCl buffer (pH 7.5) containing 0.1 M NaCl and bovine serum albumin (0.1 mg/ml)

Substrate. A fluorogenic peptide substrate,[10] Boc-Val-Pro-Arg-4-methylcoumaryl-7-amide (MCA), purchased from Protein Research Foundation, Minoh, Osaka, dissolved in distilled water to give a final concentration of 5 mM

Procedure. Each 50 μl of human prothrombin (A_{280} = 0.025–0.030) and staphylocoagulase (A_{280} = 0.01–0.03) is added in 0.6 ml of Tris-HCl buffer, and the mixture is preincubated in a cuvette thermostated at 37°C for 5 min. Then, 10 μl of the substrate is added and the increase of the relative fluorescence of 7-amino-4-methylcoumarin produced is read at regular time intervals, using a Hitachi fluorescence spectrophotometer, model MPF-2A, equipped with recorder. The measurements are carried out with excitation at 380 nm and emission at 440 nm. The instrument is standardized so that a 10 μM solution of 7-amino-4-methylcoumarin in 10% dimethyl sulfoxide gives 1.0 relative fluorescence unit.

Comments. The usual assay for staphylocoagulase is the clotting assay, originally described by Duthie and Haughton.[11] Fibrin clots are rapidly formed in the presence of fixed levels of staphylocoagulase and human prothrombin and the time required for clot formation is measured. However, the method is tedious and semiquantitative. The assay method using the fluorogenic peptide substrate is rapid, accurate, and simple to perform. With Boc-Val-Pro-Arg-MCA, the staphylocoagulase activity is measured up to the concentration of 0.05 μg/ml, using a known amount of highly purified human prothrombin, and the rate of hydrolysis is propor-

---

[7] S. Kawabata, T. Morita, S. Iwanaga, and H. Igarashi, *Seikagaku* **52**, 733 (1980).

[8] K. G. Mann, this series, Vol. 45, p. 123.

[9] J. P. Miletich, C. M. Jackson, and P. W. Majerus, *J. Biol. Chem.* **253**, 6908 (1978).

[10] T. Morita, H. Kato, S. Iwanaga, K. Takada, T. Kimura, and S. Sakakibara, *J. Biochem.* (*Tokyo*) **82**, 1495 (1977). Boc-Val-Pro-Arg-MCA is now commercially available from Protein Research Foundation, 476 Ina Minoh-shi, Osaka 562, Japan.

[11] E. S. Duthie and G. H. Haughton, *Biochem. J.* **70**, 125 (1958).

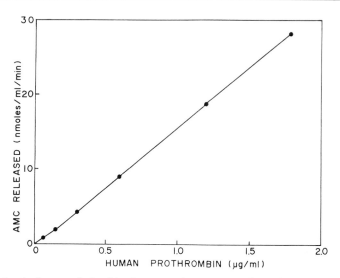

FIG. 1. A linear relationship between the concentration of staphylocoagulase–prothrombin complex and its amidase activity. The activity of the complex was measured with Boc-Val-Pro-Arg-MCA as substrate.

tional to the concentration of staphylocoagulase–prothrombin complex over at least a tenfold range (Fig. 1).

Purification[5]

Bovine Prothrombin–Sepharose-4B

About 45 ml of CNBr-activated Sepharose-4B (Pharmacia Fine Chemicals, Sweden) is swollen and washed on a sintered glass filter with 3 liters of 1 mM HCl. Bovine prothrombin (121.5 mg) purified by the method of Morita *et al.*[12] is dissolved in 0.1 M NaHCO$_3$ (pH 8.3) containing 0.5 M NaCl. The prothrombin is coupled to the gel in an Erlenmeyer's flask fixed on a rotary shaker for 3 hr at room temperature. The gel is washed with the coupling buffer until absorbance at 280 nm is negligible. Ethanolamine (1.0 M, pH 8.0) is added and mixed at room temperature by rotary shaker for 2 hr. The gel is then washed four times with 100 ml each of 0.1 M borate buffer (pH 8.5) and 0.1 M acetate buffer (pH 4.0), both containing 0.5 M NaCl. The gel is mixed with 0.05 M phosphate buffer (pH 7.4) containing 1.0 M NaCl and 0.0001% thimerosal (equilibration buffer),

[12] T. Morita, H. Nishibe, S. Iwanaga, and T. Suzuki, *J. Biochem.* (*Tokyo*) **76**, 1031 (1974).

Staphylococcus aureus (Strain st-213)
  | (1) Cultured in brain heart infusion broth at 35°C for 18 hr
  ↓ (2) Centrifuged by continuous-flow centrifuge at 14,000 rpm at the speed of 3 liters/hr
Supernatant
  ↓   Filtered through 450-nm Millipore filter disk
Culture filtrate
  | (1) NaCl concentration and pH are adjusted
  ↓ (2) Affinity chromatography on a bovine prothrombin Sepharose-4B column
1.0 *M* NaSCN eluate
  ↓   Desalted by gel filtration on a Sephadex G-25 column
Purified staphylocoagulase

FIG. 2. Flow diagram for isolation of staphylocoagulase.

and poured into a column (5 × 2.3 cm). This affinity column can be used repeatedly at least for 3 years and 50 chromatographies.

Comment. Bovine prothrombin, dissolved in 0.1 *M* NaHCO$_3$ (pH 8.5) containing 0.5 *M* NaCl, could be also coupled to Sepharose CL-6B or Sepharose-4B, which is previously activated by the procedure of March *et al.*[13]

Procedure

Purification of staphylocoagulase is carried out on culture filtrates from *S. aureus,* strain st-213, which is the reference strain of type II for coagulase typing.[14] NaCl and thimerosal are added to the culture filtrate from 450 nm Millipore filtrate disk, to give 1 *M* or 0.0001% concentration, respectively, and the solution is adjusted to pH 7.4. A flow diagram for the isolation of staphylocoagulase is shown in Fig. 2.

Step 1. The culture filtrate containing 1.0 *M* NaCl and 0.0001% thimerosal is applied to the column at a flow rate of 200 ml/hr. Ten-milliliter fractions are collected. The gel is washed with the equilibration buffer until the absorbance at 280 nm returns to zero. The chosen eluant for eluting the staphylocoagulase from the affinity column is 1.0 *M* NaSCN (Fig. 3). After elution of the adsorbed staphylocoagulase with 1.0 *M* NaSCN, the gel is equilibrated with the equilibration buffer before being used again.

Step 2. The fractions containing staphylocoagulase are pooled and applied to a Sephadex G-25 column (2.5 × 45 cm) and eluted with 0.1 *M* NH$_4$HCO$_3$ solution. Two peaks of absorbance at 280 nm emerged; the first peak is purified staphylocoagulase and the late peak contains NaSCN.

[13] S. C. March, I. Parikh, and P. Cuatrecasas, *Anal. Biochem.* **60**, 149 (1974).
[14] H. Zen-Yoji, T. Terayama, M. Benoki, and S. Kuwahara, *Jpn. J. Microbiol.* **5**, 237 (1961).

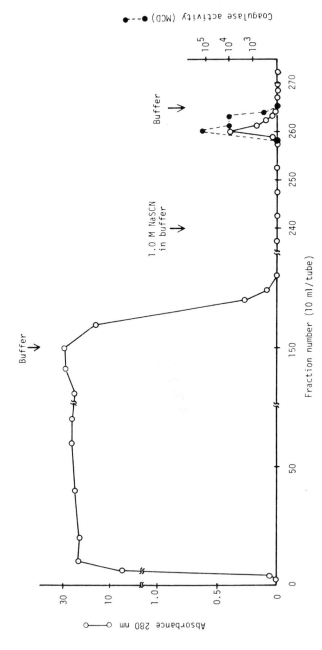

FIG. 3. Purification of staphylocoagulase by Sepharose-4B column linked with bovine prothrombin.

TABLE I

PURIFICATION OF STAPHYLOCOAGULASE FROM CULTURE FILTRATE

Step	Volume (ml)	Activity (MCD[a]/ml)	Total activity (MCD × 10⁻³)	A_{280}	Absorbance units	Specific activity (MCA/A_{280} = 1)	Yield (%)	Purification (-fold)
Culture filtrate	1500	512	768	32.5	48,750	16	100	1
Prothrombin–Sepharose eluate	50	12,800	640	0.231	11.55	55,000	83	3400
Sephadex G-25 fraction	50	12,800	640	0.113	5.65	113,000	83	7100

[a] Minimal clotting dose (MCD) is defined as a reciprocal of maximum dilution number by Method A.

TABLE II
AMINO ACID COMPOSITION OF
STAPHYLOCOAGULASE

| | Amino acid g/100 g protein | |
Amino acid	Strain st-213[a]	Strain 104[b]
Aspartic acid	13.79	13.18
Threonine	6.5	4.74
Serine	3.5	3.71
Glutamic acid	15.75	17.58
Proline	4.10	2.82
Glycine	2.08	3.38
Alanine	2.35	5.09
Half-cystine	0	1.64
Valine	5.26	6.96
Methionine	2.12	1.51
Isoleucine	4.85	6.42
Leucine	6.48	7.70
Tyrosine	8.59	3.99
Phenylalanine	3.28	4.53
Lysine	15.99	9.10
Histidine	2.57	1.89
Arginine	2.68	4.01
Tryptophan	0.84[c]	0.60

[a] H. Igarashi, T. Morita, and S. Iwanaga, *J. Biochem.* (*Tokyo*) **86**, 1615 (1979). Average or extrapolated values from 24-, 48-, and 72-hr hydrolysates.
[b] B. M. Bass, A. D. Muller, and H. C. Hemker, *Biochim. Biophys. Acta* **379**, 164 (1974).
[c] Determined by the method of R. J. Simpson, M. R. Neuberger, and T. Y. Liu, *J. Biol. Chem.* **251**, 1936 (1976).

The yields and activities during purification of staphylocoagulase are summarized in Table I.

Properties

Purity. The final preparation gives a single precipitin line with anti-crude staphylocoagulase serum and anti-purified staphylocoagulase serum, when it is tested with Ouchterlony's technique and immunoelectrophoresis. Furthermore, the purified staphylocoagulase shows a single and symmetrical peak on ultracentrifugal analysis. The value of $S_{20,w}$ is estimated to be 6.47. However, heterogeneity of the purified staphylocoagulase is demonstrated by isoelectric focusing, disc polyacrylamide-

TABLE III
PHYSICOCHEMICAL PROPERTIES OF STAPHYLOCOAGULASE

Property	Strain st-213[a]	Strain 104[b]
Molecular weight	58,000[c]	61,000
NH$_2$-terminus	Ile	Asp
pI	5.30, 5.76	4.53

[a] H. Igarashi, T. Morita, and S. Iwanaga, *J. Biochem.* (*Tokyo*) **86**, 1615 (1979).
[b] B. M. Bass, A. D. Muller, and H. C. Hemker, *Biochim. Biophys. Acta* **379**, 164 (1974).
[c] The minor components with MW 71,000 and 52,000 are also contained.

gel electrophoresis, and SDS–polyacrylamide gel electrophoresis. The NH$_2$-terminal amino acid of the staphylocoagulase is a single isoleucine.

Stability.[7] Even after the purified staphylocoagulase is allowed to stand for 24 hr at room temperature in 8 M urea, 1 M NaSCN, 6 M guanidine HCl, 10% acetic acid, or 10% acetic acid containing 8 M urea, the activity is completely restored by removing the denaturing reagents by dialysis.

Physicochemical Properties. The molecular weights of the purified staphylocoagulase estimated by SDS-gel electrophoresis are 71,000, 58,000, and 52,000. The amino acid composition of the purified staphylocoagulase is shown in Table II. It should be noted that the material isolated from the culture filtrate of *S. aureus,* strain st-213, does not contain any cystine residues (see Table II).

Specificity. Staphylocoagulase is able to interact with human prothrombin, forming a 1 : 1 complex.[4] Although the complex has a capacity for limited proteolysis of fibrinogen, resulting in releasing fibrinopeptides A and B,[2] it does not activate factor XIII, factor VIII, and factor V, unlike α-thrombin.[15] This complex also possesses a capacity for hydrolysis of synthetic substrates, such as TAME and chromogenic and fluorogenic peptide substrates. The K_m values for chromogenic peptide substrates estimated from Lineweaver–Burk plots are as follows[6]: Tosyl-Gly-Pro-Arg-pNA, K_m = 1.3 × 10^{-5} M; Z-Gly-Pro-Arg-pNA, K_m = 3.6 × 10^{-5} M; H-D-Phe-Pip-Arg-pNA, K_m = 2.0 × 10^{-5} M. The substrates, Bz-Phe-Val-Arg-pNA and H-D-Val-Leu-Lys-pNA, are not susceptible to the complex. The hydrolytic activity of the complex is not inhibited by hirudin and antithrombin III in the presence of heparin,[7,15] but is inhibited by benzamidine (K_i = 1.2 × 10^{-4} M).[7]

[15] J. P. Soulier and O. Prou-Wartelle, *Thromb. Diath. Haemorrh.* **17**, 321 (1967).

[26] Protein C

By WALTER KISIEL[1] and EARL W. DAVIE

Protein C is a vitamin K-dependent plasma glycoprotein that exhibits many of the physicochemical characteristics found in the four vitamin K-dependent blood-coagulation factors.[2,3] Inasmuch as a disorder related either to a protein C deficiency or an abnormality has not been described clinically, the precise physiological role of this protein remains unclear. Data from several *in vitro* studies, however, provide evidence that protein C may play a unique role in blood coagulation. Protein C exists in human and bovine plasma in a zymogen form. In the test tube, purified protein C is converted to a serine protease by catalytic amounts of α-thrombin.[3,4] Protein C is also activated by other proteolytic enzymes, such as pancreatic trypsin[5,6] and RVV-X, a protease from Russell's viper venom.[3,6] In contrast to the procoagulant nature of the activated forms of the vitamin K-dependent clotting factors, activated protein C (protein C_a) acts as an inhibitor of blood coagulation and markedly prolongs the kaolin–cephalin clotting time of normal plasma.[3,4,7,8] This effect has recently been ascribed to the ability of protein C_a specifically to inactivate purified preparations of blood-clotting factor V_a and factor VIII_a by limited proteolysis.[4,9-14]

Protein C was originally isolated from bovine plasma[2] and more recently from human plasma.[3] In the isolation procedures described below, dextran sulfate–agarose and preparative electrophoresis are key steps in the isolation of these proteins. Recently, a homogeneous preparation of bovine protein C has been described employing chromatography on heparin–agarose as a final purification step.[15]

[1] An Established Investigator of the American Heart Association.
[2] J. Stenflo, *J. Biol. Chem.* **251**, 355 (1976).
[3] W. Kisiel, *J. Clin. Invest.* **64**, 761 (1979).
[4] W. Kisiel, W. M. Canfield, L. H. Ericsson, and E. W. Davie, *Biochemistry* **16**, 5824 (1977).
[5] C. T. Esmon, J. Stenflo, J. W. Suttie, and C. M. Jackson, *J. Biol. Chem.* **251**, 3052 (1976).
[6] W. Kisiel, L. H. Ericsson, and E. W. Davie, *Biochemistry* **15**, 4893 (1976).
[7] E. Marciniak, *J. Lab. Clin. Med.* **79**, 924 (1972).
[8] E. Marciniak, *Science* **170**, 452 (1970).
[9] W. M. Canfield, M. Nesheim, W. Kisiel, and K. G. Mann, *Circulation* **58**, II-210 (1978).
[10] W. M. Canfield and W. Kisiel, *Circulation* **60**, II-246 (1979).
[11] M. E. Nesheim, W. M. Canfield, W. Kisiel, and K. G. Mann, *Blood* **54**, Suppl. 1, 293 (1979).
[12] F. J. Walker, P. W. Sexton, and C. T. Esmon, *Biochim. Biophys. Acta* **571**, 33 (1979).
[13] C. T. Esmon, P. C. Comp, and F. J. Walker, *in* "Vitamin K Metabolism and Vitamin K-Dependent Proteins" (J. W. Suttie, ed.), p. 72. University Park Press, Baltimore, Maryland, 1979.
[14] G. A. Vehar and E. W. Davie, *Biochemistry* **19**, 401 (1980).
[15] P. C. Comp and C. T. Esmon, *Blood* **54**, 1272 (1979).

METHODS IN ENZYMOLOGY, VOL. 80

Reagents

Normal Citrated Plasma (*Human or Bovine*). Plasma is prepared by mixing 9 volumes of blood with 1 volume of 3.8% trisodium citrate followed by centrifugation at 10,000 g for 15 min at 4°C. The plasma is stored in plastic tubes at −20°C in small aliquots.

Tris-Buffered Saline–0.1% Bovine Serum Albumin (*TBS–BSA*). 0.05 M Tris-HCl–0.1 M NaCl (pH 7.5) containing 1 mg/ml of bovine serum albumin (Sigma).

Phospholipid–Kaolin Suspension. Phospholipid (cephalin) is extracted from acetone-dehydrated bovine brain with chloroform according to Bell and Alton.[16] The chloroform-free phospholipid extract is then suspended in TBS. This stock phospholipid solution is routinely diluted 1/50 with TBS, and acid-washed kaolin (Fisher Scientific Co.) is suspended in the diluted phospholipid solution to a final concentration is 5 mg/ml. The phospholipid–kaolin suspension is stable for several hours at 37°C and is shaken each time before pipetting.

DEAE–Sephadex. DEAE–Sephadex A-50 (50 g) is allowed to swell for 1–2 days at room temperature in an excess of 0.5 M sodium phosphate (pH 6.0). Fines are removed by decantation and the ion exchanger is then resuspended several times in 0.1 M sodium phosphate (pH 6.0)–1 mM benzamidine, and stored at 4°C. After use, the DEAE–Sephadex is regenerated in a Buchner funnel by sequential washes with 0.1 M NaOH, distilled water, 0.5 M sodium phosphate (pH 6.0), and finally extensively with 0.1 M sodium phosphate (pH 6.0)–1 mM benzamidine.

Dextran Sulfate–Agarose. Dextran sulfate–agarose is prepared by a modification of the procedure described by Pepper and Prowse.[17] Sepharose-4B (500 ml) is washed with water to remove salts and resuspended in an equal volume of distilled water. Twenty grams of dextran sulfate (Pharmacia) is added to the agarose slurry and the pH adjusted to 11 with 6 M NaOH. Cyanogen bromide (100 g in 100 ml of acetonitrile) is then added with vigorous stirring and coupling is performed for 45 min. Throughout the coupling step, the pH is maintained at 11 with 6 M NaOH and the temperature is maintained between 4°C and 10°C by the addition of ice chips. After coupling, the dextran sulfate–agarose is filtered in a Buchner funnel, resuspended in 2 liters of 1 M glycine ethyl ester (pH 9.0), and stirred at room temperature for 2 hr. The dextran sulfate–agarose is then washed exhaustively with 1 M NaCl followed by 0.02 M Mes-Tris (pH 6.0) containing 2.5 mM calcium chloride, 1 mM benzamidine, and 0.02% sodium azide. The resin is then stored in this buffer at 4°C.

[16] W. N. Bell and H. G. Alton, *Nature* (*London*) **174**, 880 (1954).
[17] D. S. Pepper and C. Prowse, *Thromb. Res.* **11**, 687 (1977).

Thrombin. Human and bovine α-thrombin are isolated essentially according to Lundblad *et al.*[18] following activation of pure prothrombin by a factor X_a–Ca^{2+}–phospholipid mixture.[19] The specific activities of the thrombin preparations are 2500–3500 NIH U/mg.

Antithrombin III. Human and bovine antithrombin III are purified to electrophoretic homogeneity by heparin–agarose column chromatography according to Kurachi *et al.*[20]

Assay Methods

Protein C concentration can be assayed quantitatively by immunochemical procedures,[2] or qualitatively by its ability to prolong the kaolin–cephalin clotting time of plasma following its activation by thrombin.[3] One obvious drawback of the immunochemical method is that it requires a pure preparation of protein C for antibody production. The two-stage anticoagulant assay, described below, does not require purified protein C and appears to be a useful operational assay for evaluating the relative concentration of this protein in chromatographic eluants. It should be pointed out that with the anticoagulant assay, the prolongation of the clotting time by activated protein C is dramatically dependent on the use of plasma from which the protein C is derived.

Procedure. The biological assay for protein C is a two-stage assay in which protein C is activated in the first stage, and in the second stage an aliquot of the activation mixture is added to an assay system that measures the effect of activation protein C on the kaolin–cephalin clotting time of normal plasma. Routinely, 100 μl of a chromatographic fraction is diluted in a plastic culture tube with 900 μl of 0.1 *M* Tris-HCl (pH 8.0) containing 0.1% BSA. Thrombin (10 μg) is added and the mixture is incubated at 37°C for 30–60 min. At this time, 40 μg antithrombin III and 40 μg of heparin in 20 μl of TBS are added and the incubation is continued for an additional 15 min at 37°C to neutralize the thrombin. An aliquot of the incubation mixture is diluted in a plastic culture tube with TBS–BSA immediately before the kaolin–cephalin clotting assay. In this assay, 0.1 ml of normal plasma is incubated with 0.1 ml of kaolin–cephalin suspension in a glass culture tube (10 × 75 mm) for 3 min at 37°C. After the 3-min preincubation, 0.1 ml of the diluted test material and 0.1 ml of 25 m*M* $CaCl_2$ are added to the reaction mixture. An electric timer is started

[18] R. L. Lundblad, L. C. Uhteg, C. N. Vogel, H. S. Kingdon, and K. G. Mann, *Biochem. Biophys. Res. Commun.* **66**, 482 (1975).

[19] W. Kisiel and D. J. Hanahan, *Biochim. Biophys. Acta* **329**, 221 (1973).

[20] K. Kurachi, G. Schmer, M. A. Hermodson, D. C. Teller, and E. W. Davie, *Biochemistry* **15**, 368 (1976).

on the addition of calcium, and the time required for clot formation is recorded. Clot formation is followed by the tilt-tube method. Samples from the first stage of the assay routinely are diluted 10- to 100-fold to yield a clotting time approximately two to three times that observed for the blank sample.

Purification Steps for Bovine Protein C

Step 1. Preparation of Plasma. Bovine blood (100 liters) is collected in ten 10-liter buckets, each containing 1 liter of anticoagulant solution composed of 38 g of trisodium citrate, 100 mg of heparin (16,000 units), 100 mg of crude soybean-trypsin inhibitor, and 3.2 g of benzamidine hydrochloride. The blood is centrifuged at room temperature with a continuous-flow separator (DeLaval Model BLE 519). All subsequent steps with the plasma are performed at 4°C.

Step 2. Barium Citrate Adsorption and Ammonium Sulfate Precipitation. Eighty milliliters of 1 M BaCl$_2$ is added per liter of plasma and the mixture is stirred for 1 hr. The barium citrate precipitate is collected by centrifugation at 5000 g for 20 min and the supernatant is discarded. The barium citrate precipitate is washed twice with cold 0.1 M NaCl–5 mM BaCl$_2$–5 mM benzamidine (150 ml/liter of starting material), followed each time by centrifugation. The washed barium citrate precipitate is then suspended in 30% saturated ammonium sulfate (150 ml/liter of starting material) and stirred for 1 hr to elute adsorbed proteins. The barium citrate–sulfate and insoluble protein is collected by centrifugation at 5000 g for 15 min and discarded. The supernatant is adjusted to 65% saturation by the addition of solid ammonium sulfate, and stirring is continued for an additional 30 min. The protein precipitate is collected by centrifugation at 5000 g for 30 min, and the supernatant is discarded.

Step 3. DEAE–Sephadex A-50 Column Chromatography. The precipitate obtained in step 2 is dissolved in a minimal volume of 0.1 M sodium phosphate (pH 6.0)–10 mM EDTA–10 mM benzamidine. The protein solution is dialyzed overnight against 40 liters of 0.1 M sodium phosphate (pH 6.0)–1 mM benzamidine.

Following dialysis, the retentate is clarified by centrifugation (5000 g for 15 min) and the supernatant is applied to a DEAE–Sephadex A-50 column (5 × 50 cm) previously equilibrated with 0.1 M sodium phosphate (pH 6.0) containing 1 mM benzamidine. The column is then washed with 1 liter of 0.1 M sodium phosphate (pH 6.0)–0.15 M NaCl–1 mM benzamidine. Protein C is eluted from the column with a linear gradient of NaCl generated from 2 liters of the wash buffer and 2 liters of 0.1 M sodium phosphate (pH 6.0)–0.6 M NaCl–1 mM benzamidine. The flow rate is adjusted to 150–200 ml/hr and 20-ml fractions are collected. Under these conditions, protein C elutes at approximately 0.4 M NaCl in the descending portion of the prothrombin peak. Protein C is well resolved from factor IX, factor VII, and factor X, but contains appreciable amounts of prothrombin.

Step 4. Preparative Polyacrylamide Gel Electrophoresis. Protein C is resolved from prothrombin by preparative electrophoresis at 0–2°C in a Buchler Poly-Prep 200 apparatus (Buchler Instruments). The fractions from the DEAE–Sephadex containing protein C are pooled and made 5 mM in DFP. The pooled samples are concentrated to approximately 5 mg/ml by ultrafiltration (PM-10 membrane) and dialyzed against 4 liters of 0.025 M Tris-HCl (pH 8.0)–0.025 M glycine–0.01 M benzamidine–0.005 M EDTA for 15 hr. Quantities of approximately 50–75 mg (10–15 ml) of the retentate are electrophoresed in a 7.5% acrylamide resolving gel (100 ml)–4% acrylamide concentrating gel (20–40 ml) system at a constant current of 60 mA for about 10 hr. The resolving gel buffer initially is 0.375 M Tris-HCl (pH 8.9) and the concentrating gel buffer is 0.06 M Tris-H$_3$PO$_4$ (pH 7.2). The lower (anode) buffer is 0.4 M Tris-HCl (pH 8.0) and the upper (cathode) buffer is 0.05 M Tris–0.05 M glycine (pH 8.9). The proteins are eluted from the resolving gel with 0.1 M Tris-HCl–1 mM benzamidine at a rate of 1 ml/min. Under these electrophoretic conditions, protein C elutes well ahead of prothrombin. The fractions containing protein C are pooled, and DFP is added to a final concentration of 2 mM. The solution is then concentrated by ultrafiltration to approximately 5 mg/ml and aliquots are stored at −70°C. Aliquots are freed of DFP and benzamidine immediately before use employing a column of Sephadex G-15. Approximately 100 mg of purified protein C is obtained from 50 liters of bovine plasma by this procedure. The preparation is electrophoretically homogeneous in analytical polyacrylamide gels in the presence[21] or absence[22] of sodium dodecyl sulfate (SDS). The isolation procedure takes about 4 days.

Physicochemical Properties of Bovine Protein C

Bovine protein C has MW 54,300 as calculated from amino acid and carbohydrate composition data. It is composed of a heavy chain (M_r = 33,800) and a light chain (M_r = 20,500) held together by a disulfide bond(s). The apparent molecular weights of the heavy and light chains of protein C observed by SDS gel electrophoresis are 41,000 and 21,000, respectively. A molecular weight of 58,000 was estimated by this technique for the unreduced protein.[6]

[21] K. Weber and M. Osborn, *J. Biol. Chem.* **244**, 4406 (1969).
[22] B. J. Davis, *Ann. N.Y. Acad. Sci.* **121**, 404 (1964).

Recently, the entire amino acid sequences of the heavy and light chains of bovine protein C were reported by Stenflo and co-workers.[23,24] These studies have shown a great deal of homology between protein C and the other vitamin K-dependent clotting factors, including prothrombin, factor VII, factor IX, and factor X. The amino acid and carbohydrate compositions of bovine protein C obtained from sequence analyses are presented in Table I. Protein C contains 86% protein and 14% carbohydrate. The carbohydrate includes approximately 15 residues of hexose, 12 residues of hexosamine, and 9 residues of sialic acid distributed in 3 carbohydrate chains. These chains are linked to asparagine in the protein backbone. The light chain of protein C contains 11 residues of γ-carboxyglutamic acid, which presumably enable the protein to interact with acidic phospholipid membranes in the presence of divalent metal ions.[25]

The extinction coefficient of bovine protein C is 13.7 as determined by differential refractometry,[4] and the isoelectric point is 4.2–4.5.[26]

Isolation of Bovine Protein C_a

Protein C is converted to protein C_a by hydrolysis of a specific peptide bond in the amino-terminal region of the heavy chain. This cleavage occurs between Arg-14 and Ile-15, giving rise to protein C_a ($M_r = 52,650$) and an activation peptide ($M_r = 1650$).[6] Presumably, the new amino-terminal isoleucine residue forms an internal ion pair with Asp-198, which is adjacent to the active site Ser-199.[24]

To date, three proteases, thrombin, trypsin, and the factor X activator from Russell's viper venom (RVV-X), have been shown to activate protein C.[3–6] The proteolytic activation of protein C by trypsin appears to be nonspecific and ultimately leads to additional cleavages in both the heavy and light chains of protein C_a.[6] Proteolysis of protein C by either thrombin or RVV-X, however, appears to be limited to the Arg-14-Ile-15 peptide bond in the heavy chain, and this leads to the activation of the protein.[4] Clotting enzymes that do not activate protein C include factor XII_a, factor XI_a, factor IX_a, factor X_a, factor VII_a, and kallikrein.[27] Incubation mixtures containing protein C and either factor IX_a or factor X_a also included phospholipid and ionic calcium, while incubation mixtures of protein C and factor VII_a contained tissue factor and calcium ions. Plasmin, a serine protease involved in fibronolysis, rapidly degrades protein C without any

[23] P. Fernlund, J. Stenflo, and A. Tufvesson, *Proc. Natl. Acad. Sci. U.S.A.* **75**, 5889 (1978).
[24] P. Fernlund and J. Stenflo, *in* "Vitamin K Metabolism and Vitamin K-Dependent Proteins" (J. W. Suttie, ed.), p. 84. University Park Press, Baltimore, Maryland, 1979.
[25] G. L. Nelsestuen, W. Kisiel, and R. G. DiScipio, *Biochemistry* **17**, 2134 (1978).
[26] R. G. DiScipio and E. W. Davie, *Biochemistry* **18**, 899 (1979).
[27] W. Kisiel, unpublished data.

TABLE I

AMINO ACID AND CARBOHYDRATE COMPOSITIONS OF BOVINE PROTEIN C

| | | Protein C | |
Components	Protein C	Heavy chain	Light chain
Amino acid[a]			
Lysine	22	15	7
Histidine	11	6	5
Arginine	28	14	14
Aspartic acid	26	17	9
Threonine	17	13	4
Serine	25	15	10
Glutamic acid	25	15	10
γ-Carboxyglutamic acid	11	0	11
Proline	18	12	6
Glycine	37	21	16
Alanine	20	13	7
Half-cystine	24	7	17
Valine	31	26	5
Methionine	7	4	3
Isoleucine	16	14	2
Leucine	33	24	9
Tyrosine	13	10	3
Phenylalanine	13	5	8
Tryptophan	10	8	2
Asparagine	11	7	4
Glutamine	13	10	3
Molecular weight (protein)	46,814	28,792	18,022
Carbohydrate[b]			
Galactose	6	4	2
Mannose	9	6	3
N-Acetylglucosamine	12	8	4
N-Acetylneuraminic acid	9	6	3
Carbohydrate (%)	14	15	12
Protein (%)	86	85	88
Molecular weight (glycoprotein)	54,303	33,785	20,518

[a] Calculated from the known amino acid sequence [P. Fernlund and J. Stenflo, in "Vitamin K Metabolism and Vitamin K-Dependent Proteins" (J. W. Suttie, ed.), p. 84. University Park Press, Baltimore, Maryland, 1979]. A small increase in the protein molecular weight will be necessary on positive identification of residues 58 and 168 in the heavy chain.

[b] Assumes 2 fully sialylated sugar chains in the heavy chain of protein C and 1 fully sialylated sugar chain in the light chain. The carbohydrate structure in bovine protein C has recently been determined (A. Kobata and T. Mizuochi, unpublished data) and found to be identical to that found in bovine prothrombin [T. Mizuochi, K. Yamashita, K. Fujikawa, W. Kisiel, and A. Kobata, J. Biol. Chem. 254, 6419 (1979)].

apparent activation at an enzyme-to-substrate weight ratio of 1 : 50. As monitored by reduced SDS gel electrophoresis, the cleavage of protein C by plasmin revealed an initial rapid cleavage of the heavy chain of protein C. This yields fragments with molecular weights less than 20,000. This is then followed by a slower degradation of the light chain of protein C.[27]

Protein C_a is isolated by ion-exchange chromatography following activation by soluble thrombin or RVV-X.[4,15,28] Alternatively, a recent study utilized RVV-X immobilized on Sepharose as a convenient means of activating protein C.[29] The procedure used in our laboratory to isolate protein C_a takes advantage of the limited proteolysis of protein C by thrombin and the large difference in affinity of these reactants for SP-Sephadex. The isolation of bovine protein C_a is routinely carried out as follows. Protein C (100 mg) is incubated at 37°C with 2–4 mg of α-thrombin in the presence of 50 mM Tris-H_3PO_4 (pH 8.0) in a final volume of 40 ml. The activation is monitored by determining the amidase activity toward D-Phe-Pip-Arg-p-nitroanilide (S-2238) in temporal aliquots of the reaction mixture following neutralization of the thrombin with antithrombin III and heparin. Activation is allowed to proceed until no further increase in amidase activity is noted. This usually takes 30–60 min. Following a 1-hr incubation, the reaction mixture is made 1 mM in benzamidine and the pH of the solution is reduced to 6.5 with dilute H_3PO_4. The incubation mixture is then applied directly to an SP-Sephadex C-50 column (1.6 × 30 cm) previously equilibrated at room temperature with 50 mM sodium phosphate (pH 6.5) containing 1 mM benzamidine. After application of the sample, the column is washed with 1–2 column volumes of the equilibrating buffer. Protein C_a appears in the breakthrough peak and contains no detectable thrombin activity (as judged by its inability to clot 0.4% fibrinogen solution in 30 min at 37°C). Under these conditions, thrombin remains tightly bound to the SP-Sephadex and is eluted quantitatively from the column with 0.3 M sodium phosphate (pH 6.5). Protein C_a fractions are subsequently concentrated by ultrafiltration to about 2 mg/ml and stored as small aliquots in plastic tubes at −20°C.

Proteolytic Action of Protein C_a

Earlier work from our laboratory demonstrated that bovine protein C_a exhibited anticoagulant activity in the presence of phospholipid and calcium ions.[6] Moreover, the anticoagulant activity of protein C_a is completely inhibited by prior incubation of the enzyme with DFP.[4] This

[28] B. Dahlbäck and J. Stenflo, *Eur. J. Biochem.* **107**, 331 (1980).
[29] S. T. Steiner, G. W. Amphlett, and F. J. Castellino, *Biochem. Biophys. Res. Commun.* **94**, 340 (1980).

observation strongly suggested that the mechanism of protein C_a involved proteolytic degradation of one or more blood-clotting factors. A systematic investigation of the effects of protein C_a on several highly purified coagulation factors showed that protein C_a, in the presence of phospholipid and calcium ions, rapidly inactivated factor V_a and factor VIII_a by proteolysis.[9,14] The unactivated forms of these cofactors are also substrates for protein C_a,[12,14,30] but these reactions are at least an order of magnitude slower. This suggests that the activated forms of factor V and factor VIII may be the preferred substrates *in vivo* for protein C_a. Recently, Comp and Esmon[15] and Dahlbäck and Stenflo[28] have shown independently that protein C_a inhibits the activation of prothrombin by platelet-bound factor X_a through the destruction of the platelet receptor for factor X_a. Presumably, this platelet receptor is equivalent to a complex of factor V_a and phospholipid.

From several recent studies, a clearer picture of the molecular events associated with the inactivation of factor V_a by protein C_a is beginning to emerge. Esmon and co-workers[12] demonstrated by SDS gel electrophoresis that protein C_a degraded a slow migrating band of the factor V doublet and also degraded both the heavy and light chain of factor V_a derived from the activation of factor V by thrombin. In spite of the proteolysis, no apparent effect on factor V activity was observed when protein C_a was incubated with factor V.[12] Canfield and Kisiel,[10] using factor V_a obtained by treating factor V with a protease from Russell's viper venom (RVV-V), observed a rapid loss of factor V_a activity following treatment with catalytic amounts of protein C_a. Associated with this loss of activity was the cleavage of the light chain of factor V_a ($M_r =$ 83,000) resulting in the formation of fragments with MW 54,000 and 34,000 as determined by continuous SDS gel electrophoresis.[10] Little, if any, change in the covalent structure of the heavy chain ($M_r = 230,000$) of factor V_a was observed in this study.[30] Incubation of protein C_a with the isolated light chain of factor V_a resulted in the formation of fragments that appeared indistinguishable from those observed in the inactivation of intact factor V_a.[10] Fragments of similar size ($M_r = 54,000$ and 22,000) were observed when thrombin-activated factor V was incubated with protein C_a.[12] The precise origin of these fragments and their relationship to those seen with RVV-V activated factor V as the substrate for protein C_a is presently unclear.

Our present information on the action of protein C_a on factor VIII and factor VIII_a is derived from a recent study by Vehar and Davie.[14] These investigators demonstrated that catalytic amounts of protein C_a readily inactivated the coagulant activity of bovine preparations rich in factor VIII/vWF without any attendant loss of platelet-aggregating activity in

[30] W. M. Canfield, W. Kisiel, and E. W. Davie, in preparation.

human platelet-rich plasma. Thrombin-activated factor VIII/vWF was also rapidly inactivated by protein C_a. No cleavage of the 200,000-MW subunit of factor VIII/vWF was observed by SDS gel electrophoresis following the inactivation reaction.

The coagulant activities of purified factor VIII and thrombin-activated factor VIII were also inhibited by protein C_a.[14] The rate of inactivation of the thrombin-activated factor VIII was considerably faster than that observed for factor VIII, and both of these reactions required the presence of calcium and phospholipid.

Whether the proteolytic range of protein C_a is confined to factor V (V_a) and factor VIII (VIII_a) remains open to investigation. Preliminary work from Esmon's laboratory suggests that protein C_a may also play a significant role in fibrinolysis.[31,32] Clearly, the effect of protein C_a on the activity of factor V and factor VIII is specific and dramatic, and leads one to believe that regulation of thrombin formation by the specific inactivation of these two cofactors is a very likely role for protein C in hemostasis.

Species Specificity of Protein C_a

Bovine and human protein C_a appear to exert their maximal anticoagulant activity in the plasma from which it was derived.[3,4,7,8] The reason for this apparent species specificity is not known.

Inhibitors of Protein C_a

Protein C_a is inactivated by both diisopropyl fluorophosphate and phenylmethylsulfonyl fluoride.[4,6] Covalent binding of DFP to the heavy chain of protein C_a was shown by employing [³H]DFP.[5,6] Presumably, the inhibitor forms a covalent bond with the active-site serine residue in the heavy chain.[4] Benzamidine hydrochloride is a competitive inhibitor of protein C_a with a K_i of 7.5×10^{-4} M (S-2160 as substrate).[6]

Soybean-trypsin inhibitor, at a 10-fold molar excess to the enzyme, has no effect on its amidase activity.[4] In addition, protein C_a amidase activity is insensitive to high concentrations of antithrombin III in the presence or absence of heparin.[4]

Isolation of Human Protein C

Assay. The two-stage anticoagulant assay described earlier is used to localize human protein C in chromatographic and electrophoretic eluants.

[31] P. C. Comp and C. T. Esmon, *Circulation* **58**, II-210 (1978).
[32] P. C. Comp and C. T. Esmon, *in* "The Regulation of Blood Coagulation" (K. G. Mann and F. B. Taylor, eds.), Vol. 8, p. 583. Am. Elsevier, New York, 1980.

In the first stage of the assay, protein C is activated by human thrombin which is subsequently neutralized with a 2-fold molar excess of human antithrombin III in the presence of heparin.

Purification Procedure. With the exception of a dextran sulfate–agarose column chromatography step, protein C is isolated from human plasma essentially as described for bovine protein C. The isolation procedure is routinely carried out with 15 liters of human cryosupernatant and all steps are performed at 4°C.[3]

In contrast to the bovine molecule, human protein C elutes at a lower ionic strength than prothrombin from DEAE–Sephadex. Protein C is further purified by dextran sulfate–agarose column chromatography at pH 6 in the presence of calcium chloride. Protein C obtained from dextran sulfate–agarose is contaminated with prothrombin and protein S, which are readily separated from protein C by preparative discontinuous electrophoresis in 7.5% polyacrylamide gels. By this procedure, approximately 5 mg of purified protein C is consistently obtained from 15 liters of cryosupernatant. The product obtained is electrophoretically homogeneous and completely free of prothrombin, factor VII, and factor IX as judged by specific clotting assays. Occasionally, purified preparations of human protein C contain a small amount of factor X activity, which is subsequently removed from the preparation of immunoadsorption on a column of rabbit anti-human factor X agarose.

General Properties of Human Protein C

Human protein C is a glycoprotein composed of a heavy chain and a light chain held together by a disulfide bond(s). Human protein C appears to be more highly glycosylated than the bovine molecule and contains 77% protein and 23% carbohydrate.[3] The carbohydrate includes approximately 14 residues of galactose, 21 residues of mannose, 23 residues of glucosamine, and 12 residues of sialic acid.[3] The amino-terminal sequences of the heavy and light chains of human protein C exhibit extensive homology with the amino-terminal regions of the heavy and light chains of bovine protein C (Fig. 1).

The molecular weight of human protein C is 62,000 by SDS gel electrophoresis in the absence of reducing agent. The molecular weights of the heavy and light chains are 41,000 and 21,000, respectively, by this technique.[3] In continuous SDS electrophoresis in 10% acrylamide gels, the heavy and light chains of human protein C migrate as single bands. In the discontinuous SDS electrophoresis system[33] using 10–12.5% acrylamide gels, the heavy chain migrates as a doublet and the light chain migrates as

[33] U. K. Laemmli, *Nature (London)* **227**, 680 (1970).

Heavy chain

		5		10	↓	15		20		25		30

Human[a]: Asp -Pro- Glu- Asp -Gln- Glu -Val- Asp- Pro -Arg- Leu -Ile- Asp- Gly- Lys- Met- Thr -Arg- Arg -Gly- Asp -Ser- Pro -Trp- Gln -Val- Val -Leu-

Bovine[b]: Asp -Thr- Asn -Gln- Val -Asp- Pro -Arg- Ile- Val- Asp -Gly- Gln -Glu- Ala -Gly- Trp -Gly- Glu -Ser- Pro -Trp- Gln -Ala-

Light chain

		5		10		15		20		25		30		35		40

Human[a]: Ala -Asn- Ser -Phe- Leu -Gla- Gla -Leu- Arg -His- Ser -Ser- Leu -Gla- Arg -Gla- Cys -Ile- Gla -Gla- Ile -Cys- Asp -Phe- Gla -Gla- Ala -Leu- Gla -Ile- Phe -Gln- Asn -Val- Asp -Asp- Thr -Leu- Ala -Phe-Trp-

Bovine[b]: Ala -Asn- Ser -Phe- Leu -Gla- Gla -Leu- Arg -Pro- Gly -Asn- Val -Gla- Arg -Gla- Cys -Ser- Gla -Val- Cys -Gla- Phe- Gla -Gla- Ala -Arg- Gla -Ile- Phe -Gln- Asn -Thr- Gla -Asp- Thr -Met- Ala -Phe-Trp-

FIG. 1. Amino-terminal sequences of human and bovine protein C. Residue 18 of the heavy chain of human protein C has now been identified as methionine and not valine as previously reported [W. Kisiel, *J. Clin. Invest.* **64**, 761 (1979)]. Arrows indicate peptide bonds cleaved by thrombin in the activation process. The following observations apply to the light chain of human protein C: γ-carboxyglutamic acid (Gla) residues were not identified in this study and were presumed on the basis of homology (no PTH residues were observed in these cycles); these cycles; residues 10, 12, and 13 have now been positively identified as His, Ser, and Leu, respectively; residue 37 is tentatively identified as Thr. [a] This laboratory (W. Kisiel, unpublished data). [b] Taken from the published amino acid sequence [P. Fernlund and J. Stenflo, *in* "Vitamin K Metabolism and Vitamin K-Dependent Proteins" (J. W. Suttie, ed.), p. 84. University Park Press, Baltimore, Maryland, 1979].

a closely spaced triplet. The reason for this anomaly is presently not known. This phenomenon, however, is not observed with bovine protein C.

Purified preparations of human protein C often contain a contaminant (M_r = 38,000) as shown by SDS gel electrophoresis following reduction of the protein. This contaminant, designated β-protein C, results from proteolytic degradation of the heavy chain of protein C similar to that observed for human and bovine factor X (factor $X_{a\beta}$).[34,35] β-Protein C migrates with a slightly higher mobility than protein C from preparative discontinuous electrophoresis. Sequence analysis suggests that β-protein C arises from the proteolytic cleavage of a small peptide ($M_r \simeq 3000$) from the carboxyl-terminus of the heavy chain of intact protein C. The protease responsible for this cleavage has not been identified. Despite the liberal use of protease inhibitors, the formation of β-protein C often occurs early in the isolation procedure and may even be present in the starting plasma.[27] Protein C isolated from fresh plasma, however, usually contains little if any β-protein C, in contrast to protein C isolated from cryosupernatant or frozen plasma, which consistently yields 5–10% β-protein C.

Like the bovine molecule, human protein C exists in plasma as an inactive precursor that is readily converted to a serine protease by α-thrombin.[3] In the activation of human protein C, thrombin cleaves an Arg-Leu bond between residues 12 and 13 in the amino-terminal end of the heavy chain of the protein, releasing a small activation peptide (M_r = 1400).[3] The new amino-terminal leucine residue presumably forms an ion pair with the carboxyl group of an aspartic acid residue adjacent to the active-site serine. Human activated protein C appears to be the first example among the serine proteases characterized to date that contains an amino-terminal leucine residue. Other serine proteases usually contain either isoleucine or valine.

The extinction coefficient of human protein C, estimated from its amino acid composition, is 14.5.[3] The isoelectric point of the purified protein is 4.4–4.8.[26]

---

[34] R. G. DiScipio, M. A. Hermodson, and E. W. Davie, *Biochemistry* **16**, 5253 (1977).
[35] K. Fujikawa, K. Titani, and E. W. Davie, *Proc. Natl. Acad. Sci. U.S.A.* **75**, 3359 (1975).

[27] Factor XIII (Fibrin-Stabilizing Factor)

By L. LORAND, R. B. CREDO, and T. J. JANUS

Factor XIII is the last plasma zymogen to become activated on the coagulation cascade during the clotting of vertebrate blood.[1] The enzymatic form (called XIII_a, activated fibrin-stabilizing factor, fibrinoligase, or plasma transglutaminase) contains a cysteine active center and acts as an endo-γ-glutamine: ϵ-lysine transferase in blood coagulation. Specifically, it catalyzes the fusion of fibrin units within the existing clot network by producing intermolecular γ-glutamyl-ϵ-lysine side-chain bridges, illustrated for simple dimerization as:

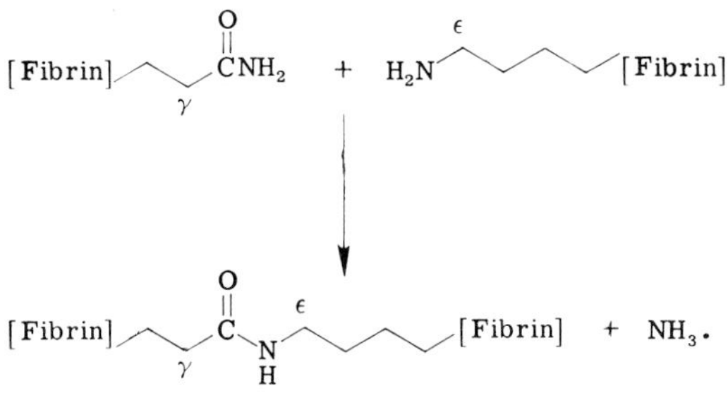

Thus, in the narrow definition of the term, this enzyme is not a proteolytic one because it does not alter the α-amide backbone of its protein substrate. With synthetic substrates, such as β-phenylpropionyl-thiocholine, however, reactions of hydrolysis, aminolysis, and O-alcoholysis can be demonstrated to proceed through a path of acylation and deacylation of the enzyme in a manner analogous to the reactions catalyzed by chymotrypsin, trypsin, or papain.[2-4]

The Physiological Pathway of Activation

Ca^{2+} ions play quite a specific and pivotal role in the activation of Factor XIII. If this zymogen, which has an a_2b_2-heterologous structure, is first acted on by thrombin (to produce XIII' = $a_2'b_2$), 10–15 mM Ca^{2+} can

[1] C. G. Curtis and L. Lorand, this series, Vol. 45, p. 177.

[2] C. G. Curtis, P. Stenberg, K. L. Brown, A. Baron, K. Chen, A. Gray, Jr., I. Simpson, and L. Lorand, *Biochemistry* **13**, 3257 (1974).

[3] P. Stenberg, C. G. Curtis, D. Wing, Y. S. Tong, R. B. Credo, A. Gray, Jr., and L. Lorand, *Biochem. J.* **147**, 155 (1975).

[4] K. N. Parameswaran and L. Lorand, *Biochemistry* **20**, 3703 (1981).

bring about the dissociation of the thrombin-modified zymogen ensemble and the unmasking of one equivalent of cysteine for each of the catalytically active a subunits (called at this stage a_2^* or XIII_a). Fibrinogen ($\leq 10^{-5}$ M) exerts a profound effect on this transition by lowering the Ca^{2+}-ion requirement to ≤ 1.5 mM, which corresponds to the concentration of free Ca^{2+} in plasma.[5] Thus the physiological regulation of the pathway for the conversion of factor XIII, occurring at pH 7.5, $\mu = 0.15$, 37°C in about 10 min, may be reconstructed as:

$$\text{Factor XIII} \xrightarrow{\text{Thrombin}} \text{XIII}' \xrightarrow[10^{-5} M \text{ fibrinogen}]{1.5 \text{ m}M \text{ Ca}^{2+}} \text{XIII}_a$$

In terms of subunit reactions, this corresponds to:

$$a_2 b_2 \xrightarrow[\substack{\searrow \\ \text{Activation} \\ \text{peptide}}]{\text{Thrombin}} a_2' b_2 \xrightarrow[10^{-5} M \text{ fibrinogen}]{1.5 \text{ m}M \text{ Ca}^{2+}} a_2^* + b_2$$

Credo et al.[6] recently showed that a peptide fragment, corresponding to residues 242–424 in the $A\alpha$ chain of human fibrinogen contains the activity of regulating the conversion of $a_2' b_2$ to a_2^* and b_2.

Establishment of the physiological pathway is significant in relation to human pathology for the correct differential diagnosis of hemorrhagic disorders of fibrin stabilization.[7] In addition to the autosomal recessive trait of hereditary factor XIII deficiency, mostly attributable to a lack of a subunits, this group of disorders is already known to comprise a variety of unrelated molecular diseases, some of which are due to difficulties in generating the a_2^* enzyme from the $a_2' b_2$ zymogen.

Details for the activation of factor XIII by thrombin and Ca^{2+} were given in this series.[1]

Thrombin-Independent Pathway

It is now known[5] that Ca^{2+} concentrations much higher than previously attempted can directly cause an activation of the $a_2 b_2$ zymogen without any prior modification by thrombin:

$$a_2 b_2 \xrightarrow{0.1 M \text{ Ca}^{2+}} a_2^0 + b_2$$

The activating effect of Ca^{2+} can be reversed by the addition of EDTA. In order to distinguish the novel form of the enzyme from the thrombin-modified $XIII_a = a_2^*$, we shall refer to it as $XIII_2^0 = a_2^0$.

[5] R. B. Credo, C. G. Curtis, and L. Lorand, Proc. Natl. Acad. Sci. U.S.A. 75, 4234 (1978).
[6] R. B. Credo, C. G. Curtis, and L. Lorand, Biochemistry 20, 3770 (1981).
[7] L. Lorand, M. S. Losowsky, and K. J. M. Miloszewski, Prog. Hemostasis Thromb. 5, 245 (1980).

James and co-workers[8] examined this mode of enzyme generation by studying the heterologous dissociation of the molecule, the unmasking of iodoacetamide-reactive cysteines of the a subunits, and the steady-state kinetics of acyl-group transfer using β-phenylpropionylthiocholine and methanol as substrates. The kinetic measurements were performed according to the protocol of Procedure D as described by Curtis and Lorand in Vol. 45 of this series (p. 183),[1] using 5,5'-dithiobis (2-nitrobenzoic acid) for monitoring the release of thiocholine. Instead of including amines, methanol was employed as nucleophile. The measured steady-state kinetic constants (K_m^{app} for β-phenylpropionylthiocholine; K_m^{app} for methanol and k_{cat} values) for the a_2^0 enzyme were indistinguishable from those obtained with the thrombin-modified a_2^*. Thus, whatever the reason for the thrombin-catalyzed release of the activation peptides may be (corresponding to 37 residues from the N-terminus of each a subunit),[9] it does not seem to be related to the unmasking of the primary catalytic site of the enzyme per se.

Chaotropic anions can accelerate the direct conversion of the a_2b_2 zymogen and can reduce the apparent Ca^{2+} requirement for the process.

The example given in Fig. 1 illustrates the methodology employed for the thrombin-independent activation of the factor XIII zymogen, by the criterion of unmasking of iodoacetamide-reactive sites.

All reagents were dissolved in 0.05 M Tris-HCl, pH 7.5, 37°C, unless otherwise specified. Hirudin (Grade II, from leeches; Sigma Chemical Company), 500 units/ml, was stored at $-10°C$, and iodo[1-^{14}C]acetamide (Amersham-Searle), 4.72 mM, 53 Ci/mol, was stored at $-10°C$. CaCl$_2$ solutions of approximately 500, 100, and 40 mM, and NaCl solutions of approximately 23, 623, 713, and 773 mM and 1 M were prepared. KCNS was dissolved to a strength of 1 M. Human factor XIII (4.33 mg/ml) was stored at 4°C in a buffer of 0.05 M Tris-HCl (pH 7.5) containing 0.001 M EDTA.

The reaction mixtures contained the following: 10 μl of 500, 100, or 40 mM CaCl$_2$ to achieve final CaCl$_2$ concentrations of 50, 10, and 4 mM, respectively, after allowance for the EDTA content of the zymogen solution; 20 μl of 23, 623, 713, or 773 mM NaCl to obtain a final ionic strength of μ = 0.7 including the final CaCl$_2$ concentration of 50, 10, 4, 0 mM, respectively; 50 μl of 1 M KCNS, in various dilutions with 1 M NaCl; 5 μl Hirudin, 500 U/ml; 5 μl of iodo[1-^{14}C]acetamide.

The 90-μl solution was brought to 37°C and the reaction was started by the addition of 10 μl of zymogen. At 10-, 20-, 30-, 40-, 50-, and 60-min

[8] T. J. James, R. B. Credo, C. G. Curtis, L. Häggroth, and L. Lorand, *Fed. Proc., Fed. Am. Soc. Exp. Biol.* **40**, 1585 (Abst. 258) (1981).
[9] T. Takagi and R. F. Doolittle, *Biochemistry* **13**, 750 (1974).

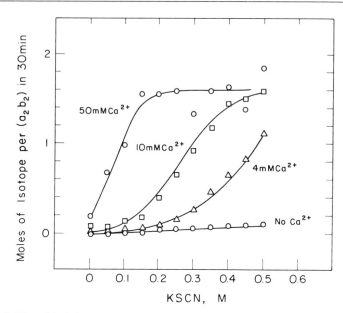

FIG. 1. Thrombin-independent alkylation of human factor XIII by iodoacetamide in the presence of KCNS and CaCl₂.

intervals, 10-μl aliquots were withdrawn and applied to Whatman 3 MM filter paper for measuring protein-bound radioactivity, as given by Curtis and Lorand (p. 186).[1]

The measured efficacies of chaotropic anions gave the approximate order of p-toluenesulfonate $\geq$ thiocyanate > iodide > bromide.

In summary, Ca^{2+}-ions alone can cause a labilization of the a_2b_2 zymogen and generate enzyme activity (a_2^o) when employed at sufficiently high concentrations (i.e., ≥ 0.1 M Ca^{2+} at 37°C, pH 7.5, $\mu > 0.3$, in a time frame of about 10 min). Chaotropic anions, such as 0.1 M p-toluenesulfonate or thiocyanate, may be used to reduce the apparent Ca^{2+} requirement for the thrombin-independent activation procedure to about 0.05 M Ca^{2+}. On the other hand, the thrombin-modified $a_2'b_2$-zymogen ensemble dissociates and can produce the a_2^* enzyme at 37°C, pH 7.5, and $\mu = 0.15$ with only 10–15 mM Ca^{2+} present. Addition of physiological concentrations of fibrinogen ($\leq 10^{-5}$ M) can reduce this further to a mere 1.5 mM Ca^{2+}.

Purification of Human Factor XIII

The method for purifying factor XIII from blood bank plasma, as described in earlier volumes of this series,[1,10] has now been further refined

[10] L. Lorand and T. Gotoh, this series, Vol. 19, p. 770.

from the point of view of separating the excess free b subunits from the heterologous zymogen ensemble.

The following procedure applies to 20 liters of outdated human plasma. If it has not been frozen, the material can be processed directly. Otherwise, the plasma may be stored frozen with the addition of 6 mM benzamidine and thawed at 37°C just prior to fractionation. All steps are carried out at 4°C in plastic containers. To 20 liters of plasma, 5 liters of saturated ammonium sulfate solution containing 1 mM EDTA are added, and the precipitate that is formed within a 1-hr period is collected by centrifugation (2400 g for 30 min). The pellet is washed with 20% saturated ammonium sulfate–1 mM EDTA, and is dissolved with 2 liters of 0.15 M KCl–1 mM EDTA, pH 7. After 6 hr, any remaining solid is collected and taken up in 1 liter of the 0.15 M KCl–1 mM EDTA solution by gentle warming (25°C, 1 hr). After centrifugation (2400 g for 30 min), the supernatants are combined and the pH is adjusted to pH 5.4 with the dropwise addition of 1 N acetic acid and stirring. Saturated ammonium sulfate (containing 1 mM EDTA) is added to 16% of saturation. Following 1 hr of precipitation and centrifugation (2400 g for 30 min), the pellet is taken up in 2 liters of buffer of 0.05 M Tris-HCl (pH 7.5)–0.2 M NaCl–1 mM EDTA. After 6 hr, an equal volume of a solution containing 0.05 M Tris-HCl (pH 7.5)–1 mM EDTA is added, followed by heating to between 53 and 56°C for a period of 3 min (by immersion into an 80°C water bath, with constant stirring and by monitoring the temperature in the protein solution). After immediate cooling to 4°C in an ice bath and centrifugation (2400 g for 30 min), in order to clear the solution of flocculated proteins, the supernatant is brought to 36% saturation by adding saturated ammonium sulfate (containing 1 mM EDTA). Precipitate formed within 1 hr is collected by centrifugation (2400 g for 30 min) and is taken up in 250 ml of 0.05 M Tris-HCl (pH 7.5)–1 mM EDTA. This step may require mild stirring overnight. The material is then dialyzed against the 0.05 M Tris-HCl–1 mM EDTA buffer of pH 7.5, cleared by centrifugation (10,000 g for 30 min) and, in an ascending mode, applied to a column of DEAE–cellulose (5 × 45 cm, Whatman DE-52, microgranular, preswollen) which was equilibrated with the same Tris–EDTA buffer. Flow rate is set to 110 ml/hr and, after washing with 2 column volumes of the equilibrating buffer, 4.8 liters of linear gradient of NaCl to 0.2 M is started using the same buffer with EDTA (i.e., one chamber of the gradient former contains 2.4 liters of buffer, the other 2.4 liters of 0.2 M NaCl in buffer). Collected fractions (13 ml) are examined for absorbancy at 280 nm and conductivity, and are also assayed for potential enzyme activity. The latter is measured by the incorporation of [14C]putrescine into N,N'-dimethylcasein, essentially as given in Procedure C in this series (p. 182)[1]; however, since no fibrinogen was present there was no need to add glycerol and the 4-min heat treatment at 56°C

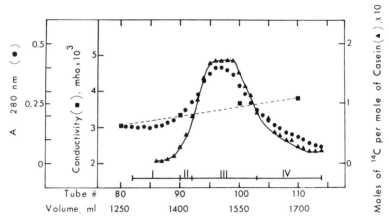

FIG. 2. Elution profile of the factor XIII peak from DEAE–cellulose.

was also omitted. For the purpose of detection of free b subunits, fractions were also subjected to nondenaturing gel electrophoresis[11] and to SDS electrophoresis.[12] The zymogen, as shown in Fig. 2, eluted with a conductivity of approximately 0.34×10^{-2} mho. In this particular example, four pools (I–IV) were separated, but in routine procedures, materials corresponding to I + II and III + IV can be combined. In the example in Fig. 2, assuming $E_{1\,cm,280\,nm}^{1\%}$ of 13.8,[13] the approximately 1 g of protein applied onto DEAE–cellulose yielded 17 mg of protein in pool I, 8 mg in II, 53 mg in III, and 18 mg in IV. Nondenaturing and denaturing (i.e., SDS) electrophoresis showed that pools I and II contained appreciable amounts of free b subunits and also some inactive polymeric substance. This leading edge of the peak could thus be used for preparing purified b subunits on further passage through Sepharose-6B.

The collected pools of material were concentrated by the addition of a saturated ammonium sulfate solution (containing 1 mM EDTA) to 36% of saturation and by centrifugation (17,000 g for 30 min). The proteins were taken up in a minimal volume of 0.55 M Tris-HCl (pH 7.5)–1 mM EDTA, and were applied to two tandem columns (2.5 × 90 cm each) of Sepharose-6B in an ascending mode, using the same Tris-EDTA solution at a flow rate of 12 ml/hr. The collected fractions were examined for absorbancy at 280 nm, potential transamidase activity in the N-N'-dimethylcasein–[^{14}C]putrescine incorporation assay, and were subjected

[11] D. Rodbard and A. Chrambach, *Anal. Biochem.* **40,** 95 (1971).
[12] K. Weber and M. Osborn, *J. Biol. Chem.* **244,** 4406 (1969).
[13] M. L. Schwartz, S. V. Pizzo, R. L. Hill, and P. A. McKee, *J. Biol. Chem.* **248,** 1395 (1973).

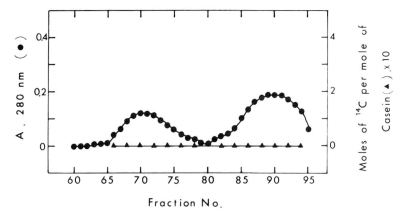

FIG. 3. Sepharose-6B chromatography of the pool I material from Fig. 2.

to nondenaturing and also to SDS–gel electrophoresis. Three illustrations are given. Figure 3 shows the results of gel filtration with DEAE–cellulose pool I, Fig. 4 with pool III, and Fig. 5 with pool IV. The electrophoretic analysis showed that the first peak in Fig. 3 corresponded to inactive polymer and that the second peak contained the free *b* carrier subunits. The zymogen activity peak in Fig. 4 comprised factor XIII, containing equal amounts of *a* and *b* subunits, whereas the follow-up peak contained only *b*. The pool IV material shown in Fig. 5, on elution

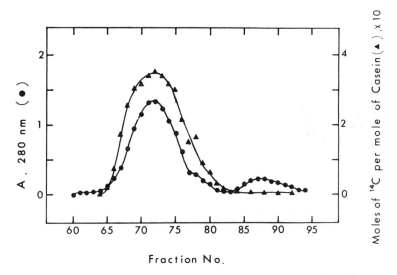

FIG. 4. Sepharose-6B chromatography of the pool III material from Fig. 2.

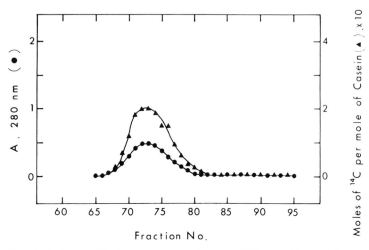

FIG. 5. Sepharose-6B chromatography of the pool IV material from Fig. 2.

from Sepharose-6B, proved to be pure factor XIII. The purified factor XIII fractions (from Figs. 4 and 5) and the free *b* subunits (from Figs. 3 and 4) were collected; each was concentrated by precipitation with 36% ammonium sulfate–1 m*M* EDTA and centrifugation as described earlier. Precipitates were dissolved in 0.05 *M* Tris-HCl (pH 7.5)–1 m*M* EDTA,

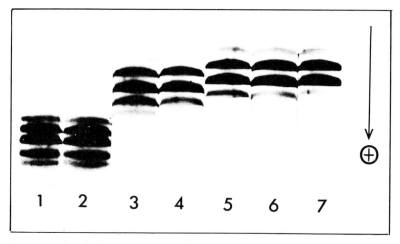

FIG. 6. Isoelectric focusing patterns of *b* subunits of a human factor XIII preparation from pooled plasma, on desialylation. Treatment with neuraminidase lasted from 1, 2, 4, 6, and 8 hr for the proteins shown in tracks 3, 4, 5, 6, and 7. The controls (in tracks 1 and 2) were not treated with the enzyme. Isoelectric focusing was performed in 5% polyacrylamide gels, containing 2% ampholine of pH 3–10 and 6 *M* urea. Approximately 20 μg of protein was applied to each gel.

and were dialyzed against the same buffer. Factor XIII was stored at 4°C (5–12 mg protein/ml); the purified b subunits at $-10°C$ (3.5 mg/ml).

Though not shown in Fig. 2, it may be mentioned that the DEAE–cellulose chromatographic procedure described can be used simultaneously for the isolation of plasma fibronectin or cold-insoluble globulin.[14] This protein emerges with the NaCl gradient as a peak starting at about 5 $\times$ 10^{-3} mho of conductance.

Microheterogeneity of the b Subunit

Isoelectric focusing showed that the b subunits, as isolated from pooled human plasma, in spite of apparent homogeneity in regard to size (i.e., in spite of showing a single band in SDS–gel electrophoresis), comprised several species with different pI's (Fig. 6). Treatment with neuraminidase (8 hr, 37°C, 64 units of enzyme/nmol of b_2) removed sialic acid entirely (about 10 mol/mol of b_2) from the protein, shifting the pI's of the remaining species in a more alkaline direction. The observed residual microheterogeneity, showing 2 major and 2 minor bands, with the desialilated material may reflect group-specific differences in the population.

[14] J. Molnar, F. B. Gelder, M. Z. Lai, G. E. Siefring, R. B. Credo, and L. Lorand, *Biochemistry* **18**, 3909 (1979).

[28] Assay of Coagulation Proteases Using Peptide Chromogenic and Fluorogenic Substrates

By Richard Lottenberg, Ulla Christensen,
Craig M. Jackson, and Patrick L. Coleman

Introduction

Amino acid chromogenic and fluorogenic substrates have been used for many years for assaying proteases. The sensitivity of the assay procedures that employ these substrates and the convenience of spectrophotometric or fluorometric measurements has led to their widespread use. Most of the early amino acid chromogenic and fluorogenic substrates are highly selective for the primary specificity-determining (P1) amino acid; thus substrates such as benzoylarginine-p-nitroanilide, for assaying trypsin-like proteases, as well as aromatic amino acid p-nitrophenyl esters for chymotrypsin-like proteases, have been extensively investigated and

METHODS IN ENZYMOLOGY, VOL. 80

employed for routine proteolytic enzyme assay. The recognition that both the selectivity of many proteases and their catalytic efficiency depend on interactions with subsite amino acids in the peptide substrate coupled with the availability of amino acid sequences around the cleavage sites in several zymogens of the coagulation and fibrinolytic systems has led to the synthesis and commercial availability of a variety of peptide chromogenic and fluorogenic substrates with much greater selectivity than the single amino acid chromogenic and fluorogenic substrates. Such increased selectivity is required because all of the proteases of these systems are trypsin-like in their primary specificity, and thus discrimination among them without exploitation of the selectivity that results from secondary binding-site interactions is virtually impossible.

A number of monographs have been published reporting procedures for specific protease assays using peptide substrates[1-3] and a comprehensive review of the literature up to 1980 on chromogenic and fluorogenic substrates for coagulation and fibrinolytic system proteases has been published.[4] A bibliography of published reports employing fluorogenic and chromogenic substrates is available from one of the manufacturers of peptide p-nitroanilide substrates.[5]

Kinetic constants that describe the hydrolysis of even the commercially available peptide chromogenic and fluorogenic substrates are relatively limited, owing in part to the fact that few of the proteolytic enzymes of the coagulation, fibrinolytic, and kinin systems are commercially available. The data presented in this report are derived from four sources: the limited literature and yet to be published work from three laboratories of the authors of this report. Several reports have listed relative rates of hydrolysis of peptide p-nitroanilide substrates by different coagulation proteases at a single substrate concentration, however such data are of very limited use in the design of assay procedures and thus have not been included in this review.

Methods

The peptide p-nitroanilide substrates used for these investigations are commercially available and were generously provided by AB Kabi Pep-

[1] M. F. Scully and V. V. Kakkar, eds., "Chromogenic Peptide Substrates: Chemistry and Clinical Usage." Churchill-Livingstone, Edinburgh and London, 1979.
[2] I. Witt, ed., "New Methods for the Analysis of Coagulation Using Chromogenic Substrates." de Gruyter, Berlin, 1977.
[3] H. R. Lijnen, D. Collen, and M. Verstraete, eds., "Synthetic Substrates in Clinical Blood Coagulation Assays." Nijhoff, The Hague, 1980.
[4] R. M. Huseby and R. E. Smith, *Semin. Thromb. Hemostasis* **6**, 175 (1980).
[5] Reference List for Enzyme Substrates, Kabi Diagnostica, Stockholm, Sweden.

tide Research (S-431 22 Molndal, Sweden), Kabi Group Inc. (One Lafayette Place, Greenwich, Connecticut), Boehringer-Mannheim, Gmbh (Postfach 120, D-8132 Tutzing, Federal Republic of Germany), Boehringer Mannheim Biochemicals (7941 Castleway Drive, Indianapolis, Indiana), and Pentapharm Ltd. (CH-4002, Basel, Switzerland). Sources of the substrates used in studies from the literature are found in the original publications.

All peptide p-nitroanilides were dissolved in deionized water that had been adjusted to pH 4 with hydrochloric acid. Substrates were stored frozen in this solution and were stable for periods longer than 1 year under these conditions. Substrate concentrations were determined from absorbance at the isosbestic wavelength for the peptide p-nitroanilide–p-nitroaniline mixtures. Extinction coefficients of 8270 liters/mol/cm in water and 8270 liters/mol/cm in solutions containing 0.01 M HEPES–0.01 M Tris–0.1 M NaCl were employed. Such a procedure was necessary as some peptide p-nitroanilide substrates are hygroscopic and thus difficult to weigh accurately.

Protease concentrations were determined by active-site titration using the procedure of Chase and Shaw.[6] Preparations of thrombin were at least 95% active by active-site titration using an extinction coefficient of 1.95 ml/mg/cm at 280 nm to determine protein concentration. Preparations of factor X_a were greater than 80% active by active-site titration using an extinction coefficient of 0.95 ml/mg/cm at 280 nm to estimate the protein concentration. All protease solutions were diluted into a buffer that contained 0.1% polyethylene glycol 6000 to prevent enzyme loss due to adsorption.

The rate of peptide p-nitroanilide hydrolysis was determined from the change in absorbance at 405 nm using an extinction coefficient for p-nitroaniline of 9920 liters/mol/cm for this reaction buffer. Data obtained by two authors (R.L. and C.M.J.) are all from solutions consisting of 0.01 M HEPES–0.01 M Tris-HCl (pH 7.8)–0.1 M NaCl–0.1% polyethylene glycol 6000. Polyethylene glycol concentrations less than 2% were without effect on the kinetic parameters being determined; at concentrations greater than 2%, K_m increased essentially linearly with polyethylene glycol concentration. Data provided by another author (U.C.) were determined in a buffer consisting of 0.05 M Tris-HCl–0.1 M NaCl, pH 8.0. All data are for a temperature of 25°C unless otherwise indicated. Measurements were made using a Cary 219 spectrophotometer at a spectral slit width of 2 nm or less. Data were transferred directly to a digital computer (DEC PDP 11/34a) and initial velocities were estimated using either the direct linear plot-based procedure of Cornish-Bowden[7] or by nonlinear least-squares

---

[6] T. Chase, Jr. and E. Shaw, *Biochemistry* **8**, 2212 (1969).
[7] A. Cornish-Bowden, *Biochem. J.* **149**, 305 (1975).

fitting to a second-order polynomial. In all but the very lowest substrate concentration mixtures for substrates with Michaelis constants less than 5 μM, less than 2% of the substrate was hydrolyzed in the reactions. Data provided by U.C. were obtained under the same conditions; however, a Beckman Model 25 spectrophotometer was employed and initial velocities were estimated from the slope of a tangent drawn to the reaction-progress curve at zero time. The substrate-concentration range investigated in the laboratories of C.M.J. and U.C. varied from 0.1 to 0.3 $\times K_m$ to at least 5, and as high as 100 $\times K_m$, depending on the actual value for the Michaelis constant and the solubility of the particular substrate. Data provided by P.L.C. were obtained using a Gemsaec centrifugal analyzer. Initial velocities were calculated from the initial linear portion of the progress curve. Substrate concentrations ranged from $0.2K_m$ to $5K_m$. All reactions were at 37°C. Data from the literature are as described in the specific reports.

Kinetic parameters were determined by unweighted nonlinear least-squares fitting of the simple Michaelis–Menten equation to the data (R.L., C.M.J.), as described by Wilkinson,[8] or a weighted linear regression to the Lineweaver–Burke transform of the Michaelis–Menten equation (U.C. and P.L.C.) as described by Cleland.[9] Substrates for which no independent values for k_{cat} and K_m are given in the tables, but which do have an estimate of k_{cat}/K_m, gave linear dependence of the velocity on substrate concentration. The estimate for k_{cat}/K_m was determined by an unweighted linear regression fit to these data. Data for all substrates were obtained for no fewer than 5 different substrate concentrations, and in some cases, as many as 20 substrate concentrations were employed. In all cases, the Michaelis–Menten equation was assumed to fit the data, although evidence for substrate activation with some of the substrates and thrombin has been observed. Values for the relative standard errors in the kinetic parameter estimates from a single experiment were approximately ±10% for K_m and ±5% for k_{cat}. When several substrate-concentration dependence data sets for thrombin and factor X_a were examined, the standard errors of the mean were approximately ±12% for both parameters. On the basis of these estimates for the "true" error, all parameters are quoted to two significant figures in the tables.

Data for Individual Proteases

Thrombin. Data obtained from the investigation of the hydrolysis of 24 peptide *p*-nitroanilides, one tosyl-arginine-nitrobenzyl ester,[10] and one

[8] G. N. Wilkinson, *Biochem. J.* **80**, 324 (1961).
[9] W. W. Cleland, *Adv. Enzymol.* **29**, 1 (1967).
[10] G. W. E. Plaut, *Haemostasis* **7**, 105 (1978).

thiobenzyl ester[11] by thrombin are given in Table I. Agreement between the kinetic parameters for the hydrolysis of benzoyl-L-arginine-p-nitroanilide determined by Takasaki et al.[12] and those reported here is very good. Limited data for thrombin for which the active enzyme concentration has been determined precludes more extensive comparisons. In general the values for both K_m and k_{cat} for bovine and human thrombins are very similar, but not identical. Until more extensive intralaboratory comparisons are available, it is impossible to determine whether these differences reflect significant differences between the enzymes from the two species or represent interlaboratory differences.

Effects of pH and solution composition on the hydrolysis of two peptide p-nitroanilides, Tos-Gly-L-Pro-L-Arg-pNA and H-D-Phe-L-Pip-L-Arg-pNA and Cbz-L-Lys-SBzl are given in Table II. The magnitude of the effects from altering buffer composition at constant pH indicate clearly the necessity for control of solution composition and the necessity for knowing solution composition when fundamental information about the properties of the particular enzyme are desired from the kinetic data. Dependence on pH is in general relatively small, between pH 7.8 and 8.5 (R. Lottenberg and C. M. Jackson, unpublished observations). Specific monovalent cation effects on the activity of thrombin have been reported and must be considered in interpreting differences in ionic strength and solution composition.[13]

Factor X_a. Kinetic parameters for the hydrolysis of 25 peptide p-nitroanilide substrates by bovine and human factor X_a are given in Table III. Interestingly, the Michaelis constants for all of the substrates investigated for factor X_a lie in the range of 0.1 mM, whereas for thrombin, values as low as 0.4 μM have been observed. Solution composition is as described above for the data provided by R. L. and C.M.J. Specific monovalent cation effects on the activity of factor X_a have been reported.[14]

Plasmin and Urokinase. Kinetic data for the hydrolysis of 16 peptide p-nitroanilide substrates, 1 thiobenzyl ester,[11] and 1 nitrobenzyl ester[10] by plasmin and urokinase are given in Table IV. Data obtained by U.C. were as described above for thrombin.

Plasminogen–Streptokinase Complex. Data obtained for the hydrolysis of H-D-Phe-L-Leu-L-Lys-pNA and Tos-Gly-L-Pro-L-Lys-pNA by the plasminogen–streptokinase complex are given in Table V. Data for six different forms of plasminogen that differ in the extent to which portions of

[11] G. D. J. Green and E. Shaw, *Anal. Biochem.* **93**, 223 (1979).
[12] S. Takasaki, K. Kasai, and S. Ishii, *J. Biochem. (Tokyo)* **78**, 1275 (1975).
[13] E. F. Workman and R. L. Lundblad, *Arch. Biochem. Biophys.* **185**, 544 (1978).
[14] C. L. Orthner and D. P. Kosow, *Arch. Biochem. Biophys.* **185**, 400 (1978).

TABLE I
THROMBIN KINETIC PARAMETER SUMMARY

Substrate	Bovine thrombin				Human thrombin			
	K_m[a]	k_cat[a]	k_cat/K_m[a]	Ref.	K_m[a]	k_cat[a]	k_cat/K_m[a]	Ref.
1. H-D-Phe-Aze-L-Arg-pNA (S-2388)	0.43	48	110	b	3[c]	120[c]	40[c]	d
2. H-D-Ile-L-Pro-L-Arg-pNa (S-2288)	1.1	74	62	b	1.5	90	60	b
3. H-D-Phe-L-Pip-L-Arg-pNA (S-2238)	1.6	95	59	b	4.4	100	23	e
4. H-D-Val-L-Pro-L-Arg-pNA (S-2234)	2.7[f]	200	71	b	13[f]	180[f]	14[f]	e
	9.0[g]	—	—	d	7[g]	150[g]	21[g]	d
	2.0	89	44	b				
5. Tos-Gly-L-Pro-L-Arg-pNA (Chromozym-TH)	4.0	100	25	b	4.2	130	31	b
	5.7[f]	200[f]	36[f]	b	8.5	130	15	e
					13[f]	220[f]	17[f]	e
6. PyrGlu-L-Pro-L-Arg-pNA (S-2366)	39	150	4.1	b	12[i]	—	—	h
7. Bz-L-Phe-L-Val-L-Arg-pNA (S-2160)	18	38	2.1	b	83	30	0.36	e
8. Tos-Gly-L-Pro-L-Lys-pNA (Chromozym-PL)	80[g]	0.068[j]		d	68[f]	45[f]	6.6[f]	e
	21	23	1.1	b	110[g]	50[g]	0.45[g]	d
25. Cbz-Lys-S-Bzl	40[l]	35[l]	0.88[l]	k				
9. Bz-L-Phe-L-Val-L-Arg-pNA (Peptide Research Foundation)	72	38	0.53	b	51	34	0.67	e
10. Bz-PyroGlu-Gly-L-Arg-pNA (S-2405)	66	28	0.44	b				
11. H-D-Val-L-Phe-L-Arg-pNA (S-2325)	130	22	0.17	b	60[f]	52[f]	0.87[f]	e

Substrate	K_m	k_{cat}	k_{cat}/K_m	Ref.	K_m	k_{cat}	k_{cat}/K_m	Ref.
(S-2302)								
13. Bz-L-Ile-L-Glu-Gly-L-Arg-pNA (S-2337) (Pip)	41	2.6	0.065	[b]	180	9.0	0.052	[e]
14. H-D-Val-L-Leu-L-Arg-pNA (S-2266)	350	19	0.054	[b]				
15. Cbz-L-Val-Gly-L-Arg-pNA (Chromozym-TRY)	73	3.2	0.045	[b]	150	4.6	0.032	[e]
	180[f]	9.0[f]	0.049	[b]	350[f]	8.7[f]	0.025[f]	[e]
16. Bz-L-Ile-L-Glu-Gly-L-Arg-pNA (S-2222)	57	0.94	0.017	[b]				
17. L-PyroGlu-Gly-L-Arg-pNA (S-2444)			0.0044	[b]	1100	3.1	0.0027	[e]
					1800[f]	6.6[f]	0.0036[f]	[e]
18. H-D-Val-L-Phe-L-Lys-pNA (S-2390)			0.0018	[b]				
19. H-D-Val-L-Leu-L-Lys-pNA (S-2251)			0.00080	[b]	1400	1.2	0.00086	[e]
					2500[n]	0.67[n]	0.00027[n]	[m]
20. PyrGlu-L-Phe-L-Lys-pNA (S-2403)			0.000033	[b]				
21. Bz-L-Arg-pNA	120[o]	0.071[o]	0.00063[o]	[b]				
	200[q]	0.12[q]	0.00058[q]	[p]				
22. Bz-L-Pro-L-Phe-L-Arg-pNA (Chromozym PK)					490	0.11	0.00022	[e]
					370[f]	0.12[f]	0.00033[f]	[e]
26. PyrGlu-L-Phe-L-Lys-pNA					1100[n]	0.07[n]	0.000064[n]	[m]
27. H-L-Ala-L-Phe-L-Lys-pNA					1000[n]	0.015[n]	0.000015[n]	[m]
28. Tos-L-Arg-p-NitroBzl	14[s]	0.037[j]		[r]				

[a] K_m (μM); k_{cat} (sec^{-1}); k_{cat}/K_m (liters/μmol/sec).
[b] R. Lottenberg and C. M. Jackson (unpublished data).
[c] 37°C, Tris, $I = 0.15$, pH 8.4.
[d] AB Kabi, manufacturer's literature.
[e] U. Christensen (unpublished data).
[f] 37°C.
[g] 37°C, Tris, $I = 0.15$, pH 8.3.
[h] Boehringer Mannheim, Gmbh, manufacturer's literature.
[i] 37°C, $I = 0.3$, pH 8.4.
[j] nM/min/NIH unit.
[k] G. D. J. Green and E. Shaw, Anal. Biochem. 93, 223 (1979).
[l] 0.1 M Tris–0.1 M NaCl, pH 8.0.
[m] D. Collen, H. R. Lijnen, F. De Cock, J. P. Durieux, and A. Loffet, Biochim. Biophy. Acta 165, 158 (1980).
[n] 0.1 M phosphate, pH 7.3.
[o] pH 8.0.
[p] S. Takasaki, K. Kasai, and S. Ishii, J. Biochem. (Tokyo) 78, 1275 (1975).
[q] 0.05 M Tris-HCl, pH 8.2.
[r] G. W. E. Plaut, Haemostasis 7, 105 (1978).
[s] 30°C, 0.1 M Tris-HCl, pH 8.4.

TABLE II
EFFECTS OF SOLUTION COMPOSITION AND pH ON THROMBIN ACTIVITY

Reaction conditions		$K_m{}^a$	$k_{cat}{}^a$	$k_{cat}/K_m{}^a$
Human THROMBIN–Tos-Gly-L-Pro-L-Arg-pNA (Chromozym-TH)				
	pH			
0.10 M Tris-HCl, 0.0 M NaCl	7.95	50	46	0.92
	8.36	43	55	1.3
	8.58	44	50	1.1
	8.76	47	34	0.72
	9.00	54	63	1.2
0.05 M Tris-HCl	pH			
0.0 M NaCl	8.58	23	32	1.4
0.375 M NaCl		11	43	3.9
0.75 M NaCl		16	52	3.3
	pH			
0.10 M Na Phosphate, 0.0 M NaCl	6.50	27	38	1.4
	7.00	14	47	3.4
	7.40	8	42	5.3
	7.74	8	49	6.1
	8.00	8	50	6.3
		8.5	72	8.5
	8.50	9	50	5.6
	pH			
0.05 M Na Phosphate, 0.0 M NaCl	8.00	6.0	52	8.7
0.15 M Na Phosphate		7.5	70	9.3
0.50 M Na Phosphate		5.0	69	13.8
	pH			
0.10 M PIPES	8.00	9	62	6.9
0.10 M Pyrophosphate	8.00	6	42	7.0
0.10 M Glycine	8.70	14	54	3.9
0.10 M Tricine	8.00	8	94	12.
0.10 M Triethanolamine	8.15	16	105	6.6
0.10 M MOPS	8.00	12	108	9.0
0.10 M HEPES	8.00	11	93	8.5
0.10 M TES	8.00	9	102	11.3
Human THROMBIN–D-Phe-L-Pip-L-Arg-pNA (S-2238)				
	pH			
0.10 M Tris-HCl, 0.0 M NaCl	7.95	18.5	23.3	1.3
	8.36	26.5	38.5	1.5
		26.	38.0	1.5

TABLE II (*continued*)

Reaction conditions	$K_m{}^a$	$k_{cat}{}^a$	$k_{cat}/K_m{}^a$
0.10 *M* Tris-HCl, 0.0 *M* NaCl	8.76	30.5	28.3	0.93
	9.00	41.5	33.6	0.81
	pH			
0.05 *M* Tris-HCl, 0.0 *M* NaCl	8.36	15	38.3	2.6
0.25 *M* Tris-HCl		27	43.5	1.6
0.50 *M* Tris-HCl		36	38.1	1.1
	pH			
0.10 *M* Na Phosphate, 0.0 *M* NaCl	6.50	4.5	35.0	7.8
	7.00	4.0	45.0	11
	7.40	4.0	50.8	13
	7.70	4.0	65.0	16
	8.00	4.0	63.3	16
	8.50	6.5	70.0	11
	pH			
0.01 *M* PIPES, 0.0 *M* NaCl	8.00	3.7	34.2	9.2
0.025 *M* PIPES		4.5	38.0	8.4
0.05 *M* PIPES		4.0	36.7	9.2
0.10 *M* PIPES		3.7	37.9	10
0.25 *M* PIPES		3.5	34.1	9.7
0.50 *M* PIPES		4.0	30.8	7.7

Human THROMBIN–Cbz-L-Lys-SBz

	$K_m{}^a$	$k_{cat}{}^a$	$k_{cat}/K_m{}^a$
	pH			
0.10 *M* Tris-HCl, 0.0 *M* NaCl	7.95	84	81	0.96
	8.15	75	90	1.2
	8.36	79	82	1.0
	8.76	78	100	1.3
	9.00	84	96	1.1
	pH			
0.10 *M* Na Phosphate, 0.0 *M* NaCl	6.50	54	38	0.70
	7.00	45	66	1.5
	7.40	34	85	2.5
	7.70	30	98	3.3
	8.00	28	107	3.8
	8.50	38	125	3.3
	pH			
0.05 *M* Na Phosphate, 0.0 *M* NaCl	8.50	42	108	2.6
0.375 *M* NaCl		24	109	4.5
0.75 *M* NaCl		29	110	3.8

[a] K_m (μM); k_{cat} (sec^{-1}); k_{cat}/K_m (liters/μmol/sec); reaction temperature 37°.

TABLE III
FACTOR X_a KINETIC PARAMETER SUMMARY

Substrate	Bovine factor X_a				Human factor X_a			
	K_m^a	k_{cat}^a	k_{cat}/K_m^a	Ref.	K_m^a	k_{cat}^a	k_{cat}/K_m^a	Ref.
10. Bz-PyroGlu-Gly-L-Arg-pNA (S-2405)	150	260	1.7	b				
13. Bz-L-Ile-L-Glu-Gly-L-Arg-pNA (S-2337) (Pip)	83	140	1.7	b	83	140	1.7	b
16. Bz-L-Ile-L-Glu-L-Gly-L-Arg-pNA (S-2222)	120[c]	210[c]	1.8[c]	b				
	140	130	0.89	b				
	290	130	0.45	b				
	300[e]	100[e]	0.33[e]	d				
	440[g]	110[g]	0.25[g]	f				
5. Tos-Gly-L-Pro-L-Arg-pNA (Chromozym-TH)	110	74	0.68	b	99	100	1.0	b
	220[c]	150[c]	0.67[c]	b				
15. Cbz-L-Val-Gly-L-Arg-pNA (Chromozym-TRY)	360	60	0.17	b				
2. H-D-Ile-L-Pro-L-Arg-pNA (S-2288)	1300	170	0.13	b				
	2000[h]	110[h]	0.055[h]	d				
12. H-D-Pro-L-Phe-L-Arg-pNA (S-2302)	700	68	0.097	b				
4. H-D-Val-L-Pro-L-Arg-pNA (S-2234)			0.070	b				
23. Bz-L-Val-Gly-L-Arg-pNA (Sigma B-4758)	620	38	0.061	b				
11. H-D-Val-L-Phe-L-Arg-pNA (S-2325)	690	27	0.039	b				
17. L-PyroGlu-Gly-L-Arg-pNA (S-2444)			0.032	b				
3. H-D-Phe-L-Pip-L-Arg-pNA (S-2238)	35	0.78	0.022	b				

	Substrate	K_m[a]	k_{cat}[a]	k_{cat}/K_m[a]	
14.	H-D-Val-L-Leu-L-Arg-pNA (S-2266)	1000	22	0.022	b
6.	PyrGlu-L-Pro-L-Arg-pNA (S-2366)	1700	24	0.014	b
8.	Tos-Gly-L-Pro-L-Lys-pNA (Chromozym-PL)			0.011	b
1.	H-D-Phe-Aze-L-Arg-pNA (S-2388)	210	2.3	0.011	b
18.	H-D-Val-L-Phe-L-Lys-pNA (S-2390)			0.0060	b
22.	Bz-L-Pro-L-Phe-L-Arg-pNA (Chromozym-PK)	190	1.1	0.0057	b
7.	Bz-L-Phe-L-Val-L-Arg-pNA (S-2160)	33	0.11	0.0033	b
		9[g]	0.27[g]	0.03[g]	f
9.	Bz-L-Phe-L-Val-L-Arg-pNA (Protein Res. Found.)	30	0.071	0.0024	b
20.	PyrGlu-L-Phe-L-Lys-pNA (S-2403)			0.0010	b
26.	PyrGlu-L-Phe-L-Lys-pNA	10000[j]	0.72[j]	0.000072[j]	i
21.	Bz-L-Arg-pNA			0.000015	b
27.	L-Ala-L-Phe-L-Lys-pNA	13000[j]	0.14[j]	0.000011[j]	i
				0.00007	b
19.	H-D-Val-L-Leu-L-Lys-pNA (S-2251)	1800[j]	0.0064[j]	0.0000036[j]	i

[a] K_m (μM); k_{cat} (sec^{-1}); k_{cat}/K_m (liters/μmol/sec).
[b] R. Lottenberg and C. M. Jackson (unpublished data).
[c] 37°C.
[d] AB Kabi, manufacturer's literature.
[e] Tris, $I = 0.25$, pH 8.3.
[f] M. J. Lindhout, B. H. M. Kop-Klaassen, and H. C. Hemker, Biochim. Biophys. Acta 533, 342 (1978).
[g] Tris-imidazole, $I = 0.15$, pH 8.2.
[h] 37°C Tris, $I = 0.15$, pH 8.4.
[i] D. Collen, H. R. Lijnen, F. De Cock, J. P. Durieux, and A. Loffet, Biochim. Biophys. Acta 165, 158 (1980).
[j] 0.1 M phosphate, pH 7.3.

TABLE IV

PLASMIN AND UROKINASE KINETIC PARAMETER SUMMARY

Substrate	Human plasmin				Urokinase			
	K_m^a	k_{cat}^a	k_{cat}/K_m^a	Ref.	K_m^a	k_{cat}^a	k_{cat}/K_m^a	Ref.
25. Cbz-Lys-SBzl	24[ee]	50[ee]	2.1[ee]	d	30[c]	60[c]	2.0[c]	b
26. PyrGlu-L-Phe-L-Lys-pNA	29	24	0.83	f	50[e]	49[e]	0.98[e]	d
8. Tos-Gly-L-Pro-L-Lys-pNA (Chromozym-PL)	150	34	0.23	f	1000	0.51	0.00051	f
	230[g]	45[g]	0.19[g]	f	1800	36	0.020	f
	280[i]	76[i]	0.27[i]	h	1600[g]	44[g]	0.027[g]	f
	290[k]	—	—	j				
	280[m]	—	—	l				
12. H-D-Pro-L-Phe-L-Arg-pNA (S-2302)	140	15	0.10	f	2100	0.6	0.00028	f
	240[g]	22[g]	0.093[g]	f				
5. Tos-Gly-L-Pro-L-Arg-pNA (Chromozym-TH)	270	27	0.098	f	50	5.3	0.11	f
	190	25	0.13	n				
27. L-Ala-L-Phe-L-Lys-pNA	530[g]	45[g]	0.084[g]	f	80[g]	9.1[g]	0.12[g]	f
19. H-D-Val-L-Leu-L-Lys-pNA (S-2251)	170[p]	15[p]	0.088[p]	o	450	0.028	0.000062	o
	120	14	0.11	f				
	240[p]	12[p]	0.050[p]	o	8300[p]	0.63[p]	0.000076[p]	o
Miniplasmin (Val442-plasmin)	130[r]	11[r]	0.085[r]	q				
	630[g]	23[g]	0.036[g]	f				
	320[t]	30[t]	0.094[t]	s				
Val442-plasmin	210[i]	26[i]	0.12[i]	h				
Plasmin B chain	280[i]	25[i]	0.093[i]	h				
	1300[i]	15[i]	0.012[i]	h				
22. Bz-L-Pro-L-Phe-L-Arg-pNA (Chromozym-PK)	200	6.6	0.033	f	360[g]	7.2[g]	0.020[g]	f
	690[g]	13[g]	0.019[g]	f			0.00065	f
14. H-D-Val-L-Leu-L-Arg-pNA (S-2266)	480[r]	12[r]	0.025[r]	f				
2. H-D-Ile-L-Pro-L-Arg-pNA (S-2288)	9000[u]	181[u]	0.020[u]	s	200	16[u]	0.080[u]	s
3. H-D-Phe-L-Pip-L-Arg-pNA	1200	19	0.016	f	1000	4.6	0.0046	f

No.	Substrate	K_m	k_{cat}	k_{cat}/K_m	Ref	K_m	k_{cat}	k_{cat}/K_m	Ref
16.	Bz-L-Ile-L-Glu-Gly-L-Arg-pNA (S-2222)	700	8.5	0.012		1300[r]	5.8[r]	0.0044[r]	[f]
15.	Bz-L-Phe-L-Val-L-Arg-pNA (S-2160)	1100[w]	9[w]	0.082[w]	[f]	670[g]	0.84	0.0013	[f]
	Val442-plasmin	1200[r]	13[r]	0.011[r]	[r] [q]	260[g]	0.5[g]	0.0019[g]	[f]
17.	L-PyroGlu-Gly-L-Arg-pNA (S-2444)	1040	8.8	0.0086	[f]	35	10.3	0.29	[f]
						54[g]	17[g]	0.32[g]	[f]
						90[x]	0.00031[y]		[s]
						60[x]	0.00013[z]		[s]
15.	Cbz-L-Val-Gly-L-Arg-pNA (Chromozym-TRY)	1400	9.1	0.0067	[f]	570	5.3	0.0094	[f]
21.	Bz-L-Arg-pNA	1400[y]	8.6[y]	0.0064[y]	[f]				
24.	H-L-Glu-Gly-L-Arg-pNA	60	0.09	0.0015	[f]				
28.	Tos-L-Arg-p-NitroBzl	78[cc]	0.31[dd]		[bb]	(HMW)200[aa]	21[aa]	0.103[aa]	[h]
						(LMW)270[aa]	16[aa]	0.061[aa]	[h]

[a] K_m (μM); k_{cat} (sec^{-1}); k_{cat}/K_m (liters/μmol/sec).

[b] G. D. J. Green and E. Shaw, *Anal. Biochem.* **93**, 223 (1979).

[c] 0.1 M Tris–0.1 M NaCl, pH 8.0.

[d] P. Coleman (unpublished data).

[e] 37°, 0.2 M Glycine/Glycinate, pH 8.5.

[f] U. Christensen (unpublished data).

[g] 37°C.

[h] R. C. Wohl, L. Summaria, and K. C. Robbins, *J. Biol. Chem.* **255**, 2005 (1980).

[i] 37°C, 0.05 M Tris–0.1 M NaCl, pH 7.4.

[j] Boehringer Mannheim Gmbh, manufacturer's literature.

[k] Tris-imidazole, I = 0.2, pH 7.5.

[l] M. J. Gallimore, E. Amundsen, A. O. Aasen, M. Larsbraaten, K. Lyngaas, and L. Svendsen, *Thromb. Res.* **14**, 51 (1979).

[m] 0.05 M Tris-HCl–0.1 M NaCl, pH 7.4.

[n] R. Lottenberg and C. M. Jackson (unpublished data).

[o] D. Collen, H. R. Lijnen, F. De Cock, J. P. Durieux, and A. Loffet, *Biochim. Biophys. Acta* **165**, 158 (1980).

[p] 0.1 M Phosphate, pH 7.3.

[q] U. Christensen, L. Sottrup-Jensen, S. Magnusson, T. E. Petersen, and I. Clemmensen, *Biochim. Biophys. Acta* **567**, 472 (1979).

[r] pH 7.8.

[s] AB Kabi, manufacturer's literature.

[t] 37°C, Tris, I = 0.15, pH 7.4.

[u] 37°C, Tris, I = 0.15, pH = 8.4.

[v] 37°C, pH 7.8.

[w] 37°C, Tris, I = 0.15, pH 8.3.

[x] 37°C, 0.05 M Tris, I = 0.05, pH 8.8.

[y] μM/min/Ploug unit.

[z] μM/min/CTA unit.

[aa] 37°C, 0.05 M Tris–0.1 M NaCl, pH 9.0.

[bb] G. W. E. Plaut, *Haemostasis* **7**, 105 (1978).

[cc] 30°C, 0.1 M Tris-HCl, pH 8.4.

[dd] μM/min/CTA unit/ml.

[ee] 37°C, 0.2 M phosphate, 0.2 M NaCl, pH 7.5.

TABLE V
PLASMINOGEN/PLASMIN-STREPTOKINASE KINETIC PARAMETER SUMMARY

Substrate	$K_m{}^a$	$k_{cat}{}^a$	$k_{cat}/K_m{}^a$	Ref.
19. H-D-Val-L-Leu-L-Lys-pNA				
(S-2251)				
Plasminogen SK	200^c	1^d		b
Glu-Plasminogen SK	320^f	39^f	0.12^f	e
Val 442-Plasminogen SK	210^f	29^f	0.14^f	e
Lys-Plasmin SK	780^f	52^f	0.067^f	e
Val 442-Plasmin SK	370^f	43^f	0.12^f	e
Plasmin B chain SK	520^f	36^f	0.070^f	e
8. Tos-Gly-L-Pro-L-Lys-pNA				
(Chromozym-PL)				
Glu-Plasminogen SK	510^f	28^f	0.055^f	e

^{a} K_m (μM); k_{cat} (sec^{-1}); k_{cat}/K_m (liters/μmol/sec).
^{b} AB Kabi, manufacturer's literature.
^{c} Tris buffer, $I = 0.05$, pH 7.4.
^{d} $V_{max} = \mu M/\text{min}/\text{CU}$.
^{e} R. C. Wohl, L. Summaria, and K. C. Robbins, *J. Biol. Chem.* **255**, 2005 (1980).
^{f} 37°C, 0.05 M Tris–0.1 M NaCl, pH 7.4.

the plasminogen molecule have been removed from the amino-terminal end are given.

Trypsin and Factor XII$_a$ (Activated Hageman Factor). Data for the hydrolysis of 16 peptide p-nitroanilides and 2 esters by trypsin, and 8 peptide p-nitroanilides and 1 thiobenzyl ester by activated Hageman factor or coagulation factor XII are given in Table VI. Unless otherwise noted in the footnotes to this table, all data for trypsin were provided by Ulla Christensen.

Kallikrein. Data from the literature for the hydrolysis of 4-peptide p-nitroanilides by various kallikreins are given in Table VII. No values for k_{cat} are available for these enzymes.

Peptide Fluorogenic Substrates. Data obtained with a variety of peptide fluorogenic substrates are summarized in Table VIII. All data are taken from the literature and no attempt has been made to convert maximum velocities reported by the authors of these papers into values for k_{cat} because insufficient data were available for determining the concentration of active enzyme in the preparations. All references to the sources for these data are given in the footnotes to the table.

Data for Other Coagulation Proteases. McRae et al.[15] have investigated the hydrolysis of a large number of peptide thioesters by bovine coagula-

[15] B. J. McRae, K. Kurachi, E. Davie, and J. C. Powers, *Biochemistry* (submitted for publication).

TABLE VI

TRYPSIN AND HAGEMAN FACTOR KINETIC PARAMETER SUMMARY

Substrate	Trypsin K_m[a]	Trypsin k_{cat}[a]	Trypsin k_{cat}/K_m[a]	Ref.	Human Hageman Fragment F K_m[a]	k_{cat}[a]	k_{cat}/K_m[a]	Ref.
5. Tos-L-Gly-L-Pro-L-Arg-pNA (Chromozym-TH)	17	69	4.1	b	210[d]	11[d]	0.051[d]	c
	20[e]	84[e]	4.2[e]	b				
3. H-D-Phe-L-Pip-L-Arg-pNA (S-2238)					170[d]	7.3[d]	0.043[d]	c
25. Cbz-Lys-SBzl	50[g]	75[g]	1.5[g]	f	1700[d]	18[d]	0.011[d]	c
	68[i]	57[i]	0.84[i]	h				
8. Tos-L-Gly-L-Pro-L-Lys-pNA (Chromozym-PL)	13	17	1.3	b				
9. Bz-L-Phe-L-Val-L-Arg-pNA (Peptide Research Foundation)	21	22	1.0	j				
17. L-PyroGlu-Gly-L-Arg-pNA (S-2444)	170	170	1.0	b	1100[d]	17[d]	0.015[d]	c
15. Cbz-L-Val-Gly-L-Arg-pNA (Chromozym-TRY)	77	72	0.93	b				
	78	75	0.96	j				
	75[l]	—	—	k				
	88[e]	84[e]	0.95[e]	b				
16. Bz-L-Ile-L-Glu-Gly-L-Arg-pNA (S-2222)	190	120	0.61	b	180[d]	23[d]	0.13[d]	c
	19[e]	270[e]	14[e]	m				
7. Bz-L-Phe-L-Val-L-Arg-pNA (S-2160)	52	30	0.58	b				
	38	28	0.74	j				
19. H-D-Val-L-Leu-L-Lys-pNA (S-2251)	21[o]	60[o]	2.9[o]	m				
	470[g]	22	0.047	m				
2. H-D-Ile-L-Pro-L-Arg-pNA (S-2288)	400[g]	7.0[a]	0.018[q]	p	400[r]	23[r]	0.058[r]	m
				(HFa)				
14. H-D-Val-L-Leu-L-Arg-pNA (S-2266)	170	37	0.22	b	310[d]	9.9[d]	0.032[d]	c
12. H-D-Pro-L-Phe-L-Arg-pNA (S-2302)	70	15	0.21	b	21[d]	14[d]	0.67[d]	c
					190[t]	15[t]	0.079[t]	s
					190[t]	15[t]	0.079[t]	s

(continued)

TABLE VI (*continued*)

Substrate	Trypsin K_m[a]	k_{cat}[a]	k_{cat}/K_m[a]	Ref.	Human Hageman fragment F K_m[a]	k_{cat}[a]	k_{cat}/K_m[a]	Ref.
22. Bz-L-Pro-L-Phe-L-Arg-pNA (Chromozym-PK)	120	13	0.11	b	710[d]	0.77[d]	0.0011[d]	c
	160[e]	20[e]	0.13[e]	b				
26. PyrGlu-L-Phe-L-Lys-pNA	200[q]	5.4[q]	0.027[q]	p				
27. L-Ala-L-Phe-L-Lys-pNA	2000[u]	4.7[u]	0.0024[q]	p				
21. Bz-L-Arg-pNA	1200[u]	2.7[u]	0.0023[u]	j				
	770[w]	2.4[w]	0.0031[w]	v				
28. Tos-L-Arg-p-NitroBzl	12[y]	140[z]		x				

[a] K_m (μM); k_{cat} (sec^{-1}); k_{cat}/K_m (liters/μmol/sec).
[b] U. Christensen (unpublished data).
[c] Y. Hojima, D. L. Tankersley, M. Miller-Andersson, J. V. Pierce, and J. J. Pisano, *Thromb. Res.* **18**, 417 (1980).
[d] 0.05 M Tris-HCl, pH 8.0.
[e] 37°C.
[f] G. D. J. Green and E. Shaw, *Anal. Biochem.* **93**, 223 (1979).
[g] 0.1 M Tris–0.1 M NaCl, pH 8.0.
[h] P. L. Coleman (unpublished data).
[i] 37°C, 0.2 M Phosphate–0.2 M NaCl, pH 7.5.
[j] R. Lottenberg and C. M. Jackson (unpublished data).
[k] Boehringer Mannheim, manufacturer's literature.
[l] Tris-HCl, pH 7.8.
[m] AB Kabi, manufacturer's literature.
[n] 37°C, Tris, I = 0.15, pH 9.0.
[o] 37°C, Tris, I = 0.15, pH 8.3.
[p] D. Collen, H. R. Lijnen, F. De Cock, J. P. Durieux, and A. Loffet, *Biochim. Biophys. Acta* **165**, 158 (1980).
[q] 0.1 M Phosphate, pH 7.3.
[r] 37°C, Tris, I = 0.15, pH 8.4.

[s] M. Silverberg, J. T. Dunn, L. Garen, and A. P. Kaplan, *J. Biol. Chem.* **255**, 7281, (1980).
[t] 0.05 M Tris-HCl–0.117 M NaCl, pH 7.8.
[u] pH 8.0.
[v] S. Takasaki, K. Kasai, and S. Ishii, *J. Biochem. (Tokyo)* **78**, 1275 (1975).
[w] 0.05 M Tris-HCl, pH 8.2.
[x] G. W. E. Plaut, *Haemostasis* **7**, 105 (1978).
[y] 30°C, 0.1 M Tris-HCl.
[z] V_{max} = μM/min/mg.

TABLE VII
KALLIKREIN KINETIC PARAMETER SUMMARY

Substrate	$K_m{}^a$	V_{max}	Ref.
22. Bz-L-Pro-L-Phe-L-Arg-pNA			
(Chromozym-PK)			
Human plasma kallikrein	140^c	—	b
12. H-D-Pro-L-Phe-L-Arg-pNA			
(S-2302)			
Human plasma kallikrein	200^e	6.8 μM/min/PEU	d
2. H-D-Ile-L-Pro-L-Arg-pNA			
(S-2288)			
Human plasma kallikrein	1000^f	1.3 μM/min/U	d
14. H-D-Val-L-Leu-L-Arg-pNA			
(S-2266)			
Porcine pancreatic kallikrein	22^g	0.008 μM/min/KU	d
Human urine kallikrein	30^g		d
Human salivary kallikrein	500^g		d

^{a} K_m (μM).
^{b} Boehringer Mannheim, manufacturer's literature.
^{c} $I = 0.15$, pH 7.9.
^{d} AB Kabi, manufacturer's literature.
^{e} 37°C.
^{f} Tris, $I = 0.15$, pH 8.4.
^{g} 0.05 M Tris, $I = 0.05$, pH 9.0.

tion factors IX_a, X_a, XI_a, thrombin, and trypsin. To date, these authors have reported the only di- and tripeptide chromogenic substrates to be hydrolyzed by factor IX_a. These dipeptide substrates, carbobenzoxy-L-Phe-L-Arg-isobutyl thioester and carbobenzoxy-L-Trp-L-Arg-isobutyl thioester, are hydrolyzed by bovine factor IX_a with k_{cat}/K_m values of 10,000–20,000 liters/mol/sec. Several tripeptide thiobenzyl esters are also hydrolyzed by factor IX_a with k_{cat}/K_m values of the order 70,000 liters/mol/sec. The original publication should be consulted for detailed information.

Discussion

Although the data base on which conclusions about peptide structure and selectivity might be made is still relatively limited, the fact that k_{cat}/K_m values are approaching values of 10^7 liters/mol/sec indicates clearly that a very high degree of selectivity is achievable. Comparison of k_{cat}/K_m values for the same substrate with different proteases further indicates that discrimination between proteases by using such substrates

TABLE VIII
FLUOROGENIC SUBSTRATES KINETIC PARAMETER SUMMARY

Substrate/enzyme	$K_m{}^a$	V_{max}	Temp.	Ref.
Cbz-L-Ala-L-Ala-L-Lys-4-methoxy-2-naphthylamide				
Plasmin	870[c]	—	37	b
D-Ala-L-Leu-L-Lys-7-amino-4-trifluoromethylcoumarin (AFC)				
Plasmin	620[e]	0.067 μM/min/CTAU[e]	25	d
L-Ala-L-Phe-L-Lys-7-amino-4-methylcoumarin (MCA)				
Plasmin	45[g]	210 μM/min/CTAU[g]	24	f
Urokinase	900[g]	100 μM/min/mg[g]	24	f
Thrombin	800[g]	9 μM/min/mg[g]	24	f
Suc-Ala-L-Phe-L-Lys-MCA				
Plasmin	400[g]	370 μM/min/CTAU[g]	24	f
Urokinase	800[g]	62 μM/min/mg[g]	24	f
MeOSuc-Ala-L-Phe-L-Lys-MCA				
Plasmin	440[g]	470 μM/min/CTAU[g]	24	f
L-Glu-Gly-L-Arg-MCA				
Urokinase	320[i]	18 μM/min/mg[i]	37	h
Boc-L-Glu-L-Lys-L-Lys-MCA				
Bovine plasmin	670[e]	1.9 μM/min/mg[e]	37	d
Plasmin	770[e]	3.7 μM/min/mg[e]	37	d
Glutaryl-Gly-L-Arg-MCA				
Urokinase	440[i]	18 μM/min/mg[i]	37	h
Gly-L-Arg-2-naphthylamide				
Plasmin	1700	0.0006 μM/min/CU	37	j
Urokinase	7500	0.000022 μM/min/CTAU	37	j
Cbz-Gly-Gly-L-Arg-MCA				
Trypsin	330[l]	3700 μM/min/μM trypsin[l]	25	k
Thrombin	110[g]	750 μM/min/mg[g]	24	f
Plasmin	450[g]	10 μM/min/CTAU[g]	24	f
Urokinase	130[g]	1600 μM/min/mg[g]	24	f
Urokinase	170[m]	0.000030 μM/min/IU[m]	25	k

Substrate / Enzyme				
Cbz-Gly-Gly-L-Arg-AFC				
Trypsin	180^l	$8200\ \mu M/min/\mu M$ trypsin^l	25	k
Urokinase	190^m	$0.000030\ \mu M/min/IU^m$	25	k
Cbz-Gly-L-Pro-L-Arg-4-methoxy-2-naphthylamide				
Thrombin	83°	—	37	n
Boc-L-Ile-L-Glu-Gly-L-Arg-MCA				
Factor X_a	160^i	$6.1\ \mu M/min/mg^i$	37	h
Cbz-L-Phe-L-Arg-MCA				
Plasma kallikrein (bov.)	240^i	$8.0\ \mu M/min/A_{280} = 1.0^i$	37	h
Urinary kallikrein (hum.)	280^i	$1.1\ \mu M/min/mg^i$	37	h
Pancreatic kallikrein (hog)	530^i	$0.3\ \mu M/min/A_{280} = 1.0^i$	37	h
Boc-L-Phe-L-Glu-L-Lys-L-Lys-MCA				
Bovine plasmin	400^i	$0.23\ \mu M/min/mg^i$	37	h
H-D-Phe-L-Pro-L-Arg-5-aminoisophthalic acid				
Thrombin	54°	—	37	p
L-Pro-L-Phe-L-Arg-MCA				
Plasma kallikrein (bov.)	610^i	$13\ \mu M/min/A_{280} = 1.0^i$	37	h
Urinary kallikrein (hum.)	220^i	$30\ \mu M/min/mg^i$	37	h
Pancreatic kallikrein (hog)	160^i	$12\ \mu M/min/A_{280} = 1.0^i$	37	h
Hageman fragment f	54^r	$k_{cat} = 1.9/sec^r$	25	q
Cbz-L-Pro-L-Phe-L-Arg-MCA				
Plasma kallikrein (bov.)	470^i	$7.2\ \mu M/min/A_{280} = 1.0^i$	37	h
Urinary kallikrein (hum.)	170^i	$7.5\ \mu M/min/mg^i$	37	h
Pancreatic kallikrein (hog)	180^i	$4.6\ \mu M/min/A_{280} = 1.0^i$	37	h
L-Val-Gly-L-Arg-2-naphthylamide				
Plasmin	900	$0.0012\ \mu M/min/CU$	37	j
Urokinase	600	$0.0000048\ \mu M/min/CTAU$	37	j
Boc-L-Val-Gly-L-Arg-2-naphthylamide				
Plasmin	1600	$0.0024\ \mu M/min/CU$	37	j
Urokinase	1100	$0.0000030\ \mu M/min/CTAU$	37	j
Boc-L-Val-L-Leu-L-Lys-MCA				
Bovine plasmin	250^e	$1.5\ \mu M/min/mg^e$	37	d
Plasmin	770^e	$3.7\ \mu M/min/mg^e$	37	d

(continued)

TABLE VIII (*continued*)

Substrate/enzyme	K_m^a	V_{max}	Temp.	Ref.
D-Val-L-Leu-L-Lys-AFC				
Plasmin	560[m]	0.088 μM/min/CTAU[m]	25	k
H-D-Val-L-Leu-L-Lys-5-aminoisophthalic acid				
Plasmin	250[t]	—	37	s
Boc-L-Val-L-Pro-L-Arg-MCA				
Bovine thrombin	21[i]	78 μM/min/mg[i]	37	h
Boc-L-Val-Ser-Gly-L-Arg-MCA				
Urokinase	330[i]	9.6 μM/min/mg[i]	37	h

[a] $K_m (\mu M)$.

[b] S. A. Clavin, J. L. Bobbit, R. T. Shuman, and E. L. Smithwick, Jr., *Anal. Biochem.* **80**, 355 (1977).

[c] 0.05 *M* Tris-HCl, pH 8.0.

[d] H. Kato, N. Adachi, Y. Ohno, S. Iwanaga, K. Takada, and S. Sakakibara, *J. Biochem. (Tokyo)* **88**, 183 (1980).

[e] 0.05 *M* Tris-HCl–0.15 *M* NaCl, pH 7.4.

[f] P. A. Pierzchala, C. P. Dorn, and M. Zimmerman, *Biochem. J.* **183**, 555 (1979).

[g] 0.05 m*M* TES-NaOH–20% dimethyl sulfoxide, pH 7.5.

[h] S. Iwanaga, T. Morita, H. Kato, T. Harada, N. Adachi, T. Sugo, I. Maruyama, K. Takada, T. Kimura, and S. Sakakibara, *in* "Kinins-II, Biochemistry, Pathophysiology, and Clinical Aspects" (S. Fujii, H. Moriya, and T. Suzuki, eds.), p. 147. Plenum, New York, 1979.

[i] 0.05 *M* Tris-HCl–0.1 *M* NaCl–0.01 *M* calcium chloride, pH 8.0.

[j] W. Nieuwenhuizen, G. Wijngaards, and E. Groeneveld, *Anal. Biochem.* **83**, 143 (1977).

[k] R. E. Smith, E. R. Bissell, A. R. Mitchell, and K. W. Pearson, *Thromb. Res.* **17**, 393 (1980).

[l] 0.45 *M* TES, pH 8.0.

[m] 0.045 *M* TES, pH 8.0.

[n] G. A. Mitchell, P. M. Hudson, R. M. Huseby, S. P. Pochron, and R. J. Gargiulo, *Thromb. Res.* **12**, 219 (1978).

[o] 0.25 *M* Glycine/Glycinate–0.03 *M* NaCl–2 m*M* potassium EDTA.

[p] G. A. Mitchell, R. J. Gargiulo, R. M. Huseby, D. E. Lawson, S. P. Pochron, and J. A. Sehuanes, *Thromb. Res.* **13**, 47 (1978).

[q] Y. Hojima, D. L. Tankersley, M. Miller-Andersson, J. V. Pierce, and J. J. Pisano, *Thromb. Res.* **18**, 417 (1980).

[r] 0.05 *M* Tris-HCl, pH 8.0.

[s] D. E. Lawson, G. A. Mitchell, and R. M. Huseby, *Thromb. Res.* **14**, 323 (1979).

[t] 0.05 *M* Glycine/Glycinate–0.15 *M* NaCl, pH 8.0.

is readily obtainable. In mixtures of proteases, assay using more than one substrate and then solving the set of simultaneous equations that relates the total velocity to that contributed by each enzyme can be done for those mixtures for which the kinetic constants are known for the individual components.[16]

The values for k_{cat} of 100–300 sec^{-1} and the large molar absorptivity for p-nitroaniline of 13,300 liters/mol/cm at 380 nm, its absorption maximum, make extremely sensitive assays readily available. It can be anticipated that the increase in sensitivity provided from fluorescence monitoring of the fluorogenic substrates may increase the sensitivity considerably, although the limited amount of information about k_{cat} for the various fluorogenic leaving groups precludes making quantitative statements at this time.

It must be noted that many of the very good substrates for one enzyme, where "good" is judged from a large k_{cat}/K_m and a large k_{cat}, are very poor substrates for one of the other enzymes by these criteria. However, if consideration of the values for K_m are made, these "poor" (low values for k_{cat}) substrates are very frequently good apparent competitive inhibitors of the enzymes by which they are hydrolyzed very slowly. This can greatly facilitate the assay of one protease in the presence of other proteases as described above; however, use of these substrates in coupled assay systems, in which one protease is being generated from its zymogen by another protease, must therefore be made with caution and one must take into account the consequences of this observation.

No consideration of the design of assay procedures using these substrates is given here, since this topic is covered extensively elsewhere.[17-19] It must be noted here, however, that the solubility of some of the chromogenic and fluorogenic substrates is relatively low when one considers the Michaelis constants for the substrates and particular proteases. This is particularly true in solutions of high ionic strength and in the presence of agents such as heparin, which may interact and initiate substrate precipitation.

[16] C. M. Jackson, *Thromb. Haemostasis* **41**, 458 (1979).
[17] A. Fersht, "Enzyme Structure and Mechanism." Freeman, San Francisco, California, 1977.
[18] A. Cornish-Bowden, "Fundamentals of Enzyme Kinetics." Butterworth, London, 1979.
[19] R. D. Allison and D. L. Purich, this series, Vol. 63, p. 3.

Section III

The Plasmin System

[29] Human Plasminogen

By Francis J. Castellino and James R. Powell

Plasminogen is a component of the fibrinolytic system and is the plasma-protein precursor of the serine protease,[1,2] plasmin. Although plasmin functions in fibrinolysis and fibrinogenolysis,[3] plasmin has been shown to be active in other systems. For example, plasmin is capable of digesting factor XII_a into factor XII_a fragments,[4] as well as activating several complement zymogens (e.g., C1,[5] C3,[6] and C5[7]). Also, plasmin is capable of inactivation of C1 inhibitor.[8] Additionally, plasmin has been shown to substitute for factor D in the alternate pathway.[9] Plasmin has also been implicated in digesting proaccelerin,[10] antihemophilic factor,[10] ACTH,[11] glucagon,[11] and γ-globulin.[12]

Regarding nonphysiological substrates, plasmin possesses tryptic-like specificity. This enzyme hydrolyzes proteins and peptides at lysyl- and arginyl-bonds.[13] Plasmin possesses esterolytic activity toward basic amino acid esters[14] and amidolytic activity toward basic amino acid amides.[15] Several p-nitrophenolate esters are also good plasmin substrates.[16,17]

Nomenclature of Plasminogen Multiple Forms

Native plasminogen is a single-chain protein[18] of MW 90,000–94,000,[19,20] with multiple isoelectric forms.[21] Each affinity-chromatogra-

[1] L. Summaria, B. Hsieh, W. R. Groskopf, K. C. Robbins, and G. H. Barlow, J. Biol. Chem. 242, 5046 (1967).
[2] W. R. Groskopf, B. Hsieh, L. Summaria, and K. C. Robbins, J. Biol. Chem. 244, 3590 (1969).
[3] S. V. Pizzo, M. Schwartz, R. L. Hill, and P. A. McKee, J. Biol. Chem. 247, 646 (1972).
[4] A. P. Kaplan and K. F. Austen, J. Exp. Med. 133, 696 (1971).
[5] O. D. Ratnoff and G. B. Naff, J. Exp. Med. 125, 337 (1967).
[6] P. A. Ward, J. Exp. Med. 125, 337 (1967).
[7] C. M. Arroyave and H. J. Müller-Eberhard, J. Immunol. 111, 536 (1973).
[8] P. C. Harpel, J. Clin. Invest. 49, 58 (1970).
[9] V. A. Brade, A. Nicholson, A. Bitter-Suermann, and U. Hadding, J. Immunol. 113, 1735 (1974).
[10] D. Allagile and J. P. Soulier, Sem. Hop. 32, 355 (1956).
[11] I. A. Mirsky, G. Perisutti, and N. C. Davis, J. Clin. Invest. 38, 14 (1959).
[12] C. A. Janeway, E. Merler, F. S. Rogen, S. Salmon, and J. D. Crain, N. Engl. J. Med. 278, 919 (1967).
[13] P. Wallen and S. Iwanaga, Biochim. Biophys. Acta 154, 414 (1968).
[14] M. J. Weinstein and R. F. Doolittle, Biochim. Biophys. Acta 258, 577 (1972).
[15] U. Christensen and S. Mullertz, Biochim. Biophys. Acta 334, 187 (1974).
[16] T. Chase and E. Shaw, Biochemistry 8, 2212 (1969).
[17] R. M. Silverstein, Thromb. Res. 3, 729 (1973).
[18] G. H. Barlow, L. Summaria, and K. C. Robbins, J. Biol. Chem. 224, 1138 (1969).
[19] E. E. Rickli and W. I. Otavsky, Biochim. Biophys. Acta 295, 381 (1973).

phy variant of human plasminogen contains glutamic acid as its amino-terminal residue. Consequently, native plasminogen is referred to as Glu-plasminogen 1 and Glu-plasminogen 2, representing the first and second variants, respectively, eluted from affinity-chromatography columns. During the activation of plasminogen, or during preparation of the protein in the absence of proteolytic inhibitors, an amino-terminal peptide can be released from Glu-plasminogen, as a result of a cleavage of the Lys^{76}-Lys^{77} peptide bond. The resulting plasminogen is termed Lys^{77}-plasminogen 1 or Lys^{77}-plasminogen 2, depending on whether affinity variant 1 or variant 2 is under consideration. Lower-molecular-weight human-plasminogen variants have been described. Extensive elastase digestion of plasminogen results in formation of a molecule of MW $\approx$ 38,000, as result of cleavage of the Val^{441}-Val^{442} peptide bond.[22] This form of plasminogen is termed Val^{442}-plasminogen 1, when derived from affinity variant 1, or Val^{442}-plasminogen 2, when derived from affinity variant 2. In addition, another low-molecular-weight form of human plasminogen has been isolated, of MW 52,000, resulting from a cleavage at position Val^{353}-Val^{354} to give Val^{354}-plasminogens 1 and 2.[23]

In addition, human plasminogen is known to possess multiple isoelectric forms, which can be isolated by isoelectric-focusing techniques.[21] Glu-plasminogen 1 possesses multiple forms of pI values of 6.2, 6.3, 6.4, and 6.6 · Glu-plasminogen 2 possesses two isoelectric forms with pI values of 6.4 and 6.6. On the other hand, Lys^{77}-plasminogen 1 contains three major isoelectric forms of pI 6.7, 7.2, and 7.5; whereas Lys^{77}-plasminogen 2 contains three major forms of pI values 7.5, 7.8, and 8.1.[21]

Purification of Plasminogen

Glu-Plasminogen 1 and Glu-Plasminogen 2

The two major variants of human Glu-plasminogen are prepared by affinity chromatography on Sepharose-lysine by a modification[24] of the procedure of Deutsch and Mertz.[25]

The affinity-chromatography resin is prepared as follows. A quantity

[20] B. N. Violand and F. J. Castellino, *J. Biol. Chem.* **251**, 3906 (1976).
[21] L. Summaria, F. Spitz, L. Arzadon, I. G. Boreisha, and K. C. Robbins, *J. Biol. Chem.* **251**, 3963 (1976).
[22] L. Sottrup-Jensen, H. Claeys, M. Zajdel, T. E. Peterson, and S. Magnusson, *Prog. Chem. Fibrinolysis Thrombolysis* **3**, 191 (1978).
[23] J. R. Powell and F. J. Castellino, *Biochem. Biophys. Res. Commun.* **102**, 46 (1981).
[24] W. J. Brockway and F. J. Castellino, *Arch. Biochem. Biophys.* **151**, 194 (1972).
[25] D. G. Deutsch and E. T. Mertz, *Science* **170**, 1095 (1970).

of 60 g of CNBr, in 300 ml of H_2O, is stirred for 10 min, at room temperature, with 300 ml of settled Sepharose-4B, previously washed with H_2O. The pH is maintained at 11 by manual addition of $4 N$ NaOH, until the pH change has essentially ceased. The resin is then added to a solution (300 ml) consisting of $0.1 M$ phosphate–$1.0 M$ L-lysine, pH 9.2, at 4°C. The mixture is gently stirred at this temperature for 18 hr. After this, the resin in equilibrated with $0.1 M$ phosphate, containing 0.02% NaN_3, pH 8.0.

Our usual starting material for the preparation of human plasminogen is human plasma Cohn Fraction III. Plasminogen can be readily extracted from this material by suspending 80 g of material in 350 ml of a buffer containing $0.05 M$ phosphate–$0.1 M$ NaCl–$0.002 M$ EDTA–10,000 KIU Trasylol (FBA Pharmaceuticals), pH 7.4, and stirring the mixture at 4°C for 2 hr. The suspension is then subjected to centrifugation at 9000 rpm for 15 min, at 4°C. The supernatant is added to 100 ml (settled volume) of the resin and gently stirred for 2 hr, at 4°C. The resin is then washed, at room temperature, by filtration, with a buffer consisting of $0.3 M$ phosphate–10 KIU/ml Trasylol, pH 8.0, until the filtrate has an absorbance at 280 nm equal to the buffer. At this point, the resin is packed into a chromatography column (2.5 cm in diameter) and equilibrated with a buffer consisting of $0.1 M$ phosphate–10 KIU/ml Trasylol, pH 8.0. The plasminogen is then batch-eluted with 250 ml of $0.1 M$ phosphate–$0.015 M$ ε-aminocaproic acid (ε-ACA)–10 KIU/ml Trasylol, pH 8.0. Fractions (10 ml) are collected at a flow rate of 150 ml/hr. The protein-containing fractions are dialyzed overnight against two changes of 2 liters of $0.1 M$ phosphate, pH 8.0, at 4°C. The protein pool is passed over a column (2.5 × 25 cm) of Sepharose-lysine, and plasminogen is eluted from this column, at room temperature, after application of linear gradient consisting of 250 ml of $0.1 M$ phosphate–10 KIU/ml Trasylol (pH 8.0) as the starting buffer, and 250 ml of $0.1 M$ phosphate–10 KIU/ml Trasylol–12 mM ε-ACA (pH 8.0) as the limit buffer. Fractions of 5 ml are collected at a flow rate of approximately 75 ml/hr. Plasminogen is eluted in 2 well-resolved fractions.[24] Each fraction is separately pooled, dialyzed extensively against H_2O at 4°C, and lyophilized. The material is then stored at $-20°C$, conditions under which the proteins are stable for at least 3 months. The overall yield is greater than 90%.

The above procedure requires no substantial modifications when plasma is utilized as the starting material. In this case, a quantity of 10,000 KIU of Trasylol should be immediately added to 300 ml of fresh plasma, and the procedure initiated at the point of treatment (above) of the Cohn Fraction III extract. When other species of mammalian plasminogens are desired, again no major modifications of the above procedure are required.

Lys^{77}-Plasminogen 1 and Lys^{77}-Plasminogen 2

These plasminogen forms will result when Cohn Fraction $III_{2,3}$ is utilized as the starting material for the human plasminogen preparation.[21] All steps in the purification procedure are as previously noted. In our hands, we find that the Lys^{77}-plasminogen forms will also be produced when proteolytic inhibitors are omitted from the Cohn Fraction III extraction procedure or from whole plasma.

Lys^{77}-plasminogen 1 and Lys^{77}-plasminogen 2 can also be produced from the appropriate parent Glu-plasminogen affinity variant by treatment with plasmin. This can be accomplished in the following fashion. A quantity of 0.3 ml (0.9 mg/ml) of activator-free plasmin[20] in 0.05 M Tris-HCl–0.1 M L-lysine (pH 8.0) is added to 3 ml (11.5 mg/ml) of Glu-plasminogen 1 or Glu-plasminogen 2, dissolved in the same buffer at 30°C for 2 hr. In order to remove the plasmin initially added, insolubilized soybean-trypsin inhibitor is added to the mixture. The plasmin-free Lys^{77}-plasminogen preparation is then extensively dialyzed against 0.1 M phosphate (pH 8.0) at 4°C, adsorbed to a Sepharose-lysine column as described, and extensively washed with this buffer at room temperature. Lys^{77}-plasminogen is then removed from the column by elution with 0.1 M phosphate–0.05 M ε-ACA (pH 8.0) at room temperature. The protein pool is then dialyzed extensively against H_2O at 4°C, and lyophilized. The protein is stable under these conditions for at least 3 months.

Val^{442}-Plasminogen 1 and Val^{442}-Plasminogen 2

These lower-molecular-weight forms of plasminogen are prepared by extensive digestion of affinity-chromatography variants 1 or 2 of Glu-plasminogen or Lys^{77}-plasminogen.[26]

A solution of plasminogen, at 1.5–3.0 mg/ml, in 100 ml of 0.05 M Tris-HCl–0.1 M NaCl (pH 8.0) containing 20,000 KIU of Trasylol is prepared. To this solution is added a quantity of elastase (Sigma Chemical Co.) to yield a plasminogen : elastase ratio of 50 : 1 (mol/mol). The digestion is allowed to proceed at room temperature for 5 hr. After this time, the reaction is terminated by adjusting the solution to 1 mM in DFP and allowing the incubation to proceed for 30 additional min. The mixture is then extensively dialyzed against 0.1 M phosphate, pH 8.0, at 4°C. The protein pool thus obtained is applied to a column (2.5 × 20 cm) of Sepharose-lysine, previously equilibrated with 0.3 M phosphate, pH 8.0, at room temperature. Continued elution with approximately 300 ml of this buffer results in elution of highly purified Val^{442}-plasminogen (other peptide fragments produced adsorb to the column and can be eluted with ε-aminocaproic acid-containing buffers).[26]

[26] J. R. Powell and F. J. Castellino, *J. Biol. Chem.* **255**, 5329 (1980).

Isoelectric Forms of Glu-Plasminogen 1, Glu-Plasminogen 2,
Lys ^{77}-Plasminogen 1, and Lys ^{77}-Plasminogen 2

Each major affinity chromatography-resolved variant of Glu-plasminogen or Lys-plasminogen can be resolved into its isoelectric forms by preparative isoelectric focusing. We utilize a LKB 110-ml column and LKB ampholines. The column is maintained at 4°C by means of an external circulating temperature-control system. The electrode solutions consist of 0.4 ml H_3PO_4–26 ml H_2O–26 g sucrose for the anode, and 0.5 g NaOH in 50 ml H_2O for the cathode. After the anode solution has been added, a linear sucrose gradient is applied. The starting solution consists of 2.82 ml ampholines (pH ranges of 5–8 and 3–10 are most commonly employed) in 40% sucrose–2.87 ml H_2O–21 g sucrose. The limit solution consists of 1.15 ml ampholines in 40% sucrose–51.1 ml H_2O–2.75 g sucrose. The gradient is layered directly above the anode solution by means of a peristaltic pump. Each gradient chamber is immersed in an ice bath prior to engaging the pump. After approximately 30 ml of the gradient has emerged, 15–20 mg of the appropriate plasminogen solution is dissolved in the dense solution; then the remainder of the gradient is applied. After this, a few extra milliliters of the limit solution is added, followed by the cathode solution. Current is then applied and the voltage constantly adjusted during the run such that the product of kilovolts × milliamperes never exceeds 3.0. Run times of 60–70 hr are employed. At the conclusion of the run, the column is drained in 1-ml fractions and each tube monitored for absorbance at 280 nm. The appropriate tubes are pooled and each pool passed over a small Sepharose-lysine column in order to adsorb the plasminogen. In this manner, excess ampholines and sucrose are conveniently removed. Each isoelectric form can then be eluted from the column as described, dialyzed against H_2O at 4°C, and lyophilized.

Assay Method

Assays of plasminogen first require its conversion to plasmin by common activators.[20] The conversion of human plasminogen to human plasmin is accomplished as a result of cleavage of the Arg^{560}-Val^{561} peptide bond.[27] The enzyme is then assayed based on its ability to hydrolyze α-N-tosyl-L-arginine methyl ester (TAME), utilizing a recording pH-stat to titrate the amount of acid, α-N-tosyl-L-arginine, liberated with time.[28] Assays that rely on the spectrophotometric determination of the plasmin-catalyzed release of p-nitroanilide from the substrate H-D-valyl-L-leucyl-L-lysine-p-nitroanilide dihydrochloride[29] (S-2251) and on the determination of the

[27] K. C. Robbins, L. Summaria, B. Hsieh, and R. J. Shah, *J. Biol. Chem.* **242**, 2333 (1967).
[28] F. J. Castellino and J. M. Sodetz, this series, Vol. 45, p. 273.
[29] Kabi Diagnostica Bulletin: 52251.

plasmin-catalyzed release of p-nitrophenolate from N^{α}-Cbz-L-lysine-p-nitrophenyl ester[30] have also been fruitfully employed. Other assay methods for plasmin are available, such as the casein,[31] azocasein,[32] and fibrin plate assay.[33]

Determination of the Active Plasmin Concentration

The concentration of active plasmin is determined by active-site titration with p-nitrophenyl-p'-guanidinobenzoate (NPGB).[34] A quantity of 0.04 ml of plasmin is added to 1 ml of 50 μM NPGB (the NPGB is prepared as a 0.04 M solution in DMF and diluted to 50 μM with 0.05 M veronal, pH 8.3). The p-nitrophenolate released is quantitated by measuring the absorbance at 410 nm on a Cary 219 spectrophotometer. The concentration of active plasmin is calculated using a $\epsilon_{1\,cm,410\,nm}^{1M}$ of 16,595 for p-nitrophenolate.[35]

TAME Assay

In this assay, we routinely utilize a potentiometric method. Our equipment consists of a Radiometer PHM 26 pH meter, Radiometer-type TTT11 titrator, Radiometer TTA-31 titration assembly, Radiometer ABU-11 Auto Burette, and REC-51-S-Servograph Recorder. Constant temperature is maintained (usually at 22°C) with a Hapke Model FF circulating water bath. The syringe is filled with carefully standardized NaOH (0.02–0.03 M). For determination of the specific activity of plasmin the substrate is first diluted with a solution of 0.1 M NaCl (pH 8.0) to a final concentration of 0.15 M. One milliliter of this solution is placed in the pH-stat vessel and allowed to equilibrate at 22°C. The pH is then adjusted to 8.0 by automatic addition of NaOH from the burette. A quantity of 0.02 ml of plasmin is added (containing 40–60 μg of enzyme) and the reaction rate followed for 2–5 min. The data is converted to micromoles of TAME hydrolyzed per minute per milligram of plasmin.

One unit of plasmin is defined as the amount of plasmin that will hydrolyze 1 μmol of TAME min^{-1}. Under these assay conditions, the V_{max} of Lys-plasmin is 16.5 μmol of TAME cleaved min^{-1} mg^{-1} of active plasmin. Therefore, one unit of plasmin activity corresponds to 0.061 mg of plasmin in this assay.

[30] R. M. Silverstein, Thromb. Res. **3**, 729 (1973).
[31] K. C. Robbins and L. Summaria, this series, Vol. 19, p. 184.
[32] Our procedure for this method can be found in L. A. Schick and F. J. Castellino, Biochemistry **12**, 4315 (1973).
[33] T. Astrup and S. Mullertz, Arch. Biochem. Biophys. **40**, 346 (1952).
[34] T. Chase and E. Shaw, this series, Vol. 19, p. 20.
[35] T. Chase and E. Shaw, Biochem. Biophys. Res. Commun. **29**, 508 (1967).

S-2251 Assay

A quantity of 0.4 ml of a buffer consisting of 0.1 M Tris-HCl–0.2 M NaCl (pH 7.4) is added to a cuvette containing 0.4 ml of 2.0–3.0 mM S-2251, dissolved in water. The temperature is allowed to equilibrate at 37°C in a Cary 219 spectrophotometer. Plasmin (10 μl containing approximately 1.0 μg of enzyme) is then added to the cuvette and the absorbance change (ΔA) is recorded at 405 nm for 1–2 min. The V_{max} of Lys[77]-plasmin (variant 1 or 2) with this substrate is approximately 15 μmol of S-2251 cleaved min^{-1} mg^{-1} of active plasmin.[23] Thus, 1 unit of Lys[77]-plasmin, defined as the amount of plasmin that will cleave 1 μmol of S-2251 min^{-1}, corresponds to 0.067 mg of enzyme.

The V_{max} of Val[442]-plasmin (variant 1 or 2) is 40.3 μmol of S-2251 cleaved min^{-1} mg^{-1} of active plasmin. Correspondingly 1 unit of Val[442]-plasmin represents 0.025 mg of enzyme.

With all plasmin assays, the activity of the enzyme can be related to International Standard of plasmin. A human plasmin reference standard, containing 8.0 IU ml^{-1}, is available through the WHO International Laboratory for Biological Standards, Holly Hill, Hampstead, London, NW3 6RB, England.

Properties of Human Plasminogen

Purity

In polyacrylamide-gel electrophoretic systems at pH 4.3, or pH 3.2 in 8 M urea, human Glu-plasminogen 1 and Glu-plasminogen 2 exhibit single bands, with significant mobility differences. At pH 9.5, each variant of Glu-plasminogen shows 3–5 bands, owing to the separation of the subforms at this pH. Single bands with near identical mobilities are observed for Glu-plasminogen 1 and 2, when examined by reduced SDS–gel electrophoresis. Similar electrophoretic properties are found for each affinity variant of Lys[77]-plasminogen, except that the higher pI values of Lys[77]-plasminogen are reflected.

When examined by ultracentrifugal sedimentation-velocity analysis, in 0.1 M Tris-HCl (pH 7.8), a single, apparently symmetrical peak is obtained for each variant of Glu-plasminogen, with a $S^\circ_{20,w}$ value of 4.8 S.[36]

Chemical Properties

The complete amino acid sequence of human plasminogen (mixture of affinity-chromatography variants 1 and 2) has been determined.[22,37–40] This sequence is given in Scheme I.[41]

[36] B. N. Violand, J. M. Sodetz, and F. J. Castellino, *Arch. Biochem. Biophys.* **170**, 300 (1975).

[37] B. Wiman, *Eur. J. Biochem.* **1**, 39 (1973).

1   10   20

NH_2 -E - P - L - D - D - Y - V - N - T - Q - G - A - S - L - F - S - V - T - K - K -

                  30   40

Q - L - G - A - G - S - I - E - E - C - A - A - K - C - E - E - D - E - E - F -

                  50   60

T - C - R - A - F - Q - Y - H - S - K - E - Q - E - C - V - I - M - A - E - N -

                  70   80

R - K - S - S - I - I - R - M - R - D - V - V - L - F - E - K - K - V - Y - L -

                  90   100

S - E - C - K - T - G - D - G - K - N - Y - R - G - T - M - S - K - T - K - N -

                  110   120

G - I - T - C - Q - K - W - S - S - T - S - P - H - R - P - R - F - S - P - A -

                  130   140

T - H - P - S - E - G - L - E - E - N - Y - C - R - N - P - D - N - D - P - Q -

                  150   160

G - P - W - C - Y - T - T - D - P - E - K - R - Y - D - Y - C - D - I - L - E -

                  170   180

C - E - E - E - C - M - H - C - S - G - E - N - Y - D - G - K - I - S - K - T -

                  190   200

M - S - G - L - E - C - Q - A - W - D - S - Q - S - P - H - A - H - G - Y - I -

                  210   220

P - S - K - F - P - N - K - N - L - K - K - N - Y - C - R - N - P - D - R - E -

                  230   240

L - R - P - W - C - F - T - T - D - P - N - K - R - W - E - L - C - D - I - P -

                  250   260

R - C - T - T - P - P - P - S - S - G - P - T - Y - Q - C - L - K - G - T - G -

                  270   280

E - N - Y - R - G - N - V - A - V - T - V - S - G - H - T - C - Q - H - W - S -

                  290   300

A - Q - T - P - H - T - H - N - R - T - P - E - N - F - P - C - K - N - L - D -
                                     |
                                   CHO

                  310   320

E - N - Y - C - R - N - P - D - G - K - R - A - P - W - C - H - T - T - N - S -

                  330   340

Q - V - R - W - E - Y - C - K - I - P - S - C - D - S - S - P - V - S - T - E -

                  350   360

E - L - A - P - T - A - P - P - E - L - T - P - V - V - Q - D - C - Y - H - G -
                |
                CHO

                  370   380

D - G - Q - S - Y - R - G - T - S - S - T - T - T - T - G - K - K - C - Q - S -

---

[38] B. Wiman, *Eur. J. Biochem.* **76**, 129 (1977).

[39] B. Wiman and P. Wallen, *Eur. J. Biochem.* **36**, 25 (1973).

[40] K. C. Robbins, I. G. Boreisha, L. Arzadon, and L. Summaria, *J. Biol. Chem.* **250**, 4044 (1975).

[41] There are conflicting literature reports as to whether an additional residue of Gln should be inserted in sequence position 32. We have not included this residue in this report. If in fact it does exist, a renumbering would be necessary with Ala[32] becoming Ala[33] and an appropriate adjustment made at all other sequence positions.

$\overset{390}{}\hspace{5cm}\overset{400}{}$

W - S - S - M - T - P - H - R - H - Q - K - T - P - E - N - Y - P - N - A - G -

$\overset{410}{}\hspace{5cm}\overset{420}{}$

L - T - M - N - Y - C - R - N - P - D - A - D - K - G - P - W - C - F - T - T -

$\overset{430}{}\hspace{5cm}\overset{440}{}$

D - P - S - V - R - W - E - Y - C - N - L - K - K - C - S - G - T - E - A - S -

$\overset{450}{}\hspace{5cm}\overset{460}{}$

V - V - A - P - P - P - V - V - L - L - P - N - V - E - T - P - S - E - E - D -

$\overset{470}{}\hspace{5cm}\overset{480}{}$

C - M - F - G - D - G - K - G - Y - R - G - K - R - A - T - T - V - T - G - T -

$\overset{490}{}\hspace{5cm}\overset{500}{}$

P - C - Q - D - W - A - A - Q - E - P - H - R - H - S - I - F - T - P - E - T -

$\overset{510}{}\hspace{5cm}\overset{520}{}$

N - P - R - A - G - L - E - K - N - Y - C - R - N - P - D - G - N - V - G - G -

$\overset{530}{}\hspace{5cm}\overset{540}{}$

P - W - C - Y - T - T - N - P - R - K - L - Y - D - Y - C - D - V - P - Q - C -

$\overset{550}{}\hspace{5cm}\overset{560}{}$

A - A - P - S - F - D - C - G - K - P - Q - V - E - P - K - K - C - P - G - R -

$\overset{570}{}\hspace{5cm}\overset{580}{}$

V - V - G - G - C - V - A - H - P - H - S - W - P - W - Q - V - S - L - R - T -

$\overset{590}{}\hspace{5cm}\overset{600}{}$

R - F - G - M - H - F - C - G - G - T - L - I - S - P - E - W - V - L - T - A -

$\overset{610}{}\hspace{5cm}\overset{620}{}$

A - H* - C - L - E - K - S - P - R - P - S - S - Y - K - V - I - L - G - A - H -

$\overset{630}{}\hspace{5cm}\overset{640}{}$

Q - E - V - N - L - E - P - H - V - Q - E - I - E - V - S - R - L - F - L - E -

$\overset{650}{}\hspace{5cm}\overset{660}{}$

P - T - R - K - D - I - A - L - L - K - L - S - S - P - A - V - I - T - D - K -

$\overset{670}{}\hspace{5cm}\overset{680}{}$

V - I - P - A - C - L - P - S - P - N - Y - V - V - A - D - R - T - E - C - F -

$\overset{690}{}\hspace{5cm}\overset{700}{}$

I - T - G - W - G - E - T - Q - G - T - F - G - A - G - L - L - K - E - A - Q -

$\overset{710}{}\hspace{5cm}\overset{720}{}$

L - P - V - I - E - P - N - K - V - C - N - R - Y - E - F - L - N - G - R - V -

$\overset{730}{}\hspace{5cm}\overset{740}{}$

Q - S - T - E - L - C - A - G - H - L - A - G - G - T - D - S - C - Q - G - D -

$\overset{750}{}\hspace{5cm}\overset{760}{}$

S* - G - G - P - L - V - C - F - E - K - D - K - Y - I - L - Q - G - V - T - S -

$\overset{770}{}\hspace{5cm}\overset{780}{}$

W - G - L - G - C - A - R - P - N - K - P - G - V - Y - V - R - V - S - R - F -

$\overset{790}{}$

V - T - W - I - E - G - V - M - R - N - N - COOH

SCHEME I

The disulfide pairings are believed (not all have been unambiguously established) to involve the following Cys residues: 30–54; 34–42; 83–161; 104–144; 132–156; 165–242; 168–296; 186–225; 214–237; 255–332; 276–315; 304–327; 357–434; 378–417; 406–429; 461–540; 482–523; 511–535; 547–665; 557–565; 587–603; 679–747; 710–726; and 737–765. The disulfide bridges between Cys^{557} and Cys^{565} and Cys^{547}-Cys^{665} are responsible for the covalent linkage of the two plasmin chains. In addition, the active-site serine residue is located at position 741,[42] and the active-center histidine is located at position 602.[43]

The carbohydrate sequence of human plasminogen has also been established, and differences exist between the two affinity-chromatography variants.[44-46] Glu-plasminogen 1 possesses two sites of glycosylation: one at Asn^{288} and the other at Thr^{345}. The sequences are:

$$Sia \xrightarrow{\alpha2,6} Gal \xrightarrow{\beta1,4} GlcNAc \xrightarrow{\beta1,2} Man \searrow^{\alpha1(6,3)}$$
$$Man \xrightarrow{\beta1,4} GlcNAc \xrightarrow{\beta1,4} GlcNAc\text{-}Asn^{288}$$
$$(\pm) Sia \xrightarrow{\alpha2,6} Gal \xrightarrow{\beta1,4} GlcNAc \xrightarrow{\beta1,2} Man \nearrow^{\alpha1(3,6)}$$

and

$$Sia \xrightarrow{\alpha2,3} Gal \xrightarrow{\beta1,3} GalNAc \longrightarrow Thr^{345}$$
$$\uparrow^{\alpha2,6}$$
$$(\pm) Sia$$

On the other hand, Glu-plasminogen 2 only contains the Thr-based oligosaccharide. The residues labeled ($\pm$)Sia are not present on all molecules.

From the above information, the complete compositions of human Glu-plasminogen, human Lys^{77}-plasminogen and human Val^{442}-plasminogen are listed in Table I.

Physical Properties

Based on the complete compositions of Glu-plasminogen 1 and Glu-plasminogen 2, in Table I, the calculated molecular weights of these single-chain proteins are 91,500 and 89,300, respectively. The two variants of Lys^{77}-plasminogen have MW 83,200 and 81,000, respectively, whereas Val^{442}-plasminogen has MW 38,380 for each variant.

A summary of some reported physical properties of Glu-plasminogen (Glu-Pg), and Lys^{77}-plasminogen (Lys-Pg) is presented in Table II.

The hydrodynamic properties of Glu-plasminogen are dramatically altered when certain antifibrinolytic amino acids such as ϵ-aminocaproic acid (ϵ-ACA) and trans-4-aminomethylcyclohexane-1-carboxylic acid (t-AMCHA) are added. These altered properties are summarized in Table III. The properties of Lys^{77}-plasminogen are not significantly altered by these amino acids.

[42] W. R. Groskopf, L. Summaria, and K. C. Robbins, J. Biol. Chem. 244, 3590 (1969).
[43] K. C. Robbins, P. Bernabe, L. Arzadon, and L. Summaria, J. Biol. Chem. 248, 1631 (1973).
[44] M. L. Hayes and F. J. Castellino, J. Biol. Chem. 254, 8768 (1979).
[45] M. L. Hayes and F. J. Castellino, J. Biol. Chem. 254, 8772 (1979).
[46] M. L. Hayes and F. J. Castellino, J. Biol. Chem. 254, 8777 (1979).

TABLE I
CHEMICAL COMPOSITION OF HUMAN PLASMINOGENS

Residue	Glu-Pg [a] (mol/mol)	Lys⁷⁷-Pg [b] (mol/mol)	Val⁴⁴²-Pg [c] (mol/mol)
Asp	36	32	13
Asn	40	38	14
Thr	60	57	22
Ser	56	50	18
Glu	56	45	22
Gln	28	24	12
Pro	70	69	33
Gly	60	57	34
Ala	37	31	20
Val	46	41	31
Met	10	8	3
Ile	21	17	11
Leu	42	38	25
Phe	20	16	11
Tyr	30	28	10
Trp	19	19	9
His	23	22	9
Lys	47	41	17
Arg	42	38	19
Cys	48	44	18
Sia	$2-4^d$ $(1-2)^e$	$2-4^d$ $(1-2)^e$	0
Gal	3^d $(1)^e$	3^d $(1)^e$	0
GalNAc	1^d $(1)^e$	1^d $(1)^e$	0
GlcNAc	4^d $(0)^e$	4^d $(0)^e$	0
Man	3^d $(0)^e$	3^d $(0)^e$	0

[a] Glu-Plasminogen. The amino acid compositions presume no differences between affinity variants 1 and 2. The values are calculated from the amino acid sequence.

[b] Lys⁷⁷-Plasminogen. The amino acid compositions presume no differences between affinity variants 1 and 2. The values are calculated from the amino acid sequence.

[c] Val⁴⁴²-Plasminogen. The amino acid compositions presume no differences between affinity variants 1 and 2. The values are calculated from the amino acid sequence.

[d] The carbohydrate composition refers to affinity-chromatography variant 1.

[e] The carbohydrate composition refers to affinity-chromatography variant 2.

TABLE II
PHYSICAL PROPERTIES OF HUMAN PLASMINOGEN[a]

Parameter	Glu-Pg	Glu-Pm	Lys^{77}-Pg	Lys^{77}-Pm
$S^{\circ}_{20,w}$ (S)[b]	5.71	5.68	4.73	4.80
f/f_{min} (translational)[c]	1.41	1.40	1.50	1.50
ρ_h (nsec)[c]	262.0	—	—	—
ρ_h/ρ_0 (rotational)[c]	3.69	—	—	—
$\bar{v}$ (ml/g)[d]	0.713	0.713	0.714	0.714

[a] Affinity variants 1 or 2.
[b] Ref. 37. The buffer employed was 0.1 M (in Cl$^-$) Tris-HCl–0.1 M NaCl, pH 7.8, at 20°C.
[c] Values in 0.1 M phosphate, pH 7.5 at 25°C. Taken from F. J. Castellino, W. J. Brockway, J. K. Thomas, H.-T. Liao, and A. B. Rawitch, *Biochemistry* **12**, 2787 (1973).
[d] Calculated from the data of Tables I and V.

Binding of Ligands to Plasminogen

The binding isotherms of ε-ACA and t-AMCHA to human plasminogen (combination of affinity variants 1 and 2) have been determined.[47,48] For ε-ACA, one strong binding site, of K_D 9 × 10^{-6} M, and approximately five weaker sites of average K_D 5 × 10^{-3} M were found, with Glu-plasminogen.[47] Utilizing Lys^{77}-plasminogen, one strong ε-ACA site of K_D 3.5 × 10^{-5} M, a slightly weaker site of K_D 2.6 × 10^{-4} M, and four weaker sites of average K_D 1 × 10^{-2} M were obtained. With t-AMCHA, a single strong site, of K_D 1.1 × 10^{-6} M and four weaker sites of K_D 7.5 × 10^{-4} M were found for Glu-plasminogen; whereas with Lys^{77}-plasminogen one strong site of K_D 2.2 × 10^{-6} M, a second weaker site of K_D 3.6 × 10^{-5} M, and two to three weaker sites of K_D 1 × 10^{-3} M were obtained.[48]

As can be seen from Table III, binding of ε-ACA to Glu-plasminogen results in a dramatic change in the conformation of the Glu-plasminogen molecule. The concentration of ligand required to produce the midpoint of the conformational alteration in Glu-plasminogen is 2.4 × 10^{-3} M for ε-ACA, and 6 × 10^{-4} mM for t-AMCHA.[49] These values closely correspond to the K_D values for binding to the weak class of sites on Glu-plasminogen, strongly suggesting the weak ligand sites as those relevant to producing the conformational alteration in the protein. Several other analogues of ε-ACA also produce similar alterations in the conformation of human Glu-plasminogen. These agents, as well as their respective concentrations required to produce one-half of the conformational change, are listed in Table IV.[49] As the number of carbon atoms between one COO$^-$

[47] G. Markus, J. L. DePasquale, and F. C. Wissler, *J. Biol. Chem.* **253**, 727 (1978).
[48] G. Markus, R. L. Priore, and F. C. Wissler, *J. Biol. Chem.* **254**, 1211 (1979).
[49] B. N. Violand, R. Byrne, and F. J. Castellino, *J. Biol. Chem.* **253**, 5395 (1978).

TABLE III
PHYSICAL PROPERTIES OF HUMAN PLASMINOGEN IN THE
PRESENCE OF ADDED ε-ACA[a]

Parameter	Glu-Pg	Glu-Pm	Lys-Pg	Lys-Pm
$S^{\circ}_{20,\omega}$ (S)[b]	4.80	4.83	4.65	4.82
f/f_{min} (translational)[c]	1.75	1.75	1.50	1.50
ρ_h (nsec)[c]	1.56	—	—	—
ρ_h/ρ_0[c]	2.20	—	—	—

[a] Affinity variants 1 or 2.
[b] Ref. 37. The buffer employed was 0.1 M (in Cl⁻) Tris-HCl–0.1 M ε-ACA, pH 7.8.
[c] Values in 0.1 M phosphate–0.1 M ε-ACA, pH 7.5. Taken from F. J. Castellino, W. J. Brockway, J. K. Thomas, H.-T. Liao, and A. B. Rawitch, *Biochemistry* **12**, 2787 (1973).

and NH_3^+ in these ligands is increased—in proceeding from 4-aminobutyric acid to 6-aminohexanoic acid (ε-ACA)—the efficacy of these agents in producing the conformational alteration increases, after which a loss in efficiency is seen on further progression to 8-aminooctanoic acid. The most effective agent so far discovered is t-AMCHA.

Mechanism of Activation of Human Plasminogen to Plasmin

The activation of single-chain Glu-plasminogen to 2-chain disulfide-linked Lys⁷⁷-plasmin by urokinase has been established.[20,50] The activation proceeds as follows:

$$
\begin{array}{ccc}
& \overset{\text{(S-S)}_2}{\boxed{}} & \\
NH_2\text{-Glu-Pg} \xrightarrow[(1)]{UK} & NH_2\text{-Glu-H} & Val^{561}\text{-L} \\
(3) \downarrow Pm & (2) \downarrow Pm & \\
& \overset{\text{(S-S)}_2}{\boxed{}} & \\
NH_2\text{-Lys}^{77}\text{-Pg} \xrightarrow[(4)]{UK} & NH_2\text{-Lys}^{77}\text{-H} & Val^{561}\text{-L}
\end{array}
$$

In the first step of the activation (1), the activator, urokinase (UK) catalyzes cleavage of the Arg⁵⁶⁰-Val⁵⁶¹ peptide bond, from Glu-Pg, resulting in formation of Glu-Pm. Glu-Pm contains a heavy chain, derived from the original amino-terminus of Glu-Pg (NH₂-Glu-H), of MW 66,300 (affinity variant 1) or 64,000 (affinity variant 2), disulfide linked to a light chain (Val-L), of MW 25,200. Next, autolytic cleavage of the Lys⁷⁶-Lys⁷⁷ peptide bond, by plasmin (Pm), results in formation of Lys⁷⁷-plasmin (2). Also, during the course of activation, Pm is capable of catalyzing cleavage

---
[50] L. Summaria, L. Arzadon, P. Bernabe, and K. C. Robbins, *J. Biol. Chem.* **250**, 3988 (1975).

TABLE IV
CONCENTRATION OF α-, ω-AMINO ACIDS
REQUIRED TO PRODUCE ONE-HALF OF THE
CONFORMATIONAL ALTERATION $\{C\}$ 0.5 IN
HUMAN GLU-PLASMINOGEN[a]

Amino acid	$\{C\}$ 0.5 (mM)
4-Aminobutyric acid	20.7
5-Aminopentanoic acid	4.9
6-Aminohexanoic acid	2.4
7-Aminoheptanoic acid	17.4
8-Aminooctanoic acid	32.5
L-Lysine	26.0
t-aminomethylcyclohexane-1-carboxylic acid	0.6

[a] Evaluated by titration of the change in $S^{\circ}_{20, w}$ of human Glu-plasminogen 2 with these amino acids. Taken from Ref. 49.

of the same Lys^{76}-Lys^{77} peptide bond, from Glu-Pg (3), resulting in formation of Lys^{77}-plasminogen (NH_2-Lys^{77}-Pg). This latter plasminogen is readily activated to Lys^{77}-plasmin, by urokinase (4).

It has been shown that steps (1), (2), and (3) are accelerated by the presence of the amino acids listed in Table IV.[49] The concentration dependence of this acceleration suggests that saturation of the weak class of ligand sites (which reflects the production of conformational alteration) is responsible for its production. Step 4 is not accelerated by addition of this class of ligands.[51]

The activation of human plasminogen by the bacterial protein streptokinase involves different considerations. Since streptokinase is not a protease, the first step in the activation requires formation of the actual plasminogen activator, the streptokinase–plasmin stoichiometric complex (SK–Pm), or the streptokinase–human plasminogen stoichiometric complex (SK–Pg); each complex contains an active site.[52] Once formed, the activator complex activates human plasminogen in a similar fashion to urokinase.[52,53]

[51] H. Claeys and J. Vermylen, *Biochim. Biophys. Acta* **342**, 351 (1974).
[52] For a review of the evolution of this mechanism, see F. J. Castellino, *Trends Biochem. Sci.* **4**, 1 (1979), and references therein.
[53] M. Gonzalez-Gronow and F. J. Castellino, *J. Biol. Chem.* **252**, 2175 (1977).

[30] Human Plasmin

By KENNETH C. ROBBINS, LOUIS SUMMARIA, and ROBERT C. WOHL

Plasmin, a serine protease with trypsin-like specificity, is formed from plasminogen in both isolated, purified systems and the plasma milieu, by a variety of activators: (*a*) urokinase-like from urine and malignant or transformed cells, and (*b*) nonurokinase from plasma, tissues, and bacteria. The preparation and properties of plasmin have been described in previous volumes of this series[1-3] and in reviews on plasminogen and plasmin.[4,5] Various forms of human plasmin have been described, prepared, and isolated, namely, Glu^1-plasmin,[6-8] Lys^{77}-plasmin,[1-3] Val^{442}-plasmin,[9,10] and Val^{561}-plasmin.[11] Glu^1-plasmin is prepared from the native Glu^1-zymogen in the presence of plasmin inhibitors, Lys^{77}-plasmin is prepared from either the native zymogen or the plasmin-degraded Lys^{77}-zymogen, and Val^{442}-plasmin is prepared from the elastase-degraded Val^{442}-zymogen. Val^{561}-plasmin, the light (B) chain of Glu^1-plasmin, Lys^{77}-plasmin, and Val^{442}-plasmin, is prepared by partial reduction of the two interchain-disulfide bonds connecting the heavy (A) chain and light (B) chain of the larger plasmins. Also, the affinity-chromatography forms of each of the plasmins containing the heavy (A) chains with lysine-binding sites can be prepared.[12] Recombinant plasmin can be prepared from the isolated sulfhydryl forms of the heavy (A) chain and light (B) chain.[13]

[1] K. C. Robbins and L. Summaria, this series, Vol. 19, p. 184.
[2] K. C. Robbins and L. Summaria, this series, Vol. 45, p. 257.
[3] F. J. Castellino and J. M. Sodetz, this series, Vol. 45, p. 273.
[4] K. C. Robbins, *Handb. Exp. Pharmacol.* [*N.S.*] **46**, 317 (1978).
[5] K. C. Robbins, *in* "Textbook of Hemostasis and Thrombosis" (R. W. Colman, J. Hirsch, V. J. Marder, and E. W. Salzman, eds.). Lippincott, Philadelphia, Pennsylvania, 1982 (in press).
[6] L. Summaria, L. Arzadon, P. Bernabe, and K. C. Robbins, *J. Biol. Chem.* **250**, 3988 (1975).
[7] B. N. Violand and F. J. Castellino, *J. Biol. Chem.* **251**, 3906 (1976).
[8] L. Summaria, I. G. Boreisha, L. Arzadon, and K. C. Robbins, *J. Biol. Chem.* **252**, 3945 (1977).
[9] R. C. Wohl, L. Summaria, and K. C. Robbins, *J. Biol. Chem.* **255**, 2005 (1980).
[10] J. R. Powell and F. J. Castellino, *J. Biol. Chem.* **255**, 5329 (1980).
[11] L. Summaria and K. C. Robbins, *J. Biol. Chem.* **251**, 5810 (1976).
[12] L. Summaria, F. Spitz, L. Arzadon, I. G. Boreisha, and K. C. Robbins, *J. Biol. Chem.* **251**, 3693 (1976).
[13] L. Summaria, I. Boreisha, R. C. Wohl, and K. C. Robbins, *J. Biol. Chem.* **254**, 6811 (1979).

Preparation of Glu[1]-Plasmin[8]

Glu[1]-plasmin can be prepared from native human Glu[1]-plasminogen by activation with urokinase in a 25% glycerol–0.04 M Tris–0.02 M lysine–0.08 M NaCl–0.003 M EDTA buffer (pH 9.0), containing 0.03 M leupeptin (Ac-L-Leu-L-Leu-L-Argal) for 20 hr, at 25°C. The concentrations of plasminogen, urokinase, and leupeptin/ml were 7 mg, 29,000 IU, and 13 mg, respectively. Preparations can be prepared with about 60–80% active sites. They are always stored at −20°C.

Preparation of Lys[77]-Plasmin and Val[442]-Plasmin

Lys[77]-plasmin and Val[442]-plasmin are prepared from the human zymogens[9,10] by methods previously described[1,2] in a 25% glycerol-Tris buffer at pH 9.0, either at 25°C for 20 hr or at 30°C for 45 min (also see Chapter 29 in this volume). The enzyme preparations contain over 95% active sites, with specific proteolytic activities (casein) of 30 IU/mg protein and 60 IU/mg protein, respectively.

Preparation of Val[561]-Plasmin [Light (B) Chain] and Heavy (A) Chain

Val[561]-plasmin can be readily prepared from human Lys[77]-plasmin. Glu- or Lys[77]-plasminogen at a concentration of 15 mg/ml is activated with urokinase (molar ratio of plasminogen to urokinase is 1000 : 1) to Lys-plasmin in the 25% glycerol-Tris buffer at pH 9.0. Leupeptin was added to the plasmin solution at a concentration of 15 mg/ml (0.04 M). 10 ml of the mixture at a concentration of 7.5 mg plasmin/ml was reduced with either 0.1 M 2-mercaptoethanol for 20 min at 20°C or 0.05 M dithioerythritol for 3 hr at 2°C, under nitrogen. The preparation, containing the sulfhydryl chains, can be alkylated with 0.1 M sodium iodoacetate at 2°C to prepare the carboxymethylated derivatives.[14] The mixture of heavy (A) chains and light (B) chains, in either their sulfhydryl or carboxymethylated forms can be passed through a L-lysine-substituted–Sepharose affinity-chromatography column equilibrated (and eluted) with 0.1 M phosphate buffer (pH 7.4) containing 0.0002 M leupeptin, at 2°C.[11,13] The heavy (A) chain was adsorbed, whereas the light (B) chains (Val[561]-plasmin) passed through the column and collected in glycerol (final concentration of 25%). The Val[561]-plasmin fractions were pooled and adjusted to 0.04 M Tris–0.02 M lysine–0.08 M NaCl–0.003 M EDTA with solid salts, and adjusted to pH 9.0 with 1 M NaOH. The heavy (A) chain was eluted with 0.1 M

---

[14] W. R. Groskopf, B. Hsieh, L. Summaria, and K. C. Robbins, *J. Biol. Chem.* **244**, 359 (1969).

phosphate–0.5 M ϵ-aminocaproic acid–0.003 M dithioerythritol (pH 7.4) at 2°C, and collected in glycerol (final concentration of 25%). The fractions were pooled and adjusted with salts and NaOH to pH 9.0 (similarly to the Val^{561}-plasmin preparation). The extinction coefficients, $E_{1cm}^{1\%}$, for the heavy (A) chain and Val^{561}-plasmin were 17.0 and 16.0, respectively. The heavy (A) chain and Val^{561}-plasmin contained 9–10 cysteine residues/mol protein and 3–4 cysteine residues/mol protein, respectively. In addition to the two interchain disulfide bonds, additional disulfide bonds were cleaved in both chains. A Val^{561}-plasmin–streptokinase complex can be prepared from the equimolar plasminogen (plasmin)–streptokinase complex by the same methodology,[13] or by mixing Val^{561}-plasmin with an equimolar amount of streptokinase. The isolated Val^{561}-plasmin–SK complex contained 6 cysteine residues/mol protein; the $E_{1cm}^{1\%}$ of this preparation was 12.5. Val^{561}-plasmin preparations contain 10 to 30% active sites, with a specific proteolytic activity (casein) of about 3 IU/mg protein, 10% of the specific activity of Lys^{77}-plasmin. The addition of an equimolar amount of streptokinase to Val^{561}-plasmin increases the active sites to >90%, indicating that streptokinase stabilizes the three-dimensional active center structure of this light (B) chain similar to the stabilization of the active center structure by the heavy (A) chain in the intact plasmin molecule. The isolated preformed Val^{561}-plasmin–streptokinase complex always has >90% active sites.

Preparation of Recombinant Lys^{77}-Plasmin and Lys^{77}-Plasmin–Streptokinase Complex

The plasmin heavy (A) and light (B) chains (sulfhydryl forms) were recombined by mixing 140 ml of the light (B)-chain preparation (21 mg) with 70 ml of the heavy (A)-chain preparation (42 mg), and dialyzing the mixture against 300 ml of 25% glycerol–0.04 M Tris–0.02 M lysine–0.08 M NaCl–0.003 M EDTA buffer (pH 9.0) at 4°C for 72 hr with stirring. The plasmin–streptokinase complex-derived chains (sulfhydryl forms) were recombined by mixing 90 ml of the light (B) chain–streptokinase complex preparation (36 mg) with 65 ml of the heavy (A) chain (26 mg) (equimolar amounts) and dialyzing by the same procedure as described above for the preparation of recombinant Lys^{77}-plasmin. The preparations were concentrated 10 times by dialysis for 18 hr at 4°C against 25% Carbowax (polyethylene glycol, 20 M), dissolved in the dialysis buffer, to a final protein concentration of about 3 mg/ml. The preparations were repurified by the lysine-Sepharose affinity chromatography method to remove excess light (B) chains and light (B) chain–streptokinase complexes that were not adsorbed. The adsorbed and eluted enzyme, or enzyme–streptokinase complex, was further purified by precipitation with

$(NH_4)_2SO_4$, 150 mg/ml, the bulk of the heavy (A) chains remain in solution. The final products had no contaminating light (B) chain or light (B) chain–streptokinase complex. Recombinant Lys^{77}-plasmin (R-Lys-Pln) had 41% active sites, without or with streptokinase, and recombinant Lys^{77}-plasmin–streptokinase complex (R-Lys-Pln–SK) had 68% active sites. R-Lys-Pln had a specific activity of ~15 IU/mg protein. It is a mixture of active plasmin and either a recombinant heavy (A) chain–light (B) chain without proteolytic activity, or a recombinant hybrid with a denatured light (B) chain. The R-Lys-Pln–SK complex is probably a mixture of active complex and a recombinant inactive covalently linked disulfide hybrid complex between asymmetric interchain–interchain, interchain–intrachain, or intrachain–intrachain sulfhydryl groups. The recombinant enzymes resemble their parents both in their physical properties and in their enzymatic functions.

Plasmin Kinetics

Steady-state amidase and esterase kinetic parameters, the apparent Michaelis constants (K_m) and the catalytic rate constants (k_{cat}), have been determined for the various types of plasmins, and plasmin–streptokinase activator complexes, on a variety of synthetic substrates, at different pH's and temperatures.[9,10,13,15–20] A summary of data obtained in our laboratories with H-D-valyl-L-leucyl-L-lysyl-p-nitroanilide (H-D-Val-Leu-Lys-pNA) and N^α-Cbz-L-lysine-p-nitrophenyl ester (CBZ-Lys-pNPE) are summarized in Table I; the amidase parameters for preparations of high- and low-molecular-weight urokinase (UK-HMW and UK-LMW) are included. Similar K_m and k_{cat} values were obtained with BAEE for Lys^{77}-plasmin and Val^{442}-plasmin.[18] Higher K_m values were obtained for these enzymes with TAME, namely 770 and 4000 μM, respectively.[10,15] With the two latter substrates, the k_{cat} values for the two enzymes were similar to those found in Table I, obtained with the $para$-nitroanilide and $para$-nitrophenyl ester substrates. With Bz-Phe-Val-Arg-pNA, higher K_m values and similar k_{cat} values were found for these two enzymes.[18,19] L-Lysine, ϵ-aminocaproic acid, and $trans$-4-aminomethylcyclohexanecarboxylic acid[1] increase both the K_m values and k_{cat} values.[21]

[15] J. M. Sodetz and F. J. Castellino, *Biochemistry* **11**, 3167 (1972).
[16] U. Christensen, *Biochim. Biophys. Acta* **397**, 459 (1975).
[17] R. C. Wohl, L. Arzadon, L. Summaria, and K. C. Robbins, *J. Biol. Chem.* **252**, 1141 (1977).
[18] U. Christensen, L. Sottrup-Jensen, S. Magnusson, T. E. Petersen, and I. Clemmensen, *Biochim. Biophys. Acta* **567**, 472 (1979).
[19] U. Christensen and H.-H. Ipsen, *Biochim. Biophys. Acta* **569**, 177 (1979).
[20] U. Christensen, *Biochim. Biophys. Acta* **570**, 324 (1979).
[21] U. Christensen, *Biochim. Biophys. Acta* **526**, 194 (1978).

TABLE I
AMIDASE AND ESTERASE KINETIC PARAMETERS OF PLASMINS
AND PLASMINOGEN ACTIVATORS

	Substrate			
	H-D-Val-Leu-Lys-pNA[a]		CBZ-Lys-pNPE[b]	
Enzyme preparations	K_m (μM)	k_{cat} (sec^{-1})	K_m (μM)	k_{cat} (sec^{-1})
Plasmins				
1. Glu^1-Pln	—	—	13.5	22.3
2. Lys^{77}-Pln	210	25.7	15.0	22.3
3. Val^{442}-Pln	280	25.5	—	—
4. Val^{561}-Pln	1260	14.5	83.0	2.1
5. Lys^{77}-Plg^{*c}	—	—	42.0	2.5
Activators				
a. SK				
1. Glu^1-Plg–SK	320	39.1	19.4	21.8
2. Lys^{77}-Pln–SK	780	51.9	33.3	22.3
3. Val^{442}-Plg–SK	210	29.0	—	—
4. Val^{442}-Pln–SK	370	43.2	—	—
5. Val^{561}-Pln–SK	520	36.3	18.0	21.8
	H-Glu-Gly-Arg-pNA[d]			
b. UK				
1. UK-HMW	200	20.8		
2. UK-LMW	270	16.5	24.0	11.4

[a] At pH 7.4 and 37°C.
[b] At pH 6.0 and 30°C.
[c] Zymogen with active site.
[d] At pH 9.0 and 37°C.

Plasminogen-Activation Kinetics

Mathematical models for studying plasminogen-activation kinetics have been developed.[9,22] The general equation (Ref. 9, Appendix, Equation 16) will accommodate all plasminogen-activator species. Steady-state kinetic parameters of activation of several well-characterized human plasminogens (i.e., the Glu^1, Lys^{77}, and Val^{442} forms) by highly purified activator species, namely, streptokinase complexes with various activated forms of human plasminogen, urokinase forms with different molecular weights derived from urine and kidney cells in tissue culture, staphylokinase, and trypsin (see Table I) were determined. Activation rates were measured at pH 6.0 and 30°C by determining the initial velocity of CBZ-Lys-pNPE ester hydrolysis by the plasmin generated, and at

[22] R. C. Wohl, L. Summaria, L. Arzadon, and K. C. Robbins, *J. Biol. Chem.* **253**, 1402 (1978).

TABLE II
KINETICS OF ACTIVATION PARAMETERS OF HUMAN PLASMINOGENS WITH
STREPTOKINASE-ACTIVATOR SPECIES AT pH 7.4 AND 37°C

| | Plasminogen | | | | | |
| | Glu^1 form | | Lys^{77} form | | Val^{442} form | |
Activators	K_{plg} (μM)	k_{plg} (min^{-1})	K_{plg} (μM)	k_{plg} (min^{-1})	K_{plg} (μM)	k_{plg} (min^{-1})
1. Glu-Plg–SK	0.12	8.97	0.11	52.39	—	—
2. Val^{442}-Plg–SK	0.25	25.53	0.11	8.91	—	—
3. Lys^{77}-Pln–SK	0.23	8.97	—	—	—	—
4. Val^{442}-Pln–SK	1.94	28.71	—	—	—	—
5. Val^{561}-Pln–SK	0.15	25.33	0.13	55.08	0.20	64.20
6. SK	0.12	8.21	0.11	44.45	0.11	19.20

pH 7.4 and 37°C by determining the initial velocity of the H-
D-Val-Leu-Lys-p NA hydrolysis by the plasmin generated. The apparent
Michaelis constants (K_{plg}) and catalytic rate constants (k_{plg}) of activation
for the different zymogen forms and streptokinase-activator species at pH
7.4 and 37°C are summarized in Table II. With Lys^{77}-plasminogen as the
substrate, the K_{plg} values were found to be similar at pH's 6.0 and 7.4, but
the k_{plg} values were found to be ~2-fold higher at pH 7.4 for streptokinase
(Lys^{77}-plasminogen–streptokinase) and Glu^1-plasminogen–streptokinase,
but were similar at both pH's for Val^{561}-plasmin–streptokinase.[9] Steady-
state kinetic parameters of activation with both Glu^1-plasminogen and
partially degraded Lys^{77}-plasminogen as the substrates with the
urokinase-activator species are summarized in Table III. There are great
differences in the molecular and functional characteristics of urokinase
preparations. The K_{plg} and k_{plg} values for Val^{442}-plasminogen with
UK-HMW were 1.37 μM and 22.2 min^{-1}, respectively. Similar data have
been reported on the urokinase-catalyzed conversion of both Glu^1- and
Lys^{77}-plasminogens[23,24] at pH 7.8 and 25°C with BAEE-coupled hydrolysis of plasmin. Since this conversion was found to obey Michaelis–
Menten rate equations, it was postulated that these data fit the established
molecular mechanism of conversion of Glu^1-plasminogen through Glu^1-
plasmin to Lys^{77}-plasmin, and not through Lys^{77}-plasminogen. The relative catalytic efficiencies of activation, or second-order rate constants,
k_{plg}/K_{plg}, of native Glu^1-plasminogen with a variety of different strep-
tokinase- and urokinase-activator species, and including staphylokinase

[23] U. Christensen, *Biochim. Biophys. Acta* **481**, 638 (1977).
[24] U. Christensen and S. Müllertz, *Biochim. Biophys. Acta* **480**, 275 (1977).

TABLE III

KINETICS OF ACTIVATION PARAMETERS OF HUMAN PLASMINOGENS WITH
UROKINASE-ACTIVATOR SPECIES AT pH 7.4 AND 37°C

| | Plasminogen | | | |
| | Glu[1] form | | Lys[77] form | |
Activators[a]	K_{plg} (μM)	k_{plg} (min^{-1})	K_{plg} (μM)	k_{plg} (min^{-1})
1. UK-HMW (U)[b]	1.72	47.06	1.90	50.41
2. UK-HMW (U)	4.08	58.79	5.65	78.48
3. UK-HMW (TC)[b]	1.11	46.94	1.25	78.24
4. UK-LMW (U)[b]	2.64	63.00	1.67	22.74
5. UK-LMW (U)	1.72	47.06	3.69	62.45
6. UK-LMW (TC)	7.43	65.45	3.16	41.78

[a] U, urine; TC, tissue culture.
[b] Homogeneous protein.

and trypsin, are summarized in Table IV. Human kallikrein, plasmin, and thrombin will not activate native Glu[1]-plasminogen in this system. The lower limit of our kinetics of activation measurements is a second-order rate constant of ~1.5 × 10^2 M^{-1} sec^{-1}.

Kinetic studies have been carried out on several human variant abnormal plasminogens (Chicago I, II, and III).[25,26] These isolated plasminogens can be completely activated to plasmin in a 25% glycerol-Tris buffer (pH 9.0) system. These plasmins have the same steady-state amidase parameters as normal plasmin. However, in the equimolar plasminogen (Glu- and Lys-)–streptokinase complexes, the K_m values are normal but the k_{cat} values are decreased 2- to 3-fold, indicating impaired active sites in these complexes. The steady-state kinetic parameters of activation of three variant plasminogens—Chicago I, II, and III (Glu[1] and Lys[77] forms)—with several different activator species, are summarized in Table V. These data lead to the conclusion that in Chicago I and II the basic defect is in the binding properties of the activators to the variant plasminogens, and not in the formation of plasmin, and also the active sites formed in the plasminogen moieties of the zymogen–streptokinase complexes have been impaired. However, in Chicago III, the basic defect is in the cleavage of the Arg^{560}-Val peptide bond to form plasmin. Interpretation of the data was possible only in terms of a homogeneous population of

[25] R. C. Wohl, L. Summaria, and K. C. Robbins, *J. Biol. Chem.* **254**, 9063 (1979).
[26] R. C. Wohl, J. Chediak, S. Rosenfeld, L. Summaria, and K. C. Robbins, submitted for publication.

TABLE IV
RELATIVE CATALYTIC EFFICIENCIES OF
ACTIVATION, OR SECOND-ORDER RATE
CONSTANTS, k_{plg}/K_{plg}, BY DIFFERENT
ACTIVATOR SPECIES WITH GLU [1]-PLASMINOGEN
AT pH 7.4 AND 37°C

Activators	k_{plg}/K_{plg} $(\mu M^{-1} min^{-1})$
1. Val[561]-Pln–SK	169
2. Val[442]-Plg–SK	102
3. Glu[1]-Plg–SK	75
4. SK	68
5. UK-HMW	27
6. UK-LMW	24
7. Staphylokinase	1
8. Trypsin	0.1

TABLE V
KINETICS OF ACTIVATION PARAMETERS OF VARIANT HUMAN PLASMINOGENS (CHICAGO I,
II, AND III) WITH STREPTOKINASE- AND UROKINASE-ACTIVATOR SPECIES AT
pH 7.4 AND 37°C

Activators	Zymogen substrates	Kinetic parameters			
		Glu [1] form		Lys [77] form	
		K_{plg} (μM)	k_{plg} (min^{-1})	K_{plg} (μM)	k_{plg} (min^{-1})
1. Val[561]-Pln–SK	N[a]	0.15	25.33	0.13	55.08
	I	22.01	25.40	0.08	33.26
	II	17.57	25.92	0.08	34.61
	III	1.32	8.35	—	—
2. SK	N	0.12	8.21	0.12	52.38
	I	0.98	9.13	0.14	22.78
	II	0.36	8.85	0.08	22.72
	III	5.26	6.49	—	—
3. Glu[1]-Plg–SK	N	0.12	8.97	—	—
	I	0.30	8.90	—	—
	II	0.11	9.00	—	—
	III	—	—	—	—
4. UK-HMW	N	1.72	47.06	1.90	52.40
	I	8.30[b]	47.05	1.28	19.87
	II	7.47[b]	47.05	0.43	11.13
	III	0.24	0.28	—	—

[a] Normal plasminogen from large plasma pool, >4000 donors.
[b] K_D, dissociation constants for I and II are 1.71 μM and 1.52 μM, respectively.

molecules in each of the three individuals studied. The urokinase data with these variant plasminogens point to an activation mechanism identical with that proposed for streptokinase—namely, the formation of a plasminogen–urokinase complex analogous to the plasminogen–streptokinase complex that will activate a second plasminogen molecule. A dissociation constant, K_D ~1.6 μM was calculated for the urokinase activation of plasminogens Chicago I and II. A dissociation constant of 5 × 10^{-11} M has been reported for the human plasmin–streptokinase complex from measurements of bovine plasminogen-activator activity generated in mixtures of human plasmin and streptokinase.[27]

Kinetics of Reaction between Human Lys[77]-Plasmin and Val[442]-Plasmin with Human α_2-Plasmin Inhibitor

The reaction between Lys[77]-plasmin and Val[442]-plasmin with α_2-plasmin inhibitor proceeds in a two-step process: first, a very rapid reversible second-order reaction (k_1) followed by a slower irreversible first-order reaction (k_2).[28-31] The k_1 and k_2 values for Lys[77]-plasmin were determined to be ~3 × 10^7 M^{-1} sec^{-1} and ~4.2 to ~6.5 × 10^{-3} sec^{-1}, respectively.[28-30] The k_1 and k_2 values for Val[442]-plasmin were determined to be ~6.5 × 10^5 M^{-1} sec^{-1} and ~4.2 × 10^{-3} sec^{-1}, respectively.[31]

[27] S. A. Cederholm-Williams, F. De Cock, H. R. Lijnen, and D. Collen, *Eur. J. Biochem.* **100**, 125 (1979).
[28] U. Christensen and I. Clemmensen, *Biochem. J.* **163**, 389 (1977).
[29] U. Christensen and I. Clemmensen, *Biochem. J.* **175**, 635 (1978).
[30] B. Wiman and D. Collen, *Eur. J. Biochem.* **84**, 573 (1978).
[31] B. Wiman, L. Boman, and D. Collen, *Eur. J. Biochem.* **87**, 143 (1978).

[31] Streptokinase and Staphylokinase

By KENNETH W. JACKSON, NAOMI ESMON, and JORDAN TANG

The process of fibrinolysis is performed by the serum protease plasmin, which is normally present in the blood as a precursor, plasminogen. Plasminogen is proteolytically activated to plasmin by its activators, which are specific proteases produced by tissues, such as urokinase and tissue activator.[1,2] There are also nonprotease plasminogen activators—streptokinase and staphylokinase—which are extracellular proteins produced by certain strains of *Streptococcus hemolyticus* and *Staphylococcus aureus,* respectively.

[1] W. F. White, G. H. Barlow, and M. M. Mozen, *Biochemistry* **5**, 2160 (1966).
[2] T. Astrup, *Blood* **11**, 781 (1956).

Because streptokinase is used for thrombolytic treatment, many studies on its purification and properties have been carried out. Unlike urokinase or tissue activator, streptokinase activates plasminogen by the formation of a strongly associated one-to-one molar complex.[3,4] As a result, proteolytic activity is generated in the plasminogen moiety of this complex.[5] There is good evidence that peptide-bond cleavage of plasminogen is not essential and occurs after active-center formation.[5,6] Hence the activation mechanism is unique in that the formation of the plasminogen activator requires no proteolysis. This is in contrast to the more commonly known zymogen-activation mechanisms. In the absence of specific inhibitors, the streptokinase–plasminogen complex rapidly converts to the streptokinase–plasmin complex. This latter complex effectively activates plasminogen to plasmin, a reaction not catalyzed by plasmin itself. It is not understood how streptokinase binding results in the generation of proteolytic-activator activity in the plasminogen moiety. Nor is it known how streptokinase binding alters the active center of plasmin to confer on it the enhanced specificity to activate plasminogen. Previous chapters in this series on streptokinase[7,8] described in detail some assay methods, as well as isolation and purification procedures. This chapter emphasizes the structure–function relationships of streptokinase. Some preliminary discussion on this aspect has been reported previously.[9,10]

Knowledge about staphylokinase is relatively limited, even though several successful purifications of this protein have been reported.[11-15] These results show that staphylokinase has a molecular weight in the range of 16,000–22,500 and exists in several forms of different isoelectric

[3] J. Zybles, W. F. Blatt, and H. Jensen, *Proc. Soc. Exp. Biol. Med.* **102**, 755 (1957).

[4] D. L. Kline and J. B. Fishman, *J. Biol. Chem.* **236**, 2807 (1961).

[5] K. N. N. Reddy and G. Markus, *J. Biol. Chem.* **247**, 1683 (1972).

[6] D. K. McClintock and P. H. Bell, *Biochem. Biophys. Res. Commun.* **43**, 694 (1971).

[7] F. B. Taylor, Jr. and R. H. Tomar, this series, Vol. 19, p. 807.

[8] F. J. Castellino, J. M. Sodetz, W. J. Brockway, and G. E. Siefring, Jr., this series, Vol. 45, p. 244.

[9] K. W. Jackson and J. Tang, *Thromb. Res.* **13**, 693 (1978).

[10] K. W. Jackson, N. Esmon and J. Tang, in "The Regulation of Coagulation" (K. G. Mann and F. B. Taylor, eds.), p. 515. Elsevier/North-Holland Publ., Amsterdam and New York, 1980.

[11] C. H. Lack and K. L. A. Glanville, this series, Vol. 19, p. 706.

[12] F. K. Vesterberg and O. Vesterberg, *J. Med. Microbiol.* **5**, 441 (1972).

[13] S. Arvidson, R. Ericksson, T. Holme, R. Mollby, T. Wadstrom, and O. Vesterberg, *Contrib. Microbiol. Immunol.* **1**, 406 (1973).

[14] B. Kowalska-Loth and K. Zakrzewski, *Acta Biochim. Pol.* **22**, 327 (1975).

[15] M. Nick, J. Brückler, W. Schaeg, K. D. Hasche, and H. Blobel, *Zentralbl. Bakteriol., Parasitenkd., Infektionskr. Hyg., Abt. 1: Orig., Reihe* A **237**, 160 (1977).

points. Some of these pI forms may be due to proteolytic interconversion.[16] Like streptokinase, staphylokinase activates human plasminogen but not the bovine zymogen.[17] However, the specificity of the two complexes differs in that the staphylokinase–human plasminogen complex cannot activate bovine plasminogen.[18] The activation of human plasminogen by staphylokinase also appears to be due to the formation of a complex with plasminogen resulting in the generation of catalytic activity in the zymogen moiety much like that for streptokinase.[14] However, considerable uncertainty exists in this area and it is obvious that more work is needed. In the following, we describe the methods of *S. aureus* culture, staphylokinase purification, and some of its properties. Some of these results have been reported previously.[10]

Assay Method

Streptokinase assays utilizing fibrin plates and synthetic esters have been described previously.[7,8] We describe below a two-stage assay for streptokinase and staphylokinase. This assay was adapted from the method developed by Radcliffe for plasma-plasminogen activator.[19] It has been used in our laboratory to measure the rate of plasminogen activation.

Principle

Streptokinase or staphylokinase forms a complex with plasminogen. The complex catalyzes the conversion of plasminogen to plasmin in the first stage of the assay. NaCl is added to a sufficient concentration to stop further plasminogen activation.[20] The amount of plasmin generated in the first stage is measured in the second stage using the plasmin-specific substrate, D-Val-Leu-Lys-p-nitroanilide (S2251). The hydrolytic product p-nitroanilide is measured spectrophotometrically.

In the streptokinase assay, Triton X-100 can be included in the first stage, since it enhances the streptokinase activity. In the staphylokinase assay, it is best to use the Lys form of human plasminogen, because the staphylokinase exhibits a serious lag in activation rate when human Glu-plasminogen is used.[21] (For review on different forms of human plasminogen, see Ref. 22.)

[16] T. Makino, *Biochim. Biophys. Acta* **522,** 267 (1978).
[17] F. M. Davidson, *Biochem. J.* **76,** 56 (1960).
[18] G. Markus and W. C. Werkheiser, *J. Biol. Chem.* **239,** 2637 (1964).
[19] R. Radcliffe and T. Heinze, submitted for publication.
[20] R. Radcliffe and T. Heinze, *in* "The Regulation of Coagulation" (K. G. Mann and F. B. Taylor, eds.), p. 551. Elsevier/North-Holland Publ., Amsterdam and New York, 1980.
[21] N. Esmon and J. Tang, unpublished results.
[22] K. C. Robbins and L. Summaria, this series, Vol. 45, p. 257.

Reagents

A. Streptokinase stock solution. 10 mg streptokinase in 1 ml of 50% (v/v) glycerol in H_2O. This solution is kept at $-20°C$. Before use, the stock solution is diluted to a concentration of 1 $\mu g/ml$ using 0.5% bovine serum albumin (BSA) in 0.05 M Tris-chloride buffer, pH 7.5.

B. 0.05 M Tris-chloride buffer, pH 7.5

C. Plasminogen solution. Glu-plasminogen is prepared using a lysine-affinity column as described previously.[23] A stock solution, stored frozen, is diluted to a concentration of 0.5 mg/ml with 0.05 M Tris-chloride–0.5% BSA (pH 7.5) before use.

D. Plasmin substrate stock solution. The synthetic tripeptide D-Val-Leu-Lys-p-nitroanilide (Kabi S2251) is dissolved in water to a concentration of 5 mg/ml.

E. Concentrated NaCl. 1.77 M in 0.032 M Tris-chloride, pH 7.5.

F. Substrate solution. Before assay, mix 2 parts of reagent D with 3 parts of reagent E.

G. Acetic acid: 0.2 M

Procedure

To the bottom of plastic test tubes (10 × 75 mm) the following solutions are added: 10 μl of a streptokinase sample to be assayed, 20 μl of buffer (B), and 20 μl of the plasminogen solution (C). The tubes are incubated at 37°C for 15 min. At this point, 30 μl of the substrate solution (F) is added. The second-stage incubation is carried out at 37°C for 10 min and 0.4 ml of acetic acid (G) is added to terminate incubation. The optical density at 405 nm is determined for this solution.

Comments

It is important to observe the development of yellow color during the second stage. Once experience is gained in visual judgment of the color intensity, the incubation time of the second stage can be shortened or prolonged to produce an optimal color intensity for spectrophotometric quantitation. The results are expressed as ΔOD per minute.

The assay for streptokinase is generally quantitative and has a linear response to the amount of streptokinase added. The dilution of streptokinase stock solution with buffer containing albumin is essential to achieve quantitative dilution. Albumin can be replaced by 10^{-4} M Triton X-100.

The assay for staphylokinase, however, is not linear. This is due to a

---

[23] R. H. Tomar and F. B. Taylor, *Biochem. J.* **125**, 793 (1971).

lag in the plasminogen-activation time course by staphylokinase.[12] The presence of ε-aminocaproic acid (EACA) or the use of Lys-plasminogen reduces but does not always eliminate the lag. This is believed to result from the presence of variable ratios of the monomer and dimer forms of staphylokinase, which exhibit different kinetic properties toward plasminogen.[10,21] With an unknown percentage of the monomer, the lag cannot be strictly predicted or controlled.

Structure and Function of Streptokinase

The tentative amino acid sequence of streptokinase is shown in Fig. 1. The detailed evidence for this structure will be published elsewhere.[24] Several features in this sequence may be pertinent to the function of streptokinase. The NH_2-terminal region, about 80 residues, of streptokinase is homologous to the corresponding region of several serine proteases.[9] This homology was found to be particularly strong in three areas: residues 1–7, 12–23, and 49–72. It is interesting that several residues important to the catalytic activity of the serine proteases are located in these regions. The N-terminal residue of most of the serine proteases is important, since it is involved in a salt linkage to Asp^{194}, thus forming the "oxyanion hole" essential for catalysis.[25] The other two regions are near the active-center residues His^{57} and Asp^{102}. The significance of this homology, which appears to diminish for the rest of the molecule, is not clear. Several functional hypotheses based on this structural homology have been advanced.[9] Another interesting structural feature is a possible internal 2-fold homology between the N- and C-terminal halves of the streptokinase molecule. This is suggested from the following evidence: (1) Some sequence homology from the N- and C-terminal halves of streptokinase has been observed; (2) the secondary structure of the two halves of the streptokinase molecule have a similar distributions of α-, β- and random structures[21]; (3) staphylokinase is only one-half the size of streptokinase and appears to require dimerization to function (see following section).

Staphylokinase

S. aureus Culture

A suitable strain of S. aureus was chosen based on its high yield of fibrinolytic activity when the centrifuged culture supernatants were tested for plasminogen activation on canine fibrin plates.[23] Although preliminary

[24] K. W. Jackson and J. Tang, in preparation.
[25] R. Huber and W. Bode, in "Regulatory Proteolytic Enzymes and Their Inhibitors" (S. Magnusson et al., eds.), p. 15. Pergamon, Oxford, 1978.

```
      1                5                 10                15                20
NH₂- Ile  -Ala- Gly -Pro- Glu -Trp- Leu -Leu- Asp -Arg- Pro -Ser- Val -Asn- Asn -Ser- Gln -Leu- Val- Val-
     21               25                30                35                40
     Ser -Val- Ala -Gly- Thr -Val- Glu -Gly- Thr -Asn- Gln -Asp- Ile -Ser- Leu -Lys- Phe -Phe- Glu- Ile-
     41               45                50                55                60
     Asp -Leu- Thr -Ser- Arg -Pro- Ala -His- Gly -Gly- Lys -Thr- Glu -Gln- Gly -Leu- Ser -Pro- Lys- Ser-
     61               65                70                75                80
     Lys -Pro- Phe -Ala- Thr -Asp- Ser -Gly- Ala -Met- Ser -His- Lys -Leu- Glu -Lys- Ala -Asp- Leu- Leu
     81               85                90                95                100
     Lys -Ala- Ile -Gln- Glu -Gln- Ile -Leu- Ala -Asn- Val -His- Ser -Asn- Ile -Xxx- Ile -Phe[ Asx, Asx
    101              105               110               115               120
     Asx ,Asx, Thr ,Glx, Glx ,Gly, Tyr ]Glu- Val -Ile- Asp -Phe- Ala -Xxx- Asp -Ala- Thr -Leu- Thr- Asp
    121              125               130               135               140
     Arg -Val- Arg -Pro- Tyr -Lys- Glu -Lys- Pro -Ile- Gln -Asn- Gln -Ala- Lys -Ser- Val -Asp- Val- Glu-
    141              145               150               155               Asp
     Tyr -Thr- Val -Gln- Phe -Thr- Pro  Asp  Asn -Pro- Asx -Asp- Asp -Phe- Arg -Pro- Gly -Leu- Lys  Leu
    161              165               170               175               180
     Thr -Lys- Leu -Leu- Lys -Thr- Leu -Ala- Leu -Gly- Asp -Thr[ Ser ,Ile] Glu -Leu- Leu -Ala- Gln- Ala-
    181              185               190               195               200
     Gln -Ser- Ile -Leu- Asn -Lys- Asn -His- Pro -Gly- Tyr -Thr- Ile -Tyr- Glu -Arg- Thr -Ile- Leu- Pro-
    201              205               210               215               220
     Met -Asp- Gln -Glu- Phe -Thr- Tyr -Arg- Val -Lys- Asn -Arg- Glu -Gln- Ala -Tyr- Arg -Ile- Asn- Lys-
    221              225               230               235               240
     Lys -Ser- Gly -Leu- Asn -Glu- Glu -Ile- Asn -Asn- Thr -Asp- Leu -Ile- Ser -Leu- Glu -Tyr- Lys- Tyr-
    241              245               250               255               260
     Val -Leu- Lys -Lys- Gly -Glu- Lys -Pro- Tyr -Asp- Pro -Phe- Asp -Arg- Ser -His- Leu -Lys- Leu- Phe-
    261              265               270               275               280
     Thr -Ile- Lys -Tyr- Val -Asp- Val -Asp- Thr -Asn- Glu -Leu- Leu -Lys- Ser -Glu- Gln -Leu- Leu- Thr-
    281              285               290               295               300
     Ala -Ser- Glu -Arg- Asn -Leu- Asp -Phe- Arg -Asp- Leu -Tyr- Asp -Pro- Arg -Asp- Lys -Ala- Lys- Leu
    301              305               310               315               320
     Leu -Tyr- Asn -Asn- Leu -Asp- Ala -Phe- Gly -Ile- Met -Asp- Tyr -Thr- Leu -Thr- Gly -Lys- Val- Glu-
    321              325               330               335               340
     Asp -Asn- His -Asp- Asp -Thr- Asn -Arg- Ile -Ile- Thr -Val- Tyr -Met- Gly -Lys- Arg -Pro- Glu- Gly-
    341              345               350               355               360
     Glu -Asn- Ala -Ser- Tyr -His- Leu -Ala- Tyr -Asp- Lys -Asp- Arg -Tyr- Thr -Glu- Glu -Glu- Arg- Glu-
    361              365               370               375
     Val -Tyr- Ser -Tyr- Leu -Arg- Tyr -Thr- Gly -Thr- Pro -Ile- Pro -Asp- Asn -Pro- Asp -Asp- Lys- COO
```

FIG. 1. The tentative amino acid sequence of streptokinase. The sequence between positions 99 and 107 has not yet been determined. The residues in this bracket represent those determined by amino acid analysis of the peptide 46–122 that were not accounted for by sequence analysis of its *S. aureus* protease-cleavage fragments. The bracket at positions 173–174 represents the residues determined to be present in the peptide 161–175 by subtraction from the sequence information obtained from its *S. aureus* protease-cleavage fragments. The amino acids in positions 96 and 114 (labeled Xxx) have not been conclusively determined. Sequence analysis of positions 148 and 160 indicated that both Asp and Leu were present in each position.

screening can be performed by growth of the organism directly on fibrin plates, this screening method produces a large number of false positive results due to high production of proteases with general specificity. The strain is stored in aliquots at $-70°C$ and fresh cultures started as needed.

Cultures are grown in 10-liter batches of brain-heart infusion medium with slow stirring and light aeration. After overnight incubation, the bacteria are removed by centrifugation at $3000-4000\ g$. Azide or Cetrimide (mixed alkyl-chain form, 0.04%) is then added to the supernatant. Unfortunately, 75°C does not effectively kill this strain. Cetrimide cannot be directly added to the culture to kill the bacteria because this interferes with the future adsorption of STK to the affinity gel (see following section).

Purification

Step 1. The culture supernatant is brought to pH 4 by the slow addition of glacial acetic acid. This precipitates a small amount of protein that is not staphylokinase. The supernatant is then adsorbed batchwise at room temperature to SP-Sephadex C-25 (Pharmacia), which had been previously equilibrated with 0.05 M citrate buffer (pH 4). 1.5 ml of gel per 100 ml of culture supernatant is more than adequate to adsorb all the activity present. After adsorption, the gel is collected on a coarse sintered glass funnel and washed with approximately 500 ml of the equilibration buffer to remove precipitated protein. Staphylokinase is stable for at least 2 weeks when stored in this form at 4°C, allowing stockpiling of this material for combined processing. Alternatively, a column can be poured (2 × 95 cm for 300 ml of gel) immediately and washed at room temperature with several column volumes of equilibration buffer until no protein is observed in the eluate. Less than 10% of the starting activity can be found in the breakthrough. Alpha-hemolysin is removed by this step. Stepwise elution is performed with 0.15 M NaCl–0.05 M Tris–0.02% azide (pH 7.5) and the staphylokinase peak is located by canine fibrin plate assay.[23] The pooled product is dialyzed against 0.05 M NH₄HCO₃ and lyophilized to concentrate it.

Step 2. A DIP-canine plasmin–Sepharose column is prepared according to the method of Cuatrecasas.[26] Canine plasminogen is purified according to the procedure for human plasminogen.[22] Plasmin is made by reacting streptokinase with plasminogen at a ratio of 1 : 50 for 2 hr at 37°C. At this time, 1 M DFP in isopropanol is added to a final concentration of 10 mM. After a minimum of 2 hr in the cold, this is coupled to CNBr-activated Sepharose at 1 mg protein/ml of Sepharose-4B gel. Before use, the affinity gel is washed with starting and elution buffer to remove any

---

[26] P. Cuatrecasas, *J. Biol. Chem.* **245**, 3059 (1967).

unreacted DFP and streptokinase that may be adsorbed. The affinity gel is reusable, although DFP reaction is repeated before each use as a precaution against additional active plasmin having been regenerated. The lyophilized product from the SP column is suspended in a minimal volume of 0.05 M Tris, pH 7 (~20 ml/10 liters of culture) and any precipitate that may be present removed. This is slowly applied to a column of the affinity gel (1 × 13 cm) at room temperature and washed with more than 10 column volumes of 0.1 M Tris–0.02% azide, pH 7. STK is then eluted slowly with a buffer containing 0.1 M Na_2CO_3–0.0125 M $NaHCO_3$–0.02% azide, pH 11. Activity is located and the product dialyzed and lyophilized as described above.

Step 3. The affinity-column product is resuspended in 0.05 M Tris (pH 7.5) and applied to a G-75 (superfine) column equilibrated in the same buffer. In general, the staphylokinase elutes at a position equivalent to MW ≈ 20,000. However, a peak of activity is also present at a volume corresponding to about MW 40,000. Evidence indicates that this is a dimer form of the protein.[10,21]

The final product gives rise to a single band on SDS–gel electrophoresis.[21] Approximately 1 mg of staphylokinase is obtained from 10 liters of culture. It is difficult to estimate precisely the extent of purification for this procedure due to the lack of linearity of the assay discussed above. Our best approximation, however, indicates a minimum of 1000-fold overall purification with respect to ethanol-precipitable protein in the crude supernatant.

Properties

As described above, staphylokinase has MW ≈ 20,000. Its monomer and dimer forms can be separated on Sephadex G-75 as well as on native polyacrylamide-gel electrophoresis.[10,21] On urea–polyacrylamide gel electrophoresis, the dimer dissociates to give a mobility identical to that of the monomer.[21] Staphylokinase is stable in solutions containing up to 4% SDS and to temperatures up to 100°C. Thus it is possible to recover it in active form after SDS–polyacrylamide gel electrophoresis. It is also stable from pH 2 (at least several hours) to pH 11 (at least several days).

The activity of staphylokinase is insensitive to treatment with DFP and other protease inhibitors. Kinetic evidence indicates the plasminogen activator is the staphylokinase–plasminogen complex.[14] Our evidence suggests that plasminogen complexes with staphylokinase dimer.[10,21] Staphylokinase dimer activates human plasminogen (both Lys and Glu forms) more rapidly than monomer.[10,21] The activation of human Lys-plasminogen by staphylokinase dimer is linear with time, whereas human Glu-plasminogen produces a severe lag.[10,21]

[32] Human α$_2$-Antiplasmin[1]

By BJÖRN WIMAN

Introduction

For many years it was believed that α$_2$-macroglobulin and α$_1$-protease inhibitor were the major inhibitors of plasmin in plasma.[2] However, a few years ago a new and very efficient plasmin inhibitor in plasma was identified by several groups independently of each other.[3-6] Müllertz[7] and also Collen and co-workers[8] suggested the existence of plasmin in complex with an unknown inhibitor in urokinase-activated plasma. This complex was later isolated and their suggestions verified.[3,4]

Aoki and von Kaulla[9] studied an inhibitor of plasminogen activation, which later turned out to be a very efficient inhibitor of plasmin.[5] They showed that this inhibitor has affinity for plasminogen and this was used as step 4 in their five-step purification procedure.[5] They were also the first to show that the inhibitor forms a very stable stoichiometric 1 : 1 complex with plasmin that is devoid of enzymatic activity.[5]

Saldeen and co-workers[6] also demonstrated a fibrinolysis inhibitor with affinity for plasminogen that also turned out to be identical with α$_2$-antiplasmin.

Name

Different names have been used to designate this plasmin inhibitor in plasma, for example: primary plasmin inhibitor,[3] α$_2$-plasmin inhibitor,[5] antiplasmin,[4] primary fibrinolysis inhibitor,[6] and α$_2$-antiplasmin.[10] The international committee on thrombosis and hemostasis, subgroup on inhibitors of fibrinolysis, proposed the name α$_2$-antiplasmin for the fast-acting plasmin inhibitor in plasma.[10] In accordance with this proposal we will use the name α$_2$-antiplasmin throughout this communication.

[1] In part supported by the Swedish Medical Research Council (Project No. 05193).
[2] A. Rimon, Y. Shamash, and B. Shapiro, J. Biol. Chem. 241, 5102 (1966).
[3] S. Müllertz and I. Clemmensen, Biochem. J. 159, 545 (1976).
[4] D. Collen, Eur. J. Biochem. 69, 209 (1976).
[5] M. Moroi and N. Aoki, J. Biol. Chem. 251, 5956 (1976).
[6] L. Bagge, I. Björk, T. Saldeen, and R. Wallin, Forensic Sci. 7, 83 (1976).
[7] S. Müllertz, Biochem. J. 143, 273 (1974).
[8] D. Collen, F. DeCock, and M. Verstraete, Thromb. Res. 7, 245 (1975).
[9] N. Aoki and K. N. von Kaulla, Am. J. Physiol. 220, 1137 (1971).
[10] Subgroup on Inhibitors, International Committee on Thrombosis and Haemostasis, Thromb. Haemostasis 39, 524 (1978).

Assay

Different methods have been described for the specific determination of α_2-antiplasmin, including immunochemical methods and procedures for measuring α_2-antiplasmin activity.

Immunochemical Assay

Electroimmunoassay is the most frequently used immunochemical method for the determination of the α_2-antiplasmin concentration, in which monospecific antisera produced in rabbits or in goats is used.[11,12] The method works well in purified samples as well as in plasma or serum samples. It has to be kept in mind, however, that the method also measures inactive forms of α_2-antiplasmin (e.g., when complexed to plasmin). This is no problem under normal physiological conditions when no or very small amounts of plasmin–α_2-antiplasmin complex has formed. However, in patients with an extensive activation of the fibrinolytic system (e.g., patients undergoing thrombolytic treatment or patients with intravascular coagulation and secondary fibrinolysis), falsely high values will be obtained.

Functional Assay

Several procedures for measuring α_2-antiplasmin activity have been described. They all measure a decrease in plasmin activity after addition of an α_2-antiplasmin sample to a specified amount of plasmin, using either synthetic substrates or clot-lysis methods.[3,5,6,11–13] Procedures using the synthetic-peptide substrate D-Val-Leu-Lys-Nan (S-2251) are most frequently and conveniently used.[11–13]

Reagents

Buffer. 0.1 M sodium phosphate buffer, pH 7.3
Substrate Solution. D-Valine-L-Leucine-L-lysine-p-nitroanilide (D-Val-Leu-Lys-Nan, S-2251, Kabi Diagnostica, Stockholm, Sweden), 6 mM dissolved in the phosphate buffer
Plasmin. A stable and soluble plasmin preparation is absolutely essential for a reliable result. Plasminogen is dissolved in 0.1 M sodium phos-

[11] B. Wiman and D. Collen, *Eur. J. Biochem.* **78**, 19 (1977).
[12] D. Collen, J. Edy, and B. Wiman, in "Chromogenic Peptide Substrates: Chemistry and Clinical Usage" (M. F. Scully and V. V. Kakkar, eds.), p. 238. Churchill-Livingstone, Edinburgh and London, 1980.
[13] A.-C. Teger-Nilsson, P. Friberger, and E. Gyzander, *Scand. J. Clin. Lab. Invest.* **37**, 403 (1977).

phate buffer (pH 7.3) containing 1 mM 6-aminohexanoic acid and 25% glycerol, to a concentration of about 20 mg/ml. Activation is performed by addition of streptokinase to a final concentration of 10.000 units/ml, and the sample is subsequently incubated for 2 hr at 0°C. The exact plasmin concentration is determined by active-site titration with p-nitrophenyl-p'-guanidinobenzoate, as described by Chase and Shaw,[14] with the exception that the veronal buffer preferably should also contain 10 mM 6-aminohexanoic acid. The plasmin stock solution is stored frozen at $-80°C$ in aliquots, under which conditions it is stable for many years.[15,16] Prior to use it is diluted to a concentration of 10.0 μM with ice-cold 0.1 M sodium phosphate buffer (pH 7.3) containing 25% glycerol and kept in an ice bath throughout the experiments.

Procedure

Ten microliters of 10 μM plasmin solution is mixed in a cuvette with 50 μl α₂-antiplasmin solution (diluted to a concentration of about 1 μM, which equals the concentration in normal plasma) and immediately diluted with 890 μl of 0.1 M phosphate buffer (pH 7.3) at 25°C, followed by the addition of 50 μl of 6 mM substrate (D-Val-Leu-Lys-Nan, final concentration 0.3 mM). The absorbance at 410 nm (or 405 nm) is spectrophotometrically recorded at 25°C for about 1 min. Blanks are run with buffer instead of α₂-antiplasmin solution. Plasmin of a final concentration 100 nM is expected to give an increase of A_{410} at 25°C of 0.39 A/min.[17] If pure or partially purified α₂-antiplasmin samples are assayed, a straight relationship is obtained between the decrease in ΔA_{410} and the concentration of α₂-antiplasmin (provided that plasmin is still in excess). The α₂-antiplasmin concentration (C_{AP}) in nmol/liter can thus be calculated from the formula:

$$(\Delta A_{blank} - \Delta A_{sample}) \times 100 \times f/0.39$$

where ΔA is the change in absorbance at 410 nm/min without α₂-antiplasmin (blank) and with α₂-antiplasmin (sample) and f is the dilution factor in the cuvette. When testing plasma samples, the best results are obtained when less than 50% of the plasmin is inactivated. Samples containing high α₂-antiplasmin concentration should thus be diluted to fulfill these conditions. Alternatively, a standard curve can be constructed by

[14] T. Chase, Jr., and E. Shaw, this series, Vol. 19, p. 20.
[15] B. Wiman, *Eur. J. Biochem.* **76**, 129 (1977).
[16] B. Wiman and D. Collen, *Eur. J. Biochem.* **84**, 573 (1978).
[17] B. Wiman, *Thromb. Res.* **17**, 143 (1980).

using different amounts (varying between 0 and 75 μl) of citrated pooled plasma (at least 10 healthy donors). The unknown samples can thereafter be expressed as percentage of the plasma pool taking the value with 50 μl as 100%.

At normal and high concentrations of α_2-antiplasmin the accuracy of the method is good, but at low concentrations an overestimation is likely to occur, since other inhibitors will then be relatively more important.[18] Furthermore plasma samples containing heparin should not be used, since a small but significant amount of the plasmin in this case will bind to antithrombin III. Such antifibrinolytic substances as 6-aminohexanoic acid and tranexamic acid strongly influence the reaction between plasmin and α_2-antiplasmin and should therefore be avoided.[13,16,19]

Purification

α_2-Antiplasmin has been shown to interact weakly with the lysine-binding sites in plasminogen.[5,11,16] This can be used for affinity-chromatographic purification of the inhibitor, and if followed by DEAE–Sephadex and concanavalin A–Sepharose chromatography, a pure α_2-antiplasmin preparation is obtained in good yield (30–40%).[11] Alternatively, affinity chromatography can be carried out on a fragment from plasminogen containing the three NH$_2$-terminal triple-loop structures (LBS I) coupled to Sepharose.[20] The material obtained from this step is already 80–90% pure, and the major contaminant, fibrinogen or high-molecular-weight (HMW) fibrinogen-degradation products, are easily removed by gel filtration on Ultrogel AcA 44.

Procedure A[11]

Reagents

Buffer. 0.04 M sodium phosphate buffer, pH 7.0

Plasminogen–Sepharose. Human plasminogen (free from substances containing amino groups) is dissolved in 0.1 M sodium phosphate buffer (pH 7.3) to a final concentration of 5–10 mg/ml and subsequently added to CNBr-activated Sepharose 4B[21] (1 ml of settled gel/ml of plasminogen solution). The coupling is usually completed in a few hours at 5°C with gentle stirring, but if left at 5°C overnight all reactive groups are destroyed

---

[18] D. Collen, N. Semeraro, P. Telesforo, and M. Verstraete, *Br. J. Haematol.* **39**, 101 (1978).
[19] U. Christensen and I. Clemmensen, *Biochem. J.* **163**, 389 (1977).
[20] B. Wiman, *Biochem. J.* **191**, 229 (1980).
[21] R. Axén, J. Porath, and S. Ernback, *Nature (London)* **214**, 1302 (1967).

and treatment with NH_2OH is then unnecessary. The gel is washed with 0.1 M sodium phosphate buffer (pH 7.3) containing 0.04% (w/w) sodium azide and stored in this solution at 5°C.

DEAE–Sephadex A-50. The ion exchanger (Pharmacia, Uppsala, Sweden) is swollen in 0.04 M phosphate buffer containing 0.04% (w/w) sodium azide and stored in this way at 5°C. Prior to use it is washed on a Büchner funnel with the phosphate buffer without sodium azide.

Concanavalin A–Sepharose. Concanavalin A–Sepharose (Pharmacia, Uppsala, Sweden) is prepared according to the procedure described by the manufacturers. Plasminogen-depleted plasma is obtained by stirring outdated human citrated plasma with 100 ml lysine-Sepharose per liter of plasma at 5°C for 30 min. The slurry is subsequently filtered through a Büchner funnel.

Purification Procedure

Affinity Chromatography on Plasminogen–Sepharose. Cohn fraction I is precipitated from 1 liter of plasminogen-depleted plasma by the addition of 70% ethanol to a final concentration of 10% while simultaneously lowering the temperature from 0°C to −3°C. After gentle stirring for 1 hr and centrifugation, the supernatant is mixed with 250 ml of plasminogen–Sepharose gel. The suspension is stirred gently at 0°C for 1 hr and the plasminogen–Sepharose is collected on a Büchner funnel. Washing is carried out with about 5 liters of cold 0.04 M sodium phosphate buffer (pH 7.0), and the gel is subsequently packed in a column (20 cm² × 12 cm) and washing is continued until the absorbance at 280 nm is below 0.05. Elution is then performed with the phosphate buffer containing 10 mM 6-aminohexanoic acid. A 200- to 400-fold purification of α₂-antiplasmin with a yield of about 50% is typically obtained in this step.

DEAE–Sephadex Chromatography. The protein peak obtained from the plasminogen–Sepharose column is applied to a DEAE–Sephadex A-50 column (5 × 5 × 10 cm) equilibrated with 0.04 M sodium phosphate buffer, pH 7.0. Elution is performed with a linear gradient to 0.4 M NaCl in this buffer. α₂-Antiplasmin is eluted between 0.15 and 0.2 M NaCl in the second protein peak. The material from this step is about 70% pure and the yield from the starting material is about 40%.

Chromatography on Concanavalin A–Sepharose. The α₂-antiplasmin peak from the DEAE–Sephadex column is applied without further treatment to a concanavalin A–Sepharose column (5 cm² × 7 cm) equilibrated with 0.04 M phosphate buffer, pH 7.0. Washing with phosphate buffer is performed until A_{280} has returned to the baseline and elution of α₂-antiplasmin is subsequently obtained by 0.02 M α-methylmannoside in phosphate buffer. The purified α_2-antiplasmin is dialyzed in the cold against several changes of distilled water and lyophilized.

Procedure B[20]

Reagents

Buffer. 0.04 M sodium phosphate buffer, pH 7.0

LBSI–Sepharose. Plasminogen is dissolved in 0.1 M NH$_4$HCO$_3$ to a final concentration of about 10 mg/ml and aprotinin is added to a final concentration of 10 μM (60 mg/liter). Elastase-Sepharose containing about 5 mg of porcine pancreatic elastase (Sigma, St Louis, Missouri) per milliliter of gel is added to give an enzyme/substrate ratio of 1/100 (w/w) and digestion is carried out at 25°C overnight with gentle stirring. The mixture is filtered through a Büchner funnel and the filtrate is passed through a lysine-Sepharose column (200 ml bed volume for 500 mg of digested material), and washing is continued with the equilibration buffer. The bound fragments are eluted with 50 mM 6-aminohexanoic acid in 0.1 M NH$_4$HCO$_3$ and subsequently chromatographed on a Sephadex G-75 column (800 ml bed volume for a 500-mg digest) equilibrated with 0.1 M NH$_4$HCO$_3$. Two peaks are obtained, the first of which represents the three NH$_2$-terminal triple-loop structures of plasminogen (LBSI). The material is dissolved after lyophilization in 0.1 M sodium phosphate buffer (pH 7.3) at a concentration of about 5 mg/ml and coupled to CNBr-activated Sepharose 4B in a similar way as described for plasminogen–Sepharose.

Ultrogel AcA 44. Ultrogel AcA 44 is obtained from LKB (Stockholm, Sweden) and used according to the procedure given by the manufacturers.

Starting Material. Plasminogen-depleted plasma can be used without further treatment, although fibrinogen interferes slightly with the purification procedure. However, since we normally use the plasma for purification of several other proteins, we have frequently used plasminogen-depleted supernatant Cohn fraction I or plasminogen-depleted supernatant after polyethylene glycol precipitation (final concentration 6%) of the plasma.

Purification Procedure

Affinity Chromatography on LBSI–Sepharose. About 1 liter of plasminogen-depleted plasma is stirred with 50–100 ml of LBSI–Sepharose at 0°C for 30 min and subsequently filtered through a Büchner funnel. Alternatively, the plasminogen-depleted plasma or supernatant after ethanol or polyethylene glycol precipitation is slowly passed (about 1 liter/hr) through the LBSI–Sepharose packed in an ordinary Büchner funnel. In this way the yield seems to be somewhat increased. The LBSI–Sepharose is washed on the Büchner funnel with about 2–3 liters of $0.04 M$ sodium phosphate buffer (pH 7.3) and then transferred to a column; washing is continued with the phosphate buffer containing 0.5 M NaCl until A_{280} is below 0.05. Elution of α_2-antiplasmin is performed with 10 mM 6-aminohexanoic acid in phosphate buffer and the protein peak is subsequently dialyzed at 0°C against several changes of distilled water. A small precipitate of fibrinogen is occasionally removed by centrifugation and the clear supernatant is thereafter lyophilized. This material is about 80–90% pure α_2-antiplasmin obtained in a yield of about 60%.

Gel Filtration on Ultrogel AcA 44. The major contaminant (fibrinogen or HMW fibrinogen degradation products) can easily be removed by gel filtration on Ultrogel AcA 44. The α_2-antiplasmin is eluted in the major second peak, which is dialyzed against distilled water in the cold and lyophilized. The obtained preparation is a homogeneous protein according to several chromatographic and electrophoretic criteria and fully active when titrated against plasmin.

Different Forms of α₂-Antiplasmin

α_2-Antiplasmin is a glycoprotein that migrates as an α_2-globulin[3–5] cently reported to contain a form of the inhibitor that does not bind to plasminogen–Sepharose.[22] Using LBSI–Sepharose, which is much more efficient in binding α_2-antiplasmin than plasminogen–Sepharose, we have been able to confirm that a form of α_2-antiplasmin with less affinity for the lysine-binding sites in plasminogen may exist even in unfractionated plasma.[20] This form of the inhibitor, constituting 25–40% of the α_2-antiplasmin antigen in plasma, has not yet been purified. It still is an active antiplasmin because it forms a complex with plasmin under highly competitive conditions, as evidenced by crossed immunoelectrophoresis (using an antiserum against α_2-antiplasmin) on a LBSI–Sepharose treated plasma sample after addition of plasmin to a final concentration of 1 μM.[20]

Properties

Physicochemical Properties

α_2-Antiplasmin is a glycoprotein that migrates as an α_2-globulin[3–5] on electrophoresis. Its molecular weight (MW) has been determined as 65,000–70,000 by both sedimentation equilibrium analysis and sodium

[22] I. Clemmensen, *in* "Physiological Inhibitors of Blood Coagulation and Fibrinolysis" (D. Collen, B. Wiman, and M. Verstraete, eds.), p. 131. Elsevier/North-Holland Publ., Amsterdam and New York, 1979.

TABLE I
AMINO ACID COMPOSITION OF DIFFERENT
α_2-ANTIPLASMIN PREPARATIONS

Amino acid[a]	A[b]	B[c]	C[d]
Aspartic acid	40.9	41.3	38.6
Threonine	25.3	23.4	24.8
Serine	41.1	43.3	39.1
Glutamic acid	79.3	80.3	75.9
Proline	41.2	40.6	38.0
Glycine	30.3	31.6	29.8
Alanine	33.5	32.2	29.9
Cysteine	5.2	5.5	5.6
Valine	21.4	23.8	30.0
Methionine	13.2	12.6	19.0
Isoleucine	9.7	11.8	10.1
Leucine	73.0	75.1	75.7
Tyrosine	5.2	4.9	6.8
Phenylalanine	34.8	32.7	30.7
Lysine	23.0	18.2	21.7
Histidine	10.3	10.1	14.0
Arginine	21.6	22.2	21.3
Tryptophan	7.4	7.6	8.1

[a] Figures are given as micromoles of amino acid per 70 mg of protein.
[b] Material obtained by procedure A (from Ref. 11).
[c] Material obtained by procedure B.
[d] Adapted from the result reported by Moroi and Aoki.[5]

dodecyl sulfate–polyacrylamide gel electrophoresis on reduced as well as nonreduced samples.[5,11]

The sedimentation constant was found to be 3.45 S,[5,11] and the partial specific volume was calculated from the amino acid and carbohydrate compositions as 0.718 ml/g.[11] The Stokes radius and the frictional ratios were calculated as 5.0 nm and 1.8 respectively, indicating a very asymmetric or highly hydrated molecule.[11] This is in agreement with gel-filtration data, which have shown MW $\approx$ 90,000.[6,23] Circular dichroism studies have demonstrated a molecule with about 16% α-helix, 19% β-structure, and 65% random coil.[24]

The amino acid composition reported by Moroi and Aoki[5] is in excellent agreement with that obtained for our preparations[11,20] (Table I).

[23] L. Bagge, I. Björk, T. Saldeen, and R. Wallin, *Thromb. Haemostasis* **39**, 97 (1978).
[24] B. Wiman, T. Nilsson, and I. Sjöholm, *Prog. Chem. Fibrinolysis Thrombolysis* **5**, 302 (1981).

TABLE II
NH2-TERMINAL AMINO ACID SEQUENCES OBTAINED
FROM INTACT α2-ANTIPLASMIN (a), DEGRADED
α2-ANTIPLASMIN (b), THE 8000-MW PEPTIDE (c), AND
6 PURIFIED CNBr FRAGMENTS (d–i)

a. Asn-Gln-Glu-Gln-Val-
b. Leu-Val-Asn-Gln-Glu-Val-Gly-Gly-Gln-
c. Met-Ser-Leu-Ser-Gly-Phe-
d. Leu-Gly-Asn-Gln-X-Pro-
e. Ser-Phe-Val-Val-Leu-Val-
f. Gln-Ala-Phe-Val-Tyr-Val-Leu
g. Leu-Ala-X-Arg-Tyr-
h. Glu-Glu-Asp-Tyr-Pro-Gln-Phe-Gly
i. Tyr-Leu-Gln-Lys-Gly-Phe-Pro-Leu-Lys-Glu-Asp-Phe-

Edman degradation yielded 1 mol of NH2-terminal asparagine per mole of protein, and carboxypeptidase Y digestion indicated leucine as COOH-terminal amino acid.[11,25] Carbohydrate analysis revealed about 11–14% carbohydrate:[5,11] 10 mol sialic acid, 30 mol hexose, and 7 mol glucosamine per mole of α2-antiplasmin.[11] The single-chain protein seems to be stabilized by 3 disulfide bridges,[24] 2 of which can be very easily reduced and S-carboxy-methylated under nondenaturing conditions.[24] This results in a fully active molecule, as measured both by activity measurement and by kinetic analysis.

The absorption coefficient, $A_{1\,cm}^{1\%}$, has been determined at 6.70[11] or 7.05.[5] α2-Antiplasmin prepared according to the method of Moroi and Aoki[5] is reported to be very unstable. Thus it quite rapidly loses its activity in solution and lyophilization results in complete inactivation. However, our preparations[11,20] are very stable provided that the pH is kept above 6.0. A solution of α2-antiplasmin kept at 20°C in 0.1 M sodium phosphate buffer (pH 7.3) containing 0.04% (w/w) sodium azide retained over 90% of the activity after 14 days.[11] Furthermore, no activity is lost on lyophilization. Indeed, lyophilization seems to be the best way of storing α2-antiplasmin, since repetitive freezing and thawing of pure preparations in solution result in a significant decrease in activity.

Amino Acid Sequence

So far only minor parts of the α2-antiplasmin molecule have been sequenced, including direct sequences of intact and partially degraded inhibitor, the 8000-MW peptide cleaved during the reaction with plasmin[25] (see later), and 6 purified CNBr fragments[26] (Table II). A seventh CNBr

25 B. Wiman and D. Collen, *J. Biol. Chem.* 254, 9291 (1979).
26 R. Lijnen, D. Collen, and B. Wiman, unpublished results.

TABLE III
NH$_2$-TERMINAL SEQUENCES OF THE TRYPTIC PEPTIDES
ISOLATED FROM A CNBr FRAGMENT WITH A PRESUMABLY
BLOCKED NH$_2$-TERMINUS

1. (Ser$_2$, Thr, Pro, Ile, Leu, Phe) -Leu-Phe-Glu-Asp-
   (Thr, Pro, Gly) -Phe-Val-Asn-Ser-Val-Arg
2. Glu-Gln-Gln-Asp-Ser-Pro-Gly-Asn-Lys
3. Ser-Phe-Leu-Gln-Asn-Leu-Lys
4. Gly-Phe-Pro-Arg
5. Gly-Asp-Lys
6. Leu-Phe-Gly-Pro-Asp-Leu-Lys
7. Leu-Ala-Arg
8. Glu-Leu-Lys
9. Asn-Pro-Asn-Pro-Ser-Ala-Pro-Arg
10. Leu-Val-(Leu, Pro$_2$) HSer

fragment has been purified, but due to unknown reasons no NH$_2$-terminal amino acid could be determined in this peptide. Nevertheless, all its tryptic peptides were purified and almost completely sequenced[26] (Table III). Peptide CB7-T1 could not be sequenced directly, but the data were obtained from chymotryptic peptides of this particular tryptic peptide. So far, only one small stretch has been found that shows clear homology with antithrombin III and α_1-antitrypsin (Table IV).[26]

Reaction with Enzymes

In purified systems, α_2-antiplasmin is capable of reacting with and forming stable enzymatically inactive complexes with many enzymes such as plasmin[16,19] (very fast), trypsin[16,19] (fast), chymotrypsin[29] (intermediate), kallikrein[30] (slow), factor X$_a$[30] (slow), urokinase[5] (very slow), and tissue-plasminogen activator[31] (very slow). Nevertheless, inhibition of plasmin is presumably its only physiologically important reaction.[32]

Kinetics. The reaction between plasmin and α_2-antiplasmin can be divided into two steps, a very fast reversible second-order reaction followed

[27] T. E. Petersen, G. Dudek-Wojciechowska, L. Sottrup-Jensen, and S. Magnusson, *in* "The Physiologic Inhibitors of Blood Coagulation and Fibrinolysis" (D. Collen, B. Wiman, and M. Verstraete, eds.), p. 43. Elsevier/North-Holland Publ., Amsterdam and New York, 1979.
[28] M. C. Owen, M. Lovier, and R. W. Carrell, *FEBS Lett.* **88,** 234 (1978).
[29] T. Nilsson and B. Wiman, unpublished results.
[30] M. Moroi and N. Aoki, *J. Biochem. (Tokyo)* **82,** 969 (1977).
[31] B. Wiman and P. Wallén, unpublished results.
[32] J. Edy and D. Collen, *Biochim. Biophys. Acta* **484,** 423 (1977).

TABLE IV

HOMOLOGIES BETWEEN A CB FRAGMENT FROM α_2-ANTIPLASMIN[a] AND
STRETCHES IN ANTITHROMBIN III[b] AND α_1-ANTITRYPSIN[c]

α_2-Antiplasmin	Y - L - Q - K - $\boxed{G - F}$ - P - $\boxed{L - K - E}$ - D - F -
Antithrombin III	$R_{316}\boxed{I}$ - E - D - $\boxed{G - F}$ - S - $\boxed{L - K - E}$ - Q - $\boxed{L}$ -
α_1-Antitrypsin	S - $\boxed{I}$ - T - G - T - Y - D - $\boxed{L - K}$ - S - Y - $\boxed{L}$ -

[a] Sequence i in Table II.
[b] From Ref. 27.
[c] From Ref. 28.

by a slower irreversible first-order transition, and schematically represented as:

$$P + A \underset{K_{-1}}{\overset{K_1}{\rightleftharpoons}} PA \overset{K_2}{\longrightarrow} PA'$$

where P is plasmin, A is α_2-antiplasmin, and PA is the complex.[16,19] Some of the kinetic constants for the reactions between α_2-antiplasmin and some enzymes are summarized in Table V.[16,29,33] The very high rate constant in its reaction with plasmin is dependent on an interaction between one of the so-called lysine-binding sites in the plasmin A chain and a complementary site in α_2-antiplasmin.[16,29,34] Thus 6-aminohexanoic acid in concentrations enough to block all the lysine-binding sites in plasmin decreases the rate of the plasmin–α_2-antiplasmin reaction about 100-fold.[16] A similar rate constant is also obtained with low-molecular-weight (LMW) plasmin, which lacks lysine-binding sites.[33] Plasmin substrates, such as D-Val-Leu-Lys-Nan, also slow down the plasmin–α_2-antiplasmin reaction, and a 50% reduction is obtained at a substrate concentration (0.38 mM) that is similar to the K_m with plasmin.[16] This shows that the active center of plasmin also plays a role in the fast formation of the reversible complex.

Structural Changes during Its Reaction with Plasmin. The reaction between plasmin and α_2-antiplasmin leads to the formation of a very stable enzyme–inhibitor complex that cannot be dissociated by reducing or denaturing agents.[5,16,25,35] The complex is very easily degraded proteolytically.[16,25] Therefore, it is absolutely essential to ascertain inhibitor excess in mechanistic studies of the complex formation. We have shown that a peptide with MW $\approx$ 8000, originating from the COOH-terminal portion of α_2-antiplasmin, can be isolated from the complex in the presence of

[33] B. Wiman, L. Boman, and D. Collen, *Eur. J. Biochem.* **87**, 143 (1978).
[34] U. Christensen and I. Clemmensen, *Biochem. J.* **175**, 635 (1978).
[35] M. Moroi and N. Aoki, *Biochim. Biophys. Acta* **482**, 412 (1977).

TABLE V
RATE CONSTANTS FOR THE INACTIVATION OF DIFFERENT
PROTEOLYTIC ENZYMES BY α_2-ANTIPLASMIN

Enzyme	Rate constant (M^{-1} sec^{-1})	Ref.
Plasmin I	3.8×10^7	16
Plasmin II	1.8×10^7	16
LMW plasmin	6.5×10^5	33
Trypsin	1.8×10^6	16
Chymotrypsin	1.0×10^5	29

sodium dodecyl sulfate.[25] A new NH$_2$-terminal amino acid (methionine), in addition to those in the parent molecules, has been demonstrated in intact complexes between plasmin and α_2-antiplasmin.[36] Methionine is also the NH$_2$-terminal amino acid of the 8000-MW peptide (Table II), indicating that a peptide bond in the COOH-terminal portion of the inhibitor is cleaved as a result of its reaction with plasmin.[25,36] Furthermore, the plasmin–α_2-antiplasmin complex is dissociated in 1.5 M NH$_4$OH yielding up to 0.2 mol of active plasmin per mole of complex and a modified form of the inhibitor, which has lost a peptide from its COOH-terminal end. Carboxypeptidase Y digestion of this modified form of the inhibitor indicated the new COOH-terminal dipeptide, Ala-Leu.[25] The reactive site in α_2-antiplasmin may thus be a Leu-Met peptide bond, which is somewhat surprising in view of the specificity of plasmin. On the other hand, if chymotrypsin reacts with the same site as plasmin, it may still be correct. The data obtained until now indicate that the stabilizing bond in the plasmin–α_2-antiplasmin complex is an ester bond between the hydroxyl group of the active-site seryl residue in plasmin and the specific carbonyl group of the leucyl residue in the reactive site of α_2-antiplasmin.[36]

Preparation of Plasmin–α_2-Antiplasmin Complex

Plasmin–α_2-antiplasmin complex can be prepared either from purified reagents or from plasma after activation of the plasminogen or after addition of plasmin. Starting from purified reagents, a suitable procedure is as follows:

Plasminogen and α_2-antiplasmin are both dissolved to final concentrations of 10 mg/ml in 0.1 M phosphate buffer, pH 7.3. This gives approximately 25% molar excess of α_2-antiplasmin as compared to plasminogen. Urokinase is added to a final concentration of 5000 CTA units/ml and the mixture is incubated at 25°C for 2 hr to obtain complete activation of the

---

[36] B. Wiman, T. Nilsson, and D. Collen, *Protides Biol. Fluids* **28**, 379 (1980).

plasminogen. The mixture is thereafter subjected to gel filtration on Ul-trogel AcA 44 equilibrated with 0.1 M NH$_4$HCO$_3$, containing 0.25 μM aprotinin. The major first peak consisting of plasmin–α₂-antiplasmin complex is lyophilized. If Glu-plasminogen has been used a complex containing Glu-plasmin is obtained.

Preparation of the plasmin–α₂-antiplasmin complex from plasma can be performed after activation of the plasma by urokinase (500 IU/ml) for 30 min at 25°C, or preferably after addition of plasmin (final concentration 1 μM) to plasminogen-depleted plasma at 5°C and incubation for several minutes. After addition of aprotinin to a final concentration of 1 μM, adsorption of the plasmin complexes is performed on lysine-Sepharose (100 ml/liter plasma) in a batch procedure. The lysine-Sepharose is collected on a Büchner funnel, washed with 0.1 M NH$_4$HCO$_3$ containing 0.25 μM aprotinin, and subsequently packed in a column. Washing is continued until A_{280} is less than 0.1, and elution is then performed with 0.1 M NH$_4$HCO$_3$ containing 0.25 μM aprotinin and 50 mM 6-aminohexanoic acid. The protein peak, after concentration by ultrafiltration, is passed through an Ultrogel AcA 44 column, equilibrated with 0.1 M NH$_4$HCO$_3$, containing 0.25 μM aprotinin. Three peaks are obtained, the second of which is the plasmin–α₂-antiplasmin complex. The first and third peaks are plasmin in complex with α₂-macroglobulin and aprotinin, respectively.[3,24]

Physiological Role

Turnover studies with radiolabeled α₂-antiplasmin in humans indicated a half-life of 2.6 days and a synthetic rate of 1.4 mg/kg/day. The half-life of plasmin–α₂-antiplasmin complex was found to be about 0.5 day.[37] The plasma concentration of α₂-antiplasmin normally is 1.0 ± 0.2 μM (70 mg/liter). It seems to be a weak acute-phase reactant. Decreased values have been reported in patients with liver disease and with severe intravascular coagulation.[38] In patients undergoing thrombolytic therapy, the inhibitor may be immediately but temporarily exhausted, resulting in a severe fibrinogenolysis.[39] A few patients with α₂-antiplasmin deficiency have been described. They seem to suffer from severe lifelong bleeding tendencies.[40] α₂-Antiplasmin plays an important role in the regulation of fi-

[37] D. Collen and B. Wiman, *Blood* **53**, 313 (1979).
[38] D. Collen and B. Wiman, *Blood* **51**, 563 (1978).
[39] D. Collen and M. Verstraete, *Thromb. Res.* **14**, 631 (1979).
[40] N. Aoki, H. Saito, T. Kamiya, K. Koie, Y. Sakata, and M. Kobakura, *J. Clin. Invest.* **63**, 877 (1979).

brinolysis.[41-44] The very high reaction rate of the plasmin–α_2-antiplasmin reaction is highly dependent on free active-site and lysine-binding sites in the plasmin molecules.[16,33] This is not the case with the fibrinolytically active plasmin molecules[45,46] formed at the fibrin surface by activation of fibrin-bound plasminogen by the tissue-plasminogen activator.[47] These plasmin molecules seem thus to be protected from the action of α_2-antiplasmin. This hypothesis explains in a plausible way the *in vivo* selective degradation of fibrin by plasmin.

[41] B. Wiman, in "Fibrinolysis: Current Fundamental and Clinical Concepts" (P. J. Gaffney and S. Balkov-Ulutin, eds.), p. 47. Academic Press, New York, 1978.
[42] B. Wiman and D. Collen, *Nature (London)* **272**, 549 (1978).
[43] B. Wiman and D. Collen, in "The Physiological Inhibitors of Blood Coagulation and Fibrinolysis" (D. Collen, B. Wiman, and M. Verstraete, eds.), p. 177. Elsevier/North-Holland Publ., Amsterdam and New York, 1979.
[44] D. Collen and B. Wiman, *Prog. Chem. Fibrinolysis Thrombolysis* **4**, 11 (1979).
[45] B. Wiman and P. Wallén, *Thromb. Res.* **10**, 213 (1977).
[46] B. Wiman, H. R. Lijnen, and D. Collen, *Biochim. Biophys. Acta* **579**, 142 (1979).
[47] P. Wallén, P. Kok, and M. Rånby, in "Regulatory Proteolytic Enzymes and Their Control" (S. Magnusson, M. Ottesen, B. Foltman, K. Danø, and H. Neurath, eds.), p. 127. Pergamon, Oxford, 1978.

[33] A Coupled Photometric Assay for Plasminogen Activator

By Patrick L. Coleman and George D. J. Green

Introduction

Plasminogen activators are important in fibrinolysis, cell transformation, metastasis, and inflammatory responses,[1] and in some systems are under hormonal control.[2,3] The great interest in understanding the many roles of plasminogen activator has led to the evolution of several different types of assays for the enzyme designed to meet the particular objectives of the investigators. Among those designed for assay with living cells are the ^{125}I-labeled fibrin-coated tissue culture plates,[4,5] the fibrin–agar over-

[1] J. K. Christman, S. C. Silverstein, and G. Acs, in "Proteinases in Mammalian Cells and Tissues" (A. J. Barrett, ed.), p. 91. Elsevier/North-Holland Publ., Amsterdam and New York, 1977.
[2] W. H. Beers, S. Strickland, and E. Reich, *Cell* **6**, 387 (1975).
[3] S. C. Seifert and T. D. Gelehrter, *Proc. Natl. Acad. Sci. U.S.A.* **75**, 6130 (1978).
[4] J. C. Unkeless, A. Tobia, L. Ossowski, J. P. Quigley, D. B. Rifkin, and E. Reich, *J. Exp. Med.* **137**, 85 (1973).
[5] J. C. Taylor, D. W. Hill, and M. Rogolsky, *Exp. Cell Res.* **73**, 422 (1972).

lay,[5,6] the casein–agar overlay,[7] and fluorescein guanidinobenzoate methods.[8] These are nondestructive to the cells, allowing their recovery after the analysis. ^{125}I-labeled plasminogen has been used for direct analysis of the activation step.[9] Various investigators have employed casein– and fibrin–agar overlay techniques for location of plasminogen activators in polyacrylamide gels.[10,11] These methods are especially useful in screening the distribution of molecular weight forms in many samples. Recently, radioimmunoassays for plasminogen activators[12] have been developed, and when combined with the molecular weight analyses on polyacrylamide gels, have led to important correlations of classes of plasminogen activators with cell type.[13] Several of these methods have been applied to routine enzyme analysis[4,14,15]; however, none of them are as rapid and sensitive as might be desired for use in protein purification or kinetic analysis.

Coleman and Green[16] have recently reported a sensitive photometric assay that is complete in less than 2 hr. Thiobenzyl benzyloxycarbonyl-L-lysinate (Z-Lys-SBzl),[17] a general thiolester substrate for trypsin-like serine proteases,[18] has been substituted for fibrin as the plasmin substrate in a two-step coupled assay.

Materials

Urokinase (B grade) and 5,5'-dithiobis(2-nitrobenzoic acid) (DTNB) were obtained from Calbiochem-Behring (LaJolla, California). Bovine serum albumin and soybean trypsin inhibitor (B grade) were obtained from Sigma Chemical Co. (St. Louis, Missouri). Z-Lys-SBzl was purchased from Peninsula Laboratories (San Carlos, California).[19] All other

[6] P. Jones, W. Benedict, S. Strickland, and E. Reich, *Cell* **5**, 323 (1975).
[7] A. R. Goldberg, *Cell* **2**, 95 (1974).
[8] W. F. Mangel, D. C. Livingston, J. R. Broklehurst, H.-Y. Liu, G. A. Peltz, J. F. Cannon, S. P. Leytus, J. A. Wehrly, B. L. Salter, and J. L. Mosher, *Cold Spring Harbor Symp. Quant. Biol.* **40**, 669 (1979).
[9] J. Unkeless, K. Danø, G. M. Kellerman, and E. Reich, *J. Biol. Chem.* **249**, 4295 (1974).
[10] A. Granelli-Piperno and E. Reich, *J. Exp. Med.* **147**, 223 (1978).
[11] C. Failly-Crepin and J. Uriel, *Biochimie* **61**, 567 (1979).
[12] D. Vetterlein, T. E. Bell, P. L. Young, and R. Roblin, *J. Biol. Chem.* **255**, 3665 (1980).
[13] R. Roblin and P. L. Young, *Cancer Res.* **40**, 2706 (1980).
[14] K. Danø and E. Reich, *Biochim. Biophys. Acta* **566**, 138 (1979).
[15] P. L. Coleman, C. Kettner, and E. Shaw, *Biochim. Biophys. Acta* **569**, 41 (1979).
[16] P. L. Coleman and G. D. J. Green, *Ann. N.Y. Acad. Sci.* **370**, 617 (1981).
[17] List of abbreviations: BSA, bovine serum albumin; DTNB, 5,5'-dithiobis(2-nitrobenzoic acid); HTC, hepatoma tissue culture; STI, soybean-trypsin inhibitor; Z-Lys-SBzl, thiobenzyl benzyloxycarbonyl-L-lysinate.
[18] G. D. J. Green and E. Shaw, *Anal. Biochem.* **93**, 223 (1979).
[19] Z-Lys-SBzl is also commercially available from BaChem (Torrance, California) and Vega Chemicals (Tucson, Arizona).

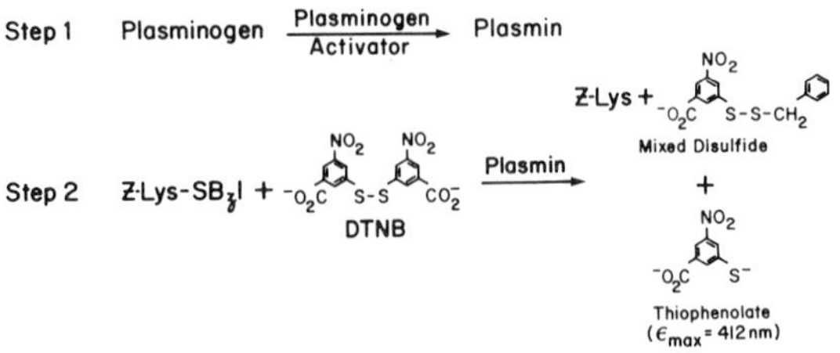

FIG. 1. The two-step reaction scheme for the assay of plasminogen activator. In step 1, plasminogen is activated to plasmin by human urokinase or the plasminogen activators from HeLa, HTC cells, or other sources. In step 2 the resulting plasmin is assayed via its reaction with Z-Lys-SBzl to form Z-lysine and benzyl mercaptan, which reacts instantly and nonenzymatically with DTNB to form the mixed disulfide and the colored thiophenolate anion.

chemicals were of reagent grade. Plasminogen was purified from fresh-frozen plasma using lysine–agarose affinity chromatography,[20] exhaustively dialyzed against 1.0 mM HCl, lyophilized, and stored at $-20°C$. Plasminogen activators from hepatoma tissue culture (HTC) cell extract and HeLa cell serum-free conditioned medium were obtained as previously described.[3,15]

Assay Method

The assay consists of two discrete steps, as illustrated in Fig. 1. In step 1, plasminogen is activated to plasmin, which is quantified in step 2. In each step, conditions are optimized for that particular reaction, and in step 2 care is also taken to minimize further plasminogen activation.

Step 1 Reagents

Glycine–BSA. 1.0 M Na glycinate–5 mg/ml bovine serum albumin, pH 8.5; stored frozen and diluted 5-fold prior to use.

BSA–HCl. 1.0 mg/ml bovine serum albumin in 1.0 mM HCl, titrated to pH 3.0 by addition of 1.0 M HCl; stored at 4°C.

Urokinase. 1.0 Ploug unit/μl in BSA–HCl, stored frozen and serially diluted with BSA–HCl to stock solutions of 0.2 to 2 × 10^{-4} Ploug unit/μl immediately prior to use (1 Ploug unit = 2.7 × 10^{-13} mol).

Plasminogen. Stored frozen as the lyophilized, salt-free protein; weighed immediately prior to use and dissolved into BSA–HCl to 10 μM. Concentration is checked by absorbance at 280 nm (1.0 mg/ml = 1.7 A_{280}, Ref. 21).

[20] D. G. Deutsch and E. T. Mertz, *Science* 170, 1095 (1970).
[21] K. C. Robbins and L. Summaria, this series, Vol. 19, p. 184.

The final assay volume of 50 μl is typically achieved by addition of 40 μl of glycine–BSA, 5 μl urokinase, and 5 μl plasminogen. Controls for urokinase and plasminogen substitute an equal volume of BSA–HCl. Step 1 is normally initiated by the addition of plasminogen to the mixture already equilibrated to 37°C. Other plasminogen activators, such as those from HeLa or hepatoma tissue culture (HTC) cells, have been substituted for urokinase, as have trypsin and streptokinase. The rate of reaction is independent of the final glycine concentration (20–1600 mM) and BSA concentration (0.2–1.0 mg/ml), thus additions to the incubation mixture may be made at the expense of the added volume of glycine–BSA without modifying the concentrations of the stocks. After 45 min, step 1 is terminated by dilution with the step 2 reagents.

Step 2 Reagents

Triton X-100. 0.1% (w/v) in H_2O stored at room temperature

DTNB. 22 mM 5,5′-dithiobis(2-nitrobenzoic acid) in 50 mM Na_2HPO_4 (a buffer with sufficient basic form to titrate the free acid of DTNB is necessary to dissolve it); stored for weeks at 4°C or for months frozen

Z-Lys-SBzl. 20 mM Z-Lys-SBzl in H_2O; same storage as DTNB

Pi–NaCl. 200 mM Na phosphate plus 200 mM NaCl, pH 7.50, stored at 4°C or room temperature

STI. 1 mg/ml soybean trypsin inhibitor in 1.0 mM HCl, titrated to pH 3.0, stored at 4°C

The step 2 (color reaction) reagent is made by the addition of 1 part each of the Triton X-100, DTNB, and Z-Lys-SBzl stock solutions to 100 parts of Pi–NaCl (at 37°C) just prior to use. Step 2 is initiated by the addition of 950 μl of the color reagent to the 50-μl step 1 mixture. The color reaction is normally allowed to proceed for 60 min, whereupon it is terminated by addition of 100 μl of the STI solution. Absorbance at 412 nm may be measured immediately or within 4 hr (if kept in subdued light).

Step 2 measures plasmin by its hydrolysis of Z-Lys-SBzl to benzyloxycarbonyl-L-lysine plus benzyl mercaptan, which immediately reacts with excess DTNB to form the mixed disulfide and the thiophenolate anion, $E_{412\ nm} = 14{,}150\ cm^{-1}\ M^{-1}$ (Ref. 22). Figure 2 demonstrates a typical standard curve.

Step 2 is optimized for maximal plasmin activity and minimal urokinase activity. Plasminogen activation is reduced 20-fold simply by the dilution with color reagent. pH 7.50 is on the broad optimum for plasmin activity, whereas urokinase is at 50% maximal activity. There is

[22] P. W. Riddles, R. L. Blakeley, and B. Zerner, *Anal. Biochem.* **94**, 75 (1979).

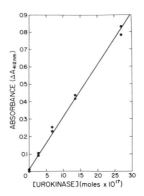

FIG. 2. Standard curve for urokinase-catalyzed activation of plasminogen. Assay conditions are as described in the text. The line was drawn based on a linear regression fit to the data (slope = $29.5 \times 10^{-3} A_{412}/10^{-17}$ mol of urokinase; intercept = $18 \times 10^{-3} A_{412}$; $r^2 = 0.995$). Precision of duplicate values varies from $\pm 10\%$ at 3×10^{-17} mol to $\pm 5\%$ at 30×10^{-17} mol of urokinase. The urokinase-free blank (0.140) was subtracted from all experimental values. The plasminogen-free and the urokinase-free, plasminogen-free blanks were equivalent at 0.130.

no effect of up to 250 mM NaCl on plasmin activity, whereas 200 mM inhibits 65% of the plasminogen activator or esterase activities of urokinase. Finally, Triton X-100 increases plasmin's k_{cat} 50% and has no effect on urokinase. By all these effects combined, there is less than 1% plasminogen activation in step 2 compared with step 1.

Remarks

The use of two-step rather than one-step coupled procedures allows for greater flexibility in experimental design and quantitative treatment of the resulting data. Since both urokinase and plasmin are trypsin-like serine proteases, it is expected that modifiers of the former would also modify the activity of the latter. Such effects would be difficult to control if plasmin concentration were constantly changing as in a one-step procedure. However, in this two-step procedure the plasmin concentration is always constant while it is being measured. Thus if there remain any effects of a diluted test compound from step 1, they can be independently tested.

The two-step procedure allows the easy calculation of kinetic constants for the plasminogen-activation reaction. This procedure has been applied in the determination of K_m and k_{cat} for plasminogen activation by urokinase and HeLa cell plasminogen activator, and in the calculation of inhibitory constants for the interactions of urokinase with Phe-Ala-

ArgCH$_2$Cl and an inhibitor from HTC cells.[16] The method has also been used in the kinetic analyses of plasminogen activation by HTC cell plasminogen activator, bovine trypsin, and streptokinase, and in monitoring the purification of HTC cell inhibitor.[23]

The Z-Lys-SBzl-based assay is the equivalent of the ^{125}I-labeled fibrin assay in sensitivity to urokinase,[16] and has advantages over the radiometric assay in ability to obtain reproducible kinetic values, ease of handling, and avoidance of radioisotopes.

There are two recent reports of one-step photometric assays for plasminogen activator relying on D-Val-Leu-Lys-p-nitroanilide. The end-point method of Overwien et al.[24] requires 2 hr and is sensitive to about 10^{-14} mol of urokinase (about 4×10^{-2} Ploug unit). The continuous method of Wohl et al.[25] is complete in a few minutes and uses $1-2 \times 10^{-8}$ mol of urokinase. It has also been applied to the kinetic analysis of plasminogen activation by kallikrein, trypsin, streptokinase, and staphylokinase.

The two-step method described herein is sensitive to 3×10^{-17} mol of urokinase (about 10^{-4} Ploug unit) as a limit of detection. Quantitative measurement requires 10^{-16} mol of urokinase or the equivalent of about 3×10^{-4} Ploug unit in other activators. Kinetic constants obtained from the two-step assay[16] compare quite closely with those reported using the continuous one-step coupled assay.[25]

Some cautions must be observed in utilizing this procedure, since it substitutes a synthetic substrate for the natural substrate. It is possible that some plasminogen activators might need to bind to a fibrin matrix in order to achieve full activity.[26,27] Some serum fibrinolytic inhibitors behave differently in the thiol esterase and ^{125}I-labeled fibrinolytic assays,[23] suggesting a requirement for fibrin.

Owing to the presence of DTNB, it is possible that there could be interference by free sulfhydryl groups. This is not a problem with dilute serum, plasma, or HTC cell extracts, but it could be a problem with high concentrations of thiol agents. The DTNB concentration in step 2 is about 200 μM. A 5-μl test solution in step 1 that contains 0.1 mM thiol would be only 5 μM in step 2. Since it would react instantaneously, it should not interfere with normal levels of plasminogen activation catalyzed by $1-2 \times 10^{-16}$ mol of urokinase (0.5 to 1×10^{-3} Ploug unit).

[23] P. L. Coleman and T. D. Gelehrter, unpublished observations.

[24] B. Overwien, C. Neumann, and C. Sorg, *Hoppe-Seyler's Z. Physiol. Chem.* **361,** 1251 (1980).

[25] R. C. Wohl, L. Summaria, L. Arzadon, and K. C. Robbins, *J. Biol. Chem.* **253,** 1402 (1978).

[26] S. Thorsen, P. Glas-Greenwalt, and T. Astrup, *Thromb. Diath. Haemorrh.* **28,** 65 (1972).

[27] R. Radcliffe, *Circulation* **62,** Suppl. III, 334 (Abstr. No. 1287) (1980).

Z-Lys-SBzl is not specific for plasmin.[15] In fact, urokinase can hydrolyze it effectively. (Under the conditions described here, 2×10^{-16} mol of urokinase will yield $\Delta A_{412} = 0.0005$.) Thus it is possible that unknown proteases in a test solution containing a plasminogen activator might hydrolyze the thiol ester. This is routinely tested for by assays in the presence of all ingredients but plasminogen.

In conclusion, this two-step photometric assay is a highly sensitive and reliable method for the measurement of low levels of plasminogen activators or their inhibitors.

Acknowledgment

P.L.C. was supported by a National Cancer Institute Grant (CA 22729) to Dr. Thomas D. Gelehrter (University of Michigan Medical School). G.D.J.G. was supported by funding from the Department of Energy to Dr. Elliott Shaw (Brookhaven National Laboratory). The authors are most grateful to Drs. Gelehrter and Shaw for financial support, encouragement, and constructive suggestions.

[34] A New Active-Site Titrant of Serine Proteases

By WALTER F. MANGEL, D. CAMPBELL LIVINGSTON,
JOHN R. BROCKLEHURST, JOHN F. CANNON, STEVEN P. LEYTUS,
STUART W. PELTZ, GARY A. PELTZ, and HO-YUAN LIU

Serine proteases are involved in a wide variety of biological reactions.[1] Sensitive assays for these enzymes are needed because in many cases they are present in very small amounts. Also, these enzymes degrade themselves at rates that are concentration dependent. From these points of view, active-site titrants are especially useful in assaying serine proteases. Good active-site titrants react quickly with an enzyme, resulting in the formation of an enzymatically inactive, stable acyl-enzyme intermediate and the release of a stoichiometric amount of product that can be detected at low concentrations.[2]

Esters of guanidinobenzoate are excellent active-site titrants of many serine proteases.[3] For example, p-nitrophenyl-p-guanidinobenzoate (NPGB)[4] rapidly acylates trypsin with the release of a stoichiometric

[1] E. Reich, D. Rifkin, and E. Shaw, eds., "Proteases and Biological Control." Cold Spring Harbor Lab., Cold Spring Harbor, New York, 1975.
[2] F. J. Kézdy and E. T. Kaiser, this series, Vol. 19 [1].
[3] P. L. Coleman, H. G. Latham, Jr., and E. N. Shaw, this series, Vol. 45 [2].
[4] List of abbreviations: NPGB, p-nitrophenyl-p-guanidinobenzoate; FDE, 3',6'-bis (4-guanidinobenzoyloxy)-5-(N'-4-carboxyphenyl)thioureidospiro [isobenzofuran-1(3H), 9'-

amount of p-nitrophenol.[5] Deacylation of the acyl-enzyme intermediate is extremely slow. Also, NPGB is sufficiently soluble and stable in aqueous buffers, so that active-site titrations can be performed at substrate concentrations above the K_m of most enzymes and in the absence of a high background. The major problem with NPGB is that p-nitrophenol can be detected by absorbance at concentrations no lower than 10^{-6} M. Elmore and his colleagues[6] increased the usefulness of p-guanidinobenzoate esters as active-site titrants by substituting 4-methylumbelliferone for p-nitrophenol. With this coumarin derivative, the product of an active-site titration fluoresces, providing an increase in sensitivity of about 2 orders of magnitude.

Fluorescein is an even better fluorophore than 4-methylumbelliferone.[7] Fluorescein has an extinction coefficient 6-fold greater than 4-methylumbelliferone and a quantum yield 4-fold greater. Furthermore, fluorescein absorbs maximally at 491 nm where the efficiency of the excitation monochromator and the xenon lamp output are 6-fold greater than at 325 nm, the absorption maximum for 4-methylumbelliferone. Also, 4-methylumbelliferone is quite sensitive to photolytic decomposition.

The molecule, 3′,6′-bis(4-guanidinobenzoyloxy)-5-(N'-4-carboxyphenyl)thioureidospiro[isobenzofuran-1(3H),9′-[9H]xanthen]-3-one, abbreviated FDE for Fluorescein DiEster, was designed as a fluorogenic active-site titrant of serine proteases (Fig. 1A).[7] It is an analog of NPGB, in which a fluorescein derivative (Fig. 1C) is substituted for p-nitrophenol. FDE is nonfluorescent because the fluorescein moiety is in the lactone state. In the presence of a serine protease, however, one of the guanidinobenzoyl esters is hydrolyzed, yielding 5-(N'-4-carboxyphenyl)-thioureido-3′-(4-guanidinobenzoyloxy)-6′-hydroxyspiro[isobenzofuran-1(3H),9′-[9H]xanthen]-3-one, hereafter abbreviated FME for Fluorescein-MonoEster (Fig. 1B). FME should be highly fluorescent, because the fluorescein moiety is in the quinone state. By substituting chromophores, the detectability of the product in an active-site titration can be increased 5–6 orders of magnitude, from 10^{-6} M for p-nitrophenol to 10^{-11}–10^{-12} M for fluorescein.[8] Since FDE is an ester of guanidinobenzoate, the kinetic

---

[9H]xanthen]-3-one; FME, 5-(N'-4-carboxyphenyl)thioureido-3′-(4-guanidinobenzoyloxy)-6′-hydroxyspiro[isobenzofuran-1(3H), 9′[9H]xanthen]-3-one; FUE, 5-(N'-4-carboxyphenyl)thioureido-3′,6′-dihydroxyspiro[isobenzofuran-1(3H), 9′-[9H]xanthen]-3-one; PBS, phosphate-buffered saline.

[5] T. Chase and E. Shaw, this series, Vol. 19 [2].

[6] G. W. Jameson, D. V. Roberts, R. W. Adams, W. S. A. Kyle, and D. T. Elmore, *Biochem. J.* **131**, 107 (1973).

[7] D. C. Livingston, J. R. Brocklehurst, J. F. Cannon, S. P. Leytus, J. A. Wehrly, S. W. Peltz, G. A. Peltz, and W. F. Mangel, *Biochemistry* **20**, 4298 (1981).

[8] The "8000" spectrofluorometer sold by SLM, Urbana, Illinois, can resolve a spectrum of 10^{-12} M fluorescein.

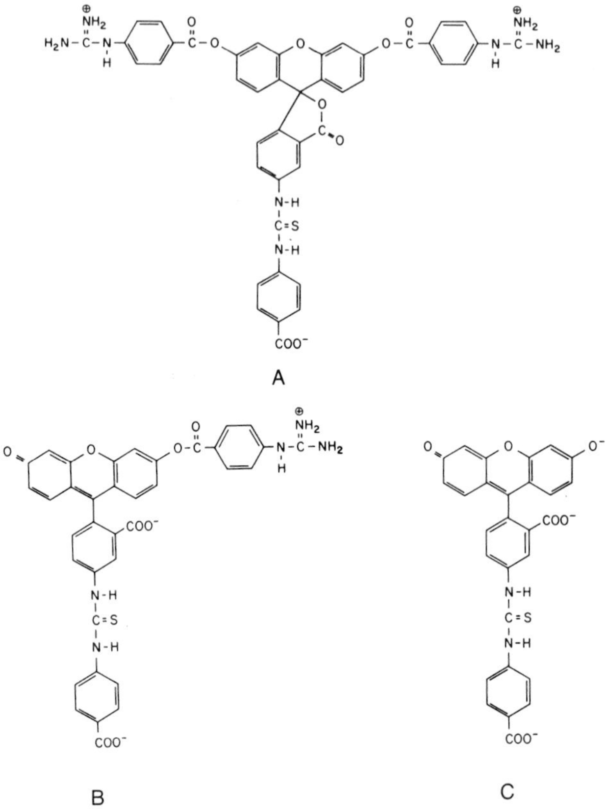

FIG. 1. The structures of (A) FDE, (B) FME, and (C) FUE in PBS at pH 7.2. (A) 3',6'-bis(4-guanidinobenzoyloxy) - 5 - (N'-4-carboxyphenyl)thioureidospiro[isobenzofuran-1(3H), 9'-[9H]xanthen]-3-one. (B) 5-(N'-4-carboxyphenyl)thioureido-3'-(4-guanidinobenzoyloxy)-6'-hydroxyspiro[isobenzofuran-1(3H),9'-[9H]xanthen]-3-one. (C) 5-(N'-4-carboxyphenyl)-thioureido-3',6'-dihydroxyspiro[isobenzofuran-1(3H),9'-[9H]xanthen]-3-one.

properties exhibited in an active-site titration of serine proteases are similar to those exhibited by NPGB (Table I).

Materials

The synthesis, purification, and characterization of FDE have been described.[7] NPGB was purchased from ICN. Stock solutions of 1 mM FDE and 10 mM NPGB in redistilled N,N'-dimethylformamide (DMF) are stored in the dark at 4°C. PBS contains 0.137 M NaCl–2.68 mM KCl–8.08 mM Na$_2$HPO$_4$–1.47 mM KH$_2$PO$_4$–0.91 mM CaCl$_2$–0.49 mM MgCl$_2$, at pH 7.2. Fraction II dog plasminogen and plasmin were purified as

TABLE I
CHARACTERISTICS OF FDE AND NPGB IN PBS, pH 7.2[a]

	FDE Trypsin	NPGB Trypsin	FDE Urokinase	FDE Plasmin
Kinetic constants				
k_2 (sec^{-1})	1.05	.370	0.112	0.799
k_3 (sec^{-1})	1.66×10^{-5}	1.66×10^{-5}	3.64×10^{-4}	6.27×10^{-6}
K_s (M)	3.06×10^{-6}	1.49×10^{-7}	2.71×10^{-5}	6.04×10^{-6}
$K_{m(app)}$ (M)	4.85×10^{-11}	6.66×10^{-12}	8.78×10^{-8}	4.74×10^{-11}
k_2/K_s (M^{-1} sec^{-1})	3.48×10^5	2.45×10^6	4.13×10^3	1.28×10^5

	FDE	NPGB
Other parameters		
$k_{(spont)}$ (sec^{-1})	5.1×10^{-6}	2.0×10^{-6}
Saturation (M)	1.3×10^{-5}	1.7×10^{-4}

[a] D. C. Livingston, J. R. Brocklehurst, J. F. Cannon, S. P. Leytus, J. A. Wehrly, S. W. Peltz, G. A. Peltz, and W. F. Mangel, *Biochemistry* **20**, 4298 (1981).

described previously.[7,9] Stock solutions contained 5–20 mg/ml in PBS. Reference standard human urokinase was purchased from Leo Pharmaceuticals. Stock solutions contained 2240 Ploug units/ml in PBS. Bovine pancreatic trypsin, three times crystallized, was purchased from Worthington. Stock solutions contained 1 mg/ml in 1 mM HCl. Human thrombin—1.86 mg/ml in 0.1 M sodium phosphate (pH 6.8), with 0.9 M NaCl—was a gift from Dr. Robert Rosenberg. Although stock solutions of enzymes are prepared by weight and checked by absorbance, aliquots of each stock solution are titrated with NPGB, or FDE to determine the concentration of active sites. All proteins and enzymes are stored in 100-μl aliquots at $-20°$C. If dilutions of stock solutions have to be made for an assay, they are done in PBS immediately prior to use. The model cell line used is a clonal subline (RT4-71-D2) of a neurotumor (RT4) induced by the subcutaneous injection of ethylnitrosourea into newborn BDIX rats.[10] The cell culture medium was Dulbecco's Modified Eagle's medium with 10% fetal calf serum, 100 units/ml penicillin, and 100 μg/ml streptomycin. Microwells (6.4-mm-diameter tissue culture dishes) were purchased from Costar.

[9] S. P. Leytus, G. A. Peltz, H.-Y. Liu, J. F. Cannon, S. W. Peltz, D. C. Livingston, J. R. Brocklehurst, and W. F. Mangel, *Biochemistry* **20**, 4307 (1981).
[10] M. Imada, N. Sueoka, and D. B. Rifkin, *Dev. Biol.* **66**, 109 (1978).

Fluorescence Measurements

Active-site titrations with FDE are monitored in a Perkin-Elmer MPF 44-A fluorescence spectrophotometer connected to a Perkin-Elmer Universal Digital Readout meter and to a Hewlett-Packard Model 7015 $X-Y$ recorder with time base. The excitation and emission wavelengths are 491 and 514 nm, respectively. The bandwidth of both monochromators is 4 nm. The fluorescence spectrophotometer is standardized several times each day using a polymethacrylate block embedded with rhodamine B (Perkin-Elmer) so that the relative fluorescence is comparable in different experiments. We do this by adjusting the digital readout to 180 fluorescence units using the fine sample sensitivity knob, under conditions in which the excitation wavelength is set at 491 nm, the emission wavelength is set at 580 nm, and the coarse sample sensitivity knob is set at 3. To detect extremely low concentrations of fluorescein, the amplifier can easily be boosted further. Fluorescence measurements are performed in 3125B standard disposable spectrovette cuvettes purchased from Evergreen Scientific Company.

Basic Active-Site Titration Procedure

The basic procedure for active-site titrations with FDE is described using urokinase and plasmin as examples (Fig. 2). Active-site titrations are performed at room temperature in PBS, pH 7.2. Immediately prior to performing an assay, 5–25 μl of FDE from a 1 mM stock solution in DMF are diluted into 5 ml of PBS. One-milliliter of this solution plus 10–20 μl of enzyme buffer is placed in a cuvette in the fluorometer and the baseline fluorescence quickly traced with the recorder at a speed of 0.5 sec/inch. The relative fluorescence units on the Universal Digital Readout meter for the baseline are noted or the meter is zero suppressed. To another 1-ml aliquot of FDE in a cuvette, 10–20 μl of enzyme are added, the solution mixed, and the cuvette quickly placed in the fluorometer. The recorder is switched on with the time axis set at 100 sec/inch, and the increase in fluorescence is traced until a steady state is reached (Fig. 2). At the end of the titration, the relative fluorescence units on the Universal Digital Readout meter are again noted. The relative fluorescence units of the baseline and of a point near the end of the titration are then used to define the ordinate scale. Extrapolation of the steady-state line to the ordinate yields the relative fluorescence units (ΔF), which are directly proportional to the concentration of active sites.[11]

[11] M. L. Bender, M. L. Begue-Canton, R. L. Blakely, L. J. Brubacher, J. Feder, C. R. Gunter, F. J. Kézdy, J. V. Killheffer, Jr., T. H. Marshal, C. G. Miller, R. W. Roeske, and J. K. Stoops, *J. Am. Chem. Soc.* **88**, 5890 (1966).

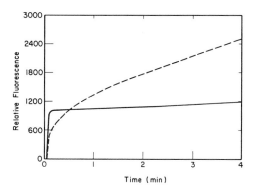

Time (min)

FIG. 2. The kinetics of hydrolysis of FDE by urokinase (dashed line) and plasmin (solid line). A 1.0-ml solution of $5 \times 10^{-6} M$ FDE in PBS containing 44 Ploug units of urokinase, $9.15 \times 10^{-9} M$ as determined by an FDE active-site titration of a stock solution, was incubated at room temperature and the kinetics of hydrolysis of FDE monitored. A 1.0-ml solution of $5 \times 10^{-6} M$ FDE in PBS containing $8.99 \times 10^{-9} M$ plasmin, whose concentration was determined by an NPGB active-site titration of a stock solution, was incubated at room temperature and the kinetics of hydrolysis of FDE monitored. The concentration of active sites can be determined from the data as follows: extrapolation of the steady-state lines to the ordinate for urokinase and plasmin yield a ΔF of 1008. With the fluorometer standardized as described in the text, the slope of a standard curve is $1.12 \times 10^{11} \Delta F/M$. Dividing the extrapolated ΔF by the slope of the standard curve yields a concentration of active sites of $9.00 \times 10^{-9} M$.

The conversion of relative fluorescence units to the concentration of active sites is accomplished with a standard curve. One-milliliter solutions of 1 μM FDE in PBS are incubated with plasmin or trypsin whose concentrations range from $(1.11–6.54) \times 10^{-8} M$. The molar concentrations of these proteases were determined by NPGB active-site titrations of stock solutions. The magnitude of the fluorescence burst after a 2-min incubation minus the magnitude of the fluorescence of a similar solution not containing enzyme is defined as ΔF. A graph of ΔF versus the protease concentration yields a straight line intersecting the origin. The slope of this line is used to convert ΔF to the molar concentration of active sites. An example of this calculation is described in the legend to Fig. 2. Standard curves are most easily prepared with plasmin or trypsin. Once acylated, their rates of deacylation are so slow that the almost horizontal steady-state lines do not have to be extrapolated to zero time in order to obtain a ΔF.

A Variation of the Basic Assay Procedure

Active-site titrations of enzymes whose rates of deacylation are so slow the steady-state lines appear horizontal (e.g., trypsin and plasmin) can be performed in volumes much less than 1 ml. For example, reaction mixtures can be incubated in 10–20 μl volumes and then, after 2 min, diluted to 1 ml with PBS before reading the change in fluorescence (ΔF). This variation conserves substrate. It also allows a titration to be performed with a low background and at a high substrate concentration, so that the burst will be completed quickly. Any background fluorescence in an FDE preparation becomes lower relative to the signal, upon dilution, compared to no dilution. If the protein concentration is high, however, nonactive site-induced hydrolysis of FDE by nucleophilic amino acid groups may become appreciable. This rate is proportional to the protein concentration and will increase the slope of the steady-state line. Thus if an impure enzyme is being assayed at a high protein concentration, a titration profile should be obtained, regardless of the rate of deacylation, so that the ΔF can be obtained by extrapolation of the steady-state line to zero time.

Use of FDE to Assay the Kinetics of Plasminogen Activation

A two-step assay using FDE has been developed to monitor the kinetics of activation of plasminogen to plasmin by urokinase and by a transformed cell-associated plasminogen activator (Fig. 3).[9,12] For example, 0.05-ml solutions of PBS containing 5.6 Ploug units of human urokinase and 0.5–5.72 mg/ml of fraction II dog plasminogen are first incubated at room temperature. The amount of plasmin formed after 30 and 60 sec is then determined by withdrawing 0.02-ml aliquots at these times, adding the aliquots to 1.0 ml of PBS containing 5 μM FDE, incubating for 2 min, and measuring the change in fluorescence (ΔF). The kinetics are shown in Fig. 3A. The transformed cell-associated plasminogen activator is assayed by first incubating 0.05-ml aliquots of PBS containing 0.2–1.3 mg/ml of fraction I or fraction II dog plasminogen in microwells containing 20,000 cells. After 30 min, the aliquots are removed from the microwells and incubated in cuvettes with 0.05-ml aliquots of PBS containing 2 μM FDE. Five minutes later, 0.90 ml of PBS is added and the change in fluorescence measured (ΔF). The data are shown in the form of a direct linear plot in Fig. 3B.

The two-step assay procedure to monitor the kinetics of plasminogen activation not only yields data that are more easy to interpret but also is much more flexible than a coupled, double-rate assay in which plasminogen, plasminogen activator, and a plasmin substrate are incubated together.[13,14] The kinetics are linear, as opposed to complex, and directly

[12] H.-Y. Liu, G. A. Peltz, S. P. Leytus, C. Livingston, J. Brocklehurst, and W. F. Mangel, Proc. Natl. Acad. Sci. U.S.A. 77, 3796 (1980).
[13] U. Christensen and S. Mullertz, Biochim. Biophys. Acta 480, 275 (1977).
[14] R. C. Wohl, L. Summaria, and K. C. Robbins, J. Biol. Chem. 255, 2005 (1980).

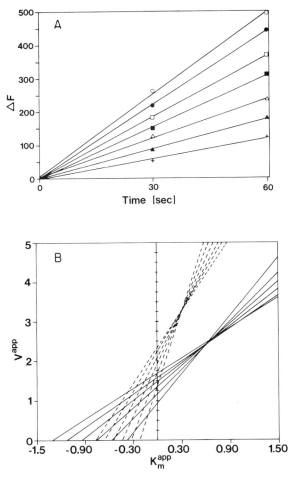

FIG. 3. The kinetics of activation of plasminogen by urokinase (A) and a direct linear plot of the initial rate of plasminogen activation by transformed cells as a function of the plasminogen concentration (B). (A) Activation reactions were performed at room temperature in 0.05 ml of PBS containing 5.6 Ploug units of human urokinase and fraction II dog plasminogen at concentrations of (+) 0.50 mg/ml, (▲) 0.75 mg/ml, (△) 1.12 mg/ml, (■) 1.70 mg/ml, (□) 2.55 mg/ml, (●) 3.82 mg/ml, and (○) 5.72 mg/ml. At the indicated times, 0.02-ml aliquots were withdrawn and added to 1.0-ml solutions of PBS containing 5 μM FDE. After 2 min, the change in fluorescence (ΔF) was measured. (B) Sixteen hours prior to the assay, 0.10-ml aliquots containing 20,000 RT4-71-D2 transformed rat cells in culture medium were placed into microwells. After this seeding period, the medium was removed and the cells were washed three times with PBS. To each microwell, 0.05-ml aliquots of PBS containing 0.2–1.3 mg/ml of fraction I (solid lines) or fraction II (dashed lines) dog plasminogen were added. After 30 min, the aliquots were removed from the microwells and incubated in cuvettes with 0.05-ml aliquots of PBS containing 2 μM FDE. Five minutes later, 0.90 ml of PBS was added and the change in fluorescence (ΔF) measured. The dimension of the initial velocity is ΔF units/min/10,000 cells and of the $K_{m(app)}$ is mg/ml.

yield the initial velocity. The first step of the assay, the activation reaction, can be performed under a variety of physiological conditions. The second step of the assay is designed solely to titrate the amount of plasmin formed in the first step. During the second step, continued activation of plasminogen by urokinase is efficiently quenched by the dilution. In part, this is because the urokinase concentration is decreased 51-fold. Furthermore, since the $K_{m(app)}$ for plasminogen and urokinase is similar to the K_s for FDE and urokinase, quenching occurs on dilution because the FDE concentration becomes much greater than the plasminogen concentration (Table I). Thus urokinase will react preferentially with FDE rather than with plasminogen, resulting in its inhibition by acylation. A coupled assay in which FDE is present during the activation is not used because a low concentration of FDE relative to the plasminogen concentration would have to be present in order for FDE not to inhibit activation. This concentration of FDE would be well below the K_s for FDE and plasmin, and thus the plasmin produced would not be titrated immediately.

The ability to assay the kinetics of plasminogen activation, before deleteriously high concentrations of plasmin are formed, illustrates the utility of FDE (Table II). The flexibility in the assay conditions should allow the factors that influence the rate and extent of plasminogen activation to be characterized quantitatively.

General Comments on the Assay Conditions

PBS was chosen as the standard buffer because of its widespread use with eukaryotic cells. Buffers that are potentially nucleophilic, such as Tris-HCl at a pH where the amino groups are unprotonated, should be avoided because they will hydrolyze FDE. As do all esters, FDE spontaneously hydrolyzes in aqueous buffers. The first-order rate constant for spontaneous hydrolysis of FDE in PBS (pH 7.2) is 5.1×10^{-6} sec^{-1} (half time = 38 hr) (Table I). At pH 8.3, the rate of spontaneous hydrolysis increases 5-fold. The rate of spontaneous hydrolysis of NPGB increases 10-fold, from pH 7.2 to 8.3. To minimize spontaneous hydrolysis, several precautions are employed. FDE in DMF is diluted into PBS immediately prior to performing an active-site titration. The pH of the PBS is 7.2, the lowest pH below which the fluorescence intensity of fluorescein begins to decrease dramatically. Fortunately, pH 7.2 is within the physiological range, so that proteases and their physiological substrates and inhibitors can be assayed. As with most assays, no more than 5% of the substrate should be hydrolyzed.

Enzymes Titrated by FDE

The kinetic constants exhibited in an active-site titration of trypsin, urokinase, and plasmin are shown in Table I. In addition, the constants for

TABLE II

MICHAELIS–MENTEN KINETIC PARAMETERS FOR THE ACTIVATION OF FRACTION I AND II DOG PLASMINOGEN

Enzyme	$K_{m(app)}$ (μM)		V_{max}		k_{cat} (sec^{-1})	
	Fraction I	Fraction II	Fraction I	Fraction II	Fraction I	Fraction II
Human urokinase[a]	31.7	19.2	$\dfrac{0.13 \text{ pmol plasmin}}{(\text{min})(10{,}000 \text{ cells})}$	$\dfrac{0.18 \text{ pmol plasmin}}{(\text{min})(10{,}000 \text{ cells})}$	1.98	1.86
Transformed rat cell[a]	6.6	3.4				
E. coli outer membrane preparations[b]		31.5		$\dfrac{0.024 \text{ pmol plasmin}}{(\text{min})(\mu g \text{ membrane protein})}$		
Bovine trypsin[c]	86	39			0.069	0.053

[a] S. P. Leytus, G. A. Peltz, H.-Y. Liu, J. F. Cannon, S. W. Peltz, D. C. Livingston, J. R. Brocklehurst, and W. F. Mangel, *Biochemistry* **20**, 4307 (1981).

[b] S. P. Leytus, L. K. Bowles, J. Konisky, and W. F. Mangel, *Proc. Natl. Acad. Sci. U.S.A.* **78**, 1485 (1981).

[c] S. P. Leytus and W. F. Mangel, unpublished observations.

trypsin and NPGB are listed, to indicate that FDE and NPGB are similar substrates kinetically. FDE has also been shown to be a good active-site titrant for human thrombin.[15] With FDE, the kinetics of activation of plasminogen to plasmin have been characterized with urokinase, trypsin, a transformed cell-associated plasminogen activator, and a plasminogen activator associated with *E. coli* outer membranes (Table II).

[15] S. W. Peltz and W. F. Mangel, unpublished observations.

Acknowledgment

This investigation was supported by Grant CA 25633 from the National Institutes of Health awarded to W.P.M.

Section IV

Enzymes Involved in Blood Pressure Regulation

[35] Renin

By EVE E. SLATER

Renin (EC 2.4.99.19) is an endopeptidase belonging to the aspartyl protease family. Renal renin is in all likelihood first synthesized as a prohormone,[1,2] by modified smooth muscle cells of the glomerular afferent arteriole, where it is stored in granules and released into the bloodstream in response to a variety of physiological stimuli.[3] Once released into the circulation, renin acts on its substrate, an α-globulin, to cleave a leucyl-leucine bond[4] (leucylvaline in humans[5]) located near the amino terminus, thereby generating the decapeptide angiotensin I, which in turn is converted to angiotensins II and III, resulting in potent vasoconstriction and aldosterone release (Fig. 1). Renal renin has been shown to participate in normal cardiovascular homeostasis,[6] in compensatory response to heart failure and cirrhosis,[7] and in the pathogenesis of renovascular hypertension in humans.[8,9] To date, renal renin has been purified from hog,[10,11] rat,[12] dog,[13] and human.[14-16]

Distribution

An enzyme possessing identical enzymatic and similar physical properties to renal renin has been isolated and purified from mouse submaxil-

[1] K. Poulsen, J. Vuust, S. Lykkegaard, A. H. Nielsen, and T. Lund, *FEBS Lett.* **98**, 135 (1979).

[2] N. Yokosawa, N. Takahashi, T. Inagami, and D. L. Page, *Biochim. Biophys. Acta* **569**, 211 (1979).

[3] S. Oparil and E. Haber, *N. Engl. J. Med.* **291**, 389 and 446 (1974).

[4] L. T. Skeggs, J. R. Kahn, K. E. Lentz, and N. P. Shumway, *J. Exp. Med.* **106**, 439 (1957).

[5] D. A. Tewksbury, R. A. Dart, and J. Travis, *Biochem. Biophys. Res. Commun.* **99**, 1311 (1981).

[6] J. Sancho, R. Re, J. Burton, A. C. Barger, and E. Haber, *Circulation* **53**, 400 (1976).

[7] L. Watkins, J. A. Burton, E. Haber, J. R. Cant, T. W. Smith, and A. C. Barger, *J. Clin. Invest.* **57**, 1606 (1976).

[8] H. Goldblatt, J. Lynch, R. F. Hanzal, and W. W. Summerville, *J. Exp. Med.* **59**, 347 (1934).

[9] H. Gavras, H. R. Brunner, E. D. Vaughn, Jr., and J. H. Laragh, *Science* **180**, 1369 (1973).

[10] T. Inagami and K. Murakami, *J. Biol. Chem.* **252**, 2978 (1977).

[11] P. Corvol, C. Devaux, T. Ito, P. Sicard, J. Ducloux, and J. Menard, *Circ. Res.* **41**, 616 (1977).

[12] T. Matoba, K. Murakami, and T. Inagami, *Biochim. Biophys. Acta* **526**, 560 (1978).

[13] V. J. Dzau, E. E. Slater, and E. Haber, *Biochemistry* **18**, 5224 (1979).

[14] E. E. Slater and H. V. Strout, Jr., *J. Biol. Chem.* **256**, 8164 (1981).

[15] H. Yokosawa, L. A. Holladay, T. Inagami, E. Haas, and K. Murakami, *J. Biol. Chem.* **225**, 3498 (1980).

[16] F. X. Galen, C. Deveaux, T. Guyenne, J. Menard, and P. Corvol, *J. Biol. Chem.* **254**, 4848 (1979).

METHODS IN ENZYMOLOGY, VOL. 80

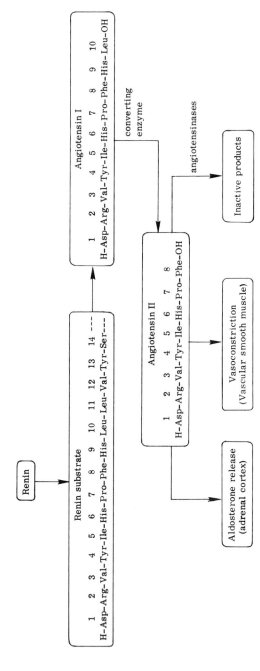

Fig. 1. A simplified schema of the renin–angiotensin–aldosterone cascade, whereby the enzyme renin acts upon its substrate to generate angiotensin I. Reprinted with permission from "Scientific American's Medicine: Principles and Progress for Practicing Physicians," p. VII-15. Scientific American, Inc.. New York, 1978.

lary gland, where its functional role has yet to be determined.[17] Its antibody cross-reacts with mouse renal renin. A large percentage of plasma renin with immunological similarity to renal renin is present in inactive form.[18] It can be activated by exposure to cold, acid, or a variety of proteolytic enzymes. A small part of this renin is thought to be prorenin; it can be activated by trypsinization.[2] The remainder is likely to represent renin bound to inhibitor proteins; its physiological role is unknown. An inactive renin that cross-reacts with antibody to renal renin is demonstrable in human amniotic fluid.[19] Although renin activity has been shown to be present in uterine extracts, its similarity to renal renin has not been clarified. A membrane-bound form of renin has been identified in rat kidney.[20] Last, renin activity that resembles renal renin in enzymatic and immunological properties has been identified both biochemically and by immunohistochemistry in nerve cells of hog,[21] rat,[22] mouse,[23,24] and human.[25] Conclusions regarding the biochemical similarity to the renal enzyme of the several renins described above await their complete purification and characterization, as well as definition of their functional roles.

Methods of Purification

Mouse submaxillary renin, and renal renins from hog and man, have been the most completely characterized to date. The purification procedure for each will be described.

Mouse Submaxillary Renin

Purification of renin from mouse submaxillary gland was achieved in 1972 by Cohen *et al.*[17] based on the relatively high concentration of renin in this organ.

Step 1. Homogenization. For each 25 g of submaxillary gland tissue obtained from adult male Swiss-Webster mice (approximately 150 mice),

---

[17] S. Cohen, J. M. Taylor, K. Murakami, A. M. Michelakis, and T. Inagami, *Biochemistry* **11**, 4286 (1972).

[18] E. E. Slater and E. Haber, *N. Engl. J. Med.* **301**, 429 (1979).

[19] T. T. Guyenne, F. X. Galen, C. Devaux, P. Corvol, and J. Menard, *Hypertension* **2**, 465 (1980).

[20] K. Nishimura, F. Alhenc-Gelas, A. White, and E. G. Erdös, *Proc. Natl. Acad. Sci. U.S.A.* **77**, 4975 (1980).

[21] S. Hirose, H. Yokosawa, T. Inagami, and R. J. Workman, *Brain Res.* **191**, 489 (1980).

[22] S. Hirose, H. Yokosawa, and T. Inagami, *Nature (London)* **274**, 392 (1978).

[23] M. R. Celio, D. L. Clemens, and T. Inagami, *Biomed. Res.* (in press).

[24] M. C. Fishman, E. A. Zimmerman, and E. E. Slater, *Science* (in press).

[25] E. E. Slater, R. Defendini, and E. A. Zimmerman, *Proc. Natl. Acad. Sci. U.S.A.* **77**, 5478 (1980).

add 100 ml of cold deionized water and homogenize for 3 min in a Waring blender. After centrifugation for 10 min at 4°C and 16,000 g, decant the supernatant, stir the residue with 90 ml of cold distilled water, and recentrifuge. Combined supernatants contain approximately 5 g of protein.

Step 2. Ammonium Sulfate Fractionation. To 9 volumes of extract, add 1 volume of stock streptomycin solution (1.46 g of streptomycin sulfate, pH 9.0, in a final volume of 20 ml). Allow the mixture (final pH 6.8–7.1) to stand at 4°C overnight, centrifuge for 5 min, and discard residue. Dissolve solid ammonium sulfate (56 g/100 ml of supernatant) in the supernatant, let stand at 0°C for 30 min, centrifuge as described in step 1, and discard supernatant. Suspend the precipitate in 20 ml of deionized water and dialyze overnight at 4°C against five changes of deionized water, 2 liters each. This dialyzed preparation contains 2.4–2.8 g of protein in approximately 110 ml. All of the remaining chromatographic steps are performed at 4°C.

Step 3. Gel Filtration. Concentrate the dialyzed protein by Amicon PM-30 to a final volume of approximately 40 ml, and centrifuge at 20,000 g for 3 hr. Divide the supernatant equally and apply sequentially to a column of Sephadex G-100 or AcA 34 (5 × 90 cm) in 0.01 M sodium acetate (pH 5.9)–0.1 M NaCl, at a flow rate of 20 ml/hr, collecting 10-ml fractions.

Step 4. DEAE Batch Concentration. Combine the renin-containing fractions from both chromatographies and concentrate by Amicon PM-30 to approximately 20 ml. Adjust the pH to 4.5 by ultrafiltration with 0.1 M sodium acetate (pH 4.5), followed by 0.01 M sodium acetate. Reconcentrate to 20 ml and centrifuge at 2500 rpm for 10 min. Apply the supernatant (400–500 mg of protein) to a column of DE-52 cellulose (1.5 × 10 cm) previously equilibrated to 0.01 M sodium acetate (pH 4.5), and elute with the same buffer. Afterward, adjust the pH of the eluate (300–350 g of protein) to 5.4 with 0.05 M sodium acetate by ultrafiltration, and concentrate to approximately 15 ml.

Step 5. CM-Cellulose Ion Exchange. Apply the sample to a column of CM-52 cellulose (1.5 × 15 cm), previously equilibrated to 0.05 M sodium acetate (pH 5.4), and wash. Then elute with a 250-ml total volume salt gradient, prepared by allowing 0.05 M sodium acetate (pH 5.4) containing 0.15 M NaCl to flow into a 125-ml constant-volume mixing chamber filled with equilibration buffer. Maintain flow at 0.2 ml/min and collect 6-ml fractions.

Step 6. DEAE–Cellulose Ion Exchange. Pool the renin-containing peaks (approximately 150 mg of protein) and concentrate to approximately 15 ml as described above. Adjust the pH to 7.5 with 0.02 M Tris-HCl, and reconcentrate to 15 ml. Apply this sample to a column of DE-52 cellulose

(1.5 × 15 cm) equilibrated with 0.02 M Tris-HCl (pH 7.5), and wash. Then elute as described in step 5 using a 0–0.3 M sodium chloride gradient.

Step 7. CM-Cellulose Ion Exchange. Pool the renin-containing peaks (approximately 60–75 mg of protein), concentrate to approximately 10 ml, adjust pH to 5.4 with 0.05 M acetate (pH 5.4), and reconcentrate to 15 ml. Apply the sample to a CM-cellulose column and then elute as described in step 5. Five renin-containing peaks, denoted A through E, will result. Final yields are 12–15 mg of protein from fraction A and 3–4 mg from fraction C; specific activity is 6.67 and 6.00 mol angiotensin I/mg/hr; yield is 11 and 2%, respectively.

Hog Renal Renin

Because of the exceedingly low concentration of renin in kidney and its rapid inactivation by co-eluted proteases, complete purification of renal renin from the hog was not accomplished until the successful application of pepstatin affinity chromatography and inclusion of protease inhibitors throughout the entire protocol. Later, the large-scale purification of hog renal renin was achieved,[10,11] and homogeneous renal renin from rat[12] and from dog[13] was obtained similarly. The protocol described below for the purification of porcine renal renin is taken from Inagami and Murakami.[10] The reader is referred to the protocol of Corvol *et al.*[11] as an alternative method.

Step 1. Extraction. Thaw frozen hog kidneys (6 kg) at room temperature for 2 hr, dissect cortex away from fat and medulla, and mince in an electric meat grinder. Freeze-dry, pulverize in a blender, stir for 30 min in 6 liters of ethyl ether to remove fat, and air-dry at room temperature. Extract the dry powder (approximately 960 g) with 16 liters of 30% (v/v) ethylene glycol monoethyl ether–water mixture containing the protease inhibitors phenylmethylsulfonyl fluoride (2 mM)–sodium tetrathionate (5 mM)–EDTA (5 mM)–diisopropylphosphofluoridate (0.05 mM). Stir for 1 hr at neutral pH and 4°C, and remove the residue by centrifugation.

Step 2. DEAE Batch Concentration. Absorb renin activity in the supernatant to DE-52 cellulose (1 kg) at neutral pH, collect the cellulose by filtration, wash once with 4 liters of 5 mM acetic acid (pH 5.3), and elute renin activity by two successive applications of 4 liters of 0.1 M sodium acetate–0.2 M NaCl (pH 4.8) containing phenylmethylsulfonyl fluoride (0.1 mM)–sodium tetrathionate (0.25 mM)–EDTA (0.25 mM).

Step 3. Ammonium Sulfate Fractionation. Precipitate renin activity by the addition of solid ammonium sulfate to 75% saturation. Collect precipitate containing approximately 28 g of protein by centrifugation, dissolve in

400 ml of 0.1 M Tris-HCl buffer (pH 7.5) containing the diluted protease inhibitors in step 2, and dialyze overnight against two changes of 8 liters of 0.01 M Tris-HCl (pH 7.5) with the same inhibitors.

Step 4. Pepstatin Affinity Chromatography. The pooled extract from 18 kg of frozen hog kidney prepared by steps 1–3 above is adjusted to pH 5.2, and the precipitate thus obtained is removed by centrifugation. The supernatant is immediately adjusted to pH 6.5 and applied to the affinity column (5 × 7.5 cm) at a flow rate of 250 ml/hr. The method of coupling pepstatin (Peptide Institute, Inc., Osaka, Japan) to aminohexyl-agarose with an approximate incorporation of 2.8 μmol/ml wet gel is described elsewhere.[26] After application of the extract, wash the column exhaustively with 0.02 M sodium acetate–1.0 M NaCl, pH 5.2. Elute renin by the stepwise addition of 0.1 M acetic acid (pH 3.2) at a flow rate of 60 ml/hr, and quickly neutralize the pH of the resultant fractions by pumping 2 M Tris-HCl (pH 8.5) into each tube (flow rate, 3 ml/hr). Collect 21-ml fractions. Approximately 167 mg of protein with a specific activity of 9.3 μg angiotensin I/μg protein/hr at pH 6.0 is obtained by this step. Regenerate the affinity column by washing with approximately 800 ml of 6.5 M urea, followed by 800 ml of 0.1 M borate, pH 9.5. This permits the gel to be recycled for up to three experiments.

Step 5. Gel Filtration. Concentrate the renin-containing fractions to approximately 25 ml by pressure filtration on an Amicon PM-10 filter, apply to a Sephadex G-150 column (5 × 85 cm) equipped with a reverse-flow plunger, and elute with 0.01 M sodium pyrophosphate–0.1 M NaCl (pH 6.5) at a rate of 24 ml/hr, collecting 8-ml fractions. An approximate 10-fold increase in specific activity results.

Step 6. DEAE–Cellulose Ion Exchange. Pool and concentrate renin-containing fractions (approximately 17 mg of protein), and equilibrate with 0.08 M sodium phosphate (pH 6.2) on an Amicon PM-10. Apply to a DE-52 column (0.9 × 40 cm) previously equilibrated with the same buffer, and elute with a linear concentration gradient between 170 ml each of the equilibration buffer and that buffer containing 0.15 M KCl at a flow rate of 24 ml/hr. Collect 2.5-ml fractions. Approximately 3 mg of renin of specific activity 260 μg angiotensin I/μg protein results.

Step 7. CM–Cellulose Ion Exchange. Pool and concentrate the renin-containing fractions and equilibrate to 0.02 M sodium acetate, pH 5.02. Apply to a CM-cellulose column (0.9 × 17 cm) and elute with a linear concentration gradient generated between 125 ml each of the initial buffer and the same buffer containing 0.1 M KCl at a flow rate of 45 ml/hr. Collect 1.5-ml fractions. Two milligrams of pure renin with a specific

---

[26] K. Murakami and T. Inagami, *Biochem. Biophys. Res. Commun.* **62**, 757 (1975).

activity of 267 μg angiotensin I/μg or 2000 Goldblatt units (GU)/mg protein and the result of a 133,000-fold purification is obtained.

Human Renal Renin[14]

Step 1. Extraction. Obtain cadaver kidneys (10 kg) from patients without renal disease and within 24 hr of death, and dissect to obtain cortex (3 kg). After five freeze–thaw cycles, homogenize by Waring blender in 4 liters of 5 mM ammonium acetate (pH 5.25) containing the following protease inhibitors: 5 mM EDTA–5 mM sodium tetrathionate–2 mM phenylmethylsulfonylfluoride. Retain these inhibitors during the remaining purification steps in 0.25, 0.25, and 0.1 mM concentrations, respectively. Centrifuge the homogenate for 20 min at 15,000 g, discard the precipitate, and repeat the centrifugation once. All succeeding steps are performed at 4°C; protein determinations are reported according to the method of Lowry *et al.*[27] and renin activity is determined by the angiotensin I radioimmunoassay of Haber *et al.*,[28] using an excess of renin-suppressed human plasma substrate. One Goldblatt unit (GU) of human renin generates 7120 ng angiotensin I/hr at pH 7.4.

Step 2. DEAE Batch Concentration. Dilute the supernatant, which contains between 55 and 70 g of protein 1 : 4 with chilled deionized water, and adjust pH to 7.5 with 5 N NaOH. Accomplish batch concentration by the addition of 660 ml of wet DE-52 cellulose previously equilibrated with 5 mM K phosphate buffer, pH 7.5. Stir gently for 2 hr, collect the gel by filtration, wash with 6 liters of equilibration buffer, and elute with 5 liters of 0.1 M K phosphate, pH 6.6. Approximately 15 g of protein are present at this step.

Step 3. Acidification, Salt Fractionation, and Ammonium Sulfate Fractionation. Adjust the pH to 3.1 with 6N HCl, add 5% NaCl (200 g), and stir for 30 min. Centrifuge as in step 1 above, and discard the pellet. Precipitate the supernatant by the addition of 65% ammonium sulfate. Resuspend the pellet in 1500 ml of 5 mM sodium acetate, pH 5.5. Dialyze overnight, employing five changes of 6 liters per change, recentrifuge as above, and discard the pellet. After this step 7–7.5 g of protein with a specific activity of 0.6–0.8 GU/mg remain, which provides a 2- to 3-fold purification and is required to allow binding of human renin to the pepstatin resin.

Step 4. Pepstatin Affinity Chromatography. Apply and reapply the

---

[27] O. H. Lowry, N. J. Rosebrough, A. L. Farr, and R. J. Randall, *J. Biol. Chem.* **193**, 265 (1951).

[28] E. Haber, T. Koerner, L. B. Page, B. Kliman, and A. Purnode, *J. Clin. Endocrinol. Metab.* **29**, 1349 (1969).

dialyzed sample (1500 ml) directly to a 150-ml column (2.4 × 35 cm) of aminohexyl Sepharose pepstatin (Peptide Institute, Inc., Osaka, Japan) coupled with an incorporation of 3.0–4.3 μmol/ml by the method of Murakami and Inagami,[26] and equilibrated to 5 mM sodium acetate, pH 5.5. Renin enzymatic activity will be almost completely bound by this procedure. Wash the column with 300 ml of equilibration buffer, followed by 1 M NaCl in 5 mM sodium acetate (pH 5.5), followed by equilibration buffer. Elute renin by application of a linear gradient established with 1 liter of 0.05 M Tris-HCl (pH 7.5) in the mixing chamber and 1 liter of 0.05 M Tris-HCl (pH 7.5)–0.25 M lithium bromide in the reservoir at a flow rate of 30 ml/hr. Collect 10-ml fractions, pool the renin-containing fractions (160–180 mg of protein), and concentrate to approximately 100 ml by pressure filtration on an Amicon PM-30. Specific activity at this step ranges from 20 to 40 GU/mg with recovery from 80 to 100% enzymatic activity. Complete separation from cathepsin D protease activity is not always accomplished by application of lithium bromide gradient elution as compared to the previously reported acid gradient elution.[29] Nonetheless, the improved yields obtained by the former method have prompted its selection. The method of gel regeneration that permits its repeated use is described under Hog Renal Renin Purification, step 4.

Step 5. Gel Filtration. Divide the concentrated sample into two equal portions and chromatograph sequentially on a previously calibrated Sephadex G-100 column (190 × 4 cm) in 0.1 M sodium acetate (pH 5.5) at a flow rate of 30 ml/hr. Collect 10-ml fractions. An approximate 10-fold purification results from this step. Pool the renin-containing peaks, concentrate by Amicon PM-30 to approximately 25 ml, and dialyze against 10 mM sodium acetate, pH 4.9 (for step 6) or 0.05 M K phosphate–0.15 M NaCl, pH 7.2 (for step 6B).

Step 6. CM-Cellulose Ion Exchange. Bind the concentrated sample to a previously equilibrated CM-52 cellulose column (19 × 1.5 cm) in 10 mM sodium acetate, pH 4.9. Elute renin activity by a gradient (200 ml total volume) from equilibration buffer to 0.1 M sodium acetate–0.15 M NaCl, pH 5.5. Collect 4-ml fractions. This procedure results in an approximate 1.5-fold purification, and effective separation of renin from cathepsin D protease activity.

Step 7. [D-leu[6]] Affinity Chromatography. Concentrate the renin-containing fractions to approximately 10 ml, dialyze against 10 mM sodium acetate (pH 5.5), and apply to a Sepharose-4B column (1.5 × 6 cm) to which has been coupled the synthetic octapeptide renin inhibitor [D-leu[6]] (Bachem) with an incorporation of 450–500 nmol/ml wet gel ac-

---

[29] E. E. Slater, R. C. Cohn, V. J. Dzau, and E. Haber, *Clin. Sci. Mol. Med.* **55**, 117 (1978).

cording to the method of Poulsen *et al.*[30] Elute by application of a 400-ml total gradient from 10 mM sodium acetate (pH 5.5) to 200 mM sodium acetate (pH 7.5), at a flow rate of 15–20 ml/hr. Collect 5-ml fractions.

Step 6B. Antibody-Affinity Chromatography. After obtaining potent antibodies to renin, the immunoglobulin fraction is purified by DEAE–cellulose ion-exchange chromatography at 0.03 M K phosphate (pH 7.0) and gel filtration on AcA 34 (1.5 × 90 cm) in 0.1 M Tris-HCl–0.20 M NaCl, pH 8.0. This fraction, which retains >90% inhibitory potency, is coupled to 50 g of cyanogen bromide-activated Sepharose-4B, as described by Livingston elsewhere in this series.[31] Renin from step 6 after dialysis is applied and reapplied several times to the antirenin antibody–Sepharose-4B column (1.5 × 12 cm), to which are applied, in stepwise fashion, 100 ml each of: 0.1 M sodium acetate (pH 4.5), 0.05 M sodium acetate (pH 3.5), 1 M acetic acid (pH 2.7), and 1 M acetic acid (pH 2.4). Flow rate is 30 ml/hr. Adjust to a pH of 6.0, each 5-ml eluted sample, by the immediate addition of 3 M potassium phosphate with vortexing. Renin enzymatic activity recovered from this step is 70% or more. Experience in applying antibody-affinity chromatography at an earlier stage of purification has not been obtained.

Renin purified using either final step (7 or 6B) should achieve a specific activity of approximately 1000 GU/mg for an overall purification of 500,000-fold. Overall yield is 1 mg of protein and 6–7% enzymatic activity. Stability in 0.02 M potassium phosphate (pH 7.0–7.5), at −70°C for over 1 year has been demonstrated. The properties of the renin purified as just described will be discussed subsequently.

A protocol that can be used alternatively is described by Yokosawa *et al.*[15] It comprises purification through step 5 of Haas, previously detailed in this series,[32] followed by hemoglobin–Sepharose chromatography to remove cathepsin D protease activity, aminohexyl Sepharose–pepstatin affinity chromatography, Sephadex G-100 gel filtration, CM-cellulose ion-exchange, and DEAE–cellulose ion-exchange chromatographies. Pure human renin has also been obtained after a 40-fold purification of a rare renin-producing tumor.[16]

Methods of Assay

Renin can be measured by bioassay as detailed in Vol. 2 of this series, where activity is expressed in Goldblatt units (1 GU equals the amount of renin required to raise dog blood pressure by 30 mm Hg).[32]

[30] K. Poulsen, J. Burton, and E. Haber, *Biochemistry* **12**, 3877 (1973).
[31] C. M. Livingston, this series, Vol. 24, p. 723.
[32] E. Haas, this series, Vol. 2, p. 124.

The indirect measurement of renin is accomplished by angiotensin I radioimmunoassay, where renin activity is expressed as nanograms angiotensin I per milliliter per hour; the protocols of Haber et al.,[28] Sealey et al.,[33] and Poulsen and Jorgensen,[34] being the most widely used. In our experience, the antibody-trapping method of Poulsen and Jorgensen, detailed below,[34] provides the most sensitive and accurate angiotensin I determination, by minimizing proteolytic destruction of the newly formed angiotensin I. To minimize angiotensin I generation by nonrenin proteases that are activated by acidification, we recommend that the method of radioimmunoassay selected is one performed at pH 7.2–7.4 in the presence of protease inhibitors.

To assay renin in tissue extracts, exogenous substrate must be provided. Its source is plasma with negligible endogenous renin activity obtained from nephrectomized animals or human subjects receiving oral contraceptives to increase substrate concentration and placed on a high-sodium diet and fluorinef (0.1 mg) for several days. Alternatively, synthetic tetradecapeptide substrate, purchased commercially, can be used. Although human renin can act readily on substrate from a variety of sources, subprimate renin is incapable of cleaving the human substrate. Total renin activity—that is, the sum of active and inactive renin—can be determined as described elsewhere[35,36] by methods designed to inactivate or remove plasma proteins. Since as much as 80% of total renin activity can derive from so-called inactive renin, which is not known to have physiological effect, these methods should be avoided as clinical assays.[18] Similarly, manipulations that release inactive renin, such as repeated freezing and thawing, should be avoided.

A rapid fluorimetric assay can be used for qualitative determination of renin activity of tissue samples after acid protease activity has been removed.[37,38] One such method will be described here.

The availability of pure renin from mouse submaxillary gland, hog kidney, and human kidney has enabled the development of direct radioimmunoassay for these renins. The method of direct radioimmunoassay for mouse submaxillary renin will be outlined below.[39] The limited

[33] J. E. Sealey, J. H. Laragh, J. Gerten-Banes, and R. M. Aceto, "Hypertension Manual," p. 621. Yorke Medical Books, New York. (1974).

[34] K. Poulsen and J. Jorgensen, J. Clin. Endocrinol. Metab. **39**, 816 (1974).

[35] A. F. Lever, J. I. S. Robertson, and M. Tree, Biochem. J. **91**, 346 (1964).

[36] S. L. Skinner, Circ. Res. **20**, 391 (1967).

[37] A. Reinharz and M. Roth, Eur. J. Biochem. **7**, 334 (1969).

[38] F. X. Galen, C. Devaux, P. Grogg, J. Menard, and P. Corvol, Biochim. Biophys. Acta **523**, 485 (1978).

[39] A. M. Michelakis, H. Yoshida, J. Menzie, K. Murakami, and T. Inagami, Endocrinology **94**, 1101 (1974).

sensitivity of the human renin assay currently precludes its clinical application.[19,40]

Angiotensin I Radioimmunoassay by Antibody Trapping of Poulsen and Jorgensen[34]

Reagents

Tris–albumin buffer. 0.2 M Tris-HCl (pH 7.5) containing 0.3% bovine serum albumin

3 M Tris–EDTA buffer. 3 M Tris-HCl (pH 7.2) measured in a 10-fold dilution, containing 200 mM EDTA

Radioimmunoassay buffer. 0.08 M sodium barbital buffer (pH 8.6) containing 0.01% merthiolate or sodium azide and 0.3% bovine serum albumin

Dextran-coated charcoal. 6.3 g of charcoal (Norit A) and 0.125 g of Dextran T 70 (Pharmacia), suspended in 100 ml of radioimmunoassay buffer

Unlabeled angiotensin I. Purchase and dilute in Tris–albumin buffer to concentrations between 0 and 2.4 ng per 5–75 μl aliquot.

Labeled angiotensin I. Purchase or react 5 μg with 1 mCi ^{125}I by the chloramine-T method.[41] Remove unreacted iodide by chromatography on a 5 × 70-mm column of CG-400 anion-exchange resin in 0.05 M sodium acetate–HCl (pH 5.0), or alternate ion-exchange resin. Resultant specific activity is 25–50 μCi/mg. Dilute in radioimmunoassay buffer and store in small aliquots. On the day of use, dilute in radioimmunoassay buffer to about 5000 cpm/ml (100–200 pg of [^{125}I]angiotensin I/ml).

Procedure

Plasma (5–75 μl) in 3 mM EDTA (in a volume appropriate to maintain pH at 7.4), or alternatively, renin-containing sample and substrate, is added to 10–50 μl of antiangiotensin serum diluted to allow 40–50% binding of tracer in the radioimmunoassay at 4°C, and then incubated for 3 hr at 37°C. Stop the reaction by placing samples in ice water, and add 1000 μl of labeled angiotensin I (approximately 5000 cpm/ml) in barbital buffer, and leave for 18 hr at 4°C. Then separate free and antibody-bound angiotensin with 100–150 μl charcoal, which is previously placed in the plastic caps of each tube and mixed simultaneously in all samples by

[40] S. Hirose, R. J. Workman, and T. Inagami, *Circ. Res.* **45**, 275 (1979).
[41] W. M. Hunter and F. C. Greenwood, *Nature (London)* **194**, 495 (1962).

turning the rack repeatedly upside down during 30 sec. Then follow quickly by centrifugation at 3000 g for 15 min at 4°C. Decant and count.

Direct Radioimmunoassay for Mouse Submaxillary Renin of Michelakis et al.[39]

Reagents

Unlabeled pure mouse submaxillary renin. Purification described in the preceding section.

Antibody to mouse submaxillary renin. Rabbits may be immunized by a variety of schedules. Michelakis *et al.* used 1 mg of renin A mixed with Freund's adjuvant and injected into the toepads of a rabbit each week; high-titer antibody production occurred at the fifth week. An antibody dilution that results in approximately 50% displacement in total binding of the tracer is used.

Labeled renin. 5 μg of renin A is iodinated with 1–2 mCi of ^{125}I using the chloramine-T method.[41] Labeled renin is freed from unbound ^{125}I by gel filtration; specific activity is 100–200 mCi/mg.

Goat antirabbit gamma globulin

Normal rabbit serum

Procedure

To 100 μl of standard containing 0 to 2000 pg of renin, or 5–100 μl of sample, add 100 μl of antibody and 100 μl of tracer containing 3000–6000 cpm. Adjust the final volume to 1.0 ml by 0.05 M phosphate buffer (pH 7.4) containing 0.5% bovine serum albumin. Incubate for a minimum of 90 hr at 4°C. Add 100 μl of previously titered goat antirabbit gamma globulin antiserum and 100 μl of normal rabbit serum, and incubate for 24 hr at 4°C. Centrifuge at 3000 rpm for 25 min, decant thoroughly, and count the precipitate.

Fluorometric Assay for Renin Activity of Galen et al.[38]

Reagents

N-acetyltetradecapeptide (Bachem)

Fluorescamine

0.02 M citrate–phosphate buffer, pH 6.0

0.01 M citrate–phosphate buffer, pH 7.3

Dioxane

Procedure

React renin-containing samples with 28 nM N-acetyltetradecapeptide in a final volume of 1 ml of 0.02 M citrate–phosphate buffer (pH 6.0) for 1 hr at 37°C. Stop the reaction by immersion in boiling water and raise pH to 7.3 by the addition of 1 ml of 0.1 M citrate–phosphate, pH 7.3. At room temperature add 0.2 ml of Fluram solution, prepared daily by dissolving 30 mg in 100 ml of dioxane. Read fluorescence at 395–495 nm in a spectrophotofluorimeter. In control experiments verify that the blank value for nonincubated media is less than 10% of the incubated mixture. Express results either in relative fluorescence or in nanomoles of tetrapeptide generated per hour of incubation. In this case a standard curve of tetrapeptide (between 1 and 4 nmol) is determined during the same experiment.

Properties

Various properties of renins from several species are summarized in Table I. The amino acid composition of these renins is presented in Table II.

Stability

Pure renin from several species has been shown to be stable under conditions of neutral pH at 4°, −20°, and −70°C.

Kinetics

The Michaelis constants for renins from a number of sources and under a variety of conditions have been reported in Volume 19 of this series.[42] The determinations for pure renin are presented as follows with the substrate used included in parentheses:

Hog renal renin	33 μM (octapeptide)[10]
	7.7 μM (tetradecapeptide)[11]
Human renin	6.8 μM (N-acetyltetradecapeptide)[16]
	1.54 μM (pure human substrate)[14]

Specificity/Inhibition/Active Site

Renin specificity is limited to the leucyl-leucine peptide bond of angiotensinogen and the tetradecapeptide renin substrate: H-Asp-Arg-Val-

---

[42] R. R. Smeby and F. M. Bumpus, this series, Vol. 19, p. 699.

TABLE I
PROPERTIES OF PURIFIED RENIN

Property	Mouse submaxillary[a]	Hog renal		Human renal		
		1[b]	2[c]	3[d]	4[e]	5[f]
Molecular weight						
Sedimentation equilibrium	36,000–37,000	36,500	36,800	—	40,000	—
Gel filtration	43,000	42,000	—	40,000	42,000	—
SDS–gel electrophoresis	—	—	—	50,000	—	50,000
4–12% Polyacrylamide-gel electrophoresis	—	—	—	38,000–42,000	—	—
Isoelectric point	5.4, 5.6	5.2	5.2	5.1, 5.4, 5.6	5.7	5.3, 5.6
Glycoprotein	?	+	+	+	+	+
Optimum pH	—	5.5–7.0	—	6.5	6.0	5.5–6.0
Extinction coefficient $E^{1\%}_{1\,cm}$	10.5, 10.8	9.1	—	—	—	—
Diffusion constant, $D_{20,w}$	—	—	—	—	$8.23 \pm 0.40 \times 10^{-7e}$	—
Friction ratio, f/f_0	—	—	—	—	1.06	—
Axial ratio, a/b	—	—	—	—	2.5	—

[a] From Cohen et al.[17]
[b] From Inagami and Murakami.[10]
[c] From Corvol et al.[11]
[d] From Galen et al.[16] (obtained from renin-secreting tumor).
[e] From Yokosawa et al.[15]
[f] From Slater and Strout.[14]

TABLE II
AMINO ACID ANALYSIS

Amino acid	Human renal renin (Ref. 14)[a]	Human renal renin (Ref. 15)[b]	Hog renin (Ref. 10)	Mouse submaxillary renin (Ref. 17)
Alanine	34	23	16	14
Arginine	13	10	9	9
Aspartic acid	35	36	26	27
Half-cysteine	4	4	3	4
Glutamic acid	50	35	31	27
Glycine	33	44	34	35
Histidine	9	6	5	8
Isoleucine	12	14	12	15
Leucine	32	27	30	33
Lysine	25	16	11	12
Methionine	4	8	4	7
Phenylalanine	15	20	17	17
Proline	15	18	18	18
Serine	28	30	32	33
Threonine	23	28	26	27
Tryptophan	—	—	5	5
Tyrosine	10	20	16	17
Valine	20	17	34	35
Galactosamine	—	—	—	—
Glucosamine	—	—	3.2	—

[a] Based on MW 50,000.
[b] Based on MW 40,000.

Tyr-Ile-His-Pro-Phe-His-*Leu-Leu*-Val-Tyr-Ser-OH[4]; the only apparent exception appears to be the recently reported amino-terminal sequence for human angiotensinogen: H-Asp-Arg-Val-Tyr-Ile-His-Pro-Phe-His-*Leu-Val*-Ile-His-Thr-Glu.[5] Although renin has been classified as an acid protease by characterization of its active site,[43] it is unique by virtue of its neutral pH requirements.

Renin is inhibited by pepstatin (for human renin: $K_1 = 1.3 \times 10^{-10}$ M,[44] $K_1 = 0.3 \times 10^{-7}$ [45] M), by bromoisocaproyl peptides as described elsewhere in this series,[46,47] by aliphatic diazo compounds as will be

[43] T. Inagami, K. Misono, and A. M. Michelakis, *Biochem. Biophys. Res. Commun.* **56**, 503 (1974).
[44] M. M. McKown, R. J. Workman, and R. I. Gregerman, *J. Biol. Chem.* **249**, 7770 (1974).
[45] P. Corvol, C. Devaux, and J. Menard, *FEBS Lett.* **34**, 189 (1973).
[46] T. Inagami and Y. Suketa, this series, Vol. 46, p. 234.
[47] Y. Suketa and T. Inagami, *Biochemistry* **14**, 3188 (1975).

discussed below,[48] and competitively by a series of synthetic substrate analogs as summarized by Burton et al.[49]

Detailed characterization of the active site of mouse submaxillary renin has been performed by Misono and Inagami.[50] By analogy to acid proteases, the renin active site was shown to contain two different catalytically essential carboxyl residues, each of which was selectively modified by two different types of inactivators: diazoacetyl methyl ester derivatives of glycine, norleucine, and leucine in the presence of cupric ion; and 1,2-epoxy-3-(p-nitrophenoxy)propane. Additional characterization demonstrated that 2 tyrosyl and 1 arginyl residue are probably located within the active site. Modification of histidyl, tryptophanyl, and amino groups did not affect renin activity. Reactions with iodoacetic acid and N-ethylmaleimide revealed that renin has a partially buried free sulfhydryl group, not essential for enzyme activity.

[48] M. M. McKown and R. I. Gregerman, *Life Sci.* **16**, 71 (1975).
[49] J. Burton, K. Poulsen, and E. Haber, *Biochemistry* **14**, 3892 (1975).
[50] K. S. Misono and T. Inagami, *Biochemistry* **19**, 2616 (1980).

[36] Human Plasma Carboxypeptidase N

By Thomas H. Plummer, Jr., and Ervin G. Erdös

Human plasma carboxypeptidase N was discovered by Erdös and Sloane[1] as a kininase that inactivated bradykinin and kallidin. It has since been proposed to serve as a regulator of action of various physiologically active peptides such as anaphylatoxins C3a,[2] C4a,[3] and C5a,[2] fibrinopeptides,[4] and plasmin-degradation products of fibrin[5] by removal of a C-terminal basic amino acid. Carboxypeptidase N has both peptidase[6] and esterase[7] activity and can be assayed conveniently in the spectrophotometer by either activity. Co^{2+} has been used to activate the peptidase activity.[6] Recent studies have shown a pronounced enhancement of

[1] E. G. Erdös and E. M. Sloane, *Biochem. Pharmacol.* **11**, 585 (1962).
[2] V. A. Bokisch and H. J. Müller-Eberhard, *J. Clin. Invest.* **49**, 2427 (1970).
[3] J. P. Gorski, T. E. Hugli, and H. J. Müller-Eberhard, *Proc. Natl. Acad. Sci. U. S. A.* **76**, 5299 (1979).
[4] A. C. Teger-Nilsson, *Acta Chem. Scand.* **22**, 3171 (1968).
[5] M. Belew, B. Gerdin, G. Lindeberg, J. Porath, T. Saldeen, and R. Wallin, *Biochim. Biophys. Acta* **621**, 169 (1980).
[6] E. G. Erdös, E. M. Sloane, and I. M. Wohler, *Biochem. Pharmacol.* **13**, 893 (1964).
[7] E. G. Erdös, H. Y. T. Yang, L. L. Tague, and N. Manning, *Biochem. Pharmacol.* **16**, 1287 (1967).

peptidase activity with a penultimate alanine,[8,9] which may explain the more rapid inactivation of C3a than of C5a by this enzyme. Additional assays are sensitive multistep procedures using hydrolysis of protamine[10] or extraction of benzoyl (Bz)-Gly from Bz-Gly-Lys reaction mixtures[11] or biological assays.[12]

Methods of Assay

Reagents

Esterase activity. Bz-Gly-argininic acid,[7] 1.6 mM (26.91 mg/50 ml) in water (stable for at least 1 week at 4°C)

Peptidase activity. Bz-Ala-Lys,[9] 2.00 mM (32.14 mg/50 ml) in water (stable for at least 2 weeks at 4°C)

Furylacryloyl (FA)-Ala-Lys,[9] 1.00 mM (18.23 mg/50 ml) in water (stable for at least 2 weeks at 4°C)

Tris-HCl, 0.10 M (pH 7.60), containing 1.0 M NaCl; 12.11 g of Tris(hydroxymethyl)aminomethane, 58.45 g NaCl in 900 ml of water adjusted to pH 7.60 with HCl and diluted to 1 liter

HEPES, 0.10 M (pH 7.75) containing 0.5 M NaCl. 23.83 g of HEPES, 29.23 g of NaCl in 900 ml of water adjusted to pH 7.75 with NaOH and diluted to 1 liter

Procedure. On the day of use, the Bz-substrates or FA-Ala-Lys are prepared by mixing equal volumes of substrate solution with Tris-HCl stock or HEPES stock, respectively. Solutions are equilibrated at 37°C. For the assay, 3 ml of substrate is pipetted into 1-cm light path cuvettes in a cuvette holder equilibrated at 37°C. Water is added to the reference cuvette in the same microliters as the samples to follow, in 10 or 100 μl, depending on the stage of purification. For fractions of very high activity, 10 μl of a dilution is used. For kinetic studies the concentration of the substrate or the light path of the cuvette can be varied (see Chapter 37 in this volume). The spectrophotometer recorder is balanced to give a ΔA of 0.1 OD full scale. Bz-substrates are followed by increase in absorbance at 254 nm and FA-Ala-Lys by the decrease in absorbance at 336 nm. FA-Ala-Lys is preferred for carboxypeptidase N solutions at low levels of purity due to negligible interference from proteins in the sample aliquot at

[8] G. Oshima, J. Kato, and E. G. Erdös, *Arch. Biochem. Biophys.* **170**, 132 (1975).

[9] T. J. McKay, A. W. Phelan, and T. H. Plummer, Jr., *Arch. Biochem. Biophys.* **197**, 487 (1979).

[10] N. C. Corbin, T. E. Hugli, and H. J. Müller-Eberhard, *Anal. Biochem.* **73**, 41 (1976).

[11] A. Koheil and G. Forstner, *Biochim. Biophys. Acta* **524**, 156 (1978).

[12] E. G. Erdös, A. G. Renfrew, E. M. Sloane, and J. R. Wohler, *Ann. N.Y. Acad. Sci.* **104**, 222 (1963).

336 nm. In microcuvettes the esterase activity of 5 μl of serum or plasma can be determined with Bz-Gly-argininic acid in 1 mM concentration.[7]

CoCl$_2$ can be used to enhance the rate of hydrolysis[6] with either peptidase substrate. In studies where different plasma samples are being compared, preincubation at 37°C for at least 30 min with 0.1 mM CoCl$_2$ is preferred.[6]

Activity Units. One activity unit is the amount of enzyme required to catalyze the hydrolysis of 1 μmol of substrate per minute at 37°C. Units can be determined by using the $\Delta\epsilon$ M^{-1} cm^{-1} of 360 at 254 nm for Bz-substrates[13] and of -1300 at 336 nm for FA-Ala-Lys.[14] The $\Delta\epsilon$ M^{-1} cm^{-1} of Bz-Gly-argininic acid is higher than that of the peptide analog and is above 700. It should be determined for commercially available substrates by comparing the absorption of substrate to that of the split products (Bz-Gly and argininic acid). Comparative rates of hydrolysis for substrates can be determined as micromoles per minute per milligram of protein and are as follows: Bz-Gly-argininic acid, 160.6; Bz-Ala-Lys, 156.0 and FA-Ala-Lys, 40.2. The lower rate of hydrolysis for FA-Ala-Lys as compared to Bz-Ala-Lys is compensated for by its higher $\Delta\epsilon$ M^{-1} cm^{-1}.

Purification

Principle. A two-step chromatographic procedure for purifying carboxypeptidase N from outdated pooled human plasma has been described.[15] The steps are outlined in Table I.[15] It consists of ion-exchange chromatography on a column of DE-52 cellulose followed by affinity chromatography on a column of p-aminobenzoyl-L-arginine–Sepharose-6B. Enzyme purified 2600-fold was eluted from the resin by the competitive inhibitor guanidinoethylmercaptosuccinic acid (GEMSA).[17]

Materials

DE-52 cellulose (Whatman) was cycled in 0.5 N HCl, water, 0.5 N NaOH, water, and 0.01 M Tris-HCl, pH 7.0. The resin was then titrated to pH 7.0 with 2 N HCl. A prewash of 1.0 M NaCl is included for previously used resin.

p-NH$_2$-Bz-L-Arg was synthesized and coupled to Sepharose-6B according to the methods outlined by Plummer and Hurwitz.[15] Amino acid

[13] J. E. Folk, K. A. Piez, W. R. Carroll, and J. A. Gladner, *J. Biol. Chem.* **235**, 2272 (1960).
[14] T. H. Plummer, Jr. and M. T. Kimmel, *Anal. Biochem.* **108**, 348 (1980).
[15] T. H. Plummer, Jr. and M. Y. Hurwitz, *J. Biol. Chem.* **253**, 3907 (1978).
[16] O. H. Lowry, N. J. Rosebrough, A. L. Farr, and R. J. Randall, *J. Biol. Chem.* **193**, 265 (1951).
[17] T. J. McKay and T. H. Plummer, Jr., *Biochemistry* **17**, 401 (1978).

TABLE I
PURIFICATION OF HUMAN PLASMA CARBOXYPEPTIDASE N[a]

Purification step	Volume (ml)	Total units[b]	Protein[c] (mg/ml)	Units/mg protein	Yield (%)	Purification (fold)
Dialyzed plasma	2043	4060	54.4	0.04	100	1
DE-52 column chromatography	1540	2129	1.54	0.89	52	22
Affinity chromatography[d]	70	1198	0.164	104	30	2608

[a] Taken from Plummer and M. Y. Hurwitz.[15]
[b] One unit equals 1 μmol of hippuryl-L-argininic acid cleaved per minute at 37°C.
[c] Protein was measured by the Lowry[16] method.
[d] The medium used was p-aminobenzoyl-L-arginine–Sepharose-6B.

analysis of several preparations indicated a substitution of 12–18 μmol of arginine per milliliter of settled gel.

GEMSA can be prepared as previously described.[17] An alternative one-step method of synthesis is as follows: Bromosuccinic acid (Aldrich, 20 mmol, 3.94 g) in 50 ml of H_2O, was adjusted to pH 7.5 with 9.5 N NaOH in a 25°C, jacketed vessel maintained in a nitrogen atmosphere. S-2-Aminoethylisothiouronium bromide hydrobromide (Aldrich, 19.4 mmol, 5.45 g), dissolved in 9.70 ml of 2 N NaOH, was added. The pH was maintained at 7.5 with 5.0 N NaOH. After 24 hr, the pH was adjusted to 2.4 with 4 N HCl and the material was applied to a 2.5 × 42-cm column of Dowex 50-X8 (Bio-Rad, Aminex Q150S) in the H^+ form. The column was developed at 50°C at a flow rate of 29.6 ml cm^{-2} hr^{-1} with 0.1 N HCl for 1 hr, 0.2 M pyridine formate (pH 3.25) for 8 hr, followed by a gradient to 0.4 M pyridine (1000 ml constant-volume elution reservoir). The first major Sakaguchi positive peak at pH 3.5 was DTNB negative and was pooled and lyophilized as product. The material inhibited carboxypeptidase N with a K_i of approximately 3 μM and was sufficiently pure to use as a column eluant.

Procedure for Preparation of Plasma. Pooled, outdated human plasma (2000 ml) was centrifuged at 13,000 g for 30 min. The supernatant was dialyzed in an Amicon DC-2 with an H1P-10 hollow-fiber cartridge for 12 hr against 16 liters of 0.005 M Tris-HCl (pH 7.0), containing 0.2 mM each of diisopropylphosphorofluoridate and phenylmethanesulfonyl fluoride. The dialyzed plasma was centrifuged again at 17,000 g for 12 min.

Ion-Exchange Chromatography. Previously prepared DE-52 cellulose was packed under 8 psi nitrogen pressure to form a column 5.0 × 41 cm. The column was equilibrated in 0.01 M Tris-HCl at a flow rate of 18 ml cm^{-2} hr^{-1} for 6 hr, and then the dialyzed, clarified plasma was applied at the same speed. Following sample addition, the column was washed with equilibration buffer for 3.5 hr. Carboxypeptidase N was eluted by a linear gradient of NaCl from 0.0 to 0.25 M in equilibration buffer. The gradient chambers contained 6.20 liters each. Effluent was collected in the last half of the gradient in 25-ml fractions. A peak of carboxypeptidase N activity was observed immediately after the blue ceruloplasmin band as the NaCl concentration approximates 0.135 M. Protein was detected by absorbance at 280 nm and carboxypeptidase N was detected by spectrophotometric assay of 100-μl aliquots. Fractions with enzyme activity at least one-fourth that of the maximum in the activity peak were pooled and concentrated to approximately 150 ml with an Amicon DC1A concentrator with a H1P-100 hollow-fiber cartridge. The concentrate was dialyzed for 4 hr against 2 liters of 0.01 M sodium phosphate (pH 7.0) containing 0.1 M NaCl and 1% butanol.

Affinity Chromatography. *p*-Aminobenzoyl-L-arginine–Sepharose-6B was packed in a column (3.8 × 25 cm) and equilibrated in 0.01 M sodium phosphate (pH 7.0) containing 0.1 M NaCl and 1% butanol. The dialyzed concentrate from the DE-52 step was applied and the column was developed at a flow rate of 8.4 ml cm^{-2} hr^{-1}. Fractions of 24 ml were collected. The buffer was automatically changed at 8 hr and 12.5 hr to equilibration buffer containing 0.001 M GEMSA and 0.3 M L-arginine-HCl, respectively. Protein and enzyme were detected as before. After every chromatographic run, the column was flushed with equilibration buffer containing 2 M guanidine.

Enzyme Concentration. Enzyme is recovered in 30% yield after the two chromatography steps. The purified enzyme from affinity chromatography averages 10–12 mg in 72–96 ml of buffer (3 or 4 fractions). Dilute enzyme can be concentrated on an Amicon ultrafilter with a XM-50 or YM-10 membrane to 3–4 ml with a 75–95% recovery of activity followed by dialysis to remove GEMSA.

An alternative method is to dialyze the enzyme pool against two changes of 2 liters each of 0.01 M sodium phosphate (pH 7.0) containing 1% butanol. The enzyme was then applied to a column (0.9 × 27 cm) of DE-52 at a flow rate of 41.5 cm^{-2} hr^{-1}. After adsorption to the column, the enzyme was eluted with 0.01 M sodium phosphate (pH 7.0) containing 0.20 M NaCl and 1% butanol. Fractions of 1.3 ml were collected. The main enzyme peak (80% of the material applied) was detected by absorbance at 280 nm and was eluted in 4 fractions. This material has been concentrated to as much as 8 mg/ml with quantitative recovery on an Amicon ultrafilter with a YM-10 membrane. Some evidence has accumulated from gel-electrophoresis studies of a number of preparations that less active, protease-degraded carboxypeptidase N is eluted after this major peak and is removed from preparations by this step. The amount of this material is minor and varies with each preparation. However, the new bands on gel electrophoresis appear to be identical to those generated by mild protease treatment of purified carboxypeptidase N.[15]

Enzyme Quantitation. Purified enzyme can be approximated by the method of Lowry *et al.*[16] using crystalline bovine serum albumin as a standard or by using the $E_{1\,cm}^{1\%}$ (280 nm) value of 11.9.[15]

Enzyme Stability. Purified solutions of carboxypeptidase N have remained stable for several months at 4°C, as evidenced by gel-electrophoresis patterns of native protein.[15] Solutions in 0.01 M sodium phosphate buffer containing 0.1 M NaCl have also been frozen in dry ice and stored for several months at -70°C with less than a 10% loss in activity.

Native carboxypeptidase N is very sensitive to degradation by proteolytic enzymes with a concomitant change in subunit size and associa-tion.[15] Since plasmin and plasminogen have an affinity for arginine–Sepharose derivatives,[18] including the p-aminobenzoyl-L-arginine–Sepha-rose-6B medium,[15] it is quite possible that instabilities observed in earlier preparations[8,19] were due to trace contaminations of plasmin(-ogen). Elution with GEMSA has been shown not to elute plasmin(-ogen) from the recommended agarose derivative.[15]

Physical Properties

Molecular Weight. The molecular weight of native carboxypeptidase N has been estimated to be between 270,000[2] and 280,000.[19] The enzyme is a zinc metalloenzyme[15] and is a tetramer of 2 each of a "heavy" subunit and "light" subunit.[8,19] The molecular weight of the large subunit has been estimated to be 83,000–98,000.[8,15] The smaller subunit of enzyme prepared by affinity chromatography appears to consist of two related derivatives having MW $\approx$ 49,000 and 55,000.[15] The 55,000-MW subunit is easily converted to the 49,000-MW subunit by mild protease treatment.

Preliminary studies using dimethylmaleic anhydride to dissociate na-tive carboxypeptidase N have provided evidence that most, if not all, of the enzymatic activity of the aggregated enzyme is associated with the smaller subunits.[20] The 280,000-MW enzyme is more stable at 37°C than its subunits,[8] thus it should be more stable in circulating blood.[8] The exact function of the large subunit will have to be determined, but it aggregates into a larger polymer under the same conditions that release the enzymati-cally active subunit as a monomer. This tendency to aggregation may be related to the amino acid composition as discussed below.

Amino Acid Composition. The amino acid composition of native car-boxypeptidase N has been reported.[15] The composition is noteworthy by a very high leucine content that approximates 16%. This observation and the higher than average content of aromatic and hydrophobic residues suggest that hydrophobic regions may facilitate aggregation. Preliminary amino acid analyses of reduced carboxymethylated heavy and light chains[20] indicate that the high leucine content is restricted to the heavy chain and increases to nearly 20% of the total composition.

Carbohydrate Composition. Carbohydrate analyses demonstrated the presence of significant amounts of glucosamine, mannose, galactose, fu-

---

[18] M. W. C. Hatton and E. Regoeczi, *Biochim. Biophys. Acta* **379**, 504 (1975).
[19] G. Oshima, J. Kato, and E. G. Erdös, *Biochim. Biophys. Acta* **365**, 344 (1974).
[20] T. H. Plummer and M. T. Kimmel, *Fed. Proc., Fed. Am. Soc. Exp. Biol.* **40**, 1603 (1981).
[21] N. Sharon and H. Lis, *Chem. Eng. News* **59**, 20 (1981).

cose, and sialic acid.[15] This content is typical for plasma proteins[21,22] and is consistent with asparaginyl-linked complex carbohydrate chains. The carbohydrate is primarily attached to the larger subunit of carboxypeptidase N, representing 28% of its weight.

Properties

Human serum carboxypeptidase N has two forms in isoelectric focusing with pI values of 3.8 and 4.3. This difference is attributed to differences in sialic acid content of two forms of carboxypeptidase N.[11] Adding trypsin to serum of human or rat or pig increases the activity of carboxypeptidase N by 36–85%, probably by cleaving it to lower-molecular-weight active subunits.[23] The enzyme is inhibited by metal sequestering agents such as EDTA or o-phenanthroline, peptide fragments of bradykinin, ϵ-NH$_2$-n-caproic acid and δ-NH$_2$-n-valeric acid.[24] Of the newer types of inhibitors,[17] GEMSA was already mentioned; it has a K_i value of about 10^{-6} M.

The K_m of bradykinin with carboxypeptidase N was $4 \times 10^{-7} M$.[25] The K_m of Bz-Gly-Lys and Bz-Gly-argininic acid was 1.4×10^{-3} and 1×10^{-4} M. Of other substrates, FA-Ala-Lys and -Arg had a K_m of $3 \times 10^{-4} M$.[13] Carboxypeptidase N originates from liver,[8] thus its activity in plasma is depressed by cirrhosis of the liver. The level of enzyme activity in blood is increased by neoplasia such as Hodgkin's disease.[26] Two cases of genetically determined low levels of carboxypeptidase N have also been reported.[27] In addition, the plasma enzyme level increases in pregnancy,[24] while Dengue fever and dextran shock[10] lower it in blood.

Acknowledgments

Some of the experiments described here were supported in part by NIH USPHS Grants HL 16320, HL 14187, and by a contract from the Department of the Navy (N00014-75-C-0807).

[22] J. R. Clamp, in "The Plasma Proteins" (F. Putnam, ed.), p. 163. Academic Press, New York, 1975.
[23] L. Jeanneret, M. Roth, and J.-P. Bargetzi, Hoppe-Seyler's Z. Physiol. Chem. 357, 867 (1976).
[24] E. G. Erdös and H. Y. T. Yang, Handb. Exp. Pharmacol. [N.S.] 25, 289 (1970).
[25] R. Zacest, S. Oparil, and R. C. Talamo, Aust. J. Exp. Biol. Med. Sci. 52, 601 (1974).
[26] E. G. Erdös, I. M. Wohler, M. I. Levine, and M. P. Westerman, Clin. Chim. Acta 11, 39 (1965).
[27] K. P. Mathews, P. M. Pan, N. J. Gardner, and T. E. Hugli, Ann. Intern. Med. 93, 443 (1980).

[37] Human Peptidyl Dipeptidase (Converting Enzyme, Kininase II)

By TESS A. STEWART, JOHN A. WEARE, and ERVIN G. ERDÖS

Angiotensin I-converting enzyme was described first as the enzyme in plasma that cleaves His-Leu from angiotensin I and thereby converts it to angiotensin II.[1] Kininase II, found in the kidney[2] and human blood,[3] inactivates bradykinin by releasing Phe-Arg. The identity of the two enzymes, one that liberates a hypertensive peptide, and the other that inactivates a hypotensive peptide, was shown later.[4] It was also shown that this peptidyl dipeptidase (dipeptidyl carboxypeptidase: EC 3.4.15.1) cleaves C-terminal dipeptides from a variety of substrates, including the opioid enkephalin peptides.[5,6] Converting enzyme (CE), as it is commonly called, has been widely studied because its inhibitors showed promise in clinical trials in treating hypertensive patients.[5,6]

The enzyme cleaves substrates of the general structure R_1—R_2—R_3OH between R_1 and R_2. R_1 is a protected amino acid or a peptide. R_2 can be any amino acid but Pro, and R_3 has to have a free C-terminus and cannot be Glu or Asp.[6]

Assay

No single assay method can be used with every source of enzyme (e.g., tissue extracts with low enzymic activity, highly purified enzyme preparations, or human blood with very low activity) equally well. Of the newer methods, one that employs a substrate in the spectrophotometer[7] at 334 nm and one that utilizes substrates that are fluorogenic,[8] show promise. The hydrolysis of naturally occurring substrates can be followed by detecting the products separated by electrophoresis, chromatography, bioassay,[6] or by high-pressure liquid chromatography.[9] A frequently used

[1] L. T. Skeggs, W. H. Marsh, J. R. Kahn, and N. P. Shumway, *J. Exp. Med.* **99**, 275 (1954).
[2] E. G. Erdös and H. Y. T. Yang, *Life Sci.* **6**, 569 (1967).
[3] H. Y. T. Yang and E. G. Erdös, *Nature (London)* **215**, 1402 (1967).
[4] H. Y. T. Yang, E. G. Erdös, and Y. Levin, *Biochim. Biophys. Acta* **214**, 374 (1970).
[5] E. G. Erdös, *Fed. Proc., Fed. Am. Soc. Exp. Biol.* **38**, 2774 (1979).
[6] E. G. Erdös, *Handb. Exp. Pharmacol.* [N.S.] **25**, Suppl., 428 (1979).
[7] P. Bünning, B. Holmquist, and J. F. Riordan, *Biochem. Biophys. Res. Commun.* **83**, 1442 (1978).
[8] A. Carmel and A. Yaron, *Eur. J. Biochem.* **87**, 265 (1978).
[9] S. Wilk and M. Orlowski, *Biochem. Biophys. Res. Commun.* **90**, 1 (1979).

assay is the method of Cushman and Cheung,[10] which involves the spectrophotometric determination of the hippuric acid after hydrolysis of Bz-Gly-His-Leu and extraction of the hippuric acid product with ethyl acetate. Another assay is based on coupling His-Leu to o-phthaldialdehyde after the release of the dipeptide and measuring an increase in fluorescence.[11]

We assayed the peptidase activity of the enzyme with a recording UV spectrophotometer by following the hydrolysis of Bz-Gly-Gly-Gly, Bz-Gly-His-Leu, or Bz-Gly-Phe-Arg. The latter two substrates represent the C-terminal dipeptide of angiotensin I and bradykinin.[6,12,13] The enzyme in tissue homogenates or biological fluids with low specific activities was assayed by extracting ^{3}H-Bz-Gly from the reaction mixture containing ^{3}H-Bz-Gly-Gly-Gly substrate.[14] High ionic strength (e.g., 0.6 M Na$_2$SO$_4$) in the incubation mixture increases the rate of hydrolysis 3- to 6-fold. In order to correct for interference by contaminating peptidases, it is advisable to use a low concentration of a specific inhibitor such as 10^{-7} M captopril (D-3-methyl-2-thiopropanoylproline).

Methods of Assay

1. Direct Spectrophotometric Assay of Bz-Gly-Dipeptide[13] *Substrates*

Reagents

Substrate 1: Bz-Gly-Gly-Gly, 0.01 M (146.7 mg/50 ml) in water, pH adjusted to 7 with 1 N NaOH (stable for 6 months at $-20°$C)

Substrate 2: Bz-Gly-His-Leu, 0.01 M (214.8 mg/50 ml) in water, pH adjusted to 7 with 1 N NaOH (stable for at least 1 month at $-20°$C; may require heating to redissolve)

Substrate 3: Bz-Gly-Phe-Arg, 0.01 M (241.3 mg/50 ml) in water (stable for at least 1 month at $-20°$C)

Buffer: HEPES–NaOH, 0.20 M, pH 8.0, containing 0.20 M NaCl (47.66 g of HEPES, 11.69 g of NaCl in 900 ml of water adjusted to pH 8.0 with NaOH and diluted to 1 liter); stable at 4°C

Procedure. Assays are performed with a recording UV spectrophotometer equipped with a heated cuvette holder set at 37°C. The 1-ml reaction mixture contains: 100 μl of substrate (final concentration = 10^{-3} M), 500 μl of buffer, and 400 μl of enzyme plus water. The reference

[10] D. W. Cushman and H. S. Cheung, *Biochem. Pharmacol.* **20**, 1637 (1971).
[11] J. Friedland and E. Silverstein, *Am. J. Clin. Pathol.* **68**, 225 (1977).
[12] T. A. Stewart, J. A. Weare, and E. G. Erdös, *Peptides* **2**, 145 (1981).
[13] H. Y. T. Yang, E. G. Erdös, and Y. Levin, *J. Pharmacol. Exp. Ther.* **177**, 291 (1971).
[14] J. W. Ryan, A. Chung, C. Ammons, and M. L. Carlton, *Biochem. J.* **167**, 501 (1977).

cuvette contains 100 μl of substrate and 900 μl of water. The reaction is started by adding substrate to enzyme samples and buffer previously equilibrated at 37°C. The recorder is set for full-scale deflection of 0.1 absorbance unit, and the zero suppression is set to bring the absorbance readings on scale. Hydrolysis is followed by an increase in absorbance at 254 nm. Using a 1-cm path-length cell limits the substrate concentration (and protein concentration), which can be used while keeping the slit width below 2 mm. Substrate concentrations may be raised to $5 \times 10^{-3} M$ by using a 0.1-cm path-length cell.

2. *Radioassay with* ^3H-Bz-Gly-Gly-Gly [14]

Reagents

^{3}H-Bz-Gly-Gly-Gly stock: Tritiated substrate (200 mCi/mmol randomly labeled) from New England Nuclear was purified by ion-exchange chromatography on Cellex-D to homogeneity as verified by high-voltage paper electrophoresis. Labeled substrate can also be purchased from Ventrex Co., 217 Read St., Portland, Maine. Diluted in aliquots and stored at $-80°C$ at a concentration of 40–60 μCi/ml; stable for at least 1 year.

^{3}H-Bz-Gly-Gly-Gly substrate: 0.015 M in water (220 mg unlabeled Bz-Gly-Gly-Gly in 40 ml of water, pH adjusted to 7 with 1 N NaOH, add 1 ml of tritiated stock solution, dilute to 50 ml). About 2 μCi/ml (50 μl $\simeq 10^5$ cpm). Stable for at least 6 months at $-20°C$.

Buffer 1: HEPES–NaOH, 0.3 M, pH 8.0, containing 0.3 M NaCl (71.49 g of HEPES, 17.53 g of NaCl in 900 ml of water, adjusted to pH 8.0 with NaOH, and diluted to 1 liter). Stable at 4°C.

Buffer 2: HEPES–NaOH, 0.30 M, pH 8.0, containing 0.3 M NaCl and 1.8 M Na$_2$SO$_4$ (35.75 g of HEPES, 8.77 g of NaCl, 127.84 g of Na$_2$SO$_4$ in 450 ml of water, dissolved by heating, adjusted to pH 8.0 with NaOH, diluted to 500 ml). Stable at room temperature for 2 weeks.

Procedure. Incubations are carried out at 37°C in disposable 13 × 100-mm glass tubes. Each 150-μl assay mixture contains: 50 μl of buffer 1 or 2, 50 μl of enzyme sample plus water, and 50 μl of substrate (added last at 0°C). Buffer 2 accelerates the reaction about 3- to 6-fold with Bz-Gly-Gly-Gly. After an appropriate incubation time ($\simeq 10\%$ substrate utilization), the reaction is stopped with 1 ml of 0.1 M HCl containing 1 M NaCl. One milliliter of ethyl acetate, which extracts the ^{3}H-Bz-Gly product, is added, the tubes mixed for 5 sec, and 200 μl of the organic phase (upper layer) placed in 10 ml of ACS scintillation fluid (Amersham). The samples are counted in a liquid scintillation beta counter. Specific CE activity can be confirmed by adding 15 μl of 10^{-6} M captopril (SQ 14225) to control tubes.

The activity present is calculated as follows: 57% (8 of 14 hydrogens) of the random ^{3}H label is present on the Bz-Gly product; 91% of the product is distributed in the organic phase. Total counts (10 μl of substrate = 150 nmol, about 2 × 10^4 cpm) are determined. The activity in nanomoles of Bz-Gly-Gly-Gly hydrolyzed per minute per milliliter is

$$\frac{750 \text{ nmol [cpm (sample)} - \text{cpm (sample} + \text{SQ 14425)]}}{(0.57)(0.91)(\text{total cpm})(\text{sample volume [ml])(incubation time [min])}}$$

$$\times \left(1 - \frac{\text{blank cpm}}{\text{substrate cpm}}\right)^{-1}$$

Activity Units

One unit (U) of CE activity is the amount of enzyme required to catalyze the hydrolysis of 1 μmol of substrate per minute at 37°C. In the spectrophotometric assay, units can be determined by using the following $\Delta \epsilon M^{-1}$ cm^{-1} for the Bz-Gly-dipeptide substrates: 360 (Bz-Gly-Gly-Gly), 520 (Bz-Gly-His-Leu), 570 (Bz-Gly-Phe-Arg).

Purification Procedure

CE has been purified from a variety of animal sources and the animal enzymes have been very well characterized.[6] The human enzyme, however, has proven to be much more elusive.[12] Purification of soluble human CE has recently been reported,[15] but the concentration of the enzyme in serum is low. The detailed procedure given here has been used for the large-scale purification of membrane-bound human lung and kidney CE.[12] We obtained 13 mg of homogeneous enzyme from 15.8 kg of lung tissue, and 29 mg of enzyme from 5.8 kg of kidney. The overall results of each step in the purification of the lung enzyme are shown in the table. About half the enzyme content of the cadaver lung goes into solution after homogenization; the kidney enzyme is less soluble. The yield after the first three steps of purification reflects the recovery of the bound enzyme. The decrease in activity after tryptic extraction (step 4) is probably due to the release of peptides that inhibit the enzyme. The inhibitory activity was removed during the subsequent chromatography steps.

Tissue Extraction

Human lungs were obtained from cadavers within 8 hr after death and stored at −20°C until use. No loss of enzymatic activity was detected for

[15] J. J. Lanzillo, R. Polsky-Cynkin, and B. L. Fanburg, *Anal. Biochem.* **103**, 400 (1980).

PURIFICATION OF HUMAN LUNG CONVERTING ENZYME

Procedure	Protein[a] (mg)	Units[b]	Specific activity (U/mg)	Purification factor	Yie (%
1. Homogenate	1,280,000	8000	0.0063	1	100
2. Supernatant, 5000 g	645,000	5870	0.0091	1.5	73
3. Acid precipitate (pH 5.2)	128,000	2250	0.0176	2.8	28
4. Tryptic extract	70,400	530	0.0075	1.2	(
5. DE-52 Batch	11,400	840	0.074	12	1(
6. DE-52, pH 8.0	3350	1490	0.44	70	19
7. DE-52, pH 7.0	1000	1090	1.09	170	14
8. Sephacryl S-300					
Fraction A	140 } 200	620 } 1070	4.4 } 5.3	830	13
Fraction B	60	450	7.5		
9. Hydroxyapatite					
Fraction A	35 } 39.1	470 } 610	13.4 } 15.6	2500	7
Fraction B	4.1	140	34		
10. Phenyl-Sepharose					
Fraction A	9.0	280	31	—	—
Overall[c]	13.1	420	32	5100	9

[a] Protein determined by the method of Lowry.
[b] One unit is 1 μmol of Bz-Gly-Gly-Gly hydrolyzed per minute at 37°C under standard assa conditions.
[c] Sum of fraction A from phenyl-Sepharose and fraction B from hydroxyapatite columns.

up to 6 months of storage. Thawed lungs were cut into small (5–10 g) pieces and washed with 0.15 M NaCl. The tissue was then homogenized in 0.02 M KH_2PO_4/K_2HPO_4 (pH 7.8) containing 0.25 M sucrose for 2 × 2 min using a Waring commercial blender set at maximum speed. The homogenate was centrifuged in a Sorvall RC-5B refrigerated centrifuge at 5000 g for 20 min. A buffer-to-tissue ratio of 3 : 1 (v/w) allowed the processing of approximately 0.5 kg of tissue (total volume, 2 liters) per run using a Sorvall Type GS-3 rotor. For smaller-scale purifications, a higher buffer-to-tissue ratio will improve the yield at this step. The precipitate, which contained about 30% of the total activity, was discarded.

The supernatant was adjusted to pH 5.2 with 20% acetic acid and centrifuged at 15,000 g for 30 min. The supernatant was discarded. The pellet was resuspended in a minimal volume (approximately 500 ml/kg tissue) of 0.01 M K_2HPO_4 and the pH adjusted to 8.0 with 1 N NaOH. The suspensions were stored at −20°C until they were pooled for solubilization.

Solubilization

Suspensions were thawed, pooled, and the pH adjusted to 8.5 with 1 N NaOH. After preincubation at 37°C for 30 min, trypsin (Trypsin 1 : 250, Difco Laboratories) was added in a ratio of 1 part trypsin to 100 parts (Lowry) protein. This mixture was incubated at 37°C for 2 hr, during which the pH dropped to approximately 7.8. The digested suspension was acidified to pH 5.2 with 20% acetic acid and centrifuged at 15,000 g for 30 min. The resulting pellet, which had less than 10% of the total activity, was discarded. The pH of the supernatant was readjusted to 8.0 with 1 N NaOH. The acid-precipitation and trypsin-extraction steps were done according to Nishimura et al.[16]

DEAE — Cellulose Batch Step

The trypsin extract was concentrated 4-fold in an Amicon Hollow-Fiber apparatus using an H1P-10 cartridge, and dialyzed in the same apparatus against deionized water until the conductivity dropped below 0.5 mΩ^{-1}. Precycled Whatman DE-52 cellulose was added (5 g/g protein), the pH adjusted to 7.0, and the mixture stirred for 1 hr at 4°C. The cellulose was drained on a coarse 3-liter glass funnel with a water aspirator and washed with 10 volumes of deionized water. The drained cellulose was resuspended in deionized water and the conductivity raised to 7 mΩ^{-1} with solid NaCl (about 0.2 M) before elution on the glass funnel.

DEAE — Cellulose Column, pH 8.0

The enzyme eluted in the DEAE–cellulose batch step was concentrated and dialyzed as before. Precycled Whatman DE-52 cellulose was added (5 g/g protein) and the mixture stirred for 1 hr at 4°C. After draining the cellulose on a 1-liter glass funnel and washing with 10 volumes of 10^{-3} M KH_2PO_4/K_2HPO_4 (pH 8.0), the suspension was packed into a 5-cm column (5 × 11 cm). The column was developed at a rate of 100 ml/hr with a 0–0.2 M linear gradient of NaCl in the above buffer. The gradient chambers contained 500 ml each. Six-milliliter fractions were collected. Enzymatic activity was eluted between 1.4 and 3.8 mΩ^{-1}.

DEAE — Cellulose Column, pH 7.0

Active fractions from the preceding step were pooled, dialyzed, and bound to Whatman DE-52 cellulose as before. The cellulose was washed

[16] K. Nishimura, N. Yoshida, K. Hiwada, E. Ueda, and T. Kokubu, Biochim. Biophys. Acta **483**, 398 (1977).

with 5 mM potassium phosphate (pH 7.0), drained on a glass funnel, and placed in a column 2.5 × 13 cm. The column was eluted at 60 ml/hr with 2 liters of a 0–0.2 M linear gradient of NaCl in 1 mM potassium phosphate buffer, pH 7.0. Six-milliliter fractions were collected. Enzymatic activity was eluted between 1.8 and 4.9 mΩ^{-1}.

Sephacryl S-300 Column

Active fractions from the DEAE–cellulose column step were pooled and concentrated with the Amicon Hollow-Fiber apparatus (H1P-10 cartridge) and with an Amicon ultrafiltration cell with a YM-10 membrane to a final volume of 20 ml. This concentrate was pumped at 40 ml/hr onto a Sephacryl S-300 column (5 × 90 cm) equilibrated with 0.1 M Tris-HCl (pH 7.8) containing 1 M NaCl and 10% (v/v) glycerol. Six-milliliter fractions were collected at a flow rate of 40 ml/hr. The void volume of the column was 670 ml as determined with Blue Dextran 2000 (Pharmacia). A portion of the enzymatic activity, fraction A, was eluted between 860 and 1060 ml. The remaining activity, fraction B, eluted as a broad band largely beyond the total volume of the column from 1600 to 3100 ml. (Changing the column buffer to 0.05 M Tris-HCl, pH 7.8, increased the proportion of enzymatic activity in fraction A.)

Hydroxyapatite Column

Fractions A and B from the Sephacryl S-300 column were dialyzed and concentrated as before to 10 ml each. The concentrates were layered on a short column (5 × 4 cm) of preequilibrated Bio-Gel HTP (Bio-Rad; 1 g/mg protein), washed with 100 ml of the phosphate buffer ($10^{-3} M$), and eluted with a linear gradient of the same buffer ranging from 0.001–0.01 M. The flow rate was 100 ml/hr; 6-ml fractions were collected. Enzymatic activity was eluted between 0.1 and 0.5 mΩ^{-1}. Fraction B was pooled, concentrated to about 1 mg protein/ml, and frozen at −20°C. This step may also be performed without concentrating the enzyme by first dialyzing against 0.01 M phosphate buffer (to a conductivity of 0.7 mΩ^{-1}), and then collecting the unbound portion after passage through the hydroxyapatite column. In any case, it is important that the enzyme be quickly eluted, since it is unstable on hydroxyapatite. The yield at this step was about 80%.

Phenyl-Sepharose Column

Fraction A from hydroxyapatite chromatography was diluted 10-fold with 3 M NaCl in 0.01 M potassium phosphate (pH 7.0), and 1 g of phenyl-Sepharose was added per milligram of protein. The suspension was stirred 30 min in the cold, and poured into a column (2.5 × 10 cm). The column was developed with 1 liter of a linear gradient of 3.0–0.5 M NaCl in 0.01 M potassium phosphate (pH 7.0), at a flow rate of 40 ml/hr. Six-milliliter fractions were collected. Enzymatically active fractions were pooled, concentrated to about 1 mg protein/ml, and frozen at −20°C. Fractions A and B gave a single protein band when 10 μg was subjected to SDS–urea polyacrylamide gel electrophoresis. The specific activity of the enzyme was 81 U/mg when assayed in a Cary 118 recording spectrophotometer with Bz-Gly-His-Leu under the conditions described by Cushman and Cheung.[10] A specific activity of 40 U/mg was reported for homogeneous human serum CE.[15] Human kidney CE, purified by essentially the same procedure reported here, had a specific activity of 65 U/mg.[12]

Rabbits immunized with either purified enzyme produced antisera that formed a single precipitin line in immunodiffusion with the purified human enzyme from either lung or kidney.

Stability

When stored at −20°C at a concentration of 1 mg/ml in 0.01 M KH_2PO_4/K_2HPO_4 (pH 7.0), containing 0.5 M NaCl, this preparation retained full activity for several months. Diluted aliquots of enzyme were less stable.

Enzyme Quantitation

The concentration of purified enzyme can be calculated from absorbance at 280 nm; $\epsilon_{280\,nm}^{1\,mg/ml} = 1.6$. Estimates by the method of Lowry (see the table), using crystalline bovine serum albumin as standard, give low protein values.

Other Purification Methods

The various techniques used for purifying CE from animal organs have been summarized elsewhere.[6] Nishimura et al.[16] reported the purification of human lung CE. We adopted two steps from their technique: trypsin solubilization and acid precipitation of membranes. They obtained 0.7 mg of enzyme with a specific activity of 9.5 U/mg with Bz-Gly-His-Leu. Their preparation migrated as a single band on 7.5% polyacrylamide gels at pH 8.6.

Lanzillo and Fanburg purified human serum CE using preparative electrophoresis as the last step.[17] Recently, Lanzillo et al.[15] have described a 1500-fold purification of the serum enzyme on an immunoadsorbent affinity gel prepared by coupling antibody to baboon lung CE to CNBr-activated Sepharose-4B. About 20% of the bound activity could be recovered from the gel by elution with 2 M MgCl$_2$, pH 5.8. Subsequent hydroxylapatite and Sephadex G-200 chromatography yielded 2.6 mg of homogeneous enzyme with a specific activity of 40 U/mg with Bz-Gly-His-Leu as substrate from 48 liters of plasma.

Harris et al.[18] have purified human serum CE by affinity chromatography using an immobilized competitive inhibitor, D-cysteinyl-L-proline, as the ligand.

Properties of Human Converting Enzyme

Human lung and kidney CE consist of a single polypeptide chain with an apparent molecular weight (MW$_{app}$) of 155,000 in electrophoresis on 7.5% SDS–urea polyacrylamide gels in presence of 2-mercaptoethanol.[12] Human serum CE has MW$_{app}$ 155,000 by sucrose-density gradient ultracentrifugation, and 140,000 in polyacrylamide gels.[17,19] The sedimentation coefficients are 8.0 ± 0.1 S for human lung CE[12] and 8.17 S for human serum CE.[18,19]

The amino acid compositions of human lung and kidney CE are similar. The mole percent content of nonpolar residues[20] is 42% and is consistent with the hydrophobic behavior of the molecule. Human CE also has a high proportion of acidic residues. One-dimensional peptide maps of lung and kidney CE suggest considerable homology between the human enzymes (unpublished).

Specific antisera raised in rabbits against human lung and human kidney CE cross-reacts fully with enzyme from either source indicating immunological identity.[12]

The isoelectric point of the major form of human lung and kidney CE is 5.2 (range 5.1–5.3).[12] The lung preparation also contained more acidic forms with isoelectric points from pH 4 to 5, which differed in their sialic acid content. Human serum CE has an isoelectric point of 4.7.[19]

Carbohydrate analyses of lung and kidney CE indicate similar amounts of N-acetylglucosamine, mannose, galactose, and fucose (Weare et al., in preparation). The lung enzyme has significantly more sialic acid

[17] J. J. Lanzillo and B. L. Fanburg, Biochemistry 16, 5491 (1977).
[18] R. B. Harris, J. T. Ohlsson, and I. B. Wilson, Arch. Biochem. Biophys. 206, 105 (1981).
[19] R. B. Harris, J. T. Ohlsson, and I. B. Wilson, Anal. Biochem. 111, 227 (1981).
[20] J. Heller, Biochemistry 7, 2906 (1968).

(about 17 residues per molecule as compared to about 3 for the kidney enzyme).

The variation of activity of CE with pH is buffer dependent, and the pH optimum shifts with salt concentration. For 0.1 M HEPES–NaOH, 0.1 M NaCl, the pH optimum is 6.5; with 1.0 M NaCl the pH optimum shifts to 7.6, with Bz-Gly-Gly-Gly as substrate.

The presence of chloride ion is an absolute requirement for activity of CE with most substrates, bradykinin being an exception.[6] In addition, high ionic strength increases the activity of the enzyme with Bz-Gly-Gly-Gly.[6]

Human CE hydrolyzes a number of natural substrates and synthetic peptides.[6,12] The turnover numbers for the model substrates Bz-Gly-Gly-Gly, Bz-Gly-His-Leu, and Bz-Gly-Phe-Arg are 17,000, 4600, and 23,000 min^{-1} for human lung CE. The K_m's for these substrates are 6.2, 1.3, and 0.6 × 10^{-3} M, respectively. Bradykinin, angiotensin I, Met^5-enkephalin, and Leu^5-enkephalin are hydrolyzed at maximum rates of 500, 500, 3500, and 700 min^{-1}. Bradykinin is a better substrate than angiotensin I, with a specificity constant (k_{cat}/K_m) of 500, compared to 20 for angiotensin I. This behavior is reflected by that of the model substrate Bz-Gly-Phe-Arg, a derivative of bradykinin that has a specificity constant of 41, whereas Bz-Gly-His-Leu, representing the C-terminal end of angiotensin I, has a value of 3.5.

Inhibition

Human CE is inhibited by a variety of agents,[6] including dipeptide products of hydrolysis, endogenous peptides such as bradykinin, sequestering agents (especially EDTA), and various snake venom peptides, some of which have been tested as antihypertensive agents. Interest in the role of converting enzyme in regulating of blood pressure, especially in pathological conditions, led to a search for new potent inhibitors that are orally active,[5,6] such as captopril.

Distribution

Human CE is widely distributed in the body. It occurs in membrane-bound form in vascular endothelial cells, in epithelial cells of renal tubules and intestinal tract, and in soluble form in various body fluids.[5,6] We have determined the activity present in crude homogenates of various human tissues using the radioassay[14] previously described. The presence of endogenous inhibitor(s) made it necessary to assay low activity at high dilution in order to obtain a linear response. We found that in humans, kidney is the best source of CE[12]; the relative specific activities of the enzyme in various crude human extracts are as follows: serum, 1; lung, 30; kidney, 150; and small intestinal aspirate, 100. The enzyme is also present in high concentrations in intestinal brush border, in parts of the brain, and in seminal plasma.[5,6] Higher than normal values were described in plasma and lymph nodes of sarcoid patients and in plasma of subjects suffering from leprosy.[6,21,22]

[21] J. Lieberman, *Am. J. Med.* **59**, 365 (1975).
[22] J. Lieberman and T. H. Rea, *Ann. Intern. Med.* **87**, 422 (1977).

[38] Human Prolylcarboxypeptidase

By C. E. ODYA and E. G. ERDÖS

Prolylcarboxypeptidase (peptidylprolylamino acid hydrolase, EC 3.4.16.2; PCP) catalyzes the hydrolysis of C-terminal amino acid residues from peptides with the general structure R_1-Pro-R_2-OH, where R_1 is either a blocking group, another protected amino acid, or a peptide, and R_2 is an aromatic or aliphatic amino acid with a free carboxyl group.[1-3] PCP does not cleave substrates where proline is replaced by another amino acid. The enzyme was first named angiotensinase C because it inactivates angiotensin II. The relative rates of hydrolysis of several substrates are shown in Table I.

Assay Methods

Method A

Cbz-Pro-Phe represents the C-terminal end of the naturally occurring substrates for PCP, angiotensins II and III, and it has been used to assay PCP activity during the purification of the enzyme from human kidney. The phenylalanine released from the synthetic substrate is measured with an automatic amino acid analyzer.

[1] H. Y. T. Yang, E. G. Erdös, and T. S. Chiang, *Nature (London)* **218**, 1224 (1968).
[2] H. Y. T. Yang, E. G. Erdös, T. S. Chiang, T. A. Jenssen, and J. G. Rodgers, *Biochem. Pharmacol.* **19**, 1201 (1970).
[3] C. E. Odya, D. V. Marinkovic, K. J. Hammon, T. A. Stewart, and E. G. Erdös, *J. Biol. Chem.* **253**, 5927 (1978).

TABLE I
Substrate Specificity of Purified Human Renal
Prolylcarboxypeptidase (pH 5)[a]

Substrate	Rate of hydrolysis (%)
Cbz-Pro-Phe	100
Cbz-Pro-Ala	420
Cbz-Pro-Val	360
Cbz-Pro-Leu	160
Cbz-Pro-Ser	95
Cbz-Pro-Tyr	71
Cbz-Pro-Gly	6
Cbz-Pro-Pro; Cbz-Pro-OHPro; Pro-Phe; Pro-Tyr; Pro-Val; Cbz-Glu-Tyr; Cbz-Glu-Phe; Cbz-Phe-Pro; Bz-Gly-Phe; Bz-Gly-Arg	0
Angiotensin II	430
Angiotensin III	1300

[a] From Odya et al.,[3] with permission of the publisher.

Reagents

Sodium acetate buffer: $0.2\,M$, pH 5.0

Substrate: 10 mM Cbz-Pro-Phe, pH 5.0

Assay mixture: 0.1 ml of buffered substrate plus 0.005 to 0.2 ml of enzyme and sufficient sodium acetate buffer to give a final volume of 0.3 ml

Sulfosalicylic acid, 3%

Procedure. Begin the reaction by adding 0.1 ml of the substrate to the test tubes (Brinkmann, micro, polypropylene conical tubes, 1.5-ml capacity) that contain enzyme and buffer or buffer alone, which serves as a blank. At the end of the incubation period, 0.5–3 hr at 37°C, add 0.7 ml of 3% sulfosalicylic acid to stop the reaction. Remove precipitated protein by centrifugation at room temperature at about 10,000 g for 1 min. Filter the supernatants through 0.45-μm disposable filter units. Rinse each tube with 1 ml of 3% sulfosalicylic acid and filter as before. Combine filtrates to give a total of 2.0 ml. The samples are now ready for processing with an amino acid analyzer. We used a Beckman 121 automatic amino acid analyzer equipped with a sample injector.

Method B

Since Method A requires the use of an amino acid analyzer, it introduces substantial delay in assaying fractions for PCP activity. To overcome this limitation, a more rapid and simpler procedure has been developed.[4,5] This procedure employs ^{14}C (or ^3H) -labeled Cbz-Pro-Ala as substrate. The substrate represents the C-terminal end of the angiotensin II antagonist saralasin and is cleaved at a rate 4.2 times greater than Cbz-Pro-Phe.[3] The amount of [^{14}C]Ala or [^3H]Ala released by PCP is measured in a scintillation counter.

Reagents

Sodium acetate buffer: 0.2 M, pH 5.0, containing 0.3% neomycin sulfate

Substrate: 0.12 M Cbz-Pro-Ala, containing Cbz-Pro-[^{14}C]Ala to give 260,000 cpm/ml. In the assay using Cbz-Pro-[^3H]Ala, the initial substrate concentration was decreased to 30 mM (containing radioactive substrate to give 800,000 cpm/ml) and the reaction volumes were cut in half [5]

Assay mixture: 0.1 ml of substrate plus 0.02 to 0.1 ml of enzyme and sufficient sodium acetate buffer to give a final volume of 0.3 ml HCl, 0.2 N

Extraction solvent: chloroform:n-butanol, 4:1

Procedure. Begin the reaction by adding 0.1 ml of the substrate to test tubes containing enzyme and buffer or buffer alone, which serves as a blank. At the end of the incubation period, 0.5–18 hr at 37°C, add 0.3 ml of 0.2 N HCl to stop the reaction. Add 4.0 ml of a chloroform:n-butanol (4:1) solution to each tube and mix for 10 sec, then centrifuge at 10,000 g for 10 min to separate the phases. Pipette 0.3 ml of the top, aqueous phase, containing radioactive alanine released, into scintillation vials, add 10 ml of scintillation fluid, mix, and count in a β-counter.

Units

One unit of enzyme activity is defined as the amount that hydrolyzes 1 μmol of substrate per hour. The specific activity is expressed in units per milligram of protein.

Purification of Human Kidney Prolylcarboxypeptidase

PCP was purified to homogeneity from human kidney by the following procedure. The results obtained at each step are summarized in Table II. Enzyme activity was measured using Method A.

Preliminary Purification Steps. Normal human kidneys were obtained at autopsy within 24 hr. They were washed with ice-cold 0.9% NaCl; fat

[4] K. Kumamoto, T. A. Stewart, A. R. Johnson, and E. G. Erdös, *J. Clin. Invest.* **67**, 210 (1981).

[5] R. A. Skidgel, E. Wickstrom, K. Kumamoto, and E. G. Erdös, *Anal. Biochem.* **118** (in press, 1981).

TABLE II

PURIFICATION OF PROLYLCARBOXYPEPTIDASE OF HUMAN KIDNEY[a]

Procedure	Total protein (mg)	Volume (ml)	Total units	Protein (mg/ml)	Units/ mg of protein	Yield (%)	Purifi- cation (-fold)
1. Supernatant, homogenized kidney heated 60°C, 30 min	61,000	640	1600	95	0.026	100	1.0
2. (NH₄)₂SO₄	53,000	610	2300	87	0.043	140	1.7
3. DEAE–cellulose	38,000	510	2000	75	0.053	120	2.0
4. CM-cellulose (pH 6.0)	6600	160	1500	41	0.23	94	8.8
5. SP-Sephadex	2100	150	760	14	0.36	48	14
6. CM-cellulose (pH 6.6 to 8.5)	500	51	540	9.9	1.1	34	42
7. Hydroxyapatite	45	5	260	9.0	5.8	16	220
8. Sephadex G-100	9.5	43	290	0.22	31	18	1200

[a] From Odya et al.,[3] with permission of the publisher.

and renal pelvises were removed and stored at $-80°C$. Approximately 20 kidneys were diced with scissors, and separated into four 500-g lots. To each lot was added 1 liter of $0.02\ M$ sodium phosphate buffer (pH 6.8) that contained $0.1\ M$ NaCl and 1 mM EDTA. The resulting mixture was homogenized four times at 4°C for 1 min at top speed in a commercial Waring blender. The blender should be chilled in an ice-water bath between homogenizations. Unless stated otherwise, all remaining steps are done at 5–10°C. The homogenates are combined and centrifuged at $10,000g$ for 30 min. The supernatant is decanted and the pH adjusted to 4.5 with the addition of solid citric acid. Precipitated proteins are removed by centrifugation as before. The supernatant is then concentrated by passage through an Amicon DC-2 ultrafiltration apparatus equipped with a hollow-fiber cartridge (50,000 MW cutoff). The pH of the retentate is adjusted at room temperature to 7.0 with solid Tris base and then heated at 60°C for 30 min, to inactivate cathepsin A in the extract. Precipitated proteins are removed by centrifugation as before. Solid ammonium sulfate is added to the supernatant to 70% saturation at room temperature, then stored at 4°C overnight. The mixture is centrifuged as before and the supernatant is discarded. The precipitated proteins are dissolved in $0.05\ M$ sodium phosphate buffer (pH 6.8), diluted to 300 ml, dialyzed at 4°C against two changes of 4 liters of the same buffer, and then diluted to 610 ml.

DEAE–Cellulose Chromatography. The dialysate (610 ml) is applied to a column (5 × 45 cm) of DEAE–cellulose. The column is eluted with $0.05\ M$ sodium phosphate buffer (pH 6.8) at a flow rate of 300 ml/hr. At this pH, PCP is not adsorbed and is eluted in 20-ml fractions, fractions 31–80.

CM-Cellulose Chromatography. The eluted PCP fractions are combined and the volume reduced to 300 ml by ultrafiltration with the Amicon DC-2 apparatus. The retentate is dialyzed against $0.01\ M$ sodium phosphate buffer (pH 6.0), diluted to 510 ml, and then applied to a column (5 × 42 cm) of CM-cellulose equilibrated and eluted (at 200 ml/hr) with the same buffer. PCP is eluted in 15-ml fractions with a linear NaCl gradient (0–2%) in the same buffer. The total volume of the gradient is 6 liters. The peak of enzyme activity appears with $0.075\ M$ NaCl. Fractions between 0.0125 and $0.09\ M$ NaCl are pooled.

SP-Sephadex Chromatography. The pooled fractions are concentrated with the Amicon DC-2 apparatus to 150 ml and dialyzed against $0.01\ M$ sodium acetate buffer, pH 4.8. The dialysate (160 ml) is diluted to 250 ml and applied to a column (5 × 46 cm) of SP-Sephadex C-25. The adsorbed activity is eluted in 18-ml fractions with a linear KCl gradient (0–2%) in the sodium acetate buffer. The total volume of the gradient is 6 liters. The flow rate should be 200 ml/hr. The peak of enzyme activity appears with $0.16\ M$ KCl. Fractions eluted between 0.1 and $0.2\ M$ KCl are pooled.

CM-Cellulose Chromatography. The combined fractions are concentrated with the Amicon DC-2 apparatus to 125 ml and dialyzed against 0.01 M potassium phosphate buffer (pH 6.6). The dialysate (150 ml) is applied to a column (2.2 × 52 cm) of CM-cellulose previously equilibrated with the same buffer. PCP is adsorbed, and the column is developed (at 80 ml/hr) with a potassium phosphate gradient, 1 liter each of 0.01 M (pH 6.6) and 0.025 M (pH 8.5). The peak of enzyme activity is eluted at pH 7.1. Fractions (7 ml) that elute between pH 6.9 and 7.35 are pooled.

Hydroxyapatite Chromatography. The combined fractions are concentrated to 51 ml in an Amicon 202 stirred cell equipped with a PM-30 membrane, and dialyzed against 0.05 M potassium phosphate buffer (pH 6.5). The dialysate is applied to a column (2.2 × 12 cm) of hydroxyapatite equilibrated with the same buffer. The column is developed (at 25 ml/hr) with a linear potassium phosphate gradient (0.05–0.4 M) at pH 6.5. The total volume of the gradient is 1200 ml. The peak enzyme activity is eluted with 0.23 M potassium phosphate buffer. Fractions (5.0 ml) eluted with 0.19–0.28 M potassium phosphate buffer are pooled.

Sephadex G-100 Chromatography. The combined fractions are concentrated as before to 5.0 ml. The concentrate is applied to a column (2.7 × 160 cm) of Sephadex G-100 equilibrated and eluted (at 20 ml/hr) with 0.05 M sodium acetate buffer (pH 6.5) containing 0.1 M NaCl. After 218 ml of buffer have passed through the column, begin collecting 2.6-ml fractions. Pool the active fractions.

The reproducibility of this procedure has been tested on two occasions, when we performed all the purification steps without assaying for PCP activity until after the Sephadex G-100 step. The yield of enzyme was comparable to that obtained when we assayed for activity after each step.

Properties of Human Kidney Prolylcarboxypeptidase

The molecular weight of human kidney PCP determined by gel filtration on Sephadex G-100 is 115,000.[3] The preparation yielded a single band in polyacrylamide-gel electrophoresis. Under denaturing conditions the enzyme dissociates into subunits of MW 45,000 and 66,500. Purified PCP cleaves Cbz-Pro-Phe at a rate of 31 μmol/hr/mg protein. The enzyme is maximally active between pH 4.5 and 5.5 with Cbz-Pro-Phe and angiotensins II and III as substrates, but with physiologically active substrates Asp[1] and Asn[1]-angiotensin II and des-Asp[1]-angiotensin II or angiotensin III, at pH 7 the enzyme retained 20–63% of optimal activity. The K_m values for Cbz-Pro-Phe, angiotensin II and III are 1, 2, and 0.77 mM, respectively. PCP is stable to heating with only 20% of the activity lost after treatment at 60°C for 30 min. Treatment at 70°C for 15 min, however, results in 90% inactivation. PCP is completely inhibited by DFP and phenylmethylsulfonyl fluoride (1 mM) and pepstatin (0.3 mM). One millimolar p-chloromercuribenzoic acid or p-chloromercuriphenylsulfonic acid inhibits only 15%. At 1 mM concentrations, none of the following agents had an effect on PCP activity: iodoacetic acid, iodoacetamide, 2-mercaptoethanol, 2-mercaptoethylamine, SQ 20881 (an angiotensin I-converting enzyme inhibitor), aprotinin, EDTA, and o-phenanthroline.

Occurrence of PCP

The lysosomal fraction of the kidney is rich in PCP; however, the same fraction of the lung also has enzymic activity; 0.43 U/mg protein with Cbz-Pro-Ala substrates.[4] PCP is also present in urine and in synovial fluid. In human blood, in decreasing order, macrophages, lymphocytes, and polymorphonuclear neutrophils have PCP activity. Cultured human fibroblasts from various sources contained 2–4 times more activity than circulating blood cells,[4] while primary cultures of human umbilical vein endothelial cells were most active, more than twice than cultured human fibroblasts.[5]

Acknowledgments

Some of the experiments described here were supported in part by NIH USPHS Grants HL 16320, HL 14187, and by a contract from the Department of the Navy (N00014-75-C-0807).

[39] Human Urinary Kallikrein

By REINHARD GEIGER and HANS FRITZ

Glandular, organ, or tissue kallikrein are serine proteinases that liberate kinins from kininogens by limited proteolysis and have no or only little proteolytic activity on other proteins.[1–3] Their most characteristic features are (1) the release of kallidin (i.e., Lys-bradykinin) from the kininogens and (2) the potent blood pressure-lowering effect of minor (less than microgram) quantities *in vivo*. The latter is due to the extremely low inhibiting potential directed against glandular kallikreins in plasma. Of the plasma proteinase inhibitors, α_1-antitrypsin seems to be the only—though very slowly reacting—inhibitor of glandular kallikreins.[4] Furthermore, they are structurally strongly homologous to the pancreatic serine pro-

[1] E. G. Erdös and A. F. Wilde, *Handb. Exp. Pharmacol.* **25** (1970).
[2] E. G. Erdös, *Handb. Exp. Pharmacol.* **25**, Suppl. (1979).
[3] J. J. Pisano and K. F. Austen, *DHEW Publ.* (*NIH*) (*U.S.*) NIH 76-791 (1976).
[4] R. Geiger, U. Stuckstedte, B. Clausnitzer, and H. Fritz, *Hoppe-Seyler's Z. Physiol. Chem.* **362**, 317 (1981).

teinases, but they differ in that they are glycoproteins.[5,6] Present evidence suggests that the glandular kallikreins of one species are all very similar or even identical, at least as far as the protein is concerned. The very close structural homology of the glandular kallikreins to arginine-esterase subunits of growth factors is striking.[7]

Glandular kallikreins differ from plasma kallikreins in several respects. Therefore, both types of kallikreins will receive different EC numbers in the future (A. J. Barrett, personal communication, 1980). The most important features of the plasma kallikreins are (1) the release of bradykinin from kininogens; (2) the presence of potent and relatively rapidly reacting inhibitors in plasma in addition to the slowly reacting α_1-antitrypsin,[8] namely, the C1 inactivator and α_2-macroglobulin; and (3) the preference for factor XII (Hageman factor), in addition to high-molecular-weight (HMW) kininogen as natural substrate. In this respect, plasma kallikrein or prokallikrein is primarily acting as a clotting factor.

Glandular kallikreins are found not only in the pancreas, salivary glands, small and large intestine, kidney, and urine,[1,2] but also in plasma or serum.[9] Kallikrein-like antigens were found in humans, in addition, in amniotic fluid, sweat and bile, ascites, and seminal plasma.[10] Kallikrein occurs in pancreatic tissue and juice as prokallikrein, but in salivary tissue and juices only in the active form. In urine both latent and active forms are found. This chapter will focus on human urinary kallikrein.

Assay Methods

Assay methods differing considerably in specificity and sensitivity are available to quantitate human urinary kallikrein as well as other glandular kallikreins. For quantitation in urine or saliva, either photometric assays using synthetic substrates—preferably AcPheArgOEt or D ValLeuArgNHNp—or the radioimmunoassay (RIA) is recommended. Samples of pancreatic and duodenal juice have to be analyzed with the RIA or a biological assay after activation (pancreatic juice) and addition of suitable trypsin inhibitors (e.g., soybean Kunitz trypsin inhibitor of M_r ~

[5] H. Fritz, I. Eckert, and E. Werle, *Hoppe-Seyler's Z. Physiol. Chem.* **348**, 1120 (1967).

[6] F. Fiedler, C. Hirschauer, and E. Werle, *Hoppe-Seyler's Z. Physiol. Chem.* **356**, 1879 (1975).

[7] R. A. Bradshaw, G. A. Grant, K. A. Thomas, and A. Z. Eisen, *Protides Biol. Fluids* **28**, 119 (1980).

[8] H. Fritz, G. Wunderer, K. Kummer, N. Heimburger, and E. Werle, *Hoppe Seyler's Z. Physiol. Chem.* **353**, 906 (1972).

[9] E. Fink, J. Seifert, R. Geiger, and C. Güttel, *Adv. Biosci.* **17**, 111 (1979).

[10] K. Mann, R. Geiger, W. Göring, W. Lipp, E. Fink, B. Keipert, and H. J. Karl, *J. Clin. Chem. Clin. Biochem.* **18**, 395 (1980).

TABLE I

METHODS FOR THE DETERMINATION OF HUMAN URINARY KALLIKREIN

Method	Detection limit (ng/assay)	Interassay coefficient of variation (%)	Ref.
Biological assay (blood pressure assay)	40	19–8	a–d
Kinin-liberating assay			
Bioassay (rat uterus)	0.5	3	b, e–g
Radioimmunoassay	0.1–0.01	—	h–m
Enzyme immunoassay	3	—	n
Radioimmunoassay	0.05	4–7	o
Radial immunodiffusion	20	8.3	p
Synthetic substrates			
AcPheArgOEt	2	4	g, q, r
DValLeuArgNHNp	4	4–6	r, s

[a] E. G. Erdös and A. F. Wilde, *Handb. Exp. Pharmacol.* **25** (1970).

[b] E. K. Frey, H. Kraut, and E. Werle, "Das Kallikrein-Kinin-System und seine Inhibitoren," p. 11. Enke, Stuttgart, 1968.

[c] A. Zimmermann, R. Geiger, and H. Kortmann, *Hoppe-Seyler's Z. Physiol. Chem.* **360**, 1767 (1979).

[d] A. Arens and G. L. Haberland, *in* "Kininogenases: Kallikrein" (G. L. Haberland and J. W. Rohen, eds.), p. 43. Schattauer, Stuttgart, 1973.

[e] M. E. Webster and E. S. Prado, this series, Vol. 19, p. 681.

[f] K. Mann, R. Geiger, and E. Werle, *Adv. Exp. Med. Biol.* **70**, 65 (1976).

[g] R. Geiger, U. Stuckstedte, and H. Fritz, *Hoppe-Seyler's Z. Physiol. Chem.* **361**, 1003 (1980).

[h] R. C. Talamo, E. Haber, and K. F. Austen, *J. Lab. Clin. Med.* **74**, 816 (1969).

[i] T. L. Goodfriend and C. E. Odya, *in* "Methods of Hormone Radioimmunoassay" (B. M. Jaffe and H. R. Behrman, eds.), p. 439. Academic Press, New York, 1974.

[j] M. Stocker and J. Hornung, *Klin. Wochenschr.* **56**, Suppl. 1, 127 (1978).

[k] R. Sipilä and A. Louhija, *Biochem. Pharmacol.* **25**, 543 (1976).

[l] U. L. Hulthén and T. Borge, *Scand. J. Clin. Lab. Invest.* **36**, 833 (1976).

[m] R. C. Talamo, K. F. Austen, and E. Haber, *Methods Invest. Diagn. Endocrinol.* **2B**, 317 (1973).

[n] A. Ueno, S. Oh-Ishi, T. Kitagawa, and M. Katori, *Adv. Exp. Med. Biol.* **120A**, 195 (1979).

[o] K. Mann, R. Geiger, W. Göring, W. Lipp, E. Fink, B. Keipert, and J. Karl, *J. Clin. Chem. Clin. Biochem.* **18**, 395 (1980).

[p] N. G. Levinsky, O. OleMoiYoi, K. F. Austen, and J. Spragg, *Biochem. Pharmacol.* **28**, 2491, 1979.

[q] F. Fiedler, R. Geiger, C. Hirschauer, and G. Leysath, *Hoppe-Seyler's Z. Physiol. Chem.* **359**, 1667 (1978).

[r] R. Geiger, E. Fink, U. Stuckstedte, and B. Förg-Brey, *Adv. Biosci.* **17**, 127 (1979).

[s] R. Geiger, U. Stuckstedte, B. Förg-Brey, and E. Fink, *Adv. Exp. Med. Biol.* **120A**, 235 (1979).

22,000). Thus far, plasma samples can be analyzed only with the RIA. In addition, the presence of various kallikrein forms (active kallikrein, pro-kallikrein, kallikrein–α_1-antitrypsin complex, kallikrein adducts to other plasma constituents) in plasma has to be considered. Urine samples may also contain precursor forms of kallikrein.[11] (Further references are given in Table I.)

Biological Assay

The biological determination of glandular kallikreins is based on the reversible reduction of blood pressure in animals. The assay is performed by comparing the blood pressure-lowering effect after intravenous injection of a sample with that caused by a standard preparation. Formerly, a urine sample was used as standard according to the definition of the biological kallikrein unit (KU or KE): 1 KU is defined as the blood pressure-reducing effect caused by 5 ml of human urine taken from 50 liters of pool urine collected from healthy volunteers and dialyzed against running water for 24 hr. At present the use of a highly purified porcine pancreatic kallikrein from Bayer AG, Federal Republic of Germany, is recommended. According to our experience, the commercially available samples are carefully tested and standardized for their biological activity. The advantage of the blood pressure assay, which is based on the liberation of kallidin from the plasma kininogens, is its relative sensitivity and specificity. Amounts of about 50 ng of kallikrein cause a clearly measurable effect. Previous dialysis of crude samples (to remove histamine, etc.) and heating or aprotinin addition (to inactivate or inhibit kallikrein) are recommended as suitable controls. The assay has already been described in detail.[1,12]

Kinin-Liberating Assay

Kallikrein activity is defined in kinin-liberating assays as microgram bradykinin equivalents liberated per minute, if an uncharacterized kininogen preparation is used, or as kininogen-U per minute, whereby 1 kininogen-U corresponds to the liberation of 1 μmol of bradykinin equivalent per minute from a purified kininogen preparation.

We strongly recommend that this assay be performed with purified kininogens preferably isolated from the same species as the kallikrein. If

[11] J. Corthorn, T. Imanari, H. Yoshida, T. Kaizu, J. V. Pierce, and J. J. Pisano, *Adv. Exp. Med. Biol.* **120B**, 575 (1979).

[12] E. K. Frey, H. Kraut, and E. Werle, *in* "Das Kallikrein-Kinin-System und seine Inhibitoren," p. 11. Enke, Stuttgart, 1968.

partially purified or heat-inactivated plasma is used as kininogen source, false-negative or false-positive results may be obtained due to the following reactions. (1) Kininases present in partially purified or crude substrates degrade liberated kinins rapidly, and may be inhibited by addition of EDTA and 8-hydroxyquinoline.[13] (2) Trypsin-like enzymes in the kallikrein preparation may liberate peptides from plasma proteins of the crude substrate that potentiate the kinin effect on the isolated organ (see later).[14] (3) Traces of plasminogen activator or urokinase in the kallikrein sample may activate plasminogen in the crude substrate; plasmin thus liberated releases bradykinin from kininogens.[15] These complications can be avoided only if purified kininogens are used as kallikrein substrates.

Bioassay

In this assay the smooth muscle contractions caused by kinins are utilized. Contraction intensity depends on the type and amount of kinin, and the muscle type used.[12] Rat uterus and guinea pig ileum have been successfully employed. Whereas the ileum gives a more reliable response,[16] the uterus is clearly more sensitive to bradykinin.[13] The contraction induced by an unknown sample is compared with that of a reference sample, preferably synthetic bradykinin. A measurable response should be obtained with 1 ng[13] or 15 ng[17] of bradykinin at the uterus or ileum. If possible, the quantity of the standard bradykinin should be checked by HPLC.[18]

Radioimmunoassay

The radioimmunoassay utilizes antibodies directed against bradykinin. To obtain an antibody response in rabbits, bradykinin coupled via toluene diisocyanate[19] or ethyl carbodiimide[20] to a larger carrier protein such as thyroglobulin, ovalbumin, or other compounds[21-23] had to be applied. The assay is performed by incubating the specific antiserum with [^{125}I]Tyr[8] or [^{125}I]Tyr[5] bradykinin and known amounts of bradykinin or test samples.

[13] K. Mann, R. Geiger, and E. Werle, *Adv. Exp. Biol. Med.* **70**, 65 (1976).
[14] M. U. Samaio, M. L. Reis, E. Fink, A. C. M. Camargo, and L. J. Greene, *Anal. Biochem.* **81**, 369 (1977).
[15] C. E. Burrows, H. Z. Movat, and M. J. Soltay, *Proc. Soc. Exp. Biol. Med.* **138**, 959 (1971).
[16] M. E. Webster and E. S. Prado, this series, Vol. 19, p. 681.
[17] D. F. Elliott, E. W. Horton, and G. P. Lewis, *Biochem. J.* **78**, 60 (1961).
[18] R. Geiger and R. Hell, in preparation.
[19] R. C. Talamo, E. Haber, and K. F. Austen, *J. Lab. Clin. Med.* **74**, 816 (1969).
[20] T. L. Goodfriend and C. E. Odya, *in* "Methods of Hormone Radioimmunoassay" (B. M. Jaffe and H. R. Behrman, eds.), p. 439. Academic Press, New York, 1974.
[21] M. Stocker and J. Hornung, *Klin. Wochenschr.* **56**, Suppl. I, 127 (1978).
[22] R. Sipilä and A. Louhija, *Biochem. Pharmacol.* **25**, 543 (1976).
[23] U. L. Hulthén and T. Borge, *Scand. J. Clin. Lab. Invest.* **36**, 833 (1976).

The percent binding of radioactive labeled bradykinin in the presence of unlabeled bradykinin is measured and the concentration of bradykinin calculated from a calibration curve.[24] Detection limit of bradykinin was reported to be 0.01 ng/liter.[23] Note, however, that the various kinins and even kininogens can cross-react with the bradykinin-directed antibodies.[24] Kinin detection after separation of the various kinins by HPLC should yield more reliable results in future.

Enzyme Immunoassays

Kinin-specific antibodies are also essential for enzyme immunoassay. Instead of a radioactive marker, bradykinin is linked to an enzyme marker (e.g., β-D-galactosidase[25]), by which photometric detection is practicable. Recently, up to 3 ng of bradykinin could be detected by this assay.[25]

Radioimmunoassay

A highly sensitive and specific radioimmunoassay was recently developed for the determination of human urinary kallikrein.[10,26] The sensitivity of the assay is 0.5 μg of kallikrein per liter.[10]

Reagents

Human urinary kallikrein used for radioiodination and as reference standard was purified as described below.
Antibodies against human urinary kallikrein and rabbit γ-globulins were raised according to Mann *et al.*[10]
Kallikrein was iodinated using the chloramine-T method.[27]
Buffer: 1.36 g KH_2PO_4, 1.42 g Na_2HPO_4, 5.84 g NaCl, 10 g bovine serum albumin, 3.72 g EDTA, and 200 mg Merthiolate (pH 7.5) per liter.

Procedure

The radioimmunoassay for human urinary kallikrein was performed as follows: kallikrein standard samples, ^{125}I-labeled kallikrein, test samples, and goat anti-rabbit IgG were dissolved or diluted with the given buffer. 3.3 ml of rabbit serum per liter of buffer was added to the ^{125}I-labeled kallikrein solution. 0.1 ml of buffer, a solution of 0.1 ml of labeled kallikrein (approximately 20,000 cpm) and 0.1 ml of specific antiserum in an

[24] R. C. Talamo, K. F. Austen, and E. Haber, *Methods Invest. Diagn. Endocrinol.* **2B**, 317 (1973).
[25] A. Ueno, S. Oh-Ishi, T. Kitagawa, and M. Katori, *Adv. Exp. Biol. Med.* **120A**, 195 (1979).
[26] K. Shimamoto, J. Chao, and H. S. Margolius, *Circulation* **60**, 133 (1979).
[27] F. C. Greenwood, W. M. Hunter, and J. Glove, *Biochem. J.* **89**, 114 (1963).

TABLE II

CONDITIONS USED FOR THE RADIOIMMUNOASSAY OF HUMAN URINARY KALLIKREIN (HUK)[a]

Substances applies and conditions	Sample B (μl)	Blank B_0 (μl)	Non-specific binding N (μl)	Total activi T (μ
Buffer	100	200	200	—
Standard or sample	100	—	—	—
Anti-HUK antibody, diluted 1 : 32,000	100	100	—	—
[125]I-labeled HUK (0.1 ng) + rabbit serum, diluted 1 : 300	100	100	100	10
Incubated for 24 hr at 22°C				
Second antibody, diluted 1 : 30	100	100	100	—
Incubated for 8 hr at 4°C				
Centrifuged at 2000 g for 30 min				

[a] K. Mann, R. Geiger, W. Göring, W. Lipp, E. Fink, B. Keipert, and H. J. Karl, *J. Clin. Chem. C Biochem.* **18**, 395 (1980).

appropriate dilution (1 : 32,000 under routine conditions) were pipetted into disposable tubes (1.8 × 5.5 cm). The reaction mixture was incubated at 22°C for 24 hr and, in order to separate bound and free kallikrein, 0.1 ml of the second antibody (anti-rabbit γ-globulins from goats) was added. After mixing, the sample was incubated at 4°C for another 4 to 8 hr. After centrifugation at 2000 g for 30 min, the supernatant was decanted, the tube wiped with filter paper, and the radioactivity of the precipitate measured. All calibration samples were set up in triplicate (Table II). Standard curves were constructed by plotting the fraction of bound radioactivity, $(B - N)/(B_0 - N)$ (B, counts from bound material; B_0, counts read for zero dose; N, counts from nonspecifically bound material), against the kallikrein dose either in a logit–log mode or in a linear–log mode. The assays were evaluated by a computer program employing the spline approximation.[28]

Comment

Due to the similarity of the glandular kallikreins of one species,[29–31] the source of kallikrein cannot be discriminated by RIA. Determination of glandular kallikrein by the described RIA is possible, not only in urine

[28] W. Huber, K. Mann, and H. J. Karl, *J. Clin. Chem. Clin. Biochem.* **17**, 241 (1979).

[29] H. Fritz, F. Fiedler, T. Dietl, M. Warwas, E. Truscheit, H. J. Kolb, G. Mair, and H. Tschesche, in "Kininogenases: Kallikrein" (G. L. Haberland, J. W. Rohen, and T. Suzuki, eds.), p. 15. Schattauer, Stuttgart, 1977.

[30] H. Tschesche, G. Mair, G. Godec, F. Fiedler, W. Ehret, C. Hirschauer, M. Lemon, and H. Fritz, *Adv. Exp. Med. Biol.* **120A**, 245 (1979).

[31] E. Fink and C. Güttel, *J. Clin. Chem. Clin. Biochem.* **16**, 381 (1978).

samples, but in saliva, pancreatic and duodenal juice, bile, and sweat.[10] Kallikrein-like antigens seem to be present also in human amniotic fluid, seminal plasma, and ascites.[10] In plasma or serum, nonparallel binding curves prevent at present an exact quantitation of glandular kallikreins; after isolation of the glandular kallikrein from plasma, parallel binding curves were obtained.[32] Further investigations to solve this problem are in progress.

Radial Immunodiffusion

Kallikrein determination is also possible by radial immunodiffusion.[33] Concentrations between 1 and 10 mg/liter can be detected after 2- to 20-fold concentration of urine by lyophilization.

Synthetic Substrate Assays

Human urinary kallikrein is capable of hydrolyzing various synthetic substrates, primarily substituted arginine esters or amides. Hydrolyzing activity of human urine toward arginine substrates was shown to be due mainly to kallikrein[34]; the contribution of urokinase is normally negligible.[35] Due to the presence of a potent trypsin-inhibitor capacity,[36] trypsin or trypsin-like enzymes cannot occur in active form in urine. In Table III substrates used for the determination of human urinary kallikrein are compiled.

Human Urinary Kallikrein Assay Using N^α-Acetyl-L-phenylalanyl-L-arginine Ethyl Ester (AcPheArgOEt)

Principle

AcPheArgOEt is hydrolyzed by kallikrein and the liberated alcohol determined by an established photometric assay using the combined NAD–ADH system.[37] One enzyme unit is defined as the amount of enzyme that releases 1 μmol of ethanol per minute.

[32] R. Geiger, B. Clausnitzer, E. Fink, and H. Fritz, *Hoppe-Seyler's Z. Physiol. Chem.* **361,** 1795 (1980).
[33] N. G. Levinsky, O. OleMoiYoi, K. F. Austen, and J. Spragg, *Biochem. Pharmacol.* **28,** 2491 (1979).
[34] R. Geiger, E. Fink, U. Stuckstedte, and B. Förg-Brey, *Adv. Biosci.* **17,** 127 (1979).
[35] G. Claeson, L. Aurell, G. Karlsson, and P. Friberger, *in* "New Methods for the Analysis of Coagulation Using Chromogenic Substrates" (I. Witt, ed.), p. 37. de Gruyter, Berlin, 1977.
[36] H. Sumi, N. Toki, Y. Takado, and A. Takado, *J. Biochem.* (*Tokyo*) **83,** 141 (1978).
[37] I. Trautschold and E. Werle, *Hoppe-Seyler's Z. Physiol. Chem.* **325,** 48 (1961).

TABLE III
SYNTHETIC SUBSTRATES USED FOR THE
DETERMINATION OF KALLIKREIN IN HUMAN URINE

Substrate	Ref.
N^α-tosyl-L-arginine methyl ester	a
[^{3}H]-N^α-tosyl-L-arginine methyl ester	b
N^α-benzoyl-L-arginine ethyl ester	c
N^α-acetyl-L-phenylalanyl-L-arginine ethyl ester	d–f
D-valyl-L-leucyl-L-arginine-p-nitroanilide	g
L-prolyl-L-phenylalanyl-L-arginine-α-naphthyl ester	h
L-prolyl-L-phenylalanyl-L-arginine 4-methylcoumaryl-7-amide	i

[a] A. M. Sieglman, A. S. Carlson, and T. Robertson, *Arch. Biochem. Biophys.* **97**, 159 (1962).
[b] V. H. Beaven, J. V. Pierce, and J. J. Pisano, *Clin. Chim. Acta* **32**, 67 (1971).
[c] I. Trautschold and E. Werle, *Hoppe-Seyler's Z. Physiol. Chem.* **325**, 48 (1961).
[d] R. Geiger, E. Fink, U. Stuckstedte, and B. Förg-Brey, *Adv. Biosci.* **17**, 127 (1979).
[e] R. Geiger, U. Stuckstedte, and H. Fritz, *Hoppe-Seyler's Z. Physiol. Chem.* **361**, 1003 (1980).
[f] F. Fiedler, R. Geiger, C. Hirschauer, and G. Leysath, *Hoppe-Seyler's Z. Physiol. Chem.* **359**, 1667 (1978).
[g] E. Amundsen, J. Pütter, P. Friberger, M. Knos, M. Larsbraaten, and G. Claeson, *Adv. Exp. Med. Biol.* **120A**, 83 (1979)
[h] Y. Hitomi, M. Niinobe, and S. Fujii, *Clin. Chim. Acta* **100**, 275 (1980).
[i] H. Kato, N. Adache, S. Iwanaga, K. Abe, K. Takada, T. Kimura, and S. Sakaki, *J. Biochem. (Tokyo)* **87**, 1127 (1980).

Reagents

Final concentrations in the test sample are given in parentheses.
Buffer: 30.4 g of $Na_4P_2O_7 \cdot 10 H_2O$, 8.2 g of semicarbazide hydrochloride, 1.8 g of glycine, and 20–24 ml of 2 N NaOH (to adjust the pH to 8.7) per liter
NAD: 204 mg of NAD (free acid) in 10 ml of water (1 mM)
ADH: 100 mg of ADH from yeast (Boehringer Mannheim) in 3.4 ml of distilled water (0.59 mg); solution kept at 0°C
AcPheArgOEt (Bachem AG, Basel, Switzerland): 7.65 mg in 1 ml of distilled water (0.5 mM), 0–4°C

Procedure

The enzyme solution (0.1 ml), NAD (0.05 ml), distilled water (0.38 ml), and ADH (0.01 ml) are added to 1 ml of buffer solution. This mixture

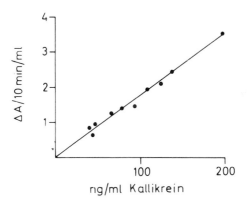

FIG. 1. Amount of kallikrein in human urine determined by RIA and compared with kallikrein activity, as determined by hydrolysis of AcPheArgOEt.

is preincubated at 25°C in a water bath for 5 min. Thereafter, 0.05 ml of AcPheArgOEt solution is added. The increase in absorbance is followed up at 366 nm in 1-min time intervals up to 10 min. From the mean ΔA/min value the enzyme units are calculated using the formula: U = (ΔA/min) × 0.455. ΔA/min values above 0.200 are not reliable.

Comments

Contamination of the solutions by alcohol has to be strictly avoided. Urine of individuals who have taken alcohol 24 hr before urine collection is not suitable for testing; in this case the initial ΔA/min decreases rapidly during the assay. This urine has to be dialyzed against running water for 24 hr before testing. Due to the high sensitivity of the assay (2 ng of kallikrein cause a ΔA/min of 0.005), urokinase does not interfere with the assay. Kallikrein activities of urine samples estimated by this method correlate well with those of the RIA (see Fig. 1) and the DValLeuArgNHN*p* assay (Fig. 2).

Human Urinary Kallikrein Assay Using D-Valyl-L-leucyl-L-arginine-*p*-nitroanilide (DValLeuArgHNH*p*)

Principle

The kallikrein-catalyzed release of *p*-nitroaniline from the chromogenic substrate DValLeuArgNHN*p* is followed photometrically at 405 nm (kinetic assay) or at a given time interval (end-point method). The end-point method is described. One enzyme unit corresponds to the amount of 1 μmol nitroaniline liberated per minute.

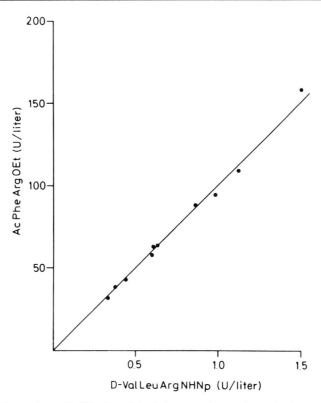

FIG. 2. Comparison of kallikrein activity in human urine, as determined by hydrolysis of AcPheArgOEt and DValLeuArgNHN*p*.

Reagents

Final concentrations in the test samples are given in parentheses.

Buffer: 48.46 g Tris in 1 liter of water, adjusted with concentrated HCl to pH 8.2 (0.4 *M*)

DValLeuArgNHN*p* (S-2266 from Kabi, Mölndal, Sweden): 5.80 mg in 10 ml of water, 0–4°C (0.1 m*M*)

Acetic acid: 100 ml of concentrated acetic acid added to 350 ml of water (4 *M*)

Procedure

Urine or enzyme sample (0.1 ml) and water (0.4 ml) are added to 0.4 ml of buffer solution. This mixture is preincubated at 37°C for 5 min. Thereafter, 0.1 ml of DValLeuArgNHN*p* solution is added and the mix-

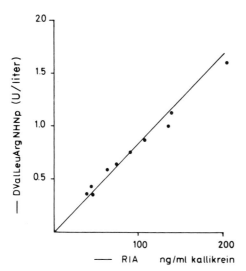

FIG. 3. Amount of kallikrein in human urine determined by radioimmunoassay and compared with kallikrein activity, as determined by hydrolysis of DValLeuArgNHNp.

ture incubated again at 37°C. After 30 min, the reaction is stopped by addition of 0.1 ml of 4 M acetic acid, and the absorbance read at 405 nm. Enzyme units are calculated after correction by the absorbance value of the urine blank sample using the formula: U = $(A_{\text{test sample}} - A_{\text{urine blank}}) \times$ 0.0036.

Comments

The amount of urokinase present in the urine of normal individuals is too low to interfere with the kallikrein assay.[35,38] Otherwise, the kallikrein inhibitor aprotinin may be added to the buffer solution in a concentration of 0.07 TIU (trypsin inhibitor units) per ml as recommended by the manufacturer of DValLeuArgNHNp.[38] In this case the difference in absorption between the test sample and the aprotinin-containing sample corresponds to the kallikrein activity of urine. Kallikrein activity in urine is then calculated by the formula: U = $(A_1 - A_2) \times 0.0036$ (A_1, absorbance of the test sample with buffer containing no aprotinin; A_2, absorbance of the test sample containing aprotinin in buffer).

Kallikrein activities of urine samples estimated by this method correlate well with those of the RIA (Fig. 3) and the AcPheArgOEt assay (see

[38] L. Svendson, *in* "New Methods for the Analysis of Coagulation Using Chromogenic Substrates" (I. Witt, ed.), p. 251. de Gruyter, Berlin, 1977.

Fig. 2). For porcine pancreatic kallikrein (often used as a reference for kallikrein activity), the following calculation factors were measured: 1 KU = 4.03 mU (substrate: DValLeuArgNHNp).[39,40]

Purification

For the purification of human urinary kallikrein, various methods are available. An optimized isolation procedure worked out in our laboratory yielding homogeneous urinary kallikrein as proved by sequence studies is described in detail.

Materials

Benzamidine hydrochloride and N^α-benzoyl-L-arginine ethyl ester were products from E. Merck, Darmstadt, Federal Republic of Germany. N^α-Benzyloxycarbonyl-L-tyrosine-p-nitrophenyl ester was purchased from Serva, Heidelberg, Federal Republic of Germany. DValLeuArg-NHNp from Kabi, Mölndal, Sweden; AcPheArgOEt synthesized according to Fiedler[41] or from Bachem AG, Basel, Switzerland; human HMW kininogen isolated according to Dittmann et al.,[42] aprotinin, the bovine basic pancreatic trypsin inhibitor (Kunitz) from Bayer AG, Leverkusen, Federal Republic of Germany; Sephacryl S-200 and DEAE–Sepharose from Pharmacia Fine Chemicals, Uppsala, Sweden.

Aprotinin–Sepharose was synthesized according to the following procedure: 4.6 g of CNBr-activated Sepharose-4B was washed and reswelled on a sintered glass filter with 1 liter of 1 mM HCl. Aprotinin (115 mg) was dissolved in 2 ml of 0.1 M NaHCO₃ (pH 8.3) containing 0.5 M NaCl. This solution was gently moved for 2 hr at room temperature, excess protein was washed off, and remaining active groups were blocked by addition of 20 ml of 1 M ethanolamine. This mixture was gently stirred for another 2 hr at room temperature. The aprotinin–Sepharose was washed with 1 liter of 0.1 M NaHCO₃ (pH 8.3) containing 0.5 M NaCl, and stored in phosphate buffer (0.025 M Na₂HPO₄–0.025 M KH₂PO₄–1 M NaCl–0.2% sodium azide, pH 7.0) at 4°C until use.

Phosphate buffer: 3.55 g of Na₂HPO₄, 3.40 g of KH₂PO₄, and 43.83 g of NaCl (pH 7.0) per liter.

Formate buffer: 4.6 g of formic acid, 5.84 g of NaCl, and concentrated ammonia (to achieve a pH of 8.0) per liter.

[39] T. Dietl, C. Huber, R. Geiger, S. Iwanaga, and H. Fritz, Hoppe-Seyler's Z. Physiol. Chem. **360**, 67 (1979).

[40] B. Förg-Brey and H. Fritz, personal communication.

[41] F. Fiedler, R. Geiger, C. Hirschauer, and G. Leysath, Hoppe-Seyler's Z. Physiol. Chem. **359**, 1667 (1978).

[42] B. Dittmann, A. Steger, R. Wimmer, and H. Fritz, Hoppe-Seyler's Z. Physiol. Chem. **362**, 919 (1981).

Procedure

Human male urine was collected from healthy volunteers. Sodium azide was preadded to the urine bottles (120 g of sodium azide per 60 liters of urine) to prevent bacterial growth. The urine was dialyzed against running water for 24 hr and lyophilized. The urine powder thus obtained was used as starting material for the isolation procedure.

Extraction of the Crude Urine Powder

The crude urine powder from 50 liters of urine was suspended in 150 ml of phosphate buffer. The suspension was gently stirred for 60 min at 4°C and centrifuged (8000 g) for 30 min. The supernatant (120 ml) was subjected to gel filtration.

Gel Filtration on Sephacryl S-200

Aliquot portions (30 ml) of the supernatant were applied to a water-cooled Sephacryl S-200 column (60 × 4.3 cm) equilibrated with phosphate buffer. The column was eluted at a flow rate of 50 ml/hr, and fractions of 12 ml were collected. The kallikrein-containing fractions of four runs were combined and concentrated to 10 ml by ultrafiltration using an Amicon UM-10 membrane. As the kallikrein activity was eluted in one symmetrical peak (Fig. 4) (tested by RIA, synthetic substrate assays, and kinin-releasing activity), normally only one of the described kallikrein assays may be used.

Affinity Chromatography on Aprotinin–Sepharose

The concentrated kallikrein solution was applied to an aprotinin–Sepharose column (15 × 1 cm) equilibrated with phosphate buffer at room temperature. The column was washed with 50 ml of the phosphate buffer before elution of kallikrein was performed with the same buffer containing 1 M benzamidine hydrochloride. The flow rate was adjusted to 4 ml/hr, and 4-ml fractions were collected. The first 12 benzamidine-containing fractions (beginning with the strong rise in absorption) were combined, separated from benzamidine hydrochloride, and concentrated by repeated ultrafiltration (five times; Amicon UM-10 membrane) against formate buffer at 4°C.

Gradient Elution Chromatography on DEAE–Sepharose

The kallikrein solution (5 ml) thus obtained was applied to a cooled (4°C) DEAE–Sepharose column (15 × 1.5 cm) equilibrated with formate buffer. A flow rate of 15 ml/hr was adjusted and 3-ml fractions were

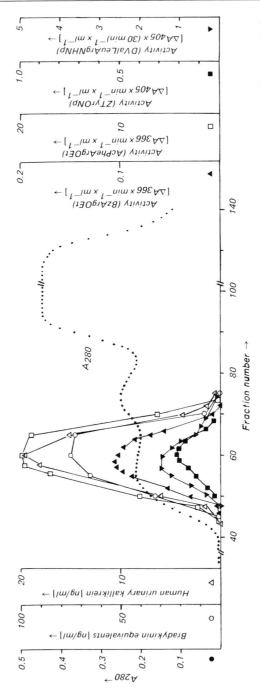

FIG. 4. Fractionation of the urine powder extract on Sephacryl S-200. Kallikrein was assayed by RIA (△—△) and by various synthetic substrates. Protein concentration is given as absorption at 280 nm (●—●), O—O, Bradykinin equivalents measured by the kinin-liberating assay at the rat uterus; ▲—▲, BzArgOEt; □—□, AcPheArgOEt; ■—■, ZTyrONp; ▼—▼, ᴅVal-LeuArgNHNp. (For abbreviations, see Table IX.)

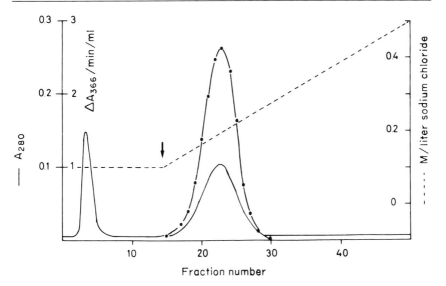

FIG. 5. Chromatography of human urinary kallikrein on DEAE–Sepharose. Kallikrein activity was measured using AcPheArgOEt as substrate (●—●). Protein concentration is given as absorption at 280 nm.

sampled. The column was developed with a linear salt gradient formed from 120 ml each of 0.1 M NaCl and 0.5 M NaCl in formate buffer (Fig. 5).

Kallikrein activity was eluted between 0.13 and 0.25 M NaCl concentration. The kallikrein-containing fractions were combined, concentrated to 1.5 ml by ultrafiltration (Amicon UM-10 membrane), and kept frozen at −30°C. The purification procedure is summarized in Table IV. The reason for the increase in kallikrein activity observed by ourselves and others[16,43] after the gel-filtration step is not yet clear. Removal of a kallikrein inhibitor seems unlikely because we could not find inhibitory activity in any of the fractions tested. More likely is activation of a prokallikrein as claimed by Corthorn *et al.*[11] So far we have not found any evidence for either possibility.

Stability and Purity Criteria

When human urinary kallikrein was kept frozen at −30°C, the preparation was stable for more than 12 months. After addition of small amounts of sodium azide (0.02%, w/v) to the kallikrein solution, the activity remained stable at 4°C for more than 6 months, and at 20°C for more than 14

[43] Y. Matsuda, K. Miyazaki, H. Moriya, Y. Fujimoto, Y. Hojima, and C. Moriwaki, *J. Biochem.* (*Tokyo*) **80,** 671 (1976).

TABLE IV

STEPS FOR PURIFICATION OF HUMAN URINARY KALLIKREIN

Step	Total units of AcPheArgOEt (U)	Specific activity of AcPheArgOEt (U/mg)	Recovery (%)	Total units (kininogen- mU)	Specific activity[a] (kininogen- mU/mg)	Recovery (%)
Urine powder extract (after centrifugation)	43.4	0.008	—	47.3	0.009	—
Sephacryl S-200 eluate after concentration	1301[b]	4.6	100	931[b]	3.3	100
Aprotinin–Sepharose eluate after separation of benzamidine	749	416	58	529	294	57
DEAE–Sepharose eluate after concentration	660	1100	51	445	742	48

[a] Bradykinin equivalents measured by the kinin-liberating assay with highly purified human HMW kininogen at the rat uterus; see text.
[b] For increase related to the urine powder extract, see Procedure.

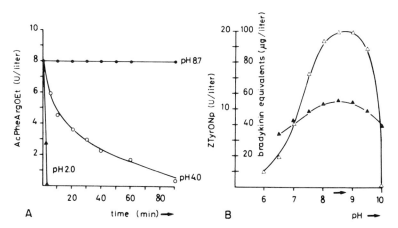

FIG. 6. (A) Influence of pH on the activity of human urinary kallikrein. Kallikrein was incubated at the given pH values; aliquot samples were removed at various time intervals and assayed. (B) pH optimum of human urinary kallikrein. Kallikrein activity was measured by kinin-liberating assay on the rat uterus (▲——▲) and hydrolysis on ZTyrONp (△——△).

days at neutral pH in 0.025 M phosphate buffer. At pH 2, an immediately irreversible loss of enzymatic activity was observed. In Fig. 6A the decrease in enzymatic activity at pH 4 is shown.

To establish the homogeneity of the kallikrein preparation, several tests were performed: Formation of the aprotinin–kallikrein complex followed by electrophoretic separation shows whether the preparation is contaminated by enzymatically inactive proteins.[44] The single-chain form of human urinary kallikrein contains isoleucine as amino-terminal residue and consists of an amino-terminal amino acid sequence closely related to that of porcine pancreatic kallikrein.[16] Thus homogeneity of the protein part is proved by quantitative end-group determination. It should be emphasized, however, that due to proteolytic cleavage(s), multiple kallikrein forms may occur (see Comments below).

Comments

Numerous alternative isolation procedures for human urinary kallikrein are available. Moriya *et al.*[43,45] isolated the enzyme (specific activity 200 KE/mg) by chromatography on DEAE–cellulose, acetone fractionation, gel filtration on Sephadex G-100, chromatography on DEAE–

[44] R. Geiger, U. Stuckstedte, and H. Fritz, *Hoppe-Seyler's Z. Physiol. Chem.* **361,** 1003 (1980).
[45] H. Moriya, J. V. Pierce, and M. E. Webster, *Ann. N.Y. Acad. Sci.* **104,** 172 (1963).

Sephadex A-50, and isoelectric focusing. Hial *et al.*[46] purified it (specific activity: 0.15 μg bradykinin equivalents per milligram and minute) by chromatography on DEAE–Sephadex A-50 and repeated gel filtration on Sephadex G-150. More simple isolation procedures were also described.[47–49] In all these cases except one,[50] the identity of the enzyme preparation with a glandular kallikrein seems to be established. However, the purity criteria given did prove that in no case was the homogeneity of the enzyme as a single-chain glycoprotein. Whereas multiple kallikrein forms due to heterogeneity of the carbohydrate portions will probably not differ in enzymatic properties,[6] this might well be the case for multiple forms produced by cleavage(s) of internal peptide bond(s),[51] either during urine collection or isolation.

Properties

Specific Activity

The mean specific activity of four preparations was 0.742 kininogen-U/mg protein, applying highly purified human HMW kininogen to the kinin-releasing assay, and 2300 KE/mg protein in the dog blood-pressure assay (see Assay Methods in this article) (protein concentration was determined by amino acid analyses). With AcPheArgOEt as substrate, a specific activity of 1100 U/mg of protein was determined; using DVal-LeuArgNHNp, 2.6 U/mg of protein were obtained.

Composition and Structure

Human urinary kallikrein has an amino acid composition similar to that of porcine pancreatic kallikrein. Values obtained in our group and related data are compiled in Table V. From the N-terminal amino acid sequence elucidated so far, a high degree of homology with other serine proteinases is already evident (Table VI).

pH Optimum

For the hydrolysis of synthetic substrates, a relatively broad pH optimum, close to pH 8.8, was found. Using kininogen as substrate, a pH optimum of 8.6 was obtained (see Fig. 6B).

[46] V. Hial, C. R. Diniz, and M. Mares-Guia, *Biochemistry* **13**, 4311 (1974).
[47] G. Porcelli, G. B. Maini-Bettolo, H. R. Croxatto, and M. DiIorio, *Ital. J. Biochem.* **23**, 44 (1974).
[48] N. B. Oza and J. W. Ryan, *Biochem. J.* **171**, 285 (1978).
[49] J. V. Pierse and K. Nustad, *Fed. Proc., Fed. Am. Soc. Exp. Biol.* **31**, 623 (1973).
[50] O. OleMoiYoi, J. Spragg, and K. F. Austen, *Proc. Natl. Acad. Sci. U.S.A.* **76**, 3121 (1979).
[51] M. Lemon, F. Fiedler, B. Förg-Brey, C. Hirschauer, G. Leysath, and H. Fritz, *Biochem. J.* **177**, 159 (1979).

TABLE V

AMINO ACID COMPOSITIONS OF HUMAN URINARY KALLIKREIN
AND PORCINE PANCREATIC KALLIKREIN

	Residues/kallikrein molecule				
Amino acid	Geiger et al.[a,b,c]	Porcelli et al.[d]	Hial et al.[e]	OleMoiYoi et al.[f]	Porcine pancreatic kallikrein[g]
Asp	23.0 ± 0.4	25	40	52.02	28
Thr	13.4 ± 0.3	20	25	25.15	15
Ser	13.7 ± 0.5	25	36	28.65	14
Glu	27.7 ± 0.3	38	50	56.80	23
Pro	14.0 ± 0.1	25	48	25.60	16
Gly	21.9 ± 0.4	30	44	40.50	22
Ala	14.1 ± 0.4	30	22	25.95	13
Cys	9.8 ± 0.2	4	10	1.42	10
Val	17.9 ± 0.6	20	21	35.50	10
Met	3.5 ± 0.1	4	2	6.42	4
Ile	7.7 ± 0.2	14	10	14.88	12
Leu	19.6 ± 0.7	20	19	43.70	20
Tyr	5.8 ± 0.2	8	14	5.60	7
Phe	9.8 ± 0.5	14	14	11.62	10
Lys	7.7 ± 0.5	27	15	14.86	10
His	8.6 ± 0.6	12	8	18.93	8
Arg	5.9 ± 0.3	8	18	9.75	3
Trp	—	—	2	—	7
Total	228	324	398		232

[a] Extrapolated to hydrolysis time zero.
[b] Without tryptophan.
[c] R. Geiger, U. Stuckstedte, and H. Fritz, *Hoppe-Seyler's Z. Physiol. Chem.* **361,** 1003 (1980).
[d] G. Porcelli, G. B. Marini-Bettolo, H. R. Croxatto, and M. DiIorio, *Ital. J. Biochem.* **23,** 44 (1974).
[e] V. Hial, C. R. Diniz, and M. Mares-Guia, *Biochemistry* **13,** 4311 (1974).
[f] O. OleMoiYoi, J. Spragg, and K. F. Austen, *Proc. Natl. Acad. U.S.A.* **76,** 3121 (1979).
[g] F. Fiedler, W. Ehret, G. Godec, C. Hirschauer, C. Kutzbach, G. Schmidt-Kastner, and H. Tschesche, *in* "Kininogenases: Kallikrein" (G. L. Haberland, J. W. Rohen, and T. Suzuki, eds.), p. 7. Schattauer, Stuttgart, 1977.

Physicochemical Data

Molecular Weight

Molecular weights and related data for human urinary kallikrein are listed in Table VII. As glycoproteins tend to show excessively high mo-

TABLE VI

N-Terminal Amino Acid Sequence of Human Urinary Kallikrein (a)[a]
as Compared to Porcine Pancreatic Kallikrein (b)[b] and Porcine Trypsin (c)[c]

7	20

(a)   Ile-Val-Gly-Gly-Trp-Glu-Cys-Glu-Gln-His-Ser-Gln-Pro-Trp-Gln-Ala-Ala-Leu-
(b)   Ile-Ile-Gly-Gly-Arg-Glu-Cys-Glu-Lys-Asn-Ser-His-Pro-Trp-Gln-Val-Ala-Ile-
(c)   Ile-Val-Gly-Gly-Tyr-Thr-Cys-Ala-Ala-Asn-Ser-Ile-Pro-Tyr-Gln-Val-Ser-Leu-

(a)   Tyr-His-Phe-Ser-Thr-Phe-Gln-Cys-Gly-Gly-Ile-Leu-Val-
(b)   Tyr-His-Tyr-Ser-Ser-Phe-Gln-Cys-Gly-Gly-Val-Leu-Val-
(c)   Asn-Ser-Gly-Ser-His-Phe-   -Cys-Gly-Gly-Ser-Leu-Ile-

---

[a] F. Lottspeich, R. Geiger, A. Henschen, and C. Kutzbach, *Hoppe-Seyler's Z. Physiol. Chem.* **360**, 1947 (1979).

[b] F. Fiedler, W. Ehret, G. Godec, C. Hirschauer, C. Kutzbach, G. Schmidt-Kastner, and H. Tschesche, *in* "Kininogenases: Kallikrein" (G. L. Haberland, J. W. Rohen, and T. Suzuki, eds.), p. 7. Schattauer, Stuttgart, 1977.

[c] M. A. Hermodson, L. H. Ericsson,   H. Neurath, and K. A. Walsh, *Biochemistry* **12**, 3146 (1973).

---

lecular weights on gel filtration,[52] the results obtained by this method must be considered with reservation. Recalculated molecular weights based on active enzyme titration and amino acid analyses are also given in Table VII. The most reliable molecular weight values of human urinary kallikrein seem to be below 30,000.

Isoelectric Points

The kallikrein preparation separated into numerous protein bands by electrofocusing with pI values of 3.75, 3.80, 3.90, 4.00, 4.05, and 4.25.[44] Similar pI values—namely, 3.80, 3.95, and 4.06[46] or 3.9, 4.0, and 4.2[43] were reported by other authors. Available data are compiled in Table VIII.

Enzymatic Properties

Substrates

Human urinary kallikrein liberates kallidin from both HMW and LMW kininogen.[53] Recently, some investigators[54,55] reported on the activation of prorenin to active renin by this kallikrein.

[52] P. Andrews, *Methods Biochem. Anal.* **18**, 1 (1970).

[53] Y. N. Han, H. Kato, S. Iwanaga, and M. Komiya, *J. Biochem. (Tokyo)* **83**, 223 (1978).

[54] J. E. Sealey, S. A. Atlas, and J. H. Laragh, *Nature (London)* **275**, 144 (1978).

[55] H. M. Derkx, H. L. Tang-Tjiong, A. J. Man in't Veld, M.P.A. Schalekamp, and M. A. D. H. Schalekamp, *J. Clin. Endocrinol. Metab.* **49**, 765 (1979).

TABLE VII
MOLECULAR WEIGHT DATA REPORTED FOR
HUMAN URINARY KALLIKREIN

Method	Molecular weight	Ref.
Gel filtration	27,000; 29,000	a,b
	25,000–30,000	c
	43,600	d
	27,000–43,000	e
	64,000	f
	45,000	g
	48,700	h
SDS electrophoresis	29,000; 45,000	f
	23,800; 29,300; 36,400	i
	37,000	j
	34,000; 41,000	k
	48,000	h
Ultracentrifugation	40,500	l
Amino acid composition	42,666	d
	35,400	j
	48,213	h
Active enzyme titration	25,500; 25,000	k

[a] Y. Matsuda, K. Miyazaki, H. Moriya, Y. Fujimoto, Y. Hojima, and C. Moriwaki, *J. Biochem.* (*Tokyo*) **80**, 671 (1976).

[b] H. Moriya, Y. Matsuda, Y. Fujimoto, Y. Hojima, and C. Moriwaki, in "Kininogenases: Kallikrein" (G. L. Haberland, J. W. Rohen, and T. Suzuki, eds.), p. 37. Schattauer, Stuttgart, 1973.

[c] J. Spragg and K. F. Austen, *Biochem. Pharmacol.* **23**, 781 (1974).

[d] V. Hial, C. R. Diniz, and M. Mares-Guia, *Biochemistry* **13**, 4311 (1974).

[e] O. OleMoiYoi, J. Spragg, S. P. Halbert, and K. F. Austen, *J. Immunol.* **118**, 667 (1977).

[f] R. Geiger, K. Mann, and T. Bettels, *J. Clin. Chem. Clin. Biochem.* **15**, 479 (1979).

[g] N. B. Oza and J. W. Ryan, *Biochem. J.* **171**, 285 (1978).

[h] O. OleMoiYoi, J. Spragg, and K. F. Austen, *Proc. Natl. Acad. Sci. U.S.A.* **76**, 3121 (1979).

[i] J. V. Pierce and K. Nustad, *Fed. Proc., Fed. Am. Soc. Exp. Biol.* **31**, 623 (1972).

[j] G.Porcelli, G. B. Marini-Bettolo, H. R. Croxatto, and M. DiIorio, *Ital. J. Biochem.* **23**, 44 (1974).

[k] R. Geiger, U. Stuckstedte, and H. Fritz, *Hoppe-Seyler's Z. Physiol. Chem.* **361**, 1003 (1980).

[l] H. Moriya, J. V. Pierce, and M. E. Webster, *Ann. N.Y. Acad. Sci.* **104**, 172 (1972).

TABLE VIII
IsoELECTRIC POINTS REPORTED FOR
HUMAN URINARY KALLIKREIN

pI Values	Ref.
3.84–4.33 (six peaks) (main peaks 4.04, 4.06, 4.13)	a
3.75, 3.80, 3.90, 4.00, 4.05, 4.25	b
3.9, 4.0, 4.2	c
3.8, 3.9, 4.09	d
3.77, 3.90, 4.04, 4.12, 4.28	e
3.80, 3.95, 4.06	f
3.8–4.4 (three peaks)	g

[a] J. V. Pierce and K. Nustad, *Fed. Proc., Fed. Am. Soc. Exp. Biol.* **31**, 623 (1972).
[b] R. Geiger, U. Stuckstedte, and H. Fritz, *Hoppe-Seyler's Z. Physiol. Chem.* **361**, 1003 (1980).
[c] Y. Matsuda, K. Miyazaki, H. Moriya, Y. Fujimoto, Y. Hoijma, and C. Moriwaki, *J. Biochem. (Tokyo)* **80**, 671 (1976).
[d] R. Geiger, K. Mann, and T. Bettels, *J. Clin. Chem. Clin. Biochem.* **15**, 479 (1977).
[e] Y. Hojiman, Y. Matsuda, C. Moriwaki, and H. Moriya, *Chem. Pharm. Bull.* **23**, 1120 (1975).
[f] Y. Hial, C. R. Diniz, and M. Mares-Guia, *Biochemistry* **13**, 4311 (1974).
[g] O. OleMoiYoi, J. Spragg, S. P. Halbert, and K. F. Austen, *J. Immunol.* **118**, 667 (1977).

Kinetic constants for the hydrolysis of synthetic substrates are compiled in Table IX. Like other serine proteinases, human urinary kallikrein is capable of hydrolyzing arginine esters but hydrolyzes p-nitroanilides, which are good substrates for trypsin[56] only more slowly. An exception is DValLeuArgNHNp, which is hydrolyzed at nearly the same rate as BzArgOEt.[41] A plausible explanation for the rapid cleavage of a tyrosine substrate, ZTyrONp, by human urinary kallikrein cannot be given yet.

Natural Inhibitors

The activity of human urinary kallikrein is strongly inhibited by aprotinin (Trasylol), the bovine basic pancreatic trypsin inhibitor (Kunitz). The dissociation constant K_i of the kallikrein–aprotinin complex was de-

[56] L. Srendsen, B. Blombäck, and P. I. Olsson, *Thromb. Res.* **1**, 267 (1972).

TABLE IX
KINETIC CONSTANTS FOR THE HYDROLYSIS OF SUBSTRATES BY
HUMAN URINARY KALLIKREIN

Substrate[a]	K_m (mM)	V_{max} μM min^{-1} mg^{-1}	Ref.
BzArgOEt	1.34	0.34	b
	1.1	7.14	c
	0.56; 0.38; 0.48	—	d
BzArgOMe	0.33; 0.19; 0.23	—	d
TosArgOMe	1.14	0.37	b
	0.4	194	e
	0.75; 0.49; 0.33	—	d
	0.67	1.9	c
AcPheArgOEt	0.06 ± 0.01	1100 ± 24	f
DValLeuArgNHNp	0.024 ± 0.003	5.67 ± 0.12	f
ZTyrONp	0.19	2.0	b
	0.007	1.4	c
HMW kininogen	0.00013	2.5	c

[a] BzArgOEt, N^α-benzoyl-L-arginine ethyl ester; BzArgOMe, N^α-benzoyl-L-arginine methyl ester; TosArgOMe, N^α-tosyl-L-arginine methyl ester; AcPheArgOEt, N^α-acetyl-L-phenylalanyl-L-arginine ethyl ester; DValLeuArgNHNp, D-valyl-L-leucyl-L-arginine-p-nitroanilide; ZTyrONp, N^α-carbobenzoxy-L-tyrosine-p-nitrophenyl ester; HMW kininogen, high-molecular-weight kininogen.
[b] Va. Hial, C. R. Diniz, and M. Mares-Guia, *Biochemistry* **13**, 4311 (1974).
[c] R. Geiger, K. Mann, and T. Bettels, *J. Clin. Chem. Clin. Biochem.* **15**, 479 (1977).
[d] Y. Matsuda, M. Miyazaki, H. Moriya, Y. Fujimoto, Y. Hojima, and C. Moriwaki, *J. Biochem. (Tokyo)* **80**, 671 (1976).
[e] O. OleMoiYoi, J. Spragg, and K. F. Austen, *Proc. Natl. Acad. Sci. U.S.A.* **76**, 3121 (1979).
[f] R. Geiger, U. Stuckstedte, and H. Fritz, *Hoppe-Seyler's Z. Physiol. Chem.* **361**, 1003 (1980).

termined to be 0.9×10^{-10} M.[44,57] Structurally homologous proteinase inhibitors of the Kunitz (aprotinin) type (bovine lung TKI, sea anemone SAI 5-II, snake venom NNV-II, snake venom HHV-II, and snail HPI$_k$[39,58]) should also inhibit human urinary kallikrein. Soybean-trypsin inhibitor (a strong inhibitor of trypsin), plasma kallikrein, and other serine proteinases,[58,59] do not inhibit human urinary kallikrein. Recently, α_1-

[57] R. Geiger, K. Mann, and T. Bettels, *J. Clin. Chem. Clin. Biochem.* **15**, 479 (1977).
[58] H. Fritz, E. Fink, and E. Truscheit, *Fed. Proc., Fed. Am. Soc. Exp. Biol.* **38**, 2753 (1979).
[59] H. Fritz, *Proc. Chem. Fibriol. Thrombol.* **3**, 285 (1978).

TABLE X

KINETIC CONSTANTS[a] FOR THE INACTIVATION OF HUMAN URINARY
KALLIKREIN BY CHLOROMETHYL KETONES (25°C, pH 7.0)[b]

Chloromethyl ketone	$10^{-4} k_2/K_i$ $(M^{-1} min^{-1})$	K_i (μM)	k_2 (min^{-1})
Ac-Phe-ArgCH$_2$Cl	0.81	56.0	0.46
Pro-Phe-ArgCH$_2$Cl	0.79	45.0	0.36
Ac-Pro-Phe-ArgCH$_2$Cl	0.34	120	0.41
Ser-Pro-Phe-ArgCH$_2$Cl	0.65	180	1.20
Phe-Ser-Pro-Phe-ArgCH$_2$Cl	4.20	21.2	0.89

[a] k_2, First-order rate constant for the inactivation reaction; K_i, dissociation constant;
[b] C. Kettner, C. Mirabelli, J. V. Pierce, and E. Shaw, *Arch. Biochem. Biophys.* **202**, 420 (1980).

TABLE XI

ACTIVATION OF HUMAN URINARY KALLIKREIN
BY AMPHIPHILES[a,b]

Amphiphile	Activity (%)
Control	100
Triton X-100	220
Triton X-45	205
Brij 35	250
Tween 20	210
Sodium dodecyl sulfate	195
Deoxycholate	170
Cholate	150
Lysolecithine	155

[a] Activation experiments were performed by the following procedure (curvette volume: 1.5 ml): 0.05 ml of human urinary kallikrein (25 ng) solution was preincubated with 0.05 ml of the amphiphile (0.02%, w/v) for 15 min at 25°C in 0.15 M sodium pyrophosphate buffer (pH 7.4), and then added (0.1 ml) to the assay mixture, which also contained the amphiphile in a final concentration of 0.01% (w/v). AcPhe-ArgOEt served as substrate.
[b] R. Geiger, U. Stuckstedte, and H. Fritz, *Hoppe-Seyler's Z. Physiol. Chem.* **361**, 1003 (1980).

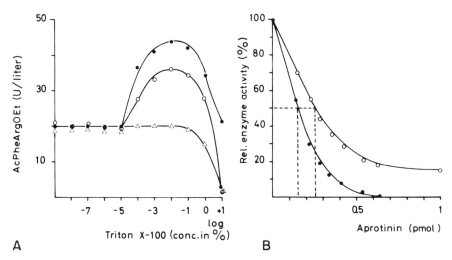

A

B

FIG. 7. (A) Time dependence of human urinary kallikrein activation by Triton X-100. ●—●, Freshly prepared kallikrein solution; O—O, the same kallikrein solution after storage at 4°C for 3 days; and for 7 days (△—△); for test conditions, see Table XI. (B) Titration of human urinary kallikrein by aprotinin. 8.25 mU/test; kallikrein with a specific activity of 1100 U/mg; substrate AcPheArgOEt. O—O, Test without Triton X-100; ●—●, test in the presence of 0.01% Triton X-100; for test conditions, see Ref. 22.

proteinase inhibitor (α_1-antitrypsin) was identified as the only progressively reacting inhibitor of human urinary kallikrein present in plasma.

Synthetic Inhibitors

Like other serine proteinases, human urinary kallikrein is also inhibited by diisopropyl fluorophosphate: at pH 7.2 and 25°C, a bimolecular velocity constant of 9 ± 2 liters mol^{-1} min^{-1} was determined. Benzamidine inhibits competitively with dissociation constants K_i of 6.42 mM [46] and 0.2 mM,[57] respectively. Enzymatic activity of human urinary kallikrein is reduced also by dioxan (K_i = 200 mM [46]). Chloromethyl ketones of arginine peptides reacting rapidly and specifically with human urinary kallikrein (see Table X) were synthesized recently.[60]

Effects of Amphiphiles

Cleavage rates of synthetic substrates (e.g., AcPheArgOEt) caused by kallikrein can be significantly increased by addition of amphiphiles to the assay system. A 1.5- to 2.5-fold increase in enzymatic activity was found in the presence of 0.01% (w/v) of amphiphile (Table XI). Human serum

[60] C. Keltner, C. Mirabelli, J. V. Pierce, and E. Shaw, *Arch. Biochem. Biophys.* **202**, 420 (1980).

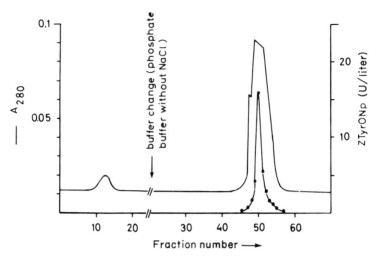

FIG. 8. Hydrophobic interaction chromatography of human urinary kallikrein on phenyl-Sepharose. The column was developed with 0.05 M phosphate buffer containing 4 M NaCl, pH 7.0 (flow rate 12 ml/hr; 1.5-ml fractions collected). After application and binding of kallikrein, the enzyme was eluted by 0.05 M phosphate buffer (without NaCl), pH 7.0. Kallikrein activity was measured using ZTyrONp as substrate (O——O).

albumin induces a similar stimulatory effect (2-fold increase). This activation holds true especially for freshly prepared kallikrein samples; no response may be observed, however, after storage of such kallikrein solutions at 4°C for a longer time (Fig. 7A). Furthermore, inactivation of kallikrein by aprotinin can be influenced by addition of amphiphiles (Fig. 7B). These observations indicate that the conformation of a hydrophobic domain present at the kallikrein molecule has an influence on its catalytic activity. Another indication for the presence of a hydrophobic domain is the high affinity of the urinary kallikrein to phenyl- or octyl-Sepharose (Pharmacia Fine Chemicals, Uppsala, Sweden). Phenyl-Sepharose especially proved to be a suitable matrix for the binding of human urinary kallikrein and may thus be used for the isolation of the enzyme (Fig. 8).

Specific Antibodies

Antibodies directed against human urinary kallikrein may depress the enzymatic activity on synthetic and natural substrates completely.[44]

[40] Porcine Glandular Kallikreins*

By Franz Fiedler, Edwin Fink, Harald Tschesche, and Hans Fritz

Nomenclature

Glandular or tissue kallikreins are serine proteinases most closely related to trypsin (for a recent review, see Fiedler[1]). Finally, they were given an EC number (3.4.21.35) of their own, as were the plasma (serum) kallikreins, a totally different group belonging to the blood-clotting enzymes. Glandular kallikreins occur mainly in pancreas and pancreatic juice (as a proenzyme activatable by trypsin), in salivary glands and saliva, and in urine.[2] In the pig, glandular kallikreins have been isolated from the pancreas, submandibular glands, and urine. Submandibular, urinary, and native pancreatic kallikrein consist of a single polypeptide chain and are called glandular α-kallikreins.[3] Porcine pancreatic kallikrein isolated as usual from pancreas homogenates incubated for some time to activate the proenzyme is obtained mainly in a form containing a split in its polypeptide chain, as pancreatic β-kallikrein. Furthermore, small amounts of the enzyme carrying an additional split in its larger chain, pancreatic γ-kallikrein, have been observed in such preparations.[3]

All these porcine glandular kallikreins occur in several forms discernible by electrophoresis or by ion-exchange chromatography, probably due to differences in their carbohydrate content (Table I).

Assay Methods

Assay methods for glandular kallikreins have been reviewed.[1] In the isolation of porcine glandular kallikreins, especially the method of Trautschold and Werle,[4] based on the hydrolysis of N^α-benzoyl-L-arginine ethyl ester (BAEE) and spectrophotometric determination of the ethanol released by means of alcohol-dehydrogenase as described below, and its adaptation to a Technicon Autoanalyzer[5] have proved useful.

---

* The physiological function of the glandular kallikrein/kinin system is unknown. Classification as blood pressure regulating enzyme is purely hypothetical.
[1] F. Fiedler, *Handb. Exp. Pharmacol.* [N.S.] 25, Suppl., 103 (1979).
[2] E. K. Frey, H. Kraut, E. Werle, R. Vogel, G. Zickgraf, and I. Trautschold, "Das Kallikrein-Kinin-System und seine Inhibitoren." Enke, Stuttgart, 1968.
[3] F. Fiedler, W. Ehret, G. Godec, C. Hirschauer, C. Kutzbach, G. Schmidt-Kastner, and H. Tschesche, *in* "Kininogenases: Kallikrein" (G. L. Haberland, J. W. Rohen, and T. Suzuki, eds.), Vol. 4, p. 7. Schattauer, Stuttgart, 1977.
[4] I. Trautschold and E. Werle, *Hoppe-Seyler's Z. Physiol. Chem.* 325, 48 (1961).
[5] F. Fiedler, this series, Vol. 45, p. 289.

TABLE I

SURVEY OF THE VARIOUS FORMS OF PORCINE GLANDULAR KALLIKREINS

Type	Source	Forms	Forms defined by	Ref.
Pancreatic prokallikrein	Pancreas	A, B	Chromatography on DEAE ion exchanger	a
Glandular α-kallikreins	Pancreas	B'	Formation from prokallikrein B	b
	Submandibular glands	A, B	SDS electrophoresis	c
	Urine	A, B	SDS electrophoresis	c
	Urine	I, II, III	Chromatography on DEAE ion exchanger	d
Pancreatic β-kallikrein	Pancreas autolyzate	A, B	Electrophoretic migration	e
	Pancreas autolyzate	A, B	Chromatography on DEAE ion exchanger	a
	Pancreas autolyzate	C	Electrophoretic migration	f
	Pancreas autolyzate	$d_1(= B)$, $d_2(= A)$, $d_3(= C?)$	Chromatography on DEAE ion exchanger	g
Pancreatic γ-kallikrein	Pancreas autolyzate	I(= A), II(= B), III	Chromatography on DEAE ion exchanger	h
	Pancreas autolyzate			i

[a] F. Fiedler and E. Werle, Hoppe-Seyler's Z. Physiol. Chem. 348, 1087 (1967).

[b] F. Fiedler, C. Hirschauer, and E. Werle, Hoppe-Seyler's Z. Physiol. Chem. 351, 225 (1970).

[c] M. Lemon, F. Fiedler, B. Förg-Brey, C. Hirschauer, G. Leysath, and H. Fritz, Biochem. J. 177, 159 (1979).

[d] G. Mair, H. Tschesche, and H. Fritz, in preparation (see this chapter).

[e] E. Habermann, Hoppe-Seyler's Z. Physiol. Chem. 328, 15 (1962).

[f] C. Kutzbach and G. Schmidt-Kastner, quoted in H. Fritz, F. Fiedler, T. Dietl, M. Warwas, E. Truscheit, H. J. Kolb, G. Mair, and H. Tschesche, in "Kininogenases: Kallikrein" (G. L. Haberland, J. W. Rohen, and T. Suzuki, eds.), Vol. 4, p. 15. Schattauer, Stuttgart, 1977.

[g] M. Zuber and E. Sache, Biochemistry 13, 3098 (1974).

[h] H. Kira, S. Hiraku, and H. Terashima, Adv. Exp. Med. Biol. 120A, 273 (1979).

[i] F. Fiedler, W. Ehret, G. Godec, C. Hirschauer, C. Kutzbach, G. Schmidt-Kastner, and H. Tschesche, in "Kininogenases: Kallikrein" (G. L. Haberland, J. W. Rohen, and T. Suzuki, eds.), Vol. 4, p. 7. Schattauer, Stuttgart, 1977.

More precise activity values are to be obtained by a pH-stat assay using the same substrate that has been described in detail previously,[5] although this method is more time consuming and requires the use of more enzyme. In this procedure, 0.1 mM thioglycolic acid added to counteract inhibition by heavy metal ions (that may originate at least in part from the calomel electrode) can be replaced by 0.1 mM ethylenediamine-tetraacetate with equal results.

The unit (U) of activity of glandular kallikrein is that amount of enzyme hydrolyzing 1 μmol of N^{α}-benzoyl-L-arginine ethyl ester per minute under the conditions of the respective assay. To distinguish the results of the various assay methods used, the method is indicated by an index to U in this chapter. Thus, U_{ADH} refers to the results of the manual spectrophotometric alcohol dehydrogenase-linked assay, U_{253} to the direct spectrophotometric assay,[4] U_{AT} to the pH-stat assay (performed on an Autotitrator), and U_{AA} to the Autoanalyzer assay (in which the unit is defined differently; see later).

Manual Spectrophotometric Assay (Alcohol Dehydrogenase-Linked)[4,6]

Reagents

Pyrophosphate–semicarbazide buffer: sodium pyrophosphate (for a 75 mM concentration in the final buffer), semicarbazide hydrochloride (75 mM), and glycine (22.5 mM) are dissolved in 220 ml of water. The pH is adjusted to 8.70 with 2 N NaOH, and the solution made up to 250 ml. This buffer can be stored in a refrigerator for about 2 weeks

BzArgOEt solution: 7.5 mM N^{α}-benzoyl-L-arginine ethyl ester hydrochloride in water (stable in a refrigerator for about 2 weeks)

NAD solution: 30 mM in water

ADH suspension: alcohol dehydrogenase from yeast (Boehringer, Mannheim, Federal Republic of Germany, about 300 U/mg), commercial suspension of 30 mg/ml

Procedure. In a 3-ml cuvette of 1 cm light path, thermostatted at 25°C, are mixed 2.40 ml of the pyrophosphate–semicarbazide buffer, 0.20 ml of BzArgOEt solution (final concentration, 0.50 mM), 0.10 ml of NAD solution (final concentration, 1 mM), and water so that the final volume after addition of the sample amounts to 3.00 ml. Twenty milliliters of ADH suspension are added with stirring. Exactly 5 min later, the sample of glandular kallikrein is added in not more than 0.28 ml of solution. Beginning 2 min later, the increase in absorbance is followed for 10 min at 366 or 340 nm. The amount of glandular kallikrein must be selected so that $\Delta A_{366}/10$ min is not higher than 0.11. Otherwise, the reaction will not be proportionate to the amount of enzyme.[6]

[6] H. Fritz, I. Eckert, and E. Werle, *Hoppe-Seyler's Z. Physiol. Chem.* **348**, 1120 (1967).

Several blanks are treated in the same way, except that the glandular kallikrein solution is replaced by water, and their average used as correction. The increase of absorbance of the blanks should not be higher than 0.007 ($\Delta A_{366}/10$ min). Higher values may indicate that the pyrophosphate–semicarbazide buffer is outdated.

Comments. This assay is a compromise in several respects, and so a number of modifications have been described. The pH control is rather poor in the above variant, the effective pH in the cuvette being only about 8.4. A more concentrated buffer,[7] however, is inconvenient to use because under refrigeration the salt crystallizes from it; however, its stability at room temperature is limited. A higher substrate concentration[7] will result in somewhat higher sensitivity and in less depletion of substrate during the reaction, but also in higher reagent blanks. In our experience a longer preincubation period[8] has not been found necessary.

Assay of Glandular Prokallikrein and Kallikrein with the Technicon Autoanalyzer [9,10]

Principle. Glandular prokallikrein is activated with trypsin during passage through a reaction coil of a Technicon Autoanalyzer. Trypsin is selectively inhibited by the addition of soybean-trypsin inhibitor (SBTI). Glandular kallikrein either already contained in the samples or formed by activation of glandular prokallikrein is assayed in the Autoanalyzer manifold described in a previous volume of this series.[5] The system is calibrated with a series of standard solutions of porcine pancreatic β-kallikrein.

Reagents

Trypsin solution: 50 mg of trypsin (Trypure Novo) is dissolved in 100 ml of 1 mM HCl. This solution is kept in an ice bath throughout

SBTI solution: 500 mg of soybean-trypsin inhibitor (pract., Serva, Heidelberg, Federal Republic of Germany) is dissolved in the 0.05 M Tris buffer described previously.[5] The solution is cleared by centrifugation

The other reagents are specified in Volume 45 of this series.[5]

Procedure. The upper half of the Autoanalyzer manifold for determination of porcine glandular kallikrein described by Fiedler[5] is modified as specified in Fig. 1. Both heating coils are contained in the same heating

---

[7] I. Trautschold, *Handb. Exp. Pharmacol.* **25**, 52 (1970).
[8] K. J. Worthington and A. Cuschieri, *Clin. Chim. Acta* **52**, 129 (1974).
[9] F. Fiedler and E. Werle, *Hoppe-Seyler's Z. Physiol. Chem.* **348**, 1087 (1967).
[10] F. Fiedler, C. Hirschauer, and E. Werle, *Hoppe-Seyler's Z. Physiol. Chem.* **351**, 225 (1970).

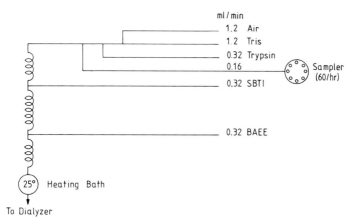

ml / min
1.2   Air
1.2   Tris
0.32 Trypsin
0.16
0.32 SBTI   Sampler (60/hr)

0.32 BAEE

25° Heating Bath

To Dialyzer

FIG. 1. Modification of the flow diagram necessary to switch from the determination of kallikrein on the Technicon Autoanalyzer (Fig. 3 of Fiedler[5]) to that of prokallikrein (plus kallikrein). Adapted from Fiedler and Werle.[9]

bath at 25°C (the temperature of 37°C given by Fiedler[5] for the first heating bath is in error). The general procedure follows the directions already given.[5] If desired, the content of already active glandular kallikrein in the samples is determined as described previously[5] in a separate run with the same samples. Glandular prokallikrein is obtained by the difference. To avoid problems with contamination of the manifold with trypsin, the determination of kallikrein is done first. After several runs, the system is rinsed with dilute alkali and the dialysis membrane changed against a new one.

Radioimmunoassay of Porcine Pancreatic Kallikrein [11]

Reagents

Buffer A: 0.015 M NaH$_2$PO$_4$–0.15 M NaCl–0.01 M ethylenediaminetetraacetate–2 g/liter bovine serum albumin–200 mg/liter Merthiolate (pH 7.4)

Buffer B: 0.015 M NaH$_2$PO$_4$–0.15 M NaCl–0.01 M ethylenediaminetetraacetate–40 g/liter bovine serum albumin–200 mg/liter Merthiolate (pH 7.4)

Buffer C: 0.015 M NaH$_2$PO$_4$–0.15 M NaCl (pH 7.4)

Iodide-125, carrier free, in NaOH solution, pH 7–11, purchased from Amersham (Code: IMS 30)

Porcine pancreatic kallikrein used for radioiodination and as standard

[11] E. Fink and C. Güttel, *J. Clin. Chem. Clin. Biochem.* **16**, 381 (1978).

in the radioimmunoassay was a highly purified neuraminidase-treated preparation of pancreatic β-kallikrein B.[5]

Rabbit anti-kallikrein serum was kindly provided by Bayer AG, Elberfeld, Federal Republic of Germany.

Precipitating antiserum. Donkey anti-rabbit IgG serum (Wellcome Company, catalog number RD 17)

Radioiodination of Porcine Pancreatic Kallikrein. A modification of the chloramine-T method of Greenwood et al.[12] is used. The following solutions are prepared in buffer C: chloramine-T, 3.0 mg/ml; sodium metabisulfite, 3.0 mg/ml; porcine pancreatic kallikrein, 0.5 mg/ml (the exact protein concentration is obtained by quantitative amino acid analysis); and potassium iodide, 10 mg/ml. The solution of Na^{125}I is diluted with buffer C to a concentration of 0.5 mCi/10 μl.

10 μl of the chloramine-T solution are added to a mixture of 5 μl porcine pancreatic kallikrein and 10 μl Na^{125}I solutions. After 5–10 sec, the reaction is terminated by adding 20 μl sodium metabisulfite solution. The reaction mixture is transferred to a Sephadex G-75 column (1.2 × 14 cm) pretreated with 1 ml of buffer B and equilibrated with buffer C; the reaction vial is then rinsed once with 50 μl of potassium iodide solution. The column is eluted with buffer C (40 ml/hr), and 0.8-ml fractions are collected into disposable polypropylene tubes each containing 50 μl of buffer B. The radioactivity of the fractions is determined in an automatic γ-counter, and the 4–6 fractions of the kallikrein peak with the highest radioactivity are pooled and diluted with buffer A to a concentration of 250,000 counts min^{-1} ml^{-1}. The calculated specific radioactivity of the [^{125}I]kallikrein is approximately 80 mCi/mg protein.

Procedure. Dilutions of kallikrein standard samples and unknowns are prepared with buffer A, dilutions of the first and second antibody with buffer B. 70 mg/liter of normal rabbit IgG are added to the [^{125}I]kallikrein solution. An anti-kallikrein serum dilution is chosen that causes a binding of approximately 50% of the [^{125}I]kallikrein under the incubation conditions of the radioimmunoassay. The incubation mixtures are set up in disposable polypropylene tubes following the scheme of Table II. Two methods of incubation can be employed: (1) Incubation under equilibrium conditions: all reagents of Table II are mixed at the same time, with the anti-kallikrein serum being the last reagent. After 24–72 hr of incubation at room temperature, bound and free antigen are separated as described below (Fig. 2, curve with open boxes). (2) Incubation under nonequilibrium conditions: all reagents except the [^{125}I]kallikrein are mixed and incubated for 24–48 hr at room temperature; [^{125}I]kallikrein is then added.

[12] F. C. Greenwood, W. M. Hunter, and J. Glover, Biochem. J. 89, 209 (1963).

TABLE II

COMPOSITION OF THE INCUBATION MIXTURES IN THE RADIOIMMUNOASSAY OF
PORCINE PANCREATIC KALLIKREIN

	Buffer A (μl)	Buffer B (μl)	Sample (standard or unknown) (μl)	Antikallikrein serum (μl)	[^{125}I]Kallikrein (μl)
N	200	100	—	—	100
B_0	200	—	—	100	100
Samples	100	—	100	100	100

After 16–24 hr further incubation at room temperature, bound and free antigen are separated. By this incubation method, an approximately 4-fold increase in sensitivity is obtained (Fig. 2, curve with filled circles).

Separation of bound and free antigen is achieved by the double-antibody method. 0.1 ml of the precipitating antiserum (diluted 1:24) is added, and after 2–3 hr of incubation at room temperature, 0.5 ml of buffer A is added. After 10-min centrifugation at 6000 g the supernatant is aspirated and the radioactivities of the precipitates are measured in an automatic γ-counter.

All calibration samples are set up in quadruplicate. Standard curves are obtained by plotting the fraction of bound radioactivity $(B - N)/(B_0 - N)$ (B, counts bound; B_0, counts bound at zero dose; N, nonspecifically bound counts), against nanograms of kallikrein in a linear–log

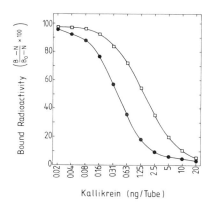

Kallikrein (ng/Tube)

FIG. 2. Radioimmunoassay of porcine pancreatic kallikrein. Standard curves obtained under different incubation conditions. □——□, Incubation of standard, tracer, and anti-kallikrein serum for 66 hr; ●——●, incubation of standard and anti-kallikrein serum for 43 hr, followed by addition of tracer and further 23-hr incubation.

mode (Fig. 2). The assays are evaluated by computer, employing the spline approximation method.[13]

Comments. Coefficients of variation in the concentration range 0.3–5.0 ng per incubation mixture were determined as: intraassay, 6–10% (quadruplicates); interassay, 6–15% (six consecutive assays). The lower detection limit was approximately 300 pg (incubation under equilibrium conditions) and 40 pg (nonequilibrium incubation conditions), respectively, corresponding to concentrations of 3.0 and 0.4 μg/liter, respectively, in the samples to be assayed. No cross-reactivity was detected for porcine trypsin, bovine trypsin and chymotrypsin, and kallikreins of guinea pig submandibular and coagulation glands. In contrast, dose–response curves of porcine submandibular and urinary kallikreins paralleled the pig pancreatic kallikrein standrad curve, indicating immunological identity of the three porcine glandular kallikreins.[11] A dose–response curve nearly parallel to the pig pancreatic kallikrein standard curve was also obtained for human urinary kallikrein.[14] The cross-reactivity calculated from the 50% intercept of the dose–response curve was approximately 4%. The radioimmunoassay proved highly useful in studies on the physiological function of glandular kallikreins. It was possible to detect glandular kallikrein in porcine serum, to demonstrate intestinal absorption of undegraded porcine pancreatic kallikrein, and to show that glandular kallikrein can be excreted by the kidney.[14]

Purification of Porcine Pancreatic α-Kallikrein B'[15]

Porcine pancreatic prokallikrein is resolved by ion-exchange chromatography into two peaks, A and B.[9,10] Native pancreatic α-kallikrein B' (so named to distinguish it from pancreatic kallikrein B from autolyzed pancreas) is obtained from partially purified pancreatic prokallikrein B[10] by spontaneous activation, most probably due to the action of some contaminating proteinase.[10]

All steps of the isolation procedure (Table III) are performed at 0–5°C, working as fast as possible in the first steps in order to minimize the premature activation of the proenzyme. Nevertheless, nearly complete activation has often been observed after the SP-Sephadex step, and such preparations are discarded. Scaling down the procedure by a factor of five appeared to lead more consistently to a high yield of prokallikrein.

---

[13] I. Marschner, F. Erhardt, and P. C. Scriba, *in* "Radioimmunoassays and Related Procedures in Medicine," Vol. I, p. 111. IAEA, Vienna, 1974.

[14] E. Fink, R. Geiger, J. Witte, S. Biedermann, J. Seifert, and H. Fritz, *in* "Enzymatic Release of Vasoactive Peptides" (F. Gross and G. Vogel, eds.), p. 101. Raven, New York, 1980.

[15] F. Fiedler and W. Gebhard, *Hoppe-Seyler's Z. Physiol. Chem.* **361,** 1661 (1980).

TABLE III
PURIFICATION OF PROKALLIKREIN B AND α-KALLIKREIN B' FROM 4.8 KG OF PORCINE PANCREAS

Purification step	Eluate volume (ml)	ml $A_{280}^{1\,cm}$	Kallikrein + prokallikrein (U_{AA})	Kallikrein (U_{AA})	Prokallikrein B (by difference) (U_{AA})	Specific activity (prokallikrein or kallikrein, respectively) [$U_{AA}/(ml\ A_{280}^{1\,cm})$]
Ammonium sulfate fractionation	1500	145,000	44,500	—	—	0.307
SP-Sephadex	4035	49,600	37,100	—	—	0.748
DEAE–Sephacel	1020	3070	7700	0	7700	2.51
Hydroxyapatite	1050	762	4850	260	4590	6.02
Sephadex G-75	108	54.2	3680	3630	(50)	67.0
DEAE–Sephacel	17	29.2	—	4000	—	137
DEAE–Sephacel[a]	21	15.4	—	2140	—	139

[a] Only three-fourths of the total material was rechromatographed in this step. Taken in part from F. Fiedler and W. Gebhard, Hoppe-Seyler's Z. Physiol. Chem. 361, 1661 (1980).

Assays for pancreatic prokallikrein and pancreatic kallikrein activity are done on the Autoanalyzer system (calibrated in the present example with pancreatic β-kallikrein standardized by another variant of the BzArgOEt assay for technical reasons[15]). Protein is determined by measuring the absorbance at 280 nm in cuvettes of 1 cm light path.

Extraction.[10] Fresh porcine pancreas is obtained from the local slaughterhouse and is transported and stored on ice for the short time until processing. The tissue is freed from fat and cut into pieces. 4.8 kg of this material is processed in four batches. Each batch of 1.2 kg is homogenized with 1.8 liters of 0.1 N acetic acid containing 0.5 g soybean-trypsin inhibitor (pract.; Serva, Heidelberg, Federal Republic of Germany) and 1.8 g ethylenediaminetetraacetate (disodium salt) in a large Waring blender (model CB-5) at medium speed, twice for 1 min with 1-min pause. The homogenate is centrifuged for 30 min at 7000 rpm and the supernatant (about 1600 ml) is decanted. One-tenth volume of 1 M sodium phosphate buffer (pH 7.0) is added, and the pH adjusted to 7.0 with 2 N NaOH at the glass electrode. The precipitate is homogenized again with 1.2 liters of the above solution and centrifuged as before. The second supernatant (about 1100 ml) is brought to pH 7.0 in the presence of sodium phosphate as before and combined with the first one.

Ammonium Sulfate Fractionation.[10] Each batch is treated separately with 176 g/liter ammonium sulfate added with stirring during 45 min. 15 min later, the precipitate is centrifuged off and discarded. To the supernatant are added 235 g/liter ammonium sulfate as before. The precipitate is transferred with a minimal amount of water and 0.25 g of soybean-trypsin inhibitor added from the centrifuge beakers to dialysis tubing (35 mm diameter). The material from all four batches is dialyzed overnight against 50 liters of distilled water.

Chromatography on SP-Sephadex.[10] To the contents of the dialysis tubings (1500 ml) are added 50 ml of 2 M ammonium acetate, and the pH is adjusted to 5.4 with 2 N acetic acid. The solution is cleared by centrifugation, if necessary, and transferred to a 14 × 56-cm column of SP-Sephadex C-50 equilibrated with 0.1 M ammonium acetate (pH 5.4). Elution is continued with the same buffer at a rate of 375 ml/hr, and fractions of 250 ml are collected in vessels each containing 1 ml of a solution of 20 mg soybean-trypsin inhibitor, and assayed for pancreatic prokallikrein activity (Fig. 3).

Chromatography on DEAE–Sephacel. The fractions containing pancreatic prokallikrein are combined and adjusted to pH 6.7 with 2 N NH₃. They are applied at a rate of 140 ml/hr to a 5 × 45-cm column of DEAE–Sephacel equilibrated with 0.1 M ammonium acetate (pH 6.7), and the column is washed with 1600 ml of the same buffer. Elution is performed

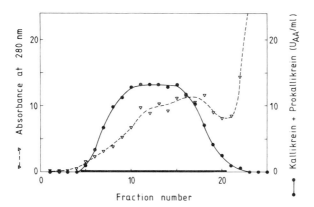

FIG. 3. Chromatography of porcine pancreatic prokallikrein on SP-Sephadex. The bar indicates the fractions pooled.

with a linear gradient, 10.5 liters, rising from 0.1 to 0.6 M ammonium acetate (pH 6.7), at a rate of 63 ml/hr. Fractions of 39 ml are collected in vessels containing 1.2 mg of soybean-trypsin inhibitor (purified by chromatography; E. Merck, Darmstadt, Federal Republic of Germany; this preparation is used from here on) in 1 ml of water, and assayed for pancreatic prokallikrein. The first peak of activity in Fig. 4 contains pancreatic prokallikrein B. The second peak is most often resolved into two: a minor, rather variable peak of pancreatic prokallikrein A and a more prominent one of pancreatic kallikrein (cf. Fig. 1 from Fiedler et al.[10]). Fractions containing pancreatic prokallikrein B are pooled and dialyzed overnight against 8 liters of water.

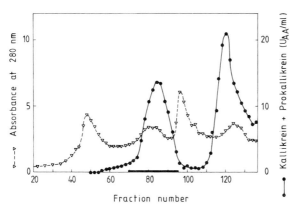

FIG. 4. Chromatography of porcine pancreatic prokallikrein on DEAE–Sephacel. The bar indicates the fractions containing prokallikrein B pooled.

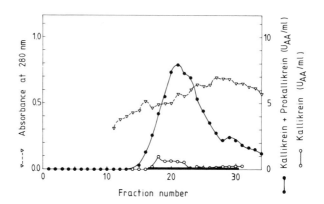

FIG. 5. Chromatography of porcine pancreatic prokallikrein B on hydroxyapatite. From Fiedler and Gebhard.[15]

Chromatography on Hydroxyapatite. The dialyzed eluate is applied to a column of Bio-Gel HT (7 cm diameter, 8 cm height) equilibrated with 0.01 M potassium phosphate (pH 7.0) containing 1 mM CaCl$_2$. The column is washed at first with 1500 ml of the equilibrating buffer, followed by a linear gradient (3600 ml) rising to 0.07 M potassium phosphate–1 mM CaCl$_2$, pH 7.0. Elution is performed at a rate of 130 ml/hr, and fractions of 75 ml (each containing 1 ml of a solution of 0.12 mg soybean-trypsin inhibitor in water) are collected. Fractions are combined as indicated by the bar in Fig. 5.

The eluate is concentrated by adsorption overnight onto a small column of DEAE–Sepharose (2.5 cm diameter, 1.2 cm height, preequilibrated with 0.05 M potassium phosphate, pH 7.0). The column is eluted with the equilibrating buffer containing 2 M KCl, and fractions of 4–5 ml are collected. Fractions containing protein according to the dye-binding assay of Bradford[16] are combined (total volume, 17 ml).

Gel Filtration. This eluate is applied to a column of Sephadex G-75 superfine (4.0 × 128 cm) equilibrated with 0.1 M ammonium acetate (pH 6.7), and eluted with the same buffer at a rate of 30 ml/hr. Fractions of 9 ml (each containing 5 μg of soybean-trypsin inhibitor in 0.1 ml of water) are collected and pooled as indicated by the bar in Fig. 6. In spite of the collection of the eluate in the presence of some soybean-trypsin inhibitor, activation of the proenzyme occurs during these operations, and enzymatically active pancreatic kallikrein B' is exclusively obtained.

Second Chromatography on DEAE–Sephacel. The pooled fractions are applied to a 0.8 × 10-cm column of DEAE–Sephacel equilibrated with 0.1M

[16] M. M. Bradford, *Anal. Biochem.* **72**, 248 (1976).

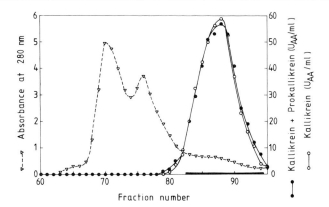

Fɪɢ. 6. Gel filtration of porcine pancreatic kallikrein B' on Sephadex G-75. From Fiedler and Gebhard.[15]

ammonium acetate, pH 6.7. After washing with about 10 column volumes of the equilibrating buffer, the material is eluted with a linear gradient (100 ml) rising to 0.6 M ammonium acetate (pH 6.7), at a rate of 3.4 ml/hr. Fractions of 2.1 ml are collected and pooled as indicated in Fig. 7.

Rechromatography on DEAE–Sephacel. The pooled fractions are diluted to a concentration of 0.1 M ammonium acetate as determined by conductivity measurements and rechromatographed under the same conditions on a 1.2 × 10-cm column of DEAE–Sephacel (Fig. 8). Porcine pancreatic kallikrein B' elutes in a single somewhat skewed peak, possibly due to the heterogeneity of its carbohydrate portion. For storage (at −20°C), the enzyme solution is dialyzed against distilled water and lyophilized.

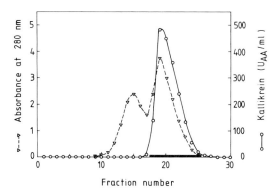

Fɪɢ. 7. Chromatography of porcine pancreatic kallikrein B' on DEAE–Sephacel. The bar indicates the fractions pooled. From Fiedler and Gebhard.[15]

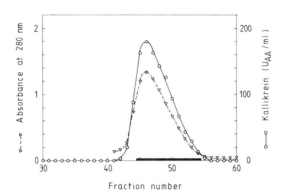

Fraction number

FIG. 8. Rechromatography of porcine pancreatic kallikrein B' on DEAE–Sephacel. The bar indicates the fractions pooled. From Fiedler and Gebhard.[15]

The total yield of porcine pancreatic kallikrein B' has been estimated to be about 12% and the purification factor to be about 10,000.[15]

Criteria of Purity. In the last purification step, the specific activity of the preparation of porcine pancreatic kallikrein B' was not increased further. The specific activity of the enzyme was constant across the whole peak and as high as that of the other porcine glandular α-kallikreins. SDS electrophoresis of the reduced enzyme yielded a single though relatively broad band. Disc electrophoresis also resulted in a single band with some structure. The amino acid composition of the preparation was very similar to that of the other glandular α-kallikreins of the pig.[15]

Purification of Porcine Pancreatic β-Kallikreins

Two-chain porcine pancreatic β-kallikrein is obtained when pancreas homogenates are subjected to some incubation step for the activation of the proenzyme. The purification of pancreatic β-kallikreins A and B has been described in a previous volume of this series.[5] For the separation of pancreatic β-kallikreins A and B, DEAE–Sepharose is much preferable to the formerly used DEAE–cellulose because of the much better flow properties of the column. A 0.9 × 35-cm column of DEAE–Sepharose preequilibrated with 0.2 M ammonium acetate (pH 6.7) is loaded with 100 mg of pancreatic β-kallikrein and eluted with a linear gradient of 300 ml, rising from 0.2 to 0.6 M ammonium acetate (pH 6.7), at a flow rate of 7.5 ml/hr. The first peak of pancreatic β-kallikrein B appears at about 150 ml. By rechromatography, pancreatic β-kallikrein B of up to 155 U_{AT}/(ml $A_{280}^{1\ cm}$) could be obtained, while a value of 141 ± 5 U_{AT}/(ml $A_{280}^{1\ cm}$) is calculated from the data of Fiedler et al.[17]

[17] F. Fiedler, C. Hirschauer, and E. Werle, *Hoppe-Seyler's Z. Physiol. Chem.* **356,** 1879 (1975).

TABLE IV

PURIFICATION OF KALLIKREIN FROM 8.45 KG OF PORCINE SUBMANDIBULAR GLANDS[a]

Purification step	Kallikrein activity		Specific activity $[U_{ADH}/(ml\ A_{280}^{1\,cm})]$	Yield (%)
	(U_{AT})	(U_{ADH})		
Neutralized supernatant	187,500			(100)
65% Acetone precipitate	134,000			71
DEAE–Cellulose	115,000	34,000	1.2	61
Affinity chromatography		20,900	95.7	38
DEAE–Sephadex		17,000	97.3	31

[a] Taken from M. Lemon, F. Fiedler, B. Förg-Brey, C. Hirschauer, G. Leysath, and H. Fritz, *Biochem. J.* **177**, 159 (1979).

A further alternative isolation procedure for porcine pancreatic β-kallikreins A and B starting from a pancreatic powder preparation has been described recently.[18] An additional minor 2-chain form, pancreatic kallikrein III, has also been obtained in this work. Pancreatic kallikreins A and B isolated recently from porcine pancreas by Moriya *et al.*[19] are most probably also the 2-chain forms, as an autolysis step was used in the purification procedure.

Purification of Porcine Submandibular Kallikrein[20]

In the following isolation procedure (Table IV), activity was assayed with a unique pH-stat method (40.5 mM BzArgOEt, pH 7.5, 37°C) or a modification of the manual spectrophotometric assay using 1 mM BzArgOEt, for technical reasons.[20]

Extraction. Submandibular glands are obtained from freshly slaughtered pigs and stored at −20°C. Frozen gland tissue (8.45 kg) is minced in a heavy-duty stainless steel mill, and the semifrozen pulp homogenized in 0.1 M acetic acid (5 liters/kg of tissue) in a Waring-type blender (1 min at maximum speed) at 5°C. The homogenate is centrifuged at 19,000 g at 20°C in a continuous centrifuge and the supernatant neutralized to pH 7 with 5 M NaOH.

Acetone Fractionation. Acetone precooled to 5°C is added with vigorous stirring to give a concentration of 35% (v/v), and left overnight at 5°C. The precipitate is removed by centrifugation (19,000 g at 20°C in a con-

[18] H. Kira, S. Hiraku, and H. Terashima, *Adv. Exp. Med. Biol.* **120A,** 273 (1979).

[19] H. Moriya, Y. Fukuoka, Y. Hojima, and C. Moriwaki, *Chem. Pharm. Bull.* **26,** 3178 (1978).

[20] M. Lemon, F. Fiedler, B. Förg-Brey, C. Hirschauer, G. Leysath, and H. Fritz, *Biochem. J.* **177,** 159 (1979); Correction: Enclosure to *Biochem. J.* **181** (No. 2, August) (1979).

tinuous centrifuge) and the concentration of acetone in the supernatant raised to 65% (v/v). Stirring is continued for 2 hr at 5°C, and the precipitate collected by centrifugation in the same way. The precipitate is dissolved by stirring overnight at 5°C in 14 liters of water that has been adjusted to pH 7.5 with ammonia solution.

DEAE–Cellulose Adsorption. DEAE–cellulose (Whatman DE 23; 300 g dry weight equilibrated with 0.1 M ammonium acetate, pH 6.7) is stirred into the solution and after 1 hr the supernatant removed by filtration through a glass sinter under reduced pressure. The DEAE–cellulose is washed with 2 × 5 liters of 0.1 M ammonium acetate (pH 6.7), and then the kallikrein eluted with 5 liters of ammonium acetate, 40 g/liter (pH 6.7), containing NaCl (50 g/liter). No further kallikrein is eluted by raising the concentration of the elution buffer to ammonium acetate, 50 g/liter (pH 6.7), containing NaCl (50 g/liter). The eluate is dialyzed twice against 10 volumes of water (pH adjusted to 7 with ammonia solution) and freeze-dried. 28.3 g of freeze-dried material is obtained from 8.45 kg of gland tissue.

Affinity Chromatography. The affinity adsorbent is prepared from guanidinated aprotinin coupled to Sepharose. 100 mg of aprotinin (basic trypsin-kallikrein inhibitor from bovine organs, Trasylol, Bayer) is guanidinated with O-methylisourea sulfate, under the conditions described by Kummel,[21] for 4 days at 4°C. The guanidinate aprotinin is isolated on a Merckogel PGM-2000 column (Merck) (1.8 × 7.5 cm), equilibrated and developed with water at a flow rate of 60 ml/hr. The guanidinated aprotinin fractions, detected by inhibition of trypsin as described by Fritz *et al.*,[22] are freeze-dried. The yield is 93 mg. Freeze-dried CNBr-activated Sepharose-4B (2 g; Pharmacia, Uppsala, Sweden) is swollen in 1 mM HCl at 4°C; 88 mg of guanidinated aprotinin is added in 12 ml of 0.1 M sodium tetraborate buffer (pH 8.0) containing 0.5 M NaCl. The mixture is allowed to react for 2 hr at room temperature (20°C), with continuous shaking. The reaction supernatant is removed from the resin by centrifugation (5000 g for 5 min) followed by decantation, and the resin suspended in 12 ml of 1 M ethanolamine (pH 9) for a further 2 hr at room temperature. The resin is then extensively washed with the borate–NaCl buffer, followed by 0.1 M triethanolamine (pH 7.8) containing 1 M NaCl. The resin is expanded to the required bed volume by mixing with Sephadex G-75.

Batches of 3–12 g of the freeze-dried kallikrein are dissolved in 250 ml of affinity-wash buffer (0.1 M triethanolamine, pH 7.8, containing 1 M NaCl) and mixed with 145 ml (bed volume) of Sepharose-linked aprotinin.

[21] J. R. Kummel, this series, Vol. 11, p. 584.
[22] H. Fritz, G. Hartwich, and E. Werle, *Hoppe-Seyler's Z. Physiol. Chem.* **345**, 150 (1966).

After decantation of the supernatant, the resin is washed with 250 ml of affinity-wash buffer and packed into a 3.5-cm-diameter column, cooled to 12°C with a water jacket. The resin is washed at a flow rate of 30–40 ml/hr until the eluate no longer contains detectable amounts of material absorbing at 260 nm (400 ml is normally sufficient). The wash buffer is replaced by eluting buffer (0.1 M triethanolamine, pH 6.45, containing 0.5 M NaCl and 0.5 M benzamidine hydrochloride). All the kallikrein is eluted in the first 300 ml. Kallikrein can be directly assayed in the eluate, in spite of the benzamidine, because dilution of the inhibitor in the assay cuvette decreases inhibition almost to zero. The combined eluate from three affinity separations is concentrated to 50 ml by ultrafiltration (Amicon UM-2 filter); the remaining benzamidine is removed by filtration through a column (2.2 × 60 cm) of Sephadex G-25, with 10 mM ammonium acetate (pH 6.7) as the buffer, and the kallikrein-containing fractions are combined and freeze-dried.

Chromatography on DEAE–Sephadex. The kallikrein (10 mg/ml of 0.1 M ammonium acetate, pH 6.7) is loaded onto a column (1.8 × 55 cm) of DEAE–Sephadex A-50 (equilibrated with 0.1 M ammonium acetate, pH 6.7, and cooled to 12°C with a water jacket). The kallikrein is eluted by means of a convex ammonium acetate gradient formed by passing 0.8 M ammonium acetate (pH 6.7) into a constant-volume mixing chamber containing 150 ml of 0.1 M ammonium acetate (pH 6.7) (Fig. 9). A peak of material absorbing strongly at 253 nm is eluted in fractions 27–35 and may be due to nucleic acids not removed by affinity chromatography. Protein is eluted in a single peak in which absorption at 280 nm corresponds well to kallikrein activity. The kallikrein is desalted through Sephadex G-25 (2.2 × 60 cm) with 10 mM ammonium acetate (pH 6.7) as the buffer, and freeze-dried.

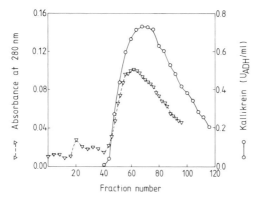

Fig. 9. Chromatography of porcine submandibular kallikrein on DEAE–Sephadex.

TABLE V

PURIFICATION OF KALLIKREIN FROM 17 LITERS OF PORCINE URINE

Purification step	Kallikrein activity (U_{ADH})	Specific activity ($U_{ADH}/ml\ A_{280}^{1\,cm}$)	Purification factor	Yield (%)
Acetone precipitation	2000	0.35	1	(100)
Sephadex G-100	1840	1.4	4	92
DEAE–Sephadex	1480	5.8	16	74
Aprotinin–Sepharose				
Buffer eluate	400	(0.4)		20
Benzamidine eluate	1000	38	108	50
DEAE–Sephadex	820			41
DEAE–Sephadex rechromatography				
Kallikrein I	330	61	174	16.5
Kallikrein II	270	67	192	13.5
Kallikrein III	60	65	186	3

Criteria of Purity. The specific activity of the preparation did not increase further on rechromatography on DEAE–Sephadex. The material eluted as a single peak with constant specific activity across the peak.[20] Submandibular kallikrein behaved as a homogeneous protein in the ultracentrifuge.[23] Disc-gel electrophoresis revealed a single, though rather broad band. On SDS electrophoresis, the preparation showed a major band of submandibular kallikrein B and a minor A form, with equal specific activities.[20] Specific activity and amino acid composition of the enzyme are very similar to those of the other porcine glandular α-kallikreins.[20]

Purification of Porcine Urinary Kallikrein[24]

All operations (Table V) are performed in the cold at about 4°C. The activity assay used was a variant of the manual alcohol dehydrogenase-linked assay with 1 mM BzArgOEt. Protein was determined in column eluates by monitoring with a flowthrough spectrophotometer, LKB Uvicord III.

Collection of Urine. Porcine urine is freshly collected from the undamaged bladder in a slaughterhouse immediately after death of the animals. After opening of the bladder with a knife, the urine is poured into a flask

[23] H. Fritz, F. Fiedler, T. Dietl, M. Warwas, E. Truscheit, H. J. Kolb, G. Mair, and H. Tschesche, *in* "Kininogenases: Kallikrein" (G. L. Haberland, J. W. Rohen, and T. Suzuki, eds.), Vol. 4, p. 15. Schattauer, Stuttgart, 1977.

[24] G. Mair. H. Tschesche, and H. Fritz, in preparation.

and cleared of solid particles and fat tissue by filtration through a coarse paper filter. The collected urine is covered with a thin layer of toluene to prevent growth of microorganisms. It is kept at 2–4°C and never stored longer than a week. The enzymatic activity ranges from 140 to 160 U_{ADH}/liter with a specific activity of about 3.5 mU_{ADH}/(ml $A_{280}^{1\,cm}$).

Acetone Fractionation. Five liters of cold urine are mixed under vigorous stirring with 4.1 liters of precooled acetone [final concentration, 45% (v/v)] in several portions. The resulting suspension is thoroughly stirred with a glass rod several times and allowed to stand overnight. The supernatant is collected from the precipitate, which is removed by decantation and/or centrifugation and discarded. The clear supernatant is treated with 7.4 liters of cold acetone in the same way as described above [final acetone concentration, 70% (v/v)]. After standing for 1 day, the precipitate is collected by centrifugation. The precipitate is kept in an open beaker for 2 hr at room temperature to allow evaporation of acetone, and then stirred several times with small portions of distilled water in order to redissolve the enzyme. Undissolved material is removed by centrifugation or filtration. The light yellow solution is dialyzed for 24 hr against two 10-liter batches of distilled water and lyophilized.

A major benefit of the acetone precipitation is the almost entire loss of the unpleasant urinary odor. The lyophilized material has a specific activity of 0.34–0.43 U_{ADH}/(ml $A_{280}^{1\,cm}$). The yield of enzyme activity is about 80%. Acetone precipitation has been used for the purification of porcine urinary kallikrein by Werle and Trautschold.[25] The fractionation limits of 50–67% (v/v) acetone given by these authors, however, resulted, in our experience, in a yield of only 60–65% of enzymatic activity, though at a somewhat higher purity of 0.57 U_{ADH}/(ml $A_{280}^{1\,cm}$).

Gel Filtration. An amount of the lyophilized material (3.5 g) is dissolved in 50 ml of 0.01 M ammonium acetate (pH 6.7) and centrifuged. The clear supernatant is applied to the top of a column (4.4 × 90 cm) of Sephadex G-100 fine, equilibrated with 0.01 M ammonium acetate (pH 6.7). The column is developed at a flow rate of 22 ml/hr, and fractions of 12 ml are collected (Fig. 10). Kallikrein activity is eluted in a single peak between $v_{rel} = 1.7–2.5$. All fractions containing activity are pooled, concentrated by ultrafiltration (Amicon Diaflow membrane UM-10 at 3 atm N_2), and lyophilized.

First Chromatography on DEAE–Sephadex. Crude kallikrein (2.5 g) from the gel-filtration step is dissolved in 40 ml of 0.1 M ammonium acetate, pH 6.7. The solution is centrifuged and applied to a column (4.4 × 35 cm) of DEAE–Sephadex A-50, equilibrated with 0.1 M ammonium

[25] E. Werle and I. Trautschold, *Ann. N.Y. Acad. Sci.* **104**, 117 (1963).

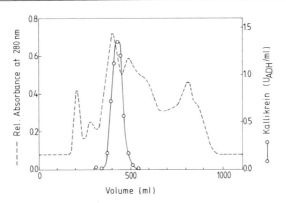

FIG. 10. Gel filtration of an extract of an acetone powder from porcine urine on Sephadex G-100.

acetate, pH 6.7. Elution is started with 100 ml of the equilibration buffer and is continued with a gradient obtained by mixing 1 liter of 0.1 M and 1.65 liters of 1.0 M ammonium acetate (pH 6.7), contained in vessels open to the atmosphere, connected with a siphon, and filled to the same level. At a flow rate of 60 ml/hr, fractions of 12 ml are collected (Fig. 11). Fractions containing enzymatic activity are pooled and concentrated by ultrafiltration. Desalting is performed on a column (3.4 × 60 cm) of Bio-Gel P2 in 0.05 M pyridine, adjusted to pH 7.5 with formic acid, at a flow rate of 40 ml/hr, and the material is lyophilized.

Affinity Chromatography. Aprotinin–Sepharose is prepared by activation of Sepharose-4B with cyanogen bromide according to Axén and Ernback,[26] and coupling of aprotinin (basic trypsin-kallikrein inhibitor from bovine organs, Trasylol, Bayer) to the activated Sepharose. 200 ml of Sepharose-4B is washed with distilled water to remove antimicrobial agents, sucked dry, and suspended in 200 ml of water. This suspension is stirred with a glass rod while adding 25 g of powdered cyanogen bromide. The reaction solution is brought to pH 11 with 10 N NaOH and kept at this pH during the reaction by the addition of further NaOH. The temperature is kept at 20–25°C by cooling with water and with pieces of ice placed directly in the reaction mixture. After the reaction has stopped, the suspension is cooled to 4°C with a greater amount of ice. Coupling is performed in a cold room at 4°C. The activation mixture is rapidly filtered and washed with 2.5 liters of cold 0.2 M sodium citrate buffer, 0.7 M in NaCl, pH 5.2. Aprotinin (200 mg in 80 ml of buffer) is added, and the suspension is stirred in a beaker with a glass rod for 2–3 hr. The reaction mixture is then allowed to stand in the cold for an additional 3 hr before it is filtered

[26] R. Axén and S. Ernback, *Eur. J. Biochem.* **18**, 351 (1971).

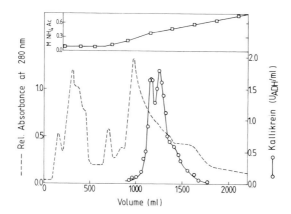

Fig. 11. Chromatography of crude porcine urinary kallikrein on DEAE–Sephadex. The inset specifies the ammonium acetate gradient used.

and washed with 1 liter of the above buffer. 100 ml of 1 M ethanolamine HCl solution (pH 8) is added and the mixture allowed to stand overnight. The gel is washed again with 1 liter of the above citrate buffer and finally with 2 liters of 0.1 M triethanolamine HCl buffer (pH 7.8) containing 1 M NaCl.

The aprotinin–Sepharose is equilibrated in 0.1 M triethanolamine HCl–0.8 M NaCl–0.2 g/liter NaN$_3$, placed in a 1.4 × 55-cm column and thoroughly washed with equilibration buffer. Crude kallikrein (1–1.5 g, 1100–1200 U$_{ADH}$) is dissolved in 10–15 ml of buffer and applied to the column, which is then washed with buffer at a flow rate of 15 ml/hr until the absorbance of the eluate at 206 nm has reached the value of the input buffer. Kallikrein is released from the affinity gel by elution with 0.3 M benzamidine HCl in 0.1 M triethanolamine HCl (pH 6.4)—0.3 M NaCl–0.2 g/liter NaN$_3$, and collected in fractions of 7–10 ml (Fig. 12). Fractions containing kallikrein activity are pooled, concentrated by ul-trafiltration, desalted on Bio-Gel P2, and lyophilized.

The binding capacity of the aprotinin–Sepharose is highly dependent on the concentration of the kallikrein solution applied. Sufficiently high concentrations can only be obtained with kallikrein preparations of specific activity not lower than 3–4 U$_{ADH}$/(ml $A_{280}^{1\,cm}$).

Second Chromatography on DEAE–Sephadex. Final purification is achieved by rechromatography on a column (1.4 × 60 cm) of DEAE–Sephadex A-50 equilibrated in 0.1 M ammonium acetate, pH 6.7. Kallikrein (30 mg) is dissolved in 3 ml of buffer, placed on top of the column, and eluted with a gradient obtained by mixing 270 ml of 0.1 M and 320 ml of 1 M ammonium acetate (pH 6.7), at a flow rate of 10 ml/hr and a fraction

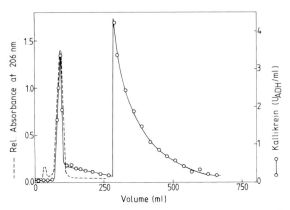

Fig. 12. Affinity chromatography of porcine urinary kallikrein on aprotinin–Sepharose.

volume of 5 ml. Activity is eluted between 0.34 and 0.57 M ammonium acetate, with a predominant peak at 0.41 M and a plateau at 0.45 M (Fig. 13). Three kallikrein fractions—I (0.36–0.42 M ammonium acetate), II (0.43–0.48 M), and III (0.49–0.56 M)—are selected.

Rechromatography of Kallikrein Fractions I, II, and III. Each kallikrein fraction is rechromatographed on DEAE–Sephadex A-50 (0.9 × 50 cm) with the same gradient as described before at a flow rate of 6 ml/hr and fraction size of 4 ml. Symmetrical peaks are obtained for each kallikrein (Fig. 14) at the same ammonium acetate concentration as in the preceding step. Fractions containing the main activity of each peak are pooled sepa-

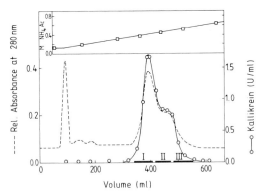

Fig. 13. Second chromatography of porcine urinary kallikrein on DEAE–Sephadex. The inset specifies the ammonium acetate gradient used. Activity was determined in this case with D-Val-Leu-Arg-*p*-nitroanilide as substrate. The bars designate the fractions pooled for urinary kallikreins I, II, and III.

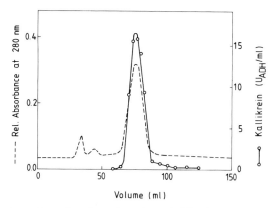

FIG. 14. Rechromatography of porcine urinary kallikrein I on DEAE–Sephadex.

rately from the side fractions, concentrated by ultrafiltration, desalted on Bio-Gel P2, lyophilized, and stored in a freezer. The distribution pattern is 50% of activity (53% by weight) for kallikrein I, 41% (38%) for kallikrein II, and 9% (9%) for kallikrein III.

Stability. At pH values between 6 and 8, the solution of urinary kallikrein is stable under refrigeration for several days. Losses in activity are below 5% after 3 days. At pH 4, losses are 7% and at pH 2, more than 85% after 3 days.[24]

Criteria of Purity. On disc electrophoresis, porcine urinary kallikreins I–III migrated toward the cathode, at a rate increasing in that order. Each form of the enzyme gave a single, but rather broad band.[24] In the ultracentrifuge, the enzyme behaved as a homogeneous protein.[23] Specific activity[20] and amino acid composition[27] are similar to those of the other porcine glandular α-kallikreins.

Properties of the Porcine Glandular Kallikreins[27a]

Molecular Properties

Heterogeneity, Molecular Weight, and Carbohydrate Content. The glandular kallikreins of the pig are all glycoproteins, and this fact is most probably the reason for their pronounced molecular heterogeneity. As compiled in Table I, several forms can be distinguished in each of the enzymes by electrophoresis and/or ion-exchange chromatography. Heterogeneity expresses itself also in the broadness of the bands observed on

[27] H. Tschesche, G. Mair, G. Godec, F. Fiedler, W. Ehret, C. Hirschauer, M. Lemon, H. Fritz, G. Schmidt-Kastner, and C. Kutzbach, *Adv. Exp. Med. Biol.* **120A**, 245 (1979).
[27a] Older data for porcine pancreatic β-kallikrein are compiled by Fiedler.[5]

TABLE
CARBOHYDRATE COMPOSITION OF PORCINE
MOLECULE, OR WEIGHT

Carbohydrate composition	Pancreatic β-kallikrein				
	A Forms				
	$d_2{}^a$	A^b	A^c	A^d	I^e
Sialic acid	0.016		0.25	0.24	
Fucose		0.51	0.41	0.80	
Hexoses	2.8 (1.55%)				3.50%
Mannose		3.6; 2.9	2.4	2.20	
Galactose		0.80; 0.70	1.8	0.76	
Hexosamines	2.4 (1.6%)				2.00%
Glucosamine		4.1 (3.9); 3.5	3.7	2.69	
Galactosamine					
Total carbohydrate	3.2%	$5.6\%^h$	5.8%	4.56%	$5.50\%^h$

[a] M. Zuber and E. Sache, *Biochemistry* **13**, 3098 (1974)
[b] F. Fiedler, C. Hirschauer, and E. Werle, *Hoppe-Seyler's Z. Physiol. Chem.* **356**, 1879 (1975).
[c] H. Moriya, Y. Fukuoka, Y. Hojima, and C. Moriwaki, *Chem. Pharm. Bull.* **26**, 3178 (1978).
[d] M. Ikekita, H. Moriya, K. Kizuki, and S. Ozawa, *Chem. Pharm. Bull.* **28**, 1948 (1980).
[e] H. Kira, S. Hiraku, and H. Terashima, *Adv. Exp. Med. Biol.* **120A**, 273 (1979).

disc or SDS electrophoresis and in the existence of numerous multiple forms with different isoelectric points (see below).

Most clearly evident have become the causes for the existence of porcine pancreatic β-kallikrein in two principal forms, A and B (Table I). These forms have the same amino acid composition and end groups,[17] and the same catalytic properties.[5] Carbohydrate analyses are compiled in Table VI. All authors agree on the much higher carbohydrate content of pancreatic β-kallikrein B. This difference in carbohydrate appears sufficient to explain the differences in the electrophoretic and chromatographic behavior. The lowest carbohydrate content is found in a recently isolated minor component, β-kallikrein C (possibly identical with pancreatic kallikrein d_3, Table I), which shows the highest electrophoretic mobility.[27b] A further minor form, β-kallikrein III, contains roughly as much carbohydrate as kallikrein A (Table VI) and also migrates like this form of the enzyme on disc electrophoresis.[18]

For the molecular weight of porcine pancreatic β-kallikrein, a newer gel-filtration result for an unresolved preparation is 31,000.[28] Newer data

[27b] F. Fiedler and C. Hirschauer, *Hoppe-Seyler's Z. Physiol. Chem.* **362**, 1209 (1981).
[28] Y. Hojima, C. Moriwaki, and H. Moriya, *Chem. Pharm. Bull.* **23**, 1128 (1975).

VI

Glandular Kallikreins (Residues per
Percent Where Indicated)

	Pancreatic β-kallikrein					
	B Forms				Form	Submandibular
$d_1{}^a$	B^b	B^c	B^d	IIe	IIIe	kallikrein^f
0.04		0.78	0.47			≥2.0
	1.23; 0.95	1.3	2.44			
3.3 (1.8%)				5.95%	3.60%	~25
	6.0; 5.3	5.0	3.92			
	1.6; 1.4	3.3	1.78			
4.3 (2.8%)				4.95%	3.31%	
	9.6 (9.2); 9.8 (9.05^a)	9.1	4.63			~14
						~0.6
4.6%	11.5%^{h}	12.6%	8.40%	10.90%^{h}	6.91%^{h}	~24%

^{f} M. Lemon, F. Fiedler, B. Förg-Brey, C. Hirschauer, G. Leysath, and H. Fritz, *Biochem. J.* **177**, 159 (1979).
^{g} F. Fiedler, W. Ehret, G. Godec, C. Hirschauer, C. Kutzbach, G. Schmidt-Kastner, and H. Tschesche, *in* "Kininogenases: Kallikrein" (G. L. Haberland, J. W. Rohen, and T. Suzuki, eds.), Vol. 4, p. 7. Schattauer, Stuttgart, 1977.
^{h} Without sialic acid.

for various forms of the enzyme are given in Table VII. The new ultracentrifugation result for the B form agrees well with the molecular weight calculated from the composition.

On SDS electrophoresis after reduction, the various forms of pancreatic β-kallikrein display their 2 chains. The decrease in the values of the sums of the apparent molecular weights of the chains of β-kallikreins B, A, and C (Table VII) is consistent with the decrease of the total carbohydrate content, respectively. The sum of the chains of β-kallikrein III resembles that of β-kallikrein A, with a similar total carbohydrate content. Since the same end groups have been found in β-kallikreins A and B,[17] the sites of cleavage and the protein moieties of the chains of the two forms of the enzyme evidently are the same. Most probably, the sites of cleavage in the polypeptide backbones of β-kallikreins C and III are also the same. Both chains of β-kallikrein B with the highest apparent molecular weights were found to contain carbohydrate,[3,27b,29] while the smaller B chain of β-kallikrein A contains none.[29] The molecular weight of the protein moiety of the B chain of β-kallikrein B calculated from the amino acid com-

[29] M. Ikekita, H. Moriya, K. Kizuki, and S. Ozawa, *Chem. Pharm. Bull.* **28**, 1948 (1980).

TABLE VII

MOLECULAR WEIGHTS AND RELATED CONSTANTS OF PORCINE GLANDULAR KALLIKREINS

	By ultracentrifuge[a]			Molecular weight (M_r) by different methods		
	$S_{20,w}$	$D_{20,w}$ ($\times 10^7$ cm² sec⁻¹)	M_r	SDS electrophoresis	Gel filtration[e]	Composition
Submandibular kallikrein A	3.1; 3.2	7.1	38,000 ± 1000	35,900 ± 700[b]		
Submandibular kallikrein B				39,600 ± 800[b]		~34,000[b]
Urinary kallikrein A	3.0; 3.1	6.6	40,000 ± 1500	36,100 ± 600[b]		
Urinary kallikrein B				39,600 ± 700[b]		
Pancreatic α-kallikrein B′				36,000[c]		
Pancreatic β-kallikrein A				{A chain: 13,000[d]; 12,600[e] {B chain: 17,100[d]; 16,300[e]	28,300	27,100[f]
Pancreatic β-kallikrein B	2.5; 2.4	7.8	28,000 ± 1000	{A chain: 13,800[e]; 12,800[e] {B chain: 20,800[e]; 19,000[e]	32,000	28,900[f]
Pancreatic β-kallikrein C				{A chain: 9900[d] {B chain: 17,100[d]		
Pancreatic β-kallikrein III				{A chain: 9700[e] {B chain: 20,000[e]	28,300	

[a] H. Fritz, F. Fiedler, T. Dietl, M. Warwas, E. Truscheit, H. J. Kolb, G. Mair, and H. Tschesche, in "Kininogenases: Kallikrein," G. L. Haberland, J. W. Rohen, and T. Suzuki, eds.), Vol. 4, p. 15. Schattauer, Stuttgart, 1977.

[b] M. Lemon, F. Fiedler, B. Förg-Brey, C. Hirschauer, G. Leysath, and H. Fritz, Biochem. J. 177, 159 (1979).

[c] F. Fiedler and W. Gebhard, Hoppe-Seyler's Z. Physiol. Chem. 361, 1661 (1980).

[d] F. Fiedler and C. Hirschauer, Hoppe-Seyler's Z. Physiol. Chem. 362, 1209 (1981).

[e] H. Kira, S. Hiraku, and H. Terashima, Adv. Exp. Med. Biol. 120A, 273 (1979).

[f] F. Fiedler, W. Ehret, G. Godec, C. Hirschauer, C. Kutzbach, G. Schmidt-Kastner, and H. Tschesche, in "Kininogenases: Kallikrein," (G. L. Haberland, J. W. Rohen, and T. Suzuki, eds.), Vol. 4, p. 7. Schattauer, Stuttgart, 1977.

position, 16,490, is quite consistent with the SDS molecular weight of the B chain of β-kallikrein A (Table VII). Evidently, there can be distinguished two types of each chain, a high-molecular-weight (index h) and a low-molecular-weight (l) one, differing only by their carbohydrate content. All four forms of pancreatic β-kallikrein that can be assembled from these types of chains do exist. β-Kallikrein B is of composition A_hB_h; A_hB_l is β-kallikrein A, A_lB_h is form III, and A_lB_l is form C.[27b]

Native single-chain pancreatic α-kallikrein B' showed on SDS electrophoresis the same apparent molecular weight as the A forms of porcine submandibular and urinary kallikreins (Table VII).

Porcine submandibular kallikrein isolated from the glands was resolved on SDS electrophoresis into a minor form A and a predominant B form of higher apparent molecular weight (Table VII), which probably differ only in their carbohydrate content. However, the absolute values of these molecular weights, as well as those of the other porcine glandular kallikreins determined by this technique, presumably are too high because of the glycoprotein nature of these enzymes.[1] In the ultracentrifuge, a similar molecular weight was obtained (Table VII). This value, however, is also certainly higher than the true one, as $\bar{\nu}$ of pancreatic β-kallikrein B with a much lower carbohydrate content (Table VI) has been used in the calculations.[23] A preliminary molecular weight calculated from the composition is distinctly lower (Table VII). Submandibular kallikreins A and B can be partly separated by ion-exchange chromatography and have the same specific activities.[20]

A preparation of porcine urinary kallikrein not resolved into the components I–III also showed on SDS electrophoresis a predominant B form and a minor A form, of the same apparent molecular weights as the two forms of submandibular kallikrein (Table VII). The weak band of urinary kallikrein A was, however, much more clearly separated from the B form than in the case of the submandibular enzyme.[20] Another cut of this preparation revealed higher amounts of the A form, and urinary kallikrein II consisted predominantly of this form.[30] Urinary kallikreins A and B showed the same specific activities and were also partially separated on ion-exchange chromatography.[20] A preparation consisting mainly of the B form displayed a similar molecular weight to that of submandibular kallikrein B, also in the ultracentrifuge (Table VII). All these values, however, are presumably too high for reasons discussed previously. The same sugars as in submandibular kallikrein (Table VI) were also found in urinary kallikrein. The galactosamine content was also much lower than that of glucosamine. Total carbohydrate amounted to as much as 30% in the most carbohydrate-rich fraction.[24]

[30] F. Fiedler, unpublished results.

TABLE VIII

CARBOHYDRATE COMPOSITION (RESIDUES PER MOLECULE) OF COMPONENTS OF PORCINE PANCREATIC KALLIKREINS A AND B RESOLVED BY ISOELECTRIC FOCUSING

| | Form and fraction no. | | | | | | | | | |
	A^a	$A\text{-}II^a$	$A\text{-}III^a$	B^a	$B\text{-}III^a$	$B\text{-}IV^a$	$B\text{-}V^a$	B^b	$B\text{-}2^b$	$B\text{-}3^b$
Isoelectric point	(Mixture)	4.13	4.24	(Mixture)	4.13	4.19	4.29	(Mixture)	3.9–4	4.1–4.2
Sialic acid	0.24	0.57	0.16	0.47	1.77	0.95	0.33	0.78	0.58	0.046
Fucose	0.80	1.47	2.35	2.44	2.93	3.14	3.21	1.3	1.7	1.2
Mannose	2.20	1.36	3.41	3.92	1.71	1.52	4.14	5.0	3.8	5.1
Galactose	0.76	1.38	0.19	1.78	1.90	1.35	0.48	3.3	1.2	0.52
Glucosamine	2.69	2.77		4.63	5.30	3.96		9.1	7.5	7.1
Total carbohydrate (%)	4.56	5.26		8.40	9.19	7.10		12.6	9.5	8.8

[a] M. Ikekita, H. Moriya, K. Kizuki, and S. Ozawa, *Chem. Pharm. Bull.* **28**, 1948 (1980).
[b] H. Moriya, Y. Fukuoka, Y. Hojima, and C. Moriwaki, *Chem. Pharm. Bull.* **26**, 3178 (1978).

Microheterogeneity and Isoelectric Points. On isoelectric focusing, porcine pancreatic β-kallikrein is resolved into numerous bands. Isoelectric points determined more recently [4.15, 4.05, 3.94, 3.85, 3.80, or 4.18, 4.11, 4.01, 3.90, 3.80, or 4.23, 4.17, 4.07, 3.92, 3.85, 3.68[31]; 4.22, 4.14, 3.97, 3.77, 3.57[28]; 4.18, 4.12, 4.01, 3.92, 3.82, 3.80 for kallikrein B and similar for the A form[19]; and 4.04, 3.92, 3.75 for β-kallikrein I (A), 4.03, 3.83, 3.67 for β-kallikrein II (B), and 3.94, 3.83 for β-kallikrein III[18]] are similar to those already compiled.[5] This microheterogeneity is most probably due to a variable content of sialic acid residues as it is eliminated by treatment with neuraminidase.[6,10,17,19,20] Indeed, the sialic acid content of isolated components increased with decreasing isoelectric point (Table VIII).

On column electrofocusing, porcine pancreatic prokallikrein B displayed a main component at pH 4.55, whereas there were two for pancreatic α-kallikrein B', at 4.32 and 4.19.[10] On gel electrofocusing, however, 6 bands were revealed in the latter preparation.[30]

Porcine submandibular kallikrein showed at least 8 (up to 10[30]) well-resolved fractions at pH 4.37, 4.19, 4.05, 3.90, 3.80, 3.61, 3.48, 3.29[32]; or 4.4, 4.2, 4.1, 3.95, 3.85, 3.65, 3.4, 3.15.[20] However, treatment with neuraminidase did not abolish, but only reduced, the microheterogeneity of this enzyme.[20]

A large number of components was also observed on gel electrofocusing of porcine urinary kallikrein.[33]

A further, different type of microheterogeneity resides in the protein portion of porcine pancreatic β-kallikrein B. Here, part of the enzyme molecules have lost the C-terminal serine residue at the cleavage site.[34]

Optical Properties. Newer values given for $A_{280}^{1\%}$ are 17.5 for porcine pancreatic β-kallikrein I (A), 16.1 for β-kallikrein II (B), and 17.5 for β-kallikrein III.[18] The molar absorbance at 280 nm for pancreatic β-kallikreins has been determined as $(50.6 \pm 1.3) \times 10^3 \, M^{-1} \, cm^{-1}$,[17] very similar to the value of $48.8 \times 10^3 \, M^{-1} \, cm^{-1}$ obtained for submandibular kallikrein.[20] The ratio A_{280}/A_{260} has been found as 1.74 for pancreatic β-,[17] as 1.77 for pancreatic α,[15] and as 1.76 ± 0.02 as submandibular[20] kallikrein.

Amino Acid Composition and Structure. The amino acid compositions of the three glandular α-kallikreins from pancreas, submandibular glands, and urine of the pig are indistinguishable within experimental limits (Table IX). The composition of pancreatic β-kallikrein B obtained by applying

[31] Y. Hojima, Y. Matsuda, C. Moriwaki, and H. Moriya, *Chem. Pharm. Bull.* **23**, 1120 (1975).

[32] F. Fiedler, B. Müller, and E. Werle, *Hoppe-Seyler's Z. Physiol. Chem.* **351**, 1002 (1970).

[33] H. Tschesche, G. Mair, B. Förg-Brey, and H. Fritz, *Adv. Exp. Med. Biol.* **70**, 119 (1976).

[34] H. Tschesche, W. Ehret, G. Godec, C. Hirschauer, C. Kutzbach, G. Schmidt-Kastner, and F. Fiedler, *Adv. Exp. Med. Biol.* **70**, 123 (1976).

TABLE IX
AMINO ACID COMPOSITION OF PORCINE GLANDULAR KALLIKREINS
(RESIDUES PER MOLECULE)

	Pancreatic β-				Pancreatic α-B'[c]	Submandibular[d]	Urinary[e]
	I(A)[a]	II(B)[a]	III[a]	B[b]			
Asp	27	27	25	28	27–28	28	28
Thr	14	15	15	15	15–16	16	15–16
Ser	12	13	12	14	14	14	14
Glu	22	23	22	23	23	23	23
Pro	16	16	15	16	16	16	16
Gly	21	22	20	22	21–22	22	22
Ala	12	12	12	13	13	13	13
Cys	12	13	12	10		10	10
Val	10	10	9	10	10	10	10
Met	4	4	4	4	4	4	4
Ile	11	11	11	12	11	11	11
Leu	19	20	18	20	21–22	21	20
Tyr	7	7	6	7	7	7	7
Phe	9	10	9	10	10	10	9
Lys	10	10	9	10	12	11	10
His	7	8	7	8	9	9	9
Arg	3	3	3	3	3	3	3
Trp	7	7	7	7		7	
Total	223	231	216	232		235	

[a] H. Kira, S. Hiraku, and H. Terashima, *Adv. Exp. Med. Biol.* **120A**, 273 (1979).
[b] F. Fiedler, W. Ehret, G. Godec, C. Hirschauer, C. Kutzbach, G. Schmidt-Kastner, and H. Tschesche, *in* "Kininogenases: Kallikrein" (G. L. Haberland, J. W. Rohen, and T. Suzuki, eds.), Vol. 4, p. 7. Schattauer, Stuttgart, 1977.
[c] F. Fiedler and W. Gebhard, *Hoppe-Seyler's Z. Physiol. Chem.* **361**, 1661 (1980).
[d] M. Lemon, F. Fiedler, B. Förg-Brey, C. Hirschauer, G. Leysath, and H. Fritz, *Biochem. J.* **177**, 159 (1979).
[e] H. Tschesche, G. Mair, G. Godec, F. Fiedler, W. Ehret, C. Hirschauer, M. Lemon, H. Fritz, G. Schmidt-Kastner, and C. Kutzbach, *Adv. Exp. Med. Biol.* **120A**, 245 (1979).

slight corrections from sequencing results[3] to the data from amino acid analyses[17] is also very similar (Table IX). Small differences in the composition of the glandular α-kallikreins might be real and due to the removal of about 3 amino acid residues at the site of cleavage in pancreatic β-kallikrein.[34a] Differences to and scatter within the data for pancreatic β-kallikreins I–III (Table IX) are most probably caused by limited precision of the latter analyses.

Porcine pancreatic β-kallikrein, submandibular and urinary kallikreins showed complete immunological identity in immunodiffusion experi-

[34a] F. Fiedler and H. Fritz, *Hoppe-Seyler's Z. Physiol. Chem.* **362**, 1171 (1981).

ments,[23] as well as in the radioimmunoassay.[35] The first 28 amino-terminal amino acid residues of submandibular kallikrein[23] and 13 of urinary kallikrein[27] are identical with those of the A chain of pancreatic β-kallikrein B. The amino acid sequence of porcine pancreatic β-kallikrein B has been determined[3,27,34] and is given in Table X. Some minor uncertainties remaining are indicated by parentheses. The sequence is highly homologous to those of other pancreatic serine proteinases, 37% of the residues being identical with those of porcine trypsin.[27] The catalytic charge-relay system of the serine proteinases, His^{57}-Asp^{102}-Ser^{195}, is also found in kallikrein. Asp^{189} is probably responsible for the specificity of the enzyme for basic amino acids, as it is in trypsin. The 5 potential disulfide bridges are presumably identical with 5 of the 6 bridges in trypsin. Among those is 1 disulfide bridge, Cys^{22}-Cys^{157}, otherwise unique to trypsin.

Peculiar features of the structure of pancreatic kallikrein are the N-terminal Ile-Ile- sequence, the C-terminal proline residue, and the extended kallikrein loop[3] at the site of cleavage between the A and B chains. The C-terminal serine residue is only found in about half of the A chains. This indicates that some further residues could also have been removed here during autolysis of the pancreas homogenates from which the enzyme had been isolated. The two probable carbohydrate binding sites are indicated in Table X. In the tentative three-dimensional structure of pancreatic kallikrein,[27] the 2 carbohydrate clusters are located in rather close proximity.

Recently, a striking degree of homology of porcine pancreatic kallikrein with mouse nerve growth factor γ-subunit and rat tonin has become evident.[34a]

Enzymatic Properties

Cleavage of Kininogen. The biological actions of the glandular kallikreins are probably all exerted via the specific release of a highly active kinin peptide from a protein precursor, kininogen. Porcine pancreatic β-kallikrein,[1,36] as well as submandibular[36,37] and urinary[36] kallikreins, liberate from bovine kininogen the decapeptide kallidin (lysyl-bradykinin, Lys-Arg-Pro-Pro-Gly-Phe-Ser-Pro-Phe-Arg). This reaction requires the cleavage of one Arg-Ser and one Met-Lys bond, while trypsin hydrolyzes the adjacent Lys-Arg bond and releases bradykinin instead.[1] The rates of kallidin release by various porcine glandular kallikreins are very

---

[35] E. Fink, T. Dietl, J. Seifert, and H. Fritz, *Adv. Exp. Med. Biol.* **120B**, 261 (1979).

[36] B. Dittmann and R. Wimmer, *in* "Current Concepts in Kinin Research" (G. L. Haberland and U. Hamberg, eds.), p. 121. Pergamon, Oxford, 1979.

[37] E. Werle, I. Trautschold, and G. Leysath, *Hoppe-Seyler's Z. Physiol. Chem.* **326**, 174 (1961).

TABLE X
AMINO ACID SEQUENCE OF PORCINE PANCREATIC β-KALLIKREIN B[a]

A Chain

16	17	18	19	20	21	22	23	24	25	26	27	28	29	30	31	32	33	34	37	38
Ile-	Ile	-Gly-	Gly	-Arg-	Glu	-Cys-	Glu	-Lys-	Asn	-Ser-	His	-Pro-	Trp	-Gln-	Val	-Ala-	Ile	-Tyr-	His	-Tyr-

40	41	41A	42	43	44	45	46	47	48	49	50	51	52	53	54	55	56	57	58	59
-Ser-	Phe	-Gln-	Cys	-Gly-	Gly	-Val-	Leu	-Val-	Asn	-Pro-	Lys	-Trp-	Val	-Leu-	Thr	-Ala-	Ala	-His-	Cys	-Lys-

61	62	63	64	65	65A	66	69[b]	70	71	72	73	74	75	76	77	78	79	80	81	82
-Asp-	Asn	-Tyr-	Glu	-Val-	Trp	-Leu-	Gly	-Arg-	His	-Asn-	Leu	-Phe-	Glu	-Asn-	Glu	-Asn-	Thr	-Ala-	Gln	-Phe-

84	85	86	87	88	89	90	91	92	93	94	95	95A	95B
-Gly-	Val	-Thr-	Ala	-Asp-	Phe	-Pro-	His	-Pro-	Gly	-Phe-	Asn-	Leu-	Ser

$$\overset{\mid}{\text{CHO}}$$

B Chain

95Y	95Z	96	97	98	99	100	101	102	103	104	105	106	107	108	109	110	111	112	113	114
Ala-	Asp	-Gly-	Lys-	Asp-	Tyr	-Ser-	His	-Asp-	Leu-	Met-	Leu-	Leu-	Arg	-Leu-	Gln	-Ser-	Pro	-Ala-	Lys	-Ile-

116	117	118	119	120	121	122	123	124	125	127	128	129	130	132	133	134	135	136	137	138
-Asp-	Ala	-Val-	Lys	-Val-	Leu	-Glu-	Leu	-Pro-	Thr	-Gln-	Glu	-Pro-	Glu	-Leu-	Gly	-Ser-	Thr	-Cys-	Glu	-Ala-

140	141	142	143	144	145	145A	145B	146	147	148	149	150	151	152	153	154	155	156	157	158
-Gly-	Trp	-Gly-	Ser	-Ile-	Glu	-Pro-	Gly	-Pro-	Asp(Asp)	Phe	-Glu-	Phe	-Pro-	Asp	-Gly-	Ile	-Gln-	Cys	-Val-	

160	161	162	163	164	165	166	167	168	169	170	171	172	173	174	175	176	177	178	179	180
-Leu-	Thr	-Leu-	Leu	-Gln-	Asn-	Thr	-Phe-	Cys(	Ala	-His-	Ala	-Asx-	Pro	-Asx)	Lys	-Val-	Thr	-Glu-	Ser	-Met-

182	183	184	184A	185	186	187	188	188A	189	190	191	192	193	194	195	196	197	198	199	200
-Cys-	Ala	-Gly-	Tyr	-Leu-	Pro	-Gly-	Gly	-Lys-	Asp	-Thr-	Cys	-Met-	Gly	-Asp-	Ser	-Gly-	Gly	-Pro-	Leu	-Ile-

202	203	204	209	210	211	212	213	214	215	216	217	218	219	220	221	221A	222	223	224	225
-Asn-	Gly	-Met-	Trp	-Gln-	Gly	-Ile-	Thr	-Ser-	Trp	-Gly-	His	-Thr-	Pro	-Cys-	Gly	-Ser-	Ala	-Asn-	Lys	-Pro-

227	228	229	230	231	232	233	234	235	236	237	238	239	240	241	242	243	244	245	246
-Ile-	Tyr	-Thr-	Lys	-Leu-	Ile	-Phe-	Tyr	-Leu-	Asp	-Trp-	Ile	(Asx)(Asx)	Thr	-Ile-	Thr	-Glu-	Asn	-Pro	

$$\overset{\mid}{\text{CHO}}$$

[a] The numbering is that of chymotrypsinogen [B. S. Hartley, *Philos. Trans. R. Soc. London, Ser. B* ., 77 (1970)]. Tentative numbering is indicated by italics. According to the alignment in M. Dayhoff, ed., "Atlas of Protein Sequence and Structure," Vol. 5, Suppl. 3, Natl. Biomed. Found., Washington, D.C., 1978, positions 37 to 41A would be numbered 35, 36, 37, 39, 40, positions 65 to 66 as 66, 67, 68; and positions 127 to 129 as 126, 127, 128. CHO, carbohydr. From H. Tschesche, G. Mair, G. Godec, F. Fiedler, W. Ehret, C. Hirschauer, M. Len H. Fritz, G. Schmidt-Kastner, and C. Kutzbach, *Adv. Exp. Med. Biol.* **120A,** 245 (1979).

[b] The sequence of residues 65A-66-69 was corrected (after eliminating overlaps) to be -Trp-Leu- on the basis of the amino acids occurring first in the sequencer run. Reinvestigation of the seque was stimulated by the current X-ray examination of crystallized porcine pancreatic kallikrein W. Bode and Z. Chen, Martinsried, Federal Republic of Germany.

similar (Table XI). This is true also for the extent of lowering of a dog's blood pressure after intravenous injection, the method used to determine the biological activity of kallikrein in biological kallikrein units,[2] KU (Table XI).

Hydrolysis of Amino Acid Esters. In several variants of the assay using BzArgOEt as substrate, the specific activities of the three porcine glandular α-kallikreins from pancreas, submandibular glands, and urine are indistinguishable from each other within experimental limits. Pancreatic β-kallikrein, however, has a distinctly different specific activity (Table XI).

At low concentrations of BzArgOEt (e.g., 0.2 mM), all three porcine α-kallikreins show a very peculiar time dependence of the hydrolysis reaction.[15,20] The fast initial hydrolysis gradually slows down in the time range of minutes until a steady rate is attained. At higher substrate concentrations the transition needs less time, and at 10 mM BzArgOEt, as used in the pH-stat assay, the fast initial period can no longer be discerned. This biphasic reaction is observed to a varying extent in the hydrolysis also of other, but not all, ester substrates by porcine α-kallikreins, especially pronounced in the case of Bz-LysOMe.[15,20] The cause of this strange phenomenon is most probably a rearrangement in the structure of the enzymes. Pancreatic β-kallikrein never showed such an effect.

Therefore, two pairs of kinetic constants pertaining, respectively, to the initial and to the stationary phase of the reaction can be determined for the hydrolysis of BzArgOEt by porcine glandular α-kallikreins. The data for the submandibular enzyme are indistinguishable within experimental precision from those for urinary kallikrein (Table XI). Initial and stationary phase evidently differ only by the higher K_m in the latter case. Under the conditions of the alcohol-dehydrogenase coupled assay with its relatively long reaction time, essentially the higher value of K_m is effective, while the time of observation determines the effective K_m in the assay of porcine α-kallikreins when the hydrolysis of BzArgOEt is followed directly at 253 nm.[4,38] This effect reconciles the disparate older values determined for K_m of submandibular and urinary kallikreins (Table XI).

Rates of hydrolysis of several N^α-acylated amino acid esters by the various porcine glandular kallikreins measured in the stationary reaction phase are also given in Table XI. While the substrate specificities of all three α-kallikreins are the same within experimental limits, pancreatic β-kallikrein generally behaves in a similar way, but shows distinct quantitative differences in the case of some of the substrates. Evidently, the primary specificity of all porcine glandular kallikreins is directed predominantly to esters of arginine, as has been demonstrated more extensively for pancreatic β-kallikrein.[39] Even the other positively charged compound, the lysine ester, is hydrolyzed at a considerably lower rate. Esters of the neutral amino acids tyrosine and methionine are very poor substrates for all these glandular kallikreins. The nature of the acyl group is of great importance for the efficiency of hydrolysis by all porcine glandular kallikreins. A tosyl residue is greatly inferior to a benzoyl residue in this

[38] G. W. Schwert and Y. Takenaka, *Biochim. Biophys. Acta* **16**, 570 (1955).
[39] F. Fiedler, G. Leysath, and E. Werle, *Eur. J. Biochem.* **10**, 419 (1969).

TABLE XI

CATALYTIC PROPERTIES OF PORCINE GLANDULAR KALLIKREINS

	Ref.	Units	Pancreatic β-kallikrein B	Pancreatic α-kallikrein B'	Submandibular kallikrein	Urinary kallikrein
Blood pressure activity (dog)		KU/U_{AT}	5.7^a; 6.37^b		5.3^a	5.6^a
Kallidin release (bovine kininogen)	a	$nmol/(min\ U_{AT})$	7.8		5.9	7.7
Specific activity (10 mM Bz-ArgOEt)		$U_{AT}/(ml\ A_{280}^{1cm})$	141^c; 155^d	176^d	166^e	168^e
Specific activity (10 mM Bz-ArgOEt)		U_{AT}/mg protein	282 ± 6^c		312^e	
Specific activity (0.5 mM Bz-ArgOEt)	d	$U_{ADH}/(ml\ A_{280}^{1cm})$	113	51	55	54
Specific activity (0.5 mM Bz-ArgOEt)	d	$U_{253}/(ml\ A_{280}^{1cm})$	127	72	75	69
Bz-ArgOEt						
K_m	f	mM	0.104 ± 0.008^f		0.60^g; 0.114^h	0.44^g; 0.125^i
k_{cat}		sec^{-1}	123 ± 8			
Initial reaction phase						
K_m	e	mM			0.15 ± 0.01	0.19 ± 0.02
V_{max}	e	$\mu mol/(min\ ml\ A_{280}^{1cm})$			169 ± 7	172 ± 6
k_{cat}	e	sec^{-1}			138 ± 6	

Stationary reaction phase		units	ref				
K_m		mM	e			0.69 ± 0.04	0.61 ± 0.03
V_{max}		μmol/(min ml $A_{280}^{1\,cm}$)	e			189 ± 7	173 ± 3
k_{cat}		sec^{-1}	e			154 ± 6	
Ac-Phe-ArgOMe (1 mM)		nmol/(min U$_{AT}$)	d, e	5780	2470	2570	2660
Bz-ArgOEt (1 mM)		nmol/(min U$_{AT}$)	d, e	955	613	698	736
Bz-ArgOMe (1 mM)		nmol/(min U$_{AT}$)	e	1120		873	955
Tos-ArgOMe (1 mM)		nmol/(min U$_{AT}$)	d, e	27.3	21.1	25.7	29.9
Bz-LysOMe (1 mM)		nmol/(min U$_{AT}$)	d, e	42.3	7.2	8.6	11.9
Ac-TyrOEt (1 mM)		nmol/(min U$_{AT}$)	d, e	2.5	2.0	2.1	1.9
Bz-MetOMe (1 mM)		nmol/(min U$_{AT}$)	e	1.8		1.2	1.2

[a] B. Dittmann and R. Wimmer, in "Current Concepts in Kinin Research" (G. L. Haberland and U. Hamberg, eds.), p. 121. Pergamon, Oxford, 1979.

[b] A. Arens and G. L. Haberland, in "Kininogenases: Kallikrein" (G. L. Haberland and J. W. Rohen, eds.), p. 43. Schattauer, Stuttgart, 1973.

[c] F. Fiedler, C. Hirschauer, and E. Werle, Hoppe-Seyler's Z. Physiol. Chem. 356, 1879 (1975).

[d] F. Fiedler and W. Gebhard, Hoppe-Seyler's Z. Physiol. Chem. 361, 1661 (1980).

[e] M. Lemon, F. Fiedler, B. Förg-Brey, C. Hirschauer, G. Leysath, and H. Fritz, Biochem. J. 177, 159 (1979).

[f] F. Fiedler, G. Leysath, and E. Werle, Eur. J. Biochem. 36, 152 (1973).

[g] I. Trautschold and E. Werle, Hoppe-Seyler's Z. Physiol. Chem. 325, 48 (1961).

[h] M. Lemon, B. Förg-Brey, and H. Fritz, Adv. Exp. Med. Biol. 70, 209 (1976).

[i] H. Tschesche, G. Mair, B. Förg-Brey, and H. Fritz, Adv. Exp. Med. Biol. 70, 119 (1976).

TABLE XII

PRELIMINARY KINETIC CONSTANTS FOR THE HYDROLYSIS OF PEPTIDE
ESTERS BY PORCINE PANCREATIC β-KALLIKREIN B (pH 9, 25°C)[a]

Substrate	K_m (mM)	k_{cat} (sec^{-1})	k_{cat}/K_m (mM^{-1} sec^{-1})
Ac-Phe-ArgOMe	0.030	920	31,000
Ac-Leu-ArgOMe			15,000
Ac-Met-ArgOMe			6000
Ac-Gly-ArgOMe	1.4	350	250
Z-Pro-Phe-ArgOMe			35,000
Ac-Leu-Met-ArgOMe			3000
Ac-Leu-Met-LysOMe			120
Ac-Phe-PheOMe			48
Bz-D,L-MetOMe[b]	20	1.5	0.07
Ac-Leu-MetOMe			3
Z-Ser-Leu-MetOMe			6

[a] F. Fiedler and G. Leysath, *Adv. Exp. Med. Biol.* **120A**, 261 (1979).
[b] pH 8; F. Fiedler, *Hoppe-Seyler's Z. Physiol. Chem.* **349**, 926 (1968).

position in arginine esters. An Ac-Phe substituent makes an especially good substrate (Table XI).

Secondary Specificity in the Hydrolysis of Ester and Peptide Substrates. This has been studied in more detail in the case of pancreatic β-kallikrein. As seen from Table XII, it is the bulk or hydrophobicity of the amino acid residue in position P_2 (in the nomenclature of Schechter and Berger[40]) that determines the substrate properties of an arginine peptide ester. Incidentally, the lysine ester is also a much worse substrate in this series.

This pronounced secondary specificity for a bulky residue, especially phenylalanine as it occurs in an equivalent position also in the natural kallikrein substrate kininogen, is seen also with peptide substrates (Table XIII). These are hydrolyzed exclusively at the arginyl bond. Substitution of the phenylalanine residue in Ac-Phe-Arg-Ser-Val-Gln by glycine reduces the specificity constant k_{cat}/K_m 730-fold. Some further minor alterations leading to Z-Gly-Arg-Ser-Gly nearly abolish hydrolysis by pancreatic kallikrein.[41] The strong secondary specificity is in sharp contrast to the properties of bovine trypsin,[41] and appears to be (besides the pronounced primary specificity for arginine) one of the reasons for the narrow proteolytic specificity of porcine pancreatic kallikrein.[1]

An influence of the leaving group is seen in the example of Ac-Phe-Arg-Ser-Val-NH$_2$, where substitution of the serine residue by an isosteric

[40] I. Schechter and A. Berger, *Biochem. Biophys. Res. Commun.* **27**, 157 (1967).
[41] F. Fiedler and G. Leysath, *Adv. Exp. Med. Biol.* **120A**, 261 (1979).

TABLE XIII

KINETIC CONSTANTS FOR THE HYDROLYSIS OF PEPTIDES BY PORCINE PANCREATIC
β-KALLIKREIN B (pH 9, 25°C)[a]

Substrate	K_m (mM)	k_{cat} (sec^{-1})	k_{cat}/K_m (mM^{-1} sec^{-1})
Bovine HMW kininogen[b]		0.94	120
Arg-Pro-Pro-Gly-Phe-Ser-Pro-Phe-Arg-Ser-Val-Gln[c]	0.075	7	93
Phe-Ser-Pro-Phe-Arg-Ser-Val-Gln[c]	0.064	6.4	100
Pro-Phe-Arg-Ser-Val-Gln	0.057	5.7	100
Phe-Arg-Ser-Val-Gln	2.2	2.8	1.3
Ac-Phe-Arg-Ser-Val-Gln	0.31	6.8	22
Ac-Phe-Arg-Ser-Val-NH$_2$	0.040	11.6	290
Ac-Phe-Arg-Abut-Val-NH$_2$	0.073	4.2	58
Ac-Phe-Arg-Ser-NH$_2$	0.21	2.1	10
Ac-Gly-Arg-Ser-Val-Gln	26	0.77	0.030

[a] F. Fiedler and G. Leysath, *Adv. Exp. Med. Biol.* **120A**, 261 (1979).

[b] $v/[E]_0$ and $v/([E]_0 \cdot [S])$; B. Dittmann and R. Wimmer, in "Current Concepts in Kinin Research" (G. L. Haberland and U. Hamberg, eds.), p. 121. Pergamon, 1979.

[c] F. Fiedler, in "Chemistry and Biology of the Kallikrein-Kinin System in Health and Disease" (J. J. Pisano and K. F. Austen, eds.), p. 93. US Govt. Printing Office, Washington, D.C. without year [1976].

α-aminobutyric acid residue leads to a distinctly worse substrate (Table XIII).

The active site of porcine pancreatic β-kallikrein recognizes about six consecutive positions in peptide substrates (Table XIII). The specificity constant for the hydrolysis of Pro-Phe-Arg-Ser-Val-Gln that represents a section of the kininogen sequence around the C-terminus of the kinin moiety[1] appears comparable to that for kallidin release from kininogen (Table XIII). Further elongation of the N-terminus is of little effect (Table XIII). The interactions of the hexapeptide with pancreatic kallikrein could thus well comprise essentially all the interactions necessary in the cleavage of the Arg-Ser bond in kininogen.[41]

The Problem of Cleavage of the Methionyl Bond in Kininogen. In contrast, the cleavage of the Met-Lys bond at the other end of the kallidin moiety is very puzzling in view of the poor hydrolysis of methionine peptide esters by porcine pancreatic kallikrein, even if they contain in P$_2$ the bulky leucine residue of kininogen (Table XII). It has been reported that a kininogen fragment of the sequence Ser-Leu-Met-Lys-bradykinin is also cleaved at the methionyl bond.[42] In view of the slow hydrolysis of the methionine esters, however, the rate of hydrolysis by porcine pancreatic β-kallikrein is expected to be very low, and was indeed found so in pre-

[42] S. Iwanaga, Y. N. Han, H. Kato, and T. Suzuki, in "Kininogeneases: Kallikrein" (G. L. Haberland, J. W. Rohen, and T. Suzuki, eds.), Vol. 4, p. 79. Schattauer, Stuttgart, 1977.

liminary experiments with the corresponding synthetic peptide.[30] So this reaction does not represent a valid model for the rapid release of kallidin from kininogen. Possibly the crucial point is the conformation of kininogen.

Inhibitors. The active site-directed reagent diisopropyl fluorophosphate inhibits all porcine glandular kallikreins at the same low rate (Table XIV); this is also true for pancreatic α-kallikrein B'.[10] Whereas the trypsin inhibitor Tos-LysCH$_2$Cl is ineffective, more extended peptide chloromethyl ketones, predominantly the arginine derivative, are inhibitory (Table XIV). Kinetic constants for the inhibition of the α-kallikreins from submandibular glands and from urine are indistinguishable, whereas pancreatic β-kallikrein also shows here some quantitative differences.

The argininal peptides leupeptin and antipain are reversible inhibitors with similar inhibition constants for all porcine glandular kallikreins (Table XIV). Inhibition constants for the basic trypsin-kallikrein inhibitor from bovine organs (Kunitz-type; aprotinin) and a number of homologous inhibitors are also compiled in Table XIV. For HHV-II and HPI$_K$, the large differences of the constants for the α-kallikreins, on the one hand, and for pancreatic β-kallikrein, on the other, are remarkable. The related trypsin inhibitor from cow colostrum, however, is not inhibitory.[43] Further high-molecular-weight inhibitors of porcine kallikreins are reviewed in Vogel.[44]

Compiled data on the inhibition of porcine pancreatic β-kallikrein by heavy metal ions and other low-molecular-weight inhibitors are to be found in the work by Fiedler.[1]

Relationship of the Various Porcine Glandular Kallikreins[1,15]

In view of the evidence from amino acid composition, partial amino acid sequences, and immunological and enzymatic behavior detailed in this chapter, there remains no doubt that native porcine pancreatic kallikrein as well as submandibular and urinary kallikreins all share the same protein moiety. Deviating properties of pancreatic β-kallikrein are easily explained by the intrachain split that this form of the enzyme has suffered during isolation. Differences in electrophoretic behavior of the various forms of the three glandular α-kallikreins are most probably due to differences in carbohydrate content. So it is striking that submandibular and urinary kallikreins B, on the one hand, and pancreatic kallikrein B', submandibular and urinary kallikreins A, on the other, show very similar

---

[43] T. Dietl, C. Huber, R. Geiger, S. Iwanaga, and H. Fritz, *Hoppe-Seyler's Z. Physiol. Chem.* **360,** 67 (1979).

[44] R. Vogel, *Handb. Exp. Pharmacol.* [N.S.] **25,** Suppl., 163 (1979).

TABLE XIV

INHIBITION CONSTANTS FOR PORCINE GLANDULAR KALLIKREINS (25°C)

Inhibitor	Ref.	Conditions	Units	Kallikrein Pancreatic β-(B)	Submandibular	Urinary
Diisopropyl fluorophosphate (DFP)	a	pH 7.2; 1 mM DFP	$k \times 10^3$ (min^{-1})	8.6 ± 2.2	$6.9 \pm 0.9;\ 8.8 \pm 2.0$	7.0 ± 1.5
Gly-Val-ArgCH$_2$Cl	b	pH 7.0	K_i (mM)	0.37 ± 0.05	0.74 ± 0.12	0.63 ± 0.04
			$k_2 \times 10^3$ (sec^{-1})	7.8 ± 0.3	16.2 ± 1.1	17.1 ± 0.5
Ala-Leu-LysCH$_2$Cl	b	pH 7.0	K_i (mM)	1.04 ± 0.12	1.95 ± 0.16	1.97 ± 0.15
			$k_2 \times 10^3$ (sec^{-1})	2.32 ± 0.10	3.62 ± 0.13	3.85 ± 0.13
		5 mM inhibitor	$t_{1/2}$ (min)	5.9 ± 0.4	4.5 ± 0.4	4.2 ± 0.3
Ala-Val-LysCH$_2$Cl	b	pH 7.0; 5 mM inhibitor	$t_{1/2}$ (min)	121 ± 13	116 ± 13	147 ± 13
Phe-Ala-LysCH$_2$Cl	b	pH 7.0; 5 mM inhibitor	$t_{1/2}$ (min)	>600	>600	>600
Tos-LysCH$_2$Cl	b	pH 7.0; 5 mM inhibitor	$t_{1/2}$ (min)	>600	>600	>600
Leupeptin	c	pH 7.8	K_i (M)	4.1×10^{-6}	1.0×10^{-5}	1.3×10^{-5}
Antipain	c	pH 7.8	K_i (M)	1.6×10^{-5}	3.9×10^{-5}	9.5×10^{-5}
Aprotinin (bovine lung)	d	pH 9.0	$K_i \times 10^9$ (M)	0.8	0.9	1.0
SAI 5-II (sea anemone)	d	pH 9.0	$K_i \times 10^9$ (M)	0.8	0.6	0.8
NNV-II (snake venom)	d	pH 9.0	$K_i \times 10^9$ (M)	1.1	1.3	1.3
HHV-II (snake venom)	d	pH 9.0	$K_i \times 10^9$ (M)	8.3	210	109
HPI$_K$ (snail)	d	pH 9.0	$K_i \times 10^9$ (M)	27.7	210	204

[a] F. Fiedler, B. Müller, and E. Werle, *Hoppe-Seyler's Z. Physiol. Chem.* **351**, 1002 (1970).
[b] F. Fiedler, C. Hirschauer, and H. Fritz, *Hoppe-Seyler's Z. Physiol. Chem.* **358**, 447 (1977).
[c] H. Fritz, B. Förg-Brey, and H. Umezawa, *Hoppe-Seyler's Z. Physiol. Chem.* **354**, 1304 (1973).
[d] T. Dietl, C. Huber, R. Geiger, S. Iwanaga, and H. Fritz, *Hoppe-Seyler's Z. Physiol. Chem.* **360**, 67 (1979).

electrophoretic properties. Synthesis of at least part of the urinary kallikrein in the kidney cannot be excluded at present. However, it appears more likely that urinary kallikrein is derived from pancreas and salivary glands by some leakage into the circulation and filtration through the kidney. The kallikrein content of the kidney[2] could well have accumulated during this process.[15]

Acknowledgment

This work has been sponsored by Deutsche Forschungsgemeinschaft, Sonderforschungsbereich 51.

Section V

Proteases of Diverse Origin and Function

[41] Cathepsin B, Cathepsin H, and Cathepsin L

By ALAN J. BARRETT and HEIDRUN KIRSCHKE

The lysosomal cysteine proteinases may well be the most active proteinases in the body. By this we mean that they hydrolyze more peptide bonds than the others, including the pancreatic serine proteinases in the intestine. The justification for this view is that whereas a 70-kg subject might be expected to eat and digest 70 g of protein per day, 200–300 g of protein will be turned over in the tissues during the same period.[1] There are now strong reasons for thinking that the turnover is mediated largely by the lysosomal enzymes. Not only do the activities of these enzymes tend to parallel the rates of protein turnover in many tissues examined,[2] but also the lysosomal system seems unique in having the capacity to handle the amount of proteolysis required.[3] Of the lysosomal enzymes, the aspartic proteinase, cathepsin D, seems to have rather a restricted activity on proteins, whereas the cysteine proteinases have broad activity.[4-6]

The lysosomal cysteine proteinases, cathepsins B, H, and L have been the subject of several recent reviews,[5-7] so the emphasis of the present chapter will be on practical aspects. Some important characteristics of the enzymes are summarized in Table I.

Assay Procedures

The ranges of sensitivity indicated for the various assay procedures are given in terms of amounts of the purified enzymes. As is explained, smaller amounts of fully active enzyme quantified by active-site titration would be needed.

---

[1] P. J. Garlick, *Compr. Biochem.* **19B**, 77 (1980).

[2] D. J. Millward, *Compr. Biochem.* **19B**, 153 (1980).

[3] D. Evered and J. Whelan, eds., "Protein Degradation in Health and Disease," Excerpta Medica, Amsterdam, 1980.

[4] A. J. Barrett, *in* "Proteinases of Mammalian Cells and Tissues" (A. J. Barrett, ed.), p. 181. North-Holland Publ., Amsterdam, 1977.

[5] A. J. Barrett, *in* "Enzyme Regulation and Mechanism of Action" (P. Mildner and B. Ries, eds.), p. 307. Pergamon, Oxford, 1980.

[6] H. Kirschke, J. Langner, S. Riemann, B. Wiederanders, S. Ansorge, and P. Bohley, *in* "Protein Degradation in Health and Disease" (D. Evered and J. Whelan, eds.), p. 15. Excerpta Medica, Amsterdam, 1980.

[7] A. J. Barrett and J. K. McDonald, "Mammalian Proteases. A Glossary and Bibliography." Academic Press, New York, 1980.

METHODS IN ENZYMOLOGY, VOL. 80

TABLE I
A COMPARISON OF THE PROPERTIES OF LYSOSOMAL CYSTEINE PROTEINASES,
CATHEPSINS B, H, AND L[a]

	Cathepsin B	Cathepsin H	Cathepsin L
Recommended test substrate (pH optimum)	Z-Arg-Arg-NMec (pH 6.0)	Arg-NMec (pH 6.8)	Z-Phe-Arg-NMec (pH 5.5)
Aminopeptidase activity	No	Yes	No
Endopeptidase activity	Moderate	Variable (see text)	High
Peptidyldipeptidase activity	Yes	No	No
Aldolase inactivation	Yes	No	Yes
Leupeptin sensitivity	High	Low	High
Z-Phe-Phe-CHN$_2$ sensitivity	Low	Low	High
pI	4.5–5.5	6.0–7.1	5.5–6.1
Concanavalin A–Sepharose	Not bound	Bound	Bound

[a] The rat and human enzymes have been found to be similar in the characteristics indicated here.

Assay of Cathepsin B

Cathepsin B has generally been assayed with Bz-DL-Arg-NPhNO$_2$[8,9] or Bz-Arg-2-NNap as substrate.[9,10] It is now clear, however, that these substrates are not only insensitive, but also susceptible to cathepsin H as well as B, so the results obtained with them are not accurate. As has been found with the serine proteinases, substrates containing longer peptide sequences are much more selective and specific than the blocked amino acid derivatives. Thus naphthylamide substrates containing the -Arg-Arg- sequence introduced by McDonald[11] are extremely sensitive to cathepsin B, and are resistant to cathepsins H[6,12] and L.[6]

The most sensitive, safe, and convenient leaving group for substrates of the cysteine proteinases so far discovered is 7-amino-4-methylcoumarin. We found Z-Phe-Arg-NMec to be an excellent substrate of cathepsin B, and thought that it was specific for this en-

[8] Standard abbreviations are used for amino acid residues and common blocking groups [*J. Biol. Chem.* **247**, 977 (1972)]. The amino acid residues should be assumed to be in the L configuration unless otherwise shown. Additional abbreviations are: -NNap, -2-naphthylamide; -NNapOMe, 2-(4-methoxy)naphthylamide; -NPhNO$_2$, -4-nitroanilide; -OPhNO$_2$, -(4-nitro)phenyl ester; -CHN$_2$, -diazomethane; -CH$_2$Cl, -chloromethane; -NMec, -7-(4-methyl)coumarylamide.

[9] A. J. Barrett, *Anal. Biochem.* **47**, 280 (1972).

[10] A. J. Barrett, *Anal. Biochem.* **76**, 374 (1976).

[11] J. K. McDonald and S. Ellis, *Life Sci.* **17**, 1269 (1975).

[12] W. N. Schwartz and A. J. Barrett, *Biochem. J.* **191**, 487 (1980).

zyme.[13] It is, indeed, unaffected by cathepsin H, but has now been discovered to be extremely sensitive to cathepsin L (see below).[14] Just as Z-Arg-Arg-NNap is specific for cathepsin B, however, so is Z-Arg-Arg-NMec, and preliminary results suggest that this will prove to be an almost ideal substrate for cathepsin B.[15] For the present, however, there is a continuing need for assays with the available naphthylamide, and the method is as follows.

Assay with Z-Arg-Arg-NNap

Principle. The substrate Z-Arg-Arg-NNap is hydrolyzed, liberating 2-naphthylamine, and this is assayed colorimetrically by coupling with a diazonium salt, Fast Garnet. The colored product is kept in solution by a nonionic detergent. A thiol reagent is required to activate the enzyme; this would normally destroy the diazonium salt, but a mercurial compound is introduced to prevent this and to stop the enzymatic reaction.[9,10]

Reagents

Buffer/activator. 88 mM KH_2PO_4–12 mM Na_2HPO_4–1.33 mM disodium EDTA (pH 6.0), freshly made 2.7 mM cysteine by addition of the free base

Substrate stock solution. Z-Arg-Arg-NNap,[16,17] dissolved in water at 20 mM, and stored at −20°C

Color reagent. To prepare the Mersalyl–Brij solution, place 2.43 g of Mersalyl acid {2-[3-(hydroxymercuri)-2-methoxypropyl]carbamoyl phenoxyacetic acid}[18] in a beaker containing a stirring bar, and dissolve the solid with 60 ml of 0.5 M NaOH. Add 0.3 g of disodium EDTA and make up to about 450 ml with water. Allow the EDTA to dissolve, then adjust the pH to 4.0 by slow addition of 1 M HCl, with stirring. Make up to 500 ml and add 500 ml of 4% (w/v) Brij 35 (polyoxyethylene lauryl ether, available from BDH Chemicals Ltd.) in water. Store the reagent in a brown bottle at room temperature.

The diazonium salt, Fast Garnet, is preferably prepared freshly by

[13] A. J. Barrett, *Biochem. J.* **187**, 909 (1980).
[14] H. Kirschke, unpublished results (1981).
[15] A. J. Barrett, D. C. Sunter, and M. A. Brown, unpublished results (1981).
[16] Available from Bachem Feinchemikalien AG, CH-4416 Bubendorf, Switzerland; Peptide Research Foundation, Minoh-Shi, Osaka, Japan; or Cambridge Research Biochemicals Ltd., Huntingdon, Cambridgeshire PE 18 6DP, England.
[17] *Warning.* 2-Naphthylamine and the naphthylamides are probably carcinogenic, and must be handled with due precautions.
[18] Obtainable from Sigma Chemical Co.

diazotization of the base. To prepare the base stock solution, take 225 mg of 4-amino-2,3-dimethylazobenzene (available from Merck-Schuchardt or United States Biochemical Corp.), crush to a powder, and dissolve in 50 ml of ethanol. Add 30 ml of $1 M$ HCl, with stirring, dilute to 100 ml with water, and store at 4°C. Place 1.0 ml of base stock solution in a test tube standing in a beaker of ice and water. Add 0.1 ml of $0.2 M$ NaNO$_2$, mix, and leave for at least 5 min, then dilute to 100 ml with Mersalyl–Brij reagent. The complete color reagent is stable in the refrigerator for 1 day.

As an alternative to diazotizing the base, one can buy the stabilized diazonium salt. Commercial preparations tend to be impure and incompletely soluble, but we have obtained satisfactory results with the fluoroborate.[19]

Naphthylamine standard. Make 20 mM 2-naphthylamine HCl (35.9 mg/10 ml) in ethanol. Store in a brown bottle at 4°C for up to 1 month. When working-strength standard is required, dilute 1 ml of the stock to 100 ml with water. Do not expose the solution to bright light, and use it the same day.

Procedure. To 0.5 ml of sample containing 10–100 ng of cathepsin B, diluted as necessary with 0.1% Brij 35, add 1.5 ml of buffer/activator. Preincubate for 5 min at 40°C. Add 50 μl of substrate stock and mix to start the reaction. After precisely 10 min, add 2.0 ml of coupling reagent and mix. Allow 10 min for color development, then read A_{520}. The color is very stable at room temperature, there being no significant change overnight.

Blanks are prepared by adding the enzyme after the color reagent. Duplicate reagent blanks, and standards prepared by replacing the enzyme with 0.5 ml of 2-naphthylamine standard, are included in each set of assays.

Calculation. The ΔA_{520} due to 100 nmol of 2-naphthylamine (0.5 ml of 0.2 mM solution) corresponds to 10 milliunits (10×10^{-3} μmol/min). The linearity of the color reaction can be checked by the preparation of a standard curve with 0.1, 0.2, 0.3, and 0.4 ml of standard. Since the $\Delta\epsilon_{520}$ of the color reaction is 28,000,[20] the 100-nmol standard should give a Δ value of 0.70.

Note. Negative interference in the assay can arise from inhibition of the enzyme or destruction of the diazonium salt. Mersalyl and chloromercuribenzoate are very powerful inhibitors of cathepsin B, so it is vital that

[19] Obtainable from Serva Feinbiochemica GmbH & Co., D-6900 Heidelberg 1, P.O. Box 105260, Federal Republic of Germany.
[20] C. G. Knight, *Biochem. J.* **189**, 447 (1980).

no traces of the assay reagents be allowed to contaminate enzyme preparations.

Destruction of the diazonium salt can be caused by azide or by thiol compounds in excess over the mercurial. If necessary, the concentration of Mersalyl acid in the coupling reagent can be doubled without any other alteration, so that as much as 10 mM thiol (5 mM dithiothreitol) can be used in the buffered activator. The possibility of interference by a component of the enzyme sample solution can be checked by running a naphthylamine standard containing the enzyme sample in addition to the normal standard.

Turbid enzyme samples such as tissue homogenates give rise to turbid final reaction mixtures that are not suitable for direct spectrophotometry. Add 4 ml of butan-1-ol to each tube, mix, and centrifuge. Read A_{520} of the clear organic phase. Standards are treated in the same way.[21]

2-Naphthylamine is by no means an ideal fluorogenic leaving group, but it is quite possible to monitor activity against Z-Arg-Arg-NMec by direct fluorimetry and this has the advantage of allowing continuous rate measurements.[11] It must be remembered, however, that the fluorescent efficiency of 2-naphthylamine is strongly pH-dependent in the acid range, and all necessary allowances should be made for this.

Assay of Cathepsin H

Cathepsin H acts both as an aminopeptidase and as an endopeptidase,[22] and can be assayed selectively by use of an unblocked substrate such as Leu-NNap, Arg-NNap, or Arg-NMec. The values of k_{cat} and k_{cat}/K_m of cathepsin H with the best substrates discovered so far are much lower than those of cathepsin B, but we have found that satisfactory assays can be made with Arg-NMec.

Assay with Arg-NMec

The principle of the assay is precisely analogous to that described below for the assay of cathepsin L with Z-Phe-Arg-NMec.

Reagents

Buffer/activator. 200 mM KH$_2$PO$_4$–200 mM NaHPO$_4$–4 mM disodium EDTA, pH 6.8. The buffer is made 40 mM cysteine (base) on the day it is to be used.

Substrate stock solution. A 1 mM solution of Arg-NMec HCl[16] in

---

[21] A. J. Barrett, *Biochem. J.* **131**, 809 (1973).

[22] H. Kirschke, J. Langner, B. Wiederanders, S. Ansorge, P. Bohley, and H. Hanson, *Acta Biol. Med. Ger.* **36**, 185 (1977).

dimethyl sulfoxide is stored at 4°C. When required, it is diluted to the working strength of 20 μM with water (i.e., 200 μl to 10 ml). *Stopping reagent, diluent and standard.* These are exactly as described for cathepsin L.

Procedure. The assay procedure is just as for cathepsin L, and units of activity are calculated in the same way. The sample should contain 40–400 ng of cathepsin H. Arg-NMec is unaffected by cathepsins B and L. Other amino acid arylamidases found in tissues tend to be inactive in the presence of EDTA and cysteine, but the possibility of positive interferences from such enzymes can be checked by use of 0.1 mM puromycin, which scarcely affects cathepsin H,[12,22] but is a powerful inhibitor of the arylamidases.[23]

Assay of Cathepsin L

Cathepsin L is a powerful proteolytic enzyme, but has much more restricted activity on synthetic substrates than cathepsin B and papain. Three synthetic substrates have been used for cathepsin L: Bz-Arg-NH₂,[24] Z-Lys-OPhNO₂,[25] and Z-Phe-Arg-NMec[25a]; the proteolytic activity is usually assayed with azocasein. Each of these substrates is also susceptible to cathepsin B, but is more sensitive to cathepsin L, and since cathepsin B can be assayed accurately with Z-Arg-Arg-NMec or Z-Arg-Arg-NNap, an appropriate correction can be introduced into the calculations of activity.

For most purposes the assay with Z-Phe-Arg-NMec (originally described for cathepsin B[13]) is most convenient, but variants of the azocasein assay are widely used at present, and a procedure is therefore given also for this substrate.

Assay with Z-Phe-Arg-NMec

Principle. The substrate, Z-Phe-Arg-NMec, which is barely fluorescent, is hydrolyzed to liberate 7-amino-4-methylcoumarin, and this is quantified by its intense fluorescence after the reaction has been stopped with monochloroacetate.

Reagents

Buffer/activator. 340 mM sodium acetate–60 mM acetic acid–4 mM disodium EDTA, pH 5.5. The buffer is made 8 mM dithiothreitol on the day it is to be used.

[23] N. Marks, R. K. Datta, and A. Lajtha, *J. Biol. Chem.* **243**, 2882 (1968).
[24] H. Kirschke, J. Langner, B. Wiederanders, S. Ansorge, and P. Bohley, *Eur. J. Biochem.* **74**, 293 (1977).
[25] A. S. Bajkowski and A. Frankfater, *Anal. Biochem.* **68**, 119 (1975).
[25a] H. Kirschke, A. A. Kembhavi, P. Bohley, and A. J. Barrett, unpublished results (1980).

Substrate stock solution. A 1 mM solution of Z-Phe-Arg-NMec HCl[16] in dimethyl sulfoxide, stored at 4°C. The working-strength substrate solution (20 μM) is prepared as required by dilution of the stock with water (i.e., 200 μl to 10 ml).

Stopping reagent. 100 mM sodium monochloroacetate–30 mM sodium acetate–70 mM acetic acid, pH 4.3

Diluent. Brij 35 (0.1%) in water

Aminomethylcoumarin standard. 7-Amino-4-methylcoumarin (1 mM) in dimethyl sulfoxide, stored at 4°C, and diluted to 0.5 μM in a 1:1 mixture of assay buffer and stopping reagent for use.

Procedure. Assays are commonly done in 10-ml polystyrene test tubes. The enzyme sample, containing 0.3–3.0 ng of cathepsin L (or 2.5–25 ng of cathepsin B), is diluted to 500 μl with diluent, and 250 μl of assay buffer is added. 1 min is allowed for activation of the enzyme and temperature equilibration in a bath at 30°C (longer preincubation can lead to loss of activity), and 250 μl of the 20 μM substrate solution is then mixed in. After exactly 10 min, 1 ml of stopping reagent is added.

The fluorescence of the free aminomethylcoumarin is determined by excitation at about 370 nm and emission at 460 nm. If the instrument is zeroed against a reaction blank prepared by adding stopping reagent before the enzyme, and set to read 1000 arbitrary units with the 0.5 μM standard, an experimental reading of 1000 units corresponds to 0.1 mU of activity in the tube.

If the fluorimeter is provided with a temperature-controlled cell and connected to a recorder, continuous rate assays are easily made. The fluorescence of aminomethylcoumarin is scarcely affected by pH in the range 4–7 (in contrast to that of 2-naphthylamine), so pH-dependence measurements pose no problem.

Z-Phe-Arg-NMec is unaffected by cathepsin H,[13] but is sensitive to cathepsin B as well as cathepsin L.

In the standard procedure, only 3.25 μg of substrate is used per tube, so the method is inexpensive. Since Z-Phe-Arg-NMec is an excellent substrate for plasma and tissue kallikreins, and for some other trypsin-like enzymes, control experiments confirming the thiol-dependence of the activity being measured may be needed.

Assay with Azocasein

Principle. Azocasein is a derivative of casein in which tyrosyl and histidyl side-chains have been coupled with diazotized sulfanilic acid or sulfanilamide in alkali. The azo-coupling confers an intense yellow color on the protein ($A_{366}^{1\%}$ about 40). Proteolytic degradation of azocasein yields peptides soluble in dilute trichloroacetic acid solution, which may be quantified by their A_{366}. The method is especially suitable for the determination of proteolytic activity in crude samples, being very resistant to interference.

The assay for cathepsin L is made rather selective by the inclusion of 3 M urea (which enhances the activity of cathepsin L, and depresses that of cathepsin B[26]), and pepstatin (which inhibits cathepsin D).

Preparation of Azocasein. The method is based on that of Charney and Tomarelli,[27,27a] except in that sulfanilamide is replaced by sodium sulfanilate. Two solutions, A and B, are prepared.

SOLUTION A. Hammarsten-grade casein (50 g) is dissolved in 1 liter of 50 mM sodium tetraborate at 4°C overnight. The pH is readjusted to 9.2 with 1 M NaOH, and solubilization of the protein completed, if necessary, by heating to 60°C. Any insoluble material is removed by low-speed centrifugation. The solution is cooled and transferred to a 2-liter beaker fitted with an efficient paddle stirrer.

SOLUTION B. First, 5 M solutions of NaOH and HCl are prepared, together with a 5 M sodium formate buffer, pH 3.0. 12 ml of 5 M NaOH is diluted to 200 ml with water; 5.0 g of sulfanilic acid is dissolved in this, followed by 2.2 g of sodium nitrite. Diazotization is started by addition of 18 ml of 5 M HCl, and allowed to proceed for exactly 2 min, with continuous stirring. 18 ml of 5 M NaOH is then added, and the solution is immediately poured into solution A, with continuation of efficient stirring. After 10 min, the precipitation of the azocasein is started by the gradual addition of 5 M sodium formate buffer to the stirred solution. When precipitation is complete (i.e., pH reaches about 4.5), and azocasein is separated by centrifugation. The sedimented protein is washed by resuspension and recentrifugation in 20, 50, 75, 96%, and absolute ethanol. The wash with absolute ethanol is repeated, and the solid allowed to dry in air. If necessary, it is reduced to a fine powder with a pestle and mortar, and drying is completed over P_2O_5. Yield is about 45 g.

Standardization of Azocasein Preparations. The degree of substitution of casein in the coupling procedure is somewhat variable, and this has the obvious result that the peptide products of proteolysis vary in $A_{366}^{1\%}$.

Less obviously, the sensitivity of the substrate to proteolysis may vary too, so it is probably not possible to establish a completely satisfactory absolute unit of activity that will be constant between batches of substrate. We prepare standardized solutions of azocasein as follows. The weight of the solid is corrected for moisture content (by drying to constant weight at 110°C) and ash content (by heating to constant weight in an

[26] S. Riemann, H. Kirschke, B. Wiederanders, A. Brouwer, E. Shaw, and P. Bohley, *Acta Biol. Med. Ger.* in press (1982).

[27] J. Charney and R. M. Tomarelli, *J. Biol. Chem.* **171**, 501 (1947).

[27a] J. Langner, A. Wakil, M. Zimmermann, S. Ansorge, P. Bohley, H. Kirschke, and B. Wiederanders, *Acta Biol. Med. Ger.* **31**, 1 (1973).

ashing oven). The solid azocasein is then allowed to dissolve in water overnight with gentle stirring to form a 6% (corrected w/v) solution, which is stored indefinitely at $-20°C$ in 20-ml portions. The $A_{366}^{1\%}$ of peptides that can be derived from the substrate is determined either by running an assay with a large amount of papain, so that no precipitation occurs in the trichloroacetic acid, or by running a "blank" assay with 1% SDS in the trichloroacetic acid, which prevents precipitation and gives a solution with A_{366} corresponding to complete digestion. In either case the total effective $A_{366}^{1\%}$ of the substrate can be determined after suitable dilution (usually to 60 ml) with 3% (w/v) trichloroacetic acid. This allows one to apply a correction factor when comparing results obtained with different batches of the substrate.

Reagents

Buffer/activator. Sodium acetate–acetic acid buffer (0.1 M, pH 5.0)– 1 μg pepstatin[28]/ml–1 mM disodium EDTA. Cysteine base (40 mM) is added freshly.

Azocasein solution. Six percent (w/v) prepared as above

Azocasein/urea solution. Mix 10 ml of 6% azocasein and 10.8 g of urea, and make up to a final volume of 30 ml (2% azocasein, 6 M urea) with assay buffer (not containing cysteine) as the urea dissolves.

Trichloroacetic acid. Three percent (w/v) aqueous solution

Procedure. The enzyme sample (0.25 ml containing 0.25–2.5 μg of cathepsin L) is mixed with 0.25 ml of assay buffer. After 5 min on the bench for activation, 0.5 ml of azocasein–urea is mixed in, and the reaction is allowed to proceed for 30 min at 40°C. Trichloroacetic acid solution (5 ml) is introduced, and the mixture is filtered (7-cm Whatman No. 1 paper). The A_{366} of the filtrate is determined. Blanks are prepared in which trichloroacetic acid is added immediately after the enzyme. The ΔA_{366} (increase in A_{366} due to enzymatic activity) is not linear with enzyme concentration, but a calibration curve may be made with a sample of cathepsin L titrated with E-64 (see later).

When facilities for microcentrifugation and microspectrophotometry are available, it may be convenient to scale down the volumes in the assay 10-fold.

Active-Site Titration of Cysteine Proteinases

The absolute molarity of solutions of purified cathepsins B and L (as well as papain) may be determined by stoichiometric titration with E-64, L-*trans*-epoxysuccinylleucylamido(4-guanidino)butane, the powerful in-

---

[28] Available from the Peptide Research Foundation (see note 11).

hibitor of cysteine proteinases discovered by Hanada and co-workers.[29,30] The synthetic analog, Ep-475 [L-*trans*-epoxysuccinylleucylamido(3-methyl)]butane, may be used in place of E-64.

Reagents

Stock E-64 solution. Pure, dry E-64[31] (about 3 mg) is dissolved in 0.1 ml of dimethyl sulfoxide, and diluted with water to a 1 mM solution (3.66 mg/10 ml). This is stable at 4°C for a few days. Working solutions of 1, 2, 3 . . . 10 μM concentration are prepared as required.

Enzyme solution. An approximately 10 μM enzyme solution (about 0.25 mg/ml) is prepared in 0.1% Brij 35.

Buffer/activator. As in the respective assay procedures already described

Procedure. To 50 μl of the buffer is added 25 μl of enzyme solution and 25 μl of 1, 2, 3, . . . 10 μM E-64. After 30 min at 30°C, each mixture is diluted to 5.0 ml by addition of 4.9 ml of 0.1% Brij 35. Samples of the dilutions (usually 10 μl) are assayed for residual activity against Z-Phe-Arg-NMec etc. A linear plot of activity against E-64 molarity reaches zero activity at the molarity of the enzyme solution, and its slope shows specific activity of the enzyme in the assay, on a molar basis (Fig. 1).

The method works well for cathepsins B and L, and for papain, which have high rates of reaction with E-64, and can be extended to cathepsin H by careful extrapolation of the upper, linear part of the curve to the abscissa. Other substrates such as Z-Arg-Arg-NNap or Bz-Arg-NNap may be used, but the method is not suitable for use with samples containing more than one enzyme reactive with E-64, or more than one enzyme active against the substrate chosen.

Staining for Activity of Cathepsins B and H

Activity of the enzymes can be located in polyacrylamide isoelectric-focusing gels by use of the appropriate 2-(4-methoxy)naphthylamide substrates,[17] Z-Arg-NNapOMe for cathepsin B, and Arg-NNapOMe for cathepsin H.[12,21]

[29] K. Hanada, M. Tamai, S. Ohmura, J. Sawada, T. Seki, and I. Tanaka, *Agric. Biol. Chem.* **42**, 529 (1978).

[30] A. J. Barrett, A. A. Kembhavi, M. A. Brown, H. Kirschke, C. G. Knight, M. Tamai, and K. Hanada, *Biochem. J.* in press (1982).

[31] E-64 was kindly supplied by, and used in collaboration with, Dr. K. Hanada, Research Laboratories of Taisho Pharmaceutical Co. Ltd., 1-403 Yoshinocho, Ohmiya, Saitama, Japan.

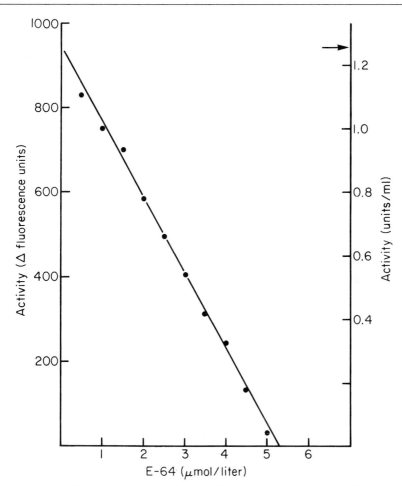

FIG. 1. A titration of human cathepsin B with E-64. For each point, 25 µl of cathepsin B solution (280 µg/ml) was reacted with an equal volume of 0–10 µM E-64 in a total volume of 100 µl. A 0.5-µl sample of each reaction mixture was then assayed for activity against Z-Phe-Arg-NMec. The measured fluorescences (left ordinate) were adjusted for the sampling factor to give the residual activities/ml indicated on the right ordinate. From this plot it can be calculated that (1) the enzyme solution had an operational molarity of 5.3 µM (53% of that expected on a protein basis), and (2) the absolute specific activity of cathepsin B in this assay is 238 units/µmol. In the same experiment (but not shown for clarity), assays were also made with Bz-DL-Arg-NNap (found 402 units/µmol) and Z-Arg-Arg-NNap (found 9160 units/µmol), as described in the text.

Reagents

Buffer/activator. A 0.1 *M* sodium phosphate buffer, pH 6.0 or 6.8 (according to the enzyme), containing EDTA and cysteine, exactly as described for the respective assay procedures above

Substrate stock solution. The substrate is dissolved in dimethyl sulf-oxide at 20 mg/ml, and stored at 4°C.

Color reagent. This is similar to that used in the spectrophotometric assay of cathepsin B with Z-Arg-Arg-NNap (see before), except that it contains no detergent, since the colored reaction product must remain insoluble. Take 1.0 ml of Fast Garnet base stock solution (see the cathepsin B assay method) and diazotize with 0.1 ml of 0.2 M NaNO$_2$ as usual. Dilute to 50 ml with Mersalyl–EDTA reagent (i.e., the solution that is normally mixed with an equal volume of 4% Brij 34 in the cathepsin B assay). The reagent is stable for 1 day.

Procedure. Assay buffer is prewarmed to 40°C, and the gel is placed in it as soon as the run is complete. After 5 min, substrate stock solution (100 μl/5 ml) is added, and the tube or dish is agitated frequently during 10 min. The buffered substrate solution is then replaced by color reagent for a further 10 min at room temperature, and the gel is stored in water. Zones of enzymatic activity acquire a crimson color in the coupling reaction, and this is stable for a week or so in the dark.

The use of this procedure with cathepsin H is illustrated in Fig. 2.

Purification of the Enzymes

We describe two types of purification methods. Method I is applicable to large amounts of frozen tissues, whereas method II is used with fresh tissue and takes advantage of a 50-fold purification factor attainable by isolation of lysosomes: it has the further advantage of separating the enzymes from inhibitors that are present in the cytosol and plasma.

Method I: Purification of Cathepsins B and H from Human Liver

We shall describe the purification of cathepsins B and H from human liver, as it has been done in Cambridge. The procedure is likely to be applicable to liver of several other species, however, with little modification. All operations, unless specified otherwise, are performed at 4°C.

Extraction, Autolysis, and Acetone Fractionation

The tissue, having been thawed at 4°C overnight, is trimmed of fat and connective tissue and cut into small pieces (about 1 cm^3). Four 250-g portions are each dispersed in 500 ml of a solution of 1% (w/v) NaCl–2% (v/v) butan-1-ol–10 mM disodium EDTA, at 4°C, by use of a Waring-type blender. The homogenate is treated with Arquad 2C-50 and centrifuged to remove particulate matter, subjected to autolysis overnight at pH 4.5 and 40°C, and then fractionated with acetone, all as described by Barrett.[21]

The 47–64% acetone precipitate is collected by centrifugation at 1500 g for 15 min at $-5°C$, redissolved in 5 mM disodium EDTA to a total volume of 50 ml, and dialyzed overnight against 5 liters of water at 4°C. The purification is then continued as described by Schwartz and Barrett.[12]

DEAE–Cellulose Chromatography

The dialyzed product of acetone fractionation (about 150 ml) is adjusted to pH 6.0 by dropwise addition of 2 M Tris-HCl buffer (pH 9.0) and applied at a flow rate of 10 cm/hr, to a column (14 × 2.6 cm, 70 cm³) of DEAE–cellulose (Whatman DE-52), equilibrated with 20 mM sodium phosphate buffer (pH 6.0) containing 1% butan-1-ol. The ion exchanger is then washed with the 20 mM buffer (about 1 bed volume) until the A_{280} falls nearly to the starting value. Cathepsins H and D pass straight through the column, and all fractions comprising the first A_{280} peak are combined. Cathepsin B is eluted stepwise with 0.2 M NaCl in the 20 mM sodium phosphate buffer.

Cathepsin B

The pool of cathepsin B from DEAE–cellulose is further purified by covalent chromatography on a column (9.4 × 2.5 cm, 50 cm³) of aminophenylmercuric acetate coupled to Sepharose, equilibrated with 50 mM sodium acetate buffer (pH 5.5) containing 0.2 M NaCl–1 mM disodium EDTA–1% butan-1-ol.[21] The enzyme pool is mixed with 0.25 volume of a 5-fold concentrated solution of the column buffer, and applied to the column at a flow rate of 20 cm/hr.

The column is washed with the buffer until the A_{280} falls close to the starting value. Cathepsin D passes unadsorbed through the organomercurial column. Cathepsin B is eluted as a sharp peak with 10 mM 2-mercaptoethanol in the column buffer. The specific activity of the enzyme purified in this way is similar to that previously determined for cathepsin B purified by other methods, and the yield is about 60 mg/kg of tissue. Multiple forms can be resolved analytically by isoelectric focusing, and preparatively by chromatography under equilibrium conditions on CM-cellulose.[21]

Cathepsin H

Gel Chromatography on Ultrogel AcA 54

The unadsorbed pool from DEAE–cellulose is concentrated to 7–8 ml by ultrafiltration, and mixed with 0.1 volume of a 10-fold stock solution of the column buffer (see later). The sample is run on a column (93 × 2.5 cm,

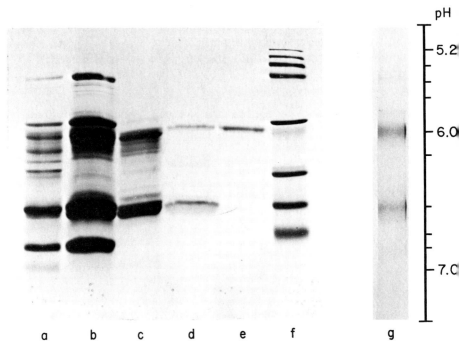

FIG. 2. SDS–polyacrylamide gel electrophoresis at various stages in the purification of cathepsin H from human liver, and isoelectric focusing of the pure enzyme. The samples, run after reduction with 2-mercaptoethanol, were taken after (a) acetone fractionation, (b) DEAE–cellulose chromatography, (c) Ultrogel AcA 54 chromatography, (d) hydroxyapatite chromatography, and (e) concanavalin A–Sepharose chromatography. A mixture of standard proteins was also run, after reduction, in lane (f): the bands correspond to phosphorylase a (100,000 M_r), transferrin (78,000), bovine serum albumin (68,000), immunoglobulin G heavy chain (50,000), carbonic anhydrase (29,000), soybean-trypsin inhibitor (21,000), cytochrome c (12,400 M_r), and aprotinin (6500). The isoelectric focusing gel (g) was stained for enzymatic activity with Arg-NNapOMe as substrate.

457 cm³) of Ultrogel AcA 54 in 20 mM KH$_2$PO$_4$–Na$_2$HPO$_4$ buffer (pH 6.0) containing 0.2 M NaCl–1 mM disodium EDTA–0.1% Brij 35. The column is run at 4 cm/hr, and cathepsin H is eluted after the major protein peaks, typically with a 10- to 15-fold increase in specific activity, completely separated from cathepsin D.

Hydroxyapatite Chromatography

The pool of cathepsin H obtained by gel filtration is dialyzed overnight against 2 liters of 20 mM KH$_2$PO$_4$–NaHPO$_4$ buffer (pH 6.0), and applied to a column (21 × 1.1 cm, 20 cm³) of hydroxyapatite equilibrated with the buffer. The column is eluted with a gradient (400 ml) increasing to 150 mM KH$_2$PO$_4$–Na$_2$HPO$_4$ (pH 6.0) at a flow rate of 7 cm/hr. Cathepsin H is eluted within a portion of the gradient corresponding to 60–85 mM phosphate. Analysis of the enzyme pool by SDS–polyacrylamide gel electrophoresis at this stage typically shows 2 major polypeptide bands, one of which corresponds to cathepsin H.

Chromatography on Concanavalin A–Sepharose

The pool of activity from hydroxyapatite is adjusted to 0.2 M NaCl–1 mM CaCl$_2$–1 mM MnCl$_2$–0.1% Brij 35, and run at 6 cm/hr on a column (6.4 × 1 cm, 5 cm³) of concanavalin A–Sepharose, equilibrated with 20 mM KH$_2$PO$_4$–Na$_2$HPO$_4$ buffer (pH 6.0), also containing 0.2 M NaCl–1 mM CaCl$_2$–1 mM MnCl$_2$–0.1% Brij 35. The column is washed with 3–4 bed volumes of starting buffer, and the enzymatic activity is eluted as a broad peak with 5 mM methyl α-D-mannoside in the same buffer. Analysis of column fractions by SDS–polyacrylamide gel electrophoresis shows that the major inactive protein contained in the sample is not adsorbed to the column under starting conditions, and thus is efficiently separated from cathepsin H (Fig. 2). Occasionally, samples from hydroxyapatite contain minor components in addition to the two referred to here. Then it is necessary to elute the concanavalin A–Sepharose column with a gradient (20 bed volumes) of 0–100 mM methyl α-D-mannoside. In either case, the fractions containing activity are combined to form the pool of purified enzyme.

The yield of cathepsin H from human liver is about one-tenth of that of cathepsin B (i.e., about 6 mg/kg).

Method II: Purification of Cathepsins B, H, and L from Rat Liver

Homogenization and Cell Fractionation

One hundred fasted male Wistar rats (300–400 g) are anesthetized with ether. The liver of each is perfused with cold 0.9% NaCl until free from blood, and placed on ice. The livers are minced with scissors and forced through a nylon sieve, at 4°C. The brei is suspended in an equal volume of 0.25 M sucrose and treated with a flowthrough, Teflon/glass homogenizer. A light mitochondrial-lysosomal fraction is prepared by differential centrifugation.[32]

---

[32] P. Bohley, H. Kirschke, J. Langner, and S. Ansorge, *FEBS Lett.* **5**, 233 (1969).

The light mitochondrial-lysosomal fraction (about 50 ml) is homogenized with 2.5 volumes of water by 10 passes of a Potter homogenizer, at 0°C. The preparation is centrifuged at 360,000 g for 45 min, and the supernatant concentrated in an Amicon Diaflo apparatus over a UM-10 membrane to 15–20 ml.

Gel Chromatography on Sephadex G-75

Purification of cathepsin L from the lysosomal lysate is much as described by Kirschke et al.,[24] but with extra precautions to prevent autolysis of the enzyme. The L20 (lysosomal, 20,000 ± 10,000 M_r) fraction of the preparation is isolated by gel chromatography on a 2-liter column of Sephadex G-75, in 10 mM potassium phosphate buffer (pH 6.9) containing 150 mM KCl. This fraction is enriched in cathepsins B, H, and L, but contains only 10–15% of the protein applied to the column.

CM-Sephadex Chromatography

The L20 fraction is concentrated once more by ultrafiltration, and dialyzed against 10 mM potassium phosphate buffer, pH 5.8. It is applied to a CM-Sephadex C-50 column equilibrated with the same buffer. The column is washed first with more of the starting buffer, and then with 200 mM KCl in the buffer, which elutes cathepsins B and H. (These enzymes can be separated by DEAE–cellulose chromatography,[33] and further purified as described for the human enzymes.) A gradient from 200 mM to 600 mM KCl then elutes cathepsin L at about 450 mM KCl.

Chromatography of Cathepsin L on Concanavalin A–Sepharose

The fractions containing cathepsin L (as a mercury derivative) are concentrated and equilibrated with 20 mM sodium phosphate buffer (pH 6.0) containing 0.2 M NaCl–1 mM CaCl₂–1 mM MnCl₂ by ultrafiltration. The enzyme sample is applied to a column (15 × 0.9 cm, 9.5 ml) of concanavalin A–Sepharose equilibrated with the same buffer. The column is eluted with 5 bed volumes of starting buffer and then with a gradient (10 bed volumes) from 0 to 100 mM methyl α-D-mannoside. In a typical experiment, 30% of the applied activity of cathepsin L is not adsorbed by concanavalin A–Sepharose (because it has lost its carbohydrate by autolysis) and 65% is eluted at 25 mM methyl α-D-mannoside. Some minor components without proteolytic activity are eluted with the starting buffer and with 5 mM methyl α-D-mannoside. The active fractions are concen-

[33] H. Kirschke, J. Langner, B. Wiederanders, S. Ansorge, and P. Bohley, Acta Biol. Med. Ger. **28**, 305 (1972).

trated and equilibrated with 0.1 M sodium acetate (pH 5.0) by ultrafiltration.

Typical yields are 6 mg each of cathepsins B and H, and 1 mg of cathepsin L from 1 kg of rat liver.

Storage of Purified Cysteine Proteinases

In the past, it has been common for cysteine proteinases to be stored in activated form, in the presence of a thiol compound. This often leads to autolytic inactivation, however, and much better stability is achieved by storage of the enzymes as their mercury derivatives. A good solution in which to store cathepsins B, H, and L is 0.5 mM HgCl$_2$–1 mM EDTA–50 mM sodium acetate buffer (pH 5.0) and 0.2 M NaCl, when they are to be kept at 4° or $-20°$C. The same buffer, but without NaCl, is suitable for freeze-drying, which usually gives good stability. If necessary, the active enzymes can be separated from the lower M_r products of autolysis by running on a small column of Ultrogel AcA 54 in the same buffer with NaCl.

Properties of the Enzymes

Cathepsins B, H, and L resemble each other in being lysosomal, having proteolytic activity dependent on an active-site cysteine residue, being irreversibly inactivated above pH 7, and having M_r's in the range of 21,000–28,000. The sequence of the N-terminus of rat liver cathepsin B shows marked similarity to that of papain,[34] and it seems probable that all three enzymes are evolutionarily homologous with papain. Nevertheless, it can be seen from Table I that there are also some marked differences between the enzymes.

Cathepsin B

Human cathepsin B appears to have a M_r of 27,500 in gel chromatography, but 24,000 in SDS–gel electrophoresis.[21] The probable explanation for this is provided by the work of Towatari et al.,[35] who found that rat liver cathepsin B can appear on SDS gels at M_r 29,000, or 25,000 plus 5000. It was shown that autolysis liberated an N-terminal fragment that contains the catalytically active Cys-29 residue analogous to Cys-25 of papain, and that several sequences in the N-terminal region, including that

[34] K. Takio, T. Towatari, M. Katunuma, and K. Titani, Biochem. Biophys. Res. Commun. **97**, 340 (1980).

[35] T. Towatari, Y. Kabata, and N. Katunuma, Eur. J. Biochem. **102**, 279 (1979).

around the cysteine, are so similar to those in papain as to leave no reasonable doubt of the evolutionary homology of the proteins.[34]

The amino acid composition of human liver cathepsin B[4] is very similar to those of the enzyme from pig parathyroid and liver[36] and from rat liver.[37] Only traces of hexosamine (3 residues/mol) were detected in the human enzyme,[4] which suggests that it has only a very small carbohydrate prosthetic group, consistent with its negligible affinity for concanavalin A–Sepharose.[12] The enzyme crystallized from rat liver, on the other hand, was found to contain 16 residues of hexosamine (including glucosamine and galactosamine)/mol,[37] but it, too, is unadsorbed by concanavalin A–Sepharose, in our hands. $A_{280\text{nm}}^{1\%}$ is 20 for human liver cathepsin B.[21]

Some kinetic constants for cathepsin B are summarized in Table II. The use of E-64 to determine the active-site molarity of cathepsin B has necessitated the reappraisal of the k_{cat} values previously calculated for the human enzyme on a protein basis.[4,13,20] The enzyme purified as described, to apparent homogeneity,[12,21] consistently shows a specific activity of 114–132 units/mg in the assay with Z-Arg-Arg-NNap, whereas the value calculated from active-site titration (for M_r 28,000) is 327 U/mg (Fig. 1). Thus it seems that our best preparations of cathepsin B have contained about 50% inactive material. This is not yet explained, but may have a bearing on the occasional observation of a doubling of activity in storage.[21] Appropriate corrections have been made to the k_{cat} values given for human cathepsin B in Table II.

Most synthetic substrates described for cathepsin B have contained arginine as the P_1 residue, and historically it was thought of as a trypsin-like enzyme. It is now clear that binding of a P_2 residue greatly increases k_{cat}/K_m values, and P_2 may well be the most important residue in the substrate, as it is for papain.[38] The specificity of the S_2 subsite of cathepsin B has yet to be clearly defined, however, since it seems to be satisfied equally well by phenylalanine (as in papain) or a basic residue. Z-Gly-Gly-Arg-NNap is a very poor substrate for cathepsin B, however.[11]

The pH optimum of cathepsin B for most substrates is about pH 6.0, and activity falls away rather sharply at neutral pH because of the irreversible inactivation of the enzyme.[9] For some protein substrates, the optimum is much lower (e.g., pH 3.5 for collagen[39]), but this is probably due to effects on the substrate.

[36] R. R. MacGregor, J. W. Hamilton, R. E. Shofstall, and D. V. Cohn, *J. Biol. Chem.* **254**, 4423 (1979).
[37] T. Towatari and N. Katunuma, *Biochem. Biophys. Res. Commun.* **83**, 513 (1978).
[38] A. Berger and I. Schechter, *Philos. Trans. R. Soc. London, Ser. B* **257**, 249 (1970).
[39] M. C. Burleigh, A. J. Barrett, and G. S. Lazarus, *Biochem. J.* **137**, 387 (1974).

TABLE II
SOME KINETIC CONSTANTS FOR CATHEPSIN B[a]

Substrate	k_{cat} (sec^{-1})	K_m (mM)	k_{cat}/K_m (10^3 sec^{-1} M^{-1})	Species	Ref.
Bz-Arg-NH$_2$	0.84	60	0.014	Rat	b
Bz-Arg-NPhNO$_2$	0.60	2.0	0.3	Cow, rabbit	c
Bz-Arg-NNap	25	4.3	5.8	Human	d
Z-Arg-Arg-NNap	178	0.19	937	Human	d,e
Z-Val-Lys-Lys-Arg-NNapOMe	68	0.14	486	Pig	f
Z-Phe-Arg-NMec	1500	0.15	9900	Human	g
Z-Arg-Arg-NMec	1340	0.39	3420	Human	g
Z-Lys-OPhNO$_2$	59	0.019	3080	Rat	g

[a] Various assumptions have been made in recalculating the data from published values (see also Ref. c). The k_{cat} values for most species are based on the assumption that 28 mg of protein represents 1 μmol of enzyme, but the k_{cat} values for rat and human cathepsin B are based on molarities determined by titration, and are about 2-fold higher than values calculated previously on the protein basis (see the text). It is very likely that the values for the other species will require similar corrections. The values given here relate to assays at pH 6.0.

[b] H. Kirschke, J. Langner, B. Wiederanders, S. Ansorge, and P. Bohley, *Eur. J. Biochem.* **74**, 293 (1977).

[c] A. J. Barrett, in "Proteinases in Mammalian Cells and Tissues" (A. J. Barrett, ed.), p. 181. North-Holland Publ., Amsterdam, 1977.

[d] C. G. Knight, *Biochem. J.* **189**, 447 (1980).

[e] J. K. McDonald and S. Ellis, *Life Sci.* **17**, 1269 (1975).

[f] R. R. MacGregor, J. W. Hamilton, R. E. Shofstall, and D. V. Cohn, *J. Biol. Chem.* **254**, 4423 (1979).

[g] H. Kirschke, A. A. Kembhavi, P. Bohley, and A. J. Barrett, unpublished results (1981).

A curious aspect of the activity of cathepsin B is its peptidyldipeptidase action; that is to say, it cleaves C-terminal dipeptides sequentially from some polypeptides, with broad specificity. We have shown this for glucagon[40] and muscle aldolase,[41] and it is this activity that leads to the inactivation of aldolase by cathepsin B (Fig. 3).

The specificity of the endopeptidase activity of cathepsin B is unclear. Two early studies with the oxidized B chain of insulin (see Ref. 4) suffered from incomplete purity of the enzyme, but showed numerous cleavage points not unlike those reported for papain (Fig. 4). There was no indication of trypsin-like specificity. Many proteins are degraded by cathepsin B,[4] but the bonds cleaved have not been identified.

[40] N. N. Aronson and A. J. Barrett, *Biochem. J.* **171**, 759 (1978).
[41] J. S. Bond and A. J. Barrett, *Biochem. J.* **189**, 17 (1980).

(a)   -Pro-Ser-Gly-Gln-Ala-Gly-Ala-Ala-Ala-Ser-

-Glu-Ser-Leu-Phe-Ile-Ser-Asn-His-Ala-Tyr

(b)   His-Ser-Gln-Gly-Thr-Phe-Thr-Ser-Asp-Tyr-Ser-Lys-Tyr-Leu-

-Asp-Ser-Arg-Arg-Ala-Gln-Asp-Phe-Val-Gln-Trp-Leu-Met-Asn-Thr

FIG. 3. The peptidyldipeptidase activity of cathepsin B. Points of cleavage (sequential from the C-terminus) in (a) the C-terminal sequence of rabbit muscle aldolase,[41] and (b) glucagon.[40]

Cathepsin H

The M_r of cathepsin H from rat liver or human liver is $28,000$[12,22]; it consists of a single polypeptide chain, contains carbohydrate, and is bound by concanavalin A–Sepharose. $A_{280\,nm}^{1\%}$ is 12.2 for the human enzyme. Cathepsin H tends to be a more basic protein than cathepsins B and L. the human liver enzyme occurs as two major forms of pI 6.0 and 6.4,[12] that from rabbit lung has several forms in the range 5.8–6.5,[42] and the rat liver enzyme has pI 7.1.[22] Rat liver cathepsin H has unusual stability to heat, but this is not seen with the human enzyme.

With the synthetic substrates tested, human cathepsin H shows greatest activity at pH 6.8, but is more active at pH 5–6 against azocasein.[12] Like the other lysosomal cysteine proteinases, cathepsin H is irreversibly inactivated above pH 7.

The most remarkable aspect of the specificity of cathepsin H is its capacity to act as an aminopeptidase, at least with synthetic substrates. This makes it unique among endopeptidases characterized so far, and has led to its being described as an "aminoendopeptidase." The most sensitive substrates yet discovered for cathepsin H are Arg-NNap and Arg-NMec; Leu-NNap is somewhat less susceptible, as is Bz-Arg-NNap. Cathepsin H has essentially no action on Z-Arg-Arg-NNap and Z-Arg-Arg-NMec, so these excellent substrates of cathepsin B can be used to quantify that enzyme in the presence of cathepsin H.[12,13] Some kinetic constants of human cathepsin H are given in Table III.

There seem to be marked species differences in the proteolytic activity

[42] H. Singh and G. Kalnitsky, J. Biol. Chem. 255, 369 (1980).

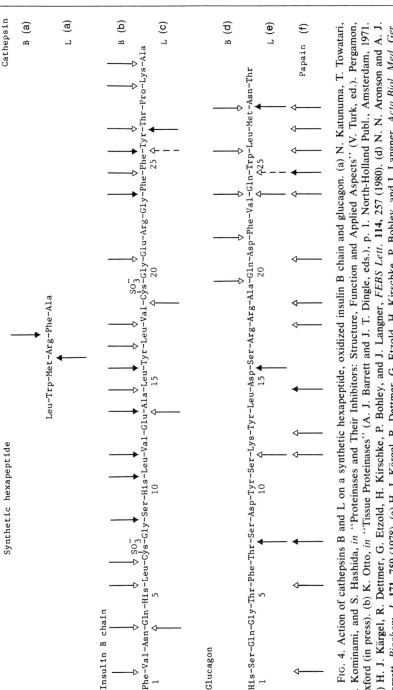

FIG. 4. Action of cathepsins B and L on a synthetic hexapeptide, oxidized insulin B chain and glucagon. (a) N. Katunuma, T. Towatari, E. Kominami, and S. Hashida, *in* "Proteinases and Their Inhibitors: Structure, Function and Applied Aspects" (V. Turk, ed.). Pergamon, Oxford (in press). (b) K. Otto, *in* "Tissue Proteinases" (A. J. Barrett and J. T. Dingle, eds.), p. 1. North-Holland Publ., Amsterdam, 1971. (c) H. J. Kärgel, R. Dettmer, G. Etzold, H. Kirschke, P. Bohley, and J. Langner, *FEBS Lett.* **114**, 257 (1980). (d) N. N. Aronson and A. J. Barrett, *Biochem. J.* **171**, 759 (1978). (e) H. J. Kärgel, R. Dettmer, G. Etzold, H. Kirschke, P. Bohley, and J. Langner, *Acta Biol. Med. Ger.* in press (1981). (f) M. J. Desmazeaud, *Biochimie* **54**, 1109 (1972).

TABLE III
KINETIC CONSTANTS OF HUMAN CATHEPSIN H [a,b]

Substrate	k_{cat} (sec^{-1})	K_m (mM)	k_{cat}/K_m (10^3 sec^{-1} M^{-1})
Arg-NNap	1.81	0.097	18.7
Bz-Arg-NNap	1.82	0.637	28.6
Arg-NMec	2.53	0.150	16.9

[a] Values of k_{cat} are based on the assumption that 28 mg of protein represents 1 μmol of enzyme, and may therefore be low: cathepsin H does not react sufficiently rapidly with E-64 to permit accurate active-site titrations, but some approximate estimates suggest that "homogeneous" preparations may be only 10% active. The values given here for Bz-Arg-NNap were obtained with the racemate, but are calculated for the L isomer on the assumption that the presence of the D isomer had no effect. The values given relate to assays at pH 6.8.

[b] W. N. Schwartz and A. J. Barrett, *Biochem. J.* **191**, 487 (1980).

of cathepsin H. Neither the rat nor the human enzyme has any action on collagen in solution, but the rat enzyme is similar to rat cathepsin B in its activity against other proteins such as azocasein, whereas the human enzyme is a very weak endopeptidase. In contrast to both of these, rabbit lung cathepsin H has strong activity against the terminal peptides of collagen, at pH 3.5.[42]

Cathepsin L

The M_r of rat liver cathepsin L is in the range of 21,000–24,000, and the pI 5.8–6.1. Despite the moderate isoelectric point, however, the enzyme is eluted long after cathepsins H and B from CM-cellulose, in a salt gradient at pH 5.8,[24] and is adsorbed by DEAE–cellulose at pH 6.5–6.8. Cathepsin L is a glycoprotein, with affinity for concanavalin A–Sepharose.[43]

Although it is the most powerful of the lysosomal proteinases, cathepsin L was discovered only relatively recently, because it has little or no activity against the conventional synthetic substrates, Bz-Arg-NPhNO$_2$

[43] H. Kirschke, H. J. Kärgel, S. Riemann, and P. Bohley, *in* "Proteinases and Their Inhibitors: Structure, Function and Applied Aspects" (V. Turk, and L. Vitale, eds.), p. 93. Pergamon, Oxford (1981).

TABLE IV
SPECIFICITY OF RAT CATHEPSIN L
Part A. Kinetic Constants for Synthetic Substrates[a]

Substrate	k_{cat} (sec^{-1})	K_m (mM)	k_{cat}/K_m (10^3 sec^{-1} M^{-1})
Bz-Arg-NH$_2$	1.68	3.0	0.56
Z-Lys-OPhNO$_2$	19.8	0.010	1980
Z-Phe-Arg-NMec	26.0	0.007	3700

Part B. Specific Activities with Protein Substrates

Substrate	pH	Unit[b]/μmol
[^{125}I]Glucagon	6.0	960.0
Azocasein	6.0	283.2
Azocasein–3 M urea	5.0	717.6
Histones	6.0	120.5
Hemoglobin	6.0	42.2
Collagen, insoluble	6.0	68.7
Collagen, insoluble	3.5	584

[a] The k_{cat} values (30°C, pH 5.5) are based on molarities determined by titration with E-64; they are about 1.5-fold higher than values calculated previously on the assumption that 24 mg of protein represented 1 μmol of cathepsin L.

[b] One unit corresponds to 1 mg of substrate degraded per minute at 37°C. The molar concentration of cathepsin L was determined by titration with E-64. Incubation conditions: 0.5–1% substrate in 10 mM phosphate buffer (pH 6.0), 0.1 M acetate buffer (pH 5.0), or 0.2 M formate buffer (pH 3.5), containing 1 mM dithiothreitol and 1 mM EDTA. Hydrolysis was determined by analysis of peptides soluble in 5% trichloroacetic acid for ^{125}I ([^{125}I]glucagon), A_{366} (azocasein), nitrogen (histones, hemoglobin) or hydroxyproline (collagen), and converted to weight of substrate degraded on the basis of the known composition of the proteins.

and Bz-Arg-NNap. Synthetic substrates that are sensitive to cathepsin L (Bz-Arg-NH$_2$, Z-Lys-OPhNO$_2$, and Z-Phe-Arg-NMec; see Table IV) are also degraded by cathepsin B, so the activities of cathepsin L were easily confused with those of cathepsin B. Nevertheless, there are some large differences in specific activity. For example, the activity of cathepsin L on Z-Phe-Arg-NMec in the assay described above is about 1200 units/μmol, whereas that of cathepsin B is 238 units/μmol. Similarly, specific activities against proteins such as azocasein and collagen are in the order of 10-fold greater than those of cathepsin B.[44] When the lysosomal cysteine proteinases are fractionated by column chromatography, the elution position of cathepsin L is usually recognized as a peak of activity against a substrate such as Z-Phe-Arg-NMec or azocasein not associated with appreciable activity against the specific substrates of cathepsins B and H. pH optima are in the range 5.0–5.5 for azocasein, ^{125}I-labeled glucagon, and Bz-Arg-NH$_2$.[24]

Many proteins are known to be degraded by cathepsin L, but the bonds cleaved have not been identified.[6] Recently, the specificity of degradation of a synthetic hexapeptide, the insulin B chain and glucagon, has been elucidated, however, The results are summarized in Fig. 4. The general impression is that cathepsin L cleaves where there are hydrophobic residues to occupy sites S$_2$ and S$_3$ (e.g., -Tyr-Leu-Asp-), but there are exceptions to this general rule. Cathepsin L does not show the peptidyldipeptidase activity seen with cathepsin B.

Inhibitors

Containing highly activated thiol (or thiolate) groups, the cysteine proteinases react readily with many thiol-blocking reagents. Among these are iodoacetate, mercuribenzoate, and 2,2'-dipyridyl disulfide. Such reagents are reactive with thiolate groups generally, but are reasonably selective for the enzyme-active sites at low pH (4–6), because low-molecular-weight thiols are almost entirely protonated under these conditions.

The aminoacyl-chloromethanes, such as Tos-Lys-CH$_2$Cl, are also general thiol-blocking reagents; thus they do not show the kind of selectivity among cysteine proteinases that they do with serine proteinases. Two classes of compounds with less reactive functional groups, whose reactions with enzymes are therefore mediated by preliminary binding to specificity subsites, are the peptidyl-diazomethanes, and the epoxy-succinyl-peptides.

The peptidyl-diazomethane (or diazomethyl ketone) group of inhibitors specific for cysteine proteinases was introduced by Shaw and co-workers.[45-47] Z-Phe-Ala-CHN$_2$ is the best diazomethane inhibitor of cathepsin B discovered so far (the apparent k_{2nd} being in the range 500–2000 $M^{-1} \cdot sec^{-1}$ at pH 6 for the rat,[48] bovine,[46] and human[30] enzymes). The rate of reaction with rat cathepsin L has been estimated as 70,000

[44] H. Kirschke, A. A. Kembhavi, A. J. Barrett, and P. Bohley, unpublished results (1981).
[45] R. Leary and E. Shaw, *Biochem. Biophys. Res. Commun.* **79**, 926 (1977).
[46] H. Watanabe, G. D. J. Green, and E. Shaw, *Biochem. Biophys. Res. Commun.* **89**, 1354 (1979).
[47] G. D. J. Green and E. Shaw, *J. Biol. Chem.* **251**, 4528 (1980).
[48] H. Kirschke and E. Shaw, *Biochem. Biophys. Res. Commun.* **101**, 454 (1981).

$M^{-1} \cdot sec^{-1}$, however. In contrast, Z-Phe-Phe-CHN$_2$ is a poor inhibitor of cathepsin B (k_{2nd} about 200 $M^{-1} \cdot sec^{-1}$), but is even better than Z-Phe-Ala-CHN$_2$ for cathepsin L. This has led to the use of Z-Phe-Phe-CHN$_2$ under defined conditions to give selective inhibition of cathepsin L in the presence of cathepsin B.[6]

Cathepsin H is only very slowly inhibited by any of the acyl diazomethanes tested so far.

The first of the epoxide inhibitors was a compound produced by *Aspergillus japonicus* discovered by Hanada and co-workers, and named E-64.[29] E-64, L-*trans*-epoxysuccinylleucylamido(4-guanidino)butane, has negligible reactivity with low-molecular-weight thiol compounds, but is extremely reactive with some cysteine proteinases, presumably because the dipeptide moiety is bound in subsites $S_1'-S_2'$, facilitating an effectively intramolecular attack of the sulfur on the epoxide. These compounds do not react with noncysteine proteinases. Hanada's group has synthesized numerous analogs of E-64, and some of these have been tested for activity against human cathepsin B and human cathepsin H in our laboratory (Table V). Compound Ep-475 is representative of the most powerful irreversible inhibitors of cathepsin B discovered so far, whereas E-64 itself is the fastest reacting with cathepsin H.

The most effective of the tight-binding but reversible inhibitors of the cysteine proteinases are the peptide aldehydes, such as leupeptin (reviewed in Ref. 49). The inhibition of human cathepsin B by leupeptin has been studied in detail by Knight[20]; it has been found to be purely competitive with K_i (apparent) being $7 \times 10^{-9} M$ (for the L form), independent of pH in the range 4.3–6.0. Leupeptin is a potent inhibitor of cathepsin L ($K_i \sim 3.7 \times 10^{-8} M$), but relatively weak for cathepsin H[42]; for the rat enzyme, the approximate K_i was $5 \times 10^{-6} M$.[50] Data on the comparative potencies of the other microbial inhibitors for the rat liver cysteine proteinases have been given by Kirschke.[6]

Little is known, as yet, of the protein inhibitors of cysteine proteinases, and very little of their action on the lysosomal enzymes. The inhibitor from chicken egg white, which we have called "cystatin,"[51] inhibits cathepsin B as well as papain ($K_i \sim 1.5 \times 10^{-9} M$ for both enzymes), and may well be closely related to the inhibitors of similar molecular weight that have been found in rat skin,[52] as well as other tissues and species.[51]

The α-cysteine proteinase inhibitor of M_r 60,000–80,000, which is

[49] H. Umezawa, this series, Vol. 45, p. 678.

[50] H. Kirschke, J. Langner, B. Wiederanders, S. Ansorge, P. Bohley, and U. Broghammer, *Acta Biol. Med. Ger.* **35**, 285 (1976).

[51] A. J. Barrett, this volume [57].

[52] M. Järvinen, *J. Invest. Dermatol.* **71**, 114 (1978).

TABLE V
ACTIVE SITE-DIRECTED INHIBITORS OF HUMAN CATHEPSINS B, H, AND L[a,b]

Compound	Abbreviation	Rate (M^{-1} sec^{-1})		
		Cathepsin B	Cathepsin H	Cathepsin L
Z-Phe-Ala-CHN$_2$		1220	—	70,000
Z-Phe-Phe-CHN$_2$		185	—	160,000
HO-Eps-Leu-amido(4-guanidino)butane	E-64	89,400	4000	96,250
HO-Eps-Leu-amido(3-methyl)butane	Ep-475	298,000	2000	206,000

[a] The apparent second-order rate constants are $k_{ss1}/[I]$ values determined under the standard assay conditions; as far as we can determine, the condition $[I] \ll K_i$ was met. Abbreviations: Eps-, *trans*-L-epoxysuccinic acid; —, not determined.

[b] A. J. Barrett, A. A. Kembhavi, M. A. Brown, H. Kirschke, C. G. Knight, M. Tamai, and K. Hanada, *Biochem. J.* in press (1982).

found in human plasma,[53,54] is probably only a weak inhibitor of cathepsin B, as opposed to papain, since no activity clearly attributable to it was detected in studies of the cathepsin B inhibitors of human plasma.[55] α_2-Macroglobulin, a plasma protein with a remarkable capacity to "trap" endopeptidases, is the major inhibitor of cathepsin B in human plasma.[55,56] α_2-Macroglobulin also binds cathepsin H,[57] incidentally dispelling any doubt that the enzyme is an endopeptidase as well as an aminopeptidase.

[53] M. Sasaki, K. Minakta, H. Yamamoto, M. Niwa, T. Kato, and N. Ito, *Biochem. Biophys. Res. Commun.* **76**, 917 (1977).
[54] H. C. Ryley, *Biochem. Biophys. Res. Commun.* **89**, 871 (1979).
[55] P. M. Starkey and A. J. Barrett, *Biochem. J.* **131**, 823 (1973).
[56] A. J. Barrett, this volume [54].
[57] C. Al-Farr and H. Kirschke, unpublished results (1976).

[42] Cathepsin G

By ALAN J. BARRETT

During 1974, several groups independently reported the existence of a chymotrypsin-like proteinase in human neutrophil leukocytes.[1-3] Subsequently, Starkey and Barrett isolated a chymotrypsin-like enzyme from human spleen, and showed that it was identical with the neutrophil enzyme.[4,5] Although chymotrypsin-like in most respects, the enzyme also differed from chymotrypsin in some ways, and clearly needed a name; for want of anything better, "cathepsin G" was suggested. Much information on cathepsin G (EC 3.4.21.20) has been presented in Refs. 6-9.

[1] A. C. Gerber, J. H. Carson, and B. Hadorn, *Biochim. Biophys. Acta* **364**, 103 (1974).
[2] R. Rindler-Ludwig, F. Schmalzl, and H. Braunsteiner, *Br. J. Haematol.* **27**, 57 (1974).
[3] W. Schmidt and K. Havemann, *Hoppe-Seyler's Z. Physiol. Chem.* **355**, 1077 (1974).
[4] P. M. Starkey and A. J. Barrett, *Biochem. J.* **155**, 255 (1976).
[5] P. M. Starkey and A. J. Barrett, *Biochem. J.* **155**, 273 (1976).
[6] K. Havemann and A. Janoff, eds., "Neutral Proteinases of Human Polymorphonuclear Leukocytes." Urban & Schwarzenberg, Munich, 1978.
[7] P. M. Starkey, *in* "Proteinases in Mammalian Cells and Tissues" (A. J. Barrett, ed.), p. 57. North-Holland Publ., Amsterdam, 1977.
[8] A. J. Barrett and J. K. McDonald, "Mammalian Proteases. A Glossary and Bibliography," see pp. 182–187. Academic Press, New York, 1980.
[9] J. Travis, P. J. Giles, L. Porcelli, C. F. Reilly, R. Baugh, and J. Powers, *in* "Protein Degradation in Health and Disease" (D. Evered and J. Whelan, eds.), p. 51. Excerpta Medica, Amsterdam, 1980.

Cathepsin G, like elastase, is a component of the azurophil granules of human neutrophil leukocytes, and it is one of the major proteins of these organelles.[10]

Assay Methods

Cathepsin G is conveniently assayed with azocasein as substrate, but such assays are, of course, nonspecific[4]; a detailed procedure has been described.[7] Other convenient, but also nonspecific substrates that have proved useful for some purposes are Boc-Tyr-OPhNO$_2$[11,12] and Bz-D,L-Phe-ONap.[4,7] For greater specificity, amide substrates such as nitroanilides or aminomethylcoumarylamides are required. The hydrophobic nature of the residues for which the enzyme is specific means that the substrates are seldom soluble at concentrations exceeding K_m; and sensitive assay procedures are therefore difficult to arrange.

Nakajima et al.[13] studied the substrate specificity of cathepsin G with a series of blocked peptide nitroanilides. As has been found with pancreatic elastase and trypsin, prolyl residues are excluded from subsite S_3 of cathepsin G. It was found that cathepsin G shows specificity of binding in at least five subsites, and when these are suitably filled, K_m values as low as those of leukocyte elastase are attainable, but still k_{cat} values are far below those of elastase.

The best synthetic substrates for cathepsin G discovered so far are blocked tetrapeptide derivatives, and these naturally tend to be expensive. This is more important for the nitroanilides, which are used in larger quantities than the fluorogenic methylcoumarylamides.

Spectrophotometric Assay Procedure

Principle. The nitroanilide substrate is cleaved to yield free 4-nitroaniline, which is quantified by its A_{410}.

---

[10] B. Dewald, R. Rindler-Ludwig, U. Bretz, and M. Baggiolini, *J. Exp. Med.* **141**, 709 (1975).

[11] Standard abbreviations are used for the amino acid residues and common blocking groups [*J. Biol. Chem.* **247**, 977 (1972)]. The amino acid residues should be assumed to be in the L configuration unless otherwise stated. Additional abbreviations are as follows: MeOSuc-, methoxysuccinyl; -NMec, 4-methyl-7-coumarylamide; -ONap, 2-naphthyl ester; -OPhNO$_2$, 4-nitrophenyl ester; Suc-, succinyl.

[12] J. C. Powers, B. F. Gupton, A. D. Harley, N. Nishino, and R. J. Whitley, *Biochim. Biophys. Acta* **485**, 156 (1977).

[13] K. Nakajima, J. C. Powers, B. M. Ashe, and M. Zimmerman, *J. Biol. Chem.* **254**, 4027 (1979).

Reagents

Suc-Ala-Ala-Pro-Phe-NPhNO$_2$ or MeOSuc-Ala-Ala-Pro-Met-NPh-NO$_2$ is dissolved at 20 mM in dimethyl sulfoxide, and the stock solution is stored at 4°C.

Buffer. 0.10 M HEPES buffer, pH 7.5

Procedure. 0.2 ml of substrate stock solution is diluted to 1.8 ml with the buffer, in a thermostatted (50°C) spectrophotometer cuvette. The reaction is started by the introduction of 25 μl of enzyme solution, and the increase in A_{410} is monitored.

Alternatively, the reaction may be carried out in a test tube, and stopped with 1 ml of soybean-trypsin inhibitor solution (200 μg/ml in water) after an appropriate interval, for reading of A_{410}.

Units of Activity. The unit of activity is hydrolysis of 1 μmol of product/min (or the Katal unit may be adopted, for which 1 nKat represents 1 nmol/sec). The amount of nitroaniline liberated is determined from its ϵ_{410} of 8800.[13,14] The substrate concentrations recommended above are below K_m, for practical reasons, so initial rates should be determined.

A temperature lower than 50°C obviously can be used for the assay, but cathepsin G shows high activity and adequate stability at this temperature.[4] Cathepsin G is poorly soluble in water unless high salt concentrations are maintained, or a nonionic detergent is present. Ideally, solutions are made in 0.5 M NaCl containing 0.1% Brij 35.

Fluorometric Assays. No fluorometric assay for cathepsin G has been described, at the time of writing, but there is little doubt that Suc-Ala-Ala-Phe-NMec would be suitable. This compound[15] is the fluorogenic analogue of a good nitroanilide substrate,[13] and has been used in another connection by Strauss *et al.*[16] An appropriate procedure for the assay of cathepsin G with the coumarylamide would be precisely analogous to that for elastase with the valyl substrate (see [44]).

Active-Site Titration

It is quite possible that a "burst-assay" with a nitrophenyl ester, such as has been described for chymotrypsin,[17] could be adapted for cathepsin G, despite its generally lower reactivity. The use of Ac-Ala-Ala-Anle-

[14] B. F. Erlanger, N. Kokowsky, and W. Cohn, *Arch. Biochem. Biophys.* **95**, 271 (1961).

[15] Available from Bachem Feinchemikalien AG, CH-4416 Bubendorf, Switzerland.

[16] A. W. Strauss, M. Zimmerman, I. Boime, B. Ashe, R. A. Mumford, and A. W. Alberts, *Proc. Natl. Acad. Sci. U.S.A.* **76**, 4225 (1979).

[17] F. J. Kezdy and E. T. Keiser, this series, Vol. 19, p. 3.

OPhNO$_2$ (in which Anle is aza-norleucine) as a titrant for cathepsin G has been briefly described.[18]

Staining for Activity

The activity of cathepsin G can be detected in electrophoresis gels by use of a naphthol ester substrate such as Bz-Phe-ONap or the acetate or chloroacetate of naphthol AS-D, liberated naphthol being coupled with a diazonium salt.[5,19]

Purification

Cathepsin G is purified from human neutrophil leukocytes by the procedure described for leukocyte elastase (this volume, [44]), but cathepsin G is eluted from the carboxymethyl-cellulose column at a higher salt concentration than the elastase.[20] The cathepsin G tends to be contaminated by traces of elastase, which are readily removed by immunoadsorption,[20] or by rechromatography on carboxymethyl-cellulose. Like leukocyte elastase, cathepsin G is well stored by lyophilization from solution in a 0.5 M pyridine acetate buffer, pH 5.5. Solutions in 0.5 M NaCl are also stable at $-20°C$ for long periods.

Properties

Cathepsin G is a very basic protein, with a remarkably high cathodal mobility in polyacrylamide-gel electrophoresis at acid pH, running even faster than chicken egg white lysozyme.[5] The M_r of cathepsin G appears to be close to 30,000 in SDS–gel electrophoresis,[5] but a value of 23,000–24,000 has been reported on the basis of compositional data.[9] The N-terminal sequence shows distinct homology with that of chymotrypsin.[9]

The specificity of action of cathepsin G on the oxidized B chain of insulin[21] and on bovine serum albumin (A. F. Bury and A. J. Barrett, unpublished results) indicates a closer similarity to porcine chymotrypsin C than to bovine chymotrypsin A, but the preference for cleavage of leucyl over phenylalanyl bonds is not seen with nitroanilide substrates.[13]

---

[18] J. C. Powers, B. F. Gupton, M. O. Lively, N. Nishono, and R. J. Whiteley, in "Neutral Proteinases of Human Polymorphonuclear Leukocytes" (K. Havemann and A. Janoff, eds.), p. 221. Urban & Schwarzenberg, Munich, 1978.

[19] R. Rindler-Ludwig, U. Bretz, and M. Baggiolini, in "Neutral Proteinases of Human Polymorphonuclear Leukocytes" (K. Havemann and A. Janoff, eds.), p. 138. Urban & Schwarzenberg, Munich, 1978.

[20] J. Saklatvala and A. J. Barrett, Biochim. Biophys. Acta 615, 167 (1980).

[21] A. M. J. Blow and A. J. Barrett, Biochem. J. 161, 17 (1977).

Cathepsin G is inhibited by phenylmethanesulfonyl fluoride and diisopropyl fluorophosphate, but the reaction with the latter is rather slow. Tosylphenylalanyl chloromethane also inhibits cathepsin G much more slowly than it does chymotrypsin, so that the inhibition often appears incomplete.[5,22] The best active site-directed irreversible inhibitor of cathepsin G described so far is Z-Gly-Leu-Phe-CH$_2$Cl, with $k_{app}/[I]$ of 51 M^{-1} sec^{-1} (Ref. 13).

Protein inhibitors of cathepsin G include lima bean and soybean-trypsin inhibitors,[5] α_1-chymotrypsin inhibitor, α_1-proteinase inhibitor, and α_2-macroglobulin.

Distribution

There is no evidence for the occurrence of cathepsin G in any human cells other than neutrophil leukocytes. Neutrophils of some other species contain "chymotrypsin-like" esterase activity, but it is not at all certain that this is due to cathepsin G. The properties of cathepsin G are very similar to those of chymase I, the chymotrypsin-like proteinase of mast cells (see the present volume, [45]), but there are differences in structure and activity of the enzymes. The comparison of human cathepsin G with rat chymase is obviously not entirely satisfactory, however, and a direct comparison within one species will be awaited with interest.

[22] G. Feinstein and A. Janoff, *Biochim. Biophys. Acta* **403**, 477 (1975).

[43] Cathepsin D from Porcine and Bovine Spleen

By TAKAYUKI TAKAHASHI and JORDAN TANG

Cathepsin D (EC 3.4.23.5) is a lysosomal carboxyl (acid) protease widely distributed in many cell types. The presence of this enzyme in relatively high amounts in mammalian spleens has long been recognized. Several purification methods were devised that resulted in pure spleen cathepsin D and in the investigation of its properties.[1-5] One of the advan-

[1] M. L. Anson, *J. Gen. Physiol.* **23**, 695 (1940).

[2] E. M. Press, R. R. Porter, and J. Cebra, *Biochem. J.* **74**, 501 (1960).

[3] H. Keilova *in* "Tissue Proteinases" (A. J. Barrett and J. T. Dingle, eds.), p. 45. North-Holland Publ., Amsterdam, 1971.

[4] J. B. Ferguson, J. R. Andrew, I. M. Voynick, and J. S. Fruton, *J. Biol. Chem.* **248**, 6701 (1973).

[5] M. Cunningham and J. Tang, *J. Biol. Chem.* **251**, 4528 (1976).

tages of studying spleen cathepsin D from pigs and cows is that large amounts of starting materials can be obtained; therefore, large-scale purifications can be carried out for the studies of the structure–function relationships of this enzyme. This chapter describes a large-scale purification scheme that has been in routine use in our laboratory for the past 3 years. It is our experience that a sufficiently pure enzyme can be obtained with this method for the structure investigation. We also describe below the isolation of polypeptide chains from porcine spleen cathepsin D as well as the interrelationships and chemical characteristics of these polypeptide chains. Some of these results have been published elsewhere.[6-8]

Assay Methods

Many methods have been devised to assay cathepsin D, and a general review has been made recently.[9] Two methods have been used in our laboratory to assay cathepsin D. The assay using hemoglobin as a substrate is essentially that of Anson's method.[10] This assay is rapid and has a reasonable sensitivity (~3000 ng of enzyme). However, because of the high substrate blank (about 0.3–0.5 OD) and narrow range of linearity (about 0.3 net OD), the accuracy is only moderate. Duplicate or triplicate assays must be carried out with several dilutions of the assay sample to achieve the best results.

The second assay utilizes renin substrate, and angiotensin I radioimmunoassay was described by Dorer et al.[11] This method, which can readily measure 30 ng of cathepsin D, is about 100 times more sensitive than hemoglobin assay, but it requires an elaborate procedure and equipment. In addition, the cathepsin D sample to be assayed must be free of renin.

Principle. Hemoglobin is degraded by the enzyme to liberate peptides soluble in a solution of trichloroacetic acid (TCA). The concentration of peptides is determined by their absorbance at 280 nm. In the assay using synthetic renin substrate, tetradecapeptide (TDP): Asp-Arg-Val-Tyr-Ile-

[6] J. S. Huang, S. S. Huang, and J. Tang, *J. Biol. Chem.* **254**, 11405 (1979).
[7] J. S. Huang, S. S. Huang, and J. Tang, *in* "Enzyme Regulation and Mechanism of Action" (P. Mildner and P. Ries, eds.), p. 289. Pergamon, Oxford, 1980.
[8] J. S. Huang, S. S. Huang, and J. Tang, *in* "Frontiers in Protein Chemistry" (T.-Y. Liu, G. Mamiya, and K. T. Yasunobu, eds.), p. 359. Am. Elsevier, New York, 1980.
[9] A. J. Barrett, *in* "Proteinases in Mammalian Cells and Tissues" (A. J. Barrett, ed.), p. 209. North-Holland Publ., Amsterdam, 1977.
[10] M. L. Anson, *J. Gen. Physiol.* **22**, 79 (1938).
[11] F. E. Dorer, K. E. Lentz, J. R. Kahn, M. Levine, and L. T. Skeggs, *J. Biol. Chem.* **253**, 3140 (1978).

His-Pro-Phe-His-Leu-Leu-Val-Tyr-Ser, angiotensin I (N-terminal, 10 residues of TDP) is released and measured by radioimmunoassay.[12]

Assay with Hemoglobin

Reagents

0.25 M Sodium formate buffer (pH 3.2)–5% bovine hemoglobin substrate. The acid-denatured protein can be obtained from a number of commercial sources. Five grams of the hemoglobin are dissolved in 100 ml of distilled water. This is best accomplished by placing the protein powder on top of the water in a beaker and very gently shaking with the hand about once every minute until all the protein powder is dissolved. Excessive stirring causes the trapping of air bubbles in viscous hemoglobin and slows its dissolving. The hemoglobin solution is stored at 4°C and can be used for at least 2 weeks. Longer storage causes an increase of blank value.

Trichloroacetic acid (TCA), 10%

Procedure. The reaction mixture contains 1.9 ml of 0.25 M sodium formate buffer (pH 3.2)–0.5 ml of hemoglobin substrate–0.1 ml of enzyme. After incubation at 37°C for 20 min, 2 ml of 10% TCA is added to precipitate proteins. After filtration through Whatman No. 50 paper, the absorbance at 280 nm is measured. The controls are carried out with incubation mixtures to which TCA is added immediately after the addition of enzyme. Duplicates are carried out for each sample including controls, and the average values used. One unit of enzyme activity is defined as the net extinction value of 1.0 of the filtrate in excess of the control. All quantitative determinations are carried out in the linear range of the assay, up to a net absorbance of 0.3.

Assay with Tetradecapeptide, TDP

Reagents

Sodium citrate buffer (0.05 M, pH 4.5) containing 0.5% lysozyme. This solution should be freshly prepared for each assay.

Renin substrate, tetradecapeptide (TDP), from Boehringer Mannheim Biochemicals, 2.5 μM. 44 μg of TDP are dissolved in 10 ml of sodium citrate buffer, pH 4.5.

Sodium chloride, 0.15 M

---

[12] E. Haber, T. Koerner, L. B. Page, B. Kliman, and A. Purnode, *J. Clin. Endocrinol. Metab.* **29**, 1349 (1969).

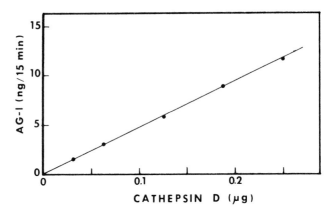

FIG. 1. Assay of bovine cathepsin D using renin substrate, TDP. The angiotensin I (AG-I) formed is determined by radioimmunoassay.

Procedure. Reaction mixtures contain in each tube 0.8 ml of sodium citrate–lysozyme buffer (pH 4.5), 0.1 ml of 2.5 μM TDP, and 0.1 ml of enzyme. After incubation for 15 min at 37°C, 0.5 ml of 0.15 M NaCl is added, and the tubes are placed in a boiling water bath for 10 min. After centrifugation, the supernatant solutions are analyzed for angiotensin I by the method of Haber *et al.*[12] Antiserum against angiotensin I can be purchased commercially from Squibb, Arlington Heights, Illinois. Appropriate corrections are made for the TDP blank (incubation of TDP in the absence of the enzyme) and the enzyme blanks. Figure 1 shows the relationship between the angiotensin I formation and the amount of cathepsin D in the total reaction mixture obtained using this assay.

Porcine Spleen Cathepsin D

Purification Procedure

Step 1. The frozen porcine spleens are thawed and freed from the connective tissues. About 2 kg of porcine spleens are diced and homogenized in a Waring blender of 1-gallon size with 2 liters of cold distilled water.

Step 2. The homogenate is centrifuged at 10,000 rpm for 30 min. The combined supernatant fractions from the centrifugation are referred to as the supernate. About 50% of cathepsin D activity is recovered in the supernate after centrifugation of porcine spleen homogenate. This recovery is not increased when the extraction of spleen homogenate is carried out with several other buffer solutions.

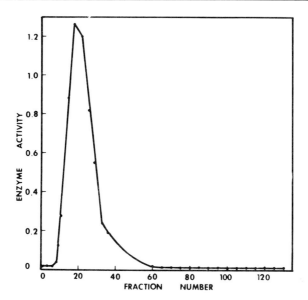

FIG. 2. Chromatographic pattern of porcine spleen cathepsin D on a column of DEAE–Sephadex A-25 (step 5, Table I).

Step 3. The pH of the supernate is adjusted to 3.7 with 5.7 N HCl, and the precipitate is removed by centrifugation at 10,000 rpm. The combined supernatant fractions are referred to as the acid supernate.

Step 4. The acid supernate is immediately dialyzed against about 12 liters of 0.05 M sodium phosphate buffer, pH 7.4. The buffer is changed at least three times during 40 hr of dialysis. The precipitate formed in this process is removed by centrifugation. The pH of the supernatant is checked and readjusted to 7.4 with 1 N NaOH if necessary.

Step 5. The supernatant is allowed to flow through a DEAE–Sephadex A-25 column equilibrated and eluted with the same buffer. Most cathepsin D activity is recovered in the breakthrough fractions (Fig. 2). It should be noted that this step is not designed to fractionate cathepsin D isozymes and only marginal purification is achieved in this step. However, it is necessary to remove the remaining apparently acidic substances, which interfere with the affinity chromatography in the following step.

Step 6. The breakthrough fractions eluted from the DEAE–Sephadex A-25 column are combined and the pH of the solution is adjusted to 3.5 with 5.7 N HCl. The solution is then immediately mixed with about 50 g (wet weight) of pepstatinyl-aminohexyl-Sepharose gels that have been washed with 0.05 M sodium acetate buffer (pH 3.5) containing 0.2 M NaCl. The gel suspension is stirred gently overnight, after which it is

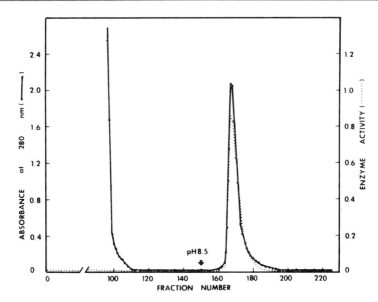

FIG. 3. Chromatographic pattern of porcine spleen cathepsin D on a column of pepstatin–Sepharose-4B (step 6, Table I).

packed on a 50-ml Plasti-pak syringe column and washed with 0.05 M sodium acetate buffer (pH 3.5) containing 0.2 M NaCl, until the absorbance of eluents at 280 nm is below 0.05. Cathepsin D is specifically eluted with 0.05 M Tris-HCl (pH 8.5) containing 0.2 M NaCl (Fig. 3). The pooled material is dialyzed against distilled water and lyophilized. This material contains nearly homogeneous cathepsin D in several isozyme forms.

Step 7. Cathepsin D isozyme mixture obtained from the affinity-column chromatography from step 6 is subjected to liquid-phase isoelectric focusing according to the method of Vesterberg,[13] employing a 110-ml column (LKB 8100). Operating temperature is maintained at 4°C. Ampholine with a pH range of 5–8 (LKB) is used in a linear sucrose gradient from 0 to 50% (w/v). The samples, 80–100 mg of protein, are focused at 400 V for 18 hr and then for 24 hr at 600 V. Column fractions of approximately 0.8 ml are collected. In a typical run, five isozymes (I–V) are separated as shown in Fig. 4. The individual isozyme fractions are pooled, dialyzed in the cold against distilled water, and lyophilized. The isozymes I–IV obtained with this procedure are usually homogeneous.

A typical scheme for large-scale purification of cathepsin D from porcine spleen is shown in Table I. It should be noted that this procedure is

[13] O. Vesterberg, this series, Vol. 22, p. 389.

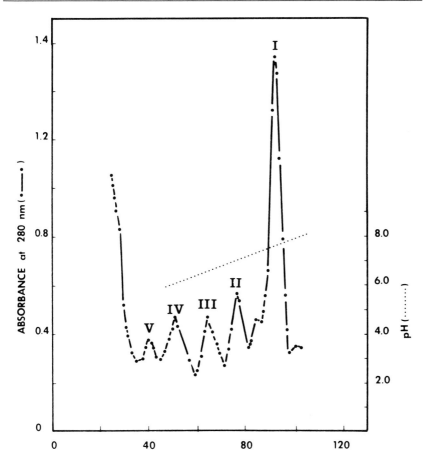

FIG. 4. Separation of porcine spleen cathepsin D in preparative isoelectric focusing.

modified from our previously published scheme.[6] The new procedure is somewhat more rapid and produces higher overall yield. However, the yield of isozymes II and V are always lower.

Isolation of High-Molecular-Weight (HMW) Cathepsin D

Porcine spleen contains a 100,000-MW (100K) cathepsin D, which can be isolated from the procedure described above. Peak V from isoelectric focusing (Fig. 4) contains HMW cathepsin D and other components. The dialyzed and lyophilized material from peak V is then further fractionated over a column of Sephadex G-100 (Fig. 5). In this procedure, a 1.8 × 160-cm column is used. Sodium acetate buffer (0.01 M), pH 5, is used to

TABLE I
PURIFICATION OF CATHEPSIN D FROM PORCINE SPLEENS

Step No.	Total protein (mg)	Total enzyme activity (units)	Specific activity (units/mg protein)	Yield (%)	Purification (-fold)
1. Homogenate	352,941	27,375	0.077	100	1
2. Supernate	123,081	13,662	0.111	49.91	1.44
3. Acid supernatant	56,265	10,184	0.181	37.20	2.35
4. Dialysis	28,255	8533	0.302	31.17	3.92
5. DEAE–Sephadex	19,093	7389	0.387	26.99	5.03
6. Pepstatin–Sepharose	88.9	3936	44.28	14.38	575.06
7. Isoelectric focusing					
Total	64.2	3527	54.94	12.88	
Isozyme I	50.0	2911	58.20	10.63	755.84
Isozyme II[a]					
Isozyme III	5.6	382	68.48	1.40	
Isozyme IV	2.0	139	67.91	0.51	
Isozyme V	6.6	95	14.35	0.35	

[a] The presence of isozyme II in this procedure is much reduced from in the previous scheme.[6] The yield of this isozyme is therefore not calculated.

equilibrate and elute the column. The protein peak eluted at the void volume contains only the HMW cathepsin D. A second peak, which is eluted in the area corresponding to MW 50,000, contains a single-chain enzyme (isozyme V), as well as a 2-chain enzyme. The 2-chain enzyme is probably the minor contamination of isozyme IV.

Enzymatic Properties

Cathepsin D isozymes I–IV have similar specific activity. HMW enzyme has a specific activity about one-tenth (hemoglobin as substrate) to one-fifth (TDP as substrate) of the others (unpublished observations). Because cathepsin D is a glycoprotein (see later), it binds to a concanavalin A–Sepharose column.[6] Cathepsin D isozymes show immunochemical identity in Ouchterlony double diffusion against an antiserum prepared from the heavy chain. HMW cathepsin D shows this immunoreactivity only after urea denaturation.[6] The molecular weights of isozymes I–IV are 50,000 (50K), as determined by gel filtration. The SDS–polyacrylamide gel electrophoresis of isozymes I–IV produces 2 bands each, a 35,000-MW (35K) heavy chain, and a 15,000-MW (15K) light chain. Isozyme V is a single-chain protein with MW 50,000. HMW isozyme, which gives a

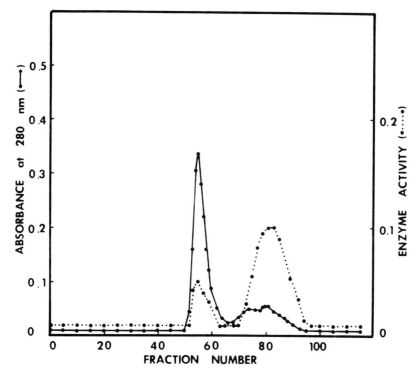

FIG. 5. Separation of 100 K cathepsin D from isozyme V on a column of Sephadex G-100. The starting material is isozyme V from isoelectric focusing (Fig. 4).

single band in SDS–gel electrophoresis, has MW ≈ 100,000. Cathepsin D isozymes have different mobilities in polyacrylamide-gel electrophoresis. This charge difference can be demonstrated for the heavy chains of these isozymes. The mobility of the light chains is the same.

Separation of Heavy Chain and Light Chain

As described above, the 50K porcine cathepsin D consists of primarily the 2-chain enzyme. The content of 50K single-chain enzyme in the affinity-chromatography product varied from trace amounts (not detectable on SDS–gel electrophoresis) to minor content (a light band in SDS–gel accounted for up to 10% of the total enzyme). This material can be used to separate the 35K heavy (H) chain from the 15K light (L) chain in a column (2 × 95 cm) of Sephadex G-75 eluted with 0.05 M sodium acetate buffer (pH 5.2) containing 6 M urea. A typical chromatogram is shown in Fig. 6. It should be noted that the H chain peak includes some minor content of

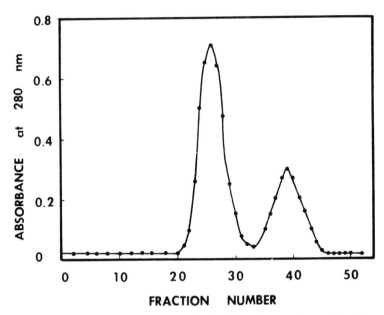

FIG. 6. Isolation of cathepsin D heavy and light chains using a column of Sephadex G-75 (1.5 × 90 cm). The elution buffer is 0.05 M sodium acetate containing 6 M urea.

50K single chain. If in the starting material the 50K band is visible on SDS–gel electrophoresis, some significant autodigestion may occur during urea denaturation. Single-chain enzyme apparently denatured slower, thus degrading the denatured H and L chains, which resulted in low yields and heterogeneous products. The proteolysis can be prevented by adding a severalfold molar excess of pepstatin prior to urea denaturation. In this procedure, 7 mg of pepstatin is dissolved in a small volume of methanol and added to a cold solution of 50 mg of cathepsin D in 10 ml of 0.05 M sodium acetate buffer, pH 5.2. After incubating in the cold while stirring for 30 min, 4 g of reagent-grade urea is added. The solution is incubated in the cold overnight before applying to a Sephadex G-75 column.

Structural Data

The structural data are summarized in Table II and Fig. 7. The amino acid compositions of porcine cathepsin isozymes, H chain, L chain, and the 100K enzyme are shown in Table III. The carbohydrate analysis indicated that isozymes have the same carbohydrate contents. Each contains 8 mannose and 4 glucosamine residues per mole. L chain contains 2 mannose and 1 glucosamine, whereas H chain contains the balance of the carbohydrates. The number of glucosamine residues is the same as in the

TABLE II
SUMMARY OF THE PROPERTIES OF CATHEPSIN D ISOZYMES FROM
BOVINE AND PORCINE SPLEENS

	pI	MW	No. of sub-units	Norleucine incor-poration[a] (residues/ mol)	Carbohydrate[b] (residues/mol)		Heavy-chain anti-serum
					Man	GlcN	
Porcine isozyme							
I	7.54	50,000	2	1	8	4	+
II	7.09	50,000	2				+
III	6.60	50,000	2				+
IV	6.19	50,000	2				+
V	5.51	�{ 50,000	1				+
		100,000	1		—	4	+
Light chain		15,000		0	2	1	±[c]
Heavy chain		35,000		1	6	3	+
Bovine isozyme							
A	6.49	46,000	1 + 2		8	4	+
B	6.04	46,000	1 + 2		8	4	+
Light chain		12,000					
Heavy chain		34,000					

[a] Data are those of amino acid analyses after reacting the native enzyme with diazo-acetyl norleucine methyl ester.
[b] Man and GlcN are mannose and glucosamine, respectively.
[c] Only trace reactivity observed.

100K enzyme, indicating that the carbohydrate is located in the 50K cathepsin D portion of the structure (Fig. 7).

The N-terminal sequence of the L chain is shown in Fig. 8. The single-chain enzyme has the same N-terminal sequence as found in the L chain. However, at least three sets of sequence could be detected from H chain prepared from a single isozyme. As shown in Fig. 7, these structural data suggest both the multiple cleavage sites between L and H chains and the structural microheterogeneity present in cathepsin D. This latter point is suggested from differences in amino acid compositions of the isozymes and also has been confirmed in peptides isolated from L chain (unpublished work).

The active-center aspartyl residues of carboxyl proteases (for review, see Ref. 14) are apparently located at the same corresponding positions in cathepsin D. Asp-32 is seen in the sequence determination (Fig. 8),

[14] J. Tang, *Mol. Cell. Biochem.* **26**, 93 (1979).

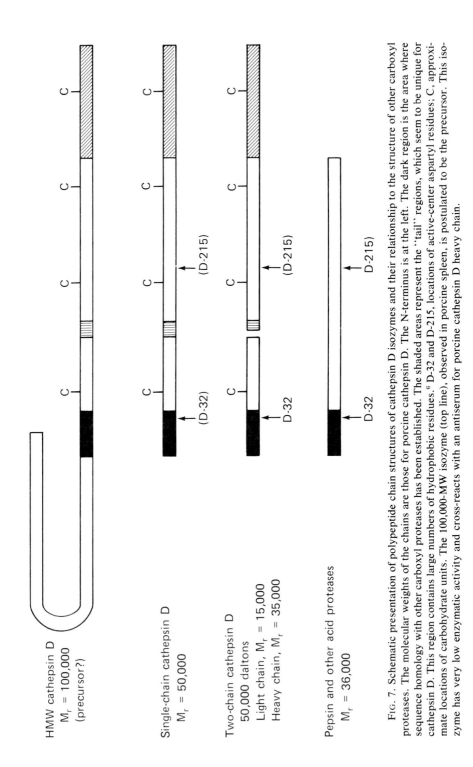

HMW cathepsin D
$M_r = 100,000$
(precursor?)

Single-chain cathepsin D
$M_r = 50,000$

Two-chain cathepsin D
50,000 daltons
Light chain, $M_r = 15,000$
Heavy chain, $M_r = 35,000$

Pepsin and other acid proteases
$M_r = 36,000$

FIG. 7. Schematic presentation of polypeptide chain structures of cathepsin D isozymes and their relationship to the structure of other carboxyl proteases. The molecular weights of the chains are those for porcine cathepsin D. The N-terminus is at the left. The dark region is the area where sequence homology with other carboxyl proteases has been established. The shaded areas represent the "tail" regions, which seem to be unique for cathepsin D. This region contains large numbers of hydrophobic residues.[6] D-32 and D-215, locations of active-center aspartyl residues; C, approximate locations of carbohydrate units. The 100,000-MW isozyme (top line), observed in porcine spleen, is postulated to be the precursor. This isozyme has very low enzymatic activity and cross-reacts with an antiserum for porcine cathepsin D heavy chain.

TABLE III
AMINO ACID COMPOSITIONS OF CATHEPSIN D ISOZYMES AND SUBUNITS

| Amino acid | Porcine isozymes[a] (residues/mol) | | | | | | Porcine isozymes I[a] (residues/mol) | | Bovine isozymes[a] (residues/mol) | | | |
	I	II	III	IV	V	100K[b]	Light chain	Heavy chain	A	B	Light chain	Heavy chain
Lysine	28	29	28	27	29	67	5	23	21	20	5	16
Histidine	10	9	10	11	10	25	5	4	8	9	3	5
Arginine	11	12	11	11	12	35	0	11	15	14	1	14
Aspartic acid	43	43	43	43	46	86	10	32	42	42	11	31
Threonine	27	27	27	27	25	72	7	19	24	24	7	17
Serine	31	31	31	30	30	83	11	20	30	30	10	20
Glutamic acid	39	40	42	44	41	74	6	32	38	38	6	32
Proline	29	30	29	30	30	30	7	22	29	29	7	22
Glycine	55	56	53	52	56	71	11	42	47	47	9	38
Alanine	25	25	25	26	27	65	3	20	23	23	3	20
Half-cystine	8	—	—	—	—	—	2	6	8	8	2	6
Valine	40	39	38	38	42	89	7	31	37	38	6	31
Methionine	11	11	10	10	11	16	1	11	11	11	1	10
Isoleucine	31	30	28	27	29	44	7	24	23	23	6	17
Leucine	40	41	39	40	42	105	7	33	37	37	7	30
Tyrosine	27	26	24	23	27	30	8	17	21	21	7	14
Phenylalanine	19	19	18	17	18	53	4	13	19	19	4	15
Tryptophan	6	—	—	—	—	—	2	5	5	5	2	3
Total	480	—	—	—	—	>945	103	365	438	365	97	342

[a] The data shown are the nearest integrals of the averages.
[b] 100K refers to the high-molecular-weight isozyme.

Residue numbers	-2 -1 1 2 3 4 5 6 7 8 9 10 11 12 13 14 15 16 17 18 19 20
Porcine cathepsin D	Gly-Pro-Ile-Pro-Glu-Val-Leu-Lys-Asn-Tyr-Met- —-Asp-Ala-Gln-Tyr-Tyr-Gly-Glu-Ile-Gly-Ile-
Bovine cathepsin D	Gly-Pro-Ile-Pro-Glu-Leu-Leu-Lys-Asn-Tyr-Met- —-Asp-Ala-Gln-Tyr-Tyr-Gly-Glu-Ile-Gly-Ile-
Gastric proteases	
Porcine pepsin	-Ala-Leu /Ile-Gly-Asp-Glu-Pro-Leu-Glu-Asn-Tyr-Leu- —-Asp-Thr-Glu-Tyr-Phe-Gly-Thr-Ile-Gly-Ile-
Bovine chymosin	Gly-Glu-Val-Ala-Ser-Val-Pro-Leu-Thr-Asn-Tyr-Leu- —-Asp-Ser-Gln-Tyr-Phe-Gly-Lys-Ile-Tyr-Leu-
Microbial proteases	
R. chinensis	Ala-Gly-Val-Gly-Thr-Val-Pro-Met-Thr-Asp-Tyr-Gly-Asn-Asp-Val/Ile-Glu-Tyr-Tyr-Gly-Gln-Val-Thr-Ile-
Penicillopepsin	Ala-Ala-Ser-Gly-Val-Ala-Thr-Asn-Thr-Pro-Thr-Ala-Asn-Asp-Glu-Glu-Tyr-Ile-Thr-Pro-Val-Thr-Ile-

Residue numbers	21 22 23 24 25 26 27 28 29 30 31 32 33 34 35 36 37
Porcine cathepsin D	Gly-Thr-Pro-Pro-Gln-Ser-Phe-Thr-Val-Phe-Asp-Thr-Gly-Ser-Ser-Asn-
Bovine cathepsin D	Gly-Thr-Pro-Pro-Gln-Ser-Phe-Thr-Val-Phe-Asp-Thr-
Gastric proteases	
Porcine pepsin	Gly-Thr-Pro-Ala-Gln-Asp-Phe-Thr-Val-Ile-Phe-Asp-Thr-Gly-Ser-Ser-Asn-
Bovine chymosin	Gly-Thr-Pro-Gln-Gln-Thr-Phe-Lys-Val-Leu-Phe-Asp-Thr-Gly-Ser-Ser-Asp-
Microbial proteases	
R. chinensis	Gly-Thr-Pro-Gly-Lys-Ser-Phe-Asp-Leu-Asn-Phe-Asp-Thr-Gly-Ser-Thr
Penicillopepsin	Gly- —-Gly-Thr-Thr-Leu-Asn-Leu-Asn-Phe-Asp-Thr-Gly-Ser-Ala-Asp-

FIG. 8. NH_2-terminal sequence of cathepsin D light chains and their homology to gastric and microbial acid proteases. The residue numbers are those of porcine pepsin. For structural information of other acid proteases, see Ref. 14 for review.

TABLE IV
PURIFICATION OF CATHEPSIN D FROM BOVINE SPLEENS

Step No.	Total proteins (mg)	Total enzyme activity (units)	Specific activity (units/mg)	Yield (%)	Purification (-fold)
1. Homogenate	362,850	42,643	0.118	100	1
2. Supernate	145,250	24,916	0.172	58.43	1.5
3. Acid supernate	66,239	19,208	0.290	45.04	2.5
4. Lyophilized powder	17,012	13,587	0.799	31.86	6.8
5. DEAE–Sephadex chromatography	9697	11,549	1.191	27.08	10
6. Pepstatin–Sepharose chromatography	93	5543	59.602	13.00	505
7. Isoelectric focusing Total	85	5211	61.306	12.22	520
Isozyme A	59	3647	61.814	8.55	524
Isozyme B	26	1564	60.154	3.67	510

whereas the presence of Asp-215 is implied from the incorporation of 1 residue of diazoacetyl norleucine methyl ester into each H chain (Table II). Together with the sequence homology to other carboxyl proteases, the relationships of these polypeptide chains are depicted in Fig. 7. Since 50K cathepsin D is larger than other carboxyl proteases (about 34K), the difference is likely to be present at the C-terminal end of cathepsin D as an extra length of polypeptide (Fig. 7). The 100K enzyme appeared in SDS–gel electrophoresis at the presence of sulfhydryl compound as a single band,[6] suggesting that it is a single polypeptide. Since 100K enzyme shows immunochemical identity with the 50K enzyme,[6] it indicates that the 50K cathepsin D must be part of 100K structure and suggests that the 100K cathepsin D may be a precursor of the other cathepsin D species. However, an alternative that a single-chain cathepsin D is covalently linked to another polypeptide chain by a nondisulfide bond cannot be excluded. When the amino acid composition of 50K enzyme (Table III) is deducted from the composition of the 100K enzyme, the difference has a composition very different from cathepsin D.

Bovine Spleen Cathepsin D

Purification Procedure

The purification procedure used for bovine spleen cathepsin D is essentially the same as described for the porcine enzyme. A typical purification scheme is shown in Table IV. As in the case of porcine enzyme

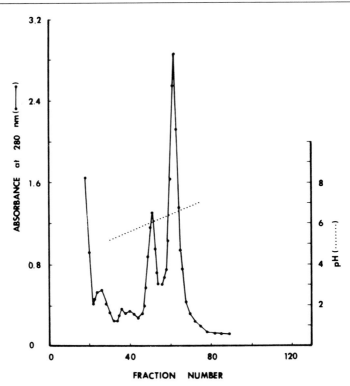

FIG. 9. Separation of bovine spleen cathepsin D isozymes in the preparative isoelectric focusing.

A   B

FIG. 10. Electrophoresis of bovine spleen cathepsin D isozymes A (left) and B (right) on SDS–polyacrylamide gel.

purification, the apparent total activity in the pepstatin–Sepharose chromatography step reduced from 11,549 units to 5543 units. As discussed in Ref. 6, the synergistic effect of sulfhydryl proteases with cathepsin D in the digestion of bovine hemoglobin is undoubtedly the reason for this apparent loss of activity.

Unlike the case of porcine spleen enzyme, bovine spleen cathepsin contains only two isozymes, A and B, which are separated in isoelectric focusing (Fig. 9). This procedure is essentially identical to that described for porcine isozymes.

Properties and Structure

The properties of bovine cathepsin D are also summarized in Table I. Both bovine isozymes contain approximately equal amounts of single-chain and 2-chain structures, as indicated in the pattern of SDS–gel electrophoresis (Fig. 10). The single chain (46K) is smaller than that in porcine enzyme. The L chain (12K) and H chain (34K) are also smaller than their porcine counterparts. Chromatography of bovine cathepsin D on $6 M$ urea on a Sephadex G-75 column as just described results in a pure L chain. However, H chain and single chain emerge in the same peak. Further separation of these two polypeptides has not been attempted. The general compositions and structural data from bovine enzymes are very similar to that described already for the porcine enzyme (Table II and Fig. 8). The bovine enzyme and porcine enzyme are indistinguishable in immunodiffusion against antiserum of porcine H chain.[7]

[44] Leukocyte Elastase

By ALAN J. BARRETT

It was recognized as early as the turn of the century that human neutrophil leukocytes are rich in neutral proteinase activity; this is now known to be largely due to the elastase (EC 3.4.21.37) and to cathepsin G (see this volume, [42]).

Much of the current interest in the enzyme stems from the work of Janoff and co-workers, who first recognized the enzyme as an elastase,[1] and then provided evidence suggesting that it might be involved in a number of important disease states, including the pulmonary emphysema associated with a hereditary deficiency of α_1-proteinase inhibitor. Useful

---

[1] A. Janoff and J. Scherer, *J. Exp. Med.* **128**, 1137 (1968).

reviews may be found in Refs. 2–5. Leukocyte elastase is one of the major proteins of the azurophil granules of human neutrophils.[6]

Assay Methods

Leukocyte elastase has been assayed with a wide variety of substrates, including elastin, other proteins, and a number of synthetic substrates. Assays with elastin are not normally to be recommended for routine use; variables introduced by the fact that the substrate is insoluble militate against precision. Thus nonionic and anionic detergents have complex effects on the results, and other proteinases sometimes act synergistically with elastase. Elastin noncovalently labeled with Congo Red has been popular for the assay of pancreatic elastase, but this has been shown to be subject to the artifact that albumin displaces the dye from the substrate, bringing it into solution without solubilization of the elastin.[7] Nevertheless, if this is borne in mind, the method can be convenient, and a procedure was described by Shotton.[8] Sensitive assays of elastinolytic activity can be made with ^{3}H-labeled elastin by the method of Takahashi et al.[9] Proteins other than elastin tend to be susceptible to cathepsin G as well as elastase, and therefore are unsuitable for the specific assay of leukocyte elastase, but assays with azocasein have been found convenient when specificity was not required.[10]

Spectrophotometric Assays

Boc-Ala-OPhNO$_2$ (Ref. 11) is a nonspecific, but sensitive, substrate that is well suited to work with purified leukocyte elastase, such as the

[2] P. M. Starkey, in "Proteinases in Mammalian Cells and Tissues" (A. J. Barrett, ed.), p. 57. North-Holland Publ., Amsterdam, 1977.

[3] K. Havemann and A. Janoff, eds., "Neutral Proteases of Human Polymorphonuclear Leukocytes." Urban & Schwarzenberg, Munich, 1980.

[4] J. Bieth, Front. Matrix Biol. 6, 1 (1978).

[5] A. J. Barrett and J. K. McDonald, "Mammalian Proteases. A Glossary and Bibliography," see pp. 165–76. Academic Press, New York, 1980.

[6] B. Dewald, R. Rindler-Ludwig, U. Bretz, and M. Baggiolini, J. Exp. Med. 141, 709 (1975).

[7] I. Banga and W. Ardelt, Biochim. Biophys. Acta 146, 284 (1967).

[8] D. M. Shotton, this series, Vol. 19, p. 113.

[9] S. Takahashi, S. Seifter, and F.-C. Yang, Biochim. Biophys. Acta 327, 138 (1973).

[10] P. M. Starkey and A. J. Barrett, Biochem. J. 155, 255 (1976).

[11] Standard abbreviations are used for the amino acid residues and common blocking groups [J. Biol. Chem. 247, 977 (1972)]. The amino acid residues should be assumed to be in the L configuration unless otherwise stated. Additional abbreviations are as follows: -CH$_2$Cl, chloromethane; MeOSuc-, methoxysuccinyl; -NMec, 4-methyl-7-coumarylamide; -NPhNO$_2$, 4-nitroanilide; -ONap, 2-naphthyl ester; -OPhNO$_2$, 4-nitrophenyl ester.

evaluation of inhibitors. Because the aryl ester is susceptible to spontaneous hydrolysis above neutral pH, assays are made at pH 6.5.[12] A more stable aryl ester is Z-Ala-ONap. Both substrates have been used successfully for leukocyte elastase.[13,14]

A much more specific, but also less sensitive substrate, is Suc-Ala-Ala-Ala-NPhNO$_2$. This is more rapidly cleaved by pancreatic than leukocyte elastase, and is hydrolyzed by some other enzymes.[15] A simple practical procedure has been described.[4] Newer spectrophotometric substrates, in which the P_1 residue is valine, promise much better selectivity for leukocyte elastase, and greater sensitivity.[16] Until such substrates become commercially available at a moderate price, however, I consider none of the spectrophotometric methods to be superior to the fluorometric method described here.

Fluorometric Assay

The most sensitive, accurate, and specific assays for leukocyte elastase available at the time of writing are those involving fluorogenic amide substrates, of which MeOSuc-Ala-Ala-Pro-Val-NMec is probably the best.[17] Importantly, it is unaffected by cathepsin G.[18]

Principle. The (almost) nonfluorescent substrate is hydrolyzed by the enzyme to liberate the intensely fluorescent 7-amino-4-methylcoumarin, which is quantified fluorometrically.

Reagents

Substrate: MeOSuc-Ala-Ala-Pro-Val-NMec[19] is dissolved in dimethyl sulfoxide as a 1 mM solution that is stored at 4°C.

Buffer: 0.20 M Tris-HCl (pH 8.5) containing 1 M NaCl

Stopping solution (if required): soybean-trypsin inhibitor (100 μg/ml) in water

[12] L. Visser and E. R. Blout, *Biochim. Biophys. Acta* **268,** 257 (1972).
[13] P. M. Tuhy and J. C. Powers, *FEBS Lett.* **50,** 359 (1977).
[14] P. M. Starkey and A. J. Barrett, *Biochem. J.* **155,** 265 (1976).
[15] J. Saklatvala, *J. Clin. Invest.* **59,** 794 (1977).
[16] H. R. Wenzel, S. Engelbrecht, H. Reich, W. Mondry, and H. Tschesche, *Hoppe-Seyler's Z. Physiol. Chem.* **361,** 1413 (1980).
[17] M. J. Castillo, K. Nakajima, M. Zimmerman, and J. C. Powers, *Anal. Biochem.* **99,** 53 (1979).
[18] K. Nakajima, J. C. Powers, B. M. Ashe, and M. Zimmerman, *J. Biol. Chem.* **254,** 4027 (1979).
[19] Available from Bachem Feinchemikalien AG, CH-4416 Bubendorf, Switzerland, or Peptide Research Foundation, Minoh-Shi, Osaka, Japan.

Procedure. 0.25 ml of the buffer is mixed with 0.50 ml of enzyme solution (containing 0.2–1.5 μg of leukocyte elastase) diluted in 0.1% Brij 35 (a nonionic detergent that prevents loss of enzyme on surfaces). The mixture is warmed to 40°C.

The stock (1 mM) substrate solution is diluted to 20 μM with water, and 0.25 ml is added to the buffered enzyme solution to start the reaction.

The reaction is monitored continuously in a fluorometer connected to a chart recorder; alternatively, the reaction is stopped after 10 min by addition of 1.0 ml of the inhibitor solution, and the fluorescence measured then.

A simple filter photofluorometer is adequate for these assays, but spectrofluorometers are more commonly used. Excellent discrimination between substrate and product is achieved by excitation at 370 nm (or in a 340–380 nm band), with monitoring of emission at 460 nm. The sensitivity of the fluorometer is adjusted initially so that a maximal reading (e.g., 1000 arbitrary units) is obtained with 0.5 μM 7-amino-4-methylcoumarin, and the reaction mixtures are read at this sensitivity.

UNITS OF ACTIVITY: The unit of activity corresponds to release of 1 μmol of product/min; alternatively, the Katal unit can be used, for which one nKat corresponds to release of 1 nmol/sec.

The substrate concentration used in this procedure is far below K_m for the substrate (0.20 mM, Ref. 17), so the rate of reaction is directly proportional to the substrate concentration. For this reason, the rate declines perceptibly if more than 10% or so of the substrate is consumed. Much greater sensitivity can be obtained in assays at higher substrate concentrations, but these become more expensive, and we have found the economical procedure just described suitable for many applications in our laboratory.

Active-Site Titration

It is often important to be able to determine the active-site molarity of an enzyme solution.[20] Shotton[8] described the use of diethyl-p-nitrophenyl phosphate for active-site titration of pancreatic elastase (with appropriate cautions), and it is very likely that this readily available, but rather nonspecific, reagent could also be used with leukocyte elastase.

A more selective procedure for the elastases (both pancreatic and leukocyte) depends on the use of Ac-Ala-Ala-Aala-OPhNO₂ (in which Aala is aza-alanine).[21]

[20] F. J. Kezdy and E. T. Kaiser, this series, Vol. 19, p. 3.
[21] J. C. Powers and B. F. Gupton, this series, Vol. 46, p. 208.

Purification

Leukocytes for use as starting material in the purification of leukocyte elastase are most conveniently obtained by leukapheresis of patients with chronic myeloid leukemia, or normal donors. Alternatively, "buffy coat" preparations from outdated blood may be used. Purulent sputum has also been described as a useful source,[22] and the enzyme has been isolated from human spleen.[10]

The following procedure has worked well in the author's laboratory, and is a modification[23] of the method of Baugh and Travis,[24] in which the crucial purification step depends on the use of aprotinin–Sepharose.

Cells (about 70 g wet weight) had been collected by leukapheresis, washed with saline by centrifugation, and stored frozen.

The cells were thawed and suspended in 9 volumes of 0.1 M Tris-HCl buffer (pH 7.5) containing 1 M MgCl$_2$–0.1% Brij 35. The mixture was homogenized (Ultra-turrax) and then centrifuged at 100,000 g for 1 hr. 95% of the proteinase activity was extracted into the supernatant. Since the supernatant was rather viscous, it was diluted with 2 volumes of 5 mM Tris-HCl buffer (pH 7.5) containing 1 M NaCl–0.1% Brij 35, to a protein concentration of 1–2 mg/ml.

Sepharose-4B was activated with CNBr[25] and coupled to aprotinin[26] in 0.1 M NaHCO$_3$ at pH 8.5. The coupling mixture contained 4 mg of aprotinin per gram of activated gel. About 85% of the aprotinin became linked to the gel, and the final preparation bound about 1 mg trypsin/ml.

The diluted extract was adsorbed batchwise with the aprotinin–Sepharose, the gel slurry was washed well with the 5 mM Tris-HCl buffer, pH 7.5 (containing 1 M NaCl–0.1% Brij 35), and the proteinases were eluted from the gel with the 50 mM glycine–HCl buffer, pH 3.3 (also containing 1 M NaCl–0.1% Brij 35). The fractions were collected in tubes containing sufficient Tris buffer to return the pH immediately to the neutral range. One-hundred milliliters of gel adsorbed about 80 mg of leukocyte proteinases and the same gel could be used repeatedly to process a large amount of material.

The eluates were dialyzed against 50 mM sodium phosphate buffer (pH 6.0), and batches of 150 mg of protein were applied to a carboxymethyl-cellulose (Whatman CM 52) column (25 ml bed volume) equilibrated with the same buffer. A linear gradient (1 liter) to 0.8 M NaCl

[22] D. Y. Twumasi and I. E. Liener, *J. Biol. Chem.* **252**, 1917 (1977).
[23] J. Saklatvala and A. J. Barrett, *Biochim. Biophys. Acta* **615**, 167 (1980).
[24] R. J. Baugh and J. Travis, *Biochemistry* **15**, 836 (1976).
[25] J. Porath, K. Aspberg, H. Drevin, and R. Axén, *J. Chromatogr.* **86**, 53 (1973).
[26] Aprotinin is obtained as Trasylol from Bayer AG, D-5600, Wuppertal 1, Federal Republic of Germany.

in the buffer eluted elastase at about 0.3 M NaCl and cathepsin G at about 0.6 M.

Elastase isolated in this way showed only protein bands attributable to the recognized iso-forms, when run in SDS–10% polyacrylamide gels, and in gels at acid pH. Moreover, no contamination by cathepsin G was detectable immunologically.

For storage, the enzyme was dialyzed into 0.5 M pyridine acetate buffer (pH 5.3) and lyophilized in ampules. In the sealed ampules stored at −20°C, the activity was stable for years.

Properties

Human leukocyte elastase appears as three major iso-forms in polyacrylamide-gel electrophoresis at acid pH,[10,27] or in isoelectric focusing.[28] There is no indication that the forms differ in catalytic activity, and minor compositional variations are assumed to be responsible for the heterogeneity; the forms are immunologically identical.

The molecular weight of leukocyte elastase is close to 30,000,[10,27] and the isoelectric points range from 8.77 to 9.15.[28] The enzyme is a glycoprotein (unlike pancreatic elastase), and amino acid analysis data have been reported.[29] The N-terminal amino acid sequence is strongly homologous with that of porcine pancreatic elastase.[29] Interestingly, though, the pancreatic elastase is generated from proelastase by tryptic cleavage of an N-terminal activation peptide, whereas no proenzyme has been detected for the leukocyte elastase.

Leukocyte elastase is a serine proteinase that shows maximal activity against most of its substrates in the region of pH 8.5. The enzyme differs from porcine pancreatic elastase in showing a preference for accommodating a valyl rather than an alanyl residue in specificity subsite S_1 when acting on synthetic substrates,[16] the insulin B chain,[30] or peptidyl chloromethanes.[31] It also differs from the pancreatic enzyme in that S_3

---

[27] J. Travis, R. Baugh, P. J. Giles, D. Johnson, J. Bowen, and C. F. Reilly, in "Neutral Proteases of Human Polymorphonuclear Leukocytes" (K. Havemann and A. Janoff, eds.), p. 118. Urban & Schwarzenberg, Munich, 1978.

[28] J. C. Taylor and J. Tlougan, Anal. Biochem. 90, 481 (1978).

[29] J. Travis, P. J. Giles, L. Porcelli, C. F. Reilly, R. Baugh, and J. Powers, in "Protein Degradation in Health and Disease" (D. Evered and J. Whelan, eds.), p. 51. Excerpta Medica, Amsterdam, 1980.

[30] A. M. J. Blow, Biochem. J. 161, 13 (1977).

[31] J. C. Powers, B. F. Gupton, M. O. Lively, N. Nishono, and R. J. Whitely, in "Neutral Proteases of Human Polymorphonuclear Leukocytes" (K. Havemann and A. Janoff, eds.), p. 221. Urban & Schwarzenberg, Munich, 1978.

does not exclude a prolyl residue.[32] Among the proteins degraded by the enzyme are such important structural proteins as elastin,[33] cartilage proteoglycan,[34] collagen of types I, II,[35] III,[36] and IV,[37] and fibronectin.[38]

Reversible inhibitors include some long-chain fatty acids,[39] polysaccharide sulfates,[40] elastatinal,[41] and elasnin.[42] Phenylmethane sulfonyl fluoride inactivates leukocyte elastase eight times more efficiently than it does pancreatic elastase, $k_{app}/[I]$ being 21 M^{-1} sec^{-1} (Ref. 43); but a far better inhibitor is MeOSuc-Ala-Ala-Pro-Val-CH$_2$Cl with $k_{app}/[I]$ 1560 M^{-1} sec^{-1} (Ref. 32).

Protein inhibitors of leukocyte elastase include soybean Kunitz inhibitor and turkey ovomucoid, inhibitors from leeches and bronchial mucus, and the plasma proteins, α_1-proteinase inhibitor and α_2-macroglobulin. In the reaction with α_1-proteinase inhibitor, a methionyl bond is cleaved, and it has been shown that the enzyme is very active against other methionyl substrates, but that the oxidized forms of methionine are resistant.[18] A striking peculiarity of the reaction of the leukocyte elastase with α_2-macroglobulin is the anomalous activation of the complexed enzyme with one of its synthetic substrates, Suc-Ala-Ala-Ala-NPhNO$_2$, by a factor of about 10-fold.[44]

There is no difficulty in raising antisera against human leukocyte elastase in rabbits, and these give precipitating reactions of complete immunological identity with the multiple forms of the enzyme.[14,45]

There is no evidence of the presence of the leukocyte type of elastase in any other type of human cell, although the amounts of the enzyme

[32] J. C. Powers, B. F. Gupton, A. D. Harley, N. Nishono, and R. J. Whitely, *Biochim. Biophys. Acta* **485**, 156 (1977).

[33] R. M. Senior, D. R. Bielefeld, and B. C. Starcher, *Biochem. Biophys. Res. Commun.* **72**, 1327 (1976).

[34] P. J. Roughley and A. J. Barrett, *Biochem. J.* **167**, 629 (1977).

[35] P. M. Starkey, A. J. Barrett, and P. M. Burleigh, *Biochim. Biophys. Acta* **483**, 386 (1977).

[36] J. E. Gadek, G. A. Fells, D. G. Wright, and R. G. Crystal, *Biochem. Biophys. Res. Commun.* **95**, 1815 (1980).

[37] M. Davies, A. J. Barrett, J. Travis, J. Sanders, and G. A. Coles, *Clin. Sci. Mol. Med.* **54**, 233 (1978).

[38] J. A. McDonald and D. G. Kelly, *J. Biol. Chem.* **255**, 8848 (1980).

[39] B. M. Ashe and M. Zimmerman, *Biochem. Biophys. Res. Commun.* **75**, 194 (1977).

[40] A. Baici, P. Salgam, K. Fehr, and A. Böni, *Biochem. Pharmacol.* **29**, 1723 (1980).

[41] G. Feinstein, C. J. Malemud, and A. Janoff, *Biochim. Biophys. Acta* **429**, 925 (1976).

[42] S. Ōmura, H. Ohno, T. Saheki, M. Yoshida, and A. Nakagawa, *Biochem. Biophys. Res. Commun.* **83**, 704 (1978).

[43] M. O. Lively and J. C. Powers, *Biochim. Biophys. Acta* **525**, 171 (1978).

[44] D. Y. Twumasi, I. E. Liener, M. Galdston, and V. Levytska, *Nature (London)* **267**, 61 (1977).

[45] K. Ohlsson and I. Olsson, *Eur. J. Biochem.* **42**, 519 (1974); K. Ohlsson, I. Olsson, M. Delshammar, and H. Schiessler, *Hoppe-Seyler's Z. Physiol. Chem.* **357**, 1245 (1976).

found in spleen and rheumatoid synovial membrane were surprisingly large in view of the scarcity of neutrophils in these tissues.[10,23] The elastase of macrophages is a metalloproteinase with quite different properties from the neutrophil enzyme. Leukocyte elastase is certainly not confined to humans, however; elastases have been detected in the polymorphonuclear cells of dog,[45] horse,[46] pig,[47] and rabbit.[48]

[46] A. Koj, J. Chudzik, and A. Dubin, *Biochem. J.* **153**, 397 (1976).
[47] M. Kopitar and D. Lebez, *Eur. J. Biochem.* **56**, 571 (1975).
[48] T. G. Cotter and G. B. Robinson, *Biochim. Biophys. Acta* **615**, 414 (1980).

[45] Mast Cell Proteases

By Richard G. Woodbury, Michael T. Everitt, and Hans Neurath

Mast cells are widely distributed in the connective tissue of most vertebrates and play a central role in the early stages of inflammation. It is generally thought that these cells help in the defense against parasites.[1] Mast cells contain dense secretory granules that, in most animal species, including humans, contain histamine, heparin, and several proteins.[2] A considerable amount of the protein of the mast cell granules of many species examined consists of serine proteases, which differ in substrate specificity from one species to another.[3] Since the activity of the mast cell granules can be detected in tissue slices using histochemical ester substrates,[3] it appears that the proteases are present in the granules in active form rather than as zymogens. Rats, as well as several other mammals examined, possess two types of mast cells. One of these, referred to as the *normal mast cell,* is found in most loose connective tissues and in the peritoneal cavity. The other one, known as the *atypical mast cell,* is present exclusively in mucosal tissues and hence is sometimes referred to as the mucosal mast cell.[4]

Although it is questionable whether normal and atypical mast cells are histologically related to each other,[5,6] it is clear that the secretory granules

[1] A. Sher, *Nature (London)* **263**, 334 (1976).
[2] D. Lagunoff and P. Pritzl, *Arch. Biochem. Biophys.* **173**, 554 (1976).
[3] H. Chiu and D. Lagunoff, *Histochem. J.* **4**, 135 (1972).
[4] R. Veilleux, *Histochemie* **34**, 157 (1973).
[5] H. R. P. Miller and R. Walshaw, *Am. J. Pathol.* **69**, 195 (1972).
[6] E. J. Ruitenberg and A. Elgersma, *Nature (London)* **264**, 258 (1976).

METHODS IN ENZYMOLOGY, VOL. 80

of each cell type contain a similar yet distinct serine protease with chymotrypsin-like esterase specificity.[7,8] In order to distinguish these enzymes from each other, that from rat normal mast cells will be called rat mast cell protease I (RMCP I) and that from atypical mast cells, rat mast cell protease II (RMCP II).

Katunuma and co-workers[9] examined both enzymes and, without a knowledge of the cellular localization of these proteases, had designated them "group-specific" intracellular proteases.

In addition to the chymotrypsin-like serine protease, normal mast cells obtained from rat peritoneum also contain an enzyme with carboxypeptidase A-like activity.[10,11]

I. Serine Protease of Peritoneal Mast Cells

Assay

Reagents

1.07 mM benzoyl tyrosine ethyl ester (BzTyrOEt) in 50% methanol
80 mM Tris-HCl buffer (pH 7.8) containing 100 mM CaCl$_2$

Procedure. This assay, which measures chymotrypsin-like esterase activity, was first described by Hummel.[12] Substrate (1.4 ml) is mixed with 1.5 ml of buffer and 0.1 ml of dilute enzyme preparation (1–10 μg) is added. The absorption change at 256 nm is recorded for about 5 min.

An enzyme unit is defined as an amount of enzyme activity that results in the hydrolysis of 1 μmol of substrate per minute at pH 7.8, 25°C.

Purification Procedure

Mast cells are collected from Sprague-Dawley rats (250–400 g) by washing the peritoneal cavities with 10 ml of ice-cold phosphate-buffered saline solution at pH 7.2 (4.1 mM Na$_2$HPO$_4$–2.1 mM KH$_2$PO$_4$–154 mM NaCl–2.17 mM KCl–0.68 mM CaCl$_2$). All subsequent steps are carried out at 4°C. Mast cells of the pooled peritoneal washes generally represent

[7] D. Lagunoff and E. P. Benditt, *Ann. N.Y. Acad. Sci.* **103,** 185 (1963).
[8] R. G. Woodbury, G. M. Gruzenski, and D. Lagunoff, *Proc. Natl. Acad. Sci. U. S. A.* **75,** 2785 (1978).
[9] N. Katunuma, E. Kominami, K. Kobayashi, Y. Banno, K. Suzuki, K. Chichibu, Y. Hamaguchi, and T. Katsunuma, *Eur. J. Biochem.* **52,** 35 (1975).
[10] R. Haas, P. C. Heinrich, and D. Sasse, *FEBS Lett.* **103,** 168 (1979).
[11] M. T. Everitt and H. Neurath, *FEBS Lett.* **110,** 292 (1980).
[12] B. C. W. Hummel, *Can. J. Biochem. Physiol.* **37,** 1393 (1959).

less than 5% of the total cell population. If desired, a preparation can be obtained that contains more than 95% mast cells by sedimentation of the cells through an isotonic solution of bovine serum albumin.[2] However, for the purification of the mast cell protease this enrichment step is unnecessary and usually results in some loss of material. The peritoneal washes are combined and the cells pelleted by centrifugation at 2300 g for 15 min. The cell pellet is resuspended in 20 ml of the phosphate-buffered saline solution, and this suspension is quickly frozen and thawed through six cycles. The mast cell granules released by cell lysis are collected along with cellular debris by centrifugation at 27,000 g for 20 min. The protease is fully active while it is associated with the granules. The pellet, containing granules, is extracted three times with 20 ml of a solution of 0.8 M potassium phosphate–2% protamine sulfate (pH 8.0). The extracts containing solubilized mast cell protease are pooled (60 ml) and centrifuged at 27,000 g for 20 min.

Affinity-Adsorption Chromatography. Hen ovoinhibitor is coupled to CNBr-activated Sepharose CL-6B by the method described by Marsh *et al.*[13] A small column (1 × 4 cm) of the Sepharose-coupled ovoinhibitor is equilibrated with a solution of 0.8 M potassium phosphate containing 1 mg/ml bovine serum albumin. The protease solution is passed slowly through the column. Approximately 5-ml fractions are collected at a flow rate of 20 ml/hr. The column is washed with equilibration buffer until no material that absorbs at 280 nm is eluted. The protease is desorbed from the ovoinhibitor–Sepharose with 25 mM formic acid containing 1 mM EDTA–1 mg/ml bovine serum albumin–20% glycerol (pH 3.0). At this point the flow rate is reduced to 10 ml/hr and 1-ml fractions are collected in tubes containing 1 ml of 0.2 M ammonium bicarbonate–1 mg/ml bovine serum albumin (pH 8.6). The eluate is monitored at 280 nm and the peak fractions are assayed for esterase activity using BzTyrOEt as substrate. The contents of fractions containing enzyme activity are pooled.

Adsorption to Barium Sulfate. The protease solution obtained by the affinity-adsorption step is diluted 2-fold with distilled water. Two grams of X-ray grade barium sulfate (Matheson, Coleman and Bell) is added slowly to the protease solution with constant stirring. After 15 min, the suspension is centrifuged at 2500 g for 10 min and the pellet is washed twice with 20 ml of 10 mM Tris-HCl buffer (pH 8.0) to remove the albumin. The protease is solubilized by washing the barium sulfate pellet with 5 ml of 10 mM Tris-HCl buffer (pH 8.0) containing 1.0 M NaCl. The purified protease is stored at $-20°C$. A summary of a typical purification of the protease from 100 rats is given in Table I.

Comment. During the purification of the mast cell protease, it is important to maintain the solutions at relatively high ionic strength in order to

[13] S. C. Marsh, I. Parikh, and P. Cuatrecasas, *Anal. Biochem.* **60**, 149 (1974).

TABLE I
PURIFICATION OF PERITONEAL MAST CELL PROTEASE[a]

Step	Volume (ml)	Total activity (BzTyrOEt units)	Total protein (mg)	Specific activity (units/mg)	Recovery (%)	Purification factor (-fold)
Crude extract supernatant	56	527	330	1.6	100	1
Affinity chromatography on ovoinhibitor–agarose	20	321	—	—	61	—
Barium sulfate adsorption	15	232	9.1	25.5[b]	44	15

[a] From the peritoneal cells of 100 rats (250 g).
[b] The preparation is known to contain approximately 50% inactive enzyme.

prevent the enzyme from adsorbing to surfaces. In solutions of low ionic strength, or in the absence of detergent, the protease adsorbs to glass, dialysis tubing, Sephadex, agar, polyacrylamide, and agarose.

Other Purification Procedures. The protease from rat peritoneal mast cells (also known as chymase) has been purified in small quantities by Yurt and Austen[14] using a procedure that includes ion-exchange chromatography on Dowex-1, gel filtration, and affinity chromatography on D-tryptophan methyl ester coupled to Sepharose.

Mast cell protease, which at the time was known as group-specific protease, was isolated by Katunuma *et al.*[9] from rat skeletal muscle and liver. This rather lengthy procedure yields small quantities of homogeneous protease with a high specific activity.

Properties

Purity. The preparation of mast cell protease (RMCP I) is homogeneous according to polyacrylamide-gel electrophoresis in the presence of sodium dodecyl sulfate and 8 M urea. Also, amino-terminal sequence analysis of the protease preparation indicates the presence of a single polypeptide chain.

Stability. At pH 7–8, the solubilized protease is quite labile even at 4°C, and most activity is lost in a few days due to autolysis. For this reason, the purification must be accomplished quickly to ensure a reasonable yield of active protease. At $-20°C$ the protease is relatively stable for several months. The protease can be freeze-dried and stored at $-20°C$ with little loss in activity. Alternatively, the enzyme can be adsorbed to barium sulfate and stored at 4°C without appreciable autolysis.

Specific Activity. Although the protease prepared as described is homogeneous, it was found that approximately 55% of the enzyme could not bind to potato chymotrypsin inhibitor immobilized on agarose. The nonbinding material also lacked esterase activity. Since the binding and nonbinding fractions of protease have identical amino acid compositions and molecular weights, it appears that a significant amount of the protease is irreversibly inactivated, most likely when it is exposed to the 25 mM formic acid solution during the affinity-chromatography step. In fact, nearly 40% of the initial esterase activity is lost during this manipulation. However, for structural studies this is of little consequence and for enzymatic studies one may readily calculate the proportion of active protease on the basis of a specific activity of fully native enzyme of 58 BzTyrOEt units/mg of enzyme.[15]

[14] R. W. Yurt and K. F. Austen, *J. Exp. Med.* **146**, 1405 (1977).
[15] M. T. Everitt and H. Neurath, *Biochimie* **61**, 653 (1979).

TABLE II
AMINO ACID COMPOSITION[a] OF RAT MAST CELL PROTEASE
(RMCP I)

Amino acid	g/100 g protein	Residues/molecule
Aspartic acid	7.4	17
Threonine[b]	6.5	16
Serine[b]	4.2	12
Glutamic acid	9.2	19
Proline	5.6	15
Glycine	5.4	21
Alanine	4.3	14
Half-cystine[c]	3.1	8
Valine	8.6	22
Methionine	2.5	5
Isoleucine[d]	5.5	13
Leucine[d]	5.1	12
Tyrosine	4.2	7
Phenylalanine	3.8	7
Histidine	4.5	9
Lysine	11.2	23
Arginine	7.0	12
Tryptophan[e]	1.2	2
Total residues		234
Molecular weight (composition)		25,800
Molecular weight (SDS electrophoresis)		26,000

[a] Duplicate samples were analyzed after 24, 48, 72, and 96 hr of hydrolysis in 6 N HCl at 110°C.
[b] Determined by extrapolation to zero-hydrolysis time.
[c] Determined as cysteic acid after performic acid oxidation.
[d] Values were obtained from analysis of 96-hr hydrolysates.
[e] Determined after alkaline hydrolysis (48 hr) of the protease.

Physical and Chemical Properties. The physical, chemical, and enzymatic properties of the protease obtained from the total cell population of the peritoneal washings are identical to those of the enzyme obtained from enriched preparations containing more than 95% mast cells. The purified protease migrates as a single polypeptide chain during electrophoresis on polyacrylamide gels in the presence of sodium dodecyl sulfate and 8 M urea, with or without the addition of 2-mercaptoethanol. The apparent molecular weight of the protease is 26,000. In the absence of urea, however, the apparent molecular weight of the protein is significantly higher and the protein migrates in a relatively broad band.

The amino acid composition (Table II) of RMCP I indicates that the protein has a relatively high content of basic residues (19%). This is consistent with the observation that the isoelectric point of the protease is at pH 9.5. The tendency of the protease to adsorb to surfaces when in solutions of low ionic strength may be explained by its highly basic nature. It has also been proposed that the relatively high content of lysine in RMCP I allows the enzyme to form strong complexes with the highly acidic heparin molecules within the mast cell granules.[16]

The amino-terminal sequence of the first 52 residues of RMCP I is known.[17] Approximately 40% of these residues are identical to those of bovine chymotrypsin A, indicating a homologous relationship between this enzyme and the mammalian serine proteases. A high degree of sequence identity (greater than 85%) is observed when the amino-terminal region of RMCP I is compared to those of the protease from rat atypical mast cells, RMCP II,[18] and human neutrophil cathepsin G[19] (Fig. 1). This sequence similarity, in addition to numerous similarities in chemical and enzymatic properties,[20] suggests that the mammalian granulocyte serine proteases may have followed a common evolutionary pathway, which has led to a distinct class of serine proteases, in the sense that the pancreatic serine proteases comprise a group that is different from the blood-coagulation enzymes.

Substrate Specificity. RMCP I has chymotrypsin-like esterase specificity. As already mentioned, the specific activity of this enzyme is 58 units/mg protein[15] using BzTyrOEt as substrate. The protease activity of RMCP I is, however, relatively low.[9] When peptides, such as glucagon and the oxidized form of the β-chain of insulin, are used as substrates, the protease preferentially cleaves peptide bonds on the carboxyl side of tyrosine and phenylalanine.[21,22] The overall protease specificity of RMCP I resembles that of human cathepsin G more nearly than that of chymotrypsin. The chemical, physical, and enzymatic similarities between the mast cell protease and cathepsin G have been discussed by Starkey.[23] However, a recent detailed investigation of the substrate specificities of both of these proteases toward 4-nitroanilide peptides clearly indicates

[16] R. G. Woodbury and H. Neurath, *FEBS Lett.* **114**, 189 (1980).
[17] R. G. Woodbury, M. Everitt, Y. Sanada, N. Katunumà, D. Lagunoff, and H. Neurath, *Proc. Natl. Acad. Sci. U. S. A.* **75**, 5311 (1978).
[18] R. G. Woodbury, N. Katunuma, K. Kobayashi, K. Titani, and H. Neurath, *Biochemistry* **17**, 811 (1978).
[19] J. Travis, personal communication.
[20] R. G. Woodbury and H. Neurath, *Proc. Life Sci.* in press (1981).
[21] K. Kobayashi, Y. Sanada, and N. Katunuma, *J. Biochem. (Tokyo)* **84**, 477 (1978).
[22] M. T. Everitt and H. Neurath, *Fed. Proc., Fed. Am. Soc. Exp. Biol.* **38**, 834 (1979).
[23] P. M. Starkey, *in* "Proteinases in Mammalian Cells and Tissues" (A. J. Barrett, ed.), p. 57. North-Holland Publ., Amsterdam, 1977.

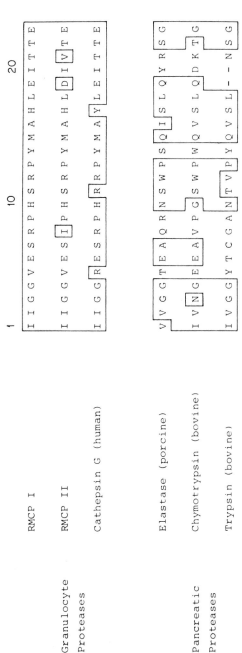

FIG. 1. The amino-terminal sequences of the granulocyte serine proteases rat mast cell protease I (RMCP I), rat mast cell protease II (RMCP II), and human neutrophil cathepsin G. A comparison of the amino-terminal sequences of these granulocyte enzymes to those of the pancreatic serine proteases elastase (porcine), chymotrypsin (bovine), and trypsin (bovine) illustrates the highly conserved structures of the granulocyte proteases, suggesting these enzymes have followed a common evolutionary pathway.

that the enzymes are quite different.[24] In this study it was observed that the best substrates of RMCP I contained hydrophobic residues not only at the primary cleavage site but also at neighboring positions that interact with the extended binding regions of the protease. The best substrate tested, succinyl-Phe-Leu-Phe-4-nitroanilide, was observed to have a k_{cat}/K_m of approximately 10^6 M^{-1} sec^{-1}, one of the highest specificity constants observed for any protease. It was also observed that, unlike other serine proteases such as chymotrypsin and cathepsin G, RMCP I will readily hydrolyze substrates that contain a prolyl residue at the P_3 subsite.

Inhibitors. RMCP I is inhibited by diisopropyl phosphorofluoridate at a rate (second-order rate constant $k_2 = 61$ liters mol^{-1} min^{-1}) which is considerably slower than that of bovine chymotrypsin ($k_2 = 17,300$ liters mol^{-1} min^{-1}).[15] The protease is also slowly inhibited by N^α-tosyl-L-phenylalanine chloromethyl ketone, and more readily so by benzoyloxycarbonyl-L-phenylalanine bromomethyl ketone. Since peptide 4-nitroanilides containing a proline at the P_3 subsite are good substrates for RMCP I, Yoshida and co-workers[24] examined the inhibitory activity of chloromethyl ketone peptides, which also contained a proline residue at the P_3 subsite. It was observed that succinyl-prolyl-leucyl-phenylalanine chloromethyl ketone was not only an excellent inhibitor of the enzyme, but also was relatively selective in its action, since it inhibited RMCP I at a rate approximately 10 times greater than that of chymotrypsin.

In addition to hen ovoinhibitor and potato chymotrypsin inhibitor I, soybean- and lima bean-trypsin inhibitors also are effective inhibitors of RMCP I. Pancreatic basic trypsin inhibitor (Kunitz) and ovomucoid, as well as cysteine, fail to inhibit the protease.

Physiological Role

Although the chymotrypsin-like protease (chymase) of rat peritoneal mast cells has been studied for over 20 years, the function of the enzyme remains obscure. Very likely it acts at some level in the inflammatory reactions associated with mast cell degranulation. There is fragmentary evidence suggesting the involvement of mast cell proteases in chemotaxis,[25] in degradation of connective tissue proteoglycans,[26] in

---

[24] N. Yoshida, M. T. Everitt, R. G. Woodbury, H. Neurath, and J. C. Powers, *Biochemistry* **19**, 5799 (1980).

[25] V. B. Hatcher, G. S. Lazarus, N. Levine, P. G. Burke, and F. J. Yost, Jr., *Biochim. Biophys. Acta* **483**, 160 (1977).

[26] H. Seppä, K. Väänänen, and K. Korhonen, *Acta Histochem.* **64**, 64 (1979).

selective degradation of basement membrane collagen (type IV),[27] as well as in the promotion of vascular permeability.[28]

Alternatively, the action of the protease following mast cell degranulation may be viewed as a pathological function, unrelated to the role of the enzyme in the normal physiological state. If mast cells function in the maintenance of homeostasis of the microenvironment without undergoing degranulation, as has been suggested,[29,30] then conceivably the protease may also have an intracellular function. Consistent with this hypothesis is the observation that the enzyme is fully active both as an esterase and as a protease when bound to the granule matrix heparin.[3,14]

II. Serine Protease of Atypical Mast Cells

Assays

Method A

Principle. This protease was first detected in rat small intestine by Katunuma and co-workers[9] by its ability to inactivate the apo form of ornithine aminotransferase, OAT (EC 2.6.1.13). In the early stages of the purification scheme this assay is much more sensitive and reliable than assays based on the esterase activity of the protease.

Reagents

Ornithine aminotransferase is isolated from rat livers and purified as described by Peraino et al.[31]
Apo ornithine aminotransferase is obtained by the method of Katunuma et al.[9]
50 mM potassium phosphate buffer, pH 8.0
0.1 mM pyridoxal phosphate in 50 mM potassium phosphate, pH 8.0

Procedure. A volume (10–100 μl) of dilute protease solution is added to 0.1 mg of apo ornithine aminotransferase in 50 mM potassium phosphate, pH 8.0 (stock solution contains 1 mg OAT/ml buffer). The volume is increased to 0.3 ml by the addition of 50 mM potassium phosphate, pH 8.0. Individual assays are carried out in duplicate for 0, 5, 15, and 30 min

[27] H. Sage, R. G. Woodbury, and P. Borstein, *J. Biol. Chem.* **254**, 9893 (1979).
[28] H. Seppä, *Inflammation* **4**, 1 (1980).
[29] K. F. Austen, *J. Immunol.* **121**, 793 (1978).
[30] J. Padawer, *Immunol. Ser.* **7**, 301 (1978).
[31] C. Peraino, L. G. Bunville, and T. N. Tahmisian, *J. Biol. Chem.* **244**, 2241 (1969).

at 37°C. Proteolytic assays are also performed using the holo form of ornithine aminotransferase as controls. At the appropriate time, each incubated solution is diluted 10-fold with 50 mM potassium phosphate buffer (pH 8.0) containing 0.1 mM pyridoxal phosphate, and is further incubated at 4°C for 30 min to allow any remaining apo ornithine aminotransferase to be converted to the holo form. Ornithine aminotransferase activity is measured by the method of Jenkins and Tsai.[32]

An enzyme unit is defined as the amount of enzyme that inactivates 50% of the ornithine aminotransferase activity in 30 min under the conditions described.

Method B

This assay measures the hydrolysis of BzTyrOEt by the method of Hummel,[12] as described earlier in this chapter.

Purification Procedure

Small intestines are removed from 50 Sprague-Dawley rats (250 g) and cut open lengthwise. The intestines are rinsed thoroughly with several changes of ice-cold 0.15 M NaCl (pH 7.2), gently scraped to remove the mucus-rich upper layer of mucosa, and then minced into small pieces. All subsequent steps are performed at 4°C. The minced intestines are homogenized in a Waring blender at moderate speed for 1–2 min in 3 volumes of 0.15 M KCl containing 5 mM benzamidine. Benzamidine is added to inhibit the activity of a small amount of a trypsin-like protease, which has also been observed by others[33] in homogenates of rat intestine. The crude homogenate is centrifuged at 15,000 g for 10 min and solid (NH$_4$)$_2$SO$_4$ is added to the supernatant containing the protease to give a solution that is 65% saturated. After 30 min, the precipitated material is collected by centrifugation at 15,000 g for 15 min and the pellet is resuspended in 200 ml of 0.15 M KCl containing 5 mM benzamidine. The suspension is centrifuged at 15,000 g for 30 min. Acetone ($-$20°C) is slowly added to the supernatant to give a concentration of 30% (v/v). The precipitate is removed by centrifugation at 15,000 g for 5 min and the supernatant is adjusted to a concentration of 55% acetone. The resulting precipitate is resuspended in 50 ml of 0.1 M Tris-HCl buffer, pH 7.8. The protease solution is incubated for 1 hr at 4°C with N^α-p-tosyl-L-lysine chloromethyl ketone (1 mM) and N^α-p-tosyl-phenylalanine chloromethyl ketone (1 mM), and sodium chloride (1 M) is then added.

[32] W. T. Jenkins and H. Tsai, this series, Vol. 17A, p. 281.
[33] R. J. Beynon and J. Kay, *Biochem. J.* **173**, 291 (1978).

Affinity-Adsorption Chromatography. Subunit C of potato chymotrypsin inhibitor that is purified by the method of Melville and Ryan[34] is coupled to Sepharose-4B as described by Marsh *et al.*[13] The inhibitor–Sepharose matrix is equilibrated with 0.1 M Tris-HCl buffer (pH 7.8) containing 1 M NaCl, and the protease-containing solution is passed slowly through the affinity-adsorption column (1.5 × 10 cm). The column contents are washed with approximately 5 bed volumes of the equilibration buffer and then with several volumes of distilled water. The protease is desorbed from the affinity adsorbant with 0.1 M acetic acid. 1-ml fractions are collected and the eluate is monitored at 280 nm. Peak fractions are assayed for protease and esterase activity and the contents of the appropriate fractions are pooled. This material is immediately dialyzed against 0.05 M Tris-HCl buffer (pH 7.8) or freeze-dried for storage at −20°C. A typical purification (Table III) of RMCP II from 50 small intestines results in about 25 mg of enzyme with more than 50% overall yield.[35]

Comments. It is essential to wash the column of Sepharose-coupled inhibitor free of the Tris buffer and NaCl, since in the presence of acetic acid, much of the desorbed protease precipitates with complete loss of activity.

The small amount of protease with trypsin-like specificity that is present in the homogenates of small intestine copurifies with RMCP II during the affinity-adsorption step. The presence of just a few micrograms of this protease results in substantial inactivation of RMCP II. The addition of N^α-*p*-tosyl-L-lysine chloromethyl ketone (1 mM) to the protease solution prior to the affinity-adsorption step serves to inhibit the contaminating protease and prevent its binding to the affinity resin.

Other Purification Procedures. This protease was first purified by Katunuma and co-workers,[9] at which time it was referred to as a group-specific protease. Although this procedure is considerably more time consuming and results in a lower overall yield of enzyme than the procedure outlined here, it does result in a homogeneous stable product with excellent activity.

Properties

Purity. Amino-terminal sequence analysis of the preparation of protease indicates that it consists of a single polypeptide chain. The preparation is also homogeneous according to polyacrylamide-gel electrophoresis in the presence of sodium dodecyl sulfate. The physical, chemical, en-

[34] J. C. Melville and C. A. Ryan, *J. Biol. Chem.* **247**, 3445 (1972).
[35] R. G. Woodbury and H. Neurath, *Biochemistry* **17**, 4298 (1978).

TABLE III

PURIFICATION OF THE PROTEASE (RMCP II) OF RAT ATYPICAL MAST CELLS[a]

Step	Total activity[b] (protease units)	Total protein (mg)	Specific activity (units/mg)	Recovery (%)	Purification factor (-fold)
Crude homogenate (supernatant)	3830	4366	0.9	100	1.0
Ammonium sulfate precipitation	2930	2264	1.4	77	1.6
Acetone precipitation	2210	590	3.8	58	4.2
Affinity adsorption	2000	25	80	52	89

[a] From the small intestines of 50 rats (250 g).
[b] Determined by measuring the ability of the protease to inactivate apo ornithine amino transferase.

zymatic, and immunological properties of RMCP II are identical to those of the protease isolated by the procedure of Katunuma *et al.*[9]

Stability. The purified protease is remarkably stable even when stored in solution at neutral pH for several weeks at 4°C, indicating the lack of significant autolysis.

Specific Activity. The activity of the purified protease (80 units/mg of protease), as measured by its ability to inactivate apo ornithine aminotransferase, is as high as that of enzyme prepared by the method of Katunuma *et al.*[9] (78 units/mg protease). Complete inactivation of the purified protease with ^{14}C-labeled diisopropyl phosphorofluoridate resulted in the incorporation of 1 mol of the ^{14}C-labeled diisopropylphosphoryl group per mole of protein, indicating that all of the protease molecules in this preparation had functional catalytic sites.[35] As measured by the hydrolysis of BzTyrOEt, the esterase activity of the homogeneous enzyme is 8.5 units/mg enzyme.

Physical and Chemical Properties. As in the case of RMCP I, RMCP II is extracted from tissue in a fully active form and thus seems to lack a zymogen precursor. The protease consists of a single polypeptide chain with a molecular weight of approximately 25,000 on the basis of the results of polyacrylamide-gel electrophoresis in the presence of sodium dodecyl sulfate. The molecular weight of the protease determined from its amino acid sequence is 24,655.[18] Table IV compares the amino acid composition obtained after acid hydrolysis of the enzyme to that calculated from the amino acid sequence. Since the enzyme has 25 basic residues and only 21 acidic ones, it is somewhat basic at neutral pH, but not as basic as RMCP I. The major structural features of RMCP II are as follows:

1. Approximately 35% sequence identity when compared to bovine chymotrypsin A.
2. The presence of the residues of the charge-relay system (Ser, Asp, His) of the active site.
3. An alanyl residue at the primary substrate binding site, identified as Asp[177] in trypsin and Ser[189] in chymotrypsin. Thus the primary binding site of RMCP II is likely to be more nonpolar when compared to those of the known serine proteases.
4. An arrangement of the three disulfide bonds of RMCP II that is unique among the known serine proteases.

Clearly, the structural data indicate a homologous relationship between RMCP II and the mammalian serine proteases. At the same time, there are structural features that distinguish this protease from other serine proteases. The significance of the various structural features of the protease with respect to enzymatic activity has been discussed previously.[18,20,24]

TABLE IV
AMINO ACID COMPOSITION[a] OF RAT ATYPICAL MAST
CELL PROTEASE (RMCP II)

Amino acid	Moles of amino acid/mole of protein	
	After hydrolysis	From sequence analysis
Aspartic acid	14.0	9
Asparagine	—	5
Threonine[b]	12.5	13
Serine[b]	13.0	13
Glutamic acid	17.2	12
Glutamine	—	5
Proline	14.8	15
Glycine	18.4	18
Alanine	16.0	16
Half-cystine[c]	5.8	6
Valine[d]	21.6	22
Methionine	4.7	5
Isoleucine[d]	16.8	18
Leucine[d]	15.6	16
Tyrosine	8.8	9
Phenylalanine	6.1	6
Histidine	8.8	9
Lysine	12.7	13
Arginine	12.0	12
Tryptophan[e]	1.7	2
Total residues	223	224
Molecular weight	24,522	24,655

[a] Duplicate samples were analyzed after 24, 48, 72, and 96 hr hydrolysis in 6 N HCl at 110°C.
[b] Determined by extrapolation to zero-hydrolysis time.
[c] Half-cystine determined as S-pyridyl-ethylcysteine.
[d] Values were obtained from analysis of 96-hr hydrolysates.
[e] Determined after 24 hr hydrolysis in 4 N methanesulfonic acid at 110°C.

A comparison of the amino acid sequence of RMCP II to those of other serine proteases has provided evidence suggesting that the ancestral protease molecule of RMCP II may have evolved some time after that of trypsin but before those of chymotrypsin and elastase.[18] This may be true also for other granulocyte serine proteases, since human neutrophil cathepsin G, RMCPI, and RMCP II show a remarkable degree of aminoterminal sequence identity (Fig. 1). If the degree of sequence identity is maintained throughout the remaining structures of RMCP I and cathepsin G, such homology would provide evidence that the granulocyte serine proteases have a common evolutionary pathway and together form a characteristic and distinct class of serine proteases.[20]

Substrate Specificity. The chymotrypsin-like esterase activity of RMCP II is significantly lower than that of commercially available (Worthington Biochemical Corporation) bovine chymotrypsin (8.5 BzTyrOEt units/mg for RMCP II as compared to 45 BzTyrOEt units/mg for bovine chymotrypsin). The protease activity of RMCP II is much lower than those of chymotrypsin and RMCP I, particularly when native proteins are used as substrates.[9,36]

Using peptide hormones as substrates, it was found that the protease specificity of RMCP II was more restrictive than that of bovine chymotrypsin.[37] The rat protease preferentially cleaved peptide bonds between hydrophobic residues. This observation is consistent with that of Yoshida *et al.*,[24] who found that the preferred 4-nitroanilide peptide substrates contained hydrophobic residues at both the primary and secondary subsites. The best substrates tested are succinyl-Phe-Pro-Phe-4-nitroanilide and succinyl-Phe-Leu-Phe-4-nitroanilide (k_{cat}/K_m = 36,000 and 30,000 M^{-1} sec^{-1}, respectively). These two substrates are also optimal for RMCP I. This similarity in substrate specificity together with the observation that RMCP II, as well as RMCP I, act on substrates containing a prolyl residue at the P_3 subsite, suggests that these two rat proteases have similar physiological functions. The major enzymatic differences between these proteases are: (1) a shift in optimal activity of RMCP II from pH 8.0 to pH 6.6 toward substrates containing a negatively charged group at the P_4 subsite; and (2) the observation that for each substrate, the catalytic rate of RMCP I is about 50-fold greater than that of RMCP II.[24] The structural features of the proteases that may explain these enzymatic differences have been discussed previously.[16,20]

Inhibitors. In contrast to RMCP I, RMCP II is not inhibited by N^α-*p*-tosyl-L-phenylalanine chloromethyl ketone. The protease is effectively inhibited by extended peptide chloromethyl ketones that contain, in addition to phenylalanine at the primary binding site, hydrophobic residues at one or more subsites.[24] The structures of inhibitors of this type are apparently analogous to those of preferred substrates. Since the substrate specificity of RMCP II (and RMCP I) shows substantive differences compared to those of other chymotrypsin-like serine proteases (e.g., cathepsin G and pancreatic chymotrypsin), it is likely that peptide chloromethyl ketone inhibitors can be designed that should prove uniquely selective for RMCP II.

---

[36] R. G. Woodbury, unpublished observation.
[37] K. Kobayashi and N. Katunuma, *J. Biochem.* (*Tokyo*) **84**, 65 (1978).

RMCP II is completely inhibited within 15 min by a 2- to 3-fold molar excess of potato chymotrypsin inhibitor, α_1-antitrypsin, hen ovoinhibitor, or lima bean trypsin inhibitor. Soybean trypsin inhibitor inactivates only 70% of the protease activity after 30 min. No inhibition of RMCP II is observed by pancreatic trypsin inhibitor.[35] The inhibition of RMCP II by various proteinase inhibitors is very similar to that observed for RMCP I,[15] again suggesting that the catalytic sites of the two proteases have similar extended binding regions.

Physiological Role

The *in vivo* function of RMCP II is unknown. The atypical (mucosal) mast cells containing the protease have long been recognized to function at some level in the expulsion of intestinal parasites.[38] On the basis of association, one may speculate that the protease also is involved in the expulsion process, perhaps by acting directly on the parasite or by mediating some other function of the mast cell. It is possible that the protease does not promote the general or widespread degradation of connective tissue proteins because those native proteins that have been examined were found to be resistant to the action of RMCP II even when high levels of enzyme are used.[9,36] Since the best synthetic substrates of RMCP II are those containing extended hydrophobic structures,[24] the resistance of native proteins to the action of the protease may be related to the fact that such hydrophobic regions are likely to be restricted to the interior of protein molecules and thus inaccessible to proteolytic attack. Such considerations suggest that the physiological action of RMCP II is highly selective and, perhaps limited to unique sites of the protein substrate.

III. Carboxypeptidase of Peritoneal Mast Cells

Assay Method

Reagents

Benzoyl-glycyl-phenylalanine (BzGlyPhe)
Benzoyl-glycyl-arginine (BzGlyArg)
50 mM Tris-HCl buffer, pH 7.5

Procedure. The assay for carboxypeptidase A activity is based on the method of Folk and Shirmer[39] using BzGlyPhe as substrate. Carboxypep-

---

[38] H. R. P. Miller and W. F. H. Jarrett, *Immunology* **20**, 277 (1971).
[39] J. E. Folk and E. W. Shirmer, *J. Biol. Chem.* **238**, 3884 (1963).

tidase B activity is measured using BzGlyArg as substrate.[40] Enzyme activities are determined spectrophotometrically as follows: extracts (10–100 μl) containing enzyme are added to 3 ml of substrate (1 mM) dissolved in 50 mM Tris-HCl buffer (pH 7.5). The initial rate of reaction is determined from the change in absorption at 254 nm.

A carboxypeptidase unit is defined as the amount of enzyme that hydrolyzes 1 μmol of substrate per min at 25°C under the conditions described.

Purification Procedure

Mast cells are obtained from Sprague-Dawley rats by washing the peritoneal cavity as described earlier for the preparation of the serine protease RMCP I. The cells are isolated by sedimentation through 30% albumin. The cell pellet consisting of at least 95% mast cells is resuspended in distilled water. Cell lysis releases the secretory granules, which are then purified by differential centrifugation.[2] Extraction of the granules with 0.8 M potassium phosphate containing 2% protamine sulfate (pH 8.0) solubilizes carboxypeptidase activity hydrolyzing BzGlyPhe, but not BzGlyArg.

In order to purify substantial amounts of mast cell carboxypeptidase,[11] the peritoneal cavities of 100 rats are lavaged as previously described in this chapter. The pooled washes are centrifuged at 2300 g for 15 min in order to collect cells and cellular debris. The pellet is resuspended in distilled water and the solution is frozen and thawed through six cycles to release mast cell granules. This suspension is centrifuged at 2300 g for 15 min and the pellet containing the granule-associated carboxypeptidase is extracted with 20 ml of 0.8 M potassium phosphate containing 2% protamine sulfate (pH 8.0).

Affinity-Adsorption Chromatography. Potato carboxypeptidase inhibitor is coupled to agarose.[13] A column (1 × 5.5 cm) of this affinity adsorbant is equilibrated with 0.8 M potassium phosphate containing 1 mg/ml bovine serum albumin (pH 8.0). The extract of granules containing solubilized carboxypeptidase A activity is passed slowly through the column. The contents of the column are washed with 20 ml of the equilibration buffer containing 1 mM phenylmethanesulfonyl fluoride and 0.1 mg/ml lima bean trypsin inhibitor. After washing the column with 2 volumes of 0.1 M NaHCO$_3$ (pH 8.0), the carboxypeptidase is eluted with 0.1 M Na$_2$CO$_3$ containing 0.5 M NaCl (pH 11.4). One-milliliter fractions are collected in tubes containing 1 ml of 0.2 M Tris-HCl buffer (pH 7.5).

[40] E. Wintersberger, D. J. Cox, and H. Neurath, *Biochemistry* **1**, 1069 (1962).

TABLE V

PURIFICATION OF MAST CELL CARBOXYPEPTIDASE A[a]

Step	Volume (ml)	Total activity (BzGlyPhe units)	Total protein (mg)	Specific activity (units/mg)	Recovery (%)	Purification factor (-fold)
Crude extract	20	81.3	245	0.33	100	1
Affinity-adsorption chromatography	10	31.5	4.18	7.54	39	23

[a] From the peritoneal cells of 100 rats.

Fractions containing significant carboxypeptidase activity are pooled. A typical purification protocol of the preparation of the enzyme from 100 rats (Table V) yields a preparation containing approximately 4 mg of protein and about 40% of the initial carboxypeptidase A activity present in the extract of granules.

Properties

Purity. The preparation of mast cell carboxypeptidase by this procedure is homogeneous on the basis of polyacrylamide-gel electrophoresis in the presence of sodium dodecyl sulfate, 8 M urea, and 2-mercaptoethanol. Also, amino-terminal sequence analysis indicates that the protein contains a single polypeptide chain.[41]

Stability. Preparations of homogeneous carboxypeptidase are relatively stable at 4°C or lower temperatures. However, during the purification, the enzyme is rapidly degraded by the mast cell serine protease (RMCP I) unless preventive measures are taken. Thus it is essential to carry out all steps quickly at 4°C and to add an inhibitor of RMCP I, e.g., lima bean-trypsin inhibitor.

Specific Activity. The purified mast cell carboxypeptidase A has a specific activity of 7.5 BzGlyPhe units/mg enzyme, which is significantly lower than that of commercially obtained (Worthington Biochemical Corporation) bovine pancreatic carboxypeptidase A (approximately 35 units/mg enzyme). The affinity-adsorption step results in a 60% loss of the original enzyme activity perhaps due, in part, to a rapid loss ($t_{1/2}$ = 1 hr) of activity when the enzyme is in the elution buffer at pH 11.4.[11] It is therefore possible that the enzyme preparation consists of both active and inactive molecules of carboxypeptidase.

Physical and Chemical Properties. The enzyme migrates during electrophoresis in polyacrylamide gels in the presence of sodium dodecyl sulfate and 2-mercaptoethanol as a single band. The apparent molecular weight is approximately 35,000.

Comparison of the amino acid composition of mast cell carboxypeptidase A and bovine carboxypeptidases A and B (Table VI) indicates a general similarity of the mast cell enzyme to both bovine enzymes. Despite its carboxypeptidase A specificity, however, the mast cell enzyme resembles more closely bovine carboxypeptidase B in its half-cystine and methionine content. In fact, the odd number of half-cystine residues of the mast cell enzyme suggests that it contains a free sulfhydryl group, which is analogous to bovine carboxypeptidase B, but unlike bovine carboxypeptidase A. A major chemical difference between the mast cell and

[41] M. T. Everitt, Doctoral Dissertation, University of Washington, Seattle (1980).

TABLE VI
AMINO ACID COMPOSITIONS[a] OF RAT MAST CELL CARBOXYPEPTIDASE AND
BOVINE PANCREATIC CARBOXYPEPTIDASES

| | | Bovine enzymes | |
Amino acid	Mast cell carboxypeptidase	Carboxypeptidase A	Carboxypeptidase B
Aspartic acid	30	29	28
Threonine[b]	19	26	27
Serine[b]	28	32	27
Glutamic acid	18	25	25
Proline	15	10	12
Glycine	19	23	22
Alanine	17	21	22
Half-cystine[c]	5	2	7
Valine	15	16	14
Methionine	7	3	6
Isoleucine[d]	20	21	16
Leucine[d]	21	23	21
Tyrosine	14	19	22
Phenylalanine	13	16	12
Histidine	9	8	7
Lysine	27	15	17
Arginine	16	11	13
Tryptophan[e]	9	7	8
Total residues	306	307	306

[a] Duplicate samples were analyzed after 24, 48, 72, and 96 hr of hydrolysis in 6 N HCl at 110°C.
[b] Determined by extrapolation to zero-hydrolysis time.
[c] Determined as cysteic acid after performic acid oxidation.
[d] Values were obtained from analysis of 96-hr hydrolysates.
[e] Determined after alkaline hydrolysis (48 hr).

bovine carboxypeptidases is the presence of nearly twice as much lysine in the mast cell enzyme, which may be involved in the binding to the highly acidic heparin and may account for the necessity of using solutions of high ionic strength to extract the enzyme from mast cell granules.

Immunofluorescent localization studies using specific antisera directed toward mast cell carboxypeptidase, with the total cell population of peritoneal washes show positive staining only of mast cell granules.[41] Similar studies using this antiserum indicated no fluorescence associated with the atypical mast cells of the small intestine.

Substrate Specificity. Purified mast cell carboxypeptidase hydrolyzes BzGlyPhe, but not BzGlyArg; hence its designation as a carboxypeptidase A. In a more detailed examination of the substrate specificity of the mast cell enzyme, it was observed that for several dipeptide substrates the kinetic parameters were similar to those of bovine carboxypeptidase A.[11]

Inhibitors. Mast cell carboxypeptidase is inhibited by 1,10-phenanthroline and by mercurials, as well as by potato carboxypeptidase inhibitor, suggesting that, in analogy with the pancreatic carboxypeptidases, the mast cell enzyme is also a metalloenzyme. Diisopropyl phosphorofluoridate, phenylmethanesulfonyl fluoride, and lima bean trypsin inhibitor are without effect.

Acknowledgment

This work was supported by a grant from the National Institutes of Health (GM-15731).

[46] γ-Subunit of Mouse Submaxillary Gland 7 S Nerve Growth Factor: An Endopeptidase of the Serine Family

By KENNETH A. THOMAS and RALPH A. BRADSHAW

Introduction

Nerve growth factor (NGF) is a hormone-like protein that is required for both maintenance of, and neurite outgrowth from, sympathetic and certain sensory neurons.[1] Cells from tissues that are innervated by NGF-responsive neurons *in vivo* secrete this substance in culture,[2] and it is presumed that these tissues are the physiologically relevant sources of the factor for its endocrine activities *in vivo*.[3] In addition, a number of exocrine tissues and their secretions—including the submaxillary gland[4] and saliva of male mice[5,6]; the prostate and semen of ox, guinea pig, sheep, and goats[7,8]; and the venom of poisonous land snakes[9]—have very high

[1] R. Levi-Montalcini and P. U. Angeletti, *Physiol. Rev.* **48**, 534 (1968).

[2] R. A. Bradshaw and M. Young, *Biochem. Pharmacol.* **25**, 1445 (1976).

[3] R. A. Bradshaw, *Annu. Rev. Biochem.* **47**, 191 (1978).

[4] S. Cohen, *Proc. Natl. Acad. Sci. U.S.A.* **46**, 302 (1960).

[5] L. J. Wallace and L. M. Partlow, *Proc. Natl. Acad. Sci. U.S.A.* **73**, 4210 (1976).

[6] R. A. Murphy, J. D. Saide, M. H. Blanchard, and M. Young, *Proc. Natl. Acad. Sci. U.S.A.* **74**, 2330 (1977).

[7] G. P. Harper, Y. A. Barde, G. Burnstock, J. R. Carstairs, M. E. Dennison, K. Suda, and C. A. Vernon, *Nature (London)* **279**, 160 (1979).

[8] G. P. Harper, *Abstr. 7th Meet., Int. Soc. Neurochem.* p. 9 (1979).

[9] R. A. Hogue-Angeletti and R. A. Bradshaw, *Handb. Exp. Pharmacol.* [N.S.] **52**, 276 (1977).

concentrations of NGF. Although the physiological importance, if any, of these sources is unknown, they have served as the source of sufficient quantities of the factor for biochemical characterization. Mouse submaxillary NGF, the best-characterized form, occurs as a multisubunit complex of MW ~140,000 and is referred to as the 7 S complex based on its sedimentation coefficient in sucrose.[10,11] It is composed of two α-subunits, each with MW 26,500, a stable β-dimer made up of 2 identical chains[12] of MW 13,250,[13] and two γ-subunits, MW ~27,500 each.[14] The complex is further stabilized by one or two bound zinc ions.[15] Other than contributing to the stability of the 7 S complex, the role of the α-subunit is unknown, with the neuronal maintenance and neurite outgrowth promoting activity residing in the β-subunit. The γ-subunit is a serine protease that cleaves both arginine and lysine substrates, although with a strong preference for the former.[16] It has been suggested that the principal function of this subunit is to process precursor(s) of the β-subunit, an activity that has been observed *in vitro*.[17] However, it remains unexplained why a catalytic process (activation of a pro-β molecule) results in the formation of a stoichiometric complex. Other activities have been observed (e.g., mitogenic stimulation of chick embryo fibroblasts[18]) that may provide a rationale for the high concentration. Furthermore, epidermal growth factor, also found in high levels in the adult male mouse submaxillary gland, is associated with a similar protease.[19] Based on partial sequence data, the enzymes show about 70% sequence identity[20] but retain absolute specificity in terms of recognizing their individual growth factor.[21]

Method of Assay

The γ-subunit catalytic activity can be measured with N^{α}-benzoyl-DL-arginine-p-nitroanilide (DL-BAPNA) (Sigma), monitoring release of the p-nitroaniline group at 410–412 nm. A 1-ml assay mixture is made with 970 μl of 0.1 M Tris-HCl (pH 8.0) at 25°C, 10 μl of 2.5 mM Na$_2$-EDTA,

---

[10] S. Varon, J. Nomura, and E. M. Shooter, *Biochemistry* **6**, 2202 (1967).

[11] S. Varon, J. Nomura, and E. M. Shooter, *Proc. Natl. Acad. Sci. U.S.A.* **57**, 1782 (1967).

[12] R. H. Angeletti, R. A. Bradshaw, and R. D. Wade, *Biochemistry* **10**, 463 (1971).

[13] R. H. Angeletti and R. A. Bradshaw, *Proc. Natl. Acad. Sci. U.S.A.* **68**, 2417 (1971).

[14] K. A. Thomas, N. C. Baglan, and R. A. Bradshaw, *J. Biol. Chem.* **256**, 9156 (1981).

[15] S. E. Pattison and M. F. Dunn, *Biochemistry* **14**, 2733 (1975).

[16] L. A. Greene, E. M. Shooter, and S. Varon, *Biochemistry* **9**, 3735 (1969).

[17] E. A. Berger and E. M. Shooter, *in* "Molecular Control of Proliferation and Differentiation" (J. Papaconstantinou and W. J. Rutter, eds.), p. 83. Academic Press, New York, 1978.

[18] L. A. Greene, J. T. Tomita, and S. Varon, *Exp. Cell Res.* **64**, 387 (1971).

[19] J. M. Taylor, S. Cohen, and W. M. Mitchell, *Proc. Natl. Acad. Sci. U.S.A.* **67**, 164 (1970).

[20] R. E. Silverman, Ph.D. Thesis, Washington University, St. Louis (1977).

[21] A. C. Server and E. M. Shooter, *J. Biol. Chem.* **252**, 165 (1976).

10 μl of 100 mM DL-BAPNA dissolved in DMSO, and 10 μl of enzyme solution. The concentration of pure enzyme can be determined from $E_{280\,nm}^{1\%,1\,cm} = 15$. The molar extinction coefficient at 410 nm of the released p-nitroaniline group is 8800 cm^{-1} M^{-1}.[22] With this assay, $K_m = 44\ \mu M$ and the specific activity is 0.38 μmol/min/mg pure γ-subunit. For routine monitoring of activity from columns, the substrate concentration can be reduced by a factor of 10, if desired. With L-BAPNA, $K_m = 24\ \mu M$ and the specific activity is 0.55 μmol/min/mg enzyme.

Purification

A number of protocols have been established for purification of the 7 S complex.[23] The γ-subunit can be readily isolated by ion-exchange chromatography following dissociation of the purified high-molecular-weight (HMW) complex[16,24] or as a by-product of the direct purification of the β-subunit[25] (2.5 S procedure).[26] The latter procedure[26] is the method most commonly used to prepare neuronally active NGF and provides the simplest route for isolating γ-subunit.

Step 1. Homogenization and Precipitation. Adult male mouse submaxillary glands (>8 weeks old), which can be obtained frozen from Pel-Freez (Rogers, Arkansas) or by excision from freshly slaughtered animals, are homogenized in 3 ml of cold water per gram of gland for 2 min in a blender at medium to high speed. This may be done in shorter segments, if necessary, to prevent the homogenate from overheating. The insoluble debris is removed by centrifugation at 10,000 g for 30 min. The supernatant is diluted with 1.5 ml of cold water for each gram of gland that is processed. A solution of 1 g of streptomycin sulfate per 8.6 ml of 0.1 M Tris-HCl (pH 7.5) is added over 5 min to a final concentration of 56 mg/g gland. The precipitation is allowed to proceed for 30–45 min on ice followed by centrifugation at 10,000 g for 30 min. The supernatant is lyophilized.

Step 2. First Gel Filtration. The extract is dissolved in cold water, centrifuged for 20 min at 10,000 g, and loaded on a column (6.5 × 100 cm) of Sephadex G-100 equilibrated in 50 mM Tris-HCl, pH 7.5. (Although the buffer solution was originally described to contain 0.5 mM EDTA, this has been found to be unnecessary. In fact, the chelator would be expected to decrease the stability of the 7 S complex form being chromatographed at this step.[27]) The column is eluted at 60 ml/hr with 12- to 15-ml fractions

[22] B. F. Erlanger, N. Kokowsky, and W. Cohen, *Arch. Biochem. Biophys.* **95**, 271 (1961).

[23] I. Jeng and R. A. Bradshaw, *Res. Methods Neurochem.* **4**, 265 (1978).

[24] L. A. Greene, E. M. Shooter, and S. Varon, *Proc. Natl. Acad. Sci. U.S.A.* **60**, 1383 (1968).

[25] I. Jeng, R. Y. Andres, and R. A. Bradshaw, *Anal. Biochem.* **92**, 482 (1979).

[26] V. Bocchini and P. U. Angeletti, *Proc. Natl. Acad. Sci. U.S.A.* **64**, 787 (1969).

[27] M. A. Bothwell and E. M. Shooter, *J. Biol. Chem.* **253**, 8458 (1978).

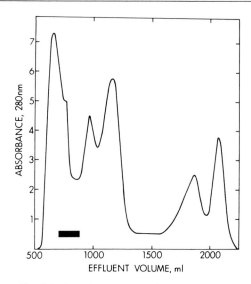

FIG. 1. Elution profile of the first chromatographic fractionation of 7 S NGF in a column (5 × 100 cm) of Sephadex G-100 equilibrated in 50 mM Tris-HCl, pH 7.5. The column was developed at 60 ml/hr and collected in 12.5-ml fractions. The sample applied was derived from 250 glands (30–35 g of tissue). Fractions were pooled as indicated by the solid bar.

collected. The NGF, as the HMW complex, follows just behind the turbid breakthrough peak but preceding the red hemoglobin fractions (Fig. 1). This position is sufficiently reproducible that, for routine work, accurate pooling can be made without absorbance measurements or bioassays. The NGF pool, about 150 ml, is dialyzed against 50 mM sodium acetate (pH 5.0) at 4°C overnight, which dissociates the 7 S complex to its constituent subunits.

Step 3. Ion-Exchange Chromatography. Any precipitate remaining after dialysis is removed by centrifugation for 20 min at 10,000 g and the supernatant, adjusted to pH 5.0 if necessary, is applied to a column (1.6 × 25 cm) of CM52-cellulose equilibrated in the same sodium acetate buffer. The column is pumped at 40 ml/hr. As shown in Fig. 2, the unbound protein in the first 300 ml, containing the impure α-subunit, is collected as a single fraction. It may be further purified on Sephadex G-100.[25] The bound protein is then eluted in 10-ml fractions with a parabolic gradient formed with three chambers containing acetate buffer with 0, 0, and 1 M NaCl. The γ-subunit, identified by its catalytic activity, is pooled. The β-subunit elutes at the end of this gradient, as shown in Fig. 2.

Step 4. Second Gel Filtration. The pool containing the γ-subunit is dialyzed, lyophilized, and dissolved in 10–20 ml of 50 mM Tris-HCl, pH 7.5. Any undissolved material is removed by centrifugation, as described

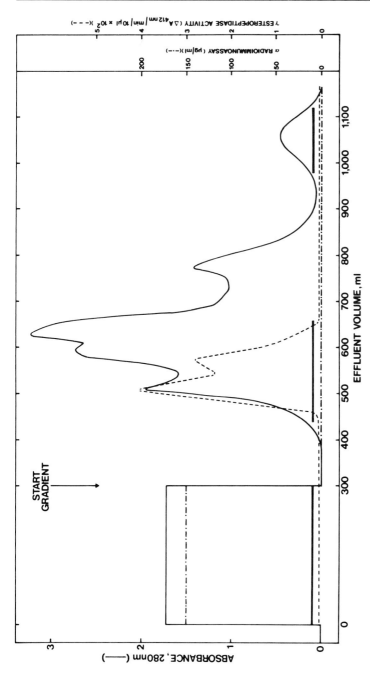

FIG. 2. Elution profile of the partially purified sample of NGF (pooled as shown in Fig. 1) on a column (1.6 × 25 cm) of CM52-carboxy-methylcellulose equilibrated in 50 mM sodium acetate, pH 5.0. The column is washed with this buffer and the first 300 ml, collected as a single fraction (designated by the first solid bar), contains the impure α-subunit. The bound protein is eluted with a parabolic gradient formed from a three-chamber reservoir containing 300 ml each of sodium acetate buffer with 0, 0, and 1 M NaCl. The flow rate is 40 ml/hr with fractions of 10 ml each. The second pool contains the γ-subunit as revealed by the amidase activity following BAPNA hydrolysis at 412 nm. The last pool contains pure β-subunit. The samples applied to this column and the one described in Fig. 3 were derived from 400 pairs of glands. Taken from Jeng et al.[25]

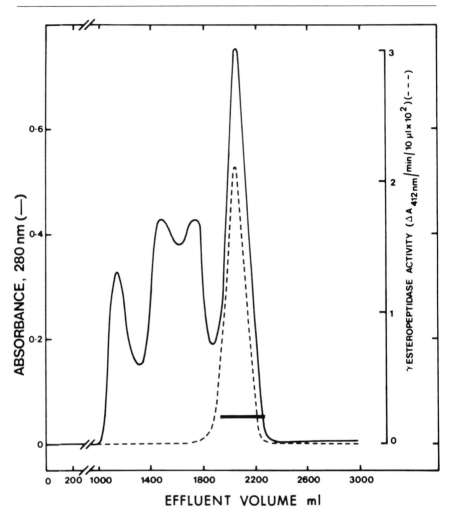

FIG. 3. Elution profile of the partially purified sample of the γ-subunit (second pool of Fig. 2) on a column (6.5 × 100 cm) of Sephadex G-100 equilibrated in 50 mM Tris-HCl, pH 7.5. The column was developed at 30 ml/hr and fractions of 12–15 ml were collected. The γ-subunit was revealed by the BAPNA assay and pooled as denoted by the solid bar. Modified from Jeng et al.[25]

above. The same Sephadex G-100 column used for step 2 can be used with the same flow rate and fraction volume. As shown in Fig. 3, the γ-subunit elutes as the last peak and is readily identified by assay. The enzyme can be dialyzed against 10 mM acetic acid and lyophilized. The impurities removed at this step are of higher molecular weight since they were found in the same gel-filtration (step 2) fraction with 7 S NGF (MW ~140,000).

FIG. 4. Elution profile of the affinity chromatography of NGF γ-subunit on a column of
benzamidine–Sepharose. The γ-subunit (20 mg) is
loaded on the column (2 ml) with a capacity of
about 15 mg of specifically bound protein/ml of
packed resin, in 0.1 M Tris-HCl, pH 7.5. The column is washed successively in the same buffer containing 1 M and 2 M NaCl. The active γ-subunit is
eluted with 50 mM sodium acetate, pH 4.0. The
column is developed at about 30 ml/hr by gravity,
collecting 3.3 ml/fraction. The peak of activity,
eluted at pH 4.0, is included in a single, or at most
two tubes. Taken from Thomas et al.[28] and reproduced by permission of the American Society of
Biological Chemists.

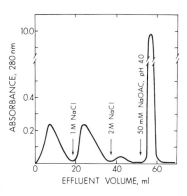

The acidification prior to the CM-cellulose separation (step 3) causes dissociation of the complex to the individual subunits, which are of considerably lower molecular weight (27,500 in the case of γ-subunit).

An alternative procedure is to dissolve the protein in 0.05 M ammonium formate (pH 4.0), and load it on a column of Sephadex G-100
equilibrated in the same buffer.[28] The potential advantage of this step is
that degradation, either autocatalytic or from other serine proteases or
carboxypeptidases, should be retarded. Furthermore, the volatile buffer
can be lyophilized directly without dialysis.

Step 5. Affinity Chromatography. To remove the remaining contaminants, affinity chromatography on benzamidine–Sepharose is utilized.[28]
The resin can be synthesized by linking p-aminocaproic acid, the spacer
arm, to agarose beads following cyanogen bromide activation. The amino
group of p-aminobenzamidine can be condensed with the free carboxyl
group of the spacer arm by the water-soluble reagent 1-ethyl-3-
(3-dimethylaminopropyl)carbodiimide to form a stable amide linkage.[29]
Alternatively, the resin can be purchased (Pierce Chemical Company,
Rockford, Illinois). The protein (20 mg) is dissolved in a few milliliters
of 0.1 M Tris-HCl (pH 7.5), centrifuged, and the supernatant loaded
onto enough resin to bind at least 1.5 times the amount of protein loaded.
The typical capacity of this resin varies from about 5 to 15 mg of bound
protein/ml of packed gel, depending on the extent of ligand substitution.
As shown in Fig. 4, following the breakthrough peak, nonspecifically
bound protein can be removed by 1 M and 2 M NaCl rinses. The γ-subunit
is eluted with 50 mM sodium acetate, pH 4.0. All of the catalytic activity
(assayed as hydrolysis of BAPNA) and 90 to 95% of the protein is recov-

[28] K. A. Thomas, R. E. Silverman, I. Jeng, N. C. Baglan, and R. A. Bradshaw, *J. Biol. Chem.* **256**, 9147 (1981).
[29] H. F. Hixson, Jr. and A. H. Nishikawa, *Arch. Biochem. Biophys.* **154**, 501 (1973).

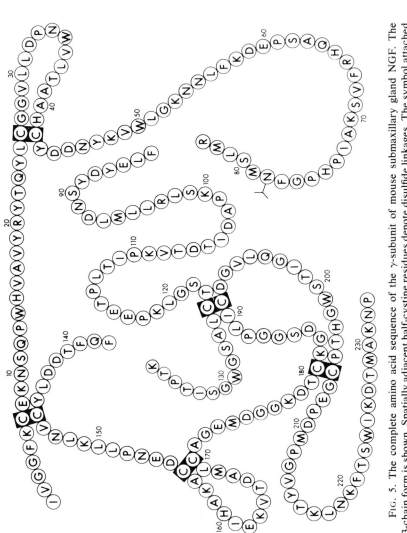

Fig. 5. The complete amino acid sequence of the γ-subunit of mouse submaxillary gland NGF. The 3-chain form is shown. Spatially adjacent half-cystine residues denote disulfide linkages. The symbol attached to Asn-78 represents a polysaccharide side-chain. Symbols for the amino acid residues are alanine (A), arginine (R), asparagine (N), aspartic acid (D), cysteine (C), glutamine (Q), glutamic acid (E), glycine (G), histidine (H), isoleucine (I), leucine (L), lysine (K), methionine (M), phenylalanine (F), proline (P), serine (S), threonine (T), tryptophan (W), tyrosine (Y), and valine (V). Taken from Thomas *et al.*[14] and reproduced by permission of the American Society of Biological Chemists.

ered in the last peak. Any enzymatic activity observed with BAPNA in any of the other peaks indicates that column overloading is likely. This can be tested by reloading the active pool on the benzamidine column after acetate elution and reequilibration with the initial buffer. Polyacrylamide-gel electrophoresis in SDS of the breakthrough and NaCl rinse pools reveal higher MW chains than the constituent γ-subunit polypeptides. The γ-subunit, purified in this fashion, appears, on electrophoresis under nondenaturing conditions, to be essentially identical to that obtained directly from the 7 S complex.[14,25,30] Further, on mixing with α- and β-subunits, it forms the characteristic HMW NGF complex.[10,11]

Properties

The complete amino acid sequence[14] of the γ-subunit (with the assumed disulfide pairing) is given in Fig. 5. All forms of the γ-subunit isolated to date contain a proteolytic cleavage following Arg-83, and some of the molecules contain a second endoproteolytic cleavage following Lys-136, resulting in a mixture of 2- and 3-chain forms of the molecule. All subunits contain the amino-terminal B_1 chain. The 2-chain forms also contain a 150-residue A chain, whereas the 3-chain forms have a C chain followed by a 97-residue B_2 chain. In addition, the carboxyl-terminal basic residues are removed from a variable portion of the B_1 and C chains *in vivo* or during purification, presumably by an endogenous exoprotease with carboxypeptidase B activity. As a result, the B_1 and C chains are 82–83 and 52–53 residues long, respectively. This heterogeneity results in multiple electrophoretic banding at three isoelectric points, 5.8, 5.6, and 5.25.[30] At these pI's, both the basic side-chains and the amino- and carboxyltermini generated by the endoproteolytic cleavages are fully charged. Thus the presence or absence of either or both of the 2 terminal basic residues, Arg-83 and Lys-136, account for the different isoelectric points. At pH 7.05, 5 major bands have been resolved.[30] At this pH, the cleavage of a peptide bond generates a net negative charge, since the newly exposed carboxyl group has a full negative charge, whereas the α-amino group, being nearer to its pK_a of about 7.8, is only partially charged. The six permutations of these endo- and exoproteolytic cleavages can account for these 5 bands, since forms 4 and 5 in Fig. 6 would, in the absence of local pK_a perturbations, be identical. The molecular weight of the γ-subunit polypeptide chains of the largest of the microheterogeneous forms is 25,900 (without the carbohydrate).

A polysaccharide side-chain has been located on Asn-78 of the aminoterminal B_1 peptide. Although it is not clear whether the polysaccharide

[30] R. W. Stach, A. C. Server, P.-F. Pignatti, A. Piltch, and E. M. Shooter, *Biochemistry* **15**, 1455 (1976).

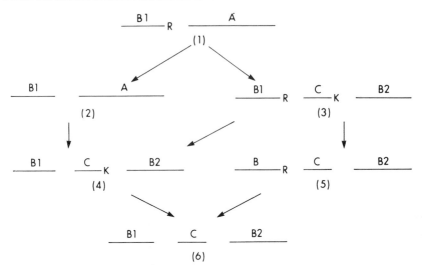

FIG. 6. Schematic diagram of the γ-subunit charge heterogeneity. Two-chain forms (1 and 2) are composed of the 82- to 83-residue amino-terminal B_1 chain followed by the 150-residue A chain. Endoproteolytic cleavage of the A chain generates the 52- to 53-residue C chain and the carboxyl-terminal 97-residue B_2 chain (forms 3–6). The chain-length variation of the B_1 and C chains results from the presence or absence of carboxyl-terminal basic residues. Those polypeptide chains containing carboxyl-terminal arginine (R) or lysine (K) residues are so indicated. The possible pathways for generating the various permutations of the endo- and exopeptidase cleavages are indicated by arrows. Taken from Thomas *et al.*[28] and reproduced by permission of the American Society of Biological Chemists.

unit is homogeneous, a tentative composition of 3 *N*-acetylglucosamine and 6 mannose residues has been obtained. Little, if any, sialic acid was detected, suggesting the polysaccharide does not make a significant contribution to the charge heterogeneity. The above carbohydrate composition would contribute an average of about 1600, making the total estimated molecular weight about 27,500. The carbohydrate structure determination, currently in progress, should clarify the size and heterogeneity of the polysaccharide component.

Substrate Specificity and Inhibition

The γ-subunit acts as an endoprotease with a primary specificity for bonds formed by the carboxyl group of arginine residues. However, the rate at which it degrades casein is only about 1% of the rate obtained with bovine pancreatic trypsin.[24] The products of the digestion of proteins or large polypeptides have not been characterized. Peptide cleavages at arginine as well as some lysine residues in a single large polypeptide have been observed, but detailed characterization was not carried out. Two

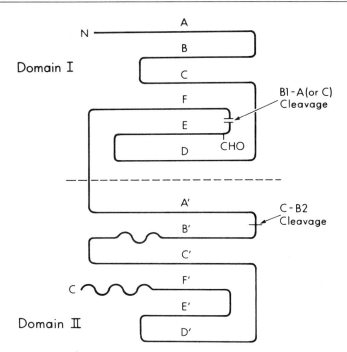

FIG. 7. Schematic representation of the presumed secondary structure of the mouse submaxillary gland NGF γ-subunit. The 2-domain structure and arrangement of β-strands found from crystal structure analysis of other serine proteases is assumed. The β-strands are sequentially lettered in each domain as they occur in the sequence with the lettering in the repeated pattern in the second domain distinguished by a prime. In the tertiary structure, the A and D and the A' and D' strands are adjacent so as to close the two β-cylinders. The two wavy lines following the B' and F' strand indicate helices. Endoproteolytic cleavages occur at turns between β-strands. The B_1 chain consists of strands A through E. The C chain is composed of strands F and A'. The B_2 chain contains strands B' through F'. The A chain, therefore, consists of strands F through F'. The carbohydrate chain is denoted as CHO. Taken from Thomas et al.[14] and reproduced by permission of the American Society of Biological Chemists.

mouse submaxillary gland proteases have similar specificity,[31] but their reported amino acid compositions[32] distinguish them from the γ-subunit.[14]

The γ-subunit has been shown to cleave amides and esters of low-molecular-weight substrates of both arginine and lysine.[16] The pH optimum for catalysis, measured with N^α-Bz-L-Arg-ethyl ester, is 8.7.[16]

In the high-molecular-weight complex (7 S NGF), the γ-subunit is inactive.[27] Since the carboxyl-terminal arginine residues of the β-subunit chains are required for formation of both the stable $\beta\gamma_2$[20] and the 7 S

[31] I. Schenkein, M. Levy, E. C. Franklin, and B. Frangione, Arch. Biochem. Biophys. 182, 64 (1977).

[32] M. Levy, L. Fishman, and I. Schenkein, this series, Vol. 19, p. 672.

complexes,[33] it has been proposed that these side-chains are positioned in the active site of the enzyme. The catalytic activity of the 7 S complex shows a lag phase caused by the dissociation of the complex in the presence of substrate. The pure γ-subunit does not show such behavior.[16,24] The free enzyme is also inhibited by bovine pancreatic trypsin inhibitor. However, other inhibitors, including soybean and lima bean trypsin inhibitors, ovomucoid, and human α_1-antitrypsin, antithrombin III, and C_1 esterase inhibitors, do not affect catalysis.[34]

The enzyme activity is abolished by diisopropyl fluorophosphate, indicating that the enzyme is a serine protease.[20] Benzamidine is reported to be an inhibitor[27] but no K_I value has been published. However, the enzyme is effectively purified using an affinity-chromatographic purification with benzamidine–Sepharose (see previous discussion).[28] Zinc ions have also been reported to inhibit the enzyme with a K_I of 8×10^{-7} M.[15] Whether or not the zinc ions in the high-molecular-weight complex might directly participate in the inhibition of the γ-subunit catalytic activity has not been determined.

Structure–Function Considerations. The γ-subunit is 10 residues longer than, and has about 40% residue identity with, bovine pancreatic trypsin. The active-site serine residue, Ser-184, and the structurally adjacent histidine and aspartic acid residues of the "charge-relay system," are conserved as in all other serine proteases. The binding-site aspartic acid residue at position 179 is in a homologous location to the aspartic acid in trypsin that binds either the competitive inhibitor benzamidine[35] or the basic residue side-chain of substrates. The γ-subunit has 5 disulfide bonds, assumed to be connected as shown in Fig. 5. These are a subset of the 6 disulfide bridges present in trypsin.

Using a schematic representation of the β-strand arrangement in the 2-domain serine protease structure as shown in Fig. 7, the two cleavages can be located between β-strands, E and F, and A' and B'. The first cleavage, present in all purified γ-subunits, is adjacent to the carbohydrate-attachment site in a region containing a 7-residue insertion relative to trypsin. Given a tertiary conformation very similar to trypsin, these cleavage sites occur at opposite sides of the substrate binding site.

[33] J. B. Moore, Jr., W. C. Mobley, and E. M. Shooter, *Biochemistry* 13, 833 (1974).
[34] A. M.-J. Au and M. F. Dunn, *Biochemistry* 16, 3958 (1977).
[35] W. Bode and P. Schwager, *J. Mol. Biol.* 98, 693 (1975).

[47] Sperm Acrosin[1]

By WERNER MÜLLER-ESTERL and HANS FRITZ

Acrosin is a spermatozoal endoproteinase with trypsin-like cleavage specificity.[2] It is found in spermatozoa of vertebrates[3] (including mammals, birds, and amphibians) and nonvertebrates (e.g., sea urchins[4]). Acrosin is located in the acrosome, a modified lysosome covering the anterior part of the sperm head.[5] The main physiological function ascribed to acrosin is the proteolysis of the zona pellucida, a glycoprotein layer surrounding the oocyte.[6] Thus penetration of the spermatozoon through the innermost investments of the ovum seems to be facilitated.

Acrosin belongs to the serine proteinase family. It selectively splits Arg-X and Lys-X bonds in natural and synthetic substrates.[7] Inactive precursors, probably zymogens, are found in the extracts of testicular, epididymal, and ejaculated spermatozoa.[8] Present evidence suggests that acrosin is a peripheral membrane protein associated with the acrosomal membranes[9,10] that exhibits some features not common to other serine proteinases.[11]

Assay Procedure

Amidase Activity Assay

The amidase activity assay of Schleuning and Fritz[2] with N^{α}-benzoyl-L-arginine-p-nitroanilide as substrate is routinely used with one minor modification: The assay buffer (0.2 M triethanolamine, pH 7.8) includes

[1] This work is supported by the Deutsche Forschungsgemeinschaft, Grant Mu 598/1-3.

[2] W. D. Schleuning and H. Fritz, this series, Vol. 45, p. 330.
[3] D. B. Morton, *in* "Proteinases in Mammalian Cells and Tissues" (A. J. Barrett, ed.), p. 445. North-Holland Publ., Amsterdam, 1977.
[4] A. E. Levine and K. A. Walsh, *J. Biol. Chem.* **225**, 4814 (1980).
[5] W. B. Schill, W. D. Schleuning, H. Fritz, V. Wendt, and N. Heimburger, *Naturwissenschaften* **62**, 540 (1975).
[6] R. Stambaugh, B. G. Brackett, and L. Mastroianni, *Biol. Reprod.* **1**, 223 (1969).
[7] H. Schiessler, W. D. Schleuning, and H. Fritz, *Hoppe-Seyler's Z. Physiol. Chem.* **356**, 1931 (1975).
[8] K. L. Polakoski and L. J. D. Zaneveld, this series, Vol. 45, p. 325.
[9] C. R. Brown and E. F. Hartree, *Hoppe-Seyler's Z. Physiol. Chem.* **357**, 57 (1976).
[10] W. Müller-Esterl and H. Fritz, *Hoppe-Seyler's Z. Physiol. Chem.* **361**, 1673 (1980).
[11] W. Müller-Esterl, B. Zippel, and H. Fritz, *Hoppe-Seyler's Z. Physiol. Chem.* **361**, 1381 (1980).

50 mM CaCl$_2$ · H$_2$O and 0.01% (w/v) Triton X-100 to obtain maximum activity of the enzyme and to prevent its adsorption on the surface of plastic cuvettes.[10] Under these conditions, 1.5–15 mU (0.04–0.4 μg) of acrosin are needed for a convenient assay procedure. Even less proteinase activity may be applied when microcuvettes are used (0.5–5 mU per assay).

Active Enzyme Staining

Principle

A simple technique is suitable for the localization of amidases in gels.[12] N^{α}-Benzoyl-D,L-arginine-β-naphthylamide serves as acrosin substrate. Coupling of the leaving group, 2-naphthylamine, to the diazo component, Fast Garnet Blue (o-aminoazotoluene diazonium salt), leads to a bright purple color of the resulting diazo dye.

Reagents

Wash solution: 13.8 g of NaH$_2$PO$_4$ · H$_2$O is dissolved in 250 ml of distilled water, adjusted to pH 7.5 with 1 M NaOH, and filled up to 1000 ml.

Dye solution: 35 mg of Fast Garnet Blue is dissolved in 40 ml of wash solution.

Substrate solution: 25 mg of N^{α}-benzoyl-D,L-arginine-β-naphthylamide is dissolved in 1 ml of dimethyl sulfoxide.

Incubation solution: To 19 ml of wash solution, 1 ml of substrate solution is added under gentle stirring.

Stability of the Solutions

Wash solution, substrate solution, and incubation solution are immediately prepared before use and maintained at 37°C.

Procedure

1. Prepare solutions during the final phase of the electrophoresis run.
2. Remove gels and place them in a suitable incubation vessel (tube, dish) containing 3–10 ml of fresh wash solution (10 min, 37°C).
3. Transfer the gels to a second incubation vessel containing 3 to 10 ml of incubation solution. Incubate gels until purple band(s) appear (10–120 min, 37°C).

[12] D. L. Garner, *Anal. Biochem.* **67**, 688 (1975).

4. Remove gels and place them in a vessel containing 7% (w/v) acetic acid to stop the reaction. Remove background stain by incubation in 7% (w/v) acetic acid for 12 hr with at least one change of 7% acetic acid.
5. Scan gels at 580 nm for densitometric analysis.

Comments

The procedure given presents a rapid and versatile technique for the localization of acrosin in various gels (polyacrylamide, agarose) after electrophoresis,[12] immunoelectrophoresis, and isoelectric focusing. Proenzymes are also detected if they are activated autoproteolytically, whereas enzyme–inhibitor complexes develop purple color very slowly, if at all. Quantitative evaluation of proteinase activities is not indicated under the conditions outlined. The color of the diazo dye is fading after 1 week of storage. Gloves should be used throughout as 2-naphthylamine is a carcinogenic aromatic amine.

Gelatin Substrate Film Technique

Principle

This technique is convenient and suitable for assessing the proteolytic (gelatinolytic) activity of individual spermatozoa.[13] Incubation of sperm cells on ultrathin gelatin films (1.5 μm) results in a disaggregation of the acrosome and hence in a liberation of acrosin. Proteolysis of the gelatin matrix leads to a visible halo formation around the anterior part of the spermatozoon. Incubation of spermatozoa with acrosin inhibitors prior to gelatin-film assay may completely prevent halo formation if the inhibitor has access to the acrosomal compartment.[14]

Reagents

Diluting buffer. 3.63 g of Tris(hydroxymethyl)aminomethane and 0.75 g of fructose are added to 90 ml of distilled water. The solution is adjusted to pH 6.9 with 1 N NaOH and filled up to 100 ml.
Incubation buffer. Same as diluting buffer, but pH adjusted to 8.1
Gelatin-film slides. Kodak fine-grain autoradiographic stripping plates AR.10 are used.[14] The film plates are exposed to daylight for 30 min. Then, film sections (0.5 × 0.5 cm) are cut off with a sharp blade. They are carefully transferred to the surface of a water bath

---

[13] P. Gaddum and R. J. Blandau, *Science* **170**, 749 (1970).
[14] V. Wendt, W. Leidl, and H. Fritz, *Hoppe-Seyler's Z. Physiol. Chem.* **356**, 1073 (1975).

(40°C, distilled water, free of dust). As the gelatin film swells it stretches out tight and flat. A slide prewarmed to 70°C is held at about 80 degrees from the horizontal, so that one end touches the (lower) gelatin layer first. The stripping layer will drape itself snugly over the glass slide, and most of the water will drain away as the slide is gradually lifted clear of the water. Avoid trapping of air bubbles between glass slide and gelatin layer. The glass slides are air-dried (1 hr, 25°C) and stored (4°C, wet chamber, free of dust) until used.

Stability

Under the conditions given, the gelatin-film slides are stable for at least 1 year. Buffer should be changed after storage for 1 month.

Procedure [15]

Preparation of spermatozoa proceeds as follows:

1. Determine sperm count of human ejaculates by light microscope 30 min to 3 hr postejaculation.
2. Adjust the sperm count to 4×10^7 spermatozoa/ml by adding an appropriate volume of diluting buffer to the semen.
3. Transfer 0.2 ml of the sperm suspension to a plastic tube and centrifuge (600 g, 10 min, 4°C).
4. Discard the supernatant and resuspend the sperm pellet in 0.2 ml of dilution buffer containing 1 mM acrosin inhibitor. Incubate for 30 min at 37°C.
5. Centrifuge (600 g, 10 min, 25°C), discard the inhibitor-containing supernatant, and wash the pellet with 10 ml of diluting buffer. Centrifuge again (600 g, 10 min, 25°C).
6. Discard the supernatant and resuspend the resultant sperm sediment in 2 ml of diluting buffer to a final sperm count of 4×10^6 spermatozoa/ml.

Steps 4 and 5 are omitted when native spermatozoa are subjected to the gelatin substrate film assay.

Gelatin lysis proceeds as follows:

1. Pipette 20 μl of incubation buffer (pH 8.1) on gelatin film (1.5 × 1.5 cm) and distribute it over the layer.
2. Preincubate the gelatin film in a closed chamber (5 min, 37°C, 94% relative humidity).

[15] W. B. Schill, M. Feifel, H. Fritz, and I. Hammerstein, *Int. J. Androl.* **4**, 25 (1981).

3. Pipette 20 μl of the washed and preincubated sperm suspension (corresponding to 8×10^4 spermatozoa) on the gelatin layer. Distribute the cell suspension evenly over the surface of the film.
4. Incubate the slides in a closed chamber (30 min to 2 hr, 37°C, 94% relative humidity).
5. Remove slides and air-dry them (10 min, 25°C) in order to stop gelatin lysis.
6. Evaluate gelatin lysis by light microscopy (40-fold magnification) and determine the diameter of the halos by an ocular micrometer (400-fold magnification).

Comments

The foregoing is a simple test suitable for the continuous observation of the inherent proteolytic activity of spermatozoa. A semiquantitative evaluation is convenient, since the mean lysis area corresponds sufficiently with the average acrosin activity of a single spermatozoon.[16] This is also a useful technique for the investigation of inhibitor diffusion through intact acrosomal membranes.[14]

Spermatozoa of different species (man, bull, boar, cock) have been successfully tested. Rapid staining of the gelatin layer with 0.05% (w/v) methylene blue is optional.

Purification Procedure

On the basis of the isolation protocol of Schleuning and Fritz,[2] a large-scale purification procedure has been developed that ensures high cumulative yield and isolation of a well-defined form of boar acrosin.[17] The following data and properties are related to boar acrosin unless otherwise stated.

Step 1. Semen Collection. Semen collected from healthy boars by an artificial vagina is diluted with an equal volume of 0.5 M sucrose to a final volume of 200–600 ml. The suspension is centrifuged at 500 g for 20 min at 4°C. (The following extraction and purification is performed at 4°C throughout if not otherwise stated.) The supernatant seminal plasma is removed, the sperm pellet washed with an equal volume of 0.25 M sucrose–0.15 M NaCl–50 mM trishydroxymethylaminomethane (pH 8.0) and the suspension again centrifuged (600 g, 20 min). The sperm sediment is shock-frozen in liquid nitrogen (-196°C) and stored at -20°C until used.

[16] V. Wendt, W. Leidl, and H. Fritz, *Hoppe-Seyler's Z. Physiol. Chem.* **356**, 315 (1975).
[17] W. Müller-Esterl, S. Kupfer, and H. Fritz, *Hoppe-Seyler's Z. Physiol. Chem.* **361**, 1811 (1980).

Step 2. Extraction Procedure. Following thawing at room temperature, the sperm pellet is extracted with an equal volume of 0.3 M acetic acid—50 mM NaCl–3 mM NaN$_3$, pH 2.6. If necessary, the pH of the resultant suspension is readjusted to 2.6 by dropwise addition of glacial acetic acid. After 15 min incubation at 25°C, the suspension is centrifuged (2000 g, 15 min) and the supernatant extract decanted. This extraction procedure is repeated three times. The opaque extracts are combined, centrifuged (30,000 g, 30 min), and the resulting clear supernatant is removed from the cell debris.

Step 3. Gel Filtration. The crude extracts of one to two ejaculates (300–400 ml) are concentrated by ultrafiltration (Amicon chamber, PM-10 membrane) to a final protein concentration of approximately 80 mg/ml. 135 g of Sephadex G-100, preswollen in 2.7 liters of 0.3 M acetic acid–50 mM NaCl–3 mM NaN$_3$ (pH 2.6), is poured into a plastic column (5 × 120 cm). A flow rate of 0.8 ml/min is adjusted and fractions of 10 ml are collected. A_{280} is monitored photometrically throughout the isolation procedure. 20 ml of the concentrated extract are applied to the column by means of an adapter. The column is developed with 0.3 M acetic acid–50 mM NaCl–3 mM NaN$_3$, pH 2.6. The eluted fractions are tested for amidase activity. Fractions comprising active acrosin are pooled and further processed.

Elution of acrosin inhibitors is monitored by an inhibitor assay using purified acrosin. Fractions containing acrosin inhibitors are pooled and processed separately.

Step 4. Hydrophobic Chromatography. An equal volume of 2 M NaCl is added dropwise to the eluate from gel filtration (600 ml). The pH of the resultant solution is adjusted to 5.5 with 2 M NaOH. 130 ml of a suspension of phenyl-Sepharose in 1 M NaCl–0.1 M sodium acetate–3 mM NaN$_3$, pH 5.5 (starting buffer) is added under gentle stirring. If necessary, additional phenyl-Sepharose is added until apparently 99% of acrosin is bound to the gel. Following incubation for at least 6 hr, the supernatant is decanted and the gel transferred to a plastic column (2.5 × 25 cm) prefilled with starting buffer. The gel is washed with 0.1 M NaCl–0.1 M sodium acetate–3 mM NaN$_3$, pH 5.5 (washing buffer) for 6 hr (flow rate 0.5 ml/min, fraction volume 8 ml). Less than 5% of the starting activity (100%) is eluted under these conditions. Then the column is developed by washing buffer including 1% (w/v) Triton X-100. Acrosin is eluted in a single sharp peak.

Step 5. Ion-Exchange Chromatography. Nine grams of CM-Sephadex C-50, preswollen in 200 ml of 0.15 M NaCl–0.1 M sodium acetate–0.1% (w/v) Triton X-100–3 mM NaN$_3$, pH 3.6 (starting buffer), is added to the active fractions of the phenyl-Sepharose eluate (530 ml) under gentle stirring. If necessary, additional CM-Sephadex is added until apparently 99% of acrosin is bound to the gel. The supernatant is decanted and the gel suspension is poured into a plastic column (2.5 × 40 cm). The column is washed with starting buffer overnight (flow rate 0.5 ml/min, fraction volume 8 ml) and then developed with a linear sodium chloride gradient ranging from 0.15 to 1.25 M (2 liters total volume applied). Acrosin is eluted in a symmetrical peak at 0.9 M NaCl.

Step 6. Affinity Chromatography. Crystalline triethanolamine HCl and NaCl is added to the combined active fractions from CM-Sephadex (740 ml) to a final concentration of 0.2 M (triethanolamine) and 1 M (NaCl), respectively. The pH of the resulting buffer is shifted to 7.8 by dropwise addition of 2 N NaOH. Then the solution is rapidly applied (flow rate 1.7 ml/min) to a plastic column (1.5 × 15 cm) filled with 9 g of benzamidine–CH-Sepharose[2] preswollen in 30 ml of 0.2 M triethanolamine–1 M NaCl–0.1% Triton X-100, pH 7.8 (starting buffer). The gel is washed with starting buffer for 6 hr (flow rate 0.8 ml/min, fraction volume 3 ml). Less than 1% of the applied activity is eluted under these conditions. The column is developed with 0.1 M glycine–0.1 M NaCl–0.1% (w/v) Triton X-100, pH 3.0. A single sharp peak of acrosin activity emerges at pH 3.4.

Step 7. Desalting and Removal of Detergents. The eluate from affinity chromatography (33 ml) is concentrated to a final volume of 10 ml (Amicon chamber, PM-10 membrane) and applied to a plastic column (2.5 × 48 cm) filled with 240 ml of a suspension of Merckogel PGM-2000 in 1 mM ammonium formate, pH 3.5 (desalting solution). The column is developed with desalting solution (flow rate 0.5 ml/min, fraction volume 5.2 ml). Acrosin is excluded from the gel and emerges as a single protein peak separated from salts. Triton X-100 is strongly retarded by the gel; hence the bulk of detergents is removed during this procedure. The active fractions are combined and lyophilized using siliconized glass vessels. A white protein powder is thus obtained.

Comments

A summary of the purification procedure is given in Table I. In a typical experiment, from one processed ejaculate containing 1.08×10^{11} spermatozoa (range $(0.31–1.78) \times 10^{11}$ cells), 4 mg of purified acrosin is obtained, with a cumulative yield of 41% (range 23.8–72.2%) and a 162-fold (range 96- to 162-fold) purification. Addition of Triton X-100 was found to enhance the recoveries in the chromatographic procedures and to stabilize the purified enzyme.[10] A critical step of the protocol is the binding of acrosin to the affinity gel at pH 7.8, as autoproteolysis proceeds

TABLE I
PURIFICATION PROCEDURE OF BOAR ACROSIN

Purification step	Volume[a] (ml)	Total protein (mg)	Protein concentration (mg/ml)	Total activity[b] (U)	Specific activity (U/mg)	Acrosin concentration[c] (mg/ml)	Recovery[d] (%)	Purification factor (-fold)	Detergent concentration [% (w/v)]
Extraction	190	1610	8.5	365[e]	0.23	0.05	—	1.0	—
Sephadex G-100	600	814	1.36	365	0.45	0.02	100	1.9	—
Phenyl-Sepharose	530	503	0.95	263	0.52	0.01	72	2.3	1.0
SP-Sephadex C-50	740	192	0.26	240	1.25	0.01	91	5.4	0.1
Benzamidine–Sepharose	33	4.3	0.13	182	42.3	0.15	76	184	0.1
Merckogel PGM-2000	51	4.0	0.08	148	37	0.08	81	161	<0.005

[a] Extraction or elution volume.
[b] Using BzArg-NHNp as substrate, 25°C.
[c] Calculated on the basis of a specific activity of 37 U/mg.
[d] For each step.
[e] As judged after gel filtration because acrosin inhibitors are present in the crude extract.

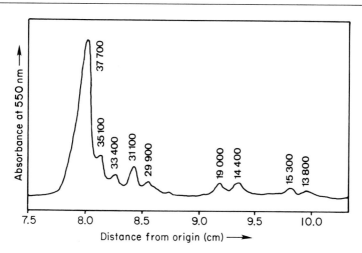

FIG. 1. Densitometric analysis of the (auto-)proteolytic fragment profile of boar acrosin. 20 μg of acrosin was reduced, carboxymethylated, and applied to a 5–20% polyacrylamide gradient gel in the presence of 0.1% sodium dodecyl sulfate.

very rapidly under these conditions. However, this purification step cannot be omitted because affinity chromatography is the most efficient purification step of the described isolation procedure (Table I).

Properties

Purity

Homogeneity of the acrosin preparation is routinely assessed by disc electrophoresis at pH 5.0,[11] revealing 1 protein band (Coomassie Brilliant Blue staining) and a corresponding amidase band (active-enzyme staining). Sodium dodecyl sulfate electrophoresis in a gradient gel[18] shows 1 predominant 38,000-MW band under reducing conditions and variable amounts (5–15%) of fragments due to (auto-)proteolysis (Fig. 1). Prior inactivation of acrosin by diisopropyl fluorophosphate efficiently reduces the amount of the low-molecular-weight fragments.[11]

Stability

Lyophilized acrosin is stable at 4°C for at least 2 years without considerable loss of activity. Purified acrosin is stable at pH 3.0 only in the presence of 0.1% Triton X-100 for at least 20 days at 4°C (with 96% of the

[18] U. K. Laemmli, *Nature* (*London*) **227**, 680 (1970).

TABLE II
AMINO ACID COMPOSITION OF BOAR, HUMAN,
AND RAM ACROSIN EXPRESSED IN
RESIDUES PER MOLECULE

Amino acid	Boar acrosin	Human[a] acrosin	Ram[b] acrosin
Asx	23	31	24
Thr	20[c]	19	19
Ser	18[c]	25–26	25
Glx	27	30–32	29
Pro	40	31–34	36
Gly	34	38	31
Ala	19	27	25
Cys	8	8	12
Val	24	22	26
Met	6[c]	4	4
Ile	17	21	18
Leu	21	22	21
Tyr	10	10	10
Phe	11	10–11	10
His	6	6–7	9
Lys	16	18	16
Arg	25	19	25
Trp	3[d]	—	3
Total	328	341–349	343

[a] R. Mindermann and W. Müller-Esterl, unpublished results.
[b] C. R. Brown and E. F. Hartree, *Biochem. J.* **175**, 227 (1978).
[c] Values extrapolated to zero time.
[d] From sequence determinations.[21]

starting activity remaining). At pH 5.5, acrosin is stable for at least 6 months at 4°C in the presence of 0.1% Triton X-100 (with 97% of the starting activity remaining).

Molecular Parameters

The *molecular weight* of boar acrosin is close to 38,000, as judged by different techniques.[11] A carbohydrate content of 8.3% has been calculated.[19] Amino acid analyses (Table II) reveal that the protein backbone comprises 328 amino acid residues corresponding to MW 35,729 for the protein moiety.[17]

[19] R. F. Klauser, Ph.D. Thesis, Technical University of Munich (1977).

TABLE III

N-Terminal Sequence of Boar Acrosin, Human Plasmin,[a] Bovine Chymotrypsin,[b] and Bovine Trypsin[c,d]

Residues 1–19 (positions 1, 5, 10, 15 marked):

	1				5					10					15				
Acrosin (boar)	Val	Val	Gly	Gly	Met	Ser	Ala	Glu	Pro	Gly	Ala	Trp	Pro	Trp	Met	Val	Ser	Leu	Gln
Plasmin (human)	Val	Val	Gly	Gly	Cys	Val	Ala	His	Pro	His	Ser	Trp	Pro	Trp	Gln	Val	Ser	Leu	Arg
Chymotrypsin (bovine)	Ile	Val	Asn	Gly	Glu	Glu	Ala	Val	Pro	Gly	Ser	Trp	Pro	Trp	Gln	Val	Ser	Leu	Gln
Trypsin (bovine)	Ile	Val	Gly	Gly	Tyr	Thr	Cys	Gly	Ala	Asn	Thr	Val	Pro	Tyr	Gln	Val	Ser	Leu	Asn

Residues 20–39 (positions 25, 30, 35 marked):

	20					25					30					35				
Acrosin	Ile	Phe	Met	Tyr	His	Asn	Asn	Arg	Arg	Tyr	His	Thr	Cys	Gly	Gly	Ile	Leu	Leu	Asn	Ser
Plasmin	Thr	Arg	Phe	Gly	—	—	—	—	—	Met	His	Phe	Cys	Gly	Gly	Thr	Leu	Ile	Ser	Pro
Chymotrypsin	Asp	Lys	Thr	Gly	—	—	—	—	—	Phe	His	Phe	Cys	Gly	Gly	Ser	Leu	Ile	Asn	Glu
Trypsin	—	—	Ser	Gly	—	—	—	—	—	Tyr	His	Phe	Cys	Gly	Gly	Ser	Leu	Ile	Asn	Ser

Residues 40–52 (positions 40, 45, 50 marked):

	40					45					50		
Acrosin	His	Trp	Val	Leu	Thr	Ala	Ala	His	Cys	Phe	Lys	Asn	Lys
Plasmin	Glu	Trp	Val	Leu	Thr	Ala	Ala	His	Cys	Leu	Glu	Lys	Ser
Chymotrypsin	Asn	Trp	Val	Val	Thr	Ala	Ala	His	Cys	Gly	Val	Thr	Thr
Trypsin	Gly	Trp	Val	Val	Ser	Ala	Ala	His	Cys	Tyr	Lys	Ser	Gly

[a] B. Wiman, *Eur. J. Biochem.* **76**, 129 (1977).
[b] D. M. Shotton and B. S. Hartley, *Nature (London)* **225**, 802 (1970).
[c] O. Mikeš, V. Holeysovsky, V. Tomašek, and F. Šorm, *Biochem. Biophys. Res. Commun.* **24**, 346 (1966).
[d] Identical residues are given in italics.

From the amino acid compositions, a striking *similarity between acrosins of different species* is obvious (Table II). A high molar percentage of proline (10.9%) and arginine (7.64%) in boar acrosin is remarkable. A polarity index of 41.2% is calculated according to Capaldi.[20] An isoelectric point of 10.5 is assessed by isoelectric focusing in an analytical gel. At 280 nm a specific extinction ($E_{280}^{1\%}$) of 16.8 is measured.

The *sequence* of 52 N-terminal amino acids of boar acrosin has been elucidated (Table III).[21] The amino-terminal sequence Val-Val-Gly-Gly typical for activated serine proteinases is found. Hence the 38,000-MW form of boar acrosin, which is the final product of the given isolation procedure, represents the active form of the proteinase. A strong structural homology to human plasmin and bovine chymotrypsin, with 48% amino acid residues being in identical positions, is obvious.

An extensive *self-association* of boar acrosin is observed at low pH after complete removal of detergents.[11] Molecular weights of the oligomers (monomer to hexamer) have been determined electrophoretically by the technique of Ferguson.[11]

Stimulation of the acrosin activity up to 260% is found in the presence of nonionic *detergents*.[10] These data correspond well with the maximum specific activity found for boar acrosin prepared in the absence (14.5 U/mg) and in the presence of detergents (38 U/mg). On the other hand, the activity of the enzyme is impaired in the presence of *phospholipid vesicles*.[10] On the basis of these results it has been concluded that acrosin has hydrophobic binding sites for amphiphiles.[10]

Present evidence suggests that acrosin is a *peripheral membrane protein*.[9,10] Immunofluorescent studies indicate that acrosin is associated with the acrosomal membranes, most probably with the inner acrosomal membrane.[3,5] An average content of 2.4×10^{-18} mol acrosin/spermatozoon (boar) is calculated. Assuming a volume of 1.5×10^{-15} liter for the acrosomal compartment,[9,22] a local acrosin concentration of 1.6 mM results (for comparison, an average concentration of 5.1 mM is found for hemoglobin in erythrocytes). Hence a powerful proteolytic potential is located at the anterior part of the spermatozoon.

[20] R. A. Capaldi and G. Vanderkooi, *Proc. Natl. Acad. Sci. U. S. A.* **69**, 930 (1972).
[21] R. Fock-Nüzel, F. Lottspeich, A. Henschen, W. Müller-Esterl, and H. Fritz, *Hoppe-Seyler's Z. Physiol. Chem.* **361**, 1823 (1980).
[22] C. R. Brown and E. F. Hartree, *J. Reprod. Fertil.* **36**, 195 (1974).

[48] Invertebrate Proteases

By ROBERT ZWILLING and HANS NEURATH

Much less is known about invertebrate proteases than about mammalian or bacterial proteases, at least partly because of the difficulties of obtaining sufficient starting material. Invertebrates are relatively small animals and hence it is not possible to obtain the starting material in kilogram quantities or from bacterial cultures. Those invertebrate species from which proteases have been isolated in significant quantities are either relatively large (e.g., crayfish), form colonies (e.g., hornet), or secrete the protease-containing fluids in quantities (e.g., cocoonase) sufficient for biochemical investigations. The invertebrates comprise 9 of the 10 phyla and, for that reason alone, are of particular importance for the study of molecular evolution. The evolutionary paths that led to the present vertebrates and invertebrates diverged from one another 0.7 to 1×10^9 years ago. The transition from the invertebrate to the vertebrate proteases, therefore, traverses a relatively large evolutionary hiatus and hence affords a wide perspective on the history of development of a protein family.

In view of this rather long span of evolution, the investigation of invertebrate proteases can serve at least two aims: (1) One can anticipate larger and perhaps more interesting changes in the structure and function of the proteases; and (2) one might discover proteases among the invertebrates that have no counterpart among the vertebrates. In what follows, we present examples for both of these possibilities. For these reasons, and also because some invertebrate proteases have been reviewed elsewhere in this series (i.e., the cocoonases in Vol. XIX; and proacrosin, cortical granule proteases, and the hatching enzyme of the sea urchin in Vol. XLV), we have restricted the present review to a few well-characterized invertebrate proteases rather than aiming at a more complete coverage of the field.

Crayfish Trypsin

Crayfish trypsin (EC 3.4.21.4) is an invertebrate serine protease that has preserved many characteristics of serine proteases of the early stage of evolution. Since this invertebrate trypsin shares some essential structural properties with bacterial trypsin rather than with bovine trypsin, it might be considered a "missing link" between prokaryotic and mammalian serine proteases.[1]

---

[1] R. Zwilling, H. Neurath, and R. G. Woodbury, *Protides Biol. Fluids* **28**, 115 (1980).

Source of Enzyme

Crayfish trypsin—together with the two other principal digestive proteases, the low-molecular-weight protease (LMWP) and crayfish carboxypeptidase—is synthesized in the hepatopancreas of decapode crustacea and secreted into the cardia, a stomach-like organ from which the enzyme is isolated.

Crayfish trypsin was reported to occur in *Astacus fluviatilis*,[2] *Astacus leptodactylus*,[3] *Cambarus affinis*,[4,5] *Carcinus maenas*,[6,7] *Homarus americanus*,[8] *Orconectes virilis*,[9] and *Penaeus setiferus*.[10]

A review of crustacean digestive proteinases has been published by DeVillez.[11]

Crayfish trypsin is present in the dark brown digestive fluid of the cardia (approximately 50 mg protein/ml in a concentration of about 3 mg/ml).

From a stock of about 1000 crayfish, 1–2 liters of cardia fluid can be collected by the following simple method.[2,12] A curved glass capillary tube is introduced through the esophagus into the cardia of the animal. Since the cardia content seems to be under a slight positive pressure, 2–3 ml will spontaneously pass into the capillary tube. After centrifugation the cardia fluid can be stored in a frozen state for several months without loss of activity.

Assay Methods

Since the substrate specificity of crayfish trypsin toward synthetic peptide as well as protein substrates is identical to that of bovine trypsin, all assay methods used for vertebrate trypsin are also applicable to crayfish trypsin.[13]

[2] G. Pfleiderer, R. Zwilling, and H.-H. Sonneborn, *Hoppe-Seyler's Z. Physiol. Chem.* **348**, 1319 (1967).

[3] R. Linke, R. Zwilling, D. Herbold, and G. Pfleiderer, *Hoppe-Seyler's Z. Physiol. Chem.* **350**, 877 (1969).

[4] R. Kleine, *Z. Vergl. Physiol.* **55**, 51 (1967).

[5] R. Kleine, *Z. Vergl. Physiol.* **56**, 142 (1967).

[6] E. J. DeVillez and K. Buschlen, *Comp. Biochem. Physiol.* **21**, 541 (1967).

[7] D. Herbold, R. Zwilling, and G. Pfleiderer, *Hoppe-Seyler's Z. Physiol. Chem.* **352**, 583 (1971).

[8] H. Brockerhoff, R. J. Hoyle, and P. C. Hwang, *J. Fish Res. Board Can.* **27**, 1357 (1970).

[9] E. J. DeVillez, *Comp. Biochem. Physiol.* **14**, 577 (1965).

[10] B. J. Gates and J. Travis, *Biochemistry* **8**, 4483 (1969).

[11] E. J. DeVillez, in ''Freshwater Crayfish'' (J. W. Avault, Jr., ed.), p. 195. Louisiana State Univ. Press, Baton Rouge, 1975.

[12] R. Zwilling, G. Pfleiderer, H.-H. Sonneborn, V. Kraft, and I. Stucky, *Comp. Biochem. Physiol.* **28**, 1275 (1969).

[13] See K. A. Walsh and P. E. Wilcox, this series, Vol. 19 [3] and [4].

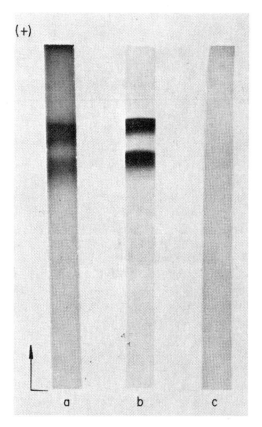

FIG. 1. Disc electrophoresis of crayfish trypsin (*A. leptodactylus*), demonstrating the two active components: (a) after incubation with N^α-benzoyl-L-arginine-β-naphthylamide stained with diazo blue; (b) stained with amido black; (c) diazo blue without the substrate (see text).

For the demonstration of tryptic activity in polyacrylamide gels, and especially for the detection of multiple forms of crayfish trypsin, the following procedure was developed.[12] Immediately after electrophoresis, the polyacrylamide gel is incubated with Bz-Arg-β-naphthylamide (5 mg in 1 ml of N,N-dimethylformamide and 9 ml of 0.05 M phosphate buffer, pH 8.8) for 15–60 min at 25°C. Hydrolysis of the substrate yields free naphthylamide, which is coupled with diazo blue B (diorthoanisidine, diazotized, 4 mg/ml of phosphate buffer), resulting in distinct bands of orange color (Fig. 1).

Ac-Phe-naphthyl ester is a still more sensitive substrate and may be used for the same purpose (red-violet color), although it is not exclusively specific for trypsin.

Purification

Methods developed for the large-scale purification of crayfish trypsin by anion-exchange chromatography and gel filtration provided a material of high purity at reasonable yield.[12] Since these methods were very time consuming, it seemed desirable to develop an efficient affinity-chromatography procedure.

Affinity procedures for the isolation of pancreatic trypsins proved to be inapplicable to crayfish trypsin, thus demonstrating marked differences between vertebrate and invertebrate trypsins. For instance, at pH 8.0, crayfish trypsin will readily bind to soybean-trypsin inhibitor coupled to CNBr–Sepharose-4B (SBTI column), but this enzyme–inhibitor complex cannot be redissolved at pH 3.0 or even below. Under similar conditions, bovine trypsin is eluted quantitatively from the affinity column in an active form. Moreover, native crayfish trypsin is completely inactivated by treatment at pH 3.0.

Initially these difficulties were overcome by eluting crayfish trypsin from the SBTI column by the application of 2 M benzamidine at pH 8. However, this procedure still is beset with several difficulties, including the need to remove benzamidine by dialysis, difficulties in photometric monitoring, and poor overall yield.

Finally, it was found that an unbuffered aqueous medium adjusted to pH 11.0 with a few drops of ammonia will assure quantitative elution of fully active crayfish trypsin from the SBTI column. Traces of ammonia are easily removed from the pure enzyme preparation by lyophilization. Both low ionic strength and high basicity of the eluting medium seem to contribute to the success of this procedure. This behavior is entirely contrary to that of bovine trypsin.

When under the above conditions, the crude digestive fluid from the cardia of A. fluviatilis is applied to a SBTI column without prior purification, an elution pattern as shown in Fig. 2 is observed.

Freeze-dried crayfish trypsin passed twice through this affinity column proved to be homogeneous, as judged by disc electrophoresis, immunological studies, and sequence analysis.[1]

Properties

Crayfish trypsin occurs in multiple forms, their number and electrophoretic mobility varying from species to species.[3,14] For instance, in A. fluviatilis 3 electrophoretic bands with distinct tryptic specificity are discernible, whereas in A. leptodactylus and in C. maenas this pattern is

[14] G. Pfleiderer and R. Zwilling, Naturwissenschaften 59, 396 (1972).

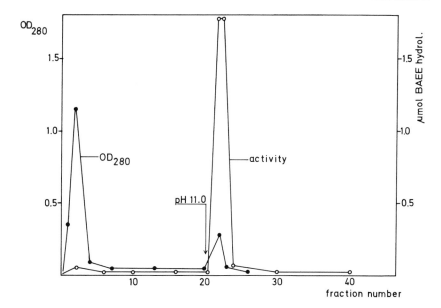

FIG. 2. Purification of crayfish trypsin from the cardia fluid of *A. fluviatilis* by affinity chromatography on CNBr-Sepharose 4B. Ligand, SBTl; elution, ammonia, pH 11.0.

different (Fig. 3A). Of these, only band II_f (*A. fluviatilis*) and band II_1 (*A. leptodactylus*) might be identical, since they are not separable by electrophoresis and are immunologically cross-reactive. The number of multiple trypsin bands is even larger when tested by the highly sensitive two-dimensional Laurell immunoelectrophoresis method, which reveals for *A. fluviatilis* 5 bands (Fig. 4A), whereas *A. leptodactylus* displays only two forms (I_1 and II_1) (Fig. 4B).

Thus in the three species, (*A. fluviatilis, A. leptodactylus,* and *C. maenas*), 11 multiple forms of trypsin are altogether present.[14] A similar situation is found for the low-molecular-weight protease (LMWP, see p. 651) of the crayfish (Fig. 3B). These observations might be related to multiple gene duplications in the crayfish genome.

Other immunological tests demonstrated that crayfish and bovine trypsin do not cross-react.[14] Even the relatively short evolutionary distance that separates the crayfish *Astacus* from the crab *Carcinus* is sufficient to abolish any detectable immunological relationships between the trypsins of these two organisms. On the other hand, the five forms of trypsin from *A. fluviatilis* and the two trypsins from *A. leptodactylus* are all mutually cross-reacting.[3]

The pH optimum for the activity of crayfish trypsin is in the range of

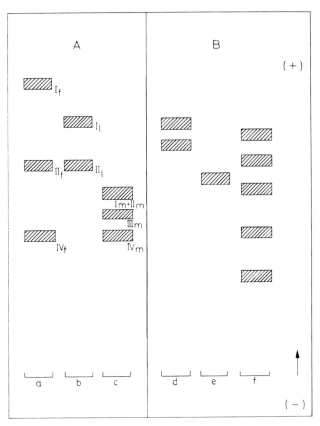

FIG. 3. Multiple forms of trypsin and low-molecular-weight protease (LMWP) from three decapode crustacea. (A) Trypsin from (a) *A. fluviatilis;* (b) *A. leptodactylus;* (c) *C. maenas.* (B) LMWP from (d) *A. fluvialitis;* (e) *A. leptodactylus;* (f) *C. maenas.* All bands shown are proteolytically active; all trypsin bands additionally are active toward benzoyl-arginyl-naphthylamide. Cellulose acetate membrane electrophoresis, pH 8.0, 50 min at 300 V and 2 mA.[14]

pH 7–8. In contrast to bovine trypsin, crayfish trypsin gradually loses activity at pH 5.0 and is immediately and irreversibly inactivated at pH 3.0. Moreover, crayfish trypsin is very resistant to self-digestion, even in the absence of $CaCl_2$. After incubation at pH 8.0 and 30°C for 2–3 days, 90–100% of the initial activity still remains.[12]

No qualitative differences between crayfish and bovine trypsin in substrate and inhibitor specificities have so far been detected. For instance, crayfish trypsin hydrolyzes the B chain of oxidized insulin in the same positions as bovine trypsin, that is, exclusively at Arg^{22}-Gly^{23}, and

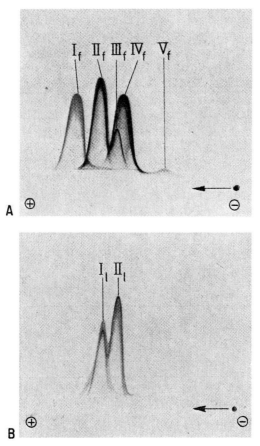

FIG. 4. Two-dimensional immunoelectrophoresis of crude digestive fluid from (A) *A. fluviatilis;* (B) *A. leptodactylus.* The samples were applied to a hole close to the cathode. First dimension: Linear electrophoretic separation in an agarose gel for 20–40 min at 250 V. Second dimension: Electrophoresis into antiserum–agarose for 12–14 hr at 100 V. Staining of tryptic precipitation lines by the chromogenic substrate benzoyl-L-arginine-β-naphthylamide.[14]

Lys^{29}-Ala^{30}. The A chain is not split by either enzyme.[15] This highly restricted tryptic specificity in combination with the marked enzymatic stability could render crayfish trypsin particularly useful for purposes of fragmentation of proteins prior to sequence analysis.

Crayfish trypsin is inactivated by all protein trypsin inhibitors. It is also inhibited by PMSF (phenylmethanesulfonyl fluoride), DFP (diiso-

[15] R. Zwilling and G. Pfleiderer, *FEBS-Symp.* **18**, 83 (1970).

TABLE I
AMINO ACID COMPOSITION OF INVERTEBRATE AND VERTEBRATE TRYPSINS

	Invertebrate	Vertebrate	
Amino acid	Crayfish[16]	Dogfish[17]	Bovine[18]
Lysine	5	5	14
Histidine	5	8	3
Arginine	2	7	2
Aspartic acid	30	24	22
Threonine	15	7	10
Serine	17	17	33
Glutamic acid	21	16	14
Proline	10	10	9
Glycine	28	28	25
Alanine	16	16	14
Half-cystine	6	12	12
Valine	18	17	17
Methionine	2	8	2
Isoleucine	15	14	15
Leucine	16	15	14
Tyrosine	11	12	10
Phenylalanine	7	1	3
Tryptophan	3	5	4
Total no. of residues	227	222	223

propyl fluorophosphate), and TLCK (tosyl-L-lysyl-chloromethyl ketone); but TPCK (tosyl-L-phenylalanyl-chloromethyl ketone), which is specific for chymotrypsin, is without effect.[14] In this connection it might be of interest that no chymotryptic activity can be detected in the digestive tract of the crayfish. Furthermore, there is no evidence for a zymogen form of the enzyme.

The amino acid composition of crayfish trypsin shows some marked differences from that of bovine trypsin (Table I).[15-18] Significantly, crayfish trypsin, like dogfish trypsin (a primitive vertebrate), contains only one-third of the lysine and one-half of the serine residues of bovine trypsin.

On the other hand, crayfish trypsin, like bacterial trypsin,[19] contains only 3 disulfide bridges, whereas vertebrate trypsins contain 6. In a de-

[16] R. Zwilling and V. Tomášek, *Nature* (*London*) **228**, 57 (1970).
[17] R. A. Bradshaw, H. Neurath, R. W. Tye, K. A. Walsh, and W. P. Winter, *Nature* (*London*) **226**, 237 (1970).
[18] K. A. Walsh and H. Neurath, *Proc. Natl. Acad. Sci. U.S.A.* **52**, 884 (1964).
[19] R. W. Olafson, L. Jurášek, M. R. Carpenter, and L. B. Smillie, *Biochemistry* **14**, No. 6, 1168 (1975).

Protein	1–122	22–157	42–58	127–232	136–201	168–182	191–220
Trypsin (bovine)		22–157	42–58	127–232	136–201	168–182	191–220
Trypsin (dogfish)		22–157	42–58	127–232	136–201	168–182	191–220
Chymotrypsin A (bovine)	1–122		42–58		136–201	168–182	191–220
Elastase (porcine)			42–58		136–201	168–182	191–220
Thrombin B (bovine)	1–122		42–58			168–182	191–220
Trypsin (crayfish)			42–58			168–182	191–220
Trypsin (Str. griseus)			42–58			168–182	191–220

FIG. 5. Acquisition of additional disulfide bridges during serine protease evolution. Position of disulfide bridges according to the chymotrypsinogen numbering system.

tailed analysis of the phylogeny of trypsin-related serine proteases, deHäen et al.[20] have pointed out the value of disulfide bridges for the investigation of distant evolutionary relationship. In crayfish trypsin we have identified 6 half-cystine residues that can be paired by homology arguments (Fig. 5). On the basis of these observations, crayfish trypsin should be placed in the evolutionary tree of the serine proteases above bacterial trypsin, and below the divergence of the vertebrate trypsins, which seemingly have acquired independently 3 additional disulfide bridges.

The amino acid sequence of crayfish trypsin is homologous to that of other serine proteases (Table II).

However, the long evolutionary distance between vertebrate and invertebrate trypsin is reflected by the observation that chymotrypsin, elastase, and even the hepatic proteases, thrombin and blood-coagulation Factor X_a, are structurally no less related to crayfish trypsin than are bovine and dogfish trypsin. At least from comparison of the amino-terminal portions of the sequences it may be concluded that the highest degree of identity exists between crayfish trypsin and bacterial trypsin,[19] rather than between crayfish trypsin and any of the vertebrate trypsins.

Crayfish trypsin and Str. griseus trypsin have in common 9 of the first 13 amino acid residues. Of the remaining 4 amino acid substitutions, 3 can be accounted for by single base changes. In the portion of the sequence discussed here, the highest degree of identity exists between crayfish trypsin and bovine Factor X_a, 10 of the first 20 residues occurring in identical positions.[21]

[20] C. deHaën, H. Neurath, and D. C. Teller, J. Mol. Biol. 92, 225 (1975).
[21] R. Zwilling, H. Neurath, L. H. Ericsson, and D. L. Enfield, FEBS Lett. 60, 247 (1975).

TABLE II

AMINO-TERMINAL SEQUENCES OF CRAYFISH TRYPSIN AND OTHER SERINE PROTEASES[a,b]

	16	17	18	19	20	21	22	23	24	25	26	27	28	29	30	31	32	33	34	35
Trypsin, crayfish	Ile	Val	Gly	Gly	Thr	Asp	Ala	Thr	Leu	Gly	Glu	Phe	Pro	Tyr	Gln	Leu	Ser	Phe	Gln	Asn
Trypsin, bovine	Ile	Val	Gly	Gly	Tyr	Thr	Cys	Gly	Ala	Asn	Thr	Val	Pro	Tyr	Gln	Val	Ser	Leu	Asn	—
Trypsin B, lungfish	Ile	Val	Gly	Gly	Tyr	Glu	Cys	Pro	Lys	His	Ser	Val	Pro	Trp	Gln	Val	Ser	Leu	—	—
Trypsin, dogfish	Ile	Val	Gly	Gly	Tyr	Glu	Cys	Pro	Lys	His	Ala	Ala	Pro	Trp	Thr	Val	Ser	Leu	Asn	—
Trypsin, *Str. griseus*	Val	Val	Gly	Gly	Thr	Arg	Ala	Ala	Gln	Gly	Glu	Phe	Pro	Phe	Met	Val	Arg	Leu	Ser	Met
Chymotrypsin A, bovine	Ile	Val	Asn	Gly	Glu	Glu	Ala	Val	Pro	Gly	Ser	Trp	Pro	Trp	Gln	Val	Ser	Leu	Gln	Asp
Elastase, porcine	Val	Val	Gly	Gly	Thr	Asp	Ala	Gln	Arg	Asn	Ser	Trp	Pro	Ser	Gln	Ile	Ser	Leu	Gln	Tyr
Thrombin B, porcine	Ile	Val	Gln	Gly	Gln	Asp	Ala	Glu	Val	Gly	Leu	Ser	Pro	Trp	Gln	Val	Met	Leu	Phe	Arg
Factor X$_a$, bovine	Ile	Val	Gly	Gly	Arg	Asp	Cys	Ala	Glu	Gly	Glu	Cys	Pro	Trp	Gln	Ala	Leu	Leu	Val	Asn

[a] From R. Zwilling, H. Neurath, L. H. Ericsson, and D. L. Enfield, *FEBS Lett.* **60**, 247 (1975).

[b] The numbering system is that of bovine chymotrypsinogen A. All positions that are occupied by identical amino acid residues in crayfish trypsin and in one or more other serine protease are enclosed by boxes.

As can be expected on functional grounds, the amino acid sequence around the "charge-relay" Asp[102] is highly conserved in crayfish trypsin (chymotrypsin numbering system):

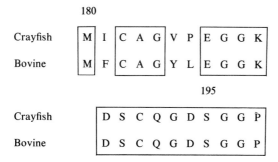

102

Crayfish	D	Y	D	L	L	D	N	D	I
Bovine	N	S	N	T	L	N	N	D	I

Similar observations can be made for the more extended sequences surrounding the catalytically active Ser[195]:

180

Crayfish	M	I	C	A	G	V	P	E	G	G	K
Bovine	M	F	C	A	G	Y	L	E	G	G	K

195

Crayfish	D	S	C	Q	G	D	S	G	G	P
Bovine	D	S	C	Q	G	D	S	G	G	P

The high degree of identity at the C-terminal portion of the sequence suggests that the protein chains of crayfish and bovine trypsin are of comparable length:

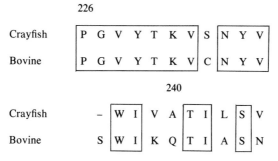

226

Crayfish	P	G	V	Y	T	K	V	S	N	Y	V
Bovine	P	G	V	Y	T	K	V	C	N	Y	V

240

Crayfish	–	W	I	V	A	T	I	L	S	V
Bovine	S	W	I	K	Q	T	I	A	S	N

Present research is directed toward the elucidation of the complete amino acid sequence of crayfish trypsin, which will allow a more detailed evaluation of the evolutionary position of crayfish trypsin in particular, and of the pattern of serine protease evolution in general.

Hornet Chymotrypsin

Most investigations of the proteolytic enzymes occurring in the digestive tract of insects were limited to a preliminary characterization, and detailed structural data are scarcely available. Probably the most thoroughly studied case is hornet chymotrypsin (EC 3.4.21.1), of which recently the complete amino acid sequence was established by Jany.[22]

Occurrence

Hornet chymotrypsin was isolated from the midgut of larvae of the two hornet species *Vespa crabro* and *Vespa orientalis*.

Assay Methods

The substrate specificity of hornet chymotrypsin is similar to that of pancreatic chymotrypsin A and B. Therefore, the assay methods described for mammalian chymotrypsins may also be used for the detection and quantitation of hornet chymotrypsin. During the purification procedure, hornet chymotrypsin activity was monitored photometrically at 405 nm using the substrate glutaryl-L-phenylalanine-*p*-nitroanilide as described by Erlanger *et al.*[23] The test mixture contains 500 μl of 4.2 mM substrate in 0.2 M Tris-HCl buffer, pH 8.2 (25°C). The reaction is started by the addition of 25 μl of enzyme solution and the liberation of *p*-nitroaniline is followed at 405 nm.[24]

Purification

Jany *et al.*[22,25] developed the following purification procedure for hornet chymotrypsin. 80–100 g of midgut material is homogenized in 200 ml of Tris-HCl buffer (10 mM, pH 7.8) and extracted for 2 hr. After centrifugation (1 hr, 40,000 g), the sediment is resuspended in 100 ml of buffer, and extracted again for 2 hr. The supernatants are combined, concentrated to 60–80 ml, and subjected to anion-exchange chromatography on DEAE–Sephadex A-50 in 10 mM Tris-HCl buffer, pH 7.8. On application of a linear NaCl gradient (0–0.4 M), a fractionation pattern as shown in Fig. 6 is observed.

This first purification step separates two chymotryptic fractions (*V. crabro* protease I and II; VCP I, II), appearing in the breakthrough peak, from a fraction of tryptic activity (*V. crabro* protease III; VCP III), which is bound to the adsorbant and is eluted only after application of the NaCl gradient. This latter fraction also displays aminopeptidase and amylase activities (Fig. 6).

[22] K. D. Jany, Habilitationsschrift, Universität Stuttgart, Germany (1980).
[23] B. F. Erlanger, A. G. Cooper, and A. J. Bendich, *Biochemistry* **3**, 1880 (1964).
[24] K. D. Jany, H. Haug, G. Pfleiderer, and J. Ishay, *Biochemistry* **17**, 4675 (1978).
[25] K. D. Jany, G. Pfleiderer, and H.-P. Molitoris, *Eur. J. Biochem.* **42**, 419 (1974).

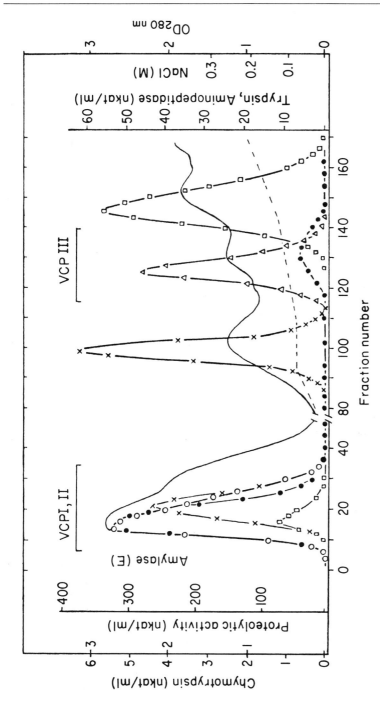

FIG. 6. Chromatography of crude extract from midgut of *V. crabro* on DEAE–Sephadex A-50, according to Jany.[22] Column, 44 × 4.5 cm; buffer, 10 mM Tris-HCl, pH 7.8. Linear NaCl gradient, 0–0.04 M. ●—●, Proteolytic activity (casein); ○—○, hydrolysis of Glu-Phe-*p*-nitroanilide; □—□, aminopeptidase activity; —, absorption at 280 nm; △—△, hydrolysis of Bz-Arg-*p*-nitroanilide; ×—×, amylase activity; -----, NaCl gradient.

After concentration and dialysis, fraction VCP I, II, containing the entire chymotryptic activity, is further purified by cation-exchange chromatography on CM-Sepharose C1-6B in McIlvaine buffer (0.1 M citric acid–0.2 M Na$_2$HPO$_4$ · 2H$_2$O, pH 6.4), diluted 1 : 3. Residual aminopeptidase and amylase activities are eluted by the buffer, and all chymotryptic activity remains bound to the gel. Fraction VCP I is eluted by the application of 70 mM NaCl and fraction VCP II, in a second step, by the application of a NaCl gradient (0.7–0.22 M) (Fig. 7).

Fraction VCP II, containing most of the active material, is next subjected to affinity chromatography on phenylbutylamine-Sepharose-4B.[26] Impurities lacking affinity for the ligand are eliminated by prolonged washing with 0.8 M NaCl in 50 mM Tris-HCl, pH 7.8. After reequilibration in 50 mM Tris-HCl (pH 7.8), the active enzyme is eluted with 0.1 M acetic acid, pH 3.0 (Fig. 8). Since the enzyme is not resistant to this treatment, a high flow rate and immediate neutralization are advisable.

The final purification of fraction VCP II requires gel filtration on Sephadex G-50, yielding about 90 mg of salt-free protein containing ~50% of the total initial activity. Fraction VCP II was the starting material for the characterization of hornet chymotrypsin and especially for the elucidation of its amino acid sequence.[22]

Purification of fraction VCP I may be performed by a procedure analogous to that given already, but the yield of this fraction was only 0.1–1.5 mg/100 g of starting material.

Properties

The fractions VCP I and II just described represent hornet chymotrypsin, whereas fraction VCP III represents a tryptic enzyme (see Fig. 6). All three fractions are homogeneous as judged by disc electrophoresis and by end-group analysis.

However, during isoelectric focusing, VCP II (*V. crabro* protease II) separates into 3 bands and VOP II (*V. orientalis* protease II) into 2 closely related bands, each of which is enzymatically active. The isoelectric points of these bands were found to be in the range of pH 9.8–10.

A rabbit antiserum against *V. orientalis* chymotrypsin (anti-VOP II) equally precipitates VOP II and VCP II. Contrarily, no immunological relationship can be found when anti-VOP II is tested against bovine chymotrypsin or the chymotrypsin-like protease from the honey bee.[22]

Hornet chymotrypsin (VCP II) exhibits similar substrate specificity as bovine chymotrypsin A. Cleavage is preferentially directed toward

[26] K. J. Stevenson and A. Landman, *Can. J. Biochem.* **49**, 119 (1971).

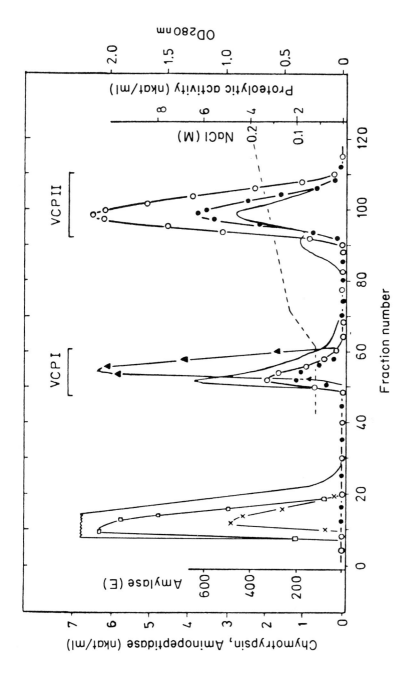

FIG. 7. Chromatography of hornet chymotrypsin (fractions VCP I and VCP II, see Fig. 6) on CM-Sepharose C1-6B.[22] Column, 37 × 2.7 cm, McIlvaine buffer (0.1 M citric acid–0.2 M Na$_2$HPO$_4$ · 2H$_2$O), 1:3 diluted, pH 6.4. ▲——▲, Carboxypeptidase A activity; other symbols as in Fig. 6.

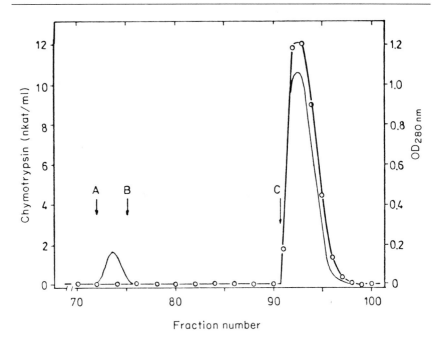

FIG. 8. Affinity chromatography of hornet protease (VCP II) on phenylbutylamine–Sepharose-4B. According to Jany.[22] Column, 14 × 3.6 cm. (A) 50 mM Tris-HCl buffer, containing 0.8 M NaCl, pH 7.8. (B) 50 mM Tris-HCl buffer, pH 7.8. (C) 0.1 M acetic acid, pH 3.0. Symbols as in Fig. 6.

aromatic and hydrophobic aliphatic amino acid side-chains, as can be seen from its action on the B chain of oxidized insulin (Fig. 9).

Additional points of cleavage at the leucine residues in positions 6, 11, 15, and 17 become manifest only after a prolonged incubation for 20 hr. Whereas bovine chymotrypsin A requires for its optimal activity a minimum substrate size of tetrapeptides,[27] VCP II hydrolyzes quantitatively tripeptides of the composition Gly-Tyr↓Gly, Gly-Phe↓Phe, or Leu-Phe↓Leu.

The molecular weight of VCP II, as calculated from the amino acid sequence, is 23,794.[22] The amino acid composition of VCP II as compared to bovine chymotrypsin A is given in Table III.

Hornet chymotrypsin (VCP II) is composed of a single peptide chain containing 221 amino acid residues.[22] There are only 3 disulfide bridges (as compared to 5 in bovine chymotrypsin and 6 in trypsin), and these are in homologous positions to those found in bacterial serine proteases (i.e., positions 42–58, 168–182, and 191–220). (See also crayfish trypsin.)

[27] K. Morihara, T. Oka, and H. Tsuzuki, *Biochem. Biophys. Res. Commun.* **35**, 210 (1964).

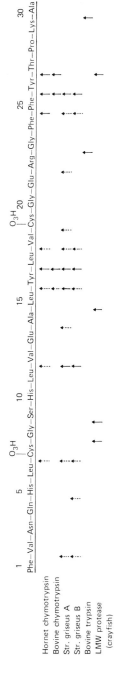

FIG. 9. Cleavage specificity of hornet chymotrypsin and other proteolytic enzymes.[22] Substrate, B chain of oxidized insulin.

TABLE III
AMINO ACID COMPOSITION OF HORNET AND BOVINE
CHYMOTRYPSIN[22]

Amino acid	Hornet chymotrypsin	Bovine chymotrypsin A
Aspartic acid	9	9
Asparagine	15	14
Threonine	12	23
Serine	15	26
Glutamic acid	5	5
Glutamine	12	9
Proline	11	7
Glycine	22	21
Alanine	11	21
Half-cystine	6	9
Methionine	0	2
Valine	24	21
Isoleucine	17	9
Leucine	18	17
Tyrosine	9	4
Phenylalanine	5	6
Histidine	6	2
Lysine	14	14
Arginine	7	3
Tryptophan	3	8
Total no. of residues	221	230

The alignment of VCP II (221 residues) with bovine chymotrypsin (230 residues) requires 16 gaps to account for the differences in the lengths of the protein chains. On this basis, VCP II shows 37% sequence identity with bovine chymotrypsin, 38% when compared to porcine elastase, and 31% when compared to bovine trypsin.

As in the case of serine proteases, the amino-terminal part of the enzyme as well as the sequences around residues essential for catalytic activity or for substrate binding are highly conserved (Fig. 10). The amino-terminal sequence of hornet chymotrypsin shown in Fig. 10 exhibits the highest degree of identity to crayfish trypsin. It is also of interest to note that the Ile-Val-Gly-Gly amino-terminal tetrapeptide sequence occurring in most serine proteases is also present in hornet chymotrypsin, and that the substitution of Asn[18] for Gly[18] is a characteristic limited to vertebrate chymotrypsins, including amphibia chymotrypsin.[28]

Although, as in other serine proteases, the amino acid residues surrounding the catalytically active Ser[195] are highly conserved in hornet

[28] W. Pies, R. Zwilling, R. G. Woodbury, and H. Neurath, *FEBS Lett.* **109**, 45 (1980).

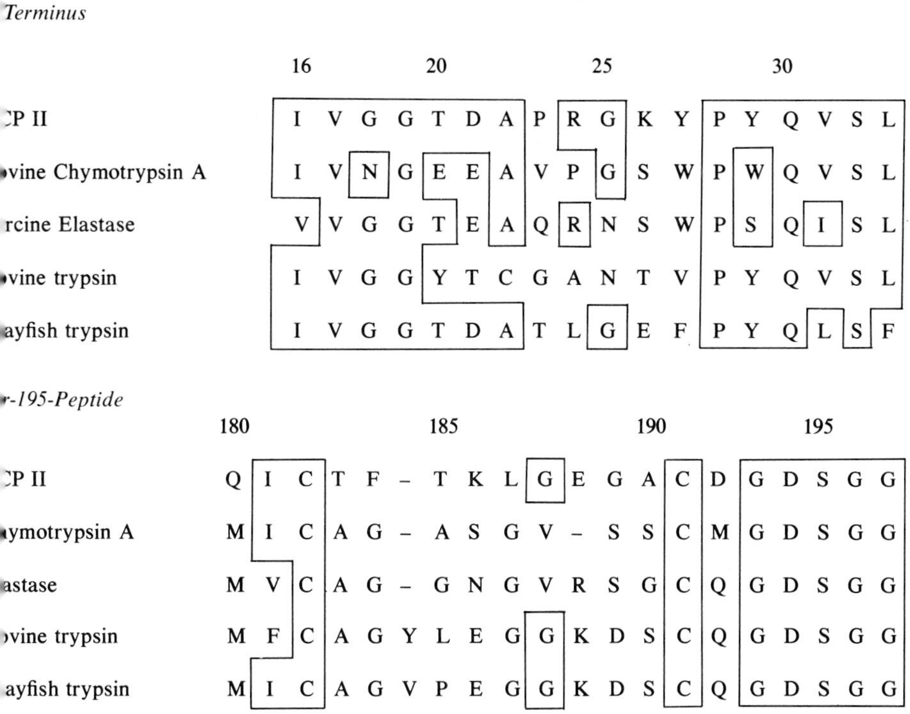

Terminus

	16				20			25				30						
CP II	I	V	G	G	T	D	A	P	R	G	K	Y	P	Y	Q	V	S	L
bovine Chymotrypsin A	I	V	N	G	E	E	A	V	P	G	S	W	P	W	Q	V	S	L
rcine Elastase	V	V	G	G	T	E	A	Q	R	N	S	W	P	S	Q	I	S	L
ovine trypsin	I	V	G	G	Y	T	C	G	A	N	T	V	P	Y	Q	V	S	L
ayfish trypsin	I	V	G	G	T	D	A	T	L	G	E	F	P	Y	Q	L	S	F

r-195-Peptide

	180					185					190					195				
CP II	Q	I	C	T	F	-	T	K	L	G	E	G	A	C	D	G	D	S	G	G
ymotrypsin A	M	I	C	A	G	-	A	S	G	V	-	S	S	C	M	G	D	S	G	G
astase	M	V	C	A	G	-	G	N	G	V	R	S	G	C	Q	G	D	S	G	G
ovine trypsin	M	F	C	A	G	Y	L	E	G	G	K	D	S	C	Q	G	D	S	G	G
ayfish trypsin	M	I	C	A	G	V	P	E	G	G	K	D	S	C	Q	G	D	S	G	G

FIG. 10. Amino-terminal sequences and Ser[195]-peptide sequences of hornet chymotrypsin and other homologous serine proteases.[22]

chymotrypsin, this is not true for the specificity-determining position 189, situated at the bottom of the substrate pocket, which is occupied by serine in bovine chymotrypsin and in porcine elastase, but is replaced by glycine in hornet chymotrypsin. Significant differences between the invertebrate and vertebrate chymotrypsins are also evident in the adjacent region from positions 183–188.

At present it is still an open question whether the chymotryptic cleavage specificity has been acquired during invertebrate evolution only once or more often.

Low-Molecular-Weight Protease

The digestive tracts of the crayfish *Astacus fluviatilis* and other decapode crustacea contain an endopeptidase with the unusually low molecular weight of 11,000.[29]

[29] G. Pfleiderer, R. Zwilling, and H. H. Sonneborn, *Hoppe-Seyler's Z. Physiol. Chem.* **348,** 1319 (1967).

Low-molecular-weight protease (LMWP) (EC 3.4.99.6) is not only remarkable for the fact that it has the lowest molecular weight of any hitherto known catalytically active protein, but that on the basis of its structural and catalytic properties, this enzyme belongs to a still unknown family of proteolytic enzymes.

Since this enzyme exhibits a new type of cleavage specificity, it is potentially a useful tool for restricted peptide degradation in conjunction with sequence analysis. Since LMWP appears to be composed of only approximately 100 amino acid residues, it is a promising candidate for the de novo synthesis of an enzyme. Moreover, this enzyme might offer the opportunity to discover a catalytic mechanism different from those already described.

Occurrence

Low-molecular-weight protease is produced in the hepatopancreas of decapode crustacea and is secreted into the stomach-like cardia of the crayfish, where it serves for digestion. LMWP can also be isolated without prior activation from the hepatopancreas, but the best source for the enzyme is the dark brown digestive fluid of the cardia. All attempts to detect in freshly collected secretions a zymogen form have failed, and, like crayfish trypsin and crayfish carboxypeptidase, the enzyme is fully active even in freshly prepared homogenates of the hepatopancreas where it is synthesized.[30]

The properties of LMWP in the species *A. fluviatilis* and *A. leptodactylus* (both crayfish), as well as in *Carcinus maenas* (crab) have been described.[2,31,32] Proteolytic enzymes with similar properties have been noticed also in other crustacea (e.g., *Orconectes virilis*,[33] *Homarus americanus*,[34] and *Panaeus setiferus*[35]). However, the distribution of LMWP in other invertebrate species is still unknown, and apparently in vertebrate animals this enzyme is entirely absent. It is therefore an open question whether this low-molecular-weight protease represents an unique evolutionary acquisition by certain invertebrate groups or whether, following a more generally occurrence, it was extinguished specifically during vertebrate evolution.

[30] R. Zwilling, G. Pfleiderer, H. H. Sonneborn, V. Kraft, and I. Stucky, *Comp. Biochem. Physiol.* **28**, 1275 (1970).

[31] R. Linke, R. Zwilling, D. Herbold, and G. Pfleiderer, *Hoppe-Seyler's Z. Physiol. Chem.* **350**, 877 (1969).

[32] D. Herbold, R. Zwilling, and G. Pfleiderer, *Hoppe-Seyler's Z. Physiol. Chem.* **352**, 583 (1971).

[33] E. J. DeVillez and J. Lau, *Int. J. Biochem.* **1**, 108 (1970).

[34] H. Brockerhoff, R. J. Hoyle, and P. C. Hwang, *J. Fish Res. Board Can.* **27**, 1357 (1970).

[35] B. J. Gates, Ph.D. Thesis, University of Georgia, Athens (1972).

Assay Methods

Since all protein substrates are readily degraded by LMWP, an effective assay procedure can be based on its proteolytic activity toward casein or azocasein. Analogous methods were given by Laskowski[36] and Reimerdes and Klostermeyer.[37] For routine measurements during the purification procedure, the following simple test is employed.

Casein Test

For each fraction to be tested for activity, an experimental value and a control value are determined.

Into small glass tubes, suited for centrifugation at low speed, are pipetted as follows:

A. Experimental value
    1. Tris-HCl buffer, 0.1 M, pH 8.0   0.2 ml
    2. Casein solution, 1%,
       in Tris-HCl buffer   0.4 ml
    3. Enzyme solution (or aliquot of the
       fraction to be tested)   0.2 ml
       Incubate for 20 min at 30°C
       Precipitate the incubation mixture
       by addition of
    4. Trichloroacetic acid, 5%   1.2 ml
                               2.0 ml

B. Control value
    1. Tris-HCl buffer, as in A (1)   0.2 ml
    2. Casein solution, as in A (2)   0.4 ml
    3. Trichloroacetic acid, as in A (4) 1.2 ml
    4. Enzyme solution, as in A (3)   0.2 ml
                               2.0 ml

Both incubation mixtures are centrifuged at low speed and in the supernatant the extinction at 280 nm of A is measured against B.

One enzyme unit is equal to the increase of extinction (A − B) for 0.001/min under the conditions given above. In order to stay in the linear range, a value of ΔE (A − B) = 0.800 should not be exceeded.

Preparation of Casein Solution

One gram of Hammarsten casein is suspended in 95 ml of Tris-HCl buffer (0.1 M, pH 8.0) in a 100-ml Erlenmeyer flask, then heated to 100°C

---

[36] M. Laskowski, this series, Vol. 2 [2].
[37] E. H. Reimerdes and H. Klostermeyer, this series, Vol. 45 [3].

in a water bath until the casein is completely solubilized. The flask is filled with buffer to 100 ml. When stored at 4°C in the refrigerator this substrate solution is good for several days.

A photometric test for LMWP based on the hydrolysis of a synthetic substrate was not known until recently. The usual synthetic substrates for trypsin or chymotrypsin (i.e., BzArgOEt, AcTyrOEt, APNE, and BANA) are not hydrolyzed by LMWP. It has been recently found that Suc-L-Ala-Ala-Ala-p-nitroanilide is a good substrate for LMWP[38] and that a sensitive spectrophotometric assay can be based on this amidase activity of the enzyme. Suc-$(Ala)_3$-p-nitroanilide was originally proposed by Bieth et al.[39] as substrate for elastase.

Spectrophotometric Test [39]

A 125 mM stock solution of Suc-$(Ala)_3$-p-nitroanilide in N-methylpyrrolidone is prepared by warming the suspension to 60°C for 10 min. This stock solution of the substrate is stable for several months if stored at 4°C in the dark.

The sample to be tested is diluted with 0.2 M Tris buffer (pH 8.0), and 2.5 ml of this solution are placed in a 1-cm cuvette. After temperature equilibration at 25°C, the reaction is started by the addition of 20 μl of the stock solution of substrate and the optical density is recorded for 2 min at 410 nm in a thermostatted spectrophotometer.

Under these assay conditions, the rate of spontaneous hydrolysis is very low and can be neglected (0.07% substrate hydrolyzed per hour in a 1 mM aqueous solution at pH 8.0 and 25°C). Nevertheless, aqueous substrate solutions should be used within 2 days. Trypsin and chymotrypsin do not interfere with this assay. Suc $(Ala)_3$-p-nitroanilide may also be used to detect LMWP activity in polyacrylamide gels (for the procedure, see Ref. 39).

Purification

LMWP occurs in the cardia extract of crayfish in a concentration of about 1 mg/ml. The digestive juice can be collected from living crayfish by the procedure described for crayfish trypsin. Since none of the commercially available protein inhibitors has any affinity for LMWP, an affinity-chromatography procedure could not be developed thus far, and purification has to rely on a combination of anion-exchange chromatography and gel filtration.

[38] R. Zwilling, H. Dörsam, H.-J. Torff, and J. Rödl, *FEBS Lett.* **127**, 75 (1981).
[39] J. Bieth, B. Spiess, and C. G. Wermuth, *Biochem. Med.* **11**, 350 (1974).

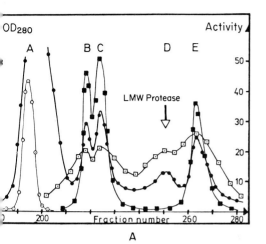
A

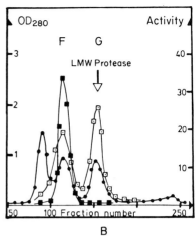
B

FIG. 11. (A) Anion-exchange chromatography of 100 ml of digestive juice of *A. fluviatilis*.[38] DEAE–Sephacel in Tris-HCl 0.01 *M* (pH 8.0), 500-ml bed volume, NaCl gradient 0.2–2.0 *M*. A, Crayfish carboxypeptidase; B, C, E, crayfish trypsin, multiple forms; D, LMWP. (B) Gel filtration of LMWP (peak D in part A). Sephadex G-50, Tris-HCl 0.01 *M* (pH 8.0)–0.5 *M* NaCl, 900-ml bed volume. F, Crayfish trypsin; G, LMWP. Symbols used in both parts: ●—●, OD_{280}; ⊙—⊙, hydrolysis of hippuryl-L-arginine, at pH 7.6 (μmol min⁻¹). ■—■, Hydrolysis of BzArgOEt (benzoyl-L-arginine ethyl ester), at pH 8.0 (μmol min⁻¹); part A (×2), part B (×10). ⊡—⊡, Caseinolytic activity (units) at pH 8.0.

After centrifugation, 100 ml of cardia extract are applied to a DEAE–Sephacel column (500-ml bed volume), equilibrated with 0.01 *M* Tris-HCl buffer, pH 8.0. Fractionation is performed by applying a NaCl gradient 0.2–2.0 *M*. (For details, see Fig. 11A.) Further purification is achieved by gel filtration on Sephadex G-50 (Fig. 11B).

Repetition of these two successive purification steps yields a LMWP preparation free of any accompanying impurities.[38] After desalting on Sephadex G-10 and lyophilization, a yield of 50–60 mg of pure LMWP from 100 ml of cardia crude extract can be expected.

The whole purification procedure is considerably facilitated by the fact that LMWP is distinguished from other accompanying proteins by a rather low isoelectric point (approx. 4.0) and its small size.

Properties

A pure preparation of *A. fluviatilis* LMWP consists of two closely related forms of the enzyme that cannot be separated by anion-exchange chromatography, but that become visible on disc electrophoresis (Fig. 12A). At the same time, disc electrophoresis testifies to the high purity of

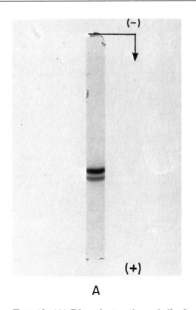

A   B

FIG. 12. (A) Disc electrophoresis[40] of purified LMWP (200 μg). The two closely related bands are multiple forms of LMWP and both hydrolyze casein. 15% polyacrylamide, tubes 6 × 130 mm; Tris-glycine buffer, pH 8.6, 6 hr at 250 V and 3 mA/gel slab; staining, amido black. (B) Laurell immunoelectrophoresis of 40 μg LMWP; 8% antiserum gel. First dimension: 250 V and 15 mA per 50 min. Second dimension: 100 V and 10 mA per 14 hr. Staining, amido black.[40]

the enzyme prepared by the method described above. A high degree of homogeneity can also be inferred from the very low background observable during automated Edman degradation of the enzyme.[38]

Two-dimensional Laurell immunoelectrophoresis shows only a single precipitation peak, thus demonstrating that the 2 protein bands are immunologically identical, as tested by this sensitive technique (Fig. 12B).

LMWP, like crayfish trypsin, occurs also in other decapode crustacea in multiple forms, varying in number and electrophoretic mobility from species to species.[32]

LMWP is not an active degradation product of crayfish trypsin or crayfish carboxypeptidase, since there is no immunological cross-reaction between any of these enzymes.[38] This possibility can also be excluded on the basis of the unique cleavage specificity and the partial amino acid sequence of the enzyme.

The optimal pH range for LMWP is around pH 8; below pH 4 (i.e., below the pI) the enzyme is irreversibly inactivated. On incubation at pH 8.0 at 30°C, LMWP is resistant to self-digestion for several days.

LMWP is clearly an endopeptidase effectively hydrolyzing casein and other protein and peptide substrates.

The synthetic substrates specific for trypsin and chymotrypsin activity (BzArgOEt, AcTyrOEt, APNE, and BANA) are not split by LMWP. Although Suc-(Ala)$_3$-p-nitroanilide and Ac-(Ala)$_3$-methyl ester are good substrates for LMWP, Cbz-Ala-Ala is not hydrolyzed. Similarly, a number of di- and tripeptides of various compositions were found to be resistant to the attack of LMWP. This suggests a minimum chain length of the substrate, a prerequisite for effective binding.

In Table IV, specific cleavage points of peptide substrates by LMWP are listed.[38] From these observations it can be inferred that LMWP liberates peptides with short aliphatic amino acid side-chains in the N-terminal position, preferentially Ala-, Gly-, Thr-, and Ser-peptides.

This finding has been confirmed by E. Krauhs et al.[39a], who applied LMWP to intact tubulin and recovered predominantly Ala-X-Y-, Gly-X-Y-, Thr-X-Y-, and Ser-X-Y-peptides.

Evidently, in contrast to many other proteolytic enzymes, the specificity of LMWP is determined by the amino acid residue providing the *amino* part of the susceptible bond. The character of the amino acid at the carboxyl side of the hydrolyzed peptide bond does not seem to have any significant influence on the cleavage specificity of this protease.

The model peptide Leu-Trp-Met-Arg-Phe-Ala, designed to cover the requirements of most proteolytic enzymes (i.e., leucine amino peptidase, carboxypeptidase, trypsin, and chymotrypsin) is not attacked at all by LMWP. Likewise, the collagen substrate Gly-Pro-Gly-Gly-Pro-Ala is not split. Among the 20 peptide bonds present in the A chain of oxidized insulin, only one is hydrolyzed. These findings suggest a new, specific, and thereby rather limited cleavage specificity for LMWP, which potentially makes the enzyme useful for sequence analysis.

LMWP is *not* inhibited by any one of the following biological trypsin, chymotrypsin, or pepsin inhibitors[40]: Polyvalent bovine pancreas inhibitor; porcine pancreas inhibitor (specific for trypsin); soybean-trypsin inhibitor; lima bean-trypsin inhibitor; ovomucoid (chicken); α_1-antitrypsin (human serum); hirudin (*Hirudo medicinalis*); and pepstatin (specific for pepsin and renin, pentapeptide from *Streptomyces*). There is evidence that LMWP degrades native soybean-trypsin inhibitor.

Side-chain modification reagents directed against Ser[195] and His[57] of serine proteases, showed that under conditions that cause complete inactivation of crayfish as well as bovine trypsin, the activity of LMWP is not

---

[39a] E. Krauhs, H. Dörsam, M. Little, R. Zwilling, and H. Ponstingl, *Anal. Biochem.* (in press).

[40] H.-J. Torff, H. Dörsam, and R. Zwilling, *Protides Biol. Fluids* **28**, 107 (1980).

TABLE IV
CLEAVAGE SPECIFICITY OF LMWP[a]

Substrate	Point of cleavage
	↓
Insulin A chain, oxidized	- Cys - Cys - Ala - Ser - 6 7 8 9
	↓
Insulin B chain,[b] oxidized	- Leu - Cys - Gly - Ser - 6 7 8 9
	↓
	- Gys - Gly - Ser - His 7 8 9 10
	↓
	- Glu - Ala - Leu - Tyr - 13 14 15 16
	↓
	- Phe - Tyr - Thr - Pro - 25 26 27 28
	↓
Mellitin	- Ile - Gly - Ala - Val - 2 3 4 5
	↓
	- Val - Leu - Thr - Thr - 8 9 10 11
	↓
Glucagon[b]	- Thr - Phe - Thr - Ser - 5 6 7 8
	↓
	- Phe - Thr - Ser - Asp - 6 7 8 9
	↓
	- Lys - Tyr - Leu - Asp - 12 13 14 15
	↓
	- Leu - Asp - Ser - Arg - 14 15 16 17
	↓
	- Arg - Arg - Ala - Gln - 17 18 19 20
	↓
Delta sleep-inducing peptide	- Ala - Gly - Gly - Asp - 2 3 4 5

[a] All cleavage peptides were separated and subsequently their composition was analyzed qualitatively and quantitatively.

[b] In insulin B chain (positions 7–9) and in glucagon (positions 6–8), two adjacent bonds are susceptible to the action of LMWP, but cleavage of one bond seems to preclude cleavage of the other.

TABLE V
AMINO ACID COMPOSITION OF LMWP[40]

Amino acid	No. of residues
Aspartic acid	13
Threonine	8
Serine	7
Glutamic acid	12
Proline	3
Glycine	8
Alanine	4
Valine	6
Half-cystine	3
Methionine	3
Isoleucine	7
Leucine	7
Tyrosine	7
Phenylalanine	3
Lysine	2
Histidine	3
Arginine	3
Tryptophan	Not determined
Total no. of residues	99

affected. LMWP is also insensitive to iodoacetic acid, EDTA, or o-phenanthroline and accordingly is neither a SH-proteinase nor a metalloenzyme.

The elucidation of the catalytic mechanism of LMWP has to await further research.

The amino acid composition of LMWP is shown in Table V. The enzyme is composed of approximately 100 amino acid residues, but the exact length of the protein chain can only be ascertained by sequence

TABLE VI
N-TERMINAL PARTIAL SEQUENCE OF LMWP[a]

Ala- Ala - Ile - Leu - Gly - Asp - Glu - Tyr - Leu - Trp -
 1   2   3   4   5   6   7   8   9   10

X^b - Gly - Gly - Val - Ile - Pro - Tyr - Thr[b] - Phe - Ala -
11   12   13   14   15   16   17   18   19   20

[a] LMWP does not share the Ile-Val-Gly-Gly-terminus characteristic of the family of serine proteases. It is likely that LMWP is a member of a new family of proteolytic enzymes occurring in invertebrates. See Ref. 38.

[b] In position 18 the recovery was low; position 11 was not identified.

analysis. Judging from the half-cystine content, the protein might contain only 1 disulfide bridge. The relatively low isoelectric point of the enzyme is consistent with the high content of aspartic and glutamic acid. The partial sequence of LMWP amino-terminal reveals no homology to known protein sequences (Table VI).

Crayfish Carboxypeptidase

In an effort to trace the evolutionary history of the pancreatic metal-lopeptidases, carboxypeptidase (EC 3.4.12.2) has been isolated from the cardia of the crayfish *Astacus fluviatilis*.[41] Comparison of the properties of the purified enzyme with those of bovine carboxypeptidases A and B indicates that the crayfish enzyme is a single molecular entity resembling in specificity bovine carboxypeptidase B, and that all three of these enzymes are the products of a common ancestral gene.

Source of Enzyme

The digestive fluid from the cardia of *A. fluviatilis* is collected by the method described for crayfish trypsin and stored in the frozen state until used. Carboxypeptidase activity is approximately 4 μmol hippuryl-L-arginine ml^{-1} min^{-1}.

Assay Methods

Enzyme assays are performed spectrophotometrically by the method of Folk and Schirmer[42] at 254 nm, using as substrate solutions 1 mM hippuryl-L-arginine (for carboxypeptidase B) or hippuryl-L-phenylalanine (for carboxypeptidase A) in 0.1 M Tris-HCl buffer (pH 8.0) containing 0.5 M NaCl, at room temperature.

One unit (U) of enzyme activity is defined as the quantity of enzyme capable of hydrolyzing 1 μmol substrate/min.

Purification

The purification of crayfish carboxypeptidase is greatly facilitated by use of affinity chromatography on a carboxypeptidase inhibitor from potatoes covalently attached to Sepharose-4B as described by Ager and Hass.[43]

---

[41] R. Zwilling, F. Jakob, H. Bauer, H. Neurath, and D. L. Enfield, *Eur. J. Biochem.* **94**, 223 (1979).
[42] J. E. Folk and E. W. Schirmer, *J. Biol. Chem.* **238**, 3884 (1963).
[43] S. P. Ager and G. M. Hass, *Anal. Biochem.* **83**, 285 (1977).

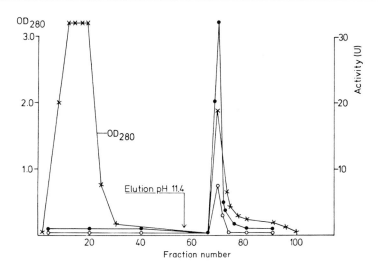

FIG. 13. Affinity chromatography on a potato inhibitor–Sepharose column of crayfish carboxypeptidase. Activities were monitored using hippuryl-L-phenylalanine as substrate for carboxypeptidase A (○) and hippuryl-L-arginine for carboxypeptidase B (●). Activity unit (U), 1 μmol of substrate hydrolyzed per minute. OD_{280} (×).

Preparation of the adsorbant proceeds as follows. 60 mg of inhibitor are coupled to 10 g of CNBr-activated Sepharose. The swollen gel product (30 ml) is transferred to a 2 × 20-cm column of the adsorbant, to which 50 ml of the cardia fluid in 0.01 M Tris-HCl buffer (pH 8.0) containing 0.5 M NaCl, is applied. After prolonged washing of the column with the initial buffer, carboxypeptidase is eluted by applying 0.01 M Na_2CO_3 (pH 11.4) containing 0.5 M NaCl (Fig. 13). The active fractions are pooled, dialyzed against 1 mM NH_4HCO_3, and lyophilized. Depending on the quality of the crude extract, up to 25 mg of freeze-dried carboxypeptidase can be obtained from 50 ml of cardia fluid.

Repetition of the affinity procedure removes additionally a minor fraction of inactive material. The final product is homogeneous, as judged by disc electrophoresis and amino-terminal analysis. Carboxypeptidase inhibitor from Russet Burbank potatoes can be prepared according to the procedure of Ryan et al.[44]

Properties

In contrast to trypsin and LMWP of crayfish, which occur in multiple forms, only a single form of carboxypeptidase could be detected in *A. fluviatilis* by affinity chromatography and by immunological tests.[41]

[44] C. A. Ryan, G. M. Hass, and R. W. Kuhn, *J. Biol. Chem.* **249**, 5495 (1974).

TABLE VII

AMINO ACID COMPOSITION OF CRAYFISH CARBOXYPEPTIDASE AND COMPARISON TO
SHRIMP AND BOVINE CARBOXYPEPTIDASES A AND B

Amino acid	Crayfish[41]	Shrimp		Bovine	
		A[45]	B[45]	A[46]	B[47]
Half-cystine	2	7	4	2	7
Aspartic acid + asparagine	55	32	45	29	28
Threonine	28	21	22	26	28
Serine	26	25	23	32	28
Glutamic acid + glutamine	34	20	33	25	27
Proline	21	15	14	10	12
Glycine	27	24	26	23	22
Alanine	26	22	22	21	22
Valine	20	18	17	16	15
Methionine	4	4	5	3	6
Isoleucine	16	13	15	20	15
Leucine	18	15	21	23	21
Tyrosine	14	16	20	19	20
Phenylalanine	11	9	11	16	13
Histidine	4	7	6	8	6
Lysine	9	4	11	15	17
Arginine	6	9	9	11	12
Tryptophan	—	3	7	7	10
M_r	35,000–36,000	30,000	34,200	35,268	35,700

The purified enzyme migrates as a single band on cellulose acetate, disc-gel, and sodium dodecyl sulfate–gel electrophoresis. The pH optimum of activity is about pH 6.5, and the isoelectric point, pH 4.0.

The specific activity of purified carboxypeptidase toward hippuryl-L-arginine is 104 μmol substrate hydrolyzed min^{-1} mg enzyme^{-1}, and 1.5 μmol min^{-1} mg^{-1} toward hippuryl-L-phenylalanine. These relationships are similar to those of porcine carboxypeptidase B. This dual specificity was confirmed by enzymatic assay of the A and B chains of oxidized bovine insulin. The A chain (terminating with -Cys-Asn) was completely resistant to crayfish carboxypeptidase, whereas the B chain (terminating with -Thr-Pro-Lys-Ala) released, after a few minutes of incubation, alanine and lysine. Since even after prolonged incubation no traces of free proline or threonine could be detected, it seems that the crayfish enzyme, like its vertebrate counterparts, is incapable of cleaving prolyl or cysteic acid bonds.

Crayfish carboxypeptidase is 10 times more sensitive than porcine carboxypeptidase to the inhibitory effects of 3-phenyllactic and

Protein	1	2	3	4	5	6	7	8	9	10	11	12	13	14	15	16	17	18	19	20	21	22	23	24	25
Crayfish							Met	Asp	Trp	X	X	Tyr	His	Asp	Tyr	Asp	Glu	Ile	Asn	Ala	Trp	Leu	Asp	X	Leu
Bovine A	Ala	Arg	Ser	Thr	Asn	Thr	Phe	Asn	Tyr	Ala	Thr	Tyr	His	Thr	Leu	Asp	Glu	Ile	Tyr	Asp	Phe	Met	Asp	Leu	Leu
Bovine B				Thr	Thr	Gly	His	Ser	Tyr	Glu	Lys	Tyr	Asn	Asn	Trp	Glu	Thr	Ile	Glu	Ala	Trp	Thr	Glu	Gln	Val
Human B				Ala	Thr	Gly	His	Ser	Tyr	Glu	Lys	Tyr	Asn	Asn	Trp	Glu	Thr	Ile	Glu	Ala	Trp	Thr		Gln	
Lungfish B								Tyr	Ser	Glu	Lys	Tyr	Asn	Thr	Trp	Asp	Glu	Ile	Ala	Ala	Trp	Thr	X	Gln	Ile

Fig. 14. Comparison of amino-terminal sequences of crayfish,[41] bovine A[46] and B,[48] human B,[49] and lung-fish B[47]-carboxypeptidases.

6-aminohexanoic acids. Inhibition by typical metal-chelating agents (i.e., ethylenediaminetetraacetate and 1,10-phenanthroline) and neutron-activation analysis indicate that, like the mammalian enzyme, crayfish carboxypeptidase is a zinc metalloenzyme.

The amino acid composition of crayfish carboxypeptidase is similar to that of pancreatic carboxypeptidases except for a higher content of acidic amino acid residues. However, the values appear to be still closer to those for shrimp carboxypeptidase B than to bovine carboxypeptidase B or A (Table VII).[41,45-47]

The amino-terminal sequence of crayfish carboxypeptidase is homologous to that of the other carboxypeptidases of known covalent structure (Fig. 14).[41,46-49] On this basis, crayfish carboxypeptidase shows an amino-terminal deletion of 6 residues as compared to bovine carboxypeptidase A, of 3 residues when compared to bovine and human carboxypeptidase B, and the addition of a methionine residue when compared to lungfish carboxypeptidase B. All enzymes have identical residues in position 12 (tyrosine), 18 (isoleucine), and, with the exception of bovine carboxypeptidase A, alanine in position 20 and tryptophan in position 21.

Acknowledgments

That portion of the work that was carried out in Heidelberg was supported by a grant from the Deutsche Forschungsgemeinschaft, Bonn–Bad Godesberg (Zw 17/7). The work in Seattle was supported by a grant from the National Institutes of Health (GM 15731). We thank Mrs. Marinette Kraemer for excellent technical assistance.

[45] B. J. Gates and J. Travis, *Biochemistry* **12**, 1867 (1973).
[46] R. A. Bradshaw, L. H. Ericsson, K. A. Walsh, and H. Neurath, *Proc. Natl. Acad. Sci. U.S.A.* **66**, 1389 (1969).
[47] G. R. Reeck and H. Neurath, *Biochemistry* **11**, 3947 (1972).
[48] K. Titani, L. H. Ericsson, K. A. Walsh, and H. Neurath, *Proc. Natl. Acad. Sci. U.S.A.* **72**, 1666 (1975).
[49] D. V. Marinkovic, J. N. Marinkovic, E. G. Erdös, and C. J. G. Robinson, *Biochem. J.* **163**, 253 (1977).

[49] Calcium-Activated Proteases in Mammalian Tissues

By LLOYD WAXMAN

The calcium-activated protease was initially described by Krebs and co-workers in the 1960s as a factor that would irreversibly activate rabbit muscle phosphorylase kinase in the presence of millimolar levels of calcium.[1,2] In the past several years this class of enzymes has received increasing attention, and studies to elucidate both their biochemical properties and possible physiological roles are currently being pursued in many laboratories.

The enzyme has been purified to homogeneity from porcine[3] and human[4] skeletal muscle, and by a very different method from bovine heart.[5] It is now clear, however, that in addition to being present in muscle, a similar protease can be found in a large number of tissues and cells, including reticulocytes and erythrocytes, brain, liver, kidney, lung, thymus, platelets, HeLa cells, and epidermal keratinocytes.[6,7] It has also been found in muscle from the horseshoe crab, but none could be detected in extracts of *E. coli*.[7] Most tissues seem to contain comparable levels of activity, although the amounts in the erythrocyte are much lower. Interestingly, the levels in rat, rabbit, and beef lung are nearly 10 times those of other tissues.[6,7]

To date, a physiological function for this class of proteases has not been resolved. Dayton *et al.*[8] have shown that the enzyme from skeletal muscle is capable of disrupting the Z band of the muscle sarcomere with the concomitant cleavage of tropomyosin and troponins T and I, but neither troponin C, nor actin, nor the heavy chain of myosin. Based on these elegant studies, the authors have suggested that the calcium-activated protease may participate in an initial step in the turnover of the contractile apparatus.[8] However, the fact that many nonmuscle cells also contain the enzyme points to a more fundamental role, perhaps as a processing enzyme or as part of a nonlysosomal degradative pathway for proteins.

A closely related problem is the unusually high concentration of calcium needed for maximal activation of the protease. Until recently, most studies have shown that this value is nearly millimolar, which is a level higher by several orders of magnitude than occurs within most cells.

[1] W. L. Meyer, E. H. Fischer, and E. G. Krebs, *Biochemistry* **3**, 1033 (1966).
[2] R. B. Huston and E. G. Krebs, *Biochemistry* **7**, 2112 (1968).
[3] W. R. Dayton, D. E. Goll, M. G. Zeece, R. M. Robson, and W. J. Reville, *Biochemistry* **15**, 2150 (1976); W. R. Dayton, W. J. Reville, D. E. Goll, and M. H. Stromer, *Biochemistry* **15**, 2159 (1976).
[4] K. Suzuki, S. Ishiura, S. Tsuji, T. Katamoto, H. Sugita, and K. Imahori, *FEBS Lett.* **104**, 355 (1979).
[5] L. Waxman, *in* "Protein Turnover and Lysosome Function" (H. L. Segal and D. J. Doyle, eds.), p. 363. Academic Press, New York, 1978.
[6] L. Waxman, *Fed. Proc., Fed. Am. Soc. Exp. Biol.* **38**, 479 (abstr.) (1979).
[7] L. Waxman, submitted for publication.
[8] W. R. Dayton, D. E. Goll, M. H. Stromer, W. J. Reville, M. G. Zeece, and R. M. Robson, *in* "Proteases and Biological Control" (E. Reich, D. B. Rifkin, and E. Shaw, eds.), p. 551. Cold Spring Harbor Press, Cold Spring Harbor, New York, 1975.

Now, however, observations in this laboratory[7] and as reported by Mellgren[9] indicate that there are two forms of calcium-activated protease. Although the relationship between the two enzymes is not clear, the new protease requires calcium in the micromolar range for activation. Clearly, this finding makes studies on the physiological role of the calcium-activated proteases all the more intriguing.

In the following discussion, I will refer to the calcium-activated protease as CAF (calcium-activated factor). It has also been called KAF (kinase-activating factor), CAP (calcium-activated protease), and CANP (calcium-activated neutral protease). The method that I shall describe for the purification of CAF has been applied to the isolation of the enzyme from rabbit skeletal muscle and lung, bovine heart and lung, and rat liver. An alternative procedure to that presented here is adequate for skeletal muscle, but is somewhat more time consuming because it requires several dialysis steps.[3] The yields, however, are comparable, although it is not known how generally applicable the latter method may be.

Assay Methods

The nonradioactive assay has been described previously.[10] α-Casein (Sigma Chemical Co.) is prepared for use as a substrate as described by Ashby and Walsh,[11] and is stored frozen as a 25-mg/ml solution in water following neutralization with NaOH. Typically, the 200-μl reaction mixture contains 100 μl of the enzyme sample, 5 mg/ml of α-casein, 50 mM Tris-HCl (pH 7.4), 3 mM 2-mercaptoethanol, and 2–5 mM CaCl$_2$. The calcium is added last to initiate the reaction, which is allowed to proceed for 20 min at 25°C. 160 μl of ice-cold 10% trichloroacetic acid is added to terminate the reaction, and the mixture is left on ice for 60 min. The precipitated protein is removed by centrifugation at top speed in a clinical centrifuge, and the absorbance of the supernatant is read at 280 nm. (If 0.5-ml cuvettes are not available, the volumes used can obviously be scaled up.) Controls are run that contain 0.1 mM EDTA (prepared from a 0.2 M stock solution neutralized to pH 7.4 with NaOH) instead of calcium.

For my experiments, I have defined a unit of proteolytic activity as that amount of enzyme that produces a change in absorbance of 0.1 in 20 min.[10] In my hands the assay is usually not linear for more than 40 min, although Dayton et al.[3] are able to obtain linearity for at least 1 hr. At higher temperatures, the linearity of the assay is quite poor. Unfortu-

[9] R. L. Mellgren, FEBS Lett. 109, 129 (1980).
[10] L. Waxman and E. G. Krebs, J. Biol. Chem. 253, 5888 (1978).
[11] C. D. Ashby and D. A. Walsh, this series, Vol. 38, p. 350.

nately, at temperatures lower than 25°C, the release of acid-soluble fragments is too slow to be practical. The radioactive assay contains the same ingredients as described above, except that 5–20 μl of enzyme sample are used, and digestion times may be shortened to 5 min. Substrate for the radioactive assay consists of α-casein that has been reductively methylated with [³H]formaldehyde (New England Nuclear) according to the method of Rice and Means.[12] The labeled casein is dialyzed exhaustively against water to remove unbound radioactivity, and the acid-soluble background counts are never more than 2% of the total. The labeled casein is diluted with cold casein such that a stock solution has a concentration of 1 mg/ml. 20 μl (20,000–40,000 cpm) of the substrate is used in a typical assay.

The radioactive assay is also carried out in a total volume of 200 μl, but the reaction is stopped by the addition of 575 μl of 10% trichloroacetic acid and 25 μl of 10% bovine serum albumin to aid the precipitation of undegraded casein. After 1 hr on ice, the denatured protein is removed by centrifugation, and 400 μl of the supernatant are counted in 4 ml of Liquiscint (National Diagnostics) or any other scintillation fluid suitable for counting aqueous tritium.

A very sensitive nonradioactive assay has been described by Mellgren et al.[13] These authors block the free amino groups of protein substrates by succinylation, and monitor the appearance of new amino groups due to proteolysis by means of reaction with fluorescamine.[14]

Purification Procedure

Although fresh tissues have always been used, frozen rabbit muscle (Pel-Freez) is a convenient source of starting material. However, because the levels of enzyme are so much higher, lung may be preferred, particularly in laboratories not equipped to process large amounts of tissue. In fact, it is possible to obtain homogeneous protease from as little as 100 g of rabbit lung.

Fresh bovine heart is obtained from a local slaughterhouse and immediately placed on ice. The tissue is carefully trimmed of fat and connective tissue with a sharp butcher's knife, and cut into strips 2–3 cm wide before passing through a meat grinder. Preparation of the tissue and all subsequent steps are carried out at 0–4°C. The ground meat is divided into 1-kg

[12] R. H. Rice and G. E. Means, J. Biol. Chem. 246, 831 (1971).
[13] R. L. Mellgren, J. H. Aylward, S. D. Killilea, and E. Y. C. Lee, J. Biol. Chem. 254, 648 (1979).
[14] P. Bohley, S. Stein, W. Dairman, and S. Udenfriend, Arch. Biochem. Biophys. 155, 213 (1973).

portions and homogenized at top speed for 1 min in a 1-gallon blender with 2.5 liters of 4 mM EDTA (prepared from a 0.2 M stock solution neutralized with NaOH to pH 7–7.4) and 15 mM 2-mercaptoethanol. The homogenate is centrifuged in a Beckman J-6 preparative centrifuge at top speed (6000–7000 g) for 40 min. The supernatant is decanted and filtered through two layers of cheesecloth to remove particulate matter.

After the pH of the extract is readjusted to neutrality with 1 M NaOH and the conductivity is lowered to 1 mS by adding ice-cold deionized water, it is poured into a small plastic pail (20-liter capacity). One liter of settled DEAE–cellulose (Whatman DE-52) is added per kilogram of tissue, and the mixture is stirred for 60 min by means of an overhead motor-driven paddle. (The resin is recycled after each use according to the manufacturer's directions. After the final water wash, the pH is adjusted to neutrality with 1 M NaOH, and the resin is equilibrated with 5 mM Tris-HCl, pH 7.4, on a Buchner funnel.) The resin is allowed to settle under gravity for 30 min, and the supernatant is siphoned off and discarded. The slurry of resin is then poured onto a Buchner funnel and washed under mild suction with 10 volumes of a buffer containing 5 mM Tris base–0.1 mM EDTA–10 mM NaCl–15 mM 2-mercaptoethanol, adjusted to pH 7.4 at room temperature (buffer A).

After the final wash, the moist resin cake is scooped into a plastic beaker and sufficient buffer A is added to make a thick slurry, which is then poured into a glass column. For extracts less than 0.5 liter, the batch procedure is not necessary, and the extract may be top-loaded onto a column that already contains the resin, followed by sufficient buffer A to remove any unbound protein. The column is then developed with a linear gradient of NaCl in buffer A to a final salt concentration of 550 mM. Generally, for preparative work, for each liter of resin, 1 liter of buffer A is placed in the mixing chamber of the gradient maker and 1 liter of buffer A plus 550 mM NaCl is placed in the other chamber. The column is most conveniently run overnight, and requires the use of a peristaltic pump. A typical elution profile is shown in Fig. 1.

Proteolytic activity is assayed with 10–20 μl if the radioactive assay is employed, or 100 μl if the nonradioactive assay is used. Generally, comparatively little activity is detected in the absence of calcium under the assay conditions described here, but it is necessary to run blanks, at least for the DEAE column, because of the presence of UV-absorbing, trichloroacetic acid-soluble compounds.

The fractions that contain calcium-activated proteolytic activity are pooled, and either dialyzed for 2 hr against buffer A, or diluted with cold deionized water such that the conductivity is below 5 mS. The protein is then top-loaded onto a column of hexylamine–Sepharose prepared ac-

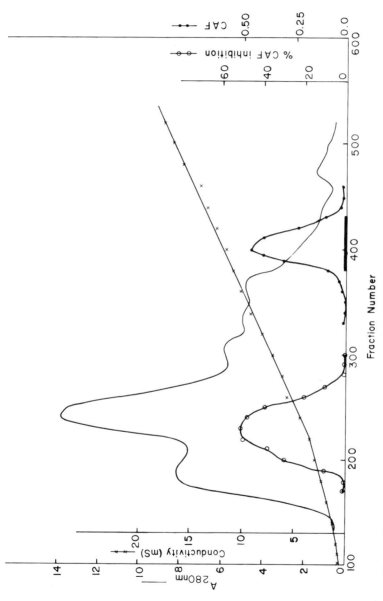

FIG. 1. Chromatography of a crude extract of bovine heart on DEAE–cellulose. A low-ionic-strength extract from five hearts was prepared as described in the text and batch-loaded onto 5 liters of DEAE–cellulose (Whatman DE-52) followed by washing with buffer A. After loading the resin into a glass column, a gradient was run from 10 mM to 550 mM NaCl (5 liters on each side), and 25-ml fractions were collected. The flow rate was 500 ml/hr. 100 μl were taken for both the protease and inhibitor assays. ——, $A_{280 nm}$; ×—×, conductivity (mS); ○—○, percent CAF inhibition; ●—●, CAF. Solid bar indicates pooled fractions.

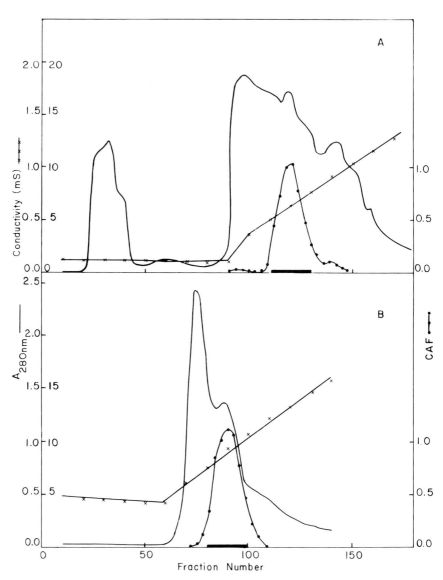

Fig. 2. Elution patterns of CAF from amino-alkyl Sepharose resins. (A) The active fractions from the DEAE–cellulose step were pooled and diluted with cold water to a conductivity of less than 5 mS and top-loaded onto a column of 6-NH₂ Sepharose (150 ml) equilibrated in buffer A. A gradient was run from 10 mM to 500 mM NaCl in buffer A (275 ml on each side), and 4.5-ml fractions were collected. The flow rate was 20 ml/hr. (B) The active fractions were taken from the 6-NH₂ step, diluted 1:2 with cold water, and applied to a 100-ml 8-NH₂ Sepharose column equilibrated in buffer A containing 50 mM NaCl. A gradient was run from 50 mM to 500 mM NaCl (250 ml on each side), and 3.5-ml fractions were collected. The flow rate was 20 ml/hr. For both columns, 40-μl fractions were taken for the protease assay. ——, $A_{280\,nm}$; ×—×, conductivity (mS); ●—●, CAF. Solid bars indicate pooled fractions.

cording to the method of Shaltiel and Er-El[15] (also available from P-L Biochemicals), which has been equilibrated in buffer A. After washing the column with 2 volumes of buffer A, a linear gradient is run from 10 mM to 500 mM NaCl. Because of the high capacity of these resins, 150–200 ml is sufficient for 5 kg of starting material. Figure 2 shows the elution pattern obtained with this resin.

The active fractions from hexylamine–Sepharose are pooled and diluted 1 : 1 with cold water, and the protein is applied to a 100-ml column of octylamine–Sepharose that is also prepared as described by Shaltiel and Er-El.[15] The column is equilibrated in buffer A plus 50 mM NaCl, and a linear gradient is run from 50 mM to 500 mM NaCl (Fig. 2).

The active fractions from the 8-NH₂ Sepharose step may be applied directly to a 50-ml column of hydroxyapatite (Bio-Rad), which has been equilibrated in 25 mM potassium phosphate (pH 7.1) containing 15 mM 2-mercaptoethanol. The column is washed with 100 ml of this buffer, and a gradient is run from 25 mM to 300 mM potassium phosphate (Fig. 3). Surprisingly, although a precipitate of calcium phosphate forms as the fractions are being assayed, the enzyme is still easily detected. The protease generally elutes between 80 and 100 mM phosphate.

The active fractions from the hydroxyapatite column are concentrated by ultrafiltration with an Amicon PM-10 membrane. The protein is applied to a 2.5 × 100-cm column of Sephadex G-150 equilibrated in buffer A (Fig. 3). The AcA-34 resin from LKB is also satisfactory for this step.

After gel filtration, the purified protease can be concentrated by ultrafiltration, or on a 5-ml column of DEAE–cellulose, which can be eluted with 2 volumes of 0.4 M NaCl. The enzyme can be stored for several months at 4°C in buffer A as long as the levels of reducing agent are maintained. It cannot be frozen without complete loss of activity, but seems to be stable in 50% glycerol at −20°C.

Comments on the Purification Procedure

The main advantage of the protocol described here is the speed with which it can be carried out. A significant amount of time is saved by eliminating the dialysis steps, and unlike earlier procedures,[2,3] this method does not require ammonium sulfate fractionation or isoelectric precipitation. The entire procedure takes 5 days, and the yields vary from 10 to 20%; 2–4 mg/kg are routinely obtained from bovine heart, and 20 mg/kg from bovine lung. A typical purification is summarized in Table I.

Although the protease is a sulfhydryl enzyme, covalent chromatography on organomercurial–Sepharose or on the glutathione-2-pyridyl di-

[15] S. Shaltiel and Z. Er-El, *Proc. Natl. Acad. Sci. U.S.A.* **20**, 778 (1973).

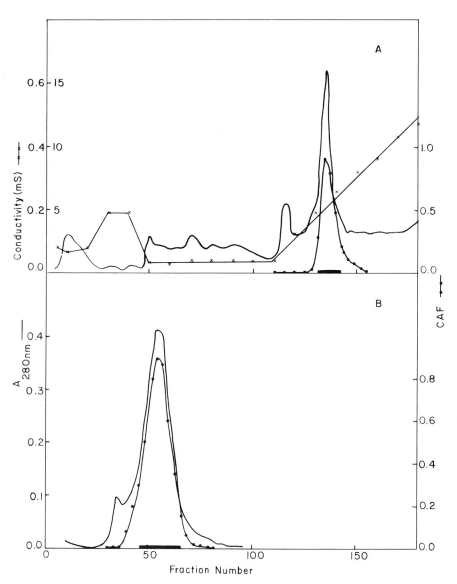

FIG. 3. Final purification steps of bovine heart CAF. (A) The pooled active fractions from the 8-NH$_2$ Sepharose step were applied directly to a 100-ml column of hydroxyapatite equilibrated in 25 mM potassium phosphate plus 15 mM 2-mercaptoethanol (pH 7.1). The column was washed with 100 ml of this buffer, and a gradient was run from 25 mM to 300 mM potassium phosphate (250 ml on each side). The flow rate was 15 ml/hr and 4-ml fractions were collected. (B) The active fractions from the hydroxyapatite step were concentrated to 5 ml on an Amicon PM-10 membrane, and applied to a column of Sephadex G-150 (2.5 × 100 cm) equilibrated in buffer A. The flow rate was 15 ml/hr and 4-ml fractions were collected. For both columns, 40-μl fractions were taken for the protease assay. ——, A_{280nm}; ×—×, conductivity (mS); ●—●, CAF. Solid bars indicate pooled fractions.

TABLE I
PURIFICATION OF BOVINE HEART CAF[a]

Step	Protein (mg)[b]	Specific activity (U/mg)	Yield (%)	Purification[c] (-fold)
Crude extract	158,400	—	—	—
DE-52	3900	60	100	1
6-NH₂ Sepharose	680	230	82	4
8-NH₂ Sepharose	200	570	66	10
Hydroxyapatite	30	1920	34	32
G-150	15	3520	32	59

[a] Starting from 5 kg of trimmed tissue (six hearts).
[b] Determined by the method of Lowry *et al.* [O. H. Lowry, N. J. Rosebrough, A. L. Farr, and R. J. Randall, *J. Biol. Chem.* **193**, 265 (1951)].
[c] The fold-purification based on the total amount of protein in the crude extract is 10,500.

sulfide derivative of Sepharose affords little additional purification. Extraction of ground tissue with buffers containing high salt or detergent does not seem to increase the yield of enzyme significantly.

In my hands, the amino-alkyl Sepharose resins have been used more than 50 times without noticeable loss of capacity. A property of these resins that has been used to advantage to purify CAF is that proteins bind more tightly with increasing length of the hydrocarbon sidearm.[15] Thus not only is it possible to obtain additional fractionation by using both 6-NH₂ and 8-NH₂ Sepharose, but based on the total amount of protein recovered, at least 10% of the protein applied to the latter resin cannot be eluted with salt. To clean the resins, it is sufficient to wash them on a sintered glass funnel with several volumes of $2\,M$ NaCl–$0.1\,M$ NaOH, and then water.

Properties of the Protease

The calcium-activated protease from heart, lung, and skeletal muscle has MW 100,000–120,000 based on gel filtration,[3⁻7] and elutes from DEAE–cellulose at 200–250 mM NaCl. It has an isoelectric point between 4.7 and 4.9, and its UV spectrum reveals the presence of no prosthetic groups. The enzymes that have been purified from these tissues all require near millimolar calcium for maximal activity, and have pH optima between 7.2 and 7.6. Magnesium cannot substitute for calcium, nor does it alter this requirement when the two cations are assayed together.[7]

CAF consists of two noncovalently associated subunits, one of MW 75,000–80,000, and a second one of MW 25,000–30,000 based on

TABLE II
POTENTIAL INHIBITORS OF THE CALCIUM-ACTIVATED PROTEASE

Reagent	Concentration (μM)	Inhibition (%)
Iodoacetamide	100	100
p-Chloromercuribenzoate	100	100
Tetrathionate	100	100
Phenylmethanesulfonyl fluoride	1000	5
Benzamidine	1000	0
Tosyl-lysyl-chloromethyl ketone	1000	10
Leupeptin	0.75	50
Antipain	1	50
E-64-C	15	50
Soybean-trypsin inhibitor	1 mg/ml	0
Trasylol	1 mg/ml	0
α_2-Macroglobulin	0.1 mg/ml	85
Trifluoperazine	100	0
Hemin	20	50

polyacrylamide-gel electrophoresis in the presence of sodium dodecyl sulfate.[3,4,7] The purified enzyme contains neither covalently bound phosphate nor carbohydrate,[7] and the amino composition is noteworthy only in the large proportion of Glx and Gly residues (13 mole %).[3,7] In addition, the amino-terminus of both subunits appears to be blocked.[7]

In some preparations, I have noticed that the enzyme contains less than stoichiometric amounts of the smaller polypeptide, but it is possible to add some of this subunit back to the purified protease and obtain enhanced activity.[7] Curiously, the purified protease from chicken skeletal muscle has been shown to contain only the 80,000 dalton subunit.[16]

Inhibitors of CAF

As summarized in Table II, CAF is very sensitive to sulfhydryl reagents such as iodoacetamide and mercurials. It is not significantly inhibited by compounds that are more specific for serine proteases, including phenylmethanesulfonyl fluoride, Trasylol, and soybean-trypsin inhibitor. Of the class of peptide aldehyde protease inhibitors that have been isolated from *Streptomyces*[17] by Umezawa and colleagues, leupeptin and antipain are the most potent. More recently, a novel group of thiol protease inhibitors that contain a reactive epoxide group have been isolated from *Aspergillus*. These compounds, as well as several synthetic analogs (e.g., E-64-C), are also good inhibitors of CAF.[18]

[16] S. Ishiura, H. Murofushi, K. Suzuki, and K. Imahori, *J. Biochem.* (*Tokyo*) **84**, 225 (1978).
[17] H. Umezawa, this series, Vol. 45, p. 678.
[18] H. Sugita, S. Ishiura, K. Suzuki, and K. Imahori, *J. Biochem.* (*Tokyo*) **87**, 339 (1980).

A kinetic analysis of the mode of inhibition by CAF by the peptide aldehydes has shown that these inhibitors are noncompetitive with respect to substrate.[19] However, the application of Michaelis–Menten theory to the case of a tightly binding inhibitor may not be valid.

Among the other compounds that have been tested, α_2-macroglobulin, which has been shown to inhibit most endoproteases,[20] will also inhibit CAF.[5] Interestingly, hemin is also an excellent inhibitor of CAF, but trifluoperazine, a drug that is known to block the action of calmodulin, has no effect on the protease.[7]

The Specific Inhibitor of CAF

Recently it has been shown that all tissues containing CAF also have a potent inhibitor of the protease.[6,7,10] The inhibitor is a protein of 270,000 daltons, and appears to be highly specific: It has no effect on any other protease that has been tested, including the lysosomal cathepsins, an ATP-stimulated protease from rabbit erythrocytes, trypsin, chymotrypsin, subtilisin, thermolysin, thrombin, papain, ficin, or bromelain.[7,10]

The existence of an inhibitor was originally suggested when it was found that CAF activity could not be detected in extracts of bovine heart, although the protease was easily measured in rabbit skeletal muscle extracts.[10] As shown in Fig. 1, the inhibitory activity may be separated from CAF on DEAE–cellulose. In all likelihood, the ability to measure calcium-stimulated proteolytic activity in crude extracts depends on the relative amounts of inhibitor and enzyme in the tissue under study.

Although the details have not yet been published, both Otsuka and Goll[21] and I have been able to purify the inhibitor protein, and studies are in progress to understand the mechanism of its interaction with CAF.

Quantitation of the Tissue Content of CAF and Its Inhibitor

Interest in the calcium-activated protease has increased greatly in the last several years, and numerous studies have attempted to correlate levels of the enzyme with nutritional state, treatment with hormones and drugs, and degenerative diseases such as muscular dystrophy. It is important, therefore, to have a reliable method to measure and quantitate changes in both protease and inhibitor levels. From early work on CAF, it was found that the protease could be precipitated by lowering the pH to 4.5–5,[2,22] while the inhibitor remained in solution. Obviously, if CAF can

[19] T. Toyo-oka, T. Shimizu, and T. Masaki, *Biochem. Biophys. Res. Commun.* **82**, 484, (1978).

[20] A. J. Barrett, P. M. Starkey, and E. A. Munn, *Bayer Symp.* **5**, 72 (1974).

[21] Y. Otsuka and D. E. Goll, *Fed. Proc., Fed. Am. Soc. Exp. Biol.* **39**, 544 (abstr.) (1980).

[22] G. I. Drummond and L. Duncan, *J. Biol. Chem.* **241**, 3097 (1966).

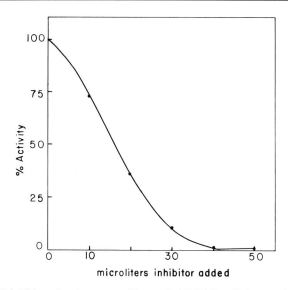

FIG. 4. CAF-inhibitor titration curve. The pooled DEAE–cellulose peak of inhibitory activity prepared from an extract of 50 g of rat liver was dialyzed against buffer A, and aliquots were assayed in increasing amounts with 5 units of partially purified CAF (see text). Following a preincubation at room temperature for 5 min, the protease assay was carried out as described in the text.

be efficiently solubilized from the precipitate, then one can accurately determine its amount. The inhibitor can then be quantitated by titrating a known amount of protease with increasing amounts of the inhibitor fraction.

For each assay, I use 5 units of CAF that has been purified through the DEAE–cellulose and gel-filtration steps. Even at this stage of purity, the levels of proteolytic activity in the absence of calcium are negligible. As can be seen in the example in Fig. 4, the titration curve of inhibitor against CAF activity is linear in the range of 25–75% inhibition, and quantitation experiments should be carried out in this region of the curve. For convenience, I have defined a unit of inhibitory activity as that amount of inhibitor that will diminish one unit of proteolytic activity by 50%.

A protocol that I have used with success is to place the tissue extract after centrifugation to remove particulate matter onto a small column of 6-NH$_2$ Sepharose (1 ml resin/g tissue) equilibrated in buffer A. The column is washed with several volumes of this buffer, followed by 2 volumes of buffer A plus 0.1 M NaCl to elute the inhibitor, and 2 volumes of buffer A + 0.4 M NaCl to elute CAF. The inhibitor fractions may be frozen at −20°C, while the CAF assays are performed. With radioactive substrate, it is not difficult to assay CAF in individual soleus muscles from 60-g rats (about 30 mg wet weight).

A Second Form of the Calcium-Activated Protease

The enzyme that has been purified from porcine muscle,[3] and bovine heart and lung,[5,7] elutes near 225 mM NaCl from DEAE–cellulose, and requires nearly 750 μM calcium for half-maximal activation. However, Mellgren[9] has recently described a calcium-activated protease in extracts of canine cardiac muscle that is similarly activated by 40 μM calcium. I have also made this observation with the calcium-activated proteolytic activity from rabbit erythrocytes.[7] In agreement with Mellgren,[9] this enzyme elutes from DEAE–cellulose at a lower ionic strength (100 mM NaCl), but appears to have a slightly lower molecular weight than the enzymes that have been purified from muscle.

Since the levels of the calcium-activated protease are very low in the erythrocyte,[6] I decided to reexamine rabbit skeletal muscle as a source for the new enzyme, because in some preparations a small amount of activity was detectable with the radioactive assay that eluted at lower ionic strength after chromatography on DEAE–cellulose (Reddy *et al.*[23] had previously made a similar observation). Indeed, as shown in Fig. 5, this tissue has a second form of the enzyme that very nearly coelutes with the inhibitor.

Note that if the fractions are first heated to 75°C for 5 min to inactivate endogenous calcium-activated proteolytic activity, the inhibitor profile shows one broad peak. However, if the fractions are assayed directly for inhibitory activity, a double peak is obtained. This difference is undoubtedly due to the contribution of the endogenous CAF activity to the assay.

In preliminary experiments, these two proteases, which I shall call type I and type II,[9] based on their elution from DEAE–cellulose, have the same substrate specificity and are inhibited by the same spectrum of inhibitors, but differ in size and elution position from all of the columns that have been used to purify the type II enzyme. Most importantly, the type I enzyme is half-maximally activated by 25 μM calcium.[7]

Clearly, the existence of two calcium-activated proteases complicates quantitation of the amounts of each in tissue extracts. In my protocol, which utilizes 6-NH$_2$ Sepharose, the two proteases are eluted together. However, by measuring the activity at 100 μM and 1 mM calcium, it is possible to assess the contributions of each protease to the total.

[23] M. K. Reddy, J. D. Etlinger, M. Rabinowitz, D. A. Fischman, and R. Zak, *J. Biol. Chem.* **250**, 4278 (1975).

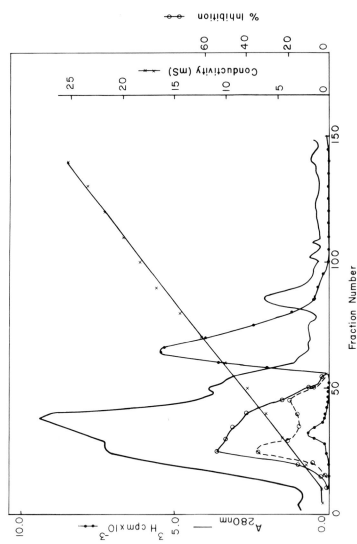

FIG. 5. Chromatography of a crude extract of rabbit skeletal muscle on DEAE–cellulose. A low-ionic-strength extract of rabbit skeletal muscle (50 g) was prepared as described in the text and top-loaded onto a 50-ml column of DEAE–cellulose equilibrated in buffer A. A gradient was run from 10 mM to 550 mM NaCl (150 ml on each side), and 2-ml fractions were collected. The flow rate was 15 ml/hr. 100-μl samples were taken for both the inhibitor and protease assays. In one series of inhibitor assays, the fractions were assayed directly (○—○). In the second series, the fractions were first heated to 75°C for 5 min (○----○). ——,

Subunit Separation

To examine the properties of the two subunits of CAF, I sought a method to separate them without having to use strong denaturants such as sodium dodecyl sulfate or guanidine hydrochloride. One approach has been to take advantage of the inactivation caused by exposure to the chaotropic agent, NaSCN.[7] The purified protease can be dialyzed overnight against 1 M NaSCN, or incubated on ice with the reagent. Since the two subunits differ in size by MW 50,000, they are easily resolved on a column of Sephadex G-100 in the presence of 0.5 M NaCSN.[7] The salt may then be dialyzed away before further experiments.

Substrate Specificity of CAF

The question of the sequence specificity of the calcium-activated proteases has not been resolved. A large number of small synthetic peptides have been tested without success.[5,24] Several well-characterized peptides are also cleaved poorly or not at all. Among those that have been examined are the A and B chains of insulin, glucagon, angiotensins I and II, and the S-peptide of ribonuclease A. It is possible that the protease will not cleave small substrates, but it will degrade a number of purified caseins and histones,[7] native tropomyosin and troponin,[8] and the regulatory subunits of the cyclic AMP-dependent protein kinases from bovine heart.[7] Although it will not degrade native chicken lysozyme, once the disulfide bonds have been reduced and the cysteine residues blocked by alkylation, and the lysine residues maleylated or succinylated, the protein becomes an excellent substrate.[5] This result indicates that denaturation of lysozyme must expose sites that are susceptible to attack by CAF. Moreover, it shows that lysine residues are not a determinant for cleavage.

Globin is also a good substrate,[7] but it is poorly soluble at alkaline pH, and this property makes kinetic studies impractical. Finally, no small synthetic chromogenic or fluorogenic substrate has been found that is facilely cleaved by CAF.

Nevertheless, a number of biologically interesting proteins have been cleaved with calcium-activated proteases of varying purity. I have found that CAF will release peptides from the α- and β-subunits of phosphorylase b kinase, but not from the γ-subunit.[7] It will also activate fructose-1,6-diphosphatase,[7] and phosphorylase phosphatase from muscle.[13] It has also been used to remove a phosphorylated site from muscle filamin,[25] and to activate a cyclic AMP-independent protein kinase from brain[26] and erythrocytes.[7]

[24] J. R. Feramisco, D. B. Glass, and E. G. Krebs, *J. Biol. Chem.* **255**, 4240 (1980).
[25] P. J. A. Davies, D. Wallach, M. C. Willingham, I. Pastan, M. Yamaguchi, and R. M. Robson, *J. Biol. Chem.* **253**, 4036 (1978).
[26] M. Inoue, A. Kishimoto, Y. Takai, and Y. Nishizuka, *J. Biol. Chem.* **252**, 7610 (1977).

Subcellular Localization of CAF

In careful studies by Reville et al.,[27] most of the calcium-activated proteolytic activity in skeletal muscle was shown to be in the soluble fraction. Because of the difficulties of homogenizing muscle without disrupting intracellular organelles, I decided to examine a soft tissue for which fractionation procedures have been well established. Unfortunately, as with bovine heart, the amount of inhibitor relative to CAF in liver is sufficiently large so as to make detection of the protease very difficult.[6] Therefore, rat kidney was chosen as the tissue for these studies. From these experiments, it was found that approximately one half of the calcium-activated protelytic activity was in the cytoplasmic fraction, and the balance was in the microsomal fraction.[7] None could be detected in any other subcellular fraction. Moreover, extraction of the microsomes with $0.2 M$ NaCl failed to solubilize any activity.

Clearly, more sophisticated techniques, such as the use of fluorescent antibodies to CAF, will be needed to clarify this issue. Hopefully, this knowledge will also be a clue to the role of CAF in the cell.

Acknowledgments

I thank Professor Edwin G. Krebs for his encouragement during the early phases of this work, and Professor Donal A. Walsh, School of Medicine, University of California at Davis, in whose laboratory most of this research was carried out. Special thanks are to Sharon Zick for many helpful discussions. My research was supported by postdoctoral fellowships from the National Institutes of Health and the Muscular Dystrophy Association of America (1977–1979), and by a grant from the NIH to Donal A. Walsh.

[27] W. J. Reville, D. E. Goll, M. H. Stromer, R. M. Robson, and W. R. Dayton, *J. Cell Biol.* **70**, 1 (1976).

[50] Proteases in *Escherichia coli*

By ALFRED L. GOLDBERG, K. H. SREEDHARA SWAMY, CHIN HA CHUNG, and FREDERICK S. LARIMORE

Although more has been learned about the regulation and selectivity of intracellular protein degradation in *Escherichia coli* than in any other cell,[1–3] the proteolytic enzymes in these organisms have only been re-

[1] A. L. Goldberg and A. C. St. John, *Annu. Rev. Biochem.* **45**, 747 (1976).
[2] M. J. Pine, *Annu. Rev. Microbiol.* **26**, 103 (1972).
[3] D. W. Mount, *Annu. Rev. Genet.* **14**, 279 (1980).

cently defined.[4-7] A detailed study of these enzymes is essential for elucidating the pathways of degradation of normal and abnormal proteins,[1,2] and the acceleration of protein breakdown when cells enter nutritionally poor conditions.[1] In these processes, cell proteins are digested completely to amino acids, and recent studies have indicated the sequential involvement of one or more endoproteases and multiple peptidases.[8-10] In addition, proteolytic enzymes must also play a key role in other important cellular processes involving only limited proteolytic cleavages, such as the maturational processing of secretory and membrane proteins,[11,12] phage morphogenesis,[3,13] breakdown of colicins,[14,15] and inactivation of certain regulatory proteins.[3,5,16,17] Finally, peptidases and perhaps proteinases are also required for the bacteria to utilize exogenous peptides,[13] although *E. coli,* unlike many gram-positive organisms, do not secrete proteases and therefore can not use polypeptides as a source of amino acids.

Proteases in E. coli

There are a growing number of reports in the literature concerning the isolation of proteases and peptidases from *E. coli.* The so-called protease I and II were identified first and purified by their ability to hydrolyze chromogenic ester substrates of chymotrypsin (e.g., N-acetylphenylalanine β-naphthyl ester) and trypsin (e.g., N-benzoyl-DL-arginine-p-nitroanilide) respectively.[18,19] These esterases are sensi-

[4] Y. E. Cheng and D. Zipser, *J. Biol. Chem.* **254**, 4698 (1979).

[5] J. W. Roberts, C. W. Roberts, and N. L. Craig, *Proc. Natl. Acad. Sci. U.S.A.* **75**, 4714 (1978).

[6] A. L. Goldberg, N. P. Strnad, and K. H. Sreedhara Swamy, *Ciba Found. Symp.* **75**, 227 (1980).

[7] K. H. Sreedhara Swamy and A. L. Goldberg, *Nature (London)* **292**, 652 (1981).

[8] R. Voellmy and A. L. Goldberg, *Nature (London)* **290**, 419 (1981).

[9] R. Voellmy, E. Rosenthal, R. Gronostajski, and A. L. Goldberg, submitted for publication.

[10] C. G. Miller, *in* "Limited Proteolysis in Microorganisms" (G. N. Cohen and H. Holzer, eds.), p. 65. US DHEW Publ. (NIH) 79-1591 (1979).

[11] W. Wickner, *Annu. Rev. Biochem.* **48**, 23 (1979).

[12] B. D. Davis and P. C. Tai, *Nature (London)* **283**, 433 (1980).

[13] C. G. Miller, *Annu. Rev. Microbiol.* **29**, 485 (1975).

[14] D. Cavard and C. Lazdunski, *Eur. J. Biochem.* **96**, 529 (1979).

[15] D. H. Watson and D. J. Sherratt, *Nature (London)* **278**, 362 (1979).

[16] G. N. Cohen and H. Holzer, eds., "Limited Proteolysis in Microorganisms," US DHEW Publ. (NIH) 79-1591 (1979).

[17] J. W. Little, S. H. Edmiston, L. Z. Pacelli, and D. W. Mount, *Proc. Natl. Acad. Sci. U.S.A.* **77**, 3225 (1980).

[18] M. Pacaud and J. Uriel, *Eur. J. Biochem.* **23**, 435 (1971).

[19] M. Pacaud and C. Richaud, *J. Biol. Chem.* **250**, 7771 (1975).

tive to diisopropyl fluorophosphate (DFP), but unlike the pancreatic proteases or the *E. coli* proteases described below, these enzymes show very little or no ability to hydrolyze protein substrates.[19–21] Furthermore, mutants lacking these two enzymes do not show any decreased ability to degrade abnormal or normal proteins to free amino acids.[10,21,22] Thus these esterases do not appear to play an essential role in these degradative processes. An additional serine esterase has been purified recently from *E. coli* using *N*-carbobenzoxy-L-alanyl-L-alanyl-L-leucine-*p*-nitroanilide, a synthetic substrate of subtilisin.[23] Its properties and amino acid composition are similar to those of an intracellular serine protease found in several species of *Bacillus*. However, its ability to hydrolyze proteins has not yet been established.

A number of proteolytic activities have been isolated that are capable of hydrolyzing polypeptides to acid-soluble material. Regnier and Thang[24,25] partially purified, from *E. coli* extracts, an activity that digests casein and that they call protease A to distinguish it from two other proteolytic peaks (B and C), which were not defined further. Subsequent studies[7] showed that protease A actually is a mixture of three different enzymes, two active against casein (Mi and Fa) and one against insulin (Pi). In addition, Cheng and Zipser[4] have purified an enzyme, named protease III, by using as a substrate, the amino-terminal fragments of β-galactosidase ("auto-α"). These polypeptides (MW $\sim$ 7000) can be generated by autoclaving β-galactosidase and can be assayed rapidly and with sensitivity by using *in vitro* complementation.[26] This assay has been used extensively to follow the degradation of nonsense fragments of β-galactosidase in intact cells.[22,26–28] However, protease III is not responsible for this degradative process *in vivo*, since mutants lacking this protease degrade such nonsense fragments as rapidly as wild-type cells.[29] As will be discussed, protease III appears identical to the periplasmic insulin-degrading enzyme, named protease Pi, that was isolated by Swamy and Goldberg[7] by a very different approach (see following discussion). After this review was completed, Regnier isolated a casein-degrading enzyme from the outer membrane of *E. coli* which was named protease IV.[29a,29b]

[20] M. Pacaud, L. Sibilli, and G. LeBrass, *Eur. J. Biochem.* **69**, 141 (1971).
[21] J. D. Kowit, W. N. Choy, S. P. Champe, and A. L. Goldberg, *J. Bacteriol.* **128**, 776 (1976).
[22] C. G. Miller and D. Zipser, *J. Bacteriol.* **130**, 34 (1977).
[23] A. Y. Strongin, D. I. Gorodetsky, and V. M. Stepanov, *J. Gen. Microbiol.* **110**, 443 (1979).
[24] P. Regnier and M. N. Thang, *C.R. Hebd. Seances Acad. Sci. Ser. D* **277**, 2817 (1973).
[25] P. Regnier and M. N. Thang, *Eur. J. Biochem.* **54**, 445 (1975).
[26] I. Zabin and M. R. Villarejo, *Annu. Rev. Biochem.* **44**, 295 (1975).
[27] J. D. Kowit and A. L. Goldberg, *J. Biol. Chem.* **252**, 8359 (1977).
[28] A. I. Bukhari and D. Zipser, *Nature (London), New Biol.* **243**, 238 (1973).
[29] Y. E. Cheng, D. Zipser, C. Cheng, and S. J. Rolseth, *J. Bacteriol.* **140**, 125 (1979).
[29a] P. Regnier, *Biochem. Biophys. Res. Commun.* **99**, 844 (1981).
[29b] P. Regnier, *Biochem. Biophys. Res. Commun.* **99**, 1369 (1981).

A systematic attempt to define the proteolytic enzymes in *E. coli* has been carried out recently by Swamy and Goldberg.[7] These workers have isolated eight soluble enzymes that hydrolyze radioactively labeled globin, casein, or insulin. All these enzymes have been purified almost to homogeneity by methods to be described. Their physical and chemical properties indicate that they are distinct endoproteases.[7] The six that degrade globin or casein are serine proteases. The other two enzymes are inactive against these substrates but rapidly hydrolyze smaller polypeptides, such as insulin or glucagon. One of these serine proteases is of particular interest, since it degrades globin and casein only in the presence of ATP. This requirement for a nucleoside triphosphate can explain the stimulation of proteolysis by ATP in crude *E. coli* extracts[30] and the energy requirement for protein breakdown in intact cells.[1,27,31,32]

For most of the proteases described thus far, clear physiological functions have not been established. One notable exception is the *recA* gene product, a protein with multiple enzymatic activities that has been purified to homogeneity by Roberts *et al.*[5] This 40,000-MW protein not only is essential for genetic recombination,[33] but also catalyzes the endoproteolytic inactivation of the phage λ-repressor during lysogenic induction of the prophage.[5,34] The purified *recA* protein cleaves the λ-repressor in the presence of ATP and a polynucleotide,[35] and also inactivates in a similar reaction the repressor of the *recA* gene (the *lexA* gene product).[17] This enzyme, however, does not digest other commonly employed protein substrates (e.g., casein or globin). Thus the *recA* product clearly differs from the ATP-dependent protease (protease La), which is very active against such polypeptides.[7-9]

A variety of observations indicate that protease La catalyzes the rate-limiting step in the breakdown of abnormal proteins *in vivo*. Studies with intact cells indicated that the initial endoproteolytic steps in this process require metabolic energy,[27] apparently because of the ATP requirement of protease La. Genetic studies have shown that mutants in *lon* gene (also called *capR* and *deg*) lead to a reduced capacity of the cells to degrade abnormal proteins *in vivo*.[28,36] Very recently, the product of the *lon* gene has been purified to near homogeneity[37] and the purified protein shown to

[30] K. Murakami, R. Voellmy, and A. L. Goldberg, *J. Biol. Chem.* **254**, 8194 (1979).
[31] K. Olden and A. L. Goldberg, *Biochim. Biophys. Acta* **542**, 385 (1978).
[32] J. Mandelstam, *Bacteriol. Rev.* **24**, 289 (1960).
[33] A. Clark, *Annu. Rev. Genet.* **7**, 67 (1973).
[34] J. W. Roberts and C. W. Roberts, *Proc. Natl. Acad. Sci. U.S.A.* **72**, 147 (1975).
[35] N. L. Craig and J. W. Roberts, *Nature (London)* **283**, 26 (1980).
[36] S. Gottesman and D. Zipser, *J. Bacteriol.* **133**, 844 (1978).
[37] B. A. Zehnbauer, E. C. Foley, G. W. Henderson, and A. Markovitz, *Proc. Natl. Acad. Sci. U.S.A.* **78**, 2047 (1981).

be an ATP-dependent protease.[37a,37b] This activity appears identical to protease La in many physical and chemical properties.[37b] Furthermore, the content of protease La (but no other *E. coli* protease) is severalfold greater in bacterial strains in which the *lon* gene is cloned on plasmid, RGC 121/pJMC40.[37b]

In bacteria as in mammalian cells, secreted and membrane proteins are synthesized as larger precursors and then undergo limited proteolytic processing.[11,12] The mechanisms of this process in *E. coli* remain controversial. Recently Zwizinski and Wickner purified an activity from the *E. coli* membrane that they termed "leader peptidase" and that catalyzes the processing of the procoat protein of phage M13.[38] Inner membrane fractions also appear active in the maturational processing of alkaline phosphatase, a periplasmic enzyme.[39] However, it is presently unclear whether one or more "signal peptidases" exist.

One important clue to the function of the other proteases is their subcellular distribution.[39a] A number of these enzymes are found within the cytoplasm (where intracellular protein degradation probably occurs), and two are localized in the periplasmic region.[7] One casein-degrading enzyme is found in both locations. The periplasmic localization suggests a possible role in the degradation of surface proteins, of signal peptides, or of exogenous polypeptides (e.g., colicins). In addition, there is appreciable proteolytic activity associated with the cell membrane even after extensive washing.[8,40,41] One protease has been demonstrated recently on the outer membrane.[29a,29b] Furthermore, in gently lysed cell homogenates, most of the ATP-stimulated protease activity can be found in the membrane fraction.[8] The relationship between the membrane-associated ATP-stimulated protease found in gently lysed extracts and the soluble ATP-dependent protease (La) remains to elucidated.

Peptidases in E. coli

E. coli can utilize exogenous peptides as a source of essential amino acids or as metabolizable substrates, and these bacteria contain specific transport systems for the uptake of small peptides.[42,43] Specific peptidases

[37a] M. Charette, G. W. Henderson, and A. Markovitz, *Proc. Natl. Acad. Sci. U.S.A.* **78**, 4728 (1981).
[37b] C. H. Chung and A. L. Goldberg, *Proc. Natl. Acad. Sci. U.S.A.* **78**, 4931 (1981).
[38] C. Zwizinski and W. Wickner, *J. Biol. Chem.* **255**, 7973 (1980).
[39] C. N. Chang, P. Model, H. Inouye, and J. Beckwith, *J. Bacteriol.* **142**, 726 (1980).
[39a] K. H. Sreedhara Swamy and A. L. Goldberg, *J. Bacteriol.* in press (1981).
[40] P. Regnier and M. N. Thang, *FEBS Lett.* **36**, 31 (1973).
[41] C. H. Chung, K. H. Sreedhara Swamy, and A. L. Goldberg, unpublished observations.
[42] A. J. Sussman and C. Gilvarg, *Annu. Rev. Biochem.* **40**, 397 (1971).
[43] J. W. Payne, *Ciba Found. Symp.* **50**, 305 (1977).

are essential for growth on defined di- or tripeptides, and this requirement has allowed the selection of mutants lacking the enzymes.[13,44] In addition, certain of these intracellular peptidases seem to play an important role in the complete degradation of intracellular proteins. In a strain lacking multiple peptidases, protein degradation during starvation for a carbon source occurs more slowly than in the wild type.[10] The slower breakdown of proteins in these strains results from the inability of these bacteria to convert peptides to free amino acids. Voellmy *et al.*[9] have also found that the crude extracts from a multiple peptidase-deficient strain hydrolyzed ^{35}S-labeled *E. coli* proteins only to large acid-soluble peptides, whereas the extracts from wild-type cells hydrolyzed them to free amino acids. Thus these peptidases appear to function in the final steps of protein degradation following the action of endoproteases.

A number of peptide hydrolases have also been isolated and purified from *E. coli,* including several aminopeptidases,[45–50] three dipeptidases,[51–53] and a dipeptidyl carboxypeptidase.[54] In addition, several peptidase-deficient mutants have been isolated in *E. coli* and in the closely related bacteria *Salmonella typhimurium.*[45,55] Most of these mutations have been mapped and their genetics have been reviewed by Miller.[13] The purification and properties of these various peptidases will not be discussed.

Nomenclature of Proteases

A novel nomenclature for the eight soluble proteases has been suggested by Swamy and Goldberg.[6,7] The obvious, overworked approaches would involve numerical or alphabetical designations. However, such names have been used already to refer to enzymes that do not hydrolyze proteins at significant rates (e.g., protease I and II), or to mixtures of several enzymes (proteases A, B, C) or to peptidases (A, B, D, M, N, P, Q) that are inactive against proteins. Since the two enzymes that degrade insulin (or similar-sized polypeptides) have distinct compartmentalizations in the cell, we suggest the name Pi for the periplasmic insulin-degrading

[44] C. G. Miller and G. Schwartz, *J. Bacteriol.* **135**, 603 (1978).
[45] S. Sarid, A. Berger, and E. Katchalski, *J. Biol. Chem.* **237**, 2207 (1962).
[46] C. S. Tsai and A. T. Matheson, *Can. J. Biochem.* **43**, 1643 (1965).
[47] V. M. Vogt, *J. Biol. Chem.* **245**, 4760 (1970).
[48] A. Yaron and A. Berger, this series, Vol. 19, p. 521.
[49] C. Lazdunski, J. Busuttil, and A. Lazdunski, *Eur. J. Biochem.* **60**, 363 (1975).
[50] L. M. Yang and R. Somerville, *Biochim. Biophys. Acta* **445**, 406 (1976).
[51] E. Haley, *J. Biol. Chem.* **243**, 5748 (1968).
[52] E. Patterson, J. S. Gatmaitan, and S. Hayman, *Biochemistry* **12**, 3701 (1973).
[53] J. L. Brown, *J. Biol. Chem.* **248**, 409 (1973).
[54] A. Yaron, D. Mlyner, and A. Berger, *Biochem. Biophys. Res. Commun.* **47**, 897 (1972).
[55] C. G. Miller and K. Mackinnon, *J. Bacteriol.* **120**, 355 (1974).

enzyme, and Ci for the cytoplasmic insulin-degrading enzyme.[39a] For the proteases that degrade globin and casein, such a simple classification is not useful. Swamy and Goldberg[6,7] therefore introduced an easily remembered designation for these enzymes: Do, Re, Mi, Fa, So, La, in the order of their elution from DEAE–cellulose. One advantage of this nomenclature is that the names can be readily applied for the names of the genetic loci that code for these proteases. When more precise information is available about their physiological or chemical specificities, a more descriptive nomenclature will be possible.

Isolation and Properties of the Proteases

Buffers

The pH of all Tris-HCl buffers is adjusted at 4°C.

Buffer A: 50 mM Tris-HCl (pH 7.8)– 10 mM MgCl$_2$–200 mM KCl

Buffer B: 10 mM Tris-HCl (pH 7.8)–5 mM MgCl$_2$

Buffer C: 50 mM Tris-HCl (pH 8.1)–10 mM MgCl$_2$

Buffer D: 50 mM Tris-HCl (pH 7.8)–5 mM MgCl$_2$–100 mM NaCl

Buffer E: 10 mM Sodium acetate, pH 5.6

Buffer F: 10 mM Tris-HCl, pH 8.4

Buffer G: 25 mM Tris-HCl (pH 7.8)–5 mM MgCl$_2$–50 mM NaCl

Buffer H: 10 mM Sodium acetate (pH 5.2)–2.5 mM MgCl$_2$

Buffer I: 20 mM Potassium phosphate, pH 7.4

Buffer J: 50 mM Tris-HCl (pH 7.8)–5 mM MgCl$_2$–50 mM NaCl

Buffer K: 10 mM Sodium acetate (pH 5.0)–2.5 mM MgCl$_2$

Buffer L: 10 mM Tris-HCl (pH 8.4)– 1 mM cysteine–20 mM NaCl–5 mM MgCl$_2$

Buffer M: 10 mM Tris-HCl (pH 8.0)–5 mM MgCl$_2$–10 mM 2-mercaptoethanol

Buffer N: 10 mM Tris-HCl (pH 7.8)–5 mM MgCl$_2$–100 mM NaCl

Buffer O: 10 mM Tris-HCl, pH 7.8

Buffer P: 10 mM Potassium phosphate, pH 6.5

Buffer Q: 50 mM Tris-HCl (pH 7.8)–5 mM MgCl$_2$–100 mM NaCl–0.25 mM ATP–10% (v/v) glycerol

Buffer R: 100 mM Potassium phosphate (pH 6.5)–10 mM 2-mercaptoethanol–1 mM EDTA–20% (v/v) glycerol

Buffer S: 10 mM Tris-HCl (pH 7.4)–1 mM 2-mercaptoethanol–1 mM EDTA–20 mM NaCl–20% (v/v) glycerol

Buffer T: 20 mM Tris-HCl (pH 7.8)–5 mM MgCl$_2$–1 mM dithiothreitol–0.1 mM EDTA–50 mM NaCl–20% (v/v) glycerol

Buffer U: 20 mM Potassium phosphate (pH 6.8)–1 mM 2-mercaptoethanol–1 mM EDTA–50 mM NaCl–20% (v/v) glycerol

Assays

Proteolytic activity is measured by following the degradation of [^{14}C]-methyl(apo)hemoglobin, [^{3}H]methyl-α-casein, [^{125}I]insulin to products soluble in 10% trichloroacetic acid (TCA). To prepare these substrates, crystalline beef hemoglobin (Sigma) is methylated with [^{14}C]formaldehyde (40–60 mCi/mmol, New England Nuclear) and bovine α-casein with [^{3}H]formaldehyde (85 mCi/mmol, New England Nuclear) according to the procedure of Rice and Means.[56] To increase its proteolytic susceptibility, heme is extracted from the hemoglobin by use of methylethylketone.[57] The resulting specific activity of ^{14}C-globin is about $(2–4) \times 10^6$ cpm/mg, whereas that of ^{3}H-casein is $(2–3) \times 10^7$ cpm/mg. [^{125}I]Insulin (New England Nuclear) is stored in the presence of 5% bovine serum albumin at $-30°$C.

Unless otherwise indicated, all the assays are carried out in buffer C in the presence and absence of 3 mM ATP. When assayed with globin or casein as substrates, each incubation mixture contains an appropriate amount of enzyme and 4–5 μg of ^{14}C-globin or 10–15 μg of ^{3}H-casein in a final volume of 0.5 ml. Assays using insulin as a substrate contain 5–15 μg of insulin (Eli Lilly Co., Indianapolis, Indiana), a trace amount of [^{125}I]-insulin (about 12,000–15,000 cpm), and the enzyme in a final volume of 0.5 ml. The incubations are performed for 1 hr at 37°C. Each reaction is stopped by the addition of 60 μl of 100% TCA and 40 μl of bovine serum albumin (30 mg/ml) as a carrier. The assay tubes are kept on ice for 30 min, and after centrifugation, 0.4 ml of acid-soluble products from [^{14}C]-globin or [^{3}H]casein hydrolysis are counted in 4 ml of Liquiscint (National Diagnostics, Parsippany, New Jersey). The products of [^{125}I]insulin hydrolysis are counted in an auto-gamma spectrometer.

In experiments with inhibitors the agents are incubated for 20–30 min at 37°C with the enzyme and buffer before the addition of substrate.

Preparation of Cell-Free Extract

Seventy-five grams (wet weight) of freshly grown, or frozen *E. coli* K12 cells (Grain Processing Co., Muscatine, Iowa) are suspended in 150 ml of buffer A at 4°C. The cell suspension is disrupted in a French press at 14,000 psi, the resulting homogenate is centrifuged at 31,000 g for 30 min, and the supernatant at 130,000 g for 3 hr. Following extensive dialysis against buffer B, the insoluble material is removed by centrifugation. This amount of cells yields 3–3.5 g of protein in the soluble extract. All the proteases to be described have been prepared from such extracts. How-

---

[56] R. H. Rice and G. E. Means, *J. Biol. Chem.* **246**, 831 (1971).
[57] F. W. J. Teale, *Biochim. Biophys. Acta* **35**, 543 (1959).

ever, protease La (the *lon* gene product) has been purified to near homogeneity by an alternative procedure[37] (see following discussion).

Fractionation of Cell-Free Extract on DEAE-Cellulose

The dialyzed cell-free extract (3 g protein) is applied to a DEAE-cellulose column (2.5 × 27 cm) equilibrated with buffer B. The column is washed with the same buffer until the A_{280} of the eluate is less than 0.05. The adsorbed proteins are then eluted with 1500 ml of a linear salt gradient (0–0.2 M NaCl). The flow rate is 100 ml/hr, and 15-ml fractions are collected. Alternate fractions are assayed for proteolytic activity against [³H]-casein and [¹²⁵I]insulin in the presence and absence of 3 mM ATP. As shown in Figs. 1 and 2, five different peaks of proteolytic activity against these substrates are found. For convenience, they are designated peaks I–V, according to their order of elution from the column.[7] Although all the casein-hydrolyzing peaks also hydrolyzed globin, the activity against casein is much higher. The activity in peak IV, but in no other peak, is stimulated dramatically (4- to 20-fold) by ATP.

To obtain sufficient material for further purification, peaks from several DEAE-cellulose columns can be pooled. The following steps are described for enzyme preparations that would correspond to starting with 250–300 g of frozen cells.

Proteases in Peak I—Do and Re

Purification of Protease Do

Peak I (i.e., the fractions that did not adsorb to DEAE-cellulose) show strong hydrolytic activity against globin and casein. To resolve this activity further, fractions under peak I are pooled and concentrated by precipitating the proteins with solid ammonium sulfate added to 95% saturation. After centrifugation at 31,000 g for 30 min, the pellet is dissolved in buffer E and dialyzed against the same buffer. After removing the precipitated material by centrifugation, the proteins are loaded on a CM-cellulose column (1.5 × 12 cm) equilibrated with buffer E. The adsorbed proteins are eluted with a 0–0.3 M linear NaCl gradient. The flow rate is 35 ml/hr and 4-ml fractions are collected. Fractions are assayed in 50 mM Tris-HCl, pH 7.8. As shown in Fig. 3, the cation exchanger resolves the activity into two distinct peaks, referred to as proteases Do and Re. Both of these peaks hydrolyze globin and casein.

The active fractions corresponding to protease Do from the CM-cellulose column are pooled, concentrated by ultrafiltration through a

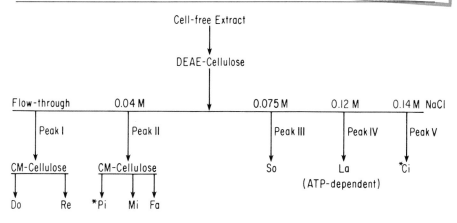

FIG. 1. Flowchart for the separation of *E. coli* proteases. *, Insulin-degrading proteases. (All others degrade globin and casein, but not insulin.)

PM-10 membrane, and dialyzed against buffer D. The dialyzed enzyme preparation is loaded on a Sepharose-6B column (2.0 × 100 cm) equilibrated with the same buffer. Fractions of 2 ml are collected at a flow rate of 20 ml/hr. Fractions containing protease Do activity are clearly separated from the nonenzymatic protein peak. The active fractions are pooled, concentrated by ultrafiltration, and stored at −30°C.

Properties of Do

Protease Do is located primarily in the cytoplasm.[39a] The enzyme has similar activity over a broad pH range from 6 to 8.5. The enzyme has an unusually large molecular weight of about 520,000, as determined by gel filtration on Sepharose-6B. Based on sucrose-density gradient centrifugation, the enzyme has a sedimentation coefficient of 16 S. In certain preparations the enzyme elutes with an apparent molecular weight of 320,000–350,000.[57a] The reason for this variability is unknown. The enzyme is inhibited by DFP (80% at 1 mM), and therefore it is a serine protease. It is not sensitive to metal chelators or sulfhydryl inhibitors.

Purification of Protease Re

Fractions exhibiting protease Re activity (Fig. 2) are combined, and precipitated by adding solid ammonium sulfate to 55% saturation. After centrifugation at 31,000 *g* for 30 min, the pellet is discarded, and the supernatant is brought to 80% saturation with ammonium sulfate. After

[57a] K. H. S. Swamy, C. H. Chung, and A. L. Goldberg, in preparation (1981).

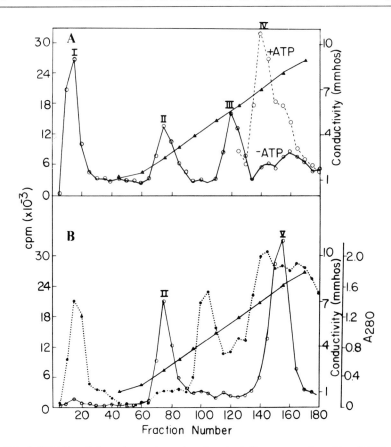

FIG. 2. DEAE–cellulose chromatography of proteolytic enzymes from *E. coli* K12. (A) Degradation of ^{3}H-casein measured in the absence (O—O) and presence (O---O) of ATP. ▲—▲, Conductivity. (B) Degradation of ^{125}I-insulin (O—O) is shown and compared with A_{280} (●···●) and the conductivity (▲—▲). The dialyzed cell-free extract is adsorbed to a DEAE–cellulose column and eluted with 1500 ml of a linear NaCl gradient (0–0.2 M NaCl). The flow rate is 100 ml/hr and 15-ml fractions are collected. Alternate fractions are assayed for proteolytic activity against ^{3}H-casein and ^{125}I-insulin in the absence and presence of 3 mM ATP.

stirring the suspension for 2 hr at 4°C, the suspension is centrifuged at 31,000 g for 30 min; the pellet is then resuspended in buffer J and dialyzed against the same buffer. The enzyme preparation is applied to an Ultrogel ACA 44 (LKB) column (1.5 × 95 cm) equilibrated with buffer J. 2-ml fractions are collected at a flow rate of 15 ml/hr. The active fractions from the Ultrogel column are then pooled, dialyzed against buffer F, and loaded

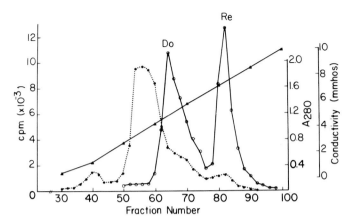

FIG. 3. CM-cellulose chromatography of proteases in peak I. The pooled peak I fractions from DEAE–cellulose are concentrated, dialyzed against buffer E, and applied to a CM-cellulose column (1.5 × 12 cm) equilibrated in buffer E. The absorbed proteins are eluted with 400 ml of 0–0.3 M linear NaCl gradient. The flow rate is 35 ml/hr and 4-ml fractions are collected. Hydrolysis of ^{3}H-casein (○—○); A_{280} (●· · ·●); conductivity (▲—▲).

on a DEAE–Sepharose column (0.5 × 9.5 cm) equilibrated with the same buffer. A linear gradient of 0 to 0.2 M NaCl is run at a flow rate of 20 ml/hr, and 1-ml fractions are collected. The peak of enzyme activity appears at about 75 mM NaCl.

Properties of Re

In intact cells protease Re is almost equally distributed between periplasm and cytoplasm.[39a] The enzyme has MW ~ 82,000 as determined by gel filtration on Sephadex G-75. Analysis of the enzyme on SDS–polyacrylamide gel electrophoresis shows a single band of MW 82,000. Thus the enzyme is composed of a single polypeptide chain.[57b]

The enzyme is active over the pH range of 7 to 8.5, with an optimal activity at pH 8.0. The enzyme is sensitive to inhibition by DFP (60% at 10 mM) and phenylmethylsulfonyl fluoride (PMSF) (60% at 10 mM). Therefore, it appears to be a serine protease, but it is also inhibited 70% by either EDTA (1 mM) or o-phenanthroline (1 mM). The enzyme is completely inhibited by L-1-tosylamido-2-phenylethylchloromethyl ketone (TPCK) (1 mM), but not by 1-chloro-3-tosylamido-7-amino-2-heptanone (TLCK).

[57b] C. H. Chung, K. H. S. Swamy, and A. L. Goldberg, in preparation (1981).

Proteases in Peak II—Pi, Mi, and Fa

Purification

The fractions in peak II from the DEAE–cellulose column contain both casein- and insulin-degrading activities. They are pooled and concentrated by precipitating the protein with solid ammonium sulfate added to 95% saturation. After centrifugation at 31,000 g for 30 min, the pellet is dissolved in buffer D and dialyzed against the same buffer. The dialyzed enzyme preparation is loaded onto a Bio-Gel A-0.5m column (2.5 × 110 cm) equilibrated with buffer D. The flow rate is 25 ml/hr, and 3-ml fractions are collected. Alternate fractions are assayed against ³H-labeled casein and [¹²⁵I]insulin. Proteolytic activities against both substrates appear in a single peak. The active fractions are pooled, dialyzed against buffer H, and applied to a CM-cellulose column (1.5 × 16 cm) equilibrated with the same buffer. After washing the column with 100 ml of buffer, the proteins are eluted with a 0 to 0.3 M linear NaCl gradient (350 ml total volume). Flow rate is 35 ml/hr, and 3-ml fractions are collected. Three peaks of proteolytic activity are resolved in this way (Fig. 4), but they can also be separated nicely with isoelectric focusing.[7] One hydrolyzes insulin, but not casein or globin, whereas the other two degrade these larger proteins but not insulin.

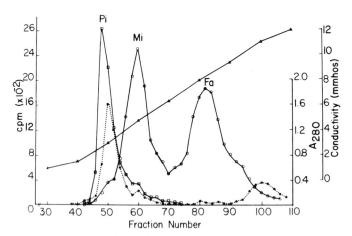

FIG. 4. CM-cellulose chromatography of proteases in peak II. The pooled Ultrogel AcA 34 fractions are dialyzed against buffer H and applied to a CM-cellulose column (1.5 × 16 cm) equilibrated in buffer H. The adsorbed proteins are eluted with 350 ml of 0–0.3 M linear NaCl gradient. The flow rate is 35 ml/hr and 3-ml fractions are collected. Hydrolysis of ¹²⁵I-insulin (□—□) and ³H-casein (○—○); A_{280} (●···●); conductivity (▲—▲).

The insulin-degrading activity has been named protease Pi (periplasmic insulin-degrading enzyme) because of its localization exclusively in the periplasmic compartment.[39a] It is identical to protease III, described by Cheng and Zipser,[4] as discussed below. The two casein-degrading enzymes are named Mi and Fa. The active fractions corresponding to protease Mi and Fa are pooled, concentrated by ultrafiltration through PM-10 membrane, dialyzed against buffer C, and stored at $-30°C$.

The protease Pi (protease III) is further purified by chromatography on hydroxyapatite. The active fractions under the peak of protease Pi are pooled, dialyzed against buffer I, and applied to a hydroxyapatite column (1.5 × 14 cm) equilibrated with the same buffer. The enzyme is eluted with a linear gradient of 20–200 mM potassium phosphate, pH 7.4 (250 ml total volume). The flow rate is 45 ml/hr, and 2.5-ml fractions are collected. The fractions with highest activity are pooled, concentrated by ultrafiltration through PM-10 membrane, dialyzed against buffer C, and stored at $-30°C$. Cheng and Zipser have purified this enzyme to homogeneity by an alternative approach that involves many additional steps.[4]

Properties of Protease Pi

This enzyme is localized in the periplasm of the cell.[39a] It hydrolyzes insulin, glucagon, and the auto-α-fragment of β-galactosidase.[26] The optimum pH for the enzyme activity is 7.5. Protease Pi has MW $\sim$ 110,000 as determined by gel filtration on Sephadex G-200. The enzyme activity is very sensitive to inhibition by EDTA (80% at 1 mM) and o-phenanthroline (50% at 1 mM). The enzyme activity is also inhibited by dithiothreitol by 60% at 5 mM but not by 2-mercaptoethanol. The enzyme is not inhibited by sulfhydryl inhibitors or inhibitors of serine proteases.

As shown by its physical properties and inhibitor sensitivity, protease Pi appears identical to protease III purified by Cheng and Zipser[4] using the auto-α-polypeptide as a substrate. In addition, the protease III-deficient mutant[29] lacks the insulin-degrading activity in peak II.[39a,58] Using oxidized insulin B chain as a substrate, Cheng and Zipser[4] have shown that protease III initially cleaves the peptide bond between tyrosine and leucine (16-17) and also cleaves between phenylalanine and tyrosine (25-26) at a slower rate.

Properties of Protease Mi and Fa

These two enzymes have different locations in the cell: protease Mi in the periplasm, protease Fa in the cytoplasm.[39a] Both enzymes degrade casein and globin. Protease Fa also degrades the auto-α-polypeptide.[58]

[58] C. H. Chung, K. H. Sreedhara Swamy, and A. L. Goldberg, unpublished observations.

Both these enzymes have MW ~ 110,000, as determined by gel filtration on Sephadex G-200.[58a]

Both Mi and Fa are serine proteases, although Fa is inhibited more strongly than Mi by DFP. At 10 mM concentration, DFP inhibits the casein-degrading activity of Mi by 55% and almost completely the activity of Fa. TPCK inhibits protease Fa (80% at 0.5 mM) but not Mi. Both are insensitive to inhibition by TLCK.

The activities of Mi and Fa are also sensitive to inhibition by metal chelators. EDTA inhibits protease Mi and Fa by 80% at 1 mM. o-Phenanthroline also inhibits Mi (63% at 1 mM) and Fa (80% at 1 mM). Sulfhydryl group-blocking agents do not affect either activity. Dithiothreitol inhibits the activity of Mi (35% at 1 mM) and Fa (67% at 1 mM), but 2-mercaptoethanol has no effect.

Protease in Peak III—So

Purification of Protease So

The activity against casein in peak III from the DEAE–cellulose column is pooled, and solid ammonium sulfate is slowly added to 45% saturation with stirring at 4°C. After additional stirring for 1 hr, the suspension is centrifuged at 31,000 g for 30 min, and the pellet is discarded. Ammonium sulfate is added again to bring the supernatant to 75% saturation, and the mixture is then stirred at 4°C for 2 hr. The pellet is collected by centrifugation at 31,000 g for 30 min, dissolved in buffer J, and then dialyzed overnight against the same buffer. The dialyzed material is applied to an Ultrogel AcA 34 column (2.5 × 120 cm) equilibrated with buffer J. Proteins are eluted with buffer J at a flow rate of 20 ml/hr, and fractions of 3 ml are collected. The fractions containing protease activity are pooled and dialyzed against buffer K at 4°C overnight. Insoluble proteins are removed by centrifugation at 31,000 g for 30 min. The supernatant is loaded on a CM-cellulose column (1.5 × 12 cm) equilibrated with buffer K. Proteins are eluted with a linear gradient of 0 to 0.3 M NaCl in buffer K. Fractions of 3 ml are collected at a flow rate of 30 ml/hr. Fractions with high protease activity are pooled, concentrated, and dialyzed against buffer L at 4°C for 8 hr.

The dialyzed protease preparation from the previous step is then applied to a DEAE–Sepharose column (1.5 × 5.6 cm) equilibrated with buffer L. Proteins are eluted with buffer L at a flow rate of 12 ml/hr, and fractions of 3 ml are collected. Protease So activity is eluted in the flow-through fractions, whereas all of the nonenzymatic proteins remain bound

[58a] K. H. S. Swamy and A. L. Goldberg, in preparation (1981).

TABLE I
PURIFICATION OF PROTEASE So

Step	Protein (mg)	Specific activity (units/mg protein)[a]	Purification factor (-fold)	Total units[a]	Yield (%)
DEAE-cellulose Peak III	369.8	0.5	1	184.9	100
Precipitates at 45-75% (NH$_4$)$_2$SO$_4$	214.1	1.2	2.4	256.9	140
Ultrogel AcA 34	61.0	4.7	9.4	286.7	155
pH precipitation	57.4	5.7	11.4	327.2	177
CM-cellulose	5.4	101.1	202.2	545.9	295
DEAE-Sepharose	0.16	1008.5	2017.0	161.4	87

[a] One unit is defined as 1 μg casein hydrolyzed per hour per milligram of protein.

to the resin. The active fractions are pooled and concentrated. These methods yield protease So with about a 2000-fold purification over that in peak III from DEAE-cellulose (Table I).

Properties of So

Protease So is a cytoplasmic enzyme of MW 140,000 as determined by gel filtration of Sephadex G-200.[39a,58b] On SDS–polyacrylamide gel electrophoresis, the enzyme shows a single band of MW 70,000. The native enzyme thus is a dimer of identical subunits. The purified enzyme appears homogeneous as determined by polyacrylamide-gel electrophoresis under denaturing and nondenaturing conditions.[58b] It is active over a pH range of 6–8, with maximal activity at 6.5. Because the enzyme is inactivated by DFP (60% at 10 mM) and PMSF (80% at 10 mM), it seems to be a serine protease. It is also inhibited by about 80% by pentamidine (1 mM). It is not sensitive to metal chelating agents or sulfhydryl inhibitors. The enzyme activity is completely inhibited by 1 mM TPCK but not by TLCK.

Protease in Peak IV—La: The ATP-Dependent Protease

Purification of Protease La

Two alternative approaches for the purification of protease La have been developed. (I) The first utilizes peak IV obtained from the DEAE-cellulose column between 0.105 and 0.15 M NaCl. This region shows

[58b] C. H. Chung and A. L. Goldberg, in preparation (1981).

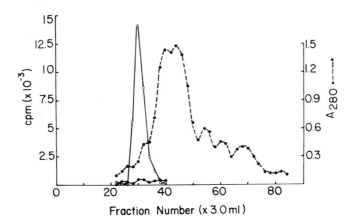

Fraction Number (x 3.0 ml)

FIG. 5. Bio-Gel P-300 column chromatography of protease La. The pooled activity from the DEAE–Sepharose column is dialyzed against buffer Q, concentrated, and loaded to a Bio-Gel P-300 column (2.6 × 45 cm) equilibrated in buffer Q. Proteins are eluted with the same buffer. Fractions of 3 ml are collected at a flow rate of 10 ml/hr. Fractions are assayed for ^{3}H-casein hydrolysis in the presence (O—O) and absence (●—●) of 3 mM ATP. A_{280} (●-----●).

appreciable casein-degrading activity in the presence of 3 mM ATP but not in its absence (Fig. 2). The fractions with high ATP-dependent activity are pooled, dialyzed against buffer M, and loaded onto a DEAE–Sepharose column (1.5 × 30 cm) equilibrated with buffer M. The proteins are eluted with a 0–0.2 M NaCl gradient in buffer M, and a symmetrical peak of ATP-dependent activity came off the column at about 0.12M NaCl concentration. Fractions with the highest activity are pooled, concentrated, and loaded onto a Bio-Gel P-300 column (2.6 × 45 cm) equilibrated with buffer Q. By this method, most of the nonenzymatic proteins can be well separated from the ATP-dependent protease activity (Fig. 5). For this purpose, we find Sephacryl S-300 is also useful. The active fractions are pooled and stored at −30°C. This approach yields a highly purified activity, which, however, is not homogeneous.

(II) As just discussed, protease La has recently been shown to be identical to the product of the lon ($CapR$) gene.[37b] Before this identification was achieved, Zehnbauer et al.[37] developed a procedure for isolating the polypeptide encoded by the lon gene, which on SDS–polyacrylamide gel electrophoresis appears to have MW 94,000. This protein shows an affinity for both DNA and ATP. Subsequent work[37a,37b] has shown that this purified material has ATP-dependent proteolytic activity. Purification of this enzyme has benefited from the use of $E. coli$ in which the lon gene has

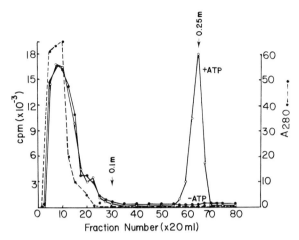

FIG. 6. Phosphocellulose column chromatography of protease La as described by Zehnbauer *et al.*[37] for the purification of *lon* gene product. Cell-free extract in buffer R is loaded onto a phosphocellulose column (2.6 × 20 cm) equilibrated with buffer R. The adsorbed proteins are eluted with 0.1–0.4 *M* linear phosphate gradient. Fractions are assayed for ³H-casein hydrolysis in the presence (○—○) and absence (●—●) of 3 m*M* ATP. A_{280} (●---●) and conductivity (×····×) are also shown. Data were kindly provided by Dr. L. A. Yeh.

been cloned on plasmid RGC 121/pJMC 40. Similar approaches, however, also work with wild-type bacteria.[58c]

Cell-free extracts are prepared by suspending 50 g of freshly grown *E. coli* K12 containing plasmid RGC 121/pJMC 40 (a strain containing the wild-type allele *lon*⁺ both on the chromosome and on the plasmids) in buffer R. The cell suspension is disrupted in a French press at 14,000 psi and the resulting homogenate is centrifuged at 48,000 *g* for 30 min. Proteins at 20 to 25 mg/ml in buffer R is then applied to a phosphocellulose (Whatman P11) column (2.6 × 20 cm) equilibrated with buffer R. The adsorbed proteins are eluted with 0.1 to 0.4 *M* linear phosphate gradient. The ATP-dependent proteolytic activity is eluted at 0.25 *M* phosphate (Fig. 6). Appropriate fractions from the phosphocellulose column are dialyzed against buffer S and applied to a DEAE–cellulose column (2.6 × 9.5 cm) equilibrated with buffer S. The column is eluted in a stepwise fashion with buffer S containing 0.075, 0.15, 0.2, 0.25, and 0.3 *M* NaCl in 5-ml fractions at a flow rate of 15 to 20 ml/hr. Fractions (0.2 *M* NaCl) containing the ATP-dependent protease activity from the DEAE–cellulose column are dialyzed against buffer T and applied to an ATP–

[58c] L. A. Yeh and L. Waxman, personal communication.

agarose (agarose-hexane-adenosine 5'-triphosphate, Type 4, PL Biochemicals) column (0.5 × 2.5 cm) equilibrated with buffer T. Fractions are eluted in buffer T containing 5 mM ATP. The fractions containing the highest enzymatic activity are pooled, dialyzed against buffer U, and applied to a DNA–cellulose (prepared according to Litman[58d]) column (0.5 × 1.5 cm) equilibrated with buffer U containing 0.3 mg bovine serum albumin per milliliter. Proteins are eluted in a stepwise fashion with buffer U containing 0.5, 0.1, 0.2, 0.3, and 0.5 M NaCl. A convenient alternative to ATP- and DNA-affinity chromatographies is gel filtration on a Sephacryl S-300, as described above.[58e] The purified ATP-dependent protease is obtained in the 0.05 M NaCl fractions and are stored at −30°C.

Properties of La

The casein- or globin-degrading activity of protease La is completely dependent on the presence of ATP and Mg^{2+} ions. Other nucleoside triphosphates (CTP, UTP, or deoxyadenosine triphosphate) can replace ATP but are less effective than ATP. Mn^{2+} and Ca^{2+} can also replace Mg^{2+}. Nonmetabolizable analogs of ATP (5'-adenylyl-β,γ-imidodiphosphate and 5'-adenylyl-β,γ-methylene diphosphate) do not support proteolytic activity, although they do bind to the enzyme, apparently at the same rate as ATP.

Protease La (*lon* protein) has MW 450,000 as determined by glycerol-density gradient centrifugation or by gel filtration on Sephacryl S-300.[37b] On SDS–polyacrylamide gel electrophoresis, the purified *lon* protein shows a single band of MW 94,000.[37,37a] Thus the native enzyme appears to be a tetramer of identical subunits. It is present in the cytoplasm of the cell,[39a] but a similar (perhaps an identical) enzyme appears to be bound to the surface membrane in intact cells.[8] Because it is sensitive to DFP, it is a serine protease. The inhibition by N-ethylmaleimide and iodoacetamide indicates that sulfhydryl groups are important for activity. The proteolytic activity is also sensitive to inhibition by vanadate, quercetin, and Diol-9, potent inhibitors of ATPases.[59] Vanadate appears to form a transition-state analog with the enzyme, as is seen with various ATPases. Therefore, ATP cleavage seems essential for the proteolytic activity. In support of this conclusion, the purified enzyme has ATPase activity that is sensitive to vanadate and that is activated by protein substrate.[59a]

Clear stoichiometry between ATP hydrolysis and protein cleavage, however, has not been demonstrated. Protease La does not show protein kinase activity and does not catalyze adenylation of itself or of casein.[59]

[58d] R. M. Litman, *J. Biol. Chem.* **243**, 6222 (1968).

[58e] C. H. Chung and A. L. Goldberg, *Proc. Natl. Acad. Sci. U.S.A.* in press (1981).

[59] F. S. Larimore, L. Waxman, and A. L. Goldberg, submitted for publication (1981).

[59a] L. Waxman and A. L. Goldberg, in preparation (1981).

Protease La in crude preparations and in the absence of glycerol (10–20%) is quite labile, and this property retarded attempts to purify it completely. The presence of organic solvents (above 1%), Triton X-100 above 0.1%, and temperatures above 37°C cause significant losses in enzyme activity. On storage at 4°C, the enzyme activity decreases about 30–40% per day. In the presence of 10 mM 2-mercaptoethanol or 3 mM ATP, these losses are reduced to less than 10%/day.[59] In addition, 3 mM ATP prevents the inactivation of the enzyme when incubated at 42°C for 1 hr. The stabilizing effects of ATP are further evidence that ATP binds directly to the protease in regulating its activity.[59]

Protease La differs in several important respects from the *recA* gene product, which cleaves certain regulatory proteins in an ATP-dependent reaction[5,17,35] (see preceding discussion). Adenosine 5'-*O*-(3-thiotriphosphate), a slowly metabolized analog of ATP, stimulates cleavage on the lambda repressor by *recA* proteins severalfold more effectively than ATP,[35] whereas it increases casein degradation by protease La only 10–20% as well as ATP. Thus proteolysis by *recA* does not seem to require ATP hydrolysis, although it contains an ATPase function. Mutants lacking a functional *recA* gene still contain protease La,[9] and thus these two ATP-dependent enzymes appear to be completely distinct proteins serving different physiological functions.

These properties differentiate this enzyme from any known proteolytic enzyme. In mammalian cells, as in *E. coli,* protein degradation requires metabolic energy,[1] and soluble extracts have been prepared from rabbit reticulocytes that degrade proteins in an ATP-dependent fashion.[60–62] An alkaline endoproteolytic activity that is stimulated 2- to 3-fold by ATP has been purified from such preparations.[63,64] It has similar molecular weight, and very similar proteases have been isolated from rat liver[65] and other tissues.[6] The ATP-dependent protease in *E. coli* differs from these enzymes in many properties.[59] For example, the mammalian enzymes are stimulated by nonmetabolizable analogs of ATP and by pyrophosphate and are not affected by vanadate.[6,63,64] Hershko, Rose, and co-workers have reported that ATP-dependent proteolysis in reticulocyte extracts[66–69]

[60] J. D. Etlinger and A. L. Goldberg, *Proc. Natl. Acad. Sci. U.S.A.* **74,** 54 (1977).
[61] A. L. Goldberg, J. D. Kowit, J. D. Etlinger, and Y. Klemes, *in* "Protein Turnover and Lysosome Function" (H. Segal and D. Doyle, eds.), p. 171. Academic Press, New York, 1978.
[62] J. D. Etlinger and A. L. Goldberg, *J. Biol. Chem.* **255,** 4563 (1980).
[63] F. S. Boches and A. L. Goldberg, *Fed. Proc., Fed. Am. Soc. Exp. Biol.* **39,** 1682 (1980).
[64] F. S. Boches, L. Waxman, and A. L. Goldberg, in preparation (1981).
[65] G. N. DeMartino and A. L. Goldberg, *J. Biol. Chem.* **254,** 3712 (1979).
[66] A. Ciechanover, H. Heller, S. Elias, A. L. Haas, and A. Hershko, *Proc. Natl. Acad. Sci. U.S.A* **77,** 1365 (1980).

involves multiple components that can be resolved by chromatography on DEAE–cellulose. They suggested that ATP does not affect a protease but instead is required for the ligation of a small peptide (identified as ubiquitin) to the protein substrates. This alteration is supposed to enhance proteolytic susceptibility. The properties of the *E. coli* protease La are not consistent with such models. The bacterial enzyme does not require ubiquitin, and its activity is not stimulated by this polypeptide (which is found in *E. coli*).

Instead, ATP cleavage appears to influence the protease directly either by an allosteric activation or by some novel involvement in the hydrolytic mechanism. Furthermore, interaction of protease La with substrates somehow promotes ATP hydrolysis.[59a] One other important property of this enzyme is that DNA stimulates both the proteolytic and ATPase activities.[58e]

Protease in Peak V—Ci

Purification of Protease Ci

The insulin-degrading activity of peak V from the DEAE–cellulose column is pooled and dialyzed overnight against buffer E at 4°C. Insoluble proteins are removed by centrifugation at 31,000 g for 30 min. The supernatant fraction is adjusted to pH 7.8 by adding 200 mM Tris-HCl (pH 8.0), and solid ammonium sulfate is then added to 40% saturation. After stirring for 2 hr at 4°C, the suspension is centrifuged at 31,000 g for 30 min, and the pellet discarded. The supernatant is brought to 65% saturation with ammonium sulfate and stirred for 2 hr at 4°C. The pellet is collected by centrifugation at 31,000 g for 30 min, dissolved in buffer N, and dialyzed against the same buffer.

The dialyzed material is applied to an Ultrogel AcA 34 column (2.5 × 110 cm) equilibrated with buffer N. Proteins are eluted with the same buffer, and 3-ml fractions are collected at a flow rate of 20 ml/hr. The active fractions from the Ultrogel column are pooled, dialyzed against buffer O overnight at 4°C, and applied to a butyl-agarose column (1.0 × 18 cm) equilibrated with the same buffer. Protein is eluted by a linear gradient of 0 to 0.2 M NaCl in buffer O, and 3-ml fractions are collected at a flow rate of 30 ml/hr. The peak of enzyme activity appears at a NaCl concentration of about 80 mM.

[67] H. Hershko, A. Ciechanover, H. Heller, A. L. Haas, and I. A. Rose, *Proc. Natl. Acad. Sci. U.S.A.* **77**, 1783 (1980).
[68] A. Ciechanover, S. Elias, H. Heller, S. Ferber, and A. Hershko, *J. Biol. Chem.* **255**, 7525 (1980).
[69] K. D. Wilkinson, M. K. Urban, and A. L. Haas, *J. Biol. Chem.* **255**, 7529 (1980).

TABLE II
PURIFICATION OF PROTEASE Ci

Step	Protein (mg)	Specific activity (units/mg protein)[a]	Purification factor (-fold)	Total units[a]	Yield (%)
DEAE–cellulose Peak V	157.3	0.5	1	991.0	100
pH precipitation	111.1	9.25	1.5	1027.7	104
Precipitates at 45–65% $(NH_4)_2SO_4$	80.8	12.3	2.0	993.4	100
Ultrogel AcA 34	35.4	23.6	3.7	835.4	84
Butyl-agarose	2.6	141.6	22.5	368.2	37
Hydroxyapatite	0.6	352.5	56.0	211.5	21

[a] One unit is defined as 1 μg insulin hydrolyzed per hour per milligram of protein.

TABLE III
PROTEASES FROM *E. coli*

Proteases	Stimulation by ATP	Inhibitors	Molecular weight
Globin and casein-degrading enzymes			
Do	−	DFP	520,000
Re	−	DFP, EDTA, o-phenan-throline, TPCK	82,000
Mi	−	DFP, EDTA, o-phenan-throline	110,000
Fa	−	DFP, EDTA, o-phenan-throline, TPCK	110,000
So	−	DFP, TPCK	140,000
La	+	DFP, N-ethyl-maleimide, EDTA	450,000
Insulin-degrading enzymes			
Pi (periplasmic)	−	EDTA, o-phenanthroline	110,000
Ci (cytoplasmic)	−	o-Phenanthroline, p-hydroxymercuri-benzoate	125,000

The fractions containing insulin-degrading activity from the butyl-agarose column are pooled, dialyzed against buffer P at 4°C overnight, and loaded on a hydroxyapatite column (1.0 × 8.5 cm) equilibrated with the same buffer. After washing the column with the same buffer, proteins are eluted with a linear gradient of 10–200 mM potassium phosphate, pH 6.5. 2-ml fractions are collected at a flow rate of 20 ml/hr. The active fractions from the column are pooled, dialyzed against buffer O, concentrated by ultrafiltration, and stored at −30°C. The summary of purification is shown in Table II.

Properties of Ci

This enzyme has MW 125,000 as determined by gel filtration on Sephadex G-200.[70] It is located in the cytoplasm of the cell[39a] and has a sharp pH optimum at 7.5. Protease Ci is inhibited by 1 mM o-phenanthroline by about 80%, but not by EDTA or EGTA. The enzyme is also completely inhibited by p-hydroxymercuribenzoate (1 mM), but is not very sensitive to other sulfhydryl inhibitors. The antibiotics bacitracin (5 μg/ml) and globomycin (5–20 μg/ml) inhibit protease activity by 40% and 70%, respectively. This activity is stimulated 2- to 4-fold by Mn^{2+} (10 mM) and Co^{2+} (1 mM); Mg^{2+} and Ca^{2+} also stimulate, but less effectively. This protease hydrolyzes insulin, glucagon, and the amino-terminal fragments of β-galactosidase (auto-α). The distinct features of Ci and other *E. coli* proteases are summarized in Table III.

Acknowledgments

These studies have been supported by grants from the National Institute of Neurological Disease and Stroke, the Juvenile Diabetes Foundation, and the Kroc Foundation. The authors are grateful to Mrs. Joanna Goldberg for her expert assistance in these studies and to Ms. Robin Levine and Ms. Maureen Rush for their assistance in the preparation of this manuscript. We have also benefited from discussions of these topics with our colleagues, L. Waxman, L. A. Yeh, and R. Voellmy. During the course of this work, Dr. Larimore held a National Research Service Award from the National Institutes of Health.

[70] C. H. Chung, K. H. S. Swamy, and A. L. Goldberg, in preparation (1981).

[51] Nonlysosomal Insulin-Degrading Proteinases in Mammalian Cells

By RICHARD J. KIRSCHNER and ALFRED L. GOLDBERG

The characterization of intracellular proteinases is essential for a complete understanding of the regulation and specificity of intracellular proteolysis. For studying such proteinases, insulin and glucagon are inexpensive substrates whose structures are well defined. We have identified two intracellular proteinases that are active at alkaline pH and digest reduced ^{125}I-insulin and ^{125}I-glucagon, but not larger polypeptides, to acid-soluble fragments.[1] Although these enzymes appear to be present in many mammalian tissues, the proteinases from human erythrocytes have been characterized most thoroughly. One of these enzymes is of unusually large molecular weight and appears to be a metalloproteinase, since it is inhibited by metal chelators.[1,2] The purification and some properties of this proteinase are described below. The second enzyme is smaller and probably a serine protease, since it is inhibited by diisopropyl fluorophosphate (DFP).[3] This enzyme can be purified manyfold, but not to homogeneity, by the same procedures used to isolate the metalloproteinase. Although in erythrocyte lysates the serine proteinase remains soluble after prolonged ultracentrifugation, in the liver an apparently identical enzyme is associated with mitochondria.[1]

The proteinases purified according to procedures to be described here appear to be responsible for the insulin hydrolysis observed at alkaline pH in soluble extracts of red cells[1,4] and of other mammalian tissues.[1,5,6] The insulin-degrading activities described in various tissues by Brush[6] and by Duckworth et al.[4] ("insulin-specific proteinase," "insulinase") are probably similar to the proteinases described herein, since in all the tissues we have studied we find one major peak of insulin-degrading activity at alkaline pH after chromatography of the soluble proteins on DEAE–cellulose. Our assay procedure, however, differs from theirs,[4,6] in that we use ^{125}I-insulin at micromolar rather than nanomolar levels, and we include a reducing agent (i.e., the A and B chains are separated). Unfortunately, the physical properties of the "insulin-specific proteinases"[4,6] are not characterized sufficiently well to make definitive comparisons with the proteinases described below. On the other hand, the soluble metalloproteinase and mitochondrial serine proteinase that we have identified are clearly distinct from the microsomal insulin-degrading metalloproteinase described in kidney[7,8] and the similar enzyme described in pancreas.[9] These mi-

[1] R. J. Kirschner and A. L. Goldberg, in preparation.

[2] R. J. Kirschner, Fed. Proc., Fed. Am. Soc. Exp. Biol. **39**, 1683 (1980).

[3] Abbreviations used are DFP, diisopropyl fluorophosphate; TCA, trichloroacetic acid; DTT, dithiothreitol; EDTA, ethylenediaminetetraacetic acid; Im buffer, buffer containing 10 mM imidazole HCl, pH 7.1; MNA, 4-methoxy-naphthylamide.

[4] W. C. Duckworth, M. A. Heinemann, and A. E. Kitabchi, Proc. Natl. Acad. Sci. U. S. A. **69**, 3698 (1972).

[5] H. Tschesche, T. Dietl, H. J. Kolb, and E. Standl, Bayer Symp. **5**, 586 (1974).

[6] J. S. Brush, Diabetes **20**, 140 (1972).

[7] M. A. Kerr and A. J. Kenny, Biochem. J. **137**, 477 (1974).

[8] P. T. Varandani and L. A. Shroyer, Arch. Biochem. Biophys. **181**, 82 (1977).

crosomal enzymes require trypsin and toluene[7,8] or detergent[9] for release from the membranes.

The use of insulin or glucagon as substrates for assaying proteinases does not imply a physiological role for the enzymes in hormone degradation. It had been suggested that *in vivo* insulin and other peptide hormones bound to or endocytosed by cells are degraded by nonlysosomal "insulinases." However, recent biochemical and physiological evidence has indicated that insulin and other peptide hormones, after binding to surface receptors, are degraded primarily within the lysosome.[10-13]

The physiological function of the metalloproteinase described here may be to hydrolyze polypeptide intermediates generated by the cytoplasmic ATP-dependent pathway for protein degradation.[14-16] In reticulocytes, this system rapidly degrades proteins with abnormal structures such as may result from mutation,[17] incorporation of amino acid analogs,[15,18] or chemical damage.[19a,19b] In addition, this ATP-dependent system hydrolyzes casein, globin, and abnormal proteins to amino acids. It is soluble, active at alkaline pH, inhibited by metal chelators, and unaffected by inhibitors of lysosomal proteinases.[15,16] The initial cleavage of proteins in this pathway is catalyzed by a soluble ATP-stimulated serine proteinase[16,20-22] that can degrade proteins to large polypeptides. We have shown that some of these products, unlike the precursor proteins, are substrates for the red cell metalloendoproteinase. The metalloendoproteinase also hydrolyzes to acid-soluble products, polypeptides such as glucagon, calcitonin, and the separate chains of insulin, but not larger molecules such as globin, casein, or albumin. Thus in many mammalian cells, this metalloproteinase may be an essential component in the concerted degradation of proteins by the soluble, ATP-dependent pathway.

---

[9] R. A. Mumford, A. W. Strauss, J. C. Powers, P. A. Pierzchala, N. Nishino, and M. Zimmerman, *J. Biol. Chem.* **255**, 2227 (1980).

[10] S. Terris, C. Hofmann, and D. F. Steiner, *Can. J. Biochem.* **57**, 459 (1979).

[11] J.-L. Carpentier, P. Gorden, P. Barazzone, P. Freychet, A. LeCam, and L. Orci, *Proc. Natl. Acad. Sci. U. S. A.* **76**, 2803 (1979).

[12] J. A. McKanna, H. T. Haigler, and S. Cohen, *Proc. Natl. Acad. Sci. U. S. A.* **76**, 5689 (1979).

[13] S. Marshall and J. M. Olefsky, *J. Biol. Chem.* **254**, 10153 (1979).

[14] A. L. Goldberg and A. C. St. John, *Annu. Rev. Biochem.* **45**, 747 (1976).

[15] J. D. Etlinger and A. L. Goldberg, *Proc. Natl. Acad. Sci. U. S. A.* **74**, 54 (1977).

[16] A. L. Goldberg, N. P. Strnad, and K. H. S. Swamy, *Ciba Found. Symp.* **75**, 227 (1980).

[17] J. D. Etlinger, K. Matsumoto, and R. F. Rieder, *Blood* **50**, 107 (1977).

[18] M. Rabinowitz and J. M. Fisher, *Biochim. Biophys. Acta* **91**, 313 (1964).

[19a] F. S. Boches and A. L. Goldberg, *Science* in press (1981).

[19b] A. L. Goldberg and F. S. Boches, *Science* in press (1981).

[20] F. S. Boches, L. Waxman, and A. L. Goldberg, submitted for publication (1981).

[21] G. N. De Martino and A. L. Goldberg, *J. Biol. Chem.* **254**, 3712 (1979).

[22] J. D. Kowit and A. L. Goldberg, *J. Biol. Chem.* **252**, 8359 (1977).

The function of the insulin-degrading serine proteinase is even less clear. Possibly it contributes to protein degradation in the mitochondrion, since this organelle recently has been shown to contain an ATP-dependent system for the breakdown of endogenous mitochondrial proteins.[23]

In addition, the metallo- and serine proteinases may function in degrading small peptides (e.g., leader sequences) released during the maturational cleavage of precursors to proteins that are secreted from the cell, inserted into membranes, or transported into mitochondria.[24]

Assay Method

Principle. The hydrolysis of radiochemically labeled proteins is measured as an increase in radioactive fragments soluble in trichloracetic acid. We use [^{125}I]insulin for most of our assays for several reasons. A single endoproteolytic cleavage of small proteins, such as the chains of insulin (MW 2300 and 3400) or glucagon (MW 3500), generally produces acid-soluble fragments. In contrast, the products of the limited proteolysis of larger proteins (e.g., casein, globin, and albumin) are generally not acid soluble. Since the iodinated tyrosines are not located at the termini of the insulin peptides, the assay is relatively insensitive to exoproteases. Furthermore,[^{125}I]insulin is hydrolyzed very slowly or not at all by two major, soluble intracellular proteinases, the ATP-stimulated proteinase[20] and the calcium-activated proteinase,[25,25a] and thus these enzymes will not complicate the measurement of insulin-degrading proteinases in extracts and in purified fractions. The assay using [^{125}I]insulin is described below, but the conditions used for hydrolyzing other radioactive proteins are identical.

Reagents

Substrate: 250 μg of nonradioactive insulin from a stock solution (5 mg/ml in 10 mM NaOH) is mixed with 1.5–2 × 10^5 cpm of [^{125}I]-insulin [1 μCi/ml, 125 μCi/μg, 50 mg/ml bovine serum albumin (BSA)] in 1.0 ml of water.

Buffer: 0.6 ml of 0.5 M sodium glycylglycine (pH 8.5) is mixed with 6 μl 1 M dithiothreitol (DTT, stored at −20°C), 12 μl 1 M MgCl$_2$, and 1.88 ml water.

Trichloroacetic acid (TCA): Dissolve 100 mg BSA in 85 ml of ice-cold water and add 15 ml of 100% TCA (454 g TCA dissolved in 190 ml H$_2$O). Store at 4°C.

[23] M. Desautels and A. L. Goldberg, *Proc. Natl. Acad. Sci. U.S.A.* in press (1982).
[24] D. F. Steiner, P. S. Quinn, S. S. Chan, J. Marsh, and H. Tager, *Ann. N.Y. Acad. Sci.* **343**, 1 (1980).
[25] L. Waxman, *in* "Protein Turnover and Lysosome Function" (H. L. Segal and D. J. Doyle, eds.), p. 363. Academic Press, New York, 1978.
[25a] L. Waxman, this volume [49].

SUMMARY OF PURIFICATION FOR ERYTHROCYTE METALLOPROTEINASE

Step	Total volume (ml)	Total protein (mg)	Total activity (units)	Specific activity (units/mg)	Relative purification (-fold)
1. Lysate	2000	2.3×10^5	84	3.6×10^{-4}	1.0
2. DEAE–Cellulose	420	750	31	0.04	110
3. 45–58% Ammonium sulfate precipitate	21	106	20	0.19	525
4. Pentyl-agarose	168	12.9	11	0.85	2360
5. Hydroxylapatite	118	2.1	8.1	3.30	9110
6. Gel filtration[a]	15	0.12	1.8	14.6	40,600

[a] The gel-filtration column was 2.5×55 cm.

Procedure. Samples containing the enzyme are mixed with 50 μl buffer at 0°C and are made up to 100 μl with water. Substrate (20 μl) is added to give mixtures that contain 5 μg of insulin in 120 μl and are 50 mM glycylglycine–1 mM DTT–2 mM MgCl$_2$. (In experiments with inhibitors the agents are incubated for 20 min at 30°C with the enzyme and buffer before the addition of substrate.) After incubation with substrate for 30 min at 37°C, reactions are stopped with 0.88 ml of TCA containing albumin as carrier, stored at 0°C for at least 15 min, and centrifuged 10 min at 1200 g. The amount of ^{125}I in 0.8 ml of the supernatant fluid is determined. Two blanks lacking enzyme (Bl 0% and Bl 100%) are treated similarly to the other samples, except that Bl 100% is not centrifuged.

This assay is linear with time for up to 90 min and with enzyme concentration so long as 60% of the radioactive substrate remains precipitable with TCA.

Definition of Activity Unit. Percent degradation (% deg) is calculated as

$$\% \ deg = \frac{cpm_{sample} - cpm_{Bl \ 0\%}}{cpm_{Bl \ 100\%} - cpm_{Bl \ 0\%}} \times 100.$$

One unit of insulin-degrading activity corresponds to 1 μmol of insulin degraded per 30 min in 120 μl under the defined conditions. 10% degradation in 30 min corresponds to 8.9×10^{-5} units of activity.

Purification Procedure

The purification of the metalloproteinase from human erythrocytes is here described, and is summarized for a typical preparation in the table. The insulin-degrading serine proteinase co-elutes from DEAE–cellulose with the metalloproteinase, but the enzymes are separated during subsequent purification steps. The serine proteinase may be purified 8000-fold using the same procedures. In related experiments, the insulin-degrading proteinases that are inhibited by metal chelators or by DFP also have been partially purified from extracts of other tissues using the protocol described below.

Preparation of Lysate. Erythrocytes from freshly drawn human blood (5 units) are allowed to settle through dextran–citrate at room temperature, the upper layer of white cells is discarded, and the red cells are washed three times with 0.9% NaCl. The saline washings and all subsequent steps in the purification procedures are at 0–4°C. The red cells are lysed with 1.2 volumes of 10 mM Tris-HCl (pH 7)–1 mM DTT, and the membranes are removed by centrifugation for 60 min at 13,000 g.

DEAE–Cellulose Chromatography. DEAE–cellulose (Whatman DE-52 equilibrated with 10 mM Tris-HCl, pH 7) is added to the centrifuged lysate (0.22 g moist DE-52 per ml lysate), and the slurry is mixed for 30 min by rotating the bottle. After filtering the slurry through a chilled Buchner funnel, the resin is washed with ice-cold 10 mM Tris-HCl (pH 7)–60 mM NaCl (2 liters/liter lysate), transferred to a column (6 cm diameter), and washed with additional 10 mM Tris-HCl (pH 7)–60 mM NaCl. A linear gradient of sodium chloride (60–400 mM) in 10 mM Tris-HCl pH 7)–0.1 mM DTT (total volume of 6 ml buffer per ml of packed resin) is run, and 5 μl of each 20-ml fraction is assayed. The insulin-degrading activities elute at about 150 mM NaCl.

Ammonium Sulfate Precipitation. The DEAE–cellulose fractions containing about 85% of the insulin-degrading activity are pooled. Ammonium sulfate is added to the pooled fractions (277 mg/ml) while stirring, and this 45% $(NH_4)_2SO_4$ solution is stirred for an additional 20 min. The precipitate collected by centrifugation (45 min at 14,700 g) is discarded. The supernatant is adjusted to 58% saturation with $(NH_4)_2SO_4$ (85 mg/ml), stirred, and centrifuged again. This 45–58% precipitate contains most of the EDTA-inhibited insulin-degrading activity and is dissolved in a small volume of 10 mM imidazole-HCl, pH 7.1 (Im buffer) before chromatography on pentyl-agarose.

Most of the insulin-degrading activity inhibitable by DFP is in the 58% $(NH_4)_2SO_4$ supernatant fluid and may be precipitated after adjusting this solution to 70% saturation with $(NH_4)_2SO_4$ (90 mg/ml).

Pentyl-Agarose Chromatography. The redissolved 45–58% precipitate is adjusted to about 45% $(NH_4)_2SO_4$ saturation by adding Im buffer or saturated $(NH_4)_2SO_4$ in Im buffer until the conductivity is 80 mmho. This sample is applied to pentyl-agarose (Sigma; recycled by washing successively with water, 70% ethanol, Im buffer, and 45% $(NH_4)_2SO_4$ in Im

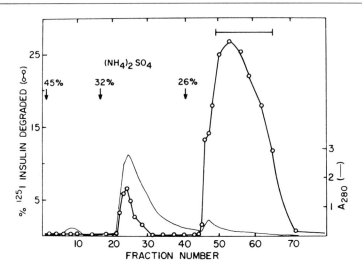

FIG. 1. Chromatography of the insulin-degrading proteinases on pentyl-agarose. The 45–58% $(NH_4)_2SO_4$ precipitate (106 mg protein) of the pooled DEAE-cellulose fractions was applied to a column containing 10 ml of pentyl-agarose equilibrated with 45% $(NH_4)_2SO_4$ in Im buffer. The protein was eluted in steps of decreased $(NH_4)_2SO_4$. O——O, Percent of [^{125}I]insulin degraded; ——, absorbance at 280 nm. The horizontal bar indicates the fractions that were pooled.

buffer) at 10 mg protein per ml of resin in a 1.5-cm-diameter column. The $(NH_4)_2SO_4$ concentration is reduced in steps (45%, 277 mg/ml; 32%, 190 mg/ml; 26%, 150 mg/ml in Im buffer). The metalloproteinase is eluted during the 26% $(NH_4)_2SO_4$ step and the serine proteinase elutes in 32% $(NH_4)_2SO_4$ (Fig. 1).

Hydroxyapatite Chromatography. The pentyl-agarose fractions containing most of the activity are pooled and dialyzed against 5 mM imidazole-HCl (pH 7.1)–0.1 mM DTT until the conductivity of the sample is similar to that of fresh buffer. Hydroxyapatite (Bio-Gel HTP, BioRad Laboratories) is equilibrated with 5 mM imidazole-HCl–5 mM phosphate (pH 6.8), and the sample (made 5 mM in phosphate by adding 100 mM phosphate, pH 6.8) is applied to a column of this resin at about 1–2 mg protein per ml of resin. On developing the column with a linear gradient of phosphate, pH 6.8 (5 mM imidazole-HCl, 5 mM phosphate to 200 mM phosphate; 20 ml of each buffer per ml of resin), the metalloproteinase elutes at about 65 mM phosphate.

Gel Filtration. The hydroxyapatite fractions containing enzyme with a specific activity greater than 3 units/mg are pooled. The proteinase is concentrated by diafiltration through a PM-10 membrane (Amicon, Lexington, Massachusetts) to less than 2 ml, and applied to a 1.5 ×

120-cm column of Ultrogel AcA 34 (LKB, Inc., Rockville, Maryland), which is equilibrated with 25 mM imidazole-acetate (pH 7.1)–10% glycerol. Fractions (3 ml) are collected at about 9 ml/hr.

Properties of the Metalloproteinase

Physical Properties. On SDS–polyacrylamide gel electrophoresis,[26] proteinase with a specific activity of 17 units/mg is homogeneous and has MW 1.1–1.15 × 10⁵ (sample reduced but not boiled). By gel filtration on Ultrogel AcA 22, the native enzyme has MW 2.8–3.1 × 10⁵. Thus the metalloproteinase appears to be composed of 2–3 subunits. The pI is 5.4.[1]

Stability. Less than 20% of the activity of the purified enzyme was lost when stored at 1.5 μg/ml, in 10 mM imidazole-HCl (pH 7.1)–5% glycerol for 5 months at 0°C. The enzyme is stable in 40% ethanol for at least 24 hr at 0°C. Activity is lost rapidly in samples incubated above 55°C, or incubated below pH 6 at 4°C.[1]

Activators and Inhibitors. The degradation of [125I]insulin is inhibited 90% by 0.1 mM EDTA or 1 mM o-phenanthroline. Over 60% of the activity lost by treatment with EDTA is restored if samples are dialyzed exhaustively and then assayed in the presence of $10^{-5} M$ ZnCl₂, CoCl₂, or MnCl₂, or $10^{-3} M$ MgCl₂ or CaCl₂.[1] This evidence suggests the enzyme is a metalloproteinase requiring a divalent cation.

The enzyme requires DTT for the hydrolysis of [125I]insulin but not of [125I]glucagon. Thus the separate A and B chains, rather than insulin, appear to be the actual substrates. The degradation of [125I]glucagon in the absence of DTT is inhibited by 10 μM HgCl₂ (85%) and 100 μM N-ethylmaleimide (84%).[1] A mixture of inhibitors from *Helix pomatia*[5] inhibited 55% at 200 μg/ml but did not inhibit at 10 μg/ml.[1] The enzyme is not inhibited by DFP, Trasylol, soybean-trypsin inhibitor, the inhibitor of the calcium-activated proteinase, α₂-macroglobulin, leupeptin, chymostatin, pepstatin, antipain, phosphoramidon, or amastatin.[1]

Substrates. [125I]insulin, [125I]glucagon, and [125I]calcitonin are hydrolyzed to acid-soluble radioactive fragments, but [³H]CH₃-casein, [¹⁴C]CH₃-globin, bovine [125I]albumin, [125I]ferritin, [125I]growth hormone, [125I]ACTH, [125I]parathyroid hormone, and [125I]epidermal growth factor are not.

Some TCA-insoluble products resulting from the proteolysis of casein or globin by the purified ATP-stimulated proteinase[20] are substrates for the metalloproteinase.[1] This observation is consistent with an intracellular role for the enzyme in the concerted hydrolysis of proteins by the ATP-dependent degradative system.

[26] U. K. Laemmli, *Nature (London)* **222**, 680 (1970).

Specificity. The enzyme has a specificity similar to that of many metal chelator-sensitive neutral proteinases of bacteria.[27] It seems to cleave on the amino-terminal side of large hydrophobic groups. After the cleavage of the insulin B chain or glucagon, only leucine, tyrosine, and phenylalanine are detected by dansylation as new amino-terminal residues. Terminal amino acids or dipeptides are not cleaved from the insulin B chain or glucagon. Thus the enzyme is an endoproteinase.[1]

Optimal pH. The rate of insulin hydrolysis is greatest from pH 8.5 to 11, whereas glucagon degradation is most rapid between pH 6.6 and 8.6.

Distribution. A metalloproteinase that degrades insulin is found in the cytosolic fraction of human erythrocytes, rabbit erythrocytes and reticulocytes, and rat skeletal muscle, liver, kidney, and brain. These enzymes are highly similar in size, net charge, and/or other chemical properties, since they all are inhibited by EDTA and behave identically when subjected to procedures (e.g., gel, ion-exchange and/or pentyl-agarose chromatography) used for purifying the erythrocyte enzyme.[1]

Properties of the Serine Proteinase

By gel filtration on Ultrogel AcA 34 the enzyme from erythrocytes has MW 1.2×10^5. It appears to be a dimer. When the purified protein was incubated with [^{3}H]DFP, a single labeled band of MW 70,000 was observed on SDS–polyacrylamide gel electrophoresis. When the enzyme is stored in 32% $(NH_4)_2SO_4$ at 4°C, less than 20% of the activity is lost in 1 month. This proteinase is inhibited 99% by DFP or $HgCl_2$ (1 mM), and about 90% by N-ethylmaleimide (1 mM), tosyl-lysyl-chloromethyl ketone (0.5 mM), or tosyl-phenylalanyl-chloromethyl ketone (0.5 mM). It is not inhibited by EDTA (10 mM), Trasylol (50 μg/ml), or soybean-trypsin inhibitor (50 μg/ml). Hydrolysis of [^{125}I]insulin or [^{125}I]glucagon is greatest at pH 7.2 to 7.8. Insulin is degraded more rapidly than glucagon.

The serine proteinase is associated with the mitochondrial/lysosomal fraction (25,000 g, 10 min) after perfused rat liver has been homogenized and subjected to differential centrifugation. The activity continues to be particulate after this fraction has been treated with digitonin (0.25 mg/mg protein), which releases lysosomal enzymes, but is soluble when the mitochondria are disrupted by freeze–thawing or sonication.[1,23] The possibility has not been eliminated that this insulin-degrading activity, like the alkaline protease ("group-specific proteinase," "chymase") of

---

[27] H. Matsubara and J. Feder, *in* "The Enzymes" (P. D. Boyer, ed.), 3rd ed., Vol. 3, p. 721. Academic Press, New York, 1971.

Katunuma *et al.*,[28] is contained within the mast cell granules that sediment with the mitochondria.[29] However, the activity we have purified from erythrocytes is clearly an enzyme of the red cell and not of mast cell origin.

Acknowledgments

These studies have been made possible by a grant from the National Institute of Neurological and Communicable Disease and Stroke, the Juvenile Diabetes Association, and the Muscular Dystrophy Association of America. Dr. Kirschner is a postdoctoral fellow of the Massachusetts Affiliate of the American Heart Association. We are grateful to Ms. Maureen Rush for assisting us in the preparation of this manuscript.

[28] N. Katunuma, E. Kominami, K. Kobayashi, Y. Banno, K. Suzuki, K. Chichibu, Y. Hamaguchi, and T. Katsunuma, *Eur. J. Biochem.* **52**, 37 (1975).
[29] R. G. Woodbury, M. Everitt, Y. Sanada, N. Katunuma, D. Lagunoff, and H. Neurath, *Proc. Natl. Acad. Sci. U.S.A.* **75**, 5311 (1978).

[52] Mammalian Collagenases

By Tim E. Cawston and Gillian Murphy

Introduction

Specific collagenases (EC 3.4.24.7) are proteinases acting at neutral pH to cleave the native helix of the interstitial collagens, at or near the same locus, approximately three-quarters of the way from the N-terminus of the molecule. They are Zn^{2+} metalloenzymes and generally require Ca^{2+} for full activity. Collagenases are found in the media from mammalian connective tissue cultures, but are rarely detectable in tissue extracts. Exceptions are tissues undergoing rapid resorption, such as involuting uterus and the specific granule fraction of human polymorphonuclear (PMN) leukocytes. Collagenases from different tissues of any one species are apparently immunologically similar, except for the PMN leukocyte enzyme.

Assay

At the present time, only assay systems employing collagen itself as a substrate are suitable for the unambiguous detection of specific collagenases. It should be noted that although collagen is relatively resistant to nonspecific proteolytic degradation, the terminal peptides are susceptible. In addition, any denatured collagen (gelatin) is rapidly degraded by a number of proteinases. It is therefore necessary to check that cleavage of the substrate helix has occurred, to give the typical three-quarter (TC_A) and one-quarter (TC_B) products. This is most conveniently carried out by conventional SDS–PAGE analysis of the products of an assay in which a fall in viscosity has been demonstrated.[1]

Type I collagen is normally used as the substrate for the assay of collagenase and is prepared in soluble form from skin and tendon.[2,3] Soluble collagen can be reconstituted into insoluble fibrils by incubating at 25–37°C at neutral pH and this provides the basis for the most widely used collagenase assays. Radiolabeled collagen fibrils are incubated with the enzyme and collagen degradation followed by the assay of radioactivity solubilized during the assay period.

If no radioactive counting facilities are available, solution assays can be used in which the initial rate of fall in viscosity of collagen is measured.[1] An assay measuring the protein-stain density of the TC_A pieces produced by collagenase degradation of collagen has also been described.[4]

Preparation of the Substrate

Acid-soluble collagen can be extracted and purified from rat skin. Skins are removed from 10 rats (weight 100–200 g), immediately cooled and scraped to remove fatty tissue, and cut into small pieces. The skin pieces are minced in an electric mincer with chips of dry ice. All subsequent procedures should be done at 0–4°C, and all solutions and equipment precooled. The minced skin is extracted three times, to remove unwanted salt-soluble collagen, with 1.5 liters of 0.9% (w/v) NaCl–0.01 M Tris-HCl (pH 7.5) for 30 min with stirring, and then filtered through cheesecloth. The tissue is then washed twice with ice-cold distilled water for 30 min, resuspended in 1 liter of 0.5 M acetic acid with 0.03% toluene, and stirred slowly overnight. The extraction is filtered through cheesecloth and the extraction repeated. The combined extracts are centrifuged at 7500 g for 2 hr at 4°C. The supernatant is dialyzed against 12 liters of 5% (w/v) NaCl in 0.1 M acetic acid to precipitate the collagen. The pellets obtained by centrifugation at 7500 g for 30 min are redissolved in 2–3 liters

[1] E. D. Harris and C. A. Vater, *in* "Collagenase in Normal and Pathological Connective Tissues" (D. E. Woolley and J. M. Evanson, eds.), p. 37. Wiley, New York, 1980.
[2] G. Chandrakasan, D. A. Torchia, and K. A. Piez, *J. Biol. Chem.* **251**, 6062 (1976).
[3] T. E. Cawston and A. J. Barrett, *Anal. Biochem.* **99**, 340 (1979).
[4] H. Turto, S. Lindy, V.-J. Uitto, O. Wegelius, and J. Uitto, *Anal. Biochem.* **83**, 557 (1977).

of 0.5 M acetic acid, dialyzed against 0.5 M acetic acid overnight, and then dialyzed against four changes of 0.02 M disodium hydrogen phosphate (20 liters) to reprecipitate the collagen. After centrifugation at 7500 g for 30 min, the pellets are redissolved in 0.5 M acetic acid (1–2 liters) and the collagen precipitated by adding solid NaCl to 5% (w/v). The precipitate is sedimented at 7500 g for 30 min, washed in 20% (w/v) NaCl (1 liter), resuspended in 0.5 M acetic acid (1 liter), and stirred slowly overnight before centrifuging at 30,000 g for 1 hr. The supernatants are combined and dialyzed against 0.1 M acetic acid (10 liters). The purified collagen can be stored frozen for long periods, but should only be frozen and thawed once. Long-term storage as a freeze-dried powder in sealed containers with silica gel is recommended. Samples can be redissolved in 0.2 M acetic acid at 4°C at a concentration of 1–2 mg/ml. The purity of the final preparation may be checked by SDS–PAGE. The exact concentration of collagen is estimated by hydroxyproline analysis,[5] assuming a hydroxyproline content of 13% (w/v).[5] The ability of the collagen to form fibrils at neutral pH with low blanks (10% of total hydroxyproline) and resistance to trypsin degradation (20% of total), may also be checked at this stage.

Radiolabeling of the Substrate

Collagen may be radioactively labeled biologically by administration of [³H]- or [¹⁴C]glycine to the animals before killing (skin[6]), or by their inclusion in culture media (calvaria[7]) and subsequent purification. More conveniently, collagen can be acetylated using [³H]- or [1-¹⁴C]acetic anhydride, after purification.[3]

Acetylation of Collagen. Collagen (250 mg) is dissolved in 0.2 M acetic acid (50 ml) at 4°C, dialyzed against 10 mM disodium tetraborate containing 200 mM CaCl₂, adjusted to pH 9.5 (2 liters, one change), and then placed in a conical flask at 4°C and slowly stirred with a large magnetic stirring bar. [1-¹⁴C]Acetic anhydride (10–30 mCi/mmol; 250 μCi) is condensed to the end of its sealed vial by placing it in dry ice. Dry dioxane (1 ml) is added to the vial and the [1-¹⁴C]acetic anhydride–dioxane mixture added to the collagen. The vial is washed with more dry dioxane (1.0 ml) and the washings are added to the collagen mixture, which is stirred for 30 min at 4°C and then dialyzed at 4°C against 50 mM Tris-HCl (pH 7.6) containing 200 mM NaCl–5 mM calcium acetate–0.03% toluene, until the radioactivity in the diffusate has fallen to background levels. The labeled collagen is diluted to 1 mg/ml with 0.2 M acetic acid and then further

[5] M. C. Burleigh, A. J. Barrett, and G. S. Lazarus, *Biochem. J.* **137**, 387 (1974).
[6] Z. Werb and M. C. Burleigh, *Biochem. J.* **137**, 373 (1974).
[7] P. B. Robertson, R. E. Taylor, and H. M. Fullmer, *Clin. Chim. Acta* **42**, 43 (1972).

diluted with unlabeled collagen solution (1 mg/ml) until the specific radioactivity is approximately 100,000 dpm/mg.

The labeled collagen can be frozen in 0.2 M acetic acid and stored at $-20°C$. Before use in the collagenase assay, it should be thawed and dialyzed at 4°C against 50 mM Tris-HCl (pH 7.6) containing 200 mM NaCl and 0.03% toluene.

Fibril Assay

Radiolabeled collagen (100 μl of 1 mg/ml solution in 50 mM Tris-HCl buffer (pH 7.6) containing 200 mM NaCl and 0.03% toluene) is incubated in microcentrifuge tubes at 35–37°C for 1 hr to allow fibril formation. 100 μl of 100 mM Tris-HCl buffer (pH 7.6) containing 15 mM CaCl$_2$ is added with the collagenase preparation, to a final volume of 300 μl. The tubes are incubated for up to 20 hr at 35–37°C before centrifuging at 10,000 g for 10 min. This sediments intact collagen fibrils, leaving digested, solubilized fragments in solution. Portions of the supernatant are counted using a suitable aqueous scintillation fluid. Controls include (*a*) buffers alone, indicating minimum soluble material; (*b*) 10 μg trypsin to indicate the maximum susceptibility to nonspecific proteinase of the substrate (due to labeling of the nonhelical regions or denaturation of the helix); (*c*) 10 μg of clostridial collagenase to totally degrade the fibril, indicating the total counts in the substrate. Units of collagenase activity are expressed as μg collagen fibril solubilized per minute.

Diffuse Fibril Assay

This assay is similar to that described above, except that the collagen is added in solution to the enzyme and buffer. A more diffuse fibril forms during the first 10 min of incubation at 37°C. The rate of collagen solubilization is generally greater than that obtained using preformed fibrils, presumably due to greater initial accessibility of the enzyme to substrate, allowing binding as the fibrils form. The assay is linear but is not suitable for enzyme preparations with a high concentration of salt, for example, culture media.[3]

Other Assays

An assay using radiolabeled collagen in solution has been described by Terato *et al.*[8] Undegraded collagen is precipitated at the end of the incubation using dioxane. A rapid procedure for determining collagenase activity

[8] K. Terato, Y. Nagai, J. Kwaniski, and S. Yamamoto, *Biochim. Biophys. Acta* **445,** 753 (1976).

involves the drying of thin films of radiolabeled collagen fibrils in microwell dishes.[9] Enzyme samples may be added to the wells, incubated, and removed for measurement of solubilized radioactivity. This method is particularly suitable for screening large numbers of samples such as culture media or column fractions. Synthetic peptide substrates resembling the sequence of amino acids around the collagen (type I) cleavage site have been described[10] but these are relatively insensitive and nonspecific.[11]

Collagenase Latency

Collagenase activity is frequently masked due to the existence of the enzyme in a latent form.[12] In serum-free culture media the enzyme can be activated by treatment with thiol-blocking reagents. Preincubation of latent enzyme with 0.1–1 mM 4-aminophenylmercuric acetate (dissolved in 0.5 M NaOH and adjusted to pH 10) at 37°C (exact concentration and length of incubation to be individually determined) may be used, or inclusion of the mercurial in the assay for longer incubations. In serum-containing culture media the presence of the ubiquitous proteinase inhibitor, α_2-macroglobulin, necessitates its inactivation before collagenase activation. Treatment of culture media with 200–500 μg/ml trypsin at 37°C for 30 min (exact conditions depending on serum concentration), followed by the addition of a 5-fold excess of soybean-trypsin inhibitor will effect both these operations. Chaotrophic ions have also been used to destroy α_2-macroglobulin and reactivate collagenase–α_2-macroglobulin complexes, but are unlikely to give quantitative or reproducible recovery of enzyme activity.

Purification

Sources

Culture Media. A large number of connective tissues (e.g., skin, uterus, calvaria, cartilage, synovium, tendon) and cells derived from these tissues (e.g., fibroblasts, chondrocytes) synthesize and secrete collagenases when placed into culture.[6,13,14] Providing that sufficient quan-

[9] B. Johnson-Wint, *Anal. Biochem.* **104**, 175 (1980).
[10] Y. Nagai, Y. Masui, and S. Sakakibara, *Biochim. Biophys. Acta* **445**, 521 (1976).
[11] J. F. Woessner, *Biochim. Biophys. Acta* **571**, 312 (1979).
[12] G. Murphy and A. Sellers, *in* "Collagenase in Normal and Pathological Connective Tissues" (D. E. Woolley and J. M. Evanson, eds.), p. 65. Wiley, New York, 1980.
[13] T. E. Cawston and J. A. Tyler, *Biochem. J.* **183**, 647 (1979).
[14] G. P. Stricklin, E. A. Bauer, J. J. Jeffrey, and A. Z. Eisen, *Biochemistry* **16**, 1607 (1977).

tities of material and facilities for large-scale culturing are available, any of these sources should provide enough collagenase for purification purposes.

Tissue Extracts. Collagenase is present in most tissues in very small amounts, and detection is beyond the limits of sensitivity of the specific assays at present available. Rapidly degrading tissues contain higher levels of enzyme, which appear to be bound to the collagen matrix. This can be assayed by monitoring the breakdown of endogenous collagen in tissue homogenates.[15,16] Collagenase can be extracted by thermal shrinkage of the collagen and the use of high calcium-ion concentrations, which seems to interfere with binding.[17] Tissue is homogenized in 0.25% (w/v) Triton X-100 and 0.01 M $CaCl_2$, and centrifuged at 6000 g for 20 min. The pellet is resuspended in 0.1 M $CaCl_2$ and 0.02 M Tris-HCl (pH 7.5), and heated at 60°C for 4 min. The pellet is centrifuged at 10,000 g for 20 min, leaving a supernatant with 65–70% of the collagenase activity detectable in the original homogenate. A further 10–15% of this activity may be recovered by a second extraction of the pellet.

Polymorphonuclear leukocyte collagenase can be efficiently extracted from either the whole granule fraction (treated with diisopropyl fluorophosphate to inactivate serine proteinases) or from the specific granule fraction[18] by the use of 0.05% (v/v) Triton X-100 in 50 mM Tris-HCl (pH 7.5) containing 150 mM NaCl, 10 mM $CaCl_2$, or 0.2 M sodium acetate buffer (pH 4.5) containing 10 mM $CaCl_2$.

Purification Procedure

Collagenase can be purified by the following procedure from the culture media of a variety of tissues or cells from different species (Scheme 1).[13] The results shown in the table are taken from the purification of collagenase from pig synovial culture media. The enzyme is assayed during the purification using the diffuse fibril assay,[3] and protein is measured using fluorescamine as described by Weigele *et al.*[19]

Ammonium Sulfate Precipitation. Pooled medium is thawed and taken to 20% saturation with $(NH_4)_2SO_4$, left overnight at 2°C, and then filtered through a Whatman 54 filter paper. $(NH_4)_2SO_4$ is then added to the filtrate to 60% saturation, stored at 2°C overnight, and then centrifuged at 22,000

[15] J. F. Woessner and J. N. Ryan, *Biochim. Biophys. Acta* **309**, 397 (1973).
[16] T. I. Morales, J. F. Woessner, D. S. Howell, J. M. Marsh, and W. J. Lemaire, *Biochim. Biophys. Acta* **524**, 428 (1978).
[17] J. G. Weeks, J. Halme, and J. F. Woessner, *Biochim. Biophys. Acta* **445**, 205 (1976).
[18] G. Murphy, U. Bretz, M. Baggiolini, and J. J. Reynolds, *Biochem. J.* **192**, 517 (1980).
[19] M. Weigele, S. L. Debernardo, J. P. Tangi, and W. Leimgruber, *J. Am. Chem. Soc.* **94**, 5927 (1972).

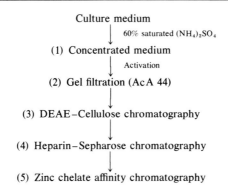

Culture medium
↓ 60% saturated (NH₄)₂SO₄

(1) Concentrated medium
↓ Activation

(2) Gel filtration (AcA 44)
↓

(3) DEAE–Cellulose chromatography
↓

(4) Heparin–Sepharose chromatography
↓

(5) Zinc chelate affinity chromatography

SCHEME 1. Flowchart for the purification of pig synovial collagenase.

g for 1 hr at 2°C. The pellet is redissolved in a minimal volume of 25 mM sodium cacodylate buffer (pH 7.2) containing 1 M NaCl–10 mM CaCl₂– 0.05% (v/v) Brij 35–0.03% toluene, dialyzed against this buffer, and then centrifuged at 40,000 g for 1 hr at 2°C and the pellet discarded. The latent collagenase is activated by treating with 4-aminophenylmercuric acetate (0.7 mM) at 37°C for 4 hr.

Gel Filtration with Ultrogel AcA 44. The activated concentrated medium (60 ml) is applied to a column of Ultrogel AcA 44 (4.4 × 86 cm) and developed by reverse flow. The collagenase is eluted as a sharp peak relatively free of contaminating proteins. The fractions containing collagenase are combined and concentrated to 5–10 ml in an ultrafiltration cell fitted with an Amicon PM-10 membrane.

PURIFICATION OF PIG SYNOVIAL COLLAGENASE [a]

	Collagenase (units)	Protein (mg)	Specific activity (units/mg)	Yield (%)	Purification factor (-fold)
Dilute medium	18,340	(524)	35	100	—
Concentrated medium	19,250	85	227	105	1
Gel filtration (AcA 44)	26,800	9.1	2940	146	12
DEAE–Cellulose	15,320	4.7	3290	84	14
Heparin–Sepharose	15,350	0.70	21,680	84	96
Zinc chelate–Sepharose	12,820	0.24	53,400	70	235

[a] Samples were taken from each collagenase pool and assayed for collagenase activity and protein. The protein determination for the dilute-medium sample was unreliable and the purification factor was calculated by taking the concentrated medium as stage 1.

DEAE–Cellulose Chromatography. The concentrated fractions from the gel-filtration column are dialyzed against 25 mM sodium cacodylate buffer (pH 7.6) containing 15 mM CaCl$_2$–0.05% Brij 35–0.03% toluene, and applied to a column of DEAE–cellulose (Grade 52) (1.0 × 9.0 cm) equilibrated with this buffer. The collagenase is unretarded and passes straight through the column under these conditions. The proteins binding to the column are eluted with a salt gradient.

Heparin–Sepharose Chromatography. Heparin–Sepharose is prepared by the method of Sakamoto *et al.*[20] The fractions containing collagenase from DEAE–cellulose chromatography are combined and dialyzed against 25 mM sodium cacodylate buffer (pH 7.5) containing 10 mM CaCl$_2$–0.05% Brij–0.03% toluene, and applied to a column of heparin–Sepharose (1.0 × 5.0 cm) equilibrated with this buffer. The sample is washed onto the column with the same buffer and then a salt gradient (0–0.5 M NaCl) applied to the column. The majority of the protein passes through the column, but the collagenase binds and is eluted at approximately 0.2–0.3 M NaCl.

Zinc Chelate Affinity Chromatography. The epoxy-activated Sepharose-6B is prepared by the method of Sundberg and Porath[21] and the zinc chelate column is prepared and charged with ZnCl$_2$ as described by Porath *et al.*[22] The fractions containing collagenase from heparin–Sepharose chromatography are combined, dialyzed against 25 mM sodium borate buffer (pH 8.0) containing 0.15 M NaCl, and then applied to the zinc chelate column (0.6 × 12 cm) equilibrated with this buffer. The fractions are loaded and washed onto the column with the borate buffer and then eluted in a stepwise manner with the following buffers: (1) 25 mM sodium cacodylate, pH 6.5; (2) 25 mM sodium cacodylate, pH 6.5, containing 0.8 M NaCl; (3) 50 mM sodium acetate buffer, pH 4.7, containing 0.8 M NaCl; (4) 50 mM EDTA adjusted to pH 7.0 containing 0.5 M NaCl. All of these buffers contain 0.05% Brij 35, 0.03% toluene, and 1 mM CaCl$_2$. The purified collagenase is eluted from the column at pH 4.7. The fractions are combined and dialyzed against 25 mM sodium cacodylate (pH 7.5) containing 0.05% Brij 35–0.03% toluene–5 mM CaCl$_2$, and stored at 4°C.

Additional Points

1. *Collagenase Stability.* The purification of collagenase from tissue culture or cell culture medium is difficult because very little enzyme protein is present and, in conventional procedures, large losses occur. The

[20] S. Sakamoto, M. Sakamoto, P. Goldhaber, and M. Glimcher, *Biochim. Biophys. Acta* **385**, 41 (1975).
[21] L. Sundberg and J. Porath, *J. Chromatogr.* **90**, 87 (1974).
[22] J. Porath, J. Carlsson, I. Olsson, and G. Belfrage, *Nature (London)* **258**, 598 (1975).

recovery of collagenase activity is dramatically improved if the detergent Brij 35 (polyoxyethylene lauryl ether) is used throughout the purification procedure. Calcium ions are also included, since the thermal stability of collagenase is much improved in the presence of Ca^{2+}. Dilute solutions of enzyme can be kept for long periods of time at 4°C in the presence of Brij 35 with relatively little loss of activity. Without detergent the enzyme is unstable during purification, and sometimes rapidly loses activity as reported for other collagenases. To ensure maximum recovery of activity, it is best to process the largest possible volume of culture medium at a time. Volumes in excess of 5 liters (containing 4–5 units collagenase/ml) give the best results.

2. The properties of collagenases do vary, depending on the species and the tissue from which it is isolated (see next section). The above procedure is used for pig synovial and rabbit bone collagenases. Human leukocyte collagenase can be purified by this procedure, but it behaves differently on some columns. For example, it does not bind to either heparin–Sepharose or the zinc chelate column. The pH of the buffer used in DEAE–cellulose chromatography can be raised for some collagenases to improve the separation. Similarly, the pH of the heparin–Sepharose chromatography buffer can be lowered if the enzyme does not bind at pH 7.5.

3. *Heparin–Sepharose Chromatography.* Similar results are obtained using CM-cellulose, but we find that substantial loss of collagenase activity occurs using this gel matrix even in the presence of detergent. This loss of activity is not observed using the heparin-substituted Sepharose.

4. *Zinc Chelate Affinity Chromatography.* Some workers have reported that losses of activity occur using this medium. If this is found, then this step can be replaced with a further gel-filtration step using Ultrogel AcA 44.

5. *Enzyme Latency.* This procedure uses a chemical activation step to ensure that all the latent collagenase is in the active form. Identical results are obtained if the enzyme is allowed to autoactivate after gel filtration.[23] Gel filtration of latent enzyme may usefully separate collagenase from other metalloproteinases.[24] The enzyme can then be activated before further purification.

Properties

Collagenase is a metalloproteinase containing zinc in the active site. The enzyme requires calcium to stabilize it at physiological temperatures. It is inhibited by chelators such as 1:10 phenanthroline and

[23] J. A. Tyler and T. E. Cawston, *Biochem. J.* **189**, 349 (1980).
[24] A. Sellers, G. Murphy, M. C. Meikle, and J. J. Reynolds, *Biochem. Biophys. Res. Commun.* **87**, 581 (1979).

ethylenediaminetetraacetic acid, and less strongly by cysteine and other thiol-containing compounds. It is strongly inhibited by dithiothreitol. Although apparently analogous to the bacterial metalloproteinases with respect to the above properties, mammalian collagenases are not inhibited by phosphoramidon. Serum, tissue extracts, and culture media from tissue contain specific collagenase inhibitors, and these have been reviewed by Murphy and Sellers.[12] The enzyme will digest collagen effectively between pH 7 and 9. It is stable, in the presence of detergent and calcium ions, between pH 6.5 and 9.0 at 4°C. Purified pig synovial collagenase can be stored at 4°C or -20°C in the presence of Brij 35 (0.05%) and Ca^{2+} ions (5 mM) at pH 7.5 for 2–3 years with little loss of activity.

Latency of Collagenase

The nature of the latency of collagenase is not yet known. A number of different activation methods of varying efficiency have been described.

1. Brief exposure to proteolytic enzymes (e.g., trypsin, plasmin, plasma kallikrein)
2. Addition of mercurials (e.g., 4-aminophenylmercuric acetate, mersalyl)
3. Addition of chaotrophic ions (e.g., SCN^-, I^-)
4. Incubation at 25–37°C for 4–24 hr to give spontaneous activation

Treatment with trypsin or 4-aminophenylmercuric acetate appear to be the most effective methods and are described in full under Assays. It is not known if all these methods activate the latent enzyme by a common mechanism or if the latent form of collagenase represents one, two, or several species. In most cases, activation is accompanied by a drop in apparent molecular weight. Two proposals have been put forward to account for the latent enzyme form: (*a*) the enzyme is a proenzyme that requires proteolytic activation; (*b*) the latent enzyme represents an enzyme–inhibitor complex from which the inhibitor can be removed either by proteolysis or by chemical treatment (mercurials, chaotrophic ions). It is not known how these two proposals fit together, but it is likely that the control of the extracellular activity of this enzyme is more complex than was previously thought. For a review of the latent forms of collagenase, see Murphy and Sellers.[12]

Specific Activity

Collagenases from many tissues and species have been identified and several have been purified to homogeneity and characterized.[13,14,25–30] It is difficult to compare these studies because the reported specific activities

[25] D. E. Woolley, R. W. Glanville, M. J. Crossley, and J. M. Evanson, *Eur. J. Biochem.* **54,** 611 (1975).

of the purified enzymes vary within the range 31.0–50,000 units/mg. A number of factors can account for these differences: the different assay conditions, the relative instability or partially purified collagenase, and the possible failure to detect and activate large amounts of latent collagenase in some purified preparations.

Substrate Specificity

The enzyme cleaves all three α-chains of the collagen triple helix at one specific point, usually at a Gly-Leu or Gly-Ile bond, about three-quarters of the way along the molecule from the N-terminus. The purified collagenases are most active against type I collagen, but also degrade type III and to a lesser extent type II.[23,26,27] Highly cross-linked insoluble forms of the interstitial collagens are much more resistant to collagenase degradation.[25] Soluble forms of type IV and the AB collagens are degraded slowly by purified collagenases at temperatures greater than 30°C. A metalloproteinase that appears to be specific for type IV collagen has been described by Liotta *et al.*,[31] but this is thought to be a separate enzyme. Metalloproteinases that degrade type IV and AB collagens have been found in some connective tissue culture media.

It is interesting to note that purified collagenases degrade gelatin very slowly; this suggests that the tertiary structure of collagen at the cleavage site is very important to facilitate the action of collagenase. Purified collagenase can also slowly cleave casein.[12]

Molecular Charge

Differences exist in the charges reported for purified collagenases. Isoelectric points of 6.9 for the rabbit tumor enzyme[27]; 6.7 for human skin procollagenase[30]; and 5.2 for pig synovial collagenase[13] have been reported. The purified collagenases also differ in their behavior on anion and cation exchangers; for example, rheumatoid synovial collagenase does not bind to QAE–Sephadex at pH 8.1, whereas the human skin enzyme binds and can be eluted with salt.[32]

[26] D. E. Woolley, R. W. Glanville, D. R. Roberts, and J. M. Evanson, *Biochem. J.* **169**, 265 (1978).

[27] P. A. McCroskery, J. F. Richards, and E. D. Harris, *Biochem. J.* **152**, 131 (1975).

[28] Z. Werb and J. J. Reynolds, *Biochem. J.* **151**, 645 (1975).

[29] S. Sakamoto, M. Sakamoto, P. Goldhaber, and M. Glimcher, *Arch. Biochem. Biophys.* **188**, 438 (1978).

[30] G. P. Stricklin, A. Z. Eisen, E. A. Bauer, and J. J. Jeffrey, *Biochemistry* **17**, 2331 (1978).

[31] L. A. Liotta, S. Abe, P. G. Robey, and G. R. Martin, *Proc. Natl. Acad. Sci. U.S.A.* **76**, 2268 (1979).

[32] D. E. Woolley, R. W. Glanville, and J. M. Evanson, *Biochem. Biophys. Res. Commun.* **51**, 729 (1973).

Molecular Size

The estimated apparent molecular sizes of active collagenases vary markedly from 23,000 to 130,000. Many of these estimates are unreliable because they have been made using gel filtration and it is known that different results are obtained using different gel matrices.[14] Collagenase can also interact with collagen[33] and heparin,[20] and these interactions account for some of the high-molecular-weight estimates. More reliable M_r values have been made on the purified collagenases using SDS–PAGE: rheumatoid synovium, 33,000[25]; human skin, 60,000[26]; human skin, 45,000 and 50,000[14]; pig synovium, 44,000[13]; mouse bone, 47,000[29]; rabbit VX$_2$ carcinoma, 34,000.[27] It is interesting to note that several workers have reported unusual protein bands using SDS–PAGE. Both Cawston and Tyler[13] and Sakamoto et al.[29] reported the presence of lower-molecular-weight bands of about 30,000, which appear to be generated from the major protein band of higher molecular weight. Stricklin et al.[14,30] have reported that latent collagenase, purified from human skin cultures, consists of two doublets after SDS–gel electrophoresis. The higher-molecular-weight doublet (MW 55,000 and 60,000) can be converted to the lower-molecular-weight doublet (MW 45,000 and 50,000) by trypsin treatment. However, no peptide has been isolated and shown to be a small cleavage product of the upper doublets. In addition, the latent collagenase can spontaneously activate without the conversion of the higher-molecular-weight doublet to the lower form. These biochemical studies must be pursued with collagenases from different species and tissues if a clear understanding of the latent enzyme, its activation mechanism(s), and its role in health and disease is to be understood.

[33] C. Fiedler-Nagy, J. W. Coffey, and R. M. Salvador, *Eur. J. Biochem.* **76**, 291 (1977).

[53] Collagenolytic Protease from Fiddler Crab (*Uca pugilator*)

By GREGORY A. GRANT, ARTHUR Z. EISEN, and RALPH A. BRADSHAW

The collagenolytic protease isolated from the hepatopancreas of the fiddler crab, *Uca pugilator,* first demonstrated in 1967,[1] is a serine protease homologous to the pancreatic serine proteases of vertebrates.[2] However, unlike that group, it is capable of cleaving the triple-helical portion of the

[1] A. Z. Eisen, *Biol. Bull.* (*Woods Hole, Mass.*) **133**, 463 (1967).
[2] G. A. Grant, K. O. Henderson, A. Z. Eisen, and R. A. Bradshaw, *Biochemistry* **19**, 4653 (1980).

polypeptide backbone of collagen without a measurable loss of helical content under physiological conditions of pH, temperature, and ionic strength.[3] This activity is characterized by cleavage at sites in a region approximately 75% of the distance from the amino-terminus of the collagen molecule, as has been found for all other animal collagenases studied.[4] Since the fiddler crab is a scavenger that feeds on animal tissues frequently containing collagen, the presence of a collagenase in the hepatopancreas of these animals is not unexpected. However, most animal collagenases are zinc metalloenzymes that require calcium for stability, have pH optima very close to neutrality, and are inhibited by such reagents as EDTA, cysteine, and o-phenanthroline.[5] These properties are often associated with other zinc-containing proteases such as thermolysin and carboxypeptidase. The crab protease, on the other hand, does not appear to have any metal requirement, has a basic pH optima, and is not inhibited by the substances mentioned above.[6] However, it is inhibited by diisopropyl fluorophosphate and soybean-trypsin inhibitor, a pattern typical of serine proteases.

The action of the crab protease is not restricted exclusively to collagen as a substrate. In addition to its collagenolytic activity, it is a good *general* protease and displays esterase and amidase activity toward synthetic low-molecular-weight substrates commonly used to assay trypsin and chymotrypsin. Furthermore, when the enzyme acts on collagen in solution, there is a marked decrease in the original intramolecular cross-linked β-component of collagen with a corresponding increase in the monomeric α-chains, in addition to the cleavage of the helical portion of the molecule.[3,6] This reaction is not commonly performed by animal collagenases, but it can be catalyzed by chymotrypsin.[7]

Crab collagenase is the first example of a serine protease that possesses significant collagenolytic activity. Other collagenolytic enzymes present in dog pancreas,[8] the fungus *Entomophthora coronata*,[9] and the insect *Hypoderma lineatum*[10,11] have been reported that may also be serine proteases. Crab collagenase, as well as these eukaryotic collagenases,

[3] A. Z. Eisen and J. J. Jeffrey, *Biochim. Biophys. Acta* **191**, 517 (1969).
[4] J. Gross, *in* "Biochemistry of Collagen" (G. N. Ramachandran and A. H. Reddi, eds.), p. 275. Plenum, New York, 1976.
[5] G. P. Stricklin, E. A. Bauer, J. J. Jeffrey, and A. Z. Eisen, *Biochemistry* **16**, 1607 (1977).
[6] A. Z. Eisen, K. O. Henderson, J. J. Jeffrey, and R. A. Bradshaw, *Biochemistry* **12**, 1814 (1973).
[7] P. Bornstein, A. H. Kang, and K. A. Piez, *Biochemistry* **5**, 3803 (1966).
[8] S. Takahashi and S. Seifler, *Isr. J. Chem.* **12**, 557 (1974).
[9] N. Hurion, H. Fromentin, and B. Keil, *Arch. Biochem. Biophys.* **192**, 438 (1979).
[10] A. Lecroisey, C. Boulard, and B. Keil, *Eur. J. Biochem.* **101**, 385 (1979).
[11] A. Lecroisey, A. DeWolf, and B. Keil, *Biochem. Biophys. Res. Commun.* **94**, 1261 (1980).

which appear to function in digestion rather than morphogenesis, represent a new group of trypsin-related serine proteases.

Methods of Assay

Crab collagenase can be assayed specifically for collagenolytic activity with a standard assay,[5,12] which quantitates the release of soluble [^{14}C]glycine-containing peptides from native reconstituted guinea pig skin collagen fibrils, or spectrophotometrically by following esterase or amidase activity with synthetic substrates.

Collagenolytic Activity. Radiolabeled collagen is purified from the skin of guinea pigs previously injected with [^{14}C]glycine according to the procedure of Gross.[13] For each assay,[12] 50 μl of a 0.4% [^{14}C]collagen solution having a specific activity of approximately 22,000 cpm/mg[5] is allowed to gel in a 6 × 50-mm tube at 37°C for at least 24 hr to allow completion of the aggregation process. After removal of excess buffer, the collagenase sample (50–100 μl) is added to the tube containing the collagen fibril and allowed to incubate at 37°C. The assay is completed by centrifuging the reaction tube at 12,000 g for 10 min and counting the supernatant in a liquid scintillation spectrometer.

This is compared to a control sample treated in an identical manner but without the presence of collagenase. Incubation of an identical gel with trypsin is used as an additional control for the presence of nonhelical or denatured collagen.

Esterase and Amidase Activity. Crab collagenolytic protease can also be assayed with a number of esterase and amidase substrates commonly used to assay trypsin and chymotrypsin such as N-benzoyl-L-tyrosine ethyl ester (BTyrEE)[14] or p-toluenesulfonyl-L-arginine methyl ester (TArgME).[15] However, an alternative spectrophotometric method using N-benzoyl-L-valylglycyl-L-arginine-p-nitroanilide (BValGlyArgNA) has the advantage of greater sensitivity over these substrates and is described here.

The assay is performed on a recording spectrophotometer by measuring the liberation of p-nitroanilide as previously described by Bundy.[16,17] BValGlyArgNA is dissolved in 0.05 M Tris buffer (pH 8.0) to a concentration of 1 mg/ml (1.7 × 10^3 M). This solution can be stored for several days at 20°C. Since the substrate slowly hydrolyzes in aqueous solution at pH

[12] Y. Nagai, C. M. Lapiere, and J. Gross, *Biochemistry* 5, 3132 (1966).
[13] J. Gross, *J. Exp. Med.* 107, 247 (1958).
[14] K. A. Walsh and P. E. Wilcox, this series, Vol. 19 [3].
[15] K. A. Walsh, this series, Vol. 19 [4].
[16] H. F. Bundy, *Anal. Biochem.* 3, 431 (1962).
[17] H. F. Bundy, *Arch. Biochem. Biophys.* 102, 416 (1963).

8.0, fresh substrate solutions should be used for precise assays. The assay is performed in substrate buffer with a substrate concentration of approximately $8.5 \times 10^5 M$ (50 μg stock solution per ml of assay mixture) by following the change in absorbance at 385 nm. The rate of increase of absorbance should be linear for several minutes, and this rate is directly proportional to the concentration of the enzyme.

One unit of activity is defined as the amount of crab protease catalyzing the hydrolysis of 1 μmol of BValGlyArgNA per minute. The molar extinction coefficient[16] of p-nitroanilide is 1.26×10^4 liters/mol, and the absorbance change accompanying the hydrolysis of 1 μmol of BValGlyArgNA per ml of assay solution is 12.6. Under the above assay conditions, crab collagenolytic protease at a concentration of 1 μg/ml in the assay cuvette results in an absorbance change of 0.012 absorbance units (min^{-1} cm^{-1}). Thus 1.0 mg of crab enzyme is equal to approximately 1 unit. In contrast, 1 mg of trypsin assayed with this substrate gives approximately 330 units.

Purification

The hepatopancreata of live fiddler crabs (*Uca pugilator*), which can be obtained from Gulf Specimen Company, Panacea, Florida, are removed immediately on arrival and stored frozen at $-80°C$. The frozen tissue can be stored without appreciable loss of enzymatic activity for several weeks. Crabs shipped over long distances during the summer months arrive in less than optimal condition and produce a significantly lower yield of enzyme. Best results are obtained with crabs either shipped during the cooler months (October through April), or obtained locally during the summer. The crab collagenolytic protease is purified from these tissues by the following method, which is a modification of that originally reported.[6] The procedure described here uses the glands from 8000 crabs, received in shipments of 1000 each, but can easily be scaled down for use with smaller amounts.

Step 1. Preparation of Acetone Powder. All procedures in the preparation of the acetone powder are performed at $-20°C$ in a walk-in freezer. Typically, the acetone powder is prepared from the glands of 2000 crabs as soon as they become available. The glands from 1000 crabs are homogenized in 10 volumes ($\sim$500 ml) of acetone in a blender for 5 min, and then poured into a 2-liter cylinder containing 1000 ml of acetone. This procedure is repeated for the second 1000 glands. The insoluble material is allowed to settle for 15–20 min and the supernatant is discarded by decantation. A second 1000 ml of acetone is added to the precipitate and, after mixing, the settling and decanting procedure is repeated. The insoluble material is then rehomogenized in 10 volumes of acetone, brought to a total volume of 1500 ml with more acetone, and centrifuged at 12,000 g for 20 min at $-20°C$. The precipitate is resuspended in 10 volumes of 1-butanol, homogenized, brought to 1500 ml with 1-butanol, centrifuged as before, and the supernatant is discarded. This procedure is repeated one more time with acetone and the supernatant again discarded. The final precipitate is combined in a single centrifuge bottle by suspending it in a small volume of acetone, centrifuging, and discarding the supernatant. The precipitate is dried in an evacuated dessicator over aluminum oxide at $-20°C$ overnight, and the powder stored at $-80°C$ in separate screw-top vials containing 8 g each. The enzyme is stable in this form for an indefinite period of time.

Step 2. First Gel Filtration. This and all subsequent procedures are performed at 4°C. Approximately 8 g of acetone powder is suspended in 50 ml of cold 0.05 M Tris-HCl (pH 7.5) by mechanical stirring in a beaker placed in an ice bath for 15–20 min. The suspension is then centrifuged at 14,000 g for 20 min, and the supernatant is decanted and saved. This procedure is repeated and the two supernatants are combined and centrifuged at 42,000 g for 1.5 hr. The supernatant is brought to 70% saturation with ammonium sulfate by the slow addition of 2.33 volumes of saturated ammonium sulfate; it is then stirred for 20 min and centrifuged at 18,000 g. The supernatant is discarded and the precipitate is dissolved in 30–40 ml of 0.05 M Tris-HCl buffer (pH 7.5), centrifuged again to remove any insoluble material, and applied to a column (5 × 150 cm) of Ultrogel Ac A 34 (LKB). The column is developed at 50–60 ml/hr with 15- to 20-ml fractions collected, and is monitored by absorbance at 280 nm. The elution profile is composed of five main peaks of absorbance and the collagenolytic activity is found coincident with the next to the last (fourth) peak. The appropriate fractions are pooled and dialyzed overnight against 100 liters of distilled water and lyophilized. At this stage of the purification, the enzyme solution also contains cellulase activity, which tends to weaken the dialysis tubing. Therefore, care should be used in handling the dialysis casing at the completion of that step in order to avoid rupture of the tubing and loss of sample. The cellulase activity is effectively separated from collagenolytic activity in step 3. The lyophilized powder can either be stored dry or dissolved in distilled water and stored at $-20°C$. Typically, the glands from 8000 crabs are processed through these first two steps and combined for the subsequent steps.

Step 3. Ion-Exchange Chromatography. The lyophilized powder from the previous step is dissolved in distilled water and centrifuged at 10,000 g for 20 min to remove any insoluble material. The sample is then applied to a column (5 × 40 cm) of DEAE–Sepharose CL-6B (Pharmacia), previously equilibrated in 0.005 M Tris-HCl (pH 7.5) containing 0.2 M NaCl.

After the sample is loaded, the column is washed with 250 ml of the same buffer and is developed with an 8-liter linear gradient composed of 4 liters of $0.005 M$ Tris-HCl (pH 7.5)–$0.2 M$ NaCl and 4 liters of $0.005 M$ Tris-HCl (pH 7.5)–$1 M$ NaCl. The column is eluted at a flow rate of 250–300 ml/hr and 20-ml fractions are collected. This chromatography results in the separation of the collagenolytic activity into two components referred to as peak 1 and peak 2, which appear in the elution profile, monitored at 280 nm, as the last two peaks eluted from the column. Both peaks contain collagenolytic activity as well as trypsin and chymotrypsin-like activity. Peak 1 contains approximately 94% of the recovered collagenolytic activity with a specific activity about four times that of peak 2, and has been the principal form of interest. Peak 2 has not been further characterized at this time. After chromatography, peak 1 is dialyzed overnight against 100 liters of distilled water and lyophilized.

Step 4. Hydroxyapatite Chromatography. The lyophilized powder from the previous step is dissolved in $0.005 M$ potassium phosphate buffer, pH 6.7 (conductivity is 570 μmho), and chromatographed on a column (2.5 × 28 cm) of spheroidal hydroxyapatite (BDH Biochemicals) equilibrated and eluted with the same buffer at 100 ml/hr. The column is monitored at 280 nm and 8- to 10-ml fractions are collected. The collagenase elutes in the void volume as a single peak, and is pooled and lyophilized. The protein remaining on the column may be removed by eluting with $0.5 M$ phosphate buffer.

Step 5. Second Gel Filtration. The lyophilized powder from the previous step is dissolved in $0.05 M$ Tris-HCl (pH 7.5) and chromatographed on a column (2.5 × 200 cm) of Sephadex G-100. The column is monitored at 280 nm, eluted at 25 ml/hr, and 4- to 5-ml fractions are collected. The enzyme elutes as a single major peak followed by several minor peaks.

This procedure produces approximately 100–150 mg of homogeneous crab protease from the glands of 8000 fiddler crabs. The enzyme is stable for several months in slightly acidic solution (pH ~6.0) at $-20°C$, and can be stored indefinitely as a lyophilized powder. Crab collagenolytic protease is also stable at pH 8.0 and 4°C for several hours. However, any disruption of the native conformation results in significant autodegradation.

Properties

Fiddler crab collagenolytic protease is a single polypeptide chain of 226 amino acid residues containing 3 disulfide bridges. The most distinctive feature of the composition is the relatively low content of basic residues. The enzyme contains only 9 basic residues (4 histidines, 4 ar-

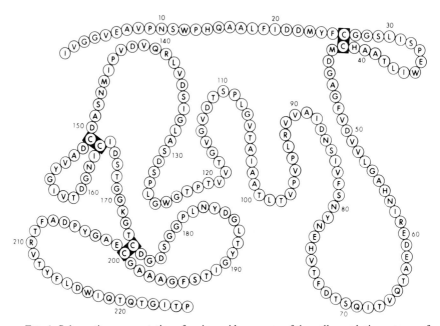

FIG. 1. Schematic representation of amino acid sequence of the collagenolytic protease of fiddler crab (*Uca pugilator*). Disulfide bonds, indicated by black boxes, were assumed from analogous pairings with other members of the family.

ginines, and 1 lysine), in contrast to 26 acidic residues, which is reflected in the acidic isoelectric point (pI 3.0). This is considerably lower than that of other serine proteases (pI 8–11), and it imparts a substantial negative charge on the enzyme at its catalytic pH optimum (pH 7.0–7.5 for collagen; pH 8.0–8.5 for esterase substrates).[6] What significance this may have in regard to its action on collagen or its physiological role as a digestive protease is not yet known.

The molecular weight of crab collagenase calculated from the amino acid sequence is 23,505. This agrees well with the molecular weight of 25,000 determined by sedimentation-equilibrium ultracentrifugation and SDS–polyacrylamide gel electrophoresis,[6] and is similar to that of trypsin and chymotrypsin.

Amino Acid Sequence. The amino acid sequence of crab collagenolytic protease is given in Fig. 1, with the degree of homology to other serine proteases presented in Table I. The residues forming the charge-relay system[18] of the active site in serine proteases (His-57, Asp-102, and Ser-195, chymotrypsin numbering), are found in corresponding regions in the crab enzyme (His-41, Asp-87, and Ser-178), and the sequences around

[18] D. M. Blow, J. J. Birktoft, and B. S. Hartley, *Nature* (*London*) **221**, 337 (1969).

TABLE I

HOMOLOGY OF CRAB COLLAGENOLYTIC PROTEASE TO OTHER
SERINE PROTEASES[a]

Enzyme	Identities/total positions	%
Chymotrypsin A (bovine)[b]	80/235	34
Chymotrypsin B (bovine)[b]	80/236	34
Trypsin (bovine)[b]	75/231	32
Elastase (porcine)[b]	75/244	31
Kallikrein (porcine)[c]	72/237	30
Group-specific protease (rat)[b]	78/237	33
Thrombin (bovine)[d]	70/260	27
Factor X_a (bovine)[d]	64/262	24
Factor IX_a (bovine)[d]	64/240	27
Protein C (bovine)[d]	68/245	28
Plasmin (human)[d]	74/244	30

[a] The comparisons were made by aligning the amino-termini and counting all positions in which at least one of the two proteins had the position occupied. Deletions were treated as nonidentities.

[b] R. G. Woodbury, N. Katunama, K. Kobayashi, K. Titani, and H. Neurath, *Biochemistry* 17, 811 (1978).

[c] H. Tschesche, G. Mair, G. Godec, F. Fiedler, W. Ehret, C. Hirschauer, M. Lemon, H. Fritz, G. Schmidt-Kastier, and C. Kutzbach, *in* "Kinins—II: Biochemistry, Pathophysiology and Clinical Aspects" (S. Fujii, H. Mariya, and T. Suzuki, eds.), p. 245. Plenum, New York, 1979.

[d] C. M. Jackson and Y. Nemerson, *Annu. Rev. Biochem.* 49, 765 (1980).

these residues are highly conserved. In addition, diisopropyl fluorophosphate selectively inhibits the enzyme through modification of Ser-178. The 6 half-cystinyl residues correspond to 6 of the 10 half-cystinyl residues in chymotrypsin. Since none of the half-cystinyl residues in the crab enzyme are present as free thiols, these residues are assumed to form disulfide bonds identical to 3 of the 5 disulfide bonds present in chymotrypsin. These bonds link Cys-26 to Cys-42, Cys-151 to Cys-164, and Cys-174 to Cys-200. These linkages are present in all other known serine proteases except the group-specific protease from rat small intestine,[19] which does not contain a disulfide bond corresponding to Cys-174 to Cys-200.

Substrate Specificity and Inhibition

In contrast to the vertebrate collagenases, the crab enzyme is a good general protease capable of extensively degrading a wide variety of

[19] R. G. Woodbury, N. Katunama, K. Kobayashi, K. Titani, and H. Neurath, *Biochemistry* 17, 881 (1978).

polypeptide substrates. The peptide bond specificity appears to be rather broad and cleaves substrates on the carboxy-terminal side of an unusually large variety of amino acid residues, including those with charged as well as hydrophobic side-chains.[20]

When polypeptide substrates are used,[20] crab collagenolytic protease appears to favor tyrosine, phenylalanine, and leucine residues, similar to chymotrypsin. Furthermore, at least one example each of highly efficient cleavages at lysine and alanine residues were observed, as well as repeated examples of slower cleavages at arginine and glutamic acid bonds. With elongated polypeptide substrates, such as bovine insulin B chain, the rate of hydrolysis catalyzed by the crab collagenase is comparable to that of chymotrypsin.

The synthetic peptide, DNP-Pro-Gln-Gly-Ile-Ala-Gly-Gln-D-Arg, possesses a sequence identical to that surrounding the vertebrate collagenase sensitive Gly-Ile bond in the α_1-chain of type I collagen, except that the amino-terminal is blocked by a dinitrophenyl group and the carboxyl-terminal residue is the stereoisomer of the naturally occurring L-arginine residue. This peptide, which was originally designed as a substrate for enzymes with collagenolytic properties, is specifically cleaved at the Gly-Ile bond[21,22] by tissue collagenases as well as certain gelatinases.[22,23] This peptide is also a good substrate for the crab collagenolytic protease, although it is totally resistant to digestion by chymotrypsin.[20] The rate of hydrolysis is 30% of the rate found with human skin fibroblast collagenase.[20] Examination of the resulting DNP-peptide products by amino acid analysis indicates that human skin collagenase cleaves the peptide exclusively at the Gly-Ile bond. However, similar examination of the peptide(s) produced by the crab enzyme reveals a different cleavage pattern. In this case, the DNP peptide products appear to be an equimolar mixture of DNP-Pro-Gln and DNP-Pro-Gln-Gly-Ile, indicating that the crab protease cleaves this peptide at the Gln-Gly and Ile-Ala bonds, rather than at the Gly-Ile bond, as seen with vertebrate skin collagenases.

The rather broad range of specificity observed for the crab collagenolytic protease when polypeptides are employed as substrates is also observed with low-molecular-weight ester and amide substrates commonly used to assay trypsin and chymotrypsin[20] (Table II). In all cases studied, however, the rate of hydrolysis is considerably less than the more selective serine proteases, trypsin and chymotrypsin. The observation

[20] G. A. Grant and A. Z. Eisen, *Biochemistry* **19**, 6089 (1980).
[21] Y. Masui, T. Tahemoto, S. Sakakibara, H. Hori, and Y. Nagai, *Biochem. Med.* **17**, 215 (1977).
[22] S. Kobayashi and Y. Nagai, *J. Biochem.* (*Tokyo*) **84**, 559 (1978).
[23] J. L. Seltzer, S. A. Adams, G. A. Grant, and A. Z. Eisen, unpublished results.

TABLE II

HYDROLYSIS OF LOW-MOLECULAR-WEIGHT SYNTHETIC SUBSTRATES BY CRAB
COLLAGENOLYTIC PROTEASE[a]

Substrate[b]	Enzyme	K_m (mM)	k_{cat} (sec^{-1})	k_{cat}/K_m (sec^{-1} M^{-1})
BArgNA	Crab protease	0.100	0.021	210
	Bovine trypsin	0.093	1.45	15,600
BValGlyArgNA	Crab protease	0.120	3.0	25,000
	Bovine trypsin	0.167	440.0	2,634,000
BTyrNA	Crab protease	0.263	0.015	57
	Bovine chymotrypsin	0.110	0.710	6500
AcAlaAlaAlaNA	Crab protease	0.044	0.002	45
	Porcine elastase	0.100	0.091	910
TArgME	Crab protease	0.233	0.810	3500
	Bovine trypsin	0.093	46.15	5,000,000
BTyrEE	Crab protease	1.11	1.09	1000
	Bovine chymotrypsin	0.270	10.20	38,000

[a] G. A. Grant and A. Z. Eisen, *Biochemistry* **19**, 6089 (1980).
[b] Abbreviations used are: BArgNA, N-benzoyl-L-arginine-p-nitroanilide; BValGly-ArgNA, N-benzoyl-L-valylglycyl-L-arginine-p-nitroanilide; BTyrNA, N-benzoyl-L-tyrosine-p-nitroanilide; AcAlaAlaAlaNA, N-acetyl-L-alanyl-L-alanyl-L-alanyl-p-nitroanilide; TArgMe, N-α-tosyl-L-arginine methyl ester; BTyrEE, N-benzoyl-L-tyrosine ethyl ester.

that the rate of hydrolysis of synthetic substrates by most serine proteases increases with increasing peptide length, due to secondary binding interactions,[24] is also observed with the crab protease. For example, the k_{cat} of the crab enzyme with N-benzoyl-L-arginyl-p-nitroanilide is approximately 150-fold less than with N-benzoyl-L-valylglycyl-L-arginyl-p-nitroanilide. This behavior is particularly common to proteases that show a low level of primary specificity.

Leupeptin, elastatinal, and chymostatin, which are protease inhibitors of microbial origin,[25] were tested for their ability to inhibit the crab protease (Table III). Of these compounds, only chymostatin, an effective inhibitor of chymotrypsin containing a carboxyl-terminal L-phenylalaninal residue essential to its specificity and inhibitory capacity, displayed significant inhibition. The corresponding inhibitors of trypsin and elastase, leupeptin and elastatinal, which contain carboxyl-terminal L-arginal and L-alaninal residues, respectively, only inhibited the crab collagenase to a

[24] J. D. Robertus, R. A. Alden, J. J. Birktoft, J. Kraut, J. G. Powers, and P. E. Wilcox, *Biochemistry* **11**, 2439 (1972).
[25] H. Umezawa, this series, Vol. 45 [55].

TABLE III
INHIBITION OF CRAB COLLAGENOLYTIC
PROTEASE[a]

Inhibitor[b]	I/E[c]	% Inhibition
DFP	200	97
Cysteine	2000	7
EDTA	1000	0
SBTI	1.4	98
TosLeuCh₂Cl	100	0
TosPheCh₂Cl	100	0
TosLysCh₂Cl	100	16
ZPheCh₂Cl	100	38
AcAlaPheCh₂Cl	100	16
Leupeptin	100	18
Elastatinal	50	15
Chymostatin	4	99

[a] Data from G. A. Grant and A. Z. Eisen.[20]
[b] Abbreviations used are: DFP, diisopropyl fluorophosphate; EDTA, ethylenediamine tetraacetic acid; SBTI, soybean-trypsin inhibitor; TosLeuCh₂Cl, tosyl-L-leucine chloromethyl ketone; TosPheCh₂Cl, tosyl-L-phenylalanine chloromethyl ketone; TosLys-Ch₂Cl, tosyl-L-lysyl chloromethyl ketone; ZPhe-Ch₂Cl, benzyloxycarbonyl-L-phenylalanine chloromethyl ketone; AcAlaPheCh₂Cl, N-acetyl-L-alanyl-L-phenylalanyl chloromethyl ketone.
[c] Molar ratio of inhibitor to enzyme.

small degree. This correlates well with the previously observed chymotrypsin-like specificity of the enzyme.[20]

Crab collagenase is not significantly inhibited by L-1-tosylamide-2-phenylethyl chloromethyl ketone, N^α-p-tosyl-L-lysine chloromethyl ketone, N^α-tosyl-L-leucine chloromethyl ketone, N-acetyl-L-alanyl-L-phenylalanine chloromethyl ketone, or N-benzyloxycarbonyl-L-phenylalanine chloromethyl ketone. Although these compounds contain the side-chains of amino acid residues that were clearly selected by the crab protease in polypeptide and synthetic substrates, none of them were capable of reacting with an active-site nucleophile. The small degree of inhibition seen with these substances (Table III) appears to be due to reversible binding at the active site rather than to covalent modification. The lack of an extended peptide chain and the presence of bulky blocking groups appears to be the most likely explanation for the poor inhibitory activity of these compounds. The enzyme is also inhibited very efficiently by soybean-trypsin inhibitor and diisopropyl fluorophosphate.

Structure–Function

All serine protease structures investigated so far show a similar pattern in their interaction with substrates, in particular polypeptide substrates. Residues 193–195 (214–216, chymotrypsin numbering) are in an extended conformation and interact with the polypeptide substrate to form a 2-stranded antiparallel β-sheet. Additionally, residues 193–195, which invariably have the sequence Ser-aromatic residue-Gly, form one side of the subsite S_1 crevice that binds the side-chain at the P_1 position in the substrate. The one known exception to this sequence is elastase, where the glycine is replaced with a valine, thus blocking the access to the S_1 crevice.[26] The primary specificity of the enzyme is determined by the S_1 crevice.

The importance of secondary substrate-enzyme interactions with the crab collagenolytic protease was shown in the kinetic experiments described above. In the crab protein, residues 194 and 196–199 (215 and 217–219 in chymotrypsin), which are all involved in secondary-subsite interaction, are changed from Tyr to Phe, and Ser-Ser-Thr to Ala-Ala-Ala-Gly. In fact, in this particular region, the crab protease much resembles elastase, both in regard to the insertion of a residue between residues 218 and 219, and with residue 215 being a phenylalanine. Both collagenase and elastase act on substrates, collagen and elastin, respectively, which have rather indistinguished features in their primary sequence, and thus tend to recognize specific conformational features in their respective substrates rather than specific residues or sequences. Thus the Ala-Ala-Ala-Gly sequence in the crab protease may well be a region of importance in the interaction of the enzyme and collagen.

In chymotrypsin and trypsin, the residue at position 189 (chymotrypsin numbering), which is serine and aspartic acid, respectively, forms the bottom of the binding pocket. In the crab collagenolytic protease, this position is occupied by a glycyl residue (172). Furthermore, the change of residue 206 (226, chymotrypsin numbering) from a glycyl to an aspartyl residue in crab collagenase is also interesting. In elastase, this position is occupied by a threonyl residue and its side-chain protrudes into the binding crevice. Although it is not possible to assess the significance of this substitution in detail without knowledge of the three-dimensional struc-

---

[26] D. M. Shotton and H. C. Watson, *Nature (London)* **225**, 811 (1970); also in this series, Vol. 19 [7].

ture, this aspartyl residue may account, at least in part, for the observed proteolysis at positively charged side-chains.

It is not known what role these structural differences relative to other serine proteases may play in the substrate-binding interactions of the crab enzyme, but they may be significant with respect to the collagenolytic activity. Its ability to attack and specifically cleave the native triple helix of collagen is a unique feature of this serine protease, and in this regard, the conformational nature of the active site and the regions involved in substrate binding are of particular interest. The structural features of the enzyme may give considerable insight into the determinants of substrate specificity of this class of enzymes as well as the mammalian metalloenzyme collagenases. In addition, investigations into the nature of the interaction of crab collagenase with collagen may lead to a better understanding of the structural features of collagen that participate in the initial cleavage of that molecule at a specific site by tissue collagenases.

Acknowledgment

This work was supported by U.S.P.H.S. Research grants AM 07284, AM 12129, and AM 13362.

Section VI

Inhibitors of Various Specificities

[54] α_2-Macroglobulin

By ALAN J. BARRETT

α_2-Macroglobulin (α_2M) is one of the major plasma proteins in humans, and homologous proteins seem to be present in all other vertebrates. The human protein has M_r 725,000, and is particularly interesting for its property of acting as a molecular trap for proteinase molecules.[1] Having been enclosed within the macroglobulin molecule, a proteinase molecule is unable to act on protein substrates, to be inhibited by protein inhibitors, or to bind antibodies. The mechanism of the trap is such that the great majority of proteinases are bound by α_2M in this way, regardless of their precise specificity or catalytic mechanism. The changed conformation of the "closed" macroglobulin molecule leads to rapid clearance from the circulation of it and the trapped proteinase, by the reticuloendothelial system. It may well be that α_2-macroglobulin represents an ancient defensive system of the body. This is suggested by the fact that α_2M is the only plasma inhibitor of a number of proteinases used by pathogens and parasites in attacking the body, and that α-macroglobulins are "acute-phase" proteins in many species; also, there is evidence that the protein has an evolutionary relationship to some of the complement components.

In all work with α_2M it is important to recognize that the native protein can exist in two quite distinct forms: the electrophoretically "slow" S form, which is the circulating form that is reactive with proteinases; and the electrophoretically "fast" F form, which is produced irreversibly from the S form by reaction with proteinases or low-molecular-weight amines.[2]

Assays

Two distinct types of assays are required in work with α_2M: determination of total α_2M (in S and F forms), and of active (S form) α_2M selectively. Total α_2M is best quantified immunologically. Active α_2M can be estimated by saturating it with trypsin, inhibiting all free trypsin with soybean trypsin inhibitor, and quantifying the bound trypsin with benzoylarginine nitroanilide as substrate.[3,4] In our hands, however, a better procedure depends on the inhibition of the activity of thermolysin on dyed

[1] A. J. Barrett and P. M. Starkey, *Biochem. J.* **133,** 709 (1973).
[2] A. J. Barrett, M. A. Brown, and C. A. Sayers, *Biochem. J.* **181,** 401 (1979).
[3] P. O. Ganrot, *Clin. Chim. Acta* **14,** 493 (1966).
[4] P. O. Ganrot, *Acta Chem. Scand.* **21,** 602 (1967).

METHODS IN ENZYMOLOGY, VOL. 80

hide powder, as described below. Other proteinases can be used, but thermolysin, a metalloproteinase, is rather selective, being inhibited by few other proteins. Equally, other protein substrates can be used, but hide powder is very convenient, and is of such large particle size that the steric inhibition by α_2M is complete.[2,5]

Thermolysin–Hide Powder Assay

Reagents

> *Substrate suspension.* Remazol brilliant blue hide powder[6] (12.5 mg/ml) in 0.6 M sucrose–0.1% Brij 35–0.1% sodium azide. Store at 4°C.
>
> *Buffer.* 0.4 M Tris-HCl (pH 7.5) containing 40 mM CaCl$_2$–0.1% Brij 35–0.1% sodium azide.
>
> *Thermolysin solution.* Thermolysin[7] is dissolved as a 1 mg/ml stock solution in the above buffer, and stored at −20°C. For use, it is diluted to 2.0 μg/ml in the same buffer with the addition of 1 mg bovine serum albumin/ml.
>
> *Stopping reagent.* 100 mM disodium EDTA.

Procedure

A 0.1-ml sample containing 2–5 μg of α_2M, or about 1 μl of plasma, diluted in 0.1% Brij 35, is mixed with 0.5 ml of buffer and 0.1 ml of thermolysin working solution. Ten minutes is allowed for reaction at room temperature, then 0.5 ml of water and 0.8 ml of substrate suspension are added and the mixture is incubated for 20 min at 37°C with continuous agitation. The reaction is stopped with 1.0 ml of the EDTA solution, and A_{595} measured.

A standard curve relating ΔA_{595} to amount of thermolysin, in the absence of α_2M, allows the decreases in ΔA_{595} due to α_2M to be expressed as micrograms of thermolysin inhibited. Standardization beyond this is not straightforward, since no active-site titrant for thermolysin is readily available. The procedure can be calibrated, however, by (*a*) making an assay of fresh plasma in which the α_2M (which may safely be assumed to be entirely S-α_2M) is also quantified immunologically, or (*b*) by making

---

[5] H. Rinderknecht, C. Carmack, and M. C. Geokas, *Immunochemistry* **12**, 1 (1975).

[6] Obtained as hide powder azure from Calbiochem, P.O. Box 12087, San Diego, California 92112, or prepared as described by H. Rinderknecht, P. Silverman, M. C. Geokas, and B. J. Haverback [*Clin. Chim. Acta* **28**, 239 (1970)], but with the amount of dye decreased to 75 mg/g hide powder.

[7] As Protease Type X from Sigma Chemical Co., St. Louis, Missouri 63118.

some assays with active-site titrated trypsin in place of thermolysin, with an acidic stopping reagent.[2]

Immunoassay for α_2M

Radial immunodiffusion plates obtainable commercially, or prepared as described by Mancini *et al.*,[8] are suitable for small numbers of analyses, although diffusion of α_2M is slow. For large numbers of analyses we find "rocket" immunoassays more convenient and economical. Rocket immunoassays are run as described by Laurell.[9] Briefly, antiserum to α_2M (commercially available) is incorporated into 1% (w/v) agarose gel at a suitable concentration determined empirically in advance (often 1–5%, v/v). The gel also contains a Tris-borate buffer (pH 8.3), composed of Tris (90 mM), boric acid (80 mM), and disodium EDTA (3 mM). The gel is cast between two glass plates (one precoated with agarose) that are separated by a 2-cm spacer. When the gel has set, the uncoated plate is slid off, and a row of 5-μl (2-mm diameter) wells is punched, 2 cm from one long edge of the gel. Standard and unknown samples are placed in the wells, and the plate is run overnight in an immunoelectrophoresis tank with large buffer reservoirs. The antigen migrates anodally from the wells, and encounters virtually immobile antibody. Migration continues so long as excess antigen keeps the complexes soluble, so the height reached by each "rocket" is directly proportional to the amount of antigen originally in the well.

This is the best method I know for determining the total concentration of α_2M, in an impure sample of the protein; yet it is not perfect, because S- and F-α_2M tend to give slightly different quantitative reactions, and the magnitude of the differences varies between antisera.[2]

Detection of α_2M in Gel Electrophoresis

The property of binding proteinases without inhibiting their activity against low-molecular-weight substrates distinguishes α_2M from other proteins, and provides a specific method for detection of the protein.[10,11]

Plasma (or another sample also containing about 2 mg α_2M/ml) is treated with papain as follows. The sample (50 μl) is mixed with 130 μl of 0.05 M sodium phosphate buffer, pH 7.4. Papain (2 μg) preactivated during 5 min in 20 μl of the same buffer containing 5 mM cysteine is then

[8] G. Mancini, A. O. Carbonara, and J. F. Heremans, *Immunochemistry* **2**, 235 (1965).

[9] C.-B. Laurell, *Scand. J. Clin. Lab. Invest.* **29**, Suppl. 124, 21 (1972).

[10] A. J. Barrett, *Biochem. J.* **131**, 809 (1973).

[11] P. M. Starkey and A. J. Barrett, *Biochem. J.* **131**, 823 (1973).

introduced, and the mixture is incubated for 60 min at 25°C before electrophoresis.

The sample is run in rate electrophoresis or pore-limit electrophoresis (see later), or immunoelectrophoresis at pH 8 (as appropriate). Free papain runs cathodally, and is lost from the gel. After the run (and, in the case of immunoelectrophoresis, development with antiserum and washing), the gel is stained for the activity of the α_2M-bound papain. The gel is immersed in 0.1 M sodium phosphate buffer (pH 6.0) containing 1.33 mM disodium EDTA–2.67 mM cysteine–0.5 mg benzyloxycarbonyl-arginine 2-(4-methoxy)naphthylamide[12,13]/ml, at 40°C. Usually, a 5- to 10-min incubation period is adequate. The gel is then transferred into the coupling reagent, prepared as follows. Mersalyl acid[13] (2.43 g) is dissolved in 60 ml of 0.5 M NaOH; 9.4 g of disodium EDTA is added, and the volume is made up to about 450 ml with water. The pH is brought down to 4.0 by dropwise addition of 1 M HCl, with continuous stirring. The volume is brought to 500 ml, and the solution is stored indefinitely in a brown bottle at room temperature. Fast Garnet GBC[14] (1 mg/ml) is dissolved in the mersalyl–EDTA reagent and the solution is used the same day. After about 10 min the red staining of parts of the gel containing methoxynaphthylamine liberated by enzymic activity is complete, and the gel is transferred to water.

Control samples without added papain are run to confirm that the activity detected is that of bound papain.

Purification

α_2M occurs in the plasma exclusively in the S form, which is active in the binding of proteins. The following procedure was developed[2] with the aim of purifying α_2M without any conversion into the inactive F form.

Blood containing acid citrate–dextrose (ACD) anticoagulant is obtained from a blood bank. (Fresh donations unsuitable for transfusion are preferable to time-expired blood.) Because the separation of α_2M from the polymeric forms of haptoglobin is extremely difficult, haptoglobin 1-1 type plasma is selected; this forms about 10% of all blood donations, and is easily identified by gel electrophoresis after saturation with hemoglobin.[15] The plasma is separated by centrifugation (10,000 g for 15 min) in plastic bottles, and can be stored until required at $-20°C$.

Plasma (250 ml) is mixed with 0.28 volume of aqueous 25% (w/v)

---

[12] Note that 2-naphthylamides are probably carcinogenic, and must therefore be handled with care.

[13] Obtainable from Sigma Chemical Co.

[14] Obtained as the fluoroborate from Serva Feinbiochemica GmbH & Co., P.O. Box 105260, D-6900, Heidelberg, Federal Republic of Germany.

[15] R. B. Halliday, *Med. Lab. Sci.* **33**, 23 (1976).

polyethylene glycol of M_r 6000, and left 30 min at 20°C. The precipitate (containing β-lipoprotein) is removed by centrifugation, and a further 0.72 volumes (on the basis of the original volume) of the 25% polyethylene glycol added to the supernatant. After 30 min the precipitate [formed at 5.5–12.5% (w/v) polyethylene glycol] is centrifuged, dissolved in 0.1 M sodium citrate buffer (pH 6.5) to an $A_{280}^{1\,cm}$ of about 50, and run on a bed (5 × 80 cm, 1570 cm³) of a 2 : 1 (v/v) mixture of LKB Ultrogel AcA 34 and AcA 22, in the same buffer, at a flow rate of 20 ml/hr.

The peak of α₂M in fractions early in the elution pattern is detected by gel electrophoresis, and fractions showing little contamination are combined, concentrated by ultrafiltration over a Sartorius type 12133 membrane under pressure of nitrogen to about 10 mg/ml, and dialyzed against water during 24–48 hr. The precipitate (largely composed of immunoglobulins) is discarded.

At this stage, the α₂M is about 95% pure (immunoglobulin A being the major contaminant), and is suitable for many purposes. Gel electrophoresis without SDS, or immunoelectrophoresis against anti(human serum proteins) serum scarcely detect contamination.

The final removal of trace contaminants is achieved by immunoadsorption. Immunoadsorbents are prepared by coupling commercial rabbit immunoglobulins against all human immunoglobulins, immunoglobulin A, α₂-pregnoglobulin, haptoglobulin, and fibrinogen, to Sepharose-4B that has been activated with CNBr essentially as described by Cuatrecasas and Anfinsen.[15a] Each gram of damp activated gel is allowed to react overnight at 20°C with 7.5–50 μl (according to titer) of the commercial immune globulin solution in 50 mM sodium phosphate buffer, pH 6.5. The adsorbents are then treated with 1% (w/v) glycine in the same buffer for 1 hr, and washed with 100 mM sodium formate buffer (pH 3.0), followed by 100 mM sodium phosphate (pH 7.0) containing 0.1% NaN₃.

The partially purified α₂M (about 100 mg in 10 ml) is passed through a series of small columns (each about 20 ml bed volume) containing the immunoadsorbents, in 0.1 M sodium phosphate buffer, pH 7.0. The unadsorbed protein is washed through with the same buffer, and reconcentrated to A_{280} about 10 by ultrafiltration. The immunoadsorbent columns are regenerated for reuse by washing with 0.1 M sodium formate buffer (pH 3.0) containing 0.5 M NaCl. The immunoadsorption process is repeated as necessary until no immunoprecipitation reaction is obtained for any of the original contaminants.

Finally, the solution of α₂M is twice dialyzed for 24 hr against 100 volumes of 20 mM sodium citrate buffer (pH 6.5) and stored at 4°C, sometimes after freeze-drying.

A typical yield of α₂M from 250 ml of citrate plasma is 200 mg (40%)

[15a] P. Cuatrecasas and C. B. Anfinsen, this series, Vol. 22, p. 345.

before immunoadsorption, and 160 mg (32%) after. The product is expected to be exclusively in the S form, as judged by rate electrophoresis or pore-limit electrophoresis (see later).

A useful new technique that may be considered for inclusion in α_2M purification schemes is affinity chromatography on a zinc-chelate column.[16]

Electrophoresis of α_2M

Two particularly useful electrophoretic techniques for α_2M are pore-limit electrophoresis in a gradient of increasing polyacrylamide concentration, and SDS-gel electrophoresis with or without reduction of disulfide bonds.[2]

Pore-Limit Electrophoresis

In this technique the molecules of α_2M or its subunits migrate through a gradient of 4–26% polyacrylamide until their progress is stopped by the decreased pore size of the gel. Gel slabs may be purchased commercially, or prepared in the laboratory.[2] Figure 1 shows the results obtained.

SDS-Gel Electrophoresis

SDS-gel electrophoresis forms the most critical electrophoretic test of the homogeneity of purified S-α_2M. In some systems, the fragments IVa and IVb resulting from cleavage of the "bait" region (see Fig. 3) are not resolved, but we have obtained good results with the 2-amino-2-methylpropane-1-diol (Ammediol)/glycine/chloride system originally described by Wyckoff et al.[17] Our modified procedure, in which we use slab gels containing 7% total acrylamide, 2.6% of this as methylenebisacrylamide, has been described fully.[2] Typical results are shown in Fig. 2B.

As is explained below, pure S-α_2M shows three bands on an SDS gel after reduction, but it should not show any component in the 75,000–100,000 M_r range: this would represent contaminants, or subunits of F-α_2M cleaved at the bait region.

Properties

Stability

α_2M is denatured below about pH 4, and separates into the noncovalently associated half-molecules. Mild reduction of the native protein (e.g.,

[16] T. Kurecki, L. F. Kress, and M. Laskowski, Sr., Anal. Biochem. 99, 415 (1979).
[17] M. Wyckoff, D. Rodbard, and A. Chrambach, Anal. Biochem. 78, 459 (1977).

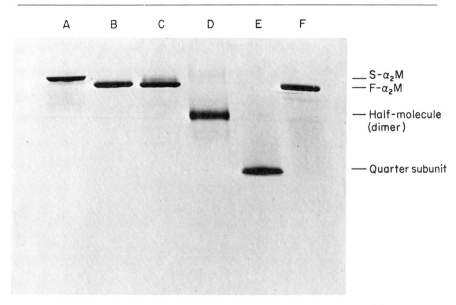

FIG. 1. Pore-limit gel electrophoresis of α_2M. The samples, run in a gradient of increasing polyacrylamide concentration, were (A) S-α_2M, (B) F-α_2M formed by saturation with trypsin, (C) F-α_2M formed by treatment with methylammonium chloride, (D) subunit (dimeric) formed by exposure of S-α_2M to pH 2.0, (E) subunit (monomeric) formed by treatment of S-α_2M with 1 mM dithiothreitol, and (F) reduced tetramer formed by treating F-α_2M (trypsin-treated) with 1 mM dithiothreitol. Reproduced with permission of the publisher from Barrett *et al.*[2]

1 mM dithiothreitol, pH 8.0) causes reversible dissociation into inactive but native quarter subunits. S-α_2M also reacts with ammonia, methylamine, and hydroxylamine above pH 7 to form F-α_2M, so the use of $(NH_4)_2SO_4$ in the purification of α_2M is undesirable.

Physiochemical Properties

The M_r of α_2M is 725,000, as determined by ultracentrifugation,[18,19] and $A_{280}^{1\%}$ is close to 9.0.[19,20] Estimates of partial specific volume range between 0.720 and 0.735,[18,21] the frictional ratio is 1.43–1.57,[18,22] and the pI 5.0–5.2.[2]

[18] J. M. Jones, J. M. Creeth, and R. A. Kekwick, *Biochem. J.* **127**, 187 (1972).
[19] P. K. Hall and R. C. Roberts, *Biochem. J.* **171**, 27 (1978).
[20] J. T. Dunn and R. G. Spiro, *J. Biol. Chem.* **242**, 5549 (1967).
[21] J.-P. Frénoy, R. Bourillon, R. Lippoldt, and H. Edelhoch, *J. Biol. Chem.* **252**, 1129 (1977).
[22] M. von Schönenberger, R. Schmidtberger, and H. E. Schultze, *Z. Naturforsch. B: Anorg. Chem., Org. Chem., Biochem., Biophys., Biol.* **13B**, 761 (1958).

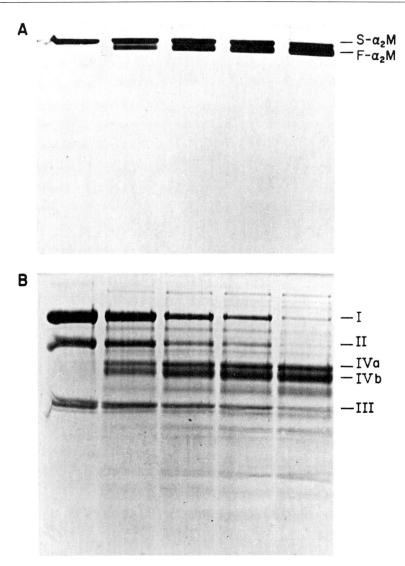

FIG. 2. Effect of reacting α_2M with increasing amounts of trypsin, examined in two gel-electrophoresis systems. The samples were prepared by reacting S-α_2M (1 mg/ml) with 0, 7.5, 15, 20, and 30 μg of trypsin/ml (from left to right). They were then run (A) without denaturation in polyacrylamide gels with the Ammediol–glycine–HCl buffer system (see the text), or (B) with reduction in SDS, on 7% polyacrylamide gels in the same buffer system. Gel (A) clearly shows the conversion of S-α_2M into F-α_2M as saturation is approached. Gel (B) shows the partial autolytic fragmentation of the subunit (I) into the components II and III in the absence of trypsin, and the replacement of this type of cleavage by scission at the bait region to give components IVa and IVb on reaction with the enzyme. Some of the minor bands running between I and IVa represent covalently linked enzyme–IVa complexes.

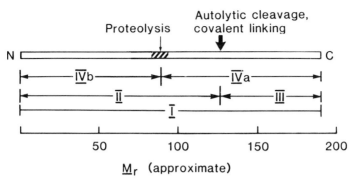

FIG. 3. Diagrammatic structure of the S-α₂M subunit. The relative positions of the parts of the subunit are deduced from the results of our peptide mapping, and lengths of the various regions of the chain correspond to the M_r values from SDS-gel electrophoresis (I, 181,000; II, 139,000; III, 68,000; IVa, 111,000; IVb, 98,000). The banded area indicates the maximum dimension of the bait region.

Composition

α₂M is glycoprotein containing 8–10% carbohydrate, including sialic acid. Data for the amino acid composition have been summarized by Harpel.[23]

Structure

α₂M consists of four identical subunits that are disulfide bonded in pairs, and the half-molecules thus formed are associated noncovalently. The subunit has recently been largely sequenced. The subunit contains a bond about two-thirds of the way from the N-terminus that tends to be cleaved spontaneously in denatured S-α₂M, especially at elevated temperatures and alkaline pH. This "autolytic" bond gives rise to the minor bands II and III in SDS-gel electrophoresis (Fig. 2B); its location is indicated in Fig. 3, and its chemical nature will be discussed later with the covalent linking reaction. There is no autolytic bond in the subunits of F-α₂M.[2]

When S-α₂M is converted to F-α₂M, it seems that the conformation of the individual subunits changes, and this results in a major increase in interactions between subunits, and a major change in the shape of the molecule.[2] Further aspects of the structure of α₂M are described below.

Binding Reactions of α₂M

The most interesting and potentially important aspect of α₂M is its remarkable binding properties. We now recognize three distinct types of

---

[23] P. C. Harpel, this series, Vol. 45, p. 639.

binding of other molecules: (*a*) the specific *trapping* of proteinases; (*b*) the covalent *linking* of proteinases and other molecules that can follow the closing of the trap; and (*c*) the noncovalent, nonsteric *adherence* reaction with a variety of proteins and other molecules, which is unrelated to proteolytic activity.

The Trapping Reaction

The "trap hypothesis" for α_2M was originally formulated[1] to explain the peculiar characteristics of the interaction of α_2M with proteinases. These have been reviewed,[24] and can be summarized as follows:

1. This type of reaction is confined to active endopeptidases, but the exact specificity and catalytic mechanism of the enzyme seem unimportant (see the table).
2. Saturation of α_2M with any proteinase prevents the reaction with any other.
3. All reactions of the bound proteinase molecule are sterically hindered.
4. The proteinase molecule is irreversibly bound.

What we suggested was that each subunit of the α_2M molecule has a region (later called the "bait" region) that is very susceptible to limited proteolysis, and that when this is cleaved by a proteinase the macroglobulin molecule changes shape in such a way as to trap the proteinase molecule within itself. The reactivity of the active site of the enzyme with substrates and inhibitors of low M_r typically remains almost normal, clearly showing that binding is not through the active site, as it would be with any normal inhibitor. Since only molecules of M_r less than about 10,000 diffuse at all freely in and out of the closed trap, however, there is severe steric hindrance of the reactivity of the trapped enzyme with protein substrates, protein inhibitors, and antibodies. (Because of its steric character, immunoinhibition of proteinases has some strong resemblances to inhibition by α_2M.[5])

Evidence for the postulated proteolytic cleavage of α_2M was almost immediately forthcoming,[25,26] and the conformational change is strongly suggested by electron microscopy[26] and other evidence.[2] The α_2M that we isolated in 1973 was not fully active, and bound only one molecule of trypsin per mole, but we agree with most other workers that up to two molecules of trypsin (and most other proteinases) can be bound in favor-

[24] P. M. Starkey and A. J. Barrett, *in* "Proteinases in Mammalian Cells and Tissues" (A. J. Barrett, ed.), p. 663. North-Holland Publ., Amsterdam, 1977.

[25] P. C. Harpel, *J. Exp. Med.* **138**, 508 (1973).

[26] A. J. Barrett, P. M. Starkey, and E. A. Munn, *in* "Proteinase Inhibitors" (H. Fritz, H. Tschesche, L. J. Green, and E Truscheit, eds.), p. 72. Springer-Verlag, Berlin, 1974.

able circumstances. The actual binding ratio in a particular experiment varies with the conditions (including concentration of reactants) because the trapping reaction can be more or less efficient.[27] It is not difficult to see that it would be possible for some molecules to fail to capture two proteinase molecules while undergoing the conformational change, and the scatter of experimentally determined binding ratios between one and two is a reflection of the difference between a trapping mechanism and one in which a definite number of actual binding sites becomes saturated.

α₂M has a far broader specificity for reaction with proteinases than any other inhibitor that shows selectivity for such enzymes, since it traps the majority of proteinases from each of the four catalytic classes (see the table). Nevertheless, some endopeptidases do not react with α₂M at an appreciable rate. In light of the trap mechanism, one would expect that some proteinases would have such restricted specificity as not to cleave the bait region, whereas others might be so small as to escape from the trap, or too large to be enclosed by it. Urokinase (M_r 33,000 or 47,000) is not too large, so it is presumably excluded by its specificity. It is doubtful whether there is any proteinase small enough ($M_r \approx 10,000$) to escape from the trap, but most of the other proteinases that are not trapped happen to have both narrow specificity and M_r greater than about 80,000, and may be too large. Plasma kallikrein and plasmin (both $M_r \approx 90,000$) are the largest proteinases yet reported to be bound. It has been found that trypsin or chymotrypsin linked to Sepharose have very little capacity to be trapped by α₂M (or bind it in any way) or even to cleave the bait region.[28] This suggests that the bait region may be internal, so that any proteinase that cleaves it must already be well within the trap.

Absolute rates of trapping of individual proteinases by α₂M have seldom been determined, but there is ample evidence that they differ by many orders of magnitude. Thus the half-time for binding of trypsin at moderate concentrations of enzyme and α₂M was calculated to be 0.05 sec,[29] as compared to 30 min or so for thrombin or plasma kallikrein.[30,31] As would be expected, the rate of reaction of any proteinase is decreased by competing substrates or inhibitors. The irreversibility of trapping by α₂M means that it will progressively dissociate complexes with other very tight but reversible inhibitors. Thus trypsin is transferred from its complex with soybean-trypsin inhibitor to α₂M,[32] and k_{off} rates can be measured in this way.[33] The α₂M acts as an efficient "sink" for all free en-

[27] G. S. Salvesen, Ph.D. Dissertation, University of Cambridge (1981).

[28] M. Burdon, *Clin. Chim. Acta* **100**, 225 (1980); A. J. Barrett, unpublished results (1979).

[29] G. S. Salvesen, C. A. Sayers, and A. J. Barrett, *Biochem. J.* **195**, 453 (1981).

[30] M. Iwamoto and Y. Abiko, *Biochim. Biophys. Acta* **214**, 402 (1970).

[31] K. D. Wuepper and C. G. Cochrane, *Proc. Soc. Exp. Biol. Med.* **141**, 271 (1972).

[32] G. Krebs and Y. Jacquot-Armand, *Biochem. Biophys. Res. Commun.* **55**, 929 (1973).

[33] M. Aubry and J. Bieth, *Biochim. Biophys. Acta* **438**, 221 (1976).

SPECIFICITY OF THE TRAPPING REACTION[a]

Serine Proteinases

Trapped
Chymotrypsin
Trypsin
Thrombin
Plasmin
Plasma kallikrein
Acrosin
Pancreatic elastase
Leukocyte elastase
Cathepsin G
Arvin (coagulant enzyme of *Agkistrodon rhodostoma*)[d]
Batroxobin (coagulant enzyme of *Bothrops atrox* venom)[e]
Schistosoma mansonii chymotrypsin-like proteinase[f]
Brinase (*Aspergillus oryzae* protease I)[g]
Serratia sp. E$_{15}$ proteinase[b]
Staphylococcus aureus "acid" proteinase[i]
Staphylococcus aureus V8 proteinase[j]
Subtilisins A and B

Not trapped
Tissue kallikrein
Urokinase
Enteropeptidase[k]
Coagulation factor XII$_a$
Complement component $\overline{C1s}$[l]

Cysteine Proteinases

Trapped
Cathepsin B
Cathepsin H[m]
Cathepsin L[n]
Calcium-dependent proteinase[o]
Papain
Ficin
Bromelain
Clostripain[p]

Aspartic Proteinases

Trapped
Chymosin
Cathepsin D
Periplaneta proteinase

Not trapped
Renin

Metalloproteinases

Trapped
Vertebrate collagenases
Leukocyte collagenase
Crotalus atrox venom proteinase
Crotalus adamanteus venom proteinase[q]
Trichophyton mentagraphytes keratinase
Pseudomonas aeruginosa elastase
Thermolysin
Bacillus subtilis metalloproteinase

Not trapped
Clostridium histolyticum collagenase[r]

(continued)

Unknown Mechanism

Trapped
Proteus vulgaris neutral proteinase
Fusiformis nodosus neutral proteinase

---

[a] Only active endopeptidases are included here. Many examples of the failure of various inactive forms of proteinases, and of zymogens, to bind to α₂M have been listed previously.[b,c] In the interests of brevity, the individual references are not given if they have been included in one of the reviews.
[b] A. J. Barrett and P. M. Starkey, *Biochem. J.* **133**, 709 (1973).
[c] P. M. Starkey and A. J. Barrett, *in* "Proteinases in Mammalian Cells and Tissues" (A. J. Barrett, ed.), p. 663. North-Holland Publ., Amsterdam, 1977.
[d] W. R. Pitney and E. Regoeczi, *Br. J. Haematol.* **19**, 67 (1970).
[e] N. Egberg, *Thromb. Res.* **4**, 35 (1974).
[f] M. Dresden, personal communication (1978).
[g] H. Kiessling, *Protides Biol. Fluids* **23**, 47 (1976).
[h] K. Miyata, M. Tsuda, and K. Tomada, *Anal. Biochem.* **101**, 332 (1980).
[i] U. Neubüser, J. Brückler, and H. Blobel, *Zentralbl. Bakteriol., Parasitenkd., Infektionskr. Hyg., Abt. 1: Orig., Reihe A* **247**, 164 (1980).
[j] G. S. Salvesen, Ph.D. Thesis, University of Cambridge (1981).
[k] D. A. W. Grant and J. Hermon-Taylor, personal communication (1979).
[l] R. Sim, personal communication (1980).
[m] W. N. Schwartz and A. J. Barrett, *Biochem. J.* **191**, 487 (1980).
[n] H. Kirschke, personal communication (1980).
[o] L. Waxman, *in* "Protein Turnover and Lysosome Function" (H. L. Segal and D. J. Doyle, eds.), p. 363. Academic Press, New York, 1978.
[p] E. Giroux and B. B.Vargaftig, *Biochim. Biophys. Acta* **525**, 429 (1978).
[q] L. F. Kress, personal communication (1979).
[r] A. J. Barrett, unpublished results (1978).

---

zyme, and the amount of released enzyme in α₂M can be quantified by its action on a synthetic substrate.

The properties of a proteinase molecule trapped by α₂M are adequately summarized by saying that all its reactions are subject to steric hindrance of access of macromolecules, M_r 10,000 being approximately the value above which molecules may be totally excluded. Action on low-M_r substrates is usually approximately normal, although minor changes in apparent K_m and V can be detected, and are attributable to the new environment of the proteinase molecule. One striking change is the activation by about 10-fold of leukocyte elastase seen with one particular substrate, succinyl-trialanine 4-nitroanilide,[34] which is explicable in terms of the effects described by Lestienne and Bieth.[35]

[34] D. Y. Twumasi, I. E. Liener, M. Galdston, and V. Levytska, *Nature (London)* **267**, 61 (1977).
[35] P. Lestienne and J. G. Bieth, *J. Biol. Chem.* **255**, 9289 (1980).

Sometimes, emphasis has been placed on the fact that proteinase–α_2M complexes often show slight proteolytic activity, e.g., coagulant activity of thrombin–α_2M, or fibrinolytic activity of plasmin–α_2M. Such activities are usually not more than 1% of the activity expected of the uninhibited enzyme, however, and may be attributable to only partial enclosure of a few of the proteinase molecules. The antigenic activity of proteinase molecules trapped by α_2M is blocked.[35a,37,37a] This is one of the clearest indications that the entire molecule is enclosed: since antigenic determinants are likely to be scattered over the whole surface of the molecule, they would not all be masked by an enzyme–α_2M interaction that was restricted to the active-site region of the proteinase. Unexpected, however, is the affinity of plasmin–α_2M for lysine–Sepharose; this seems to show that the lysine-binding site on the large molecule of plasmin remains exposed, whereas the active site is within the trap.

Usually it has been accepted, as Ganrot[3] thought, that trypsin–α_2M is completely unaffected by soybean trypsin inhibitor. In our hands, inhibition by as much as 20% is not unusual, and this should be looked for by anyone using Ganrot's assay procedure. This and other aspects of the trap mechanism have been reviewed and discussed more fully elsewhere.[24]

The Covalent Linking Reaction

The trap hypothesis implied that denaturation of the α_2M–proteinase complexes should release the proteinase molecule, possibly in an active form. To some extent this is the case, since collagenase can be liberated in active form by the action of 3 M sodium thiocyanate,[36] and a *Serratia* proteinase can be exposed with antigenic activity by 60% acetone.[37] It was pointed out by Harpel,[38] however, that when complexes of radiolabeled plasmin and α_2M are run in SDS-gel electrophoresis, radioactive components are seen that are too large for free plasmin chains. We confirmed this, and showed that with a variety of proteinases a certain proportion of the proteinase molecules (ranging from 8% for papain to 61% for trypsin) behave as if covalently linked to α_2M chains.[39] Peptide mapping of the α_2M chains labeled by covalent linking of radioactive proteinase molecules shows them to correspond to the larger derivative of cleavage at the bait region, i.e., the IVa fragment (Fig. 3). With a proteinase containing more than one polypeptide chain (such as plasmin or β-trypsin), complexes corresponding to each chain can be identified.

[35a] M. C. Geokas, J. W. Brodrick, J. H. Johnson, and C. Largman, *J. Biol. Chem.* **252**, 61 (1977).

[36] H. Birkedahl-Hansen, C. M. Cobb, R. E. Taylor, and H. M. Fulmer, *Arch. Oral Biol.* **20**, 681 (1975).

[37] K. Miyata, M. Tsuda, and K. Tomada, *Anal. Biochem.* **101**, 332 (1980).

[37a] C. A. Sayers and A. J. Barrett, *Biochem. J.* **189**, 255 (1980).

[38] P. C. Harpel, *J. Exp. Med.* **146**, 1033 (1977).

The linking reaction proves not to be confined to proteinases; that is to say, other proteins can become covalently linked, but only in the presence of an active proteinase. We have postulated that in the closing of the trap following cleavage of the bait region, a reactive species of α_2M appears, which we term α_2M*. Once formed, α_2M* reacts with a nucleophile, or rapidly decays, probably by reaction with water, the $t_{1/2}$ being about 2 min.[29]

In investigating the chemical nature of the linking reaction, we found that the linking of proteins can be inhibited by nucleophiles such as glycine and lysine at high concentration, and more efficiently by certain hydroxamic acids originally designed to inhibit the fixation of complement. Radioactive glycine and lysine themselves become bound, and we suggested that they compete with nucleophilic groups of the proteinase for reaction at an electrophilic site on α_2M*. In agreement with this, the extent of linking of partially N-acetylated derivatives of trypsin was found to be related to their remaining lysine content. The linking is probably mainly through lysine side chains, but not by a transglutaminase mechanism. Proteinase molecules that are trapped without covalent linking to α_2M are just as efficiently inhibited as those that are linked, so the covalent reaction has no obvious connection with the function of α_2M as an inhibitor.

Clues to the chemical nature of the electrophilic site of α_2M* came initially from the study of the reaction of the protein with methylamine. It has been mentioned already that α_2M reacts with NH_3 and CH_3NH_2 to produce an inactive F-α_2M very like that produced in the trapping of a proteinase.[2] Swenson and Howard[40] showed that radioactive methylamine is incorporated as a glutamylmethylamide residue into a peptide:

$$NH \cdot CH_3$$
$$|$$
$$\text{-Gly-Cys-Gly-Glu-Glu-Asn-Met-}$$

and suggested that this sequence initially contains a reactive internal pyroglutamyl residue. The reactive site was shown to be identical with the autolytic site[41] (Fig. 3). The reaction of these very small nucleophiles occurs without the need for proteolytic cleavage, unlike that with amines generally. Nevertheless, peptide mapping showed us that reaction is with the same part of the molecule.[29]

In contrast to Howard's suggestion of reactivity being due to a pyroglutamyl residue, Tack *et al.*[42] offered evidence for a reactive thiolester in complement components C3 and α_2M. C3 was shown to contain the same seven-residue sequence found in α_2M; this reacts with methylamine in the

[39] G. S. Salvesen and A. J. Barrett, *Biochem. J.* **187**, 695 (1980).
[40] R. P. Swenson and J. B. Howard, *Proc. Natl. Acad. Sci. U.S.A.* **76**, 4313 (1979).
[41] J. B. Howard, M. Vermeulen, and R. P. Swenson, *J. Biol. Chem.* **255**, 3820 (1979).

same way as the α_2M peptide, but its cysteine residue then becomes available for reaction with iodoacetate. In agreement with the suggestion that S-α_2M has a similar thiolester bond, we found the reaction with methylamine to expose a new cysteine side-chain to reactivity with [^{14}C]iodoacetate. We think that the thiolester bond in S-α_2M is sterically shielded from reaction with nucleophiles larger than CH_3NH_2, but that it is exposed in α_2M* for general reactivity.

As was mentioned earlier, the autolytic bond—a chemically unstable bond in the polypeptide chain—is at the site of covalent linking, and we suggest that attack by peptide bond nitrogen on the electron-deficient carbonyl carbon of the thiolester leads to peptide bond cleavage and the formation of the pyroglutamyl residue identified by Howard at the new N-terminus.[14] There is evidence from work with model compounds that this process occurring as an intramolecular reaction could be quite rapid.[29]

The finding that F-α_2M formed by reaction with methylamine (cleaving the thiolester bond) is not susceptible to autolytic cleavage[2,43] is consistent with this.

The Adherence Reaction

The adherence reaction of α_2M is quite unrelated to the trapping and linking reactions, and is probably mediated by quite different parts of this very large molecule. Nevertheless, it is a striking property of α_2M that may have physiological importance.

The S-α_2M molecule behaves as if "sticky" for a variety of molecules and particles. Since α_2M is a sialoglycoprotein, it is not surprising that it binds lectins (concanavalin A, phytohemagglutinin) and a number of cationic molecules. The selectivity for cations is not understood, however. A significant part of the plasma zinc is firmly bound by α_2M,[44] and the protein has affinity for a zinc-chelate medium.[16] Basic proteins bound by α_2M include cationic aspartate aminotransferase,[45] myelin basic protein,[46] histone H4,[47] and anhydrotrypsin,[37a] but many other cationic proteins are not bound.[37a,47]

[42] B. F. Tack, R. A. Harrison, J. Janatova, M. L. Thomas, and J. W. Prahl, *Proc. Natl. Acad. Sci. U.S.A.* **77**, 5764 (1980).

[43] A. J. Barrett and G. S. Salvesen, in "The Physiological Inhibitors of Blood Coagulation and Fibrinolysis" (D. Collen, B. Wiman, and M. Verstraete, eds.), p. 247. Elsevier/North-Holland Biomedical Press, Amsterdam, 1979.

[44] M. K. Song, N. F. Adham, and H. Rinderknecht, in "Proteolysis and Physiological Regulation" (D. W. Ribbons and K. Brew, eds.), p. 409. Academic Press, New York, 1976.

[45] T. R. C. Boyd, *Biochem. J.* **111**, 59 (1969).

[46] T. A. McPherson, J. J. Marchalonis, and V. Lennon, *Immunology* **19**, 929 (1970).

[47] D. B. Stoller and W. Rezuke, *Arch. Biochem. Biophys.* **190**, 398 (1978).

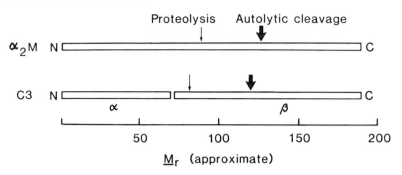

FIG. 4. Comparison of the α₂M subunit with complement component C3. Based on Sim and Sim.[55]

The other major type of affinity of α₂M is of hydrophobic type, encompassing endotoxin,[48] phenyl–Sepharose, liposomes,[49] and various cell membranes.[50] It is not yet clear whether this type of binding differs between S-α₂M and F-α₂M, and whether it may have any part to play in the rapid clearance of F-α₂M from the circulation.[51]

Binding to α₂M by the adherence mechanism is probably largely through a combination of ionic and hydrophobic reactions, and is reversible, unlike the trapping and linking reactions; it does not prevent the trapping of proteinases, and is not known to be associated with any conformational change.

Relation of α₂M to Other Proteins

The structure of α₂M is unusual, with such very large subunits being assembled as a dimer of dimers, and as far as is known the trapping reaction is a unique form of protein–protein interaction. It has recently been recognized, however, that the covalent reaction following activation by limited proteolysis is shared by the complement components C3 and C4.[42,52,53] It seems that each of these proteins has a thiolester bond that is normally reactive with methylamine, and becomes much more widely reactive after proteolytic activation. In each protein, the bond can cause autolytic cleavage of the peptide chain. Moreover, the amino acid sequence containing the reactive bond is identical in α₂M and C3.[42] Sim and Sim[52] have pointed out that the precursor polypeptide chains of C3 and C4 (before posttranslational cleavage into α and β, and α, β, and γ compo-

[48] M. Yoshioko and S. Konno, *Infect. Immun.* **1**, 431 (1970).
[49] D. V. Black and G. Gregoriadis, *Biochem. Soc. Trans.* **4**, 253 (1976).
[50] R. L. Nachman and P. C. Harpel, *J. Biol. Chem.* **251**, 4514 (1976).
[51] K. Ohlsson, *Acta Physiol. Scand.* **81**, 269 (1971).
[52] R. B. Sim and E. Sim, *Biochem. J.* **193**, 129 (1981).
[53] J. B. Howard, *J. Biol. Chem.* **255**, 7082 (1980).

nents, respectively) are of similar M_r to the α_2M subunit, and that the locations of their proteolysis-sensitive and covalent-linking/autolysis sites correspond closely to those in α_2M (Fig. 4). It thus seems that there is a strong possibility of evolutionary homology between α_2M, C3, C4, and possibly C5.

In another approach to the evolution of α_2M, Starkey and I have surveyed the chordate phyla for proteins with the characteristic properties of α_2M, using as our primary test the capacity to bind papain in a form active against low-M_r substrates.[54] The mammals, birds, reptiles, and amphibia all seem to have an α_2M-like protein of similar molecular size to human α_2M. The corresponding protein in the fish, however, has a molecular size corresponding to a half-molecule of α_2M. A study of its structure indicates that it is not simply the dimer of subunits of about 180,000 M_r, but that each precursor polypeptide chain is further cleaved just N-terminal to the proteolysis site, as it is in human C3. The protein nevertheless operates the trap mechanism, and binds methylamine covalently.

[54] P. M. Starkey and A. J. Barrett, unpublished results (1980).

[55] Human α_1-Proteinase Inhibitor

By JAMES TRAVIS *and* DAVID JOHNSON

Human plasma is known to contain at least seven proteinase inhibitors,[1,2] many of which are believed to have specific functions in the coagulation, fibrinolytic, or complement pathways. However, the major proteinase inhibitor in plasma, referred to as either α_1-proteinase inhibitor (α_1-PI) or α_1-antitrypsin, apparently acts as a general scavenger for tissue serine proteinases. This protein was first isolated by Schultze *et al.*[3] and referred to as the 3.5 S α_1-glycoprotein of plasma. It was found by Jacobsson to be the major trypsin inhibitor in plasma[4] and renamed α_1-antitrypsin in 1962.[5] However, because of the broad spectrm of inhibitory activities of this protein and, in particular, its apparent major func-

[1] N. Heimburger, H. Haupt, and H. Schwick, *Proc. Int. Res. Conf. Proteinase Inhibitors, 1st, 1970* p. 1 (1971).
[2] M. Moroi and N. Aoki, *J. Biol. Chem.* **251**, 5956 (1976).
[3] H. E. Schultze, I. Gollner, K. Heide, M. Schonenberger, and G. Schwick, *Z. Naturforsch., B: Anorg. Chem., Org. Chem., Biochem., Biophys., Biol.* **10B**, 463 (1955).
[4] K. Jacobsson, *Scand. J. Clin. Lab. Invest., Suppl.* **14**, 55 (1955).
[5] H. E. Schultze, K. Heide, and H. Haupt, *Klin. Wochenschr.* **40**, 427 (1962).

tion to control the activity of neutrophil proteinases, it was renamed α_1-proteinase inhibitor.[6]

Major interest in α_1-PI began in 1963 when Laurell and Eriksson[7] noted a deficiency in plasma levels of this protein in several patients who had developed emphysema. These individuals synthesize an aberrant form of α_1-PI that is poorly transported into the circulation. It is now known, in fact, that α_1-PI can exist in at least nine forms known as Pi variants, with the Pi_M protein present in most individuals with normal plasma levels of α_1-PI (130 mg α_1-PI/100 ml plasma[8,9]). In contrast, the Pi_Z protein is found primarily in individuals with low plasma α_1-PI levels (less than 20 mg/100 ml plasma). The difference between the M and Z proteins has been described and is primarily due to the substitution of a glutamic acid residue in the M protein by a lysine residue in the Z protein.[10,11]

The fact that low α_1-PI levels in plasma can be correlated with emphysema development has stimulated an intensive study of the role of this protein in the maintenance of a proper proteinase–proteinase inhibitor equilibrium in tissues. In fact, the protein has now been purified by at least six different procedures, using $(NH_4)_2SO_4$ fractionation, gel-filtration chromatography, ion-exchange chromatography, affinity chromatography, and preparative gel electrophoresis to isolate the protein.[3,4,6,8,12–15]

Assay Methods

Assays are based on the inhibition of proteinase activity resulting from preincubation of the enzyme with inhibitor. Since α_1-PI inhibits a variety of serine proteinases, the assay procedures actually depend on the proteinase being inhibited. Historically, trypsin has been employed for the assay of α_1-PI because it is available in high purity and easily measured using synthetic substrates. However, α_1-PI apparently functions *in vivo* as an inhibitor of polymorphonuclear leukocyte elastase.[16] Because this en-

[6] R. Pannell, D. Johnson, and J. Travis, *Biochemistry* **13**, 5439 (1974).

[7] C. B. Laurell and S. Eriksson, *Scand. J. Clin. Lab. Invest.* **15**, 132 (1963).

[8] J. O. Jeppson, C. B. Laurell, and M. Fagerhol, *Eur. J. Biochem.* **83**, 143 (1978).

[9] M. K. Fagerhol, *Ser. Haematol.* **1**, 153 (1968).

[10] A. Yoshida, J. Lieberman, L. Gaidulis, and C. Ewing, *Proc. Natl. Acad. Sci. U.S.A.* **73**, 1324 (1976).

[11] J. O. Jeppson, *FEBS Lett.* **65**, 195 (1976).

[12] I. P. Crawford, *Arch. Biochem. Biophys.* **156**, 215 (1973).

[13] I. E. Liener, O. R. Garrison, and Z. Pravda, *Biochem. Biophys. Res. Commun.* **51**, 436 (1973).

[14] P. Musiani, G. Massi, and M. Piantelli, *Clin. Chim. Acta* **73**, 561 (1976).

[15] M. Plancot, A. Delacourte, K. K. Han, M. Dautrevaux, and G. Biserte, *Int. J. Pept. Protein Res.* **10**, 113 (1977).

[16] K. Beatty, J. Bieth, and J. Travis, *J. Biol. Chem.* **255**, 3931 (1980).

zyme is not commercially available, elastase-inhibitory activity is routinely measured using porcine pancreatic elastase. The proteinase used in the assay must be free of other proteolytic enzymes, because α_1-PI would also bind to such impurities and would, therefore, give incorrect values. Additionally, cysteinyl[17] and metalloproteinases[18] have been shown to inactivate α_1-PI catalytically.

Trypsin-Inhibitory Assay

   Reagents

   Buffer: 0.01 M Tris-HCl, pH 8.0
   Substrate: Benzoyl-L-arginine ethyl ester (Mann), 0.58 mM. Dissolve 20 mg of the hydrochloride salt in 100 ml of the Tris-HCl buffer. Prepare fresh daily.
   Enzyme: Porcine trypsin. Prepare a stock solution of 1 mg enzyme in 5 ml of 2 mM HCl. This solution is stable at 4°C for at least 2 weeks.
   Procedure. Porcine trypsin (usually 0.10 ml) is dispensed into a series of 13 × 100-mm tubes, together with 0.2 ml of buffer. The inhibitor solution (up to 0.2 ml) is added and the mixture incubated at room temperature for at least 1 min. The assay is begun by adding substrate solution to a final volume of 3.0 ml. After mixing well, the sample is transferred to a 3.0-ml cuvette and the increase in absorbance at 253 nm measured for 1 min. The trypsin control, without inhibitor, should have a ΔA_{253} of 0.1–0.2 per minute. The amount of α_1-PI used should be adjusted to give a ΔA_{253} between 30 and 70% of the control. One unit of trypsin-inhibitory activity is defined as that amount which reduces the ΔA_{253} of trypsin by 1.0 absorbance units per minute relative to the trypsin control. The value of $\Delta\epsilon$ (change in molar extinction coefficient) for the complete hydrolysis of substrate is 1160. Therefore, a change in absorbance of 1.0 unit translates into the production of 2.59 μmol of product.

Elastase Inhibitory Activity

   Reagents

   Buffer: 0.2 M Tris-HCl, pH 8.0
   Substrate: Succinyl-L-alanyl-L-alanyl-L-alanyl-p-nitroanilide (Sigma). Dissolve 45 mg in 5 ml of dimethylformamide or dimethyl sulfoxide to yield a 20 mM stock solution that is stable at 4°C for several weeks.

[17] D. Johnson and J. Travis, *Biochem. J.* **163**, 639 (1977).
[18] J. Morihara, T. Suzuki, and K. Oda, *Infect. Immun.* **24**, 188 (1979).

Enzyme: Porcine pancreatic elastase, chromatographically purified (Worthington). Prepare a stock solution of 0.3 mg enzyme in 5 ml of 0.1 M Tris-HCl, pH 8.0. The enzyme is not stable below pH 4.0. *Procedure.* Porcine elastase (usually 0.5 ml) is dispensed into 13 × 100-mm tubes and 0.2 ml of buffer is added. Inhibitor solution (up to 0.2 ml) is mixed with the elastase and the mixture incubated at room temperature for at least 1 min. Buffer is added for a total of 2.95 ml and the assay is begun by adding 0.05-ml of substrate solution. After mixing well, the sample is transferred to a 3.0-ml cuvette and the increase in absorbance at 410 nm measured for 1 min. A ΔA_{410} per minute for the elastase control should be 0.1–0.2. The amount of α_1-PI used in the assay should reduce the ΔA_{410} by 30–70%, relative to the control. One unit of elastase-inhibitory activity is defined as the reduction in the ΔA_{410} of the elastase control by 1.0 absorbance unit per minute. Since the molar extinction coefficient of p-nitroaniline is 8800 at 410 nm, one unit corresponds to the production of 0.341 μmol of product.

Purification Procedure

The purification described here is a modification of the original procedure described by Pannell *et al.* in 1974,[6] and primarily involves removal of albumin from plasma by affinity chromatography on dye-bound gels.

Preparation of Cibacron Blue Sepharose

The coupling of Cibacron Blue F-3GA (Ciba-Geigy) to Sepharose follows the procedure of Bohme *et al.*[19] The coupling of the dye to Sepharose requires heating to 95°C. For this reason a cross-linked gel stable to high temperatures is used.[20]

Typically, 1 liter of washed Sepharose-4B or 6B is mixed at room temperature with 1 liter of 1 M NaOH, containing 5 g of NaBH₄. To this mixture is added, with stirring, 20 ml of epichlorohydrin and the mixture heated to 60°C for 1 hr. The cross-linked gel is then washed on a coarse Buchner filter with hot water until the washings are neutral.

The Sepharose–dye conjugate referred to as Cibacron Blue Sepharose is prepared by suspending 500 ml of cross-linked gel in an equal volume of water and warming the mixture to 60°C. The dye (5 g in 50 ml of water) is added dropwise with vigorous stirring and, after 15 min, 50 g of NaCl is added. The mixture is heated to 95°C to ensure efficient coupling, and 10 g

[19] H. J. Bohme, G. Kopperschlagaer, J. Schultz, and E. Hofmann, *J. Chromatogr.* **69**, 209 (1972).
[20] J. Porath, J. C. Janson, and T. Lass, *J. Chromatogr.* **60**, 167 (1971).

of Na_2CO_3 is added. After 30 min, the dyed gel is washed with hot water, followed by 0.05 M Tris-HCl–0.05 M NaCl (pH 8.0), until the washings are color free. Traces of blue dye may still elute during the first passage of plasma through columns of this material. However, subsequent usage usually indicates no further dye leakage. Gel prepared in this manner is capable of binding up to 40 mg of human albumin per milliliter of dye–gel conjugate.

Inhibitor Purification

Step 1. Ammonium Sulfate Fractionation. Solid $(NH_4)_2SO_4$ is added with stirring to human plasma (usually 1 liter) to bring the concentration to 0.5 saturation (213 g per liter of plasma). The solution is clarified by centrifugation at 25,000 g (10 min at 4°C), and the precipitate discarded. $(NH_4)_2SO_4$ is then added to bring the concentration to 0.8 saturation (210 g/liter). After centrifugation the precipitate is retained, dissolved in 100 ml of 0.03 M sodium phosphate buffer (pH 6.7), and dialyzed against the same buffer for 24 hr (4 liters of buffer, 4°C, four changes).

Step 2. Cibacron Blue Sepharose Fractionation. A column of Cibacron Blue Sepharose is prepared (5.0 × 100 cm) and equilibrated with 0.03 M sodium phosphate buffer, pH 6.7. The yellow solution obtained from step 1 is applied to the column. The column is then washed with equilibration buffer (225 ml/hr) and 19-ml fractions collected. The fractions containing α_1-PI are eluted in the void volume, together with transferrin, orosomucoid, pre-albumin, and traces of high-molecular-weight components. Collection of fractions may be discontinued when the A_{280} falls below 0.100. Since albumin remains tightly bound to the column (the color of the dye–albumin complex becomes a light blue), stripping of the protein may be obtained by elution with equilibration buffer containing 0.50 M KSCN. The eluted albumin is usually about 95% pure and contains only traces of albumin dimer.

Reequilibration of the column with starting buffer makes it suitable for further use. Occasionally, however, the column should be further washed with either 6 M urea or 5 M guanidine-HCl to remove tightly bound components, such as lipoproteins, before reequilibration. In our laboratory this is usually performed when the binding capacity of the column for albumin is reduced to less than 20 mg albumin/ml gel conjugate.

Step 3. Fractionation on DEAE–Cellulose. A column of DEAE–cellulose (DE-52, Whatman Chemicals) is prepared (2.5 × 30 cm) and equilibrated with 0.03 M sodium phosphate buffer, pH 6.5. The pooled fractions containing α_1-PI, obtained in step 2, are adjusted to pH 6.5 with 0.01 N HCl and charged directly onto the column. The column is washed

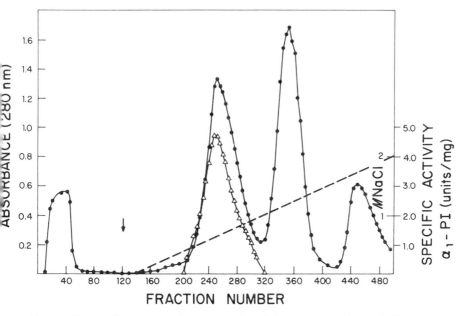

FIG. 1. Salt gradient-elution chromatography of human α_1-proteinase inhibitor on DEAE–cellulose. ●—●, Absorbance at 280 nm; △—△, inhibitory activity; -----, NaCl concentration. Arrow represents initiation of gradient.

with equilibration buffer until the A_{280} is less than 0.050. An inactive breakthrough peak containing transferrin is usually observed. A linear gradient of NaCl from 0 to 0.2 M NaCl in 2000 ml of 0.03 M sodium phosphate buffer (pH 6.5) is applied to develop the chromatogram (30 ml/hr) and 5-ml fractions collected. Three peaks are obtained by this procedure (Fig. 1). The first contains all of the α_1-PI, whereas the last two are orosomucoid (α_1-acid glycoprotein) and pre-albumin, both of which are obtained in homogeneous form.

Step 4. Gel-Filtration Chromatography. A column of Sephadex G-75 (5.0 × 100 cm) is equilibrated with 0.05 M Tris-HCl–0.05 M NaCl, pH 8.0. The pooled active fractions from step 4 are adjusted to pH 8.0 with 1 M Tris (unbuffered) and concentrated to 30 ml by ultrafiltration on a UM-10 membrane. After application to the Sephadex G-75 column, equilibration buffer is used to elute the α_1-PI containing fractions (flow rate 30 ml/hr; 5-ml fractions). Inactive protein appears in the void volume, followed by the purified inhibitor.

Normally, α_1-PI is stored frozen in the equilibration buffer, at concentrations up to 5 mg/ml, without loss of inhibitory activity.

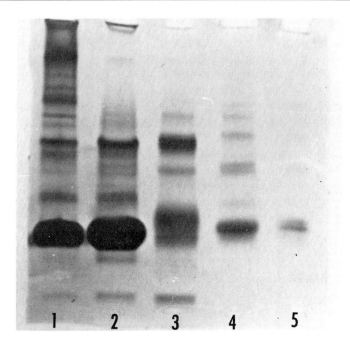

FIG. 2. Polyacrylamide-gel electrophoresis of human α_1-proteinase inhibitor at each stage of purification. (1) Whole human plasma; (2) $(NH_4)_2SO_4$ fractionation; (3) Cibacron Blue Sepharose chromatography; (4) DEAE–cellulose chromatography; (5) gel-filtration chromatography.

Figure 2 depicts the purification as noted by gel electrophoresis of fractions obtained in the steps given above. Table I indicates the recovery at each stage of the purification.

Properties

Stability

Human α_1-PI is stable for several months at 4°C in solutions containing 0.01% sodium azide. This protection is required because of the potential inactivating effect of proteinases released from bacterial contaminants. α_1-PI may also be stored frozen in solution at $-80°C$, but once thawed the protein should not be refrozen because this leads to inactivation. Lyophilized preparations of α_1-PI, dialyzed free of salts before freeze-drying, retain less than half of their activity on solubilization. However, preparations lyophilized in the presence of 0.15 M NaCl retain nearly all of their inhibitory activity after resolubilization.

TABLE I
PURIFICATION OF α_1-PROTEINASE INHIBITOR

Step	Protein (mg)	Activity units (total)	Specific activity (units/ $A_{280\,nm}$)	Recovery (%)	Purification (-fold)
1. Starting plasma (850 ml)	54,570	4529	0.083	100	1
2. $(NH_4)_2SO_4$, (0.5–0.8 saturation)	24,720	4252	0.172	94	2.1
3. Cibacron Blue Sepharose	2332	4012	1.72	89	20.7
4. DEAE–cellulose	927	2808	3.03	62	36.5
5. Sephadex G-75	350	2793	7.98	61	96

Unlike many of the low-molecular-weight inhibitors, α_1-PI is not stable below pH 5.5. This is probably because inactive polymers form near the isoelectric point. However, rapid acidification to pH 2.0 does not seem to cause irreversible inactivation of the inhibitor, since readjustment to neutral pH results in the retention of most of the inhibitory activity.[21]

Purity

The purified inhibitor exhibits a single band on gel electrophoresis at alkaline pH, as well as on electrophoresis after treatment with sodium dodecyl sulfate (reduced and nonreduced preparations). Two major and three minor bands are observed after isoelectric focusing of the Pi_M proteins in a pH 4–6 ampholyte system. It is believed that these differences are due to different degrees of sialylation of the individual protein moieties, since treatment of the purified preparation with neuraminidase shifts the pattern to focus at a higher pH value as one major fraction.[8] Immunoelectrophoresis against either anti-whole human plasma or anti-human α_1-PI gives only single precipitin lines in each case. Ouchterlony immunodiffusion against antibodies to albumin, orosomucoid, α_1-antichymotrypsin, antithrombin III, pre-albumin, group-specific components, and transferrin are all negative. Such immunological tests are particularly important in the case of orosomucoid, since, due to its high carbohydrate content, common protein stains are not tightly bound and are removed during prolonged destaining.

Physical Properties

Human α_1-PI is a glycoprotein containing about 12% carbohydrate. It has MW 53,000 determined by both sedimentation equilibrium experi-

[21] C. Glaser, L. Karic, and A. Cohen, *Biochim. Biophys. Acta* **491**, 325 (1977).

ments and gel electrophoresis after treatment with sodium dodecyl sulfate.[6,12] The protein has an $\epsilon_{280}^{1\%}$ of 5.30[6,22] and a range of isoelectric points from 4.4 to 4.8, depending on the phenotype. The determination of phenotype by isoelectric focusing has been clearly shown to be a valuable analytical tool.[23]

Human α_1-PI has the following amino acid composition:

$$Lys_{39}His_{13}Arg_7Trp_1Asp_{49}Thr_{25}Ser_{23}Glu_{54}Pro_{22}Gly_{24}Ala_{24}$$
$$\tfrac{1}{2}Cys_1Val_{27}Met_8Ile_{18}Leu_{50}Tyr_6Phe_{28}$$

The single cysteine residue is usually intermolecularly linked to either cysteine or glutathione.[24] The amino-terminal sequence of α_1-PI is as follows:

Glu-Asp-Pro-Glu-Gly-Asp-Ala-Ala-Gln-Lys-Thr-Asp-Thr-Ser

His-His-Asp-Gln-Asp-His-Pro-Thr-Phe-Asn-Lys-Ile-Thr-Pro-

Asn-Leu-Ala-Glu-Phe-Ala-Phe-Ser-Leu-Tyr-Arg-Gln-Leu-Ala

Val-Thr-Gly[25]

The carboxy-terminal sequence has also been determined:

Gly-Lys-Val-Val-Asn-Pro-Thr-Gln-Lys [26]

However, although many other fragments have been isolated and sequenced, the complete primary structure of α_1-PI has not yet been determined.[27-30]

The carbohydrate composition of α_1-PI is as follows: N-acetyl hexosamine$_{12}$hexose$_{18}$sialic acid$_7$.[8] The sequence of the carbohydrate linked to asparaginyl residues of α_1-PI through N-glycosidic linkages has been reported.[31,32] Three side chains were found, two of type A and one of type B. The A-chain composition consisted of $Man_3Gal_3(GlcNAc)_4$ and $(NeuAc)_2$, while the B chain contained $Man_3Gal_4(GlcNAc)_5$ and $(NeuAc)_3$. A and B have been found to be identical except that B contains an additional mannose-linked trisaccharide.

[22] M. Schonenberger, Z. Naturforsch., B: Anorg. Chem., Org. Chem., Biochem., Biophys., Biol. **B10**, 474 (1955).

[23] J. Pierce, J. O. Jeppson, and C. B. Laurell, Anal. Biochem. **74**, 227 (1976).

[24] C. B. Laurell, J. A. Pierce, U. Persson, and E. Thulin, Eur. J. Biochem. **57**, 107 (1975).

[25] J. Travis, D. Garner, and J. Bowen, Biochemistry **17**, 5647 (1978).

[26] J. Travis and D. Johnson, Biochem. Biophys. Res. Commun. **84**, 219 (1978).

[27] W. Hrong and J. C. Gan, Int. J. Biochem. **6**, 555 (1975).

[28] M. C. Owen, M. Lorier, and R. W. Carrell, FEBS Lett. **88**, 234 (1978).

[29] D. Johnson and J. Travis, J. Biol. Chem. **253**, 7142 (1978).

[30] R. W. Carrell, M. C. Owen, S. O. Brennan, and L. Vaughan, Biochem. Biophys. Res. Commun. **91**, 1032 (1979).

[31] L. Hodges, R. Laine, and S. K. Chan, J. Biol. Chem. **254**, 8208 (1979).

[32] T. Mega, E. Lujan, and A. Yoshida, J. Biol. Chem. **255**, 4057 (1980).

Biological Properties

Specificity

Human α_1-PI specifically inhibits serine proteinases, including pancreatic and leukocyte elastases,[33-36] pancreatic trypsin and chymotrypsin,[33] leukocyte cathepsin G,[16] thrombin,[37] plasmin,[38] acrosin,[39] kallikrein,[40] and skin and synovial collagenases.[41,42] In addition to these interactions, it has also been reported that microbial enzymes from *Aspergillus*[43] and *B. subtilus*[44] are also inactivated by α_1-PI. All of these interactions appear to be due to the formation of a 1 : 1 molar complex that is stable to acid, as well as to boiling in 1% sodium dodecyl sulfate.

Significantly, α_1-PI does not inhibit nonserine proteinases. In fact, the inhibitor is rapidly inactivated by papain and cathepsin B (thiol proteinases), as well as by an elastolytic enzyme from *Pseudomonas aeroginosa* (a metalloproteinase).[17,18] Structural studies indicate that this reaction is occurring by peptide-bond cleavage at the reactive site.

Kinetic Properties

The association rate of enzymes with α_1-PI has been determined for a number of serine proteinases.[16] Human leukocyte elastase was found to have the highest k_{ass} (6.5×10^7), followed by human chymotrypsin (5.9×10^6), human leukocyte cathepsin G (4.1×10^5), human anionic trypsin (7.3×10^4), human cationic trypsin (1.1×10^4), human plasmin (1.9×10^2), and human thrombin (4.8×10^1). It would appear from these data that the primary function of α_1-PI is to control the activity of leukocyte elastase and not trypsin, as the original name given for this protein tended to indicate. Significantly, chemical oxidation of α_1-PI markedly reduces

[33] H. C. Schwick, N. Heimburger, and H. Haupt, *Z. Gesamte Inn. Med. Ihre Grenzgeb.* **21**, 1 (1966).

[34] A. Janoff, *Am. Rev. Respir. Dis.* **105**, 121 (1972).

[35] P. D. Kaplan, C. Kuhn, and J. A. Pierce, *J. Lab. Clin. Med.* **82**, 349 (1973).

[36] R. Baugh and J. Travis, *Biochemistry* **15**, 836 (1976).

[37] N. R. Matheson and J. Travis, *Biochem. J.* **159**, 495 (1976).

[38] A. Rimon, Y. Shamash, and B. Shapiro, *J. Biol. Chem.* **241**, 5102 (1966).

[39] H. Fritz, N. Heimburger, M. Meier, M. Arnold, L. J. D. Zaneveld, and G. B. F. Schumacer, *Hoppe-Seyler's Z. Physiol. Chem.* **353**, 1953 (1972).

[40] H. Fritz, B. Brey, A. Schmal, and E. Werle, *Hoppe-Seyler's Z. Physiol. Chem.* **350**, 1551 (1969).

[41] Y. Tokoro, A. Z. Eisen, and J. T. Jeffrey, *Biochim. Biophys. Acta* **258**, 289 (1972).

[42] E. D. Harris, D. R. DiBona, and S. M. Krane, *J. Clin. Invest.* **48**, 2104 (1969).

[43] R. Bergkvist, *Acta Chem. Scand.* **17**, 2239 (1963).

[44] V. Wicher and J. Dolovich, *Immunochemistry* **10**, 239 (1973).

TABLE II

COMPARATIVE AMINO ACID SEQUENCES AT THE REACTIVE CENTER OF
PROTEINASE INHIBITORS

α_1-PI[a]	Leu- Glu-Ala- Ile- Pro-Met- Ser- Ile- Pro- Pro-Glu- Val-Lys
Antithrombin III[b]	Val- Val- Ile- Ala-Gly- Arg-*Ser*[c]-Leu- Asn-*Pro*-Asn
Garden bean (elastase)[d]	Val-Cys- Thr- Ala- *Ser*- *Ile*-Pro-*Pro*-Gln-Cys-Ile
Lima bean (trypsin)[e]	His-Cys-*Ala*-Cys- Thr- Lys- *Ser*- *Ile*-Pro-*Pro*-Gln-Cys-Arg
Lima bean (chymotrypsin)[f]	Ser-Cys- Ile-Cys- Thr- Phe- *Ser*- *Ile*-*Pro*- Ala-Gln-Cys-Val
Soybean (trypsin)[g]	Gln-Cys-*Ala*-Cys- Thr- Lys- *Ser*-Asn- *Pro*-*Pro*-Gln-Cys-Arg
Soybean (chymotrypsin)[h]	Ser-Cys- Ile-Cys- Ala-Leu- *Ser*- Tyr- *Pro*- Ala-Gln-Cys-Arg
	P_6 P_5 P_4 P_3 P_2 P_1 P'_1 P'_2 P'_3 P'_4 P'_5 P'_6 P'_7

[a] D. Johnson and J. Travis, *J. Biol. Chem.* **253**, 7142 (1978).
[b] H. Jornvall, W. W. Fish, and I. Bjork, *FEBS Lett.* **106**, 358 (1979).
[c] Italicized residues are homologous to those in α_1-PI.
[d] K. A. Wilson and M. Laskowski, Sr., *Bayer Symp.* **5**, 344 (1974).
[e] F. C. Stevens, C. Wuerz, and J. Krahn, *Bayer Symp.* **5**, 344 (1974).
[f] J. Krahn and F. C. Stevens, *Biochemistry* **11**, 1804 (1970).
[g] S. Odani and T. Ikenaka, *J. Biochem.* (*Tokyo*) **71**, 839 (1972).
[h] T. Ikenaka, S. Odani, and T. Koide, *Bayer Symp.* **5**, 325 (1974).

the k_{ass} for leukocyte elastase (to 3.1×10^4) and for pancreatic elastase (to zero).

Reactive Center

The amino acid sequence surrounding the inhibitory site of α_1-PI has been determined. This was partly obtained by forming complexes of α_1-PI with various serine proteinases, dissociating the complexes with nucleophilic reagents, and examining new amino-terminal sequences derived by proteinase cleavage of the α_1-PI molecule during the dissociation.[45] With all enzymes studied (trypsin, chymotrypsin, elastase), the new amino-terminal sequences obtained were identical but different from that of the native inhibitor. Significantly, papain cleaved α_1-PI at the same site, even though no inhibition of this enzyme was found. Further studies involving papain cleavage of denatured α_1-PI resulted in the isolation and sequencing of a peptide whose structure overlapped with that of the new sequence derived by complex dissociation.[29] This sequence of amino acids, together with sequences derived for the reactive site of other serine proteinase inhibitors, is given in Table II. Homology with other inhibitors is apparent.

There have been claims that α_1-PI may have more than one inhibitory site and that, in fact, it may be a "multiheaded" inhibitor.[46-49] This re-

[45] D. Johnson and J. Travis, *Biochem. Biophys. Res. Commun.* **72**, 33 (1976).
[46] T. Lo, A. B. Cohen, and H. James, *Biochim. Biophys. Acta* **453**, 344 (1976).

mains to be unequivocally proved, however, since nonspecific peptide-bond cleavage during the interaction of proteinases with α_1-PI may be responsible for the interpretations drawn by these authors.

Inactivators

α_1-PI inhibitory activity is readily abolished by interaction with several nonserine proteinases.[17,50,51] It is also sensitive to both chemical and enzymatic oxidants,[52-55] which convert the reactive-site methionine to a sulfoxide form. Cigarette smoke also inactivates α_1-PI, apparently by the same oxidation process.[56] It has been suggested in all of these studies that the oxidation process may occur *in vivo*, resulting in the loss of inhibitory activity in the lung and the concomitant development of emphysema due to uncontrolled proteolysis by proteinases released from phagocytic cells.

---

[47] J. Baumstark, C. Lee, and R. Luby, *Biochim. Biophys. Acta* **484**, 400 (1973).
[48] H. James and A. B. Cohen, *J. Clin. Invest.* **62**, 1344 (1978).
[49] S. Satoh, T. Kurecki, L. Kress, and M. Laskowski, Sr., *Biochem. Biophys. Res. Commun.* **86**, 130 (1979).
[50] R. W. Moskowitz and C. Heinrich, *J. Lab. Clin. Med.* **77**, 777 (1971).
[51] L. F. Kress and E. A. Paroski, *Biochem. Biophys. Res. Commun.* **83**, 649 (1978).
[52] D. Johnson and J. Travis, *J. Biol. Chem.* **254**, 4022 (1979).
[53] N. R. Matheson and J. Travis, *Biochem. Biophys. Res. Commun.* **88**, 402 (1979).
[54] H. Carp and A. Janoff, *J. Clin. Invest.* **63**, 5461 (1979).
[55] D. Johnson, *Am. Rev. Respir. Dis.* **121**, 1031 (1980).
[56] H. Carp and A. Janoff, *Am. Rev. Respir. Dis.* **118**, 617 (1978).

[56] Human α_1-Antichymotrypsin

By JAMES TRAVIS and MITSUYOSHI MORII

Human α_1-antichymotrypsin (α_1-Achy) is a plasma proteinase inhibitor that specifically inactivates serine proteinases of the chymotrypsin class. This protein was first described by Schultze *et al.* as α_1-X glycoprotein[1] but was later found to be a chymotrypsin inhibitor.[2] It is now known to specifically control the activity of chymotrypsin-like proteinases from phagocytic cells (neutrophils, basophils, tissue mast cells), including cathepsin G and "chymase."[3,4] Furthermore, it possesses the unusual fea-

[1] H. E. Schultze, K. Heide, and H. Haupt, *Klin. Wochenschr.* **40**, 427 (1962).
[2] N. Heimburger and H. Haupt, *Clin. Chim. Acta* **12**, 116 (1965).
[3] J. Travis, J. Bowen, and R. Baugh, *Biochemistry* **17**, 5652 (1978).
[4] K. Beatty, J. Bieth, and J. Travis, *J. Biol. Chem.* **255**, 3931 (1980).

ture of being a primary acute-phase protein whose plasma concentration may double within 8 hr after tissue damage.[5] These facts would suggest a primary role of α_1-Achy in regulating specific chymotrypsin-like enzymes, particularly those released during inflammatory episodes. Ryley and Brogan[6] have suggested that α_1-Achy is an important protector of membranes, since it is present in sputum from patients with chronic bronchitis at concentrations higher than could be predicted by diffusion from serum, and this has been recently confirmed.[7] However, the enzymes being regulated in this process have not yet been identified.

Assay Methods

Assays are based on the inhibition of chymotrypsin activity resulting from complex formation during preincubation with the inhibitor. Since plasma contains other inhibitors that can inactivate chymotrypsin [α_2-macroglobulin (α_2-M), α_1-proteinase inhibitor (α_1-PI)] it is not possible to assay samples containing these inhibitory activities for α_1-Achy. If trypsin is carefully added to saturate both α_2-M and α_1-PI, one could conceivably then add chymotrypsin and measure the inhibition due to α_1-Achy. Unfortunately, this does not work in practice, since proteinases other than chymotrypsin may convert α_1-Achy into an inactive form.

Chymotrypsin-Inhibitory Assay

Reagents

Buffer: 0.05 M Tris-HCl, pH 8.0

Substrate: Benzoyl-L-tyrosyl ethyl ester (Sigma). Dissolve 67.1 mg of the hydrochloride salt in 113 ml of methanol-water (63/50 v/v).

Enzyme: Bovine chymotrypsin (Worthington). Prepare a stock solution of 10 mg/ml in 0.001 M HCl, containing 0.05 M CaCl$_2$. Prepare fresh daily a 1 mg/ml solution in 0.05 M Tris-HCl–0.05 M CaCl$_2$, pH 8.0.

Procedure. Bovine chymotrypsin (10 μl) and inhibitor prepared in buffer (300 μl) are mixed in a 3.0-ml cuvette. After preincubation for 3 min at room temperature, 1.0 ml of substrate and buffer to give a final volume of 3.0 ml are added. The solution is mixed thoroughly and the increase in absorbance at 256 nm measured for 1 min. The chymotrypsin control, without inhibitor, should have a ΔA_{256} of 0.1–0.2 per minute, and the

[5] K. F. Aronsen, G. Ekelund, C. O. Kindmark, and C. B. Laurell, *Scand. J. Clin. Lab. Invest.* **29**, 127 (1972).
[6] H. C. Ryley and T. D. Brogan, *J. Clin. Pathol.* **26**, 852 (1973).
[7] R. A. Stockley and D. Burnett, *Am. Rev. Respir. Dis.* **122**, 81 (1980).

amount of α_1-Achy used should be adjusted to give a ΔA_{256} between 30 and 70% of the control. One unit of chymotrypsin-inhibitory activity is defined as 1 μmol of substrate hydrolyzed per minute of reaction. One inhibitory unit is equivalent to the inhibition of one unit of enzyme. Specific inhibitory activity is defined as units of inhibitory activity per milligram of inhibitor.

Purification Procedure

The isolation of α_1-Achy described here is a modification of the original procedure described by Travis *et al.*[8] The steps involved are primarily dependent on the adsorptive properties of Cibacron Blue Sepharose (CBS), which tightly binds human albumin and weakly binds α_1-Achy. The preparation of this dye–gel conjugate is described elsewhere in this volume.[8] All purification steps for α_1-Achy isolation are performed at 4°C.

Step 1. Ammonium Sulfate Fractionation. Pooled plasma (500 ml) is brought to 50% saturation by the addition of an equal volume of saturated ammonium sulfate solution, with constant stirring. The mixture is allowed to stand for 2 hr and centrifuged at 25,000 g for 30 min. The supernatant is retained and dialyzed against 4 liters of 0.03 M sodium phosphate buffer (pH 6.8) with five changes.

Step 2. Cibacron Blue Sepharose Fractionation. The dialyzed fraction from step 1 is applied to a column of Cibacron Blue Sepharose (5.0 × 50 cm) equilibrated with 0.03 M sodium phosphate buffer, pH 6.8. Unbound protein (including α_1-PI, which may be further purified as described elsewhere in this volume[9]) is removed by washing with the same buffer until the A_{280} of the eluate is less than 0.050. Some inhibitory activity toward chymotrypsin is lost at this step due to the elution of α_1-PI. All α_1-Achy remains bound and is subsequently eluted in a single peak by washing of the column with 0.03 M sodium phosphate buffer–0.1 M NaCl (pH 6.8), together with traces of albumin and other unidentified plasma proteins (Fig. 1).

Step 3. Rechromatography on Cibacron Blue Sepharose. The α_1-Achy containing fractions from the previous step are concentrated to 50 ml and dialyzed against 0.05 M Tris-HCl, pH 8.0 (4 liters; at least four changes). The solution is applied to a column of Cibacron Blue Sepharose (5.0 × 20 cm) equilibrated with 0.05 M Tris-HCl (pH 8.0), and washed with the equilibration buffer until the A_{280} of the eluate is less than 0.005. When the column is subsequently washed with 0.05 M Tris-HCl–0.2 M NaCl (pH 8.0), only α_1-Achy is eluted, with albumin and other contaminants remain-

[8] J. Travis, D. Garner, and J. Bowen, *Biochemistry* **17**, 5647 (1978).
[9] J. Travis, and D. Johnson, this volume [55].

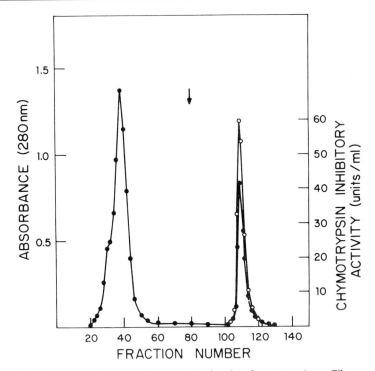

FIG. 1. Chromatography of $(NH_4)_2SO_4$ salt fraction from step 1 on Cibacron Blue Sepharose. ●—●, Absorbance at 280 nm; ○—○, inhibitory activity. Arrow indicates increase in salt concentration.

ing bound to the column (Fig. 2). By this procedure a minimum of 45 mg of homogenous α_1-Achy can be obtained. Higher levels should be attainable if plasma is used from individuals in an acute-phase condition. The yield is approximately 67% of theoretical values. However, because of the presence of other chymotrypsin inhibitors in plasma (α_1-PI, α_2-M, for example) it is not possible to calculate yields at each stage of the purification.

Polyacrylamide-gel electrophoresis patterns of fractions obtained at each step of the purification are shown in Fig. 3.

Properties

Stability

Purified α_1-Achy is unstable to both heat and pH alteration, with losses of inhibitory activity occurring above 50°C, as well as below pH 5.5.

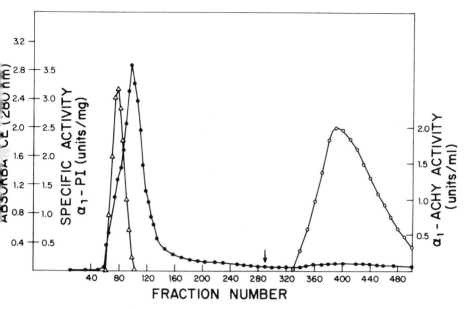

FIG. 2. Rechromatography of α_1-antichymotrypsin containing fractions, obtained from step 2, on Cibacron Blue Sepharose. ●—●, Absorbance at 280 nm; △—△, inhibitory activity due to α_1-proteinase inhibitor; ○—○, inhibitory activity due to α_1-antichymotrypsin. Arrow indicates initiation of gradient.

Inactivation also occurs above pH 10.5. α_1-Achy frozen in solution at pH 8.0 is stable indefinitely. Oxidizing agents do not affect the inhibitory activity of this protein.[4]

Purity

α_1-Achy purified by the method described above migrates as a single band in alkaline-gel electrophoresis. Under these conditions the protein has a mobility slightly slower than that of α_1-PI. When α_1-Achy is subjected to electrophoresis after boiling in 1% sodium dodecyl sulfate, a single band is detected, migrating with an apparent molecular weight of 68,000.[8]

α_1-Achy gives a single arc after immunoelectrophoresis using anti-whole human serum. Curiously, using commercial antisera to α_1-Achy (Behring Diagnostics) and α_1-Achy, a major arc with a spur can be detected. With whole plasma, two distinct lines can be seen using the antiserum to α_1-Achy. The nature of these results has not yet been established, but could represent an α_1-Achy complex, a form of desialylated inhibitor, or polymeric forms of the inhibitor in whole plasma.

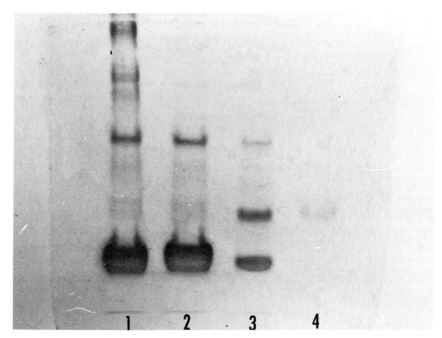

FIG. 3. Polyacrylamide-gel electrophoresis of α_1-antichymotrypsin at each stage of purification. (1) Whole human plasma; (2) $(NH_4)_2SO_4$ fractionation; (3) first Cibacron Blue Sepharose chromatography; (4) second Cibacron Blue Sepharose chromatography.

Physical Properties

The molecular weight of α_1-Achy has been reported as 65,000 and 69,000 after sedimentation-equilibrium centrifugation,[8,10] and as 68,000 after treatment with sodium dodecyl sulfate, followed by polyacrylamide-gel electrophoresis. The inhibitor is composed of a single polypeptide chain, since gel electrophoresis after prior reduction with 0.1 M mercaptoethanol in sodium dodecyl sulfate still yields a single band with MW 68,000. An extinction coefficient ($\epsilon_{280 nm}^{1\%}$) of 6.2 has been reported.[8]

The amino acid composition of α_1-Achy is as follows:

$$Lys_{27}His_9Arg_{15}Trp_4Asp_{49}Thr_{31}Ser_{32}Glu_{50}Pro_{15}Gly_{18}Ala_{34}$$
$$\tfrac{1}{2}Cys_2Val_{27}Met_{11}Ile_{23}Leu_{56}Tyr_9Phe_{26}$$

The amino-terminal sequence of α_1-Achy is as follows:

Arg-Gly-Thr-His-Val-Asp-Leu-Gly-Leu-Ala-Ser-Ala-Asp-Val
Ser-Phe-Ala-Phe-Ser-Leu-Tyr-Lys-Tyr-Leu-Val-Val-Thr-Gly

[10] N. Heimburger, H. Haupt, and H. G. Schwick, *Proc. Int. Res. Conf. Proteinase Inhibitors, 1st, 1970* p. 1 (1971).

This sequence shows a great deal of homology with that given for the amino-terminal sequence of α_1-PI,[8] and suggests a common origin for both plasma-proteinase inhibitors. The carboxy-terminus has been found to have the sequence Ser-Gly-COOH.

The carbohydrate composition of α_1-Achy is as follows: N-acetyl hexosamine$_{30-35}$hexose$_{38-40}$sialic acid$_{11-14}$.[8,10]

Biological Properties

Specificity

Human α_1-Achy inactivates pancreatic chymotrypsin, neutrophil cathepsin G, and mast cell "chymase."[3] No inhibition of pancreatic trypsin, pancreatic elastase, or neutrophil elastase has been observed. The complex that is formed is apparently a 1:1 interaction, as is found in the case of α_1-PI, and is stable to boiling in 1% sodium dodecyl sulfate.[8]

Kinetic Properties

The association rate for α_1-Achy with human pancreatic chymotrypsin is 1.0×10^4, whereas k_{ass} for neutrophil cathepsin G is 5.1×10^7. These data, when compared to the k_{ass} for α_1-PI with the same enzymes (5.4×10^6 for chymotrypsin and 4.1×10^5 for cathepsin G), clearly show that the function of α_1-Achy is to control the activity of cathepsin G, rather than chymotrypsin.[4] Significantly, oxidation of α_1-Achy has no effect on the rates of association, whereas there are significant differences when α_1-PI is oxidized.

As in the case of α_1-PI, the specificity of α_1-Achy indicates that the name of this protein is incorrect, and it seems logical that another name that correctly specifies the *in vivo* function of this protein should be considered, since it is doubtful that chymotrypsin is ever complexed with α_1-Achy under *in vivo* conditions.

[57] Cystatin, the Egg White Inhibitor of Cysteine Proteinases

By ALAN J. BARRETT

The cysteine proteinase inhibitor of chicken egg white was discovered by Fossum and Whitaker[1] as an inhibitor of ficin and papain, and further purified and characterized by Sen and Whitaker.[2] The inhibition of bovine

[1] K. Fossum and J. R. Whitaker, *Arch. Biochem. Biophys.* **125**, 367 (1968).
[2] L. C. Sen and J. R. Whitaker, *Arch. Biochem. Biophys.* **158**, 623 (1973).

cathepsin B and dipeptidyl peptidase I (cathepsin C) were described by Keilová and Tomášek,[3–5] and further work on the purification and properties of the inhibitor have been done in our own laboratory.[6] We suggest that the egg white protein be named "cystatin." Some distinctive characteristics of egg white cystatin, including $M_r \approx 13,000$, binding of inactive derivatives of the cysteine proteinases as well as of the active enzymes, and stability to heat, are shared with cysteine-proteinase inhibitors found in the cells of higher animals, and if these prove to be closely similar as now seems likely, they may well also be termed "cystatins."

Assay

Principle. The activity of papain on benzoyl-D,L-arginine 4-nitroanilide is inhibited, and the concentration of the inhibitor is expressed in terms of equivalent papain. The choice of assay substrate is important: the K_i of cystatin for papain (see later) is such that one should work with cystatin–papain complexes at $10^{-6} M$ or greater concentration, in order not to have an appreciable proportion of free enzyme in equilibrium with the complex (which would make the titration curve nonlinear). The substrate must be insensitive enough for use with such relatively high concentrations of enzyme, and must also be usable well below its K_m so as not to impair the linearity of the response. Azocasein has proved to be unsatisfactory.

The selection of the reference preparation of papain is also important because cystatin binds many inactive forms of the enzyme as well as the active enzyme. Fully active papain prepared by chromatography on an organomercurial or disulfide adsorbent[7,8] is probably most suitable, since one cannot easily determine whether the extraneous material in partially active preparations is inactive papain that would also participate in the binding of cystatin, or is truly inactive material. If pure cystatin is available, this can obviously be used to calibrate the assay.

Reagents

Substrate stock solution. Benzoyl-D,L-arginine 4-nitroanilide hydrochloride is dissolved in dimethyl sulfoxide (43.4 mg/ml, 50 mM L-isomer), and the solution stored at 4°C.

---

[3] H. Keilová and V. Tomášek, *Biochim. Biophys. Acta* **334**, 179 (1974).
[4] H. Keilová and V. Tomášek, *Collect. Czech. Chem. Commun.* **40**, 218 (1975).
[5] H. Keilová and V. Tomášek, *Acta Biol. Med. Ger.* **36**, 1873 (1977).
[6] A. Anastasi, C. A. Sayers, A. A. Kembhavi, D. C. Sunter, M. A. Brown, and A. J. Barrett, unpublished results (1981).
[7] L. A. Æ. Sluyterman and J. Wijdenes, *Biochim. Biophys. Acta* **200**, 593 (1970).

Assay buffer. Sodium phosphate buffer (pH 6.8) containing 2 mM disodium EDTA. Cysteine base (8 mM) is added freshly before use.

Stopping reagent. 1.0 M sodium formate buffer (pH 3.0) containing 0.1 M sodium monochloroacetate.

Papain solution. Fully active papain (1 thiol/mol) prepared by affinity chromatography (see preceding discussion) is stored as the mercurial derivative in 0.5 mM HgCl$_2$ and 1 mM EDTA, at a concentration of 300 μg/ml (A_{280} 0.75).

Procedure. Each assay tube contains 0.5 ml of assay buffer, 0.1 ml of papain stock, and the cystatin sample, diluted to a total volume of 2.0 ml with water. The reaction is started by the addition of 50 μl of stock substrate solution, and incubation is for 10 min at 40°C. The reaction is stopped with 1 ml of the acid chloroacetate solution, and the nitroaniline liberated by enzymatic activity is quantified by measurement of A_{410} of the solutions. A standard curve prepared with 0–100 μl of the papain solution is expected to show a linear response, and allows the papain equivalence of the cystatin solution to be determined. A standard curve in which cystatin is varied should also be linear.

Immunoassay by Radial Immunodiffusion

Cystatin is readily quantified by radial immunodiffusion as described by Mancini.[9] Antisera are raised in rabbits by intramuscular injection of purified cystatin emulsified with Freund's complete adjuvant, on two occasions, 2 weeks apart. The animals are bled 2 weeks after the second injection. The specificity of the antisera is checked by double immunodiffusion against whole egg white, side fractions from the purification, and pure cystatin. If necessary, contaminating antibodies are removed by adding predetermined amounts of whole egg white or fractions to the serum, and elimination of the immunoprecipitate. The concentration of antiserum required in the agarose gel for the assays obviously depends on the potency of the antiserum, but is typically in the order of 5% (v/v).

Purification

Egg white cystatin can be purified almost completely as a mixture of multiple forms by affinity chromatography on carboxymethyl-papain (Cm-papain) linked to Sepharose.

---

[8] B. B. S. Baines and K. Brockelhurst, *Biochem. J.* **173**, 345 (1978).
[9] G. Mancini, A. O. Carbonara, and J. F. Heremans, *Immunochemistry* **2**, 235 (1965).

Preparation of Cm-Papain—Sepharose

Papain (200 mg, twice crystallized) is activated with 2 mM cysteine and 1 mM disodium EDTA in 20 ml of 0.1 M sodium phosphate buffer (pH 6.0) during 10 min at 20°C, and is then carboxymethylated with 10 mM (final concentration) iodoacetic acid. Two hundred grams (wet weight) of Sepharose-4B is activated with CNBr (50 mg/g) as described by Axén and Ernback,[10] washed, suspended in 200 ml of 0.1 M NaHCO$_3$/NaOH (pH 9.0), and stirred overnight with the Cm-papain. The gel slurry is made 0.1 M glycine for a further 4 hr, and then washed at pH 3 (0.1 M sodium formate buffer in 0.5 M NaCl) and pH 11 (50 mM trisodium phosphate/NaOH in 0.5 M NaCl). Determination of the protein content of the washed gel by the method of Lowry et al.[11] typically shows 0.6 mg of papain bound per gram (wet weight) of gel.

Purification of the Inhibitor

The whites of 48 chicken eggs are separated from the yolks, and blended with an equal volume of 0.25% NaCl by use of a mixer-homogenizer (Silverson Machines Ltd., London SE1). The pH of the mixture is adjusted to 6.0–6.5 with 5 M sodium formate buffer (pH 3.0), and the precipitate of ovomucin removed by centrifugation (30 min at 2100 g) and discarded.

The supernatant is stirred gently overnight with 200 g of Cm-papain–Sepharose, and the adsorbent is then collected by centrifugation and washed by several cycles of resuspension in 0.5 M NaCl–0.1% Brij 35 and recentrifugation, until the A_{280} of the washings approaches zero. The Sepharose is packed into a chromatographic column (45 mM diameter), and the inhibitor eluted with 50 mM trisodium phosphate (pH 11.5) in 0.5 M NaCl. Fractions containing the single peak of protein detected by A_{280} are combined, and adjusted to pH 8 with the 3 M formate buffer.

All dialysis of cystatin is done in acetylated dialysis tubing (because the molecular size of cystatin is small enough to allow some loss through normal tubing). The tubing is immersed overnight at 20°C in a mixture of pyridine–acetic anhydride (3 : 1, v/v),[12] and then rinsed free of the reagent. The solution of cystatin eluted from the Cm-papain–Sepharose column is dialyzed in the acetylated tubing against moist polyethylene glycol, Carbowax 20M,[13] until its volume decreases to about 50 ml.

[10] R. Axén and S. Ernback, Eur. J. Biochem. 18, 351 (1971).
[11] O. H. Lowry, N. J. Rosebrough, A. L. Farr, and R. J. Randall, J. Biol. Chem. 193, 265 (1951).
[12] L. C. Craig, this series, Vol. 11, p. 870.
[13] From Raymond A. Lamb Ltd., North Acton, London NW10 6JL.

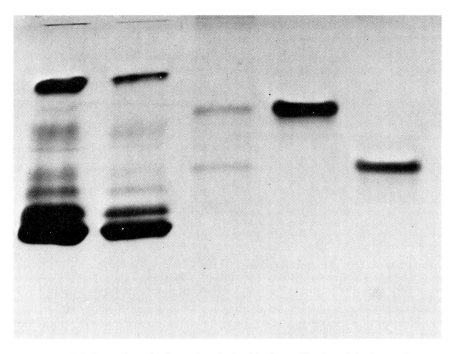

Fig. 1. Gel electrophoresis of samples obtained in the purification of the forms of cystatin.[6] The samples were, from left to right, whole egg white (56 μg protein); fraction not adsorbed by Cm-papain–Sepharose (28 μg); sample eluted from Cm-papain–Sepharose (5 μg); cystatin form I from isoelectric focusing (11 μg); and cystatin form II (10 μg). Migration was downward, toward the cathode, in a 12.5% polyacrylamide gel, as described in the text.

Gel electrophoresis of the cystatin at this stage (Fig. 1)[6] shows only trace contamination by other proteins. These are removed by gel chromatography on a column of LKB Ultrogel AcA 54 (105 × 25 cm, 515 cm³) in 50 mM Tris-HCl buffer (pH 7.8) containing 0.2 M NaCl, at a flow rate of 20 ml/hr. Assays for inhibitory activity detect cystatin in the major protein peak that emerges late in the elution profile, and the fractions containing activity are combined.

At this stage the preparation is a mixture of the multiple forms of cystatin, with very little contamination by other proteins.

If necessary, the multiple forms are separated by preparative isoelectric focusing. The solution is concentrated to 25 ml by dialysis against Carbowax as before, and dialyzed against two 1-liter portions of 1% glycine during 24 hr. It is then run in an LKB 8101 column (according to the manufacturer's instructions) in a sucrose gradient containing Am-

TYPICAL RESULTS IN THE PURIFICATION OF CYSTATIN[a]

	Protein (mg)	Cystatin (mg)	Purity (%)
Egg white (from 48 eggs, diluted to 3000 ml)	200,000	100	0.05
Affinity chromatography (Cm-papain–Sepharose)	80	34	45
Gel chromatography (Ultrogel AcA 54)	33	32	97
Isoelectric focusing			
Cystatin I	7	7	100
Cystatin II	5	5	100

[a] Protein was determined by the method of Lowry, and cystatin by radial immuno-diffusion (see the text for details).

pholines (1%, w/v) of pH range 3.5–10. The separated multiple forms in the effluent fractions are located by assays for inhibitory activity, and by electrophoresis in nondenaturing conditions.[14]

The quantitative results of a typical purification of cystatin forms 1 and 2 are summarized in the table. The yields from isoelectric focusing have been poor, in our hands, and an alternative method might be preferable. Our very recent results have indicated that chromatofocusing in the range pH 7–4 can be used with advantage in place of both AcA 54 and isoelectric focusing steps.[15]

Cystatin stores very well at $-20°C$ in the presence of sufficient sucrose or glycerol to prevent freezing, but freezing and freeze-drying usually cause partial inactivation. So long as bacterial growth is prevented with a suitable preservative, cystatin may also be stored for long periods at 4°C. Cystatin is, in fact, stable to high temperatures, surviving exposure to 80–100°C quite well.[2]

Gel Electrophoresis of Cystatin

Polyacrylamide-gel electrophoresis is done in a discontinuous 2-amino-2-methylpropane-1,3-diol (Ammediol)/glycine/HCl system as described previously.[14] Slab gels are of 12.5% total acrylamide concentration for native proteins, and 15% for SDS–protein complexes.

As is shown in Fig. 1, the two major forms of cystatin are well separated from each other and the major egg white proteins in the nondenaturing electrophoresis system. In SDS–gel electrophoresis the several forms of cystatin are found to be single-chain proteins with identical mobilities (see later).

[14] A. J. Barrett, C. A. Sayers, and M. A. Brown, *Biochem. J.* **181**, 401 (1979).

Properties

Fossum and Whitaker[1] calculated a value of 12,700 for the M_r of cystatin from their gel-chromatography results. We have found the protein (at least forms I and II) to run fractionally slower than cytochrome C (M_r 12,700) and egg white lysozyme (M_r 14,700), which would be consistent with an M_r of 13,000–15,000.[6] pI values for the two major forms, I and II, are 6.5 and 5.6, respectively. Minor forms with pI 5.3 and 5.0 are also detectable. With our antisera, the forms show precipitating reactions of complete identity.

Sen and Whitaker[2] reported an amino acid composition for what was probably a mixture of the forms of cystatin. No carbohydrate was detected. The protein probably contains two disulfide bridges. We have found[15] that the N-terminus of cystatin I is Ser-Glx-Asx-, and that the N-terminus of cystatin II is identical. There is, as yet, no indication of the chemical basis of the charge heterogeneity of cystatin; we have found the compositions of the separated forms I and II of cystatin to be closely similar to each other, and to the values given by Fossum and Whitaker.[1]

Cystatin inhibits ficin, papain, cathepsin B, and dipeptidyl peptidase I, but scarcely bromelain, trypsin, chymotrypsin, or various bacterial proteinases.[1,3,4,6] Sen and Whitaker[2] determined K_i to be $1.5 \times 10^{-8} M$ for ficin (k_{on} $4.3 \times 10^6 M^{-1}$ sec^{-1}; k_{off} 6.3×10^{-2} sec^{-1}). In contrast, we have obtained K_i values of $1.5 \times 10^{-9} M$ or below for both papain and human cathepsin B.[6]

One molecule of ficin, papain, or cathepsin B is bound per molecule of cystatin, and it is very likely that these three cysteine proteinases share the same binding site. Complexes of bovine cathepsin B are still able to bind bovine dipeptidyl peptidase I, however, so a separate site must exist for this exopeptidase.[4]

Cystatin binds ficin, papain, or cathepsin B even when the essential cysteine residue of the enzyme has been blocked with mercurial or alkylating reagents. Sen and Whitaker[2] reported evidence that the complex with carboxymethylated ficin is even more stable than that with the active enzyme. It seems clear, therefore, that there is no energetically important covalent binding. The interaction is suspected of being primarily hydrophobic.[11]

Distribution of Cystatins

The concentration of cystatin in chicken egg white is approximately 60 μg/ml. We have detected low concentrations (about 1 μg/ml) in serum of

[15] A. Anastasi and A. J. Barrett, unpublished results (1980).

both male and female chickens, so the synthesis of the protein is not confined to the female reproductive tract.

Skin and other tissues of all the mammals examined so far contain inhibitors with properties similar to egg white cystatin, and it seems possible that these are members of a single homologous family of cysteine-proteinase inhibitors.[16-19] Common characteristics of the cystatin-type inhibitors are M_r close to 13,000, inhibition of papain and cathepsin B (with interaction even when the catalytic site is blocked), inhibition of dipeptidyl peptidase I but scarcely of bromelain, pI 4.5–6.5, and stability to heat and alkali. Amino acid sequence data will be required to confirm or deny the possibility of homology, however.

[16] K. Udaka and H. Hayashi, *Biochim. Biophys. Acta* **104**, 600 (1965).
[17] M. Järvinen, *Acta Chem. Scand., Ser. B* **30**, 933 (1976).
[18] M. Järvinen, *J. Invest. Dermatol.* **72**, 114 (1978).
[19] T. Hibino, K. Fukuyama, and W. L. Epstein, *Biochim. Biophys. Acta* **632**, 214 (1980).

[58] Carboxypeptidase Inhibitor from Potatoes

By G. Michael Hass and C. A. Ryan

Although proteinaceous inhibitors of the serine proteases are widely distributed in nature, polypeptides that specifically inhibit the pancreatic carboxypeptidases have only been found in potatoes,[1] tomatoes,[2] and roundworms.[3] The inhibitors from potatoes and tomatoes are highly homologous,[4] and are believed to function both as storage proteins and as a defense mechanism by arresting digestive proteolysis of invading pests.

The availability of relatively large amounts of the carboxypeptidase inhibitor from potatoes has facilitated extensive investigations of its structure and mechanism of action. In addition, affinity chromatography utilizing immobilized inhibitor has proved to be an effective and economical means of purifying a variety of metallocarboxypeptidases that are target enzymes for the inhibitor.

Assay Procedure

The carboxypeptidase inhibitor (CPI) is usually assayed by measuring the inhibition of carboxypeptidase A activity. The substrate most com-

[1] J. M. Rancour and C. A. Ryan, *Arch. Biochem. Biophys.* **125**, 380 (1968).
[2] G. M. Hass and C. A. Ryan, *Phytochemistry* **19**, 1329 (1980).
[3] G. A. Homandberg and R. J. Peanasky, *J. Biol. Chem.* **251**, 2226 (1976).
[4] G. M. Hass and M. A. Hermodson, *Plant Physiol.* **65**, 33 (1980).

monly employed is hippuryl-L-phenylalanine,[5,6] although chloroacetyl-L-tyrosine[7] and hippuryl-DL-phenyllactate are also commonly used. A radial diffusion immunoassay for screening CPI levels in large numbers of samples has also been employed[8]; however, a source of anti-carboxypeptidase inhibitor antibodies is required for this procedure.

Reagents

> *Substrate:* Hippuryl-L-phenylalanine (1 mM) in 0.5 M NaCl–20 mM Tris-HCl (pH 7.5)
>
> *Enzyme:* Carboxypeptidase A (Worthington or Sigma) is dissolved in 5 M NaCl–0.05 M Tris-HCl (pH 7.5) at 4°C, and diluted with assay buffer in a concentration of approximately 0.1 mg/ml. The protein concentration can be estimated by absorbance using $E_{280\,\text{nm}}^{0.1\%} = 1.88$.
>
> *Inhibitor:* The stock inhibitor solution should be diluted with buffer such that an aliquot of 20–50 μl produces 25–70% inhibition.

Procedure

Substrate (3ml) is equilibrated at 25°C in a recording spectrophotometer. Controls are performed by adding enzyme alone (20 μl) and monitoring the increase in absorbance at 254 nm. The inhibitor is detected as a decrease in the rate of change in absorbance. Since equilibration between enzyme, inhibitor, and substrate is rapid, there is no need for preincubation.

Data Analysis

The amount of inhibitor present in the aliquot taken for assay is given by the following:

$$\text{CPI } (\mu\text{mol}) = \frac{\%\text{ Inhibition}}{100} \times \frac{\text{CPA } (\mu\text{g})}{34{,}500 \ \mu\text{g}/\mu\text{mol}}$$

where CPA is the amount of carboxypeptidase A present in the assay solution. This expression is valid for approximately 0–70% inhibition using hippuryl-L-phenylalanine as substrate and for 0–95% inhibition with chloroacetyl-L–tyrosine. The low K_m for hippuryl-DL-phenyllactate and low enzyme concentration used in the assays render impractical the titration of enzymatic activity with this substrate.

[5] J. E. Folk and E. W. Schirmer, *J. Biol. Chem.* **238**, 3884 (1963).
[6] C. A. Ryan, G. M. Hass, and R. W. Kuhn, *J. Biol. Chem.* **249**, 5495 (1974).
[7] G. M. Hass, H. Ako, D. G. Grahn, and H. Neurath, *Biochemistry* **15**, 93 (1976).
[8] C. A. Ryan, T. Kuo, G. Pearce, and R. Kunkel, *Am. Potato J.* **53**, 443 (1976).

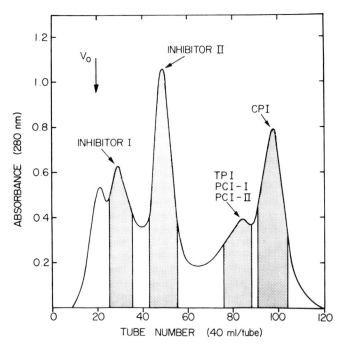

Fig. 1. Chromatography of crude inhibitor I on Sephadex G-75. (See Ryan *et al.*[6] for details.)

Purification

Method

The purification of the principal carboxypeptidase isoinhibitor has been previously described in this series.[9] Briefly, this procedure involves homogenization of Russet Burbank potato tubers in H_2O containing sodium dithionite, acidification to pH 3.0, ammonium sulfate fractionation (0–70% saturation), heat treatment (80% for 5 min), exhaustive dialysis against H_2O, and lyophilization. This partially purified inhibitor (crude inhibitor I) is resolved into several zones by gel filtration on Sephadex G-75 (Fig. 1). The first inhibitor to elute from Sephadex G-75 is inhibitor I (MW ~40,000[10]), the second is inhibitor II (MW ~20,000[11]), and in a "polypeptide peak" (fractions 77–104), there are several low-molecular-weight inhibitors of trypsin and chymotrypsin and the carboxy-

[9] Y. Birk, this series, Vol. 46, p. 736.
[10] J. C. Melville and C. A. Ryan, *J. Biol. Chem.* **249**, 5495 (1972).
[11] J. Bryant, T. R. Greene, T. Gurusaddaiah, and C. A. Ryan, *Biochemistry* **15**, 3418 (1976).

peptidase isoinhibitors. Ion exchange is utilized to purify the inhibitor further. In earlier reports phosphocellulose was employed[6,9]; however, carboxymethyl cellulose appears to give better resolution of the many proteinase inhibitors present in the polypeptide fraction and is the preferred resin. After lyophilization, polypeptide material derived from approximately 14 kg of potatoes is dissolved in 5 mM sodium citrate (pH 4.3) and chromatographed on a column (2.5 × 40 cm) of carboxymethyl cellulose (Whatman CM-52) (Fig. 2). Following elution of the breakthrough, the column is developed with a linear gradient (2 liters total) of NaCl (0–0.5 M) in equilibration buffer. Three zones exhibiting carboxypeptidase-inhibitory activity (Fig. 2) are observed and the three pools are lyophilized and desalted by gel filtration on Sephadex G-25. Isoinhibitor II, the predominant species, is apparently homogeneous at this stage, whereas additional chromatography steps are required to purify completely the minor isoinhibitors, CPI-I and CPI-III.[12]

Yield

The amount of carboxypeptidase inhibitor differs considerably among varieties of potatoes,[8] and it is recommended that a sampling of tubers be homogenized and assayed before engaging in a large-scale preparation. Russet Burbank potatoes grown in the states of Idaho and Washington typically yield 200 mg CPI-II, 25 mg CPI-I, and 25 mg CPI-III per 50 kg (fresh weight).

Purity

The homogeneity of CPI preparations can be easily assessed by polyacrylamide-gel electrophoresis[6] or by isoelectric focusing.[12] Alternatively, the absence of leucine[6] in CPI is a convenient monitor that allows detection of most protein contaminants after acid hydrolysis and amino acid analysis.

Characterization

Physical and Chemical Properties

CPI-II is an unusually stable protein easily withstanding temperatures up to 90°C and extremes of pH (pH 2–11.5) for extended time periods. Even in the tuber it is extraordinarily stable during cooking by several

[12] G. M. Hass, J. E. Derr, D. J. Makus, and C. A. Ryan, *Plant Physiol.* **64**, 1022 (1979).

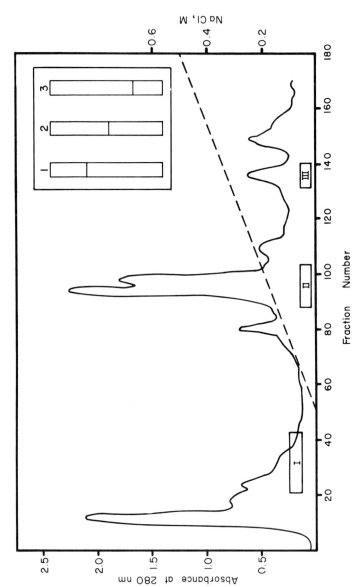

FIG. 2. Chromatography of polypeptide fraction from crude inhibitor I on carboxymethyl cellulose at pH 4.3. The inset is polyacrylamide-gel electrophoresis of the purified CPIs at pH 3.2. Migration was from the anode (top) to the cathode (bottom). ——, Absorbance at 280 nm; ---, NaCl concentration; I–III, zones containing isoinhibitors I, II, III activity. (See Hass *et al.*[12] for details.)

TABLE I
CHARACTERISTICS OF CPIs FROM POTATO TUBERS

Property[a]	CPI-I	CPI-II	CPI-III
Molecular weight	4089, 4219[b]	4204, 4333[b]	4075
Isoelectric point	4.6	5.8	6.5
N-terminus	<Glu	<Glu	His
$K_{i,app.}$ (CPA)	1.9–3.5 nM	1.5–2.7 nM	1.6–3.3 nM
$K_{i,app.}$ (CPB)	3.2–6 nM	1.1–5.5 nM	1.3–3.3 nM

[a] Data are taken from Hass et al.[12]
[b] Data given for 38- and 39-residue species (see Hass et al.[12] and Fig. 2).

methods.[13] The properties of all three of the isoinhibitor species are given in Table I. Molecular weights (~3000) estimated by gel filtration are somewhat lower than the values determined by amino acid sequence analysis, presumably because the inhibitors are rather hydrophobic, and thus are retarded during chromatography.[6,12]

The amino acid compositions of the carboxypeptidase isoinhibitors are given in Table II.[12] The isoinhibitors contain neither methionine nor leucine, and CPI-I lacks arginine as well. The extinction coefficient $E_{280 nm}^{0.1\%} = 3.0$, determined by titration of stock solutions against carboxypeptidase A, is in good agreement with that predicted from the content of aromatic amino acids.

Amino Acid Sequence

The principal form of the inhibitor (CPI-II) is a nearly equimolar mixture of the 38- and 39-amino acid residue peptides indicated in Fig. 3.[14] This amino acid sequence, determined primarily by convential procedures,[14] has been confirmed by GC–MS analysis of suitably derivatized peptide fragments.[15] The two isoinhibitors may be separated by ion-exchange chromatography, although this is not usually necessary. CPI-I differs from CPI-II at two positions ($Ser_{30} \rightarrow Ala$ and $Arg_{32} \rightarrow Gly$). CPI-III is identical to CPI-II except that it lacks the amino-terminal pyrrolidone carboxylic acid and the adjacent glutamine residue. The six residues of half-cystine of CPI are found as 3 disulfide bonds (Cys_8-Cys_{24}, Cys_{12}-Cys_{27}, and Cys_{18}-Cys_{34}).[16]

[13] D. Y. Huang, B. G. Swanson, and C. A. Ryan, *J. Food Sci.* **46**, 287 (1981).
[14] G. M. Hass, H. Nau, K. Biemann, D. T. Grahn, L. H. Ericsson, and H. Neurath, *Biochemistry* **14**, 1334 (1975).
[15] H. Nau and K. Biemann, *Biochem. Biophys. Res. Commun.* **73**, 175 (1976).
[16] T. R. Leary, D. T. Grahn, H. Neurath, and G. M. Hass, *Biochemistry* **18**, 2252 (1979).

TABLE II

AMINO ACID COMPOSITION OF CPIs

Calculated no. of amino acid residues per molecule

Amino acid	CPI-I		CPI-II		CPI-III	
	Average[a]	Integral	Average[a]	Integral	Average[a]	Integral
Asp	5.00	5	4.96	5	5.00	5
Thr	1.91	2	1.95	2	1.92	2
Ser	1.04	1	1.98	2	1.75	2
Glu	2.66	2–3	2.42	2–3	1.15	1
Pro	3.25	3	2.94	3	3.28	3
Gly	3.74	4	2.96	3	3.02	3
Ala	4.83	5	4.00	4	4.18	4
Val	1.27	1	0.98	1	0.95	1
Ile	0.85	1	0.90	1	0.95	1
Tyr	1.08	1	0.99	1	1.11	1
Phe	1.13	1	1.03	1	1.24	1
His	2.01	2	2.15	2	2.14	2
Lys	1.89	2	2.13	2	2.20	2
Arg	0.01	0	1.05	1	1.09	1
Trp	1.69	1–2	1.3	1–2	1.23	1–2
½Cys	5.75	6	5.8	6	5.85	6

[a] Taken from Hass et al.[12]

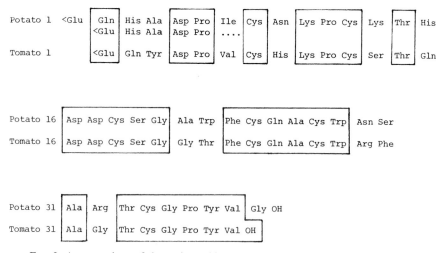

FIG. 3. A comparison of the amino acid sequences of the carboxypeptidase inhibitors from potato tubers and tomato fruit. Both 38- and 39-residue forms of CPI-II from potatoes[12] are shown. The residues enclosed in boxes are identical in both sequences. Sequence data are taken from Hass et al.[4,14]

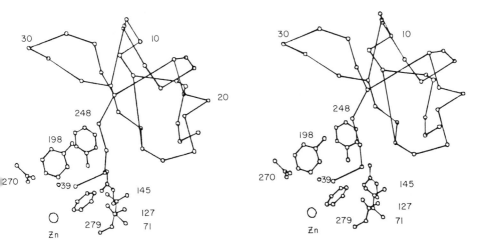

FIG. 4. Stereoview of the carboxypeptidase inhibitor in association with carboxypeptidase A as determined by X-ray diffraction at 2.5 Å resolution.[19]

The amino acid sequence of CPI demonstrates that it is homologous to the carboxypeptidase inhibitor from tomato fruit (Fig. 3), and suggests that it may be homologous to a polypeptide chymotrypsin inhibitor from potatoes,[17] as well as to several toxic peptides including viscotoxin A2 from the European mistletoe and β-purothinine from wheat,[18] although these latter comparisons are highly speculative and are based primarily on the half-cystine residues.

Three-Dimensional Structure

The structure of CPI-II has been deduced by X-ray diffraction analysis at 2.5 Å resolution of the inhibitor–carboxypeptidase A complex.[19] CPI-II has neither helical structures nor β-pleated sheets, as is seen in Fig. 4.

X-Ray Diffraction Analysis

The interaction of the carboxyl-terminal region of the inhibitor from potatoes with carboxypeptidases has been clearly indicated by X-ray crystallographic analysis of the enzyme–inhibitor complex at 2.5 Å resolution[19] (Fig. 4). That report demonstrates that the inhibitor binds like an extended substrate. The most unusual features of the inhibitor–enzyme

[17] G. M. Hass, R. Venkatakrishnan, and C. A. Ryan, *Proc. Natl. Acad. Sci. U.S.A.* **73**, 1941 (1976).

[18] D. J. Strydom, *J. Mol. Evol.* **9**, 349 (1977).

[19] D. C. Rees and W. N. Lipscomb, *Proc. Natl. Acad. Sci. U.S.A.* **77**, 4633 (1980).

interaction are that the Val_{38}-Gly_{39} peptide bond is cleaved in the complex, and that Gly_{39} appears to be trapped in the binding pocket of the enzyme.[19] The comparison of the amino acid sequences of the inhibitors from tomatoes and potatoes (Fig. 3) is of particular interest in this respect. The amino acid residue that would correspond to Gly_{39} of the inhibitor from potatoes is not found in the inhibitor from tomatoes. This observation has led to recent studies demonstrating that carboxypeptidase A rapidly removes Gly_{39} from the potato inhibitor and that this residue is not required for enzyme–inhibitor interaction.[26]

The X-ray diffraction analysis of the enzyme–inhibitor complex[19] indicates that residues 37–39 of the inhibitor interact extensively with the enzyme active site, an observation consistent with the complete homology in this region between the inhibitor from potatoes and from tomatoes (Fig. 3). Additional enzyme–inhibitor contacts involve the rings of Trp_{28}, Phe_{22}, and, to a lesser extent, His_{15}. Of these residues the only replacement observed in $His_{15} \rightarrow Gln$ (Fig. 3). The final enzyme–inhibitor contact involves the backbone of Trp_{28}-Ser_{30} of the inhibitor and Ile_{247}, Tyr_{248}, and Lys_{239} of the enzyme. Although Trp_{28} is conserved, nonconservative replacements are found at residues 29 and 30. Apparently, these replacements do not significantly affect the conformation of the polypeptide backbone in this region.[18] The structure is extremely compact with the disulfide bond joining Cys_{18} and Cys_{34} passing through a closed loop that involves the other disulfide bridges. The structure indicated that the Val_{38} and Gly_{39} bond was broken but that the Gly_{39} was trapped in the specificity pocket of the enzyme (see later). Unfortunately, a structure based on X-ray diffraction of crystals of free inhibitor is not yet available. Thus the possibility that conformation changes occur in the inhibitor other than the breaking of the Val_{38}-Gly_{39} bond on binding enzyme has not been investigated.

Inhibition Spectrum

CPI was first detected as an impurity in crystals of inhibitor I by its ability to inhibit porcine carboxypeptidase B.[1,20] The inhibitor was purified and demonstrated to affect bovine carboxypeptidase A.[6] It has now been tested against a variety of carboxypeptidases from animal, plant, fungal, and bacterial sources.[21] The association between each enzyme and CPI was monitored either by inhibition of enzymatic activity ([CPI] approx. 1–10 nM) or by the binding of enzyme to CPI–Sepharose. Enzymes that participate in the degradation of dietary protein were par-

[20] C. A. Ryan, *Biochemistry* **5**, 1592 (1966).
[21] G. M. Hass, S. P. Ager, and D. J. Makus, *Fed. Proc., Fed. Am. Soc. Exp. Biol.* **39**, 1689 (1980).

TABLE III
BINDING OF CPI TO DIGESTIVE TRACT CARBOXYPEPTIDASES AND TO
THEIR ZYMOGENS[a]

	Procedure	
Enzyme	Inhibition of enzymatic activity	Binding to CPI–Sepharose
Bovine CPDase A	+	+
Bovine CPDase B	+	+
Porcine CPDase A	+	+
Porcine CPDase B	+	+
Shrimp CPDase A	+	ND[b]
Shrimp CPDase B	+	ND
Lungfish CPDase B	+	ND
Crayfish CPDase	+	+
Cabbage looper CPDase	+	ND
Limpet CPDase	+	+
Bovine Pro-CPDase A	−	−
Bovine Pro-CPDase B	NA[c]	−
Porcine Pro-CPDase A	NA	−
Porcine Pro-CPDase B	NA	−
Lungfish Pro-CPDase B	NA	−

[a] Data are taken from Hass et al.[21]
[b] ND, not determined.
[c] NA, not applicable.

tially purified from animal species as diverse as the cow and the limpet, and all were potently affected by the inhibitor (Table III). However, several zymogens of the enzymes in this group were tested and shown not to bind immobilized inhibitor (Table III). With the exception of an enzyme from mast cells and a novel carboxypeptidase A-like enzyme from bovine placenta, all animal carboxypeptidases that were not of digestive tract origin were not affected by the inhibitor (Table IV). The inhibitor had no effect on the enzymatic activities of all plant and most microbial carboxypeptidases (Table V) tested. However, a strong association between the inhibitor and *Streptomyces griseus* carboxypeptidase has been noted,[22] and a low affinity ($K_i \sim 10^{-5}$ M) for a carboxypeptidase G_1 from an *Acinetobacter* was found as well. This inhibition spectrum not only elucidates possible evolutionary relationships among the carboxypeptidases, but also identifies enzymes whose purification might be facilitated by using immobilized inhibitor as described next.

[22] L. Gage-White, B. E. Hunt, G. M. Hass, and W. M. Awad, Jr., *Fed. Proc., Fed. Am. Soc. Exp. Biol.* **36**, 892 (1977).

TABLE IV
BINDING OF CPI TO INTRACELLULAR AND REGULATORY CARBOXYPEPTIDASES[a]

	Procedure	
	---	---
Enzyme	Inhibition of enzymatic activity	Binding of CPI–Sepharose
Mast cell CPDase (rat)	+	+
Placenta CPDase (bovine)	+	+
Cathepsin A (bovine spleen)	−	ND[b]
Catheptic CPDase B (bovine spleen)	−	ND
CPDase P (bovine kidney)	−	ND
CPDase N (human, bovine, porcine serum)	−	−
CPDase (chicken muscle)	−	ND
Kininase II	−	ND

[a] Data are taken from Hass et al.[21]
[b] ND, not determined.

Mechanism of Action

The inhibition of bovine carboxypeptidase A and porcine carboxypeptidase B is competitive with K_i values of approximately 5 and 50 nM, respectively.[6] Because CPI binds much tighter than model substrates to the carboxypeptidases and because CPI represents a new class of inhibitors, much effort has been directed at probing the mode of binding of this inhibitor.

TABLE V
BINDING OF CPI TO MICROBIAL CARBOXYPEPTIDASES[a]

	Procedure	
	---	---
Enzyme	Inhibition of enzymatic activity	Binding to CPI–Sepharose
CPDase G_1 (P. stutzeri)	+	ND[b]
CPDase G_1 (Acinetobacter sp.)	+	+
CPDase G_1 (Flavobacterium sp.)	−	−
S. griseus CPDase	+	+
Yeast protease α	−	ND
Yeast protease Y	−	ND
P. omnivorum CPDase	−	ND
S. sclerotiorum CPDase	−	ND

[a] Data are taken from Hass et al.[21]
[b] ND, not determined.

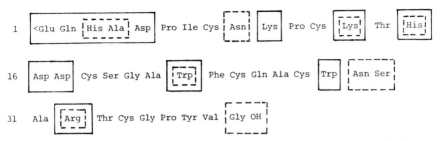

FIG. 5. Amino acid residues of the carboxypeptidase inhibitor from potatoes that do not contribute significantly to the free energy of association of the enzyme–inhibitor complex are enclosed in boxes. Evidence is based on a comparison with the sequence of the inhibitor from tomatoes (---)[4] and chemical modification studies (——).[7]

Chemical Modification Studies

The interaction in solution of CPI with target enzymes has been investigated by modifying either the inhibitor or the enzyme. First, several chemical derivatives of the inhibitor were prepared and their K_i values with respect to carboxypeptidases A and B determined to evaluate the effect of each modification on enzyme–inhibitor interaction.[7] Residues 1–5 could be removed, the side-chain carboxylate groups modified, lysines at positions 10 and 13 acetylated, or the arginine at position 31 reacted with phenylglyoxal without affecting the strength of binding to either carboxypeptidase. Tyrosine-37 and the alpha-carboxylate of glycine-39, however, appear to be involved in contact with target enzymes (Fig. 5). In all cases, parallel effects of a given modification on the binding to both carboxypeptidases were observed, indicating that similar or identical contacts between enzyme and inhibitor were utilized in both systems.[7]

A second study of CPI binding to derivatives of carboxypeptidase A revealed that the binding of the inhibitor to arsanilazocarboxypeptidase A produced spectral changes similar to those observed when typical competitive inhibitors bind to this enzyme derivative.[23] Enzymatic activity was not required for association with inhibitor, since apocarboxypeptidase bound inhibitor. This situation was analogous to the tight association between anhydrotrypsin and anhydrochymotrypsin and their respective proteinase inhibitors.[24] On the other hand, blockage of the binding pocket of carboxypeptidase A by reaction with N-bromoacetyl-N-methyl-L-phenylalanine had little effect on the free energy of association with inhibitor, whereas a similar modification of trypsin or chymotrypsin greatly weakens their interactions with proteinase inhibitors.[25]

[23] H. Ako, G. M. Hass, D. G. Grahn, and H. Neurath, *Biochemistry* **15**, 2573 (1976).
[24] H. Ako, R. J. Foster, and C. A. Ryan, *Biochemistry* **13**, 132 (1974).
[25] D. S. Ryan and R. E. Feeney, *J. Biol. Chem.* **250**, 843 (1975).

Amino Acid Sequence Comparisons

A comparison of the amino acid sequences of the carboxypeptidase isoinhibitors provides little insight into enzyme–inhibitor interactions. However, the carboxypeptidase inhibitors from potatoes and from tomatoes are clearly homologous and yet exhibit differences in 11 of 37 positions (Fig. 3).[4] The virtually indistinguishable K_i values of the two inhibitors for their target enzymes (carboxypeptidase A, carboxypeptidase B, *S. griseus* carboxypeptidase)[2] suggest that they bind through the same interactions. If this is so, nonconservative amino acid replacements probably do not occur at sites on the inhibitor that are in contact with enzyme.

Figure 5 presents the amino acid sequence of the carboxypeptidase inhibitor from potatoes, with nonconservative replacements (Fig. 3) enclosed in broken lines. Residues of the inhibitor that may be chemically modified or removed without appreciable effect on inhibitory activity are indicated in Fig. 5 by solid lines.[7] Based on these data the region from Cys_{18} to Cys_{27} and the carboxyl-terminal section of the inhibitor should be of particular importance in dictating conformation or in direct interaction with enzymes.

Affinity Chromatography

The affinity chromatography of pancreatic carboxypeptidases has been the object of several previous investigations. Although these procedures are effective in retaining carboxypeptidases from solution, the relatively high K_i values of the inhibitors employed (typically 0.1–10 mM) required high concentrations of ligand coupled to the matrix. At these concentrations of ligand, significant hydrophobic or ionic characteristics are imparted to the matrix, resulting in substantial nonspecific binding of protein during chromatography of crude enzyme preparations. In contrast to these substrate analogs, CPI binds tightly to target enzymes (K_i ~5–50 nM), a feature apparently retained after immobilization on Sepharose. Additional properties that render this inhibitor particularly suitable as a ligand in affinity chromatography include the following: (1) The inhibitor is stable to extremes of pH; (2) guanidine hydrochloride (6 M) does not irreversibly denature the inhibitor and thus may be used to wash the resin; (3) the amino groups of the inhibitor may be modified without reducing its activity[7]; thus direct coupling of CPI to CNBr-activated Sepharose is possible. This combination of tight binding and desirable characteristics has produced an affinity medium that is both effective and stable over a period of several months.

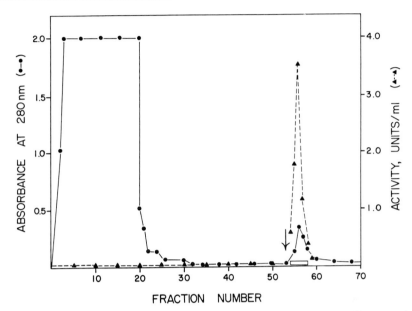

FIG. 6. Chromatography of an extract from the visceral hump of the limpet on immobilized carboxypeptidase inhibitor. The extract was applied at pH 6.0 and the enzyme was eluted (↓) at pH 10. ●—●, Absorbance at 280 nm; ▲---▲, enzyme activity. (Details are provided in Hass.[30])

Immobilized CPI has been used effectively in the purification of bovine and porcine carboxypeptidases A and B,[27] mast cell carboxypeptidase,[28] crayfish carboxypeptidase,[29] and carboxypeptidase A-like enzymes from the limpet[30] and from bovine placenta.[31] An example of the use of immobilized CPI is the partial purification of limpet carboxypeptidase shown in Fig. 6.[30] After application of the crude extract and extensive elution with 0.5 M NaCl–0.01 M MES (pH 6.0), the enzyme was eluted at pH 10.0 in buffer containing L-phenylalanine. This step produced a purification of over 600-fold with an 89% yield of enzymatic activity.

[26] G. M. Hass and C. A. Ryan, *Biochem. Biophys. Res. Commun.* **97**, 1481 (1980).

[27] S. P. Ager and G. M. Hass, *Anal. Biochem.* **83**, 285 (1977).

[28] M. T. Everitt and H. Neurath, *FEBS Lett.* **110**, 292 (1980).

[29] R. Zwilling, F. Jakob, H. Bauer, H. Neurath, and D. L. Enfield, *Eur. J. Biochem.* **94**, 223 (1979).

[30] G. M. Hass, *Arch. Biochem. Biophys.* **198**, 247 (1979).

[31] J. E. Derr-Makus and G. M. Hass, in preparation.

[59] Bull Seminal Plasma Proteinase Inhibitors

By Dana Čechová *and* Věra Jonáková

The high concentration of proteinase inhibitors in the seminal plasma of mammals, as well as the presence of acrosin, a trypsin-like enzyme associated with the acrosome of sperm, suggest that this protease-inhibitor system may play an important physiological role in fertilization. The presence of trypsin inhibitors in the sex glands and seminal plasma of mammals, including bull, was first demonstrated by Haendle *et al.*[1] in 1965. The isolation and characterization of proteinase inhibitors from the seminal plasma of man[2] and from boar,[3] and from the seminal vesicles of guinea pig,[4] have been described in an earlier volume of this series. The amino acid sequence of the boar seminal plasma inhibitor has been determined[5] and a partial structure has been reported from the inhibitor from guinea pig seminal vesicles.[6] Both inhibitors are homologous to the pancreatic secretory trypsin inhibitors of the Kazal type.[5] Seminal plasma inhibitors are relatively low-molecular-weight proteins (6000–12,000), are stable in acid, and, like the pancreatic secretory trypsin inhibitor, are "temporary inhibitors."[5]

In this article we describe procedures for the isolation and characterization of two proteinase inhibitors isolated from bull seminal plasma that differ in molecular weight.[7] Bull seminal plasma inhibitor I (BUSI I), MW 8900, exists as three strongly acidic isoinhibitors[8] A, B1, and B2; BUSI II, MW 6200, is an extremely basic protein[9] whose amino acid sequence has been determined.[10] The distribution of these inhibitors in bull sexual organs as determined by immunological methods[11] will also be described.

[1] H. Haendle, H. Fritz, I. Trautschold, and E. Werle, *Hoppe-Seyler's Z. Physiol. Chem.* **343**, 185 (1965).

[2] H. Schiessler, E. Fink, and H. Fritz, this series, Vol. 45, p. 847.

[3] H. Fritz, H. Tschesche, and E. Fink, this series, Vol. 45, p. 837.

[4] E. Fink and H. Fritz, this series, Vol. 45, p. 825.

[5] H. Tschesche, S. Kupfer, R. Klauser, E. Fink, and H. Fritz, *Protides Biol. Fluids* **23**, 255 (1976).

[6] E. Fink, Dissertation, Ph.D. Thesis, Faculty of Natural Science, University of Munich (1970).

[7] D. Čechová and H. Fritz, *Hoppe-Seyler's Z. Physiol. Chem.* **357**, 401 (1976).

[8] D. Čechová, V. Jonáková, M. Havranová, E. Sedláková, and O. Mach, *Hoppe-Seyler's Z. Physiol. Chem.* **360**, 1759 (1979).

[9] D. Čechová, V. Jonáková, E. Sedláková, and O. Mach, *Hoppe-Seyler's Z. Physiol. Chem.* **360**, 1753 (1979).

[10] B. Meloun and D. Čechová, *Collect. Czech. Chem. Commun.* **44**, 2710 (1979).

[11] L. Veselský and D. Čechová, *Hoppe-Seyler's Z. Physiol. Chem.* **361**, 715 (1980).

Assay Methods

Principle. The assay of the inhibitory activity of the proteinase inhibitors is based on the reduction of enzymatic activity obtained by preincubation of the enzyme with the inhibitor for 10 min before the addition of the substrate. Enzyme activity is conveniently measured by standard spectrophotometric methods using synthetic ester or amide substrates.[12,13]

Inhibition of the Amidase Activity of Trypsin. The routine assay of the inhibition of tryptic activity is carried out with N^α-benzoyl-DL-arginine-*p*-nitroanilide (DL-BAPA) as substrate, and the hydrolysis was monitored by absorbance measurements at 405 nm, as was described in another volume[14] of this series.

Inhibition of the Esterolytic Activity of Trypsin, Kallikrein, Acrosins, and Chymotrypsin. The esterolytic activity of trypsin, kallikrein, and acrosin on N^α-benzoyl-L-arginine ethyl ester (BAEE) was measured by the direct spectrophotometric method (253 nm) as described in the preceding volume,[15] except that we use a quartz cell without the silicone coating. The esterolytic activity of chymotrypsin was measured using N^α-acetyl-L-tyrosine ethyl ester (ATEE), as described by Schwert and Takenaka.[12] The conditions used for the measurement of the esterolytic activity of these proteinases are summarized in Table I.

Definition of Unit and Specific Activity. One inhibition unit (IU) reduces the activity of two enzyme units by 50%. One enzyme unit (U) corresponds to the hydrolysis of 1 μmol of substrate (DL-BAPA) per min ($\Delta_{405}^A = 3.32/\text{min}$). Specific activity is the number of inhibitor units (DL-BAPA) per milligram of protein.

Purification Procedures (Tables II and III)

In view of the finding[5] that the low-molecular-weight inhibitors from the seminal plasma of mammals are "temporary inhibitors," that is, slowly digested in the presence of excess trypsin, the use of affinity chromatography with immobilized trypsin columns is not recommended for their isolation. For this reason conventional methods, such as gel filtration and ion-exchange chromatography are preferred.

Desalting. Bull seminal plasma inhibitors require desalting at certain stages of the isolation procedure. This can be effected in Sephadex G-25

---

[12] G. W. Schwert and Y. Takenaka, *Biochim. Biophys. Acta* **16,** 570 (1955).
[13] H. Fritz, I. Trautschold, and E. Werle, *in* "Methoden der enzymatischen Analyse" (H. U. Bergmeyer, ed.), 2nd Engl. ed., Vol. 2, p. 1064. Verlag Chemie, Weinheim, 1970.
[14] D. Čechová, this series, Vol. 45, p. 806.
[15] W.-D. Schleuning and H. Fritz, this series, Vol. 45, p. 330.

TABLE I

CONDITIONS OF MEASUREMENT OF ENZYMATIC ACTIVITIES[a]

Enzyme	Activity[b] (U/3 ml)	Substrate	Concentration (mg/3 ml)	Wavelength (nm)	Buffer	pH
Trypsin, bovine	0.141(1)	BAEE	0.50	253	0.1 M Tris-HCl	8.0
Acrosin, bull	0.048(1)	BAEE	0.50	253	0.1 M Tris-HCl	8.0
Acrosin, boar	0.059(1)	BAEE	0.50	253	0.1 M Tris-HCl	8.0
Kallikrein, porcine	0.104(1)	BAEE	0.50	253	0.1 M Tris-HCl	8.0
Chymotrypsin, bovine	0.164(2)	ATEE	0.76	237	0.1 M phosphate	7.0

[a] B. Železná, M. Havranová, D. Čechová, and E. Sedláková, *Hoppe-Seyler's Z. Physiol. Chem.* **361**, 461 (1980).
[b] Quantity of enzyme used in each assay. One enzyme unit (U) corresponds to the hydrolysis of 1 μmol of substrate/min; i.e., (1) an increase in absorption at 253 nm of 0.385/min; (2) a decrease in optical density at 237 nm of 0.133/min.

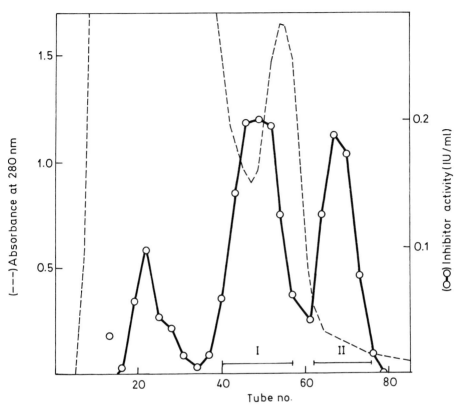

FIG. 1. Fractionation of bull seminal plasma by gel filtration. See step 2 in the text for details. Inhibitory activity (trypsin), O—O; absorbance at 280 nm, ----. Bar I, partly purified isoinhibitors of BUSI I; bar II, partly purified BUSI II. From Čechová *et al.*[9]

columns eluted by 0.2% (v/v) acetic acid. The size of the column should correspond to the sample volume and the flow rate should be kept at 10 ml/hr. For the concentration of the samples and as an alternative method of desalting (i.e., during the isolation of BUSI I), we use Diaflo membranes, Amicon UM05 for ultrafiltration.

Step 1. Preparation of Crude Concentrate. Bull ejaculates are centrifuged at 600 g at 5°C for 20 min in plastic tubes. The supernatant is then centrifuged at 5000 g at 5°C for 30 min, and the clear seminal plasma obtained processed as described under step 2 or stored at −20°C.

Step 2. Gel Filtration. Bull seminal plasma (50 ml) is applied to a column of Sephadex G-50 fine, 4.2 × 175 cm, equilibrated and eluted with 0.1 M Tris-HCl (pH 7.4) containing 0.2 M NaCl and 0.2% sodium azide. The column is eluted at 120 ml/hr and fractions of 16 ml are collected. The elution diagram is shown in Fig. 1.

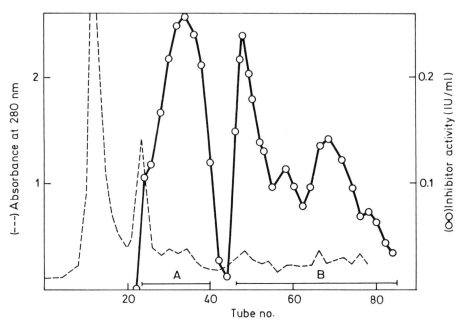

Fig. 2. Separation of fractions A and B on DEAE–Sephadex during the isolation of BUSI I isoinhibitors. See step 3 in the text for details. Inhibitory activity (trypsin), O—O; absorbance at 280 nm, ----. Bar A, fractions containing partly purified isoinhibitor A; bar B, fractions containing partly purified isoinhibitors B1 and B2. From Čechová et al.[8]

These two initial steps are the same for the isolation of isoinhibitors BUSI I (A, B1, and B2), and BUSI II. Different procedures are used for each type of inhibitor in the subsequent purification steps.

Isolation of Isoinhibitors A, B1, and B2 of BUSI I

Method A

Step 3. Chromatography on DEAE–Sephadex. The desalted and lyophilized pooled fractions I from the preceding purification step (690 mg of protein, i.e. the material obtained from five gel-filtration columns) are dissolved in the equilibrating buffer, centrifuged, and the supernatant applied to a DEAE–Sephadex A-25 column (3 × 14 cm), equilibrated with 0.05 M sodium acetate adjusted to pH 5.0. Fractions of 14 ml are collected every 15 min. The column is eluted with buffer until the absorbance at 280 nm is nearly zero. The antitryptic activity is then eluted with equilibrating buffer containing 0.5 M NaCl in two separate peaks of antitryptic activity, A and B, as shown in Fig. 2. The pooled fractions corresponding to the two peaks are treated separately in the subsequent isolation step.

Step 4. Preparation of Pure Isoinhibitors A, B1, and B2 by Chromatography on SE–Sephadex. An SE–Sephadex C-25 column (2.4 × 13 cm), equilibrated with 0.5 M sodium acetate adjusted to pH 3.0 by acetic acid, is used both for isolation of isoinhibitor A from fraction A obtained by chromatography on DEAE–Sephadex, and for the isolation of isoinhibitors B1 and B2 from fraction B. The desalted and lyophilized fraction A or B from the preceding isolation step is dissolved in the equilibrating buffer and applied to the column. The column is eluted in both cases by the equilibrating buffer until the effluent absorbance at 280 nm is nearly zero. The isoinhibitors are eluted by linear gradients of sodium chloride: isoinhibitor A in 0–0.3 M NaCl (360 ml total volume) and isoinhibitors B1 and B2 by 0–0.25 M NaCl (400 ml total volume). The fractions are collected at a rate of 9.4 ml/15 min. The elution diagrams obtained with both columns are shown in Fig. 3. The specific activity of the isoinhibitors A, B1, and B2 is 2.7 IU/mg lyophilized protein. The recoveries of the inhibitory activities from the individual isolation steps and the specific activities of the materials obtained are summarized in Table II.

Method B

Step 3a. Adsorption of BUSI I Isoinhibitors on DEAE–Sephadex. Fraction I (obtained by repeating step 2 seven times) are pooled (1900 ml), and 200 ml (settled volume) of DEAE–Sephadex A-25 in OH⁻-form is added. The mixture is stirred gently with a magnetic stirrer overnight at room temperature. The slurry is poured into a glass tube 6.2 × 13 cm; and after emergence of the liquid fraction, the column is eluted with 0.1 M Tris-HCl buffer containing 0.2 M NaCl and 0.02% sodium azide at pH 7.4 at a rate of 100 ml/hr until the absorbance of the effluent at 280 nm is nearly zero (1200 ml). The partially purified isoinhibitors are eluted as a single peak with 1 M NaCl. Fractions (12 ml) showing antitryptic activity are pooled and desalted by repeated ultrafiltration using the Diaflo UMO5 ultrafiltration membrane.

Step 4a. Separation of Isoinhibitors A, B1, and B2 by Chromatography on SE–Sephadex. The pH and conductivity of the desalted fractions possessing antitryptic activity and obtained in the preceding step are adjusted to be equivalent to the equilibrating buffer and the fractions are applied on a SE–Sephadex C-25 column (2.5 × 18 cm), equilibrated with 0.05 M sodium acetate adjusted to pH 3.0 by concentrated acetic acid. The column is eluted by 0.5 liter of the same buffer and then by a linear salt gradient 0–0.3 M NaCl in the equilibrating buffer (total volume 1000 ml). The fractions are collected at a rate of 9 ml/15 min. The isoinhibitors are eluted as sharp and symmetrical peaks of constant specific activity in the

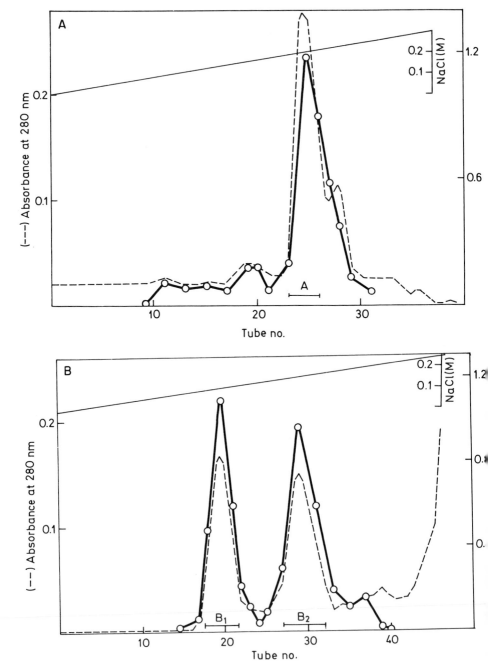

FIG. 3. Chromatography on SE–Sephadex of purified isoinhibitors of BUSI I. (A) Purification of isoinhibitor A. (B) Purification of isoinhibitors B1 and B2. See step 4 in the text for experimental details. Inhibitory activity (trypsin), O—O; absorbance at 280 nm, ----. From Čechová et al.[8]

TABLE II
PURIFICATION OF BULL SEMINAL PLASMA ISOINHIBITORS BUSI I
(A, B1 AND B2)[a]

Isolation step	Fraction	Total activity (IU)	Recovery (%)	Specific activity (mIU/mg)
	Seminal plasma (250 ml)	450	(100)	1.34[b]
Gel filtration	Fraction I	275	61	100
pH 3.5 Adjustment and desalting	Desalted fraction I	100	22	
DEAE–Sephadex	Fraction A + B	63	14	1200
SE–Sephadex	BUSI I (A, B1 + B2)	46	10	2700

[a] D. Čechová, V. Jonáková, M. Havranová, E. Sedláková, and O. Mach, *Hoppe-Seyler's Z. Physiol. Chem.* **360,** 1759 (1979).
[b] The absorbance of 1 mg was taken as 1.0 at 280 nm per 1 ml of solution in 1-cm cuvettes.

order B1, B2, and A. The advantage of this method is its simplicity. Isoinhibitor A obtained by this method contains 5% of contaminating proteins.

Isolation of Inhibitor BUSI II

Steps 1 and 2 of the isolation of BUSI II are the same as the corresponding steps of isolation of BUSI I isoinhibitors.

Step 3. Preparation of Purified Inhibitor BUSI II by Chromatography on CM–Sephadex C-25. The CM–Sephadex C-25 column (1.7 × 17 cm), is equilibrated with 0.1 M Tris-HCl buffer, pH 7.4. The desalted and lyophilized fraction II (77 mg) from step 2 (Fig. 1) is dissolved in the equilibrating buffer and applied to the column. After application of the sample, the column is eluted by the same buffer until the effluent absorbance at 280 nm is nearly zero (100 ml). BUSI II is eluted by applying a linear salt gradient (0–0.25 M NaCl) in the equilibrating buffer (total volume 400 ml). Fractions are collected at a rate of 5 ml/15 min. The homogeneous inhibitor is eluted as a sharp and symmetrical peak approximately in the middle part of the gradient. The specific activity of the preparation thus obtained is 2.1 IU/mg of lyophilized protein. The recoveries of the total inhibitory activities and the specific activities of the material obtained are summarized in Table III.

TABLE III
PURIFICATION OF BULL SEMINAL PLASMA ACROSIN INHIBITOR BUSI II[a]

Isolation step	Fraction	Total activity (IU)	Recovery (%)	Specific activity (mIU/mg)
	Seminal plasma (100 ml)	180	(100)	1.34[b]
Gel filtration	Fraction II	52	29	312
CM–Sephadex	BUSI II	42	23	2100

[a] D. Čechová,V. Jonáková, E. Sedláková, and O. Mach, *Hoppe-Seyler's Z. Physiol. Chem.* **360**, 1753 (1979).
[b] The absorbance of 1 mg was taken as 1.0 at 280 nm per 1 ml of solution in 1-cm cuvettes.

Properties

Stability. The bull seminal plasma inhibitors are stable at pH 3. BUSI I isoinhibitors, however, are less stable at pH 2 than BUSI II. BUSI I isoinhibitors are partly denatured by lyophilization in 0.2% formic acid. Inhibitors desalted in 0.2% acetic acid, lyophilized, and stored in dry state at 4°C show no decrease of activity for 6 months. Inhibitor activity is lost when purification by affinity chromatography with trypsin or chymotrypsin Sepharose-4B was attempted.

Purity. Isoinhibitors A, B1, and B2 of BUSI I purified by the last step of method A showed no evidence of contamination by other proteins as judged by various homogeneity criteria, such as electrophoresis, amino acid composition (integral values of constituent amino acids), N-terminal end-group analysis (glutamic acid as the single N-terminal amino acid). If the isoinhibitors are isolated by method B, then isoinhibitor A contains small traces of other proteins, yet below 5%.

Inhibitor BUSI II obtained from the last isolation step is homogeneous by the criteria of electrophoresis (cellulose acetate sheets or in acrylamide gel) and N-terminal end-group analysis. (In fact, no N-terminal amino acid was detected.) The inhibitor is also homogeneous as judged by its amino acid composition (integral values of constituent amino acids), and the primary structure determination (no evidence of impurities was obtained during the amino acid sequence determination).

Physical Properties. The molecular weight of BUSI I isoinhibitors determined by gel filtration is 8900. This value markedly differs from the value of 5600 determined by electrophoresis in SDS–polyacrylamide gel. However, these isoinhibitors also contain a sugar moiety that can affect the molecular weight determination. The values of isoelectric point, de-

TABLE IV
DISSOCIATION CONSTANTS OF COMPLEXES OF BUSI I
AND BUSI II WITH PROTEINASES[a]

	K_i (M)	
Proteinase	BUSI I	BUSI II
Acrosin, bull	6.5×10^{-10}	3.0×10^{-6}
Acrosin, boar	5.8×10^{-11}	7.0×10^{-6}
Trypsin, bovine	8.1×10^{-9}	5.0×10^{-7}
Kallikrein, porcine	1.6×10^{-5}	1.0×10^{-3}
Chymotrypsin, bovine	1.6×10^{-6}	2.5×10^{-5}

[a] B. Železná, M. Havranová, D. Čechová, and E. Sedláková, *Hoppe-Seyler's Z. Physiol. Chem.* **361**, 461 (1980).

termined by isoelectric focusing of the individual isoinhibitors, are 3.9 (A), 3.4 (B1), and 3.7 (B2). The specific absorptivity $A_{280}^{1\%}$ of isoinhibitor A in water is 5.4, and the same value (6.0) was obtained for both isoinhibitor B1 and B2.

The molecular weight of BUSI II determined from amino acid composition is 6200; this value is the same as that determined by electrophoresis in SDS–polyacrylamide gel. The molecular weight determined by gel filtration is 6800. The isoelectric point lies in the strongly basic pH range; the pI value determined by isoelectric focusing is 10.5. The specific absorptivity of the inhibitor in water is $A_{280}^{1\%} = 4.8$.

Inhibitory Properties. Both inhibitors from bull seminal plasma inhibit acrosin[7,16] from boar and bull spermatozoa, bovine trypsin and chymotrypsin, and hog plasmin.[7] BUSI I isoinhibitors also inhibit hog pancreatic kallikrein[16] and human granulocyte proteinases, cathepsin G, and elastase.[17]

Kinetic Properties. The dissociation constants of the complexes of BUSI I and BUSI II inhibitors with proteinases are given in Table IV.

Immunological Properties. The antibodies produced by rabbits immunized by BUSI I (isoinhibitor B1) cross-react with BUSI II, similarly, the antibodies against BUSI II cross-react with BUSI I. Antisera against both BUSI I and the BUSI II inhibitor cross-react with the boar seminal vesicle fluid and with ram seminal plasma. They do not react with rabbit seminal plasma.

[16] B. Železná, M. Havranová, D. Čechová, and E. Sedláková, *Hoppe-Seyler's Z. Physiol. Chem.* **361**, 461 (1980).
[17] M. Stančíková, D. Čechová, and K. Trnavský, *Hoppe-Seyler's Z. Physiol. Chem.* **361**, 1129 (1980).

TABLE V
AMINO ACID COMPOSITION OF BUSI I PROTEINASE
ISOINHIBITORS A, B1, B2, AND BUSI II INHIBITOR FROM BULL
SEMINAL PLASMA

Amino acid	BUSI I isoinhibitors (moles/mole)			BUSI II (moles/mole)
	A	B1	B2	
Aspartic acid	10	10	10	6
Threonine	4	4	4	2
Serine	2	2	2	3
Glutamic acid	8	9	9	5
Proline	4	4	4	2
Glycine	2	2	2	7
Alanine	4	4	4	4
Half-cystine	6	6	6	6
Valine	1	1	1	3
Methionine	1	1	1	1
Isoleucine	2	3	3	1
Leucine	0	0	0	1
Tyrosine	3	5	5	2
Phenylalanine	4	6	6	2
Tryptophan	0	0	0	0
Lysine	3	3	3	8
Histidine	2	2	2	2
Arginine	1	1	1	2
Total	57	63	63	57

Distribution in Organs. Both inhibitors, BUSI I and BUSI II, were detected by immunological techniques in seminal vesicle tissues and fluids, in bull ampullae and urethra, and on acrosomes of ejaculated and ampullar spermatozoa. BUSI II, moreover, was detected in the epididymal fluid and on the acrosomes of epididymal spermatozoa.

Amino Acid Composition. The composition of the inhibitors isolated from bull seminal plasma is given in Table V. The composition of BUSI I isoinhibitors is completely different from the composition of BUSI II. B1 and B2 isoinhibitors of BUSI I are identical in amino acid composition. Isoinhibitor A differs in the absence of 6 amino acid residues.

Amino Acid Sequence. The sequence of the basic bull seminal plasma inhibitor BUSI II is shown in Fig. 4 and compared with the sequence of the inhibitor from boar seminal plasma and partially determined structure of inhibitor from guinea pig seminal vesicles. Inhibitor BUSI II from bull seminal plasma, like the inhibitors from boar seminal plasma and guinea pig seminal vesicles, is homologous with the group of pancreatic secretory

ll		Pyr	Gly	Ala	Gln	Val	Asp	Cys	—	—	—	Ala	Glu	Phe	Lys	Asp	Pro	Lys	—
ar		Thr	Arg	Lys	Gln	Pro	Asn	Cys	—	—	—	Asn	Val	Tyr	Arg	Ser	His	Leu	—
inea Pig	Phe	Ala	Pro	Ser	Lys	Val	Asx	Ser	Cys	Arg	Pro	Asx	Ser	Asx	—	Arg	Tyr	Val	Glx	Arg

ll	Val	Tyr	Cys	Thr	Arg	His	Ser	Asp	Pro	Gln	Cys	Gly	Ser	Asn	Gly	Glu	Thr	Tyr	Gly	Asn
ar	Phe	Phe	Cys	Thr	Arg	Gln	Met	Asp	Pro	Ile	Cys	Gly	Thr	Asn	Gly	Lys	Ser	Tyr	Ala	Asn
inea Pig	Tyr	Met	Cys	Thr	Lys	(Glx	Leu	Asx	Pro	Val	[Cys,	Gly,	Thr,	Asx,	Gly,	His,	Thr]	Tyr)		

ll	Lys	Cys	Ala	Phe	Cys	Lys	Ala	Val	Met	Lys	Ser	Gly	Gly	Lys	Ile	Asn	Leu	Lys	His	Arg
ar	Pro	Cys	Ile	Phe	Cys	Ser	Glu	Lys	Gly	Leu	Arg	Asn	Gln	Lys	Phe	Asp	Phe	Gly	His	Trp
inea Pig														(Phe	Thr	Phe	Ser	His	Tyr	

ll	Gly	—	Cys	Lys
ar	Gly	His	Cys	Arg	Glu	Tyr	Thr	Ser	Ala	Arg	Ser
inea Pig	Gly	Arg)

FIG. 4. Comparison of the amino acid sequences of the bull seminal plasma inhibitor BUSI II[10]; the boar seminal plasma inhibitor,[3,5] and the partial sequence of the guinea pig seminal vesicles trypsin–plasmin inhibitor.[3,6] Based on Meloun and Čechová,[10] Tschesche *et al.*[5]

trypsin inhibitors (Kazal type).[18] (See Tschesche *et al.*[5] and Kato *et al.*[19] for recent reviews.)

Possible Biological Function. The physiological role of the BUSI I isoinhibitors, which are strong acrosin inhibitors, most likely consists in the inactivation of prematurely activated acrosin as has been suggested for trypsin acrosin inhibitor from human seminal plasma.[2] Since these isoinhibitors inhibit the granulocyte proteinases, which act as mediators of inflammatory processes, we assume that BUSI I isoinhibitors represent a mechanism for the protection of the tissues against inflammation.

The physiological role of BUSI II remains uncertain. We assume that this inhibitor, which is present in the epididymal fluid and on acrosomes of epididymal spermatozoa, is one of the basic proteins that are adsorbed in the epididymis to the plasma membrane of the sperm acrosome, thus preventing a premature acrosomal reaction.[11]

Acknowledgment

The authors are indebted to Dr. Lewis J. Greene for advice and assistance in the preparation of this manuscript.

[18] D. C. Bartelt, R. Shapanka, and L. J. Greene, *Arch. Biochem. Biophys.* **179**, 189 (1977).
[19] I. Kato, J. Schrode, K. A. Wilson, and M. Laskowski, Jr., *Protides Biol. Fluids* **23**, 235 (1976).

[60] Eglin: Elastase–Cathepsin G Inhibitor from Leeches

By Ursula Seemüller, Hans Fritz, and Manfred Eulitz

Thus far two groups of proteinase-inhibitor proteins from the leech *Hirudo medicinalis* have been isolated and characterized in detail: (1) hirudin, the thrombin-specific inhibitor[1,2]; and (2) the bdellins, inhibitors of trypsin and plasmin.[3,4] During isolation of the bdellins an antichymotryptic activity was also detected.[3] In the course of our investigations it appeared that the antichymotryptic activity is associated with a group of proteins, named eglin, which also inhibits subtilisin and the human neutral granulocytic proteinases elastase and cathepsin G.

A more detailed characterization of the eglins, including the elucidation of the primary structure of eglin c and studies concerning the inhibition mechanism, revealed that the eglins represent a new type of low-molecular-weight proteinase-inhibitor proteins.[5,6]

The eglins belong to the most potent inhibitors of the human neutral granulocytic proteinases known so far; therefore, they are of special medical interest. Uncontrolled liberation of such proteinases in the organism may enhance inflammatory processes or cause tissue degradation by unspecific proteolysis.[7,8] In a recent study on experimental endotoxemia it could be demonstrated that application of an exogenous inhibitor of granulocytic elastase and cathepsin G causes a reduction of the unspecific proteolysis.[9] Comparable results have been obtained by application of a synthetic elastase inhibitor in experimental emphysema.[10] Thus the eglins might be a useful tool for therapeutic approaches in diseases evoked by neutral granulocytic proteinases.

[1] F. Markwardt, this series, Vol. 19, p. 924.
[2] T. E. Petersen, H. R. Roberts, L. Sottrup-Jensen, and S. Magnusson, *in* "Protides of the Biological Fluids, 23rd Colloquium" (H. Peeters, ed.), p. 145. Pergamon, Oxford and New York, 1976.
[3] H. Fritz, M. Gebhardt, R. Meister, and E. Fink, *Proc. Int. Res. Conf. Proteinase Inhibitors, 1st, Munich 1970*, p. 271. de Gruyter, Berlin, 1971.
[4] K. Krejci and H. Fritz, *FEBS Lett.* **64**, 152 (1976).
[5] U. Seemüller, M. Meier, K. Ohlsson, H.-P. Müller, and H. Fritz, *Hoppe-Seyler's Z. Physiol. Chem.* **358**, 1105 (1977).
[6] U. Seemüller, Dissertation im Fachbereich Chemie und Pharmazie der Ludwig-Maximilinas-Universität, Müchen (1980).
[7] A. Janoff, J. Blondin, R. A. Sandhaus, A. Mosser, and C. Malemud, *in* "Proteases and Biological Control" (E. Reich, D. B. Rifkin, and E. Shaw, eds.), p. 603. Cold Spring Harbor Lab., Cold Spring Harbor, New York, 1975.
[8] A. J. Barrett, *Agents Actions* **8**, 11 (1978).
[9] J. Witte, Habilitationsschrift (Academic Thesis). Medizinische Fakultät der Universität München (1979).
[10] A. Janoff and R. Dearing, *Am. Rev. Respir. Dis.* **121**, 1025 (1980).

TABLE I

PHOTOMETRIC ASSAY CONDITIONS FOR DETERMINATION OF PROTEINASE ACTIVITIES

Enzyme	Enzyme conc.[a] (μM)	Substrate	Substrate conc.[a] (mM)	Wavelength (nm)	Buffer of pH 7.8
Chymotrypsin	0.27	SucPheNHNp	4.32	405	0.2 M TRA-HCl 0.1 M CaCl$_2$
Trypsin	0.13	BzArgNHNp	0.38	405	0.2 M TRA-HCl
Elastase	0.12	Suc(Ala)$_3$NHNp	1	405	0.2 M TRA-HCl
Cathepsin G	0.4	BzTyrOET	0.5	256	0.08 M Tris-HCl 0.1 M CaCl$_2$

[a] In the 3-ml test volume.

Assay

Principle. The inhibition of proteinases is measured by reduction of enzyme-catalyzed hydrolysis of synthetic peptide-*p*-nitroanilides, peptide esters, or azocasein. The substrate hydrolysis is determined utilizing the change in absorbance at 405 nm (*p*-nitroanilides), 256 nm (peptide esters), and 366 nm (azocasein), respectively.

The experimental conditions for the different enzyme assays are compiled in Table I.

Definition of Enzyme and Inhibitor Unit. One enzyme unit (U) corresponds to the hydrolysis of 1 μmol substrate/min at 25°C.

One inhibitor unit (IU) reduces the enzyme-catalyzed hydrolysis by 1 μmol substrate/min at 25°C.

Purification Procedure

Starting Material. The starting material is a lyophilized crude leech extract prepared by acetone precipitation and standardized to 100 ATU/mg (ATU, antithrombin unit).[1,11] This so-called raw hirudin exhibits inhibition activity against bovine and human thrombin, bovine trypsin and chymotrypsin, elastase and cathepsin G from human neutral polymorphonuclear granulocytes, subtilisin, human and porcine plasmin.

Step 1. Gel Filtration. The starting material is first fractionated by gel chromatography on a Sephadex G-75 column (150 × 3.8 cm), equilibrated with 0.05 *M* triethanolamine–0.4 *M* sodium chloride, pH 7.8.

A solution of 1.5 g crude leech extract, dissolved in 15 ml of the equilibration buffer, is applied to the column, which is developed with the same buffer at a rate of 52 ml/hr. Fractions of 9.7 ml are collected and aliquot samples are tested for inhibition activity against trypsin, chymotrypsin, and thrombin (Fig. 1). Three inhibitor fractions are thus separated.

Fraction I (MW ~ 38,000) contains mainly antitryptic and antiplasmin activity and only traces of inhibitors directed against chymotrypsin, subtilisin, and the granulocytic proteinases elastase and cathepsin G.

Fraction II (MW ~ 6000) contains the eglins with inhibition activity against chymotrypsin, subtilisin, and the granulocytic proteinases elastase and cathepsin G. The nature of the trypsin and/or plasmin inhibitors also present in fraction II has not been studied yet. From earlier work on leech inhibitors,[3] we assume that this inhibitory activity is probably due to the bdellins.

Intermediate fraction I/II (MW 8000–9000) contains most of the thrombin-specific inhibitor hirudin.

[11] F. Markwardt, *Naunyn-Schmiedebergs Arch. Exp. Pathol. Pharmakol.* **229**, 389 (1956).

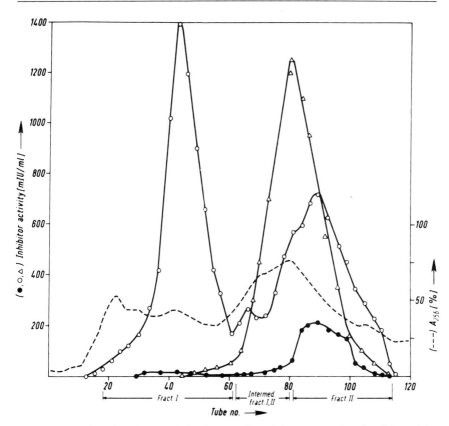

FIG. 1. Fractionation of the crude leech extract by gel chromatography. Conditions: 1.5 g of the crude leech extract, dissolved in the equilibration buffer, was applied to a Sephadex G-75 column (150 × 3.8 cm), equilibrated and developed with 0.05 M triethanolamine HCl– 0.4 M NaCl, pH 7.8, at a rate of 52 ml/hr; 9.7 ml/tube were collected. ----, Relative absorbance at 256 nm; ○—○, trypsin inhibition; ●—●, chymotrypsin inhibition; △—△, thrombin inhibition.

The pooled eluates of fraction II are concentrated and desalted by ultrafiltration using an Amicon cell with an UM-05 membrane. The salt-free protein solution is lyophilized.

Step 2. Ion-Exchange Chromatography on DEAE–Cellulose. Fraction II is further purified by ion-exchange chromatography on a DEAE–cellulose column (80 × 1.5 cm) equilibrated with 0.03 M ammonium acetate, pH 6.5. The solution of 70 mg fraction II in 1 ml equilibration buffer is adjusted to pH 6.5 and applied to the column, which is developed with the equilibration buffer at a rate of 17 ml/hr. After 12 hr the main part of the chymotrypsin-inhibiting activity has been eluted. Then the column is de-

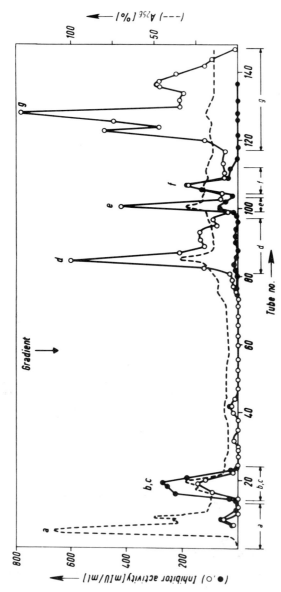

Fig. 2. Ion-exchange chromatography of fraction II on DEAE–cellulose. Conditions: 70 mg fraction II from gel chromatography, dissolved in the equilibration buffer, was applied to a DEAE–cellulose column (80 × 1.5 cm) equilibrated with 0.03 M ammonium acetate, pH 6.5. The column was developed with the equilibration buffer at a rate of 17 ml/hr until most of the antichymotryptic activity was eluted, subsequently with a linear salt gradient formed from 300 ml each of the equilibration buffer and 0.06 M NH₄Ac–0.4 M NaCl, pH 6.5; 3.7 ml/tube were collected. ----, Relative absorbance at 256 nm; ○—○, trypsin inhibition; ●—●, chymotrypsin inhibition; a–g, different inhibitor fractions eluted.

veloped further with a salt gradient (0.03 M ammonium acetate, pH 6.5, and 0.06 M ammonium acetate–0.4 M sodium chloride, pH 6.5, 300 ml each). Several trypsin and chymotrypsin-inhibiting fractions are thus separated (Fig. 2).

Eglin exists in multiple forms. The two main components, eglin b and eglin c, are eluted together in fraction IIb,c; two other forms (in fraction IIe and IIf) are eluted by the salt gradient. The small part of chymotrypsin-inhibiting activity that appears in the exclusion volume is discarded. The bdellins and hirudin (both inhibitors have low isoelectric points) are eluted with increasing salt concentration (in fraction IIe, IIf, and IIg); the same holds true for a relatively trypsin-specific inhibitor (in fraction IId), which shows neither antiplasmin nor antichymotryptic activity. Fraction IIb,c is desalted by ultrafiltration (as described for fraction II) and lyophilized.

Step 3. Ion-Exchange Chromatography on SP–Sephadex. The final purification and separation of the two isoinhibitors, eglin b and eglin c, is achieved by ion-exchange chromatography on an SP–Sephadex column (75 × 1.5 cm) equilibrated with 0.05 M ammonium acetate, pH 5.0. The solution of 50 mg fraction IIb,c in 1 ml equilibration buffer is adjusted to pH 5.0 and applied to the column, which is developed with the equilibration buffer at a rate of 17 ml/hr. After 6 hr chromatography is continued with a salt gradient (0.05 M ammonium acetate, pH 5.0, and 0.05 M ammonium acetate–0.5 M sodium chloride, pH 5.0, 300 ml each). The two isoinhibitors are eluted in two clearly separated fractions (symmetrical peaks, see Fig. 3), which are reduced to a small volume under vacuum. Desalting is performed on a Fractogel PGM 2000 column (60 × 1.5 cm), equilibrated with 0.1 M acetic acid; the salt-free inhibitory solutions are lyophilized.

Comments on the Isolation Procedure. The isolation steps starting from raw hirudin to the pure inhibitor forms eglin b and eglin c are summarized in Fig. 4.

The yields in chymotrypsin-inhibiting activity and the increase in the specific activity are compiled in Table II.

Besides conventional purification methods, we also used affinity chromatography on immobilized enzymes. However, in experiments performed with immobilized trypsin and chymotrypsin, only 25% of the chymotrypsin-inhibiting activity applied was recovered, thus we did not pursue this method.

Criteria of Purity. The homogeneity of the isoinhibitors eglin b and eglin c is proved by polyacrylamide-gel electrophoresis at pH 8.9 and pH 4.3, by SDS–polyacrylamide gel electrophoresis, and by isoelectric focusing. It is further confirmed by amino acid analysis and by detection of a single

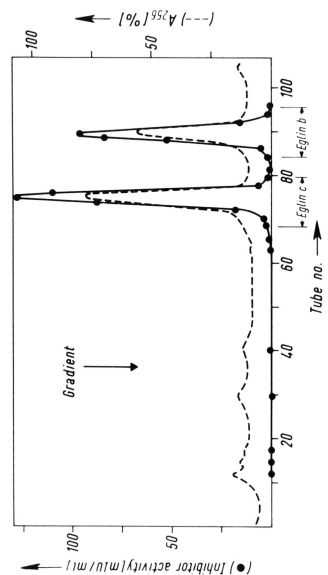

Fig. 3. Ion-exchange chromatography on SP–Sephadex. Conditions: 50 mg fraction IIb,c from DEAE–cellulose chromatography, dissolved in the equilibration buffer, was applied to an SP–Sephadex column (75 × 1.5 cm) equilibrated with 0.05 M NH$_4$Ac, pH 5.0. The column was developed with the equilibration buffer at a rate of 17 ml/hr for 6 hr, subsequently with a linear salt gradient formed from 300 ml each of the equilibration buffer and 0.05 M NH$_4$Ac–0.5 M NaCl, pH 5.0; 3.0 ml/tube were collected. ---, Relative absorbance at 256 nm; ●—●, chymotrypsin inhibition.

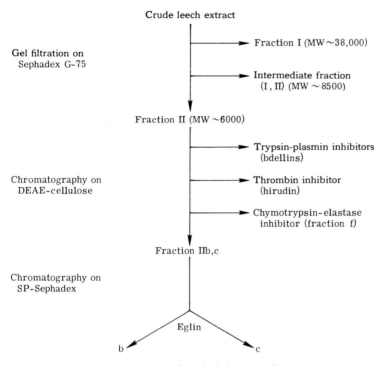

Fig. 4. Scheme of the isolation procedure.

TABLE II

Recoveries of Inhibitory Activities and Specific Inhibitory Activities after Various Isolation Steps[a]

Isolation step	Recovery (%)	Fraction	Specific inhibitory activity (mIU/mg)
		Starting material: raw hirudin	5.5
Gel filtration on Sephadex G-75	85	Fraction II	15.5
DEAE–cellulose chromatography	73	Fraction IIb,c	74
SP–Sephadex chromatography	68	Eglin b	115
		Eglin c	120

[a] Chymotrypsin inhibition was measured with SucPheNHNp as substrate.

amino acid residue for the N-terminus as well as for the C-terminus of the proteins.

Properties and Characterization

The molecular weights calculated from the amino acid composition are 8073 for eglin b and 8099 for eglin c. Determination of the molecular weight by SDS–gel electrophoresis results in a much lower value of ~6500. At present we cannot explain this difference in the molecular weight. But as the amino acid composition was confirmed by sequence analysis, the values close to 8100 are correct.

The isoelectric point of eglin b is 6.6 ± 0.3, and of eglin c, 6.45 ± 0.3 as obtained by isoelectric focusing in polyacrylamide slab gel in the presence of ampholine pH 3.5–8.0.

Stability. The eglins are highly resistant against denaturation by heat and acidic solutions. For example, no loss of inhibitory activity is observed in a solution of pH 2 at 80°C within 10 min.

Specificity. Eglin b and eglin c are identical in their inhibition characteristics. They are strong inhibitors of chymotrypsin, subtilisin, and the human neutral granulocytic proteinases elastase and cathepsin G. Inhibition constants (K_i) of the corresponding enzyme–eglin complexes were obtained by the method of Green and Work.[12] They are in the range of 10^{-10}–10^{-11} M. These are the lowest values described to date for a naturally occurring inhibitor of granulocytic proteinases.

Complex formation of eglin with chymotrypsin was demonstrated by incubation of constant amounts of inhibitor with increasing amounts of enzyme up to equimolar concentrations. The incubation was performed in $0.2\,M$ triethanolamine (pH 7.8) for 5 min. The mixtures were separated by polyacrylamide-gel electrophoresis at pH 4.3. Two distinct protein bands appeared, one corresponding to the native inhibitor, the other one to the complex. The intensity of the complex band increased with increasing enzyme concentration, whereas the intensity of the inhibitor band decreased simultaneously (Fig. 5). Equimolar concentrations of enzyme and inhibitor yielded a single band, thus demonstrating formation of a $1:1$ molar complex.

Amino Acid Composition. Eglin b and eglin c contain 70 amino acid residues. The amino acid composition is essentially the same for both inhibitors, except for one additional tyrosine in eglin c and one additional histidine in eglin b (Table III). The most striking feature of the amino acid composition is the lack of cystine and cysteine, respectively. This result was confirmed by the fact that the electrophoretic mobility, the amino

---

[12] N. M. Green and E. Work, *Biochem. J.* **54**, 347 (1952).

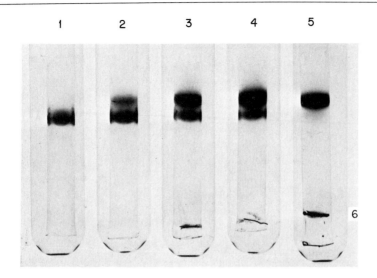

FIG. 5. Demonstration of the complex formation of eglin c with chymotrypsin by disc electrophoresis. Constant amounts of eglin c were incubated with increasing amounts of chymotrypsin in buffer solution (pH 7.8) for 5 min before the solutions were applied to polyacrylamide gels (pH 4.3). After electrophoresis, proteins were stained with Coomassie Brilliant Blue. (1) eglin c; (2–5) eglin plus increasing amounts of chymotrypsin; (6) marker dye.

acid composition, and the specific inhibition activity of reduced and carboxymethylated inhibitor did not differ from the native protein.

The lack of disulfide bridges is an uncommon structural feature distinguishing the eglins from most proteinase-inhibitor proteins described so far. At present there are only two disulfide bridge-free inhibitors known, the proteinase B inhibitor from yeast[13] and α_1-proteinase inhibitor.[14]

End-Group Analysis. N-terminus and C-terminus are identical for eglin b and eglin c. The N-terminal residue, threonine, was identified by dansylation. The C-terminal sequence, -Val-Gly-COOH, was determined with carboxypeptidase Y. Interestingly, the eglins are susceptible to degradation by carboxypeptidase Y only in the presence of 0.5% SDS.

Amino Acid Sequence of Eglin c. The primary structure of eglin c was elucidated by automatic sequential protein degradation.[15] The phenylthiohydantoin amino acids were identified by gas chromatography, thin-

[13] K. Maier, H. Müller, R. Tesch, R. Trolp, J. Witt, and H. Holzer, *J. Biol. Chem.* **254**, 12555 (1979).

[14] M. Morii, S. Odani, T. Koide, and T. Ikenaka, *J. Biochem. (Tokyo)* **83**, 269 (1978).

[15] U. Seemüller, M. Eulitz, H. Fritz, and A. Strobl, *Hoppe-Seyler's Z. Physiol. Chem.* **361**, 1841 (1980).

TABLE III
AMINO ACID COMPOSITION OF EGLIN b AND EGLIN c

Amino acid	Eglin b		Moles/ mole	Eglin c		Moles/ mole
	20 hr	72 hr		20 hr	72 hr	
Asp	6.74	6.70	7	6.76	6.70	7
Thr	4.53	4.07	5	4.52	4.08	5
Ser	3.04	2.48	3	2.93	2.31	3
Glu	7.25	7.17	7	7.16	7.11	7
Pro	6.15	6.49	6	5.9	6.22	6
Gly	5.11	5.19	5	5.18	5.27	5
Ala	1.29	1.32	1	1.16	1.17	1
Cys[a]						
Val	10.02	10.64	11	10.1	10.84	11
Met[a]						
Ile						
Leu	4.71	4.71	5	4.83	4.83	5
Tyr	3.86	3.78	4–5	5.12	4.61	5–6
Phe	4.76	4.89	5	4.93	4.93	5
His	4.07	3.47	4	2.99	3.26	3
Lys	2.34	2.30	2	2.18	2.23	2
Arg	3.76	3.71	4	2.60	3.84	4
Trp[b]	0.62		0–1	0.72		0–1
N-Terminus		Thr			Thr	
Total			70			70
Molecular weight			8073			8099

[a] After performic acid oxidation.
[b] Photometrically determined.

layer chromatography, and partly by back hydrolysis of the phenylthiohydantoin amino acids with hydrogen iodide and subsequent amino acid analysis. The eglin structure shown in Fig. 6 was determined by sequential degradation of the native protein and of the tryptic peptides. The sequence of positions 48–53 was elucidated from a chymotryptic peptide derived from limited proteolysis. This peptide was obtained by incubation of eglin with 50 mol % chymotrypsin for 90 min at pH 7.5 and room temperature. Separation of this mixture on a Sephadex G-50 column equilibrated and developed with 0.1 M acetic acid led to three main fractions: the eglin–chymotrypsin complex, native eglin c, and a peptide that could be identified as the C-terminal fragment of eglin c. Amino acid composition and sequence data of this peptide fragment, which comprises positions 46–70, reveals the cleavage of the Asp-45 to Leu-46 bond in eglin c.

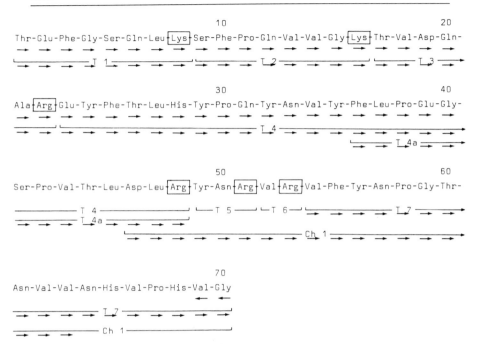

FIG. 6. Amino acid sequence of eglin c. Boxed amino acids indicate the tryptic cleavage sites; the arrow indicates the chymotrypsin-reactive bond. T1–T7, tryptic peptides; Ch 1, chymotryptic peptide. →, residues identified by automatic Edman degradation; ←, residues identified after cleavage with carboxypeptidase Y.

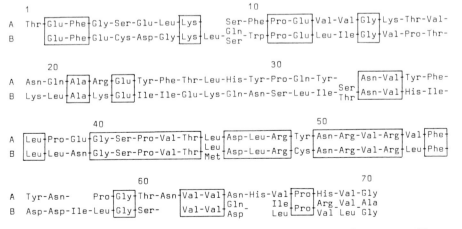

FIG. 7. Sequence homology of eglin c (A) and chymotrypsin inhibitor I from potatoes (B).

Inhibition Mechanism. Preceding investigations showed that incubation of eglin with varying amounts of chymotrypsin (5, 50, and 90 mol %) and subsequent separation of the mixtures in SDS–gel electrophoresis yields a low-molecular-weight peptide (Ch 1, see Fig. 6) in a quantity corresponding to the applied enzyme concentration. Obviously, during incubation of eglin with chymotrypsin, limited proteolysis occurs, resulting in the separation of the C-terminal peptide from the inhibitor molecule. The N-terminal part of the inhibitor remains firmly bound to the enzyme, thus forming the inactive complex. It cannot be dissociated by SDS–gel electrophoresis or by gel chromatography in acidic medium (0.1 M acetic acid). This mechanism of complex formation seems to be comparable to the mode of action of the plasma inhibitors with proteinases (e.g., α_1-proteinase inhibitor).

Sequence Homology. The structure of eglin shows no similarity to the well-characterized Kunitz- or Kazal-type inhibitors. Surprisingly, there exists a striking homology to chymotrypsin inhibitor I from potatoes[16] (Fig. 7). Furthermore, the potato inhibitor contains 1 disulfide bridge, which can be reduced and carboxymethylated without effect on the inhibitory activity of this protein.[17]

[16] M. Richardson and L. Cossius, *FEBS Lett.* **52,** 161 (1975).
[17] G. Plunkett and C. A. Ryan, *J. Biol. Chem.* **255,** 2752 (1980).

[61] The Broad-Specificity Proteinase Inhibitor 5 II from the Sea Anemone *Anemonia sulcata*

By GERT WUNDERER, WERNER MACHLEIDT, and HANS FRITZ

The sea anemone *Anemonia sulcata* contains abundant amounts of proteinase inhibitors. These are basic miniproteins of high stability that inhibit various serine proteinases such as trypsin, chymotrypsin, plasmin, plasma, and tissue kallikreins, and thus resemble aprotinin, the basic trypsin inhibitor from bovine organs. As published earlier in this series,[1] the inhibitory principle of *A. sulcata* can be separated by chromatography into at least 10 isoinhibitors that are similar in their amino acid compositions but differ in their abilities to inhibit kallikreins.[2]

The present contribution describes the amino acid sequence of the main isoinhibitor 5 II in comparison with known inhibitory polypeptides of similar structure.

[1] G. Wunderer, L. Béress, W. Machleidt, and H. Fritz, this series, Vol. 45, p. 881.
[2] G. Wunderer, K. Kummer, H. Fritz, L. Béress, and W. Machleidt, *Bayer Symp.* **5,** 277 (1974).

Amino Acid Sequence

The inhibitor 5 II was purified from sea anemones as described earlier.[1-3] The native as well as the performic acid-oxidized polypeptide was homogeneous as judged by electrophoresis, gel filtration, ion-exchange chromatography at different pHs, and end-group determination. The amino acid composition, however, departed from integer values for a few amino acids. This result can be explained now as a consequence of microheterogeneities in the amino acid sequence. Six heterogeneous positions were found in the sequence of inhibitor 5 II (see later). Obviously, the observed exchanges of charged residues must be balanced within the molecule because the corresponding inhibitory polypeptides could not be separated according to charge by electrophoresis or ion-exchange chromatography.

The amino acid sequence of inhibitor 5 II was determined by automated solid-phase Edman degradation of the reduced and carboxymethylated polypeptide up to lysine[53], which was used as an anchor point on the solid support. This sequence was completed and confirmed by sequencing of peptide fragments obtained by cleavage with *Staphylococcus aureus* protease. Details of sequence analysis will be presented elsewhere.[4] The results are consistent with partial sequences elucidated from tryptic fragments, which have been published earlier.[3] Figure 1 shows the complete amino acid sequence, including the duplicate residues found in microheterogeneous positions.[5-9]

Structural Homology and Kallikrein Inhibition

Inhibitor 5 II from *A. sulcata* is homologous to the family of aprotinin-type proteinase inhibitors, some of which are included in Fig. 1 for comparison. The inhibitor from the sea anemone shows a striking homology to the whole family, especially in regions that are invariant. Aprotinin, the most prominent representative of the family, is well characterized with respect to its three-dimensional structure[10] and its interaction with trypsin.[11] From these results it is known that the amino acid residues with the

[3] G. Wunderer, Dissertation, Technical University of Munich (1975).

[4] W. Machleidt and G. Wunderer, *FEBS Lett.* (submitted for publication).

[5] B. Kassell and M. Laskowski, Sr., *Biochem. Biophys. Res. Commun.* 20, 463 (1965).

[6] D. Čechova, V. Jonáková, and F. Šorm, *Proc. Int. Res. Conf. Proteinase Inhibitors, 1st, 1970* p. 105 (1971).

[7] Y. Hokama, S. Iwanaga, R. Tatsuki, and T. Suzuki, *J. Biochem.* (*Tokyo*) 79, 559 (1976).

[8] T. Dietl and H. Tschesche, *Bayer Symp.* 5, 254 (1974).

[9] IUPAC-IUB Commission on Biochemical Nomenclature, *J. Biol. Chem.* 243, 3557 (1968).

[10] R. Huber, D. Kukla, O. Epp, and H. Formanek, *Naturwissenschaften* 57, 389 (1970).

[11] A. Rühlmann, D. Kukla, P. Schwager, K. Bartels, and R. Huber, *J. Mol. Biol.* 77, 417 (1973).

```
                              5           10          15          20          25          30
Sea anemone 5 II    I N G D C E L P K V V G P C R A R F P R Y Y Y N S S S K R C
                                                                        L         R
Aprotinin           R P D F C L E P P Y T G P C K A R I I R Y F Y N A K A G L C
                                        G G
Cow colostrum   F Q T P P P D L C Q L P Q A R G P C K A A L L R Y F Y N S T S N A C
Cobra HHV II        R P D F C E L P A E T G L C K A Y I R S F H Y N L A A Q Q C
Cobra NNV II        R P R F C E L P A E T G L C K A R I R S F H Y N R A A Q Q C
Snail, K            R P S F C N L P A E T G P C K A S F R Q Y Y Y N S K S G G C

                              35          40          45          50          55
Sea anemone 5 II    E K F I Y G G C G G N A N N F H T L E E C E K V C G V R S
                                        R
Aprotinin           Q T F V Y G G C R A K R N N F K S A E D C M R T C G G A
Cow colostrum       E P F T Y G G C Q G N N N F E T T E M C L R I C E P P Q Q T D K S
Cobra HHV II        L Q F I Y G G C G G N A N R F K T I D E C R R T C V G
Cobra NNV II        L E F I Y G G C G G N A N R F K T I D E C H R T C V G
Snail, K            Q Q F I Y G G C R G N Q N R F D T T Q Q C Q G V C V
```

FIG. 1. Amino acid sequence of the proteinase inhibitor 5 II from A. sulcata. The above sequences are given for comparison: aprotinin[5]; proteinase inhibitor from cow colostrum,[6] from Haemachatus haemachatus venom,[7] and from Naja nivea venom[7]; and inhibitor K from snail mucus.[8] The amino acid residues are given in the one-letter code.[9] Residues identical in all sequences are printed in boldface type.

most intimate contact to trypsin in the complex are in the regions glycine[12] to arginine[17] (forming the specific pocket) and glycine[36] to arginine[39].

The sequence homology of aprotinin[5] and inhibitor 5 II from *A. sulcata* suggests a close similarity in the spatial arrangement of these two and other homologous polypeptides.

Like aprotinin, the polypeptide 5 II from the sea anemone is a strong inhibitor of tissue and plasma kallikreins, which are enzymes of much more restricted specificity than trypsin. It is reasonable to assume that the basic amino acid residues in positions 17 and 39 are responsible for the relatively strong interaction of both aprotinin and the sea anemone inhibitor with kallikreins. Exchange of both basic amino acid residues in these positions by neutral residues completely abolishes the affinity of an aprotinin-type inhibitor to kallikreins, as it has been found for the cow colostrum inhibitor.[6] If, however, a basic residue is present in position 19 (cobra HHV II),[7] inhibitory activity against kallikrein is restored. Obviously, in an aprotinin-type inhibitor the basic nature of amino acid residues in one or more of these positions, namely 17, 19, and 39, is necessary to gain enough energy for complex formation with a kallikrein. The observation that the more acidic isoinhibitors from sea anemones do not inhibit kallikreins as strongly as the more basic inhibitors do, can be explained on the basis of this hypothesis. As shown above, the inhibitor 5 II contains microheterogeneous forms in which arginine[17] and or arginine[39] are replaced by glycine. The inhibitor fraction 5 II must be regarded, therefore, as a mixture of weaker and stronger kallikrein inhibitors. In contrast, the affinity of aprotinin-type inhibitors to plasmin (and trypsin) is not significantly affected by similar amino acid exchanges in the contact regions.

Pharmacological and Biological Aspects

Due to their lower basicity (pI ≤ 9.5), the inhibitors from *A. sulcata* differ from aprotinin (pI = 10.5) in their pharmacological properties. Essentially, they neither become fixed to kidney membranes after parenteral application, nor to negatively charged cell wall surfaces. This different behavior compared to aprotinin makes them interesting candidates for medical and/or pharmacological studies.[12]

The biological function of the proteinase inhibitors for the sea anemones themselves is not clear yet. The inhibitors are diffusely distributed in the ectoderm of tentacles and body as observed by immunofluorescence; a similar distribution was found for the toxins[13] of sea anemones (T. Dietl and G. Wunderer, unpublished results). The biolog-

[12] H. Fritz, E. Fink, and E. Truscheit, *Fed. Proc., Fed. Am. Soc. Exp. Biol.* **38**, 2753 (1979).
[13] L. Béress, R. Béress, and G. Wunderer, *FEBS Lett.* **50**, 311 (1975).

ical function of the toxins is easily understood as a means for paralyzing the prey.

Proteinase inhibitors from sea anemones will maintain special interest as they represent the phylogenetically oldest aprotinin-type inhibitors known until now. They can be easily isolated in large quantities.[1] Isoinhibitors with different inhibitory specificities may provide desirable properties for therapeutic uses and for the study of structure–function relationships.

[62] Inactivation of Thiol Proteases with Peptidyl Diazomethyl Ketones

By ELLIOTT SHAW and GEORGE D. J. GREEN

Enzymes that have a catalytically essential thiol group can often be inactivated by commonly available group-specific reagents such as p-chloromercuribenzoate, N-ethylmaleimide, and iodoacetic acid (or amide). There are recognized dangers in such diagnostic procedures of both false positive and false negative reactions. The former may arise from modification of either amino acid side-chains such as the occasional reaction of N-ethylmaleimide with amino groups.[1] The need to use these general reagents at a relatively high concentration, such as $10^{-3} M$ in many cases, favors nontypical reactions. The false negative presumably arise from the nonaccesssibility or "buried" state of a catalytic group to the reagent examined. For example, cathepsin L is only partially inhibited by millimolar N-ethylmaleimide or iodoacetamide,[2] even though these are relatively small molecules.

In the study of purified enzymes, application of a number of such tests may eventually lead to a satisfactory conclusion about mechanism. *In vivo* studies, however, can rarely permit the use of such reactants, due to the widespread occurrence of thiols both in proteins and small molecular metabolites.

Affinity labeling has introduced a number of reagents for the study of proteolytic enzymes with a considerable reduction in the number of likely side reactions. However, even in this limited biochemical area there may remain an ambiguity about the mechanistic class to which a susceptible protease belongs, since both serine proteases and thiol proteases are inac-

[1] J. F. Riordan and B. L. Vallee, this series, Vol. 11, p. 541.
[2] H. Kirschke, J. Langner, B. Wiederanders, S. Ansorge, and P. Bohley, *Eur. J. Biochem.* **74**, 293 (1977).

tivated by substrate-derived chloromethyl ketones[3] or the naturally occurring peptidyl aldehydes such as antipain, leupeptin, chymostatin, and elastatinal.[4] On the other hand, a more selective class of reagents for thiol proteases has appeared in the form of peptidyl diazomethyl ketones,[5-8] which react rapidly in high dilution with thiol proteases but not other types of proteases,[7,8] and can be synthesized to satisfy the specificity of individual members of the thiol protease family.

Protein Chemistry

The reaction of a diazomethyl carbonyl group with thiol groups in proteins was clarified by the work of Buchanan and his collaborators on the mechanism of action of the antibiotic, azaserine, in inactivating glutamine-dependent aminotransferases. The alkylation of an essential —SH group in its protonated form was demonstrated.[9] A similar reaction was shown to take place in the inactivation of papain, a thiol proteinase, with Z-Phe diazomethyl ketone.[5] The pH dependence of the inactivation indicated more rapid reaction with the essential thiol at pH 6.5 than at higher pH values. This is a useful diagnostic characteristic of this type of reagent. In the cases of the other thiol proteases described below, the site of reaction has not been rigorously demonstrated.

The reaction of the diazomethyl ketones with simple thiols such as mercaptoethanol or cysteine is very slow in contrast to a protein thiol group susceptible to affinity labeling. The rate difference[8] may be at least 10^{10}! This is a dramatic demonstration of a rate enhancement obtainable by a proximity effect. It has the additional advantage of permitting the use of these reagents with impunity even in millimolar mercaptoethanol, which is often present to preserve thiol proteases from oxidative inactivation.

Synthetic Aspects and Reagent Stability

Only a limited number of reagents of this type have been synthesized, largely from blocked amino acid derivatives or peptides activated at the

[3] W. H. Porter, L. W. Cunningham, and W. M. Mitchell, *J. Biol. Chem.* **246**, 7675 (1971).
[4] H. Umezawa, "Enzyme Inhibitors of Microbial Origin." Univ. Park Press, Baltimore, Maryland, 1972.
[5] R. Leary, D. Larsen, H. Watanabe, and E. Shaw, *Biochemistry* **16**, 5857 (1977).
[6] R. Leary and E. Shaw, *Biochem. Biophys. Res. Commun.* **79**, 926 (1977).
[7] H. Watanabe, G. D. J. Green, and E. Shaw, *Biochem. Biophys. Res. Commun.* **89**, 1354 (1979).
[8] G. D. J. Green and E. Shaw, *J. Biol. Chem.* **256**, 1923 (1981).
[9] J. B. Dawid, T. C. French, and J. M. Buchanan, *J. Biol. Chem.* **238**, 2178 (1963).

carboxyl group by the mixed anhydride method and reacted with diazomethane.[5,7,8] The diazomethyl ketone group is stable to alkali and unstable to acid; therefore, methods of peptide chemistry utilizing acid deblocking conditions must be avoided. The solid reagents are apparently very stable. Some have been kept at ambient temperature for 3 years without deterioration. It is probably advisable to protect them from light, however. Since some of the peptide derivatives are not very soluble in water, it is often necessary to prepare a stock solution in an organic solvent miscible with water such as acetonitrile, dimethyl sulfoxide, or ethanol for eventual dilution in the biological system under study.

Inhibitor Syntheses

Z-Phe-Ala Diazomethyl Ketone.[7] Z-Phe-Ala (3.7 g, vacuum-dried) in tetrahydrofurane (50 ml) and N-methylmorpholine (1.25 ml) was treated at $-15°C$ with isobutyl chloroformate (1.25 ml). After 15 min, the mixture was added to freshly prepared ethereal diazomethane (250 ml) at 0°C. This solution had been obtained from N-methyl-N-nitroso-p-toluenesulfonamide (Aldrich Chemical Co.) according to the supplier's instructions. After a half hour at 0°C, the reaction was left overnight at room temperature. After successive washings with water and aqueous sodium bicarbonate, the ethereal solution was dried over magnesium sulfate and concentrated under reduced pressure. The crystalline residue was recrystallized from ethyl acetate and hexane to yield 3.0 g, mp 132–133°C decomposition.

Z-Lys Diazomethyl Ketone.[8] The $N^ε$-trifluoroacetyl derivative was prepared from Z-$N^ε$-TFA-lysine[10] by conversion to the mixed anhydride and reaction with diazomethane as described in the preceding synthesis. The product, Z-N-TFA-lysine diazomethyl ketone, mp 94–98°C (after crystallization from methanol and water), was obtained in 73% yield and deblocked with alkali. For this purpose, 0.40 g (1 mmol) was dissolved in methanol and cooled to $-15°C$. Sodium hydroxide (1 M, 2 ml) was added and the reaction mixture stirred overnight at $-15°C$. Water (30 ml) was added and the organic solvent removed under reduced pressure. The resulting oil–water mixture was extracted three times with ethyl acetate, and to the pooled extracts (35 ml), water (50 ml) was added. This mixture was stirred vigorously with dilute acetic acid added dropwise until the pH of the aqueous layer was steady at pH 6. The aqueous layer was separated and freed of dissolved ethyl acetate under reduced pressure, then lyophilized. The residual yellow solid is the acetate of the product.

Gly-Phe Diazomethyl Ketone.[8] Z-Phe-Gly-Phe was obtained in 48% yield by the reaction of Z-Phe hydroxysuccinimide ester with Gly-Phe in aqueous dioxane containing triethylamine and had mp 202–204°C.

[10] A. T. Moore, H. N. Rydon, and M. J. Smithers, *J. Chem. Soc.* p. 2349 (1966).

Z-Phe-Gly-PheCHN$_2$ was prepared from the foregoing tripeptide as described above in the synthesis of Z-Phe-AlaCHN$_2$. The glasslike product appeared to be pure by thin-layer chromatography in methylene chloride; ethanol (10:1) on silica gel, $R_f = 0.60$, 82% yield.

A chymotrypsin–Affigel conjugate was prepared by treating 5 ml of Affigel-10 (Bio-Rad) with chymotrypsin (200 mg) in 0.1 M NaHCO$_3$ (5 ml) overnight at 4°C. Residual reactivity was blocked by addition of 1 M ethanolamine, pH 8 (1 ml). After 1 hr, the gel was extensively washed with 0.1 M NaHCO$_3$ followed by 0.2 M NaCl until the absorbance at 280 nm dropped to 0.02.

Enzymatic generation of Gly-PheCHN$_2$ was carried out in solutions obtained by dilution of a stock solution of Z-Phe-Gly-PheCHN$_2$ (10^{-2} M in acetonitrile) with 9 volumes of 0.1 M Tris (pH 8), and treatment at room temperature with 0.4 ml of immobilized chymotrypsin/10 ml and with occasional shaking. Gradual disappearance of the starting material could be followed by TLC. At intervals, aliquots were removed, centrifuged, and the supernatant solution analyzed by HPLC on a column (4.2 × 300 mm) of LiChrosorb RP-18, 10 μm (E. Merck), using a linear gradient from 20 mM phosphate buffer (pH 5.6) to 50% acetonitrile in the same buffer. Only one of the peaks corresponded to the desired product, having a diazomethyl ketone spectrum (λ_{max} at 275 nm in aqueous solvents) and producing only glycine on amino acid analysis (the phenylalanine converted to a diazomethyl group is not recovered). The concentration of Gly-PheCHN$_2$ in this peak is based on the glycine concentration.

Enzyme Inactivations

If proteases present a nuisance in a biological isolation procedure, the addition of an effective thiol protease inactivator such as Z-Phe-AlaCHN$_2$ to a final concentration of 10^{-5} M is worth consideration in combination with other inhibitors, for example DFP. This will have a very rapid action near pH 6, irreversibly nullifying cathepsins B and L.

For enzyme characterization, however, more gradual responses are desired for kinetic information generally obtainable in a range of dilutions that are not enzyme saturating. For peptidyl diazomethyl ketones the useful range is often 10^{-6}–10^{-8} M.

A typical procedure involves setting up an incubation mixture containing enzyme and inhibitor in the buffer to be used for assay. Some organic solvent may be necessary to keep the inhibitor in solution. At various intervals, an aliquot is removed and assayed by addition of an appropriate substrate to permit measurement of the rate of inactivation. In the table are listed some characteristic inactivations. This table is not a complete survey since it does not include results with different combinations of

SOME CHARACTERISTIC INACTIVATIONS OF THIOL PROTEASES WITH
PEPTIDYL DIAZOMETHYL KETONES

Thiol protease	Inhibitor	Concentration (M)	pH	$t_{1/2}$ (min)	Ref.
Cathepsin B					
Beef spleen	Z-Phe-AlaCHN$_2$[a]	5×10^{-7}	5.4	18.5	b
Rat liver	Z-Phe-PheCHN$_2$[c]	10^{-4}	5.5	>15	d
Cathepsin L					
Rat liver	Z-Phe-PheCHN$_2$[c]	5×10^{-8}	5.5	15	d
Cathepsin H					
Rat liver	Z-Phe-AlaCHN$_2$	10^{-4}	5.5	>15	d
Cathepsin C	Gly-PheCHN$_2$	0.6×10^{-7}	6	12	e
Clostripain	Z-LysCHN$_2$	1.8×10^{-6}	7.8	14	e
Streptococcal proteinase	Z-Phe-AlaCHN$_2$[f]	10^{-7}	5.4	15	e
Postproline endopeptidase	Z-Ala-Ala-ProCHN$_2$[g]	4×10^{-7}	7.4	5	h

[a] 10% Acetonitrile or 0.1% dimethyl sulfoxide present in buffer.
[b] H. Watanabe, G. D. J. Green, and E. Shaw, *Biochem. Biophys. Res. Commun.* **89**, 1354 (1979).
[c] 2% Dimethyl sulfoxide present.
[d] H. Kirschke, J. Langner, S. Riemann, B. Wiederanders, S. Ansorge, and P. Bohley, *Ciba Found. Symp.* **75**, 15 (1980).
[e] G. D. J. Green and E. Shaw, *J. Biol. Chem.* **256**, 1923 (1981).
[f] 5% Acetonitrile present.
[g] 1% Ethanol present.
[h] G. D. J. Green and E. Shaw, to be submitted for publication.

enzymes and inhibitors that would permit the investigator to judge selectivity. Some of this information is available in the references cited.

There is preliminary indication that the degree of inhibition may vary according to species or tissue. Thus cathepsin B from beef spleen is moderately susceptible[7] to Z-Phe-PheCHN$_2$ in contrast to cathepsin B from rat liver (see the table).

In Vivo Use

A few experiments have demonstrated that peptidyl diazomethyl ketones can be used to inactivate thiol proteinases in cultured cells. Z-Phe-AlaCHN$_2$ at $10^{-5} M$ (in 0.2% dimethyl sulfoxide) completely inactivated thiol proteases within cultured mouse peritoneal macrophages and permitted the measurement of the role of these enzymes in protein degradation.[11] On removal of the inhibitor, the recovery of cathepsin B activity by new synthesis was observable. This inhibitor has also been used with cultured mouse bones to block the resorption induced by parathyroid hor-

[11] E. Shaw and R. T. Dean, *Biochem. J.* **186**, 385 (1980).

mone.[12] It is not certain, in this case, whether or not the inhibitor is acting intracellularly.

Z-Ala-Ala-ProCHN$_2$ is rapidly taken up by cultured mouse peritoneal macrophages with resultant inactivation of a cytoplasmic endopeptidase that cleaves at proline residues.[13]

Relationship to Thiol Protease Specificity: Selectivity

At the present time few thiol endoproteases have been identified with a clear-cut primary specificity. Clostripain is one, with a specificity reminiscent of trypsin.[3] Others, such as cathepsins B and L, appear to position substrates in a way that hydrophobic residues are located one or more positions N-terminal to the site of cleavage.[8] In any case, it appears that, as expected from results with serine proteases and peptidyl chloromethyl ketones, great variations in reactivity are encountered depending on reagent sequence, and that highly selective inactivators for thiol proteases can be expected to emerge from the extension of this approach.

Judging from an initial survey, proteases of other mechanistic families, viz. serine proteases, metalloproteases, and acid proteases, are not inactivated by this class of reagents.[8] In the presence of cupric ions, however, acid proteases may become inactivated by reactive species formed on decomposition of the reagent.[14]

Enzyme Assays

The most convenient substrates from the point of view of speed, sensitivity, and accuracy are generally those permitting spectroscopic or fluorimetric monitoring of concentration changes. Such substrates may be neither specific or necessarily reflective of proteolytic activity. However, for enzymes that have been purified by some established procedure and whose identities are thus relatively secure, these substrates are reliable indicators of function. In crude extracts, however, activities measured with any of these substrates may include the contributions of many proteases.

Esterase Assays. Cathepsin B, cathepsin L, papain, and streptococcal proteinase can be measured using Z-Lys-*p*-nitrophenyl ester. Aliquots of enzyme are added to 2.0 ml of 0.10 M acetate (pH 5.4), which also is millimolar in EDTA and 2.5 mM in mercaptoethanol. The substrate is maintained as a $10^{-2} M$ stock solution in 95% acetonitrile; 20 μl is added to

[12] J.-M. Delaisse, Y. Eeckhout, and G. Vaes, *Biochem. J.* **192**, 365 (1980).
[13] G. D. J. Green and E. Shaw, unpublished results.
[14] R. C. Lunblad and W. H. Stein, *J. Biol. Chem.* **244**, 154 (1969).

the assay cuvette and the increase at 340 nm is recorded. For clostripain, benzoyl-arginine ethyl ester is recommended.[3]

Proteolytic Assays.[2] Activity is measured by incubating test samples with a 2.5% solution of azocasein in 0.05 M phosphate buffer (pH 5.5–6.0) containing 5 mM glutathione and 5 mM EDTA at 37°C. After selected time intervals, the reaction is stopped by addition of an equal volume of 10% trichloroacetic acid. Following centrifugation, the optical density of the supernatant solution is measured at 366 nm.

[63] Inactivation of Trypsin-Like Enzymes with Peptides of Arginine Chloromethyl Ketone

By CHARLES KETTNER and ELLIOTT SHAW

Introduction

The value of affinity labels in the identification and characterization of proteases in their physiological environments has been clearly demonstrated by the use of TPCK (tos-PheCH$_2$Cl), TLCK (tos-LysCH$_2$Cl), and alanyl peptide chloromethyl ketones. These reagents distinguish among chymotryptic, tryptic, and elastolytic activities by conforming to the primary specificity requirements of the proteases.[1–3] This communication summarizes many of the results we have obtained in the development of selective-affinity labels for individual trypsin-like proteases that function in blood coagulation, fibrinolysis, and hormone processing. This group of enzymes functions by the hydrolysis of either one or two bonds of their physiological substrates, usually at arginyl residues. We have used the approach of incorporating part of the sequence of the physiological substrate in the reagent,[4] thus taking advantage of binding selectivity in both primary and secondary sites.

Synthesis and Properties of Arginine Chloromethyl Ketones

A typical synthesis of peptides of arginine chloromethyl ketones is demonstrated by the preparation of Pro-Phe-ArgCH$_2$Cl · 2HCl.[5] This pro-

[1] G. Schoellmann and E. Shaw, *Biochemistry* **2**, 252 (1963).
[2] E. Shaw, M. Mares-Guia, and W. Cohen, *Biochemistry* **4**, 2219 (1964).
[3] J. C. Powers and P. M. Tuhy, *Biochemistry* **12**, 4767 (1973).
[4] J. R. Coggins, W. Kray, and E. Shaw, *Biochem. J.* **137**, 579 (1974).
[5] C. Kettner and E. Shaw, *Biochemistry* **17**, 4778 (1978).

cedure has been applied to the synthesis of numerous other reagents with only minor modification. Following the preparation of the first intermediate, Boc-Arg(NO$_2$)CHN$_2$ and its conversion to H-Arg(NO$_2$)CH$_2$Cl · HCl with anhydrous HCl, Z-Pro-Phe-OH was added by the mixed anhydride procedure. The crystalline product, Z-Pro-Phe-Arg(NO$_2$)CH$_2$Cl, was then deprotected with anhydrous HF to yield H-Pro-Phe-ArgCH$_2$Cl · 2HCl. The addition of the dipeptide to H-Arg(NO$_2$)CH$_2$Cl rather than the sequential addition of amino acids is advantageous in that it avoids the dipeptide intermediate obtained in sequential addition. The protected tripeptide arginine chloromethyl ketones are easier to isolate and characterize than the dipeptide intermediate, whose elongation, in addition, results in low yields. Racemization of the residue penultimate to the arginyl residue is a potential problem in the coupling of the N-protected dipeptide to H-Arg(NO$_2$)CH$_2$Cl, but we have no indication that this has occurred to a significant extent.

Final products are amorphous white powders that are hygroscopic without exception. The reagents are stable in this form when stored at 4°C desiccated with P$_2$O$_5$ *in vacuo* for at least several years. In addition, they are stable up to a year in the form of frozen solutions in 10^{-3} M HCl. We have found that it is convenient to prepare a 10^{-2} M stock solution for routine use.

An extensive study of the stability of arginine chloromethyl ketones at physiological pH has not been made. However, no change in the biological activity of Phe-Ala-ArgCH$_2$Cl was observed when it was incubated in 50 mM PIPES buffer (pH 7.0) for 24 hr at 25°C. This is in contrast to the instability of Phe-Ala-LysCH$_2$Cl, which lost half of its biological activity under the same conditions. It is expected that other lysine and arginine chloromethyl ketones will behave similarly, and that at higher pH values the chloromethyl ketones will be progressively less stable. For example, 70% of the biological activity of Phe-Ala-ArgCH$_2$Cl was lost when it was incubated for 30 min at pH 9.0 and 25°C.

Preparation of H-Arg(NO$_2$)CH$_2$Cl · HCl

Boc-Arg (NO$_2$)-OH (5.0 g, 15.6 mmol) was dissolved in 60 ml of tetrahydrofurane and allowed to react with isobutyl chloroformate (2.06 ml, 15.6 mmol) in the presence of N-methylmorpholine (1.72 ml, 15.6 mmol) for 4 hr at −20°C. The mixed anhydride preparation was filtered and the filtrate was added to 120 ml of ethereal diazomethane. After stirring the reaction solution for 30 min at 0°C, 80 ml of ether was added. The product crystallized from the reaction solution in the cold and was isolated by decanting the supernatant and washing the residue with ether. The diazomethyl ketone of Boc-Arg(NO$_2$)-OH was recrystallized from methanol to yield 2.56 g, mp 150–151°C.

Boc-Arg(NO$_2$) CHN$_2$ (1.47 g, 4.30 mmol) was dissolved in a minimal volume of tetrahydrofurane and was allowed to react with ethanolic HCl (20 mmol) at room temperature until nitrogen evolution ceased. Solvent was removed by evaporation and the residue was taken up in 40 ml of 1.8 N ethanolic HCl. After stirring the solution for 30 min at room temperature, 1.34 g of an amorphous white solid was obtained by evaporating the solvent and triturating the residue with ether. This product was dried over KOH and P$_2$O$_5$ *in vacuo* and was used in subsequent reactions without further purification.

Preparation of Z-Pro-Phe-Arg(NO$_2$)CH$_2$Cl

Z-Pro-Phe-OH (0.43 g, 1.1 mmol) was allowed to react with N-methylmorpholine (0.12 ml, 1.1 mmol) in 5 ml of tetrahydrofurane and isobutyl chloroformate (0.15 ml, 1.1 mmol) for 10 min at $-20°$C. Cold tetrahydrofurane containing triethylamine (0.15 ml, 1.1 mmol) was added to the mixed anhydride preparation and the mixture was immediately added to H-Arg(NO$_2$)CH$_2$Cl · HCl (0.31 g, 1.1 mmol) dissolved in 5 ml of cold dimethylformamide. After stirring for 1 hr at $-20°$C and 2 hr at room temperature, the reaction mixture was filtered, the filtrate was evaporated to dryness, and the residue was dissolved in 3 ml of methanol. The solution was diluted to 100 ml with ethyl acetate and washed with 0.1 N HCl, 5% NaHCO$_3$, and saturated aqueous NaCl. The organic phase was briefly dried over anhydrous Na$_2$SO$_4$ and was concentrated to yield 0.33 g of crystals, mp 102–105°C.

Preparation of H-Pro-Phe-ArgCH$_2$Cl · 2HCl

Z-Pro-Phe-Arg(NO$_2$)CH$_2$Cl (0.50 g, 0.79 mmol) was treated with approximately 10 ml of anhydrous hydrogen fluoride in the presence of 1 ml of anisole using a KelF hydrogen fluoride line. After allowing the reaction to proceed for 30 min at 0°C, the hydrogen fluoride was removed by distillation and the residue was dissolved in 20 ml of water. The solution was extracted with three 20-ml portions of ether and the aqueous phase was applied to a column containing 20 ml of SP–Sephadex (C-25, H$^+$ form). The column was washed with 120 ml of water, and the product eluted with 120 ml of 0.4 N HCl. After lyophilization of the HCl solution, the residue was transferred to a small container by solution in methanol and concentration under a stream of nitrogen. The product was obtained as 0.36 g of an amorphous white solid by trituration with ether and drying over P$_2$O$_5$ and KOH. TLC indicated a single spot, R_f = 0.21 by Sakaguchi, and ninhydrin stains after development on silica gel in butanol, acetic acid, and water (4 : 1 : 1).

Kinetics of Affinity Labeling

The reactivity of the chloromethyl ketones with proteases has been determined by incubating the protease with the affinity label, removing timed aliquots, and measuring residual enzymatic activity. The initial concentration of protease was selected so that the concentration of affinity label was at least 10 times greater than that of the enzyme. Values of k_{app}, the apparent, pseudo-first-order rate constant for inactivation of the protease under observation, were determined from slopes of semilogarithmic plots of protease reactivity versus time.[6] Kinetic constants for the affinity labeling mechanism, reaction (1), are shown in Eq. (2). K_i is the dissociation constant for the reversible, substrate-like complex EI, and k_2 is the first-order rate constant for the irreversible alkylation of the protease.

$$E + I \rightleftharpoons EI \xrightarrow{k_2} \text{alkylated protease} \tag{1}$$

$$\frac{1}{k_{app}} = \frac{K_i}{k_2}\frac{1}{I} + \frac{1}{k_2} \tag{2}$$

$$\frac{k_2}{K_i} = \frac{k_{app}}{I} \quad \text{if } I \ll K_i \tag{3}$$

Bimolecular constants, k_2/K_i, were determined from Eq. (2) or estimated by the relationship in Eq. (3). The relationship in Eq. (3) is valid for most of the enzymes we have studied at levels of affinity label that give half-times of inactivation of greater than 15 min.

Determination of individual kinetic constants, K_i and k_2, are less reliable, since k_2 is determined from the intercept on the Y axis of double reciprocal plots of k_{app} versus I. Small variations in the intercept, particularly when it is close to the origin, result in large differences in K_i and k_2. Lineweaver–Burk plots[7] are a more reliable method of measuring K_i.[8] In this approach, the affinity label is incubated with the protease in the presence of substrate and initial velocities of substrate hydrolysis are measured. The substrate sufficiently protects the enzyme from inactivation so that an initial velocity can be measured before inactivation of the protease in the cuvette. (A detailed description can be found in Ref. 8.)

In typical assays, the proteases (10–15 nM) were incubated with the affinity label in 2.00 ml of 50 mM PIPES buffer (pH 7.0) at 25°C. Timed aliquots (200 μl) were removed and diluted in 0.20 M Tris buffer (pH 8.0) containing 0.20 M NaCl, Z-Lys-SBzl as a substrate,[9] and 5,5'-dithiobis-

[6] R. Kitz and I. B. Wilson, J. Biol. Chem. 237, 3245 (1962).
[7] H. Lineweaver and D. Burk, J. Am. Chem. Soc. 56, 658 (1934).
[8] C. Kettner, C. Mirabelli, J. V. Pierce, and E. Shaw, Arch. Biochem. Biophys. 202, 420 (1980).
[9] G. D. J. Green and E. Shaw, Anal. Biochem. 93, 223 (1979).

(2-nitrobenzoic acid) as a chromogen. The final volume of the assay solution was 2.00 ml and the final concentrations of substrate and chromogen were 0.10 mM and 0.33 mM, respectively. Enzymatic activity was followed for 2–3 min at 412 nm on a Beckman 5230 recording spectrophotometer using a 0.05 chart scale. For the evaluation of the reactivity of highly effective affinity labels, it was occasionally advisable to use lower concentrations of enzymes in order to maintain the concentration of affinity label in excess of that of the protease. It should be noted that under these conditions the rate of inactivation is independent of the initial concentration of protease. Protease concentrations were decreased by increasing the volume of the incubation mixture and proportionally increasing the size of the aliquot (up to 1.0 ml) removed for assay, or by increasing the sensitivity of the spectrophotometer as much as 5-fold.

In almost all cases, Z-Lys-SBzl has been an adequate substrate for the assay of purified enzymes, and in most cases it is comparable in sensitivity to peptide chromogenic substrates.[9] Among the advantages of Z-Lys-SBzl is its ease of synthesis and low cost as well as its adaptability for use with a variety of enzymes with minor modification. An exception is factor X_a, which hydrolyzes this substrate poorly. The chromogenic substrate of Kabi, S-2222, was more suitable for this purpose.

Affinity Labeling of Kallikreins

Plasma kallikrein and glandular kallikrein are two distinctive enzymes, differing in their physical properties, specificity, and physiological function. Both enzymes, however, hydrolyze the same Arg-Ser bond of kininogen, liberating the C-terminal of bradykinin (-Phe-Ser-Pro-Phe-Arg-OH).[10] Plasma kallikrein also hydrolyzes factor XII in the initiation of blood coagulation[10] and activates prorenin,[11,12] but the sequence of these substrates is unknown. Less is known about the physiological substrates of glandular kallikreins, although they are implicated in renal function[13] and hormone processing.[14] Therefore, the sequence corresponding to the C-terminal of bradykinin was used in the design of affinity labels for kallikreins.

Plasma Kallikrein. The tripeptide analog corresponding to the C-terminal of bradykinin, Pro-Phe-ArgCH$_2$Cl, readily inactivates plasma

---

[10] J. J. Pisano, *in* "Proteases and Biological Control" (E. Reich, D. B. Rifkin, and E. Shaw, eds.), p. 199. Cold Spring Harbor Lab., Cold Spring Harbor, New York, 1975.

[11] N. Yokosawa, N. Takahashi, T. Inagami, and D. L. Page, *Biochim. Biophys. Acta* **569**, 211 (1979).

[12] F. H. M. Derkx, B. N. Bouma, M. P. A. Schalekamp, and M. A. D. H. Schalekamp, *Nature (London)* **280**, 315 (1979).

[13] H. S. Margolius, R. Geller, J. J. Pisano, and A. Sjoerdsma, *Lancet* **2**, 1063 (1971).

[14] M. A. Bothwell, W. H. Wilson, and E. M. Shooter, *J. Biol. Chem.* **254**, 7287 (1979).

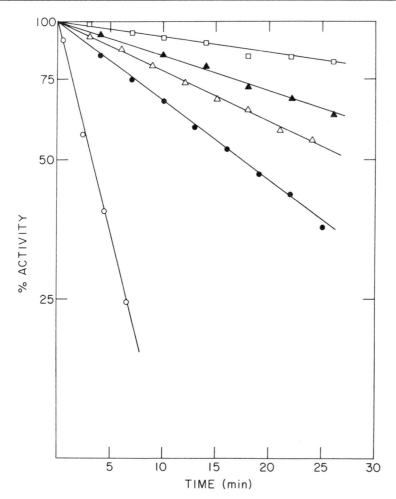

FIG. 1. Selectivity of Pro-Phe-ArgCH₂Cl in the inactivation of plasma kallikrein. The proteases were incubated at 25°C in 50 mM PIPES buffer (pH 7.0) containing Pro-Phe-ArgCH₂Cl at the indicated concentration. Inactivation of each protease was monitored by removing timed aliquots and assaying for esterase activity. O——O, Kallikrein inactivation, I = 0.20 μM; ●——●, plasmin, I = 2.0 μM; △——△, factor X_a, I = 2.0 μM; ▲——▲, thrombin, I = 15 μM; □——□, urokinase, I = 500 μM.

kallikrein at $2.0 \times 10^{-7} M$, and reacts with other trypsin-like proteases more slowly in the presence of higher concentrations of the reagent[5] (Fig. 1). Comparison of values of k_{app}/I shows that plasma kallikrein differs in its reactivity toward Pro-Phe-ArgCH₂Cl from other proteases by a factor of 48 in the case of plasmin, and by factors of 10^2-10^5 for factor X_a, thrombin, and urokinase.[5]

TABLE I
AFFINITY LABELING OF KALLIKREINS[a]

	$10^{-4} \times k_{app}/I \ (M^{-1} \ min^{-1})$[b]		
Affinity Label	Human plasma kallikrein	Human urinary kallikrein	Submaxillary gland protease (mouse)
Ac-Phe-ArgCH$_2$Cl	88	0.81	30
Pro-Phe-ArgCH$_2$Cl	120	0.79	12
Ac-Pro-Phe-ArgCH$_2$Cl	65	0.34	42
Ser-Pro-Phe-ArgCH$_2$Cl	150	0.65	78
Phe-Ser-Pro-Phe-ArgCH$_2$Cl	140	4.20	150
D-Phe-Phe-ArgCH$_2$Cl	2300	18	1500
Glu-Phe-ArgCH$_2$Cl	570	0.32	
DNS-Ala-Phe-ArgCH$_2$Cl	570		
Ala-Phe-ArgCH$_2$Cl	350	0.34	8.4
Phe-Phe-ArgCH$_2$Cl	170	0.63	55
DNS-Pro-Phe-ArgCH$_2$Cl	33	7.1	
Ala-Phe-LysCH$_2$Cl	7.4	0.020	0.19
Pro-Gly-ArgCH$_2$Cl	3.3	0.0029	0.0052
Pro-Phe-HarCH$_2$Cl	0.15	0.031	0.012

[a] Inactivation reactions were conducted at 25°C in 50 mM PIPES buffer (pH 7.0), which was 0.20 M in NaCl, except for plasma kallikrein, for which the NaCl was omitted.
[b] k_{app}/I, the ratio of the apparent, first-order rate constant for the inactivation to the concentration of the affinity label, is an estimate of k_2/K_i, the bimolecular constant for the reaction. Values shown for the first five reagents are actual values of k_2/K_i in the case of plasma and urinary kallikreins.

The reactivity of plasma kallikrein has been determined with two series of reagents. The first consists of di- through pentapeptide analogs corresponding to the sequence of kininogen to determine the influence of binding in the P$_3$–P$_5$ sites on reagent effectiveness. The affinity of the acetylated dipeptide analog, Ac-Phe-ArgCH$_2$Cl, was comparable to that of Phe-Ser-Pro-Phe-ArgCH$_2$Cl (Table I), indicating that the P$_3$–P$_5$ binding sites contribute little to the specificity of kallikrein for kininogen.

In the second series of reagents, the reactivity of plasma kallikrein was determined with chloromethyl ketones in which amino acid substituents in the Pro-Phe-ArgCH$_2$Cl sequence were replaced by other residues. The importance of the P$_1$ Arg on reagent effectiveness is clearly shown by the 50- and 800-fold greater reactivity of the arginine chloromethyl ketone than the corresponding lysine and homoarginine analogs. Similarly, the importance of the P$_2$ Phe is shown by the 36-fold greater reactivity of Pro-Phe-ArgCH$_2$Cl than Pro-Gly-ArgCH$_2$Cl. However, in P$_3$, replacement of Pro by Ala, Phe, or Glu resulted in reagents of comparable or superior reactivity, indicating that the P$_3$ residue is of minor importance in determining the specificity of plasma kallikrein.

A reagent with a P_3 D-Phe was prepared for kallikrein due to the favorable effect obtained with this substituent in a related case. D-Phe-Phe-ArgCH$_2$Cl is the most effective reagent for plasma kallikrein obtained, inactivating it with a $t_{1/2}$ of 15 min at a reagent concentration of 2.0 $\times$ 10^{-9} M; however, the selectivity of this reagent for the inactivation of plasma kallikrein is no greater than that of Pro-Phe-ArgCH$_2$Cl.

As shown in Fig. 2, D-Phe-Phe-ArgCH$_2$Cl reacts stoichiometrically

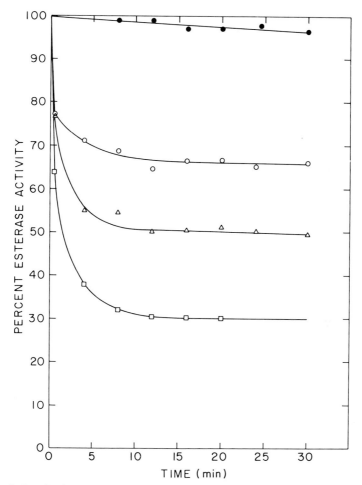

FIG. 2. Inactivation of human plasma kallikrein by D-Phe-Phe-ArgCH$_2$Cl. Plasma kallikrein (1 $\times$ 10^{-7} M) was incubated with the reagent at the indicated concentrations in 50 mM PIPES buffer, pH 7.0 (final volume 1.54 ml). Reagent concentrations were as follows: ●——●, 0; ○——○, 2.5 $\times$ 10^{-8} M; △——△, 5 $\times$ 10^{-8} M; □——□, 7.5 $\times$ 10^{-8} M. Timed aliquots (20 μl) were removed and diluted into 2.00 ml of the Z-Lys-SBzl assay solution[9] and the change in absorbance monitored on a 0.10 recorder scale.

with plasma kallikrein, inactivating it in proportion to the level of affinity label. Titration of plasma kallikrein by D-Phe-Phe-ArgCH$_2$Cl has the advantage that when coupled with a sensitive colorimetric assay such as provided by Z-Lys-SBzl,[9] considerably lower concentrations of kallikrein can be determined than is possible with the usual p-nitrophenyl guanidinobenzoate (NPGB) titration.

Human Urinary Kallikrein. The differences between the specificity of glandular and plasma kallikrein are clearly illustrated by the differences in the kinetic constants in Table I for the two enzymes.[8] In all cases, the arginine chloromethyl ketones were more reactive with plasma kallikrein than urinary kallikrein and the pattern of reactivities of the affinity labels was distinct for each protease. As expected for enzymes hydrolyzing the same bond of kininogen, reagents with a P$_1$ Arg and P$_2$ Phe were the most effective of the inhibitors tested for both proteases.

Submaxillary Gland Protease. Our interest in the specificity differences among trypsin-like proteases led us to examine the reactivity of a series of arginine chloromethyl ketones toward the arginine-specific protease[15] of the mouse submaxillary gland, which functions as the epidermal growth factor-binding protein. Reagents prepared specifically for kallikrein proved to be most effective in the inactivation of this enzyme (Table I). The pattern of reactivity of the arginine-specific protease with the affinity labels is distinct from both plasma and urinary kallikrein, but the overall results suggest that this protease is kallikrein-like in specificity. In fact, the differences observed between the arginine-specific protease and human urinary kallikrein may be due at least in part to species differences, since previous results[8] have shown considerable differences between species for glandular kallikreins. Subsequent to these observations, Bothwell *et al.*[14] reported the independent conclusion that this protease exhibited considerable kininogenase activity and was immunologically similar to other mouse glandular kallikreins. Thus this result represents an example in which the specificity of a protease for a physiological substrate was deduced from its reactivity toward peptides of arginine chloromethyl ketone.

Affinity Labeling of Factor X$_a$

Factor X$_a$, a protease common to both the intrinsic and extrinsic blood coagulation pathways, hydrolyzes two bonds in the conversion of prothrombin to thrombin. The hydrolysis sites in bovine prothrombin are[16]

---

[15] M. Boesman, M. Levy, and I. Schenkein, *Arch. Biochem. Biophys.* **175**, 463 (1976).

[16] S. Magnusson, T. E. Petersen, L. Sottrup-Jensen, and H. Claeys, *in* "Proteases and Biological Control" (E. Reich, D. B. Rifkin, and E. Shaw, eds.), p. 123. Cold Spring Harbor Lab., Cold Spring Harbor, New York, 1975.

TABLE II
REACTIVITY OF FACTOR X_a WITH AFFINITY LABELS[a]

Affinity labels	Concentrations (M)	$t_{1/2}$ (min)	$10^{-4} \times k_{app}/I$ (M^{-1} min^{-1})
DNS-Glu-Gly-ArgCH₂Cl	2.5×10^{-9}	12.8	2200
Ile-Glu-Gly-ArgCH₂Cl	2.5×10^{-8}	14.4	190
Ac-Glu-Gly-ArgCH₂Cl	2.0×10^{-8}	24.7	140
Glu-Gly-ArgCH₂Cl	2.5×10^{-7}	17.5	16
Glu-Phe-ArgCH₂Cl	2.0×10^{-7}	25.1	14

[a] Inactivation reactions of factor X_a (44 ng/ml) were conducted at 25°C in 50 mM PIPES buffer (pH 7.0), which was 0.20 M in NaCl and 1.0 mM in CaCl₂. Rates of inactivation were determined by removing at least four timed aliquots (0.50 ml) and diluting with 0.60 ml of 1.67×10^{-4} M Bz-Ile-Glu-Gly-Arg-pNA in 0.20 M Tris buffer (pH 8.0), 0.20 M in NaCl, followed by observation at 405 nm.

$$-\text{Ile-Glu-Gly-Arg}{\downarrow}\text{Ile} \quad \text{and} \quad -\text{Ile-Glu-Gly-Arg}{\downarrow}\text{Ser}-$$

Reagents corresponding to this sequence were prepared and shown to be the most effective affinity labels for bovine factor X_a (Table II).[17]

Several aspects of the specificity of factor X_a are unique. Addition of the dansyl group (DNS) to Glu-Gly-ArgCH₂Cl increased the reactivity of the reagent 140-fold, yielding an affinity label effective at 2.5×10^{-9} M. For other plasma proteases this change yielded only a 10- to 20-fold increase in reactivity.[18] Acetylation or the addition of Ile to Glu-Gly-ArgCH₂Cl increased the reactivity of the reagent about 10-fold, whereas for other proteases there was little effect. Reagents containing Val, Pro, and Lys in the P_2 position were at least 2 orders of magnitude less effective than Glu-Gly-ArgCH₂Cl; however, factor X_a can readily accommodate a P_2 Phe.

Of the reagents tested, DNS-Glu-Gly-ArgCH₂Cl and Ac-Glu-Gly-ArgCH₂Cl were the most selective affinity labels for factor X_a, inactivating it about 20 times more readily than human plasma kallikrein and at least 50 times more effectively than thrombin and plasmin.

Affinity Labeling of Thrombin

Thrombin, the last protease in the blood-coagulation cascade, catalyzes the conversion of fibrinogen to fibrin by the hydrolysis of two peptide bonds, liberating fibrinopeptides A and B. It also hydrolyzes the ''pro'' portion of thrombin and activates factor XIII. The amino acid

[17] C. Kettner and E. Shaw, *Thromb. Res.* **22,** 645 (1981).
[18] C. Kettner and E. Shaw, *Biochim. Biophys. Acta* **569,** 31 (1979).

sequences preceding the hydrolyzed bonds are -Gly-Val-Arg-, -Ile-Pro-Arg-, and -Val-Pro-Arg- for the A chain of fibrinogen,[19] prothrombin,[16] and factor XIII,[20] respectively. Initially, chloromethyl ketones corresponding to each of these sequences were prepared.[21] Both reagents with a penultimate Pro inactivated thrombin with a $t_{1/2}$ of 17–22 min at 7.5×10^{-8} M, whereas Gly-Val-ArgCH$_2$Cl was about 20 times less effective. The degree of selectivity in inactivation of thrombin by the reagents containing a P$_2$ Pro was demonstrated by the 20-, 120-, and 150-fold greater reactivity of Ile-Pro-ArgCH$_2$Cl with thrombin than with plasma kallikrein, factor X$_a$, and plasmin, respectively.

An earlier observation in this laboratory indicated that thrombin was more susceptible to an inhibitor with a D-amino acid in the P$_3$ position, and Bajusz et al.[22] showed that a substrate with a D-Phe in the P$_3$ position was particularly well cleaved by thrombin. Subsequently, D-Phe-Pro-ArgCH$_2$Cl was prepared and its reactivity with thrombin was determined.[23] As shown in Fig. 3, 5×10^{-10} M D-Phe-Pro-ArgCH$_2$Cl inactivates 2.3×10^{-10} M thrombin within a few minutes, whereas it was at least 3 orders of magnitude less reactive with other trypsin-like proteases. When D-Phe-Pro-ArgCH$_2$Cl is allowed to react with a molar excess of thrombin, it acts as a titrant, rapidly inactivating thrombin stoichiometrically and providing a means of determining the concentrations of solutions of thrombin as low as 2.5×10^{-9} M. (For an example of titration data, cf. Ref. 23 or compare the titration of plasma kallikrein by D-Phe-Phe-ArgCH$_2$Cl, Fig 1.)

Studies by Collen have demonstrated that D-Phe-Pro-ArgCH$_2$Cl is an effective anticoagulant in vivo.[24] No significant drop in plasma fibrinogen levels was observed when rabbits received an intravenous injection of 1.0 mg of affinity label before an intravenous infusion of 0.5 mg of thrombin over a 5-min period. Rabbits receiving only thrombin died due to massive thrombosis in the right heart before the infusion of the thrombin was complete. This effective level of D-Phe-Pro-ArgCH$_2$Cl is considerably less than its LD$_{50}$ in mice, which is greater than 50 mg/kg.

[19] S. Iwanaga, P. Wallen, N. J. Grondahl, A. Nenschen, and B. Blombach, Eur. J. Biochem. 8, 189 (1969).

[20] T. Takagi and R. F. Doolittle, Biochemistry 13, 750 (1974).

[21] C. Kettner and E. Shaw, in "Chemistry and Biology of Thrombin" (R. L. Lundblad, J. W. Fenton, II, and K. G. Mann, eds.), p. 129. Ann Arbor Sci. Publ., Ann Arbor, Michigan, 1977.

[22] S. Bajusz, E. Barabas, E. Szell, and D. Bagdy, in "Peptides: Chemistry, Structure and Biology" (R. Walter and J. Meienhofer, eds.), p. 603. Ann Arbor Sci. Publ., Ann Arbor, Michigan, 1975.

[23] C. Kettner and E. Shaw, Thromb. Res. 14, 969 (1979).

[24] D. Collen, O. Matsuo, J. M. Stassen, C. Kettner, and E. Shaw, J. Lab. Clin. Med. (submitted for publication).

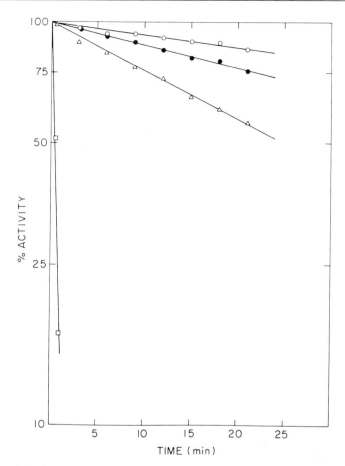

FIG. 3. Selectivity of D-Phe-Pro-ArgCH$_2$Cl in the inactivation of thrombin. The proteases were incubated with D-Phe-Pro-ArgCH$_2$Cl at the indicated concentrations at 25°C in 50 mM PIPES buffer, pH 7.0; timed aliquots were removed for measurement of residual esterase activity.[9] The initial concentrations of thrombin, plasma kallikrein, factor X$_a$, and plasmin were 0.23, 2.5, 14, and 1.8 nm, respectively. O——O, Plasmin inactivation, I = 6 × 10^{-8} M; ●——●, factor X$_a$, I = 6 × 10^{-8} M; △——△, kallikrein, I = 6 × 10^{-8} M; □——□, thrombin, I = 5 × 10^{-10} M.

Affinity Labeling of Plasmin

Plasmin hydrolyzes multiple bonds of its physiological substrate, fibrin, during the lysis of blood clots. The specificity of plasmin for fibrin *in vivo* does not appear to be highly dependent on the sequence of its substrate, but rather on the binding of plasmin and plasminogen on the sur-

TABLE III

SELECTIVITY OF AFFINITY LABELS IN INACTIVATION OF PLASMA PROTEASES[a]

	$10^{-4} \times k_{app}/I$ $(M^{-1}$ $min^{-1})$[b]			
Affinity label	Human plasma kallikrein	Bovine factor X_a	Bovine thrombin	Human plasmin
D-Phe-Phe-ArgCH₂Cl	2320	19.3	45	36.7
DNS-Glu-Phe-ArgCH₂Cl	770	42.1	2.62	85
Glu-Phe-ArgCH₂Cl	568	13.8	0.154	14.5
DNS-Ala-Phe-ArgCH₂Cl	566	49	—	95
Ala-Phe-ArgCH₂Cl	440	2.22	0.17	7.83
Phe-Phe-ArgCH₂Cl	171	4.7	—	2.64
Pro-Phe-ArgCH₂Cl	140	1.33	0.12	1.86
DNS-Glu-Gly-ArgCH₂Cl	141	2200	26	28.6
D-Phe-Pro-ArgCH₂Cl	47.5	27.2	69,000	4.3
DNS-Pro-Phe-ArgCH₂Cl	33.2	2.26	—	2.08
Ile-Glu-Gly-ArgCH₂Cl	28.6	190	3.03	0.45
Ala-Lys-ArgCH₂Cl	18.9	0.0158	—	0.44
Ile-Leu-ArgCH₂Cl	18.8	0.385	5.2	0.317
Glu-Gly-ArgCH₂Cl	15.8	16.0	2.2	1.00
Leu-Gly-Leu-Ala-ArgCH₂Cl	8.25	27.9	15.3	3.33
Gln-Gly-ArgCH₂Cl	4.3	14	1.67	1.00
D-Val-Leu-LysCH₂Cl	3.5	—	12.0	5.45
Pro-Gly-ArgCH₂Cl	3.3	2.0	1.2	0.11
Val-Pro-ArgCH₂Cl	2.96	0.028	54.0	0.31
Phe-Ala-ArgCH₂Cl	2.85	1.37	9.0	0.10
Val-Ile-Pro-ArgCH₂Cl	2.2	2.18	73.0	0.26
Ile-Pro-ArgCH₂Cl	2.0	0.34	42.0	0.27
Gly-Val-ArgCH₂Cl	1.6	0.036	1.9	0.051
Val-Val-ArgCH₂Cl	1.47	0.037	2.1	0.034
Ac-Gly-Gly-ArgCH₂Cl	1.40	5.59	0.74	0.053
Glu-Gly-HarCH₂Cl	0.057	0.004	—	0.005

[a] Inactivation reactions were conducted at 25°C in 50 mM PIPES buffer, pH 7.0. The incubation solutions also contained 0.20 M NaCl for bovine thrombin and 0.20 M NaCl and 1.0 mM CaCl₂ for factor X_a.

[b] k_{app}/I, the bimolecular constants for the inactivation reactions, are the ratios of the apparent, pseudo-first-order rate constants for the inactivation of the proteases to the concentration of the affinity label used in the inactivation.

face of fibrin.[25] In consequence, the usual approach for the development of a selective-affinity label is not applicable to plasmin. In the course of our studies of plasma proteases, effective affinity labels for plasmin have been obtained (Table III), and several aspects of the response of plasmin are worthy of note.

[25] B. Wiman and D. Collen, *Nature (London)* **272**, 549 (1978).

The difference in the affinity of plasmin for Ala-Phe-ArgCH$_2$Cl and the Lys analog indicated that plasmin has a 2-fold greater preference for the lysine chloromethyl ketone.[5] However, this difference in affinity is relatively minor when the influence of residues in the P$_2$ and P$_3$ sites are examined. For example, there is a 400-fold difference in the reactivity of Glu-Phe-ArgCH$_2$Cl and Val-Val-ArgCH$_2$Cl.

DNS-Ala-Phe-ArgCH$_2$Cl is the most effective affinity label for plasmin, inactivating it 50% in 18 min at 4.0×10^{-8} M, and other arginine chloromethyl ketones containing a P$_2$ Phe were among the most effective reagents, indicating a preference of plasmin for this P$_2$ residue. The reactivity of plasmin is also dependent on the P$_3$ residue, as shown by the 2- to 8-fold difference in the reactivity of plasmin with Pro-Phe-ArgCH$_2$Cl and its analogs containing a P$_3$ Glu, Ala, and Phe. Finally, addition of DNS in the P$_4$ position or substitution of a D-residue in the P$_3$ position resulted in reagents with enhanced reactivity as observed for other plasma proteases.[17,18]

Affinity labels that will distinguish plasmin from other trypsin-like proteases have not been obtained. However, urokinase reacts very poorly with arginine chloromethyl ketones containing a P$_2$ Phe; therefore, such reagents readily distinguish plasmin from urokinase and probably from other plasminogen activators.

Affinity Labeling of Plasminogen Activators

Aside from the role of plasminogen activators in hemostasis, tissue and plasma activators are associated with a number of pathological processes, which include malignant cell transformation,[26,27] inflammation,[28] and the promotion of carcinogenesis.[29]

The urinary plasminogen activator, urokinase, hydrolyzes a single Arg-Val bond in the Pro-Gly-Arg-Val- sequence of plasminogen in the activation process.[30] The affinity label corresponding to this sequence, Pro-Gly-ArgCH$_2$Cl, was effective in the inactivation of urokinase at the $10^{-6} M$ level ($K_i = 68$ μM, $k_2 = 0.47$ min^{-1}).[18] Measurement of the inactivation rate, k_{app}/I, for Glu-Gly-ArgCH$_2$Cl, DNS-Glu-Gly-ArgCH$_2$Cl, and Ac-Gly-Gly-ArgCH$_2$Cl revealed that these reagents were even somewhat

[26] L. Ossowski, J. C. Unkeless, A. Tobia, J. P. Quigley, D. B. Rifkin, and E. Reich, *J. Exp. Med.* **137**, 112 (1973).

[27] J. C. Unkeless, A. Tobia, L. Ossowski, J. P. Quigley, D. B. Rifkin, and E. Reich, *J. Exp. Med.* **137**, 85 (1973).

[28] J.-D. Vassalli, J. Hamilton, and E. Reich, *Cell* **11**, 695 (1977).

[29] M. Wigler and I. B. Weinstein, *Nature* (*London*) **259**, 232 (1976).

[30] L. Sottrup-Jensen, M. Zajdel, H. Claeys, T. E. Petersen, and S. Magnusson, *Proc. Natl. Acad. Sci. U.S.A.* **72**, 2577 (1975).

more reactive than Pro-Gly-ArgCH$_2$Cl by factors of 25, 6, and 3, respectively.

The effectiveness of arginine chloromethyl ketones as affinity labels is highly dependent on binding in the S$_2$ and S$_3$ sites, resulting in differences in reactivity of up to 4 orders of magnitude. These differences in the susceptibility of urokinase suggested that the series of affinity labels would be useful in evaluating differences among plasminogen activators from various sources.

The plasminogen activator secreted by HeLa cells following exposure to the tumor promoter, phorbol myristic acetate,[28] was also examined. Because of the low concentration of protease, use was made of the coupled assay with plasminogen and ^{125}I-labeled fibrin to measure activity. Since values of k_{app}/I are difficult to obtain in the nonlinear ^{125}I-labeled fibrin assay, values of I_{50}, defined in Fig. 4, were determined for the affinity labels. The validity of this approach was established by the good correlation obtained in the case of urokinase between values of k_{app}/I and I_{50}. As shown in Fig. 4,[31] the HeLa cell plasminogen activator and urokinase are similar in their susceptibilities toward affinity labels in many respects. Reagents containing a P$_2$ Gly were the most effective for both enzymes, reagents with a P$_2$ Pro and Ala were intermediate in effectiveness, and those containing a P$_2$ Phe were the least reactive. However, distinctive differences exist between the responses of the two plasminogen activators. Most notable is the effect of adding a residue to the P$_4$ position. For example, the addition of DNS to Glu-Gly-ArgCH$_2$Cl increased its reactivity with the HeLa cell enzyme by an order of magnitude, but decreased its reactivity with urokinase.

Thus far we have not obtained an affinity label that will distinguish either the HeLa cell enzyme or urokinase from all plasma proteases. For example, the most effective inhibitor of the HeLa cell plasminogen activator, DNS-Glu-Gly-ArgCH$_2$Cl, reacts with factor X$_a$ at the 10^{-9} M level.[17] On the other hand, Ac-Gly-Gly-ArgCH$_2$Cl is 50 times more reactive with urokinase than plasmin, providing a means of distinguishing these two proteases.

Selectivity in Affinity Labeling

This section summarizes the results obtained in the affinity labeling of plasma trypsin-like enzymes by providing values of k_{app}/I for the reactivity of the affinity labels with plasma kallikrein, plasmin, factor X$_a$, and thrombin. In addition to making the data on the reactivities of the chloromethyl ketones accessible, Table III illustrates the fact that a

---

[31] P. Coleman, C. Kettner, and E. Shaw, *Biochim. Biophys. Acta* **569**, 41 (1979).

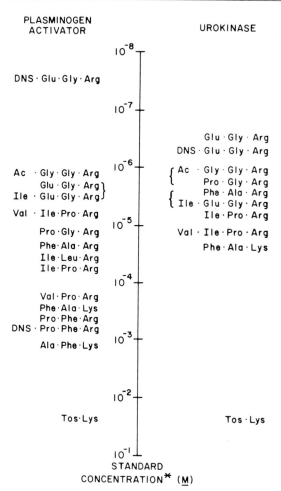

Fig. 4. The relative effectiveness of chloromethyl ketone inhibitors with plasminogen activator (from HeLa cells) and urokinase. The position of the chloromethyl ketone on the log concentration scale indicates the approximate value of its "standard concentration." Inhibitors with close to identical values are designated by brackets. (*) The concentration of inhibitor that increases the time of the midpoint in the reaction profile of ^{125}I-labeled fibrin solubilization to twice that of the uninhibited control. From Coleman et al.[31]

unique pattern of reactivity is obtained for each protease. These patterns should be useful in the identification of proteases in their physiological environments. In crude systems the reactivity of the proteases with affinity labels may be less than that obtained with purified enzymes, particularly if their physiological substrate is present; however, the relative reactivities can be expected to be characteristic (Table III).

Acknowledgments

This research was conducted under the auspices of the U.S. Department of Energy. Partial support was obtained from NIH Contract 1-Y01-HV-70021. The authors extend their gratitude to John Ruscica, Christopher Mirabelli, and Frances Brown for their excellent technical assistance, to Dr. Yale Nemerson, Mt. Sinai School of Medicine, New York City, and Jolyon Jesty, SUNY, Stony Brook, New York, for the gift of factor X_a, to Dr. John Fenton II, N.Y. Health Department, Albany, New York, for human thrombin, Dr. Jack V. Pierce, NIH, Bethesda, Maryland, for human urinary kallikrein, and to Dr. Isaac Schenkein, N.Y. University Medical Center, New York City, for submaxillary gland protease.

[64] Affinity Methods Using Argininal Derivatives

By SHIN-ICHI ISHII and KEN-ICHI KASAI

Leupeptin, which is produced by several strains of Streptomyces, has been known as a potent inhibitor for trypsin-family proteases including trypsin, plasmin, kallikrein, and urokinase.[1,2] It is a mixture of tripeptide derivatives; α-amino group of L-leucyl-L-leucyl-L-argininal is acylated by either acetyl or propionyl group. The strong inhibitory action has been explained by formation of a covalent hemiacetal adduct between an aldehyde group in the inhibitor and a serine hydroxyl group in the enzyme-active site.[3] We shall obtain useful biospecific affinity adsorbents if leupeptin or its essential moiety can be immobilized. Because the guanidino group of the terminal argininal is also indispensable for the activity, there remains no functional group suitable for immobilization reaction. In order to use leupeptin as an immobilized ligand for affinity adsorbents, appropriate modification is required.

We have developed a new method to expose an α-amino group by using thermolysin. Thermolysin was found to hydrolyze exclusively the leucyl–leucyl bond. However, the direct thermolysin digestion of leupeptin gave complex products and yield of the expected leucylargininal was very low. This would be due to the reaction of the exposed α-amino group with the aldehyde group in the same or the other molecule. Thus the aldehyde group was protected prior to thermolysin digestion and the product was handled under the protected form throughout a series of reaction steps. Finally, the aldehyde group was regenerated (see Scheme I).

[1] T. Aoyagi, S. Migita, M. Nanbo, F. Kojima, M. Katsuzaki, M. Ishizuka, T. Takeuchi, and H. Umezawa, *J. Antibiot.* **22**, 558 (1969).
[2] H. Umezawa, this series, Vol. 45, p. 678.
[3] H. Kuramochi, H. Nakata, and S. Ishii, *J. Biochem.* (*Tokyo*) **86**, 1403 (1979).

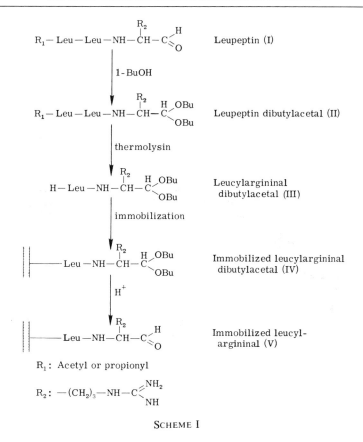

R_1: Acetyl or propionyl

R_2: $-(CH_2)_3-NH-C\underset{NH}{\overset{NH_2}{<}}$

SCHEME I

The outline of preparation of affinity adsorbents is schematically shown. The aldehyde group was protected as dibutylacetal[1] and deprotected after immobilization. Affinity adsorbents prepared by such a procedure showed strong affinity for trypsin-family proteases as had been expected. The affinity is so strong that the adsorbents are quite suitable for removal of the last trace of such proteases from their zymogen preparations. When the object of experiment is to purify the active proteases themselves, the adsorbed enzymes may be eluted with a dilute solution of leupeptin. The eluted enzymes are freed from leupeptin by inactivating the latter by $NaBH_4$ treatment.

Leucylargininal dibutylacetal (compound III) is also an important starting material for various soluble derivatives of leupeptin. For example, dansylleucylargininal prepared essentially in the same way as explained in Scheme I shows almost the same affinity as that of leupeptin toward bovine trypsin. It may be very useful as a fluorescent affinity-labeling reagent for trypsin-family proteases. Preparation and application of these derivatives will be reported in detail elsewhere.

Preparation of Affinity Adsorbents

Leupeptin Dibutylacetal (II). Leupeptin sulfate (commercially available from Protein Research Foundation, Osaka 562, Japan) was converted to chloride salt with Dowex 1, Cl⁻ form. Leupeptin chloride (2.93 g) was dissolved in 170 ml of 1-butanol and refluxed for 4 hr. To this solution, 170 ml of butylacetate and 340 ml of water were added. The upper layer was taken and dried with Na_2SO_4. After evaporation, 2.3 g of crude II was obtained. This was purified chromatographically on silica gel [column volume, 200 ml; solvent, chloroform : 1-butanol (1 : 2)], and the purified product showing a single spot on silica gel TLC (the same solvent system) was obtained in a yield of 72%.

Thermolysin Hydrolysis. II (1.05 g) was dissolved in 110 ml of 0.1 M N-ethylmorpholine-HCl buffer (pH 8.0) containing 0.02 M $CaCl_2$ and 5.5 mg of thermolysin. The mixture was incubated at 40°C. The extent of hydrolysis was checked by an appropriate method for amino-group determination such as the trinitrobenzene sulfonate colorimetry. After 72 hr, the amount of free amino group reached to that of the aldehyde group. The reaction mixture was extracted with 90, 75, and 60 ml of 1-butanol. After evaporation of the solvent, 0.75 g of III was obtained.

Immobilization of III to Epoxy-Activated Sepharose: EPOX-Leu-Argal–Sepharose. Sepharose-6B was activated with 1,4-butanediol-diglycidyl ether.[4,5] The activated wet gel was mixed with an equal volume of 0.1 M borate buffer (pH 9.5) containing III (0.1–2 μmol/1 ml gel). The suspension was shaken for 40 hr at 40°C. After washing, the gel was treated with 2% glycine in the borate buffer for 40 hr at 40°C to block remaining activated groups. The gel was suspended in 2.5 volumes of 0.2 M citrate buffer (pH 2.2), and shaken for 10 hr at 40°C to regenerate the aldehyde group.

The amount of immobilized ligand was determined as follows. The gel (ordinary 0.5 ml) was suspended in 2 ml of 0.1 M borate buffer (pH 8.0) containing 0.1% $NaBH_4$ for 1 hr at 0°C in order to reduce the argininal residue to argininol. The gel was then washed with water, water-acetone (1 : 1), and acetone successively, and dried *in vacuo*. The dried gel was hydrolyzed with 1 ml of 5.7 N HCl under reduced pressure for 24 hr at 110°C. The hydrolysate was chromatographed on a CM–cellulose (Whatman CM-52) column (0.6 × 16.8 cm), which had been equilibrated with

---

[4] L. Sundberg and J. Porath, *J. Chromatogr.* **90**, 87 (1974).

[5] J. Porath, this series, Vol. 34, p. 13.

0.1 M acetate–NaOH buffer (pH 6.0). Argininol, eluted from the column with the same buffer at about 60 ml, was quantitated by the ninhydrin reaction for guanidino group.[6] Adsorbents containing 3.2–270 nmol of ligand/ml gel were obtained.

Quantitation of arginine produced by oxidation of argininal would have been much easier. However, quantitative conversion of argininal to arginine in the Sepharose-bound ligand could not be attained.

Immobilization of III to Aminohexanoic Acid–Sepharose: AH-Leu-Argal–Sepharose. Sepharose-4B was activated with cyanogen bromide by ordinary procedure.[7] Forty milliliters of the activated gel were suspended in an equal volume of 0.2 M borate buffer (pH 9.2) containing 0.1 M 6-aminohexanoic acid, and the suspension was shaken for 10 hr at room temperature. After washing, the gel was suspended in 0.1 M ethanolamine-HCl (pH 8.5), and the suspension was shaken for 3 hr at room temperature. Thirty milliliters of the aminohexanoic acid–Sepharose thus obtained and washed were mixed with 30 ml of 3.3% aqueous solution of III. Two milliliters of 1-ethyl-3-(dimethylaminopropyl)carbodiimide were added dropwise over 3 hr. The pH of the reaction mixture was maintained at 4.3–4.5 with 2 N HCl. The suspension was shaken for an additional 8 hr at room temperature. The aldehyde group was regenerated in 10 mM HCl for 3 hr at 60°C.

The total amount of immobilized ligand was determined by amino acid analysis after acid hydrolysis of the gel (5.7 N HCl, 24 hr at 110°C). The amount of leucine was used for calculation of the ligand content. An adsorbent containing 2.3 μmol of ligand/ml gel was obtained.

Immobilization of III to Gly-Gly–Sepharose: Gly-Gly-Leu-Argal–Sepharose. Gly-Gly was coupled to Sepharose by the cyanogen bromide method as described above. Immobilization of III to this Gly-Gly–Sepharose and deprotection of the aldehyde group were carried out essentially in the same way as described above. Ligand content was determined by amino acid analysis. Immobilized Gly-Gly was 4 μmol/ml, and to this spacer 0.67 μmol of III/ml was introduced as judged from leucine content.

Properties and Application of Affinity Adsorbents

Properties of Adsorbents. The adsorbents showed very strong affinity for trypsin-family enzymes. Adsorbed enzymes were bovine and *Streptomyces griseus* trypsin, plasmin, kallikrein, urokinase, and clostripain. The importance of the aldehyde group of argininal was evident. The

[6] R. B. Conn, Jr., *Clin. Chem.* **6**, 537 (1960).
[7] P. Cuatrecasas, *J. Biol. Chem.* **245**, 3059 (1970).

Sepharose derivatives adsorbed enzymes only after the acid treatment for removal of the protecting group. The adsorbents completely lost their ability if the aldehyde group was converted to the corresponding alcohol by treatment with $NaBH_4$.

Elution of adsorbed enzymes was practically impossible by procedures usually used, such as elevation of ionic strength, lowering of pH, or addition of denaturants. Successful elution was achieved only by the addition of leupeptin to the eluant at low concentrations (e.g., 0.2 mM). However, the transfer of adsorbed enzymes to this mobile phase proceeds only slowly, so the flow rate should be low. Practically, a leupeptin-containing buffer was introduced into the column to fill the void and allowed to stand overnight without flow. Elution was started again the next day. The eluted enzyme formed a tight complex with leupeptin which could not be removed by dialysis or gel filtration. Thus leupeptin was inactivated by converting its aldehyde group to alcohol with $NaBH_4$ (e.g., 2 mM). To regain intact enzyme, it is important that the objective enzyme tolerates such reductive conditions. The enzymes that we tested were stable under these conditions, and about 90% of the activity originally applied to the column was recovered. Such an elution procedure may have two merits. First, we can expect improved efficiency in the enzyme purification, because biospecific interaction must be acting not only on the adsorption step but also on the elution step. Second, we can operate the enzyme, throughout the process of affinity chromatography, in the inhibitor-bound form in which it must be to protect from autodigestion.

Chromatography of Bovine Trypsin on AH-Leu-Argal–Sepharose. A column of AH-Leu-Argal–Sepharose (0.6 × 6.0 cm) was equilibrated with 0.05 M Tris-HCl buffer (pH 8.2) containing 0.02 M $CaCl_2$. The column temperature was maintained at 15°C. Commercial bovine trypsin was dissolved in the same buffer (1.4 mg/ml). Three ml of this solution was applied, and the column was washed with 35 ml of the same buffer at a flow rate of 5 ml/hr. A small amount of UV-absorbing material with no enzyme activity appeared in this washing process, whereas active trypsin was adsorbed. Then, the mobile phase in the column was replaced with the same buffer containing 0.2 mM leupeptin and the elution was interrupted. After standing overnight, the elution started again. A protein peak appeared in the effluent, each fraction of which was then treated with 2 mM $NaBH_4$ at 0°C for 15 min to inactivate leupeptin. It had been shown that this treatment did not affect the activity of native trypsin. Trypsin activity in this leupeptin eluate was assayed by using benzoyl-DL-arginine-p-nitroanilide (0.45 mM) as a substrate. The elution profile of activity corresponded exactly to that of protein. About 90% of the trypsin activity originally applied to the column was recovered (see Fig. 1A).

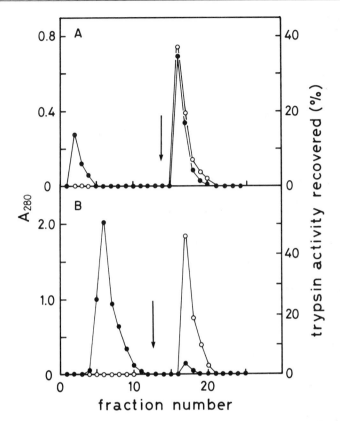

FIG. 1. Chromatography of trypsins on the immobilized leupeptin derivatives. Columns were equilibrated with 0.05 M Tris-HCl buffer (pH 8.2) containing 0.02 M CaCl₂. After addition of a sample, the column was washed with the same buffer. Adsorbed trypsin was eluted as described in the text with 0.2 mM leupeptin dissolved in the same buffer. ●—●, A_{280}; ○—○, trypsin activity. The arrow indicates addition of the leupeptin solution. (A) Bovine trypsin on AH-Leu-Argal–Sepharose. Bovine trypsin (Worthington, twice recrystallized) was dissolved in the same buffer. A_{280} of the solution was 2.2. Three milliliters of this solution were applied to a column of AH-Leu-Argal–Sepharose (0.6 × 6.0 cm). Column temperature was 15°C. Flow rate was 5 ml/hr and fractions of 3 ml were collected. (B) *S. griseus* trypsin on EPOX-Leu-Argal–Sepharose. Pronase P (Kaken Kagaku Co., Tokyo, Japan) dissolved in the same buffer (1.8 ml, total absorbance unit at 280 nm was 9.9) was applied to a column of EPOX-Leu-Argal–Sepharose (0.9 × 9.2 cm). The ligand content was 0.27 μmol/ml. Column temperature was 4°C. Fractions of 2.1 ml were collected.

Chromatography of Pronase on EPOX-Leu-Argal–Sepharose. Pronase is a mixture of various proteases including a trypsin-type enzyme produced by *S. griseus.*[8,9] Figure 1B shows its chromatography on EPOX-Leu-Argal–Sepharose (lot 1). The procedure used was essentially the same as described above. Trypsin activity was completely adsorbed and eluted with the pH 8.2 buffer containing 0.2 mM leupeptin.

Considerations on the Handling of Adsorbents. Each Sepharose derivative has both merits and demerits. The ligand immobilized by the cyanogen bromide method is unstable at alkaline pH.[10] Leakage of the ligand, and consequently of the ligand–enzyme complex, will occur if chromatography is carried out near pH 9. On the other hand, the ligand immobilized by the bisepoxide method was very stable at neutral and slightly alkaline pH. However, it was considerably unstable at acidic pH. Even by the acid treatment for deprotection of aldehyde group (pH 2.2, 10 hr at 40°C), significant release (50–70%) of the immobilized ligand was observed. On the contrary, the ligand immobilized by the cyanogen bromide method was more stable at acidic pH. In the case of Gly-Gly-Leu-Argal–Sepharose, both glycine and leucine contents did not change after the acid treatment (10 mM HCl, 3 hr at 60°C). Thus if chromatography is carried out mainly at neutral and alkaline pH, adsorbents prepared by the bisepoxide method are preferable, while at lower pH, those prepared by the cyanogen bromide method are preferable.

Quantitative Estimation of Affinity of Adsorbents. Since most of the immobilized argininal derivatives adsorbed trypsin-family enzymes too tightly, it was difficult to estimate their affinities by ordinary procedures. However, if we prepare an adsorbent of quite low ligand content and analyze its behavior toward an enzyme by frontal chromatography, we can determine the dissociation constant (K_d) of enzyme-immobilized ligand complex.[11] Thus we prepared an EPOX-Leu-Argal–Sepharose (lot 2) whose ligand content was 3.2 nmol/ml ($3.2 \times 10^{-6} M$) and determined K_d values for several enzymes under various conditions. K_d values for bovine α- and β-trypsin and *S. griseus* trypsin were 1.1, 0.33, and 0.025 μM, respectively, at 25°C, pH 8.2. K_d for bovine β-trypsin increased slightly when pH was lowered from 8.2. At pH 6, K_d value was 1 μM, and at pH 4 it was 3.2 μM, that is, about 10 times larger than at pH 8.2. Those for bovine anhydrotrypsin and DIP–trypsin were more than 500 μM. These results were consistent with our previous findings on the mechanism of leupeptin action.[3]

[8] Y. Narahashi, K. Shibuya, and M. Yanagita, *J. Biochem.* (*Tokyo*) **64**, 427 (1968).
[9] Y. Narahashi, this series, Vol. 19, p. 651.
[10] G. I. Tesser, H.-U. Fisch, and R. Schwyzer, *FEBS Lett.* **23**, 56 (1972).
[11] K. Kasai and S. Ishii, *J. Biochem.* (*Tokyo*) **84**, 1051 (1978).

Author Index

Numbers in parentheses are footnote reference numbers and indicate that an author's work is referred to although his name is not cited in the text.

H

Subject Index

A

Boc-Val-Pro-Arg-MCA
  amidase assay of staphylocoagulase, 313, 314
Boc-L-Val-L-Pro-L-Arg-MCA
  assay of thrombin, 360
Boc-L-Val-Ser-Gly-L-Arg-MCA, assay of urokinase, 360
Bothrops atrox venom, Factor V activator, *see* Thrombocytin
Bowman-Birk trypsin inhibitor, 311
Bradykinin
  activity, 173, 174
  antiserum, 175
  enzyme immunoassay, 175
  hydrolysis by converting enzyme, 459
  release, 467
Brinase, trapped by α_2-macroglobulin, 748
Bromelain, trapped by α_2-macroglobulin, 748
Bromide, inactivation of Factor XIII, 336
BTyr EE, hydrolysis by crab collagenase, 731
BTyr NA, hydrolysis by crab collagenase, 731
Bull seminal plasma inhibitor, 792–803
  amino acid composition, 802
  assay, 794
  distribution in organs, 802
  enzyme unit, 793
  function, 803
  inhibition unit, 793
  properties, 800–803
    immunological, 801
    inhibitory, 801
    kinetic, 801
  purification, 793–796
  purity, 800
  specific activity, 793
  stability, 800
Bull seminal plasma inhibitor I, 792
  isoinhibitors
    isoelectric points, 800, 801
    isolation, 796–799
    molecular weight, 800, 801
Bull seminal plasma inhibitor II, 792
  amino acid sequence, 802, 803
  isoelectric point, 801
  isolation, 799, 800
  molecular weight, 801

BVal-Gly-Arg-NA, hydrolysis by crab protease, 731
Bz-Arg-NH$_2$, substrate for cathepsin L, 557, 558
Bz-Arg-NNap, substrate for cathepsin H, 556
Bz-Gly-Gly-Gly, assay of peptidyl dipeptidase, 450–453
Bz-Gly-His-Leu, assay of peptidyl dipeptidase, 450–453
Bz-Gly-Phe-Arg, assay of peptidyl dipeptidase, 450–453
Bz-Gly-Phe, hydrolysis by mast cell carboxypeptidase, 608
Bz-Gly-Pro-Arg-p NA-HCl, *see* Chromozym TH
Bz-L-Ile-L-Glu-Gly-L-Arg-p NA, *see* S-2222 substrate, S-2337 substrate substrate for thrombocytin, 285
Bz-L-Phe-L-Val-L-Arg-p NA, *see also* S-2160 substrate
  hydrolysis
    by Factor X$_a$, 351
    by Factor XII$_a$, 355
    by thrombin, 346
    by trypsin, 355
  substrate for thrombocytin, 285
Bz-L-Pro-L-Phe-L-Arg-p NA, *see also* Chromozym-PK
  assay of kallikrein amidase activity, 159
Bz-L-Val-Gly-L-Arg-p NA, hydrolysis by Factor X$_a$, 350

C

C1
  human, 6–16
    activated form, *see* C$\overline{1}$
    activation, 15–17, 365
    assay
      hemolytic, 7–10
      procedure, 8–10
      reagents, 7, 8
    CH$_{50}$ values, 9
    dissociation, 14
    properties, 13, 14
    purification, 10–13
      euglobulin precipitation, 10, 11
      gel filtration, 11

.